GSM, GPRS AND EDGE Performance

GSM, GPRS and EDGE Performance

Evolution Towards 3G/UMTS

Second Edition

Edited by

Timo Halonen
Nokia Networks, Finland

Javier Romero and Juan Melero
Tartessos Technologies S.A. (TarTec), Spain

John Wiley & Sons, Ltd

Telephone (+44) 1243 779777

Email (for orders and customer service enquiries): cs-books@wiley.co.uk
Visit our Home Page on www.wileyeurope.com or www.wiley.com

Reprinted July 2004, November 2005, November 2006

Other Wiley Editorial Offices

John Wiley & Sons Inc., 111 River Street, Hoboken, NJ 07030, USA

Jossey-Bass, 989 Market Street, San Francisco, CA 94103-1741, USA

Wiley-VCH Verlag GmbH, Boschstr. 12, D-69469 Weinheim, Germany

John Wiley & Sons Australia Ltd, 33 Park Road, Milton, Queensland 4064, Australia

John Wiley & Sons (Asia) Pte Ltd, 2 Clementi Loop #02-01, Jin Xing Distripark, Singapore 129809

John Wiley & Sons Canada Ltd, 22 Worcester Road, Etobicoke, Ontario, Canada M9W 1L1

Wiley also publishes its books in a variety of electronic formats. Some content that appears in print may not be available in electronic books.

Library of Congress Cataloging-in-Publication Data

GSM, GPRS, and edge performance : evolution towards 3G/UMTS / edited by Timo Halonen, Javier Romero, Juan Melero.—2nd ed.
p. cm.
Includes bibliographical references and index.
ISBN 0-470-86694-2
1. Global system for mobile communications. I. Halonen, Timo, II. Romero, Javier (Romero García) III. Melero, Juan.

TK5103.483.G753 2003
621.3845′6—dc22

2003057593

British Library Cataloguing in Publication Data

A catalogue record for this book is available from the British Library

ISBN13:978-0-470-86694-8(HB)

Typeset in 10/12pt Times by Laserwords Private Limited, Chennai, India
Printed and bound in Great Britain by TJ International, Padstow, Cornwall
This book is printed on acid-free paper responsibly manufactured from sustainable forestry in which at least two trees are planted for each one used for paper production.

Contents

Acknowledgements

We would like to thank all contributors for the "world-class" material produced to support this book. This book has been the result of a collaborative effort of a truly motivated and committed team.

We would also like to thank our listed colleagues for their valuable input and comments: Heikki Annala, Harri Jokinen, Mattias Wahlqvist, Timo Rantalainen, Juha Kasinen, Jussi Reunanen, Jari Ryynanen, Janne Vainio, Kari Niemelä, Simon Browne, Heikki Heliste, Petri Gromberg, Oscar Salonaho, Harri Holma, Fabricio Velez, Jyrki Mattila, Mikko Säily, Gilles Charbit, Jukka Kivijärvi, Martti Tuulos, Lauri Oksanen, Ashley Colley, Mika Kähkölä, Per Henrik Michaelson, Rauli Jarvela, Rauli Parkkali, Kiran Kuchi, Pekka Ranta, Brian Roan, James Harper, Raul Carral, Joe Barret, Timothy Paul, Javier Munoz, Poul Larsen, Jacobo Gallango, Regina Rodríguez, Manuel Martínez, Sheldon Yau, and Art Brisebois. Special thanks to Jussi Sipola for the link performance material provided in Chapter 7.

Thanks to Mark Keenan for the support, guidance and encouragement provided during the last years. We hope he feels as proud as we do about the result of this book.

Thanks to the authors of the forewords (Mike Bamburak, Mark Austin and Chris Pearson) and the companies/institutions they represent (AT&T, Cingular and 3G Americas) for the support provided.

Many thanks to the operators (CSL, Radiolinja, Sonofon, AT&T, Cingular, Telefonica and many others) we have been closely working with since only through this tight collaboration the understanding of the technology capabilities contained in this book has been possible. Special thanks to Optus, for their collaboration, support and excellent technical team (Andrew Smith, Carolyn Coronno, Bradley Smith and the rest of the team).

Part of the studies presented in this book have been performed as the result of the cooperation agreement between Nokia and the University of Malaga. This agreement is partially supported by the "Programa de Fomento de la Investigación Técnica" (PROFIT) of the Spanish Ministry of Science and Technology. We want to thank Malaga University and the Ministry of Science and Technology for the provided support.

Thanks to the publishing team from Wiley (Mark Hammond, Sarah Hinton and Geoff Farrell), which has given an excellent support throughout this project, which has been very important to accomplish the demanding time schedule.

Finally we would like to express our loving thanks to our wives Cristina, Laila and Susanna for the patience and understanding during the holidays, weekends and late nights devoted to the writing of this book.

The authors welcome any comments and suggestions for improvements or changes that could be implemented in possible new editions of this book. The e-mail address for gathering such input is geran_book@hotmail.com.

Malaga, Spain
The editors of the "GSM, GPRS & EDGE Performance" book

Forewords

(Taken from GSM, GPRS and EDGE Performance, 1st Edition).

I have worked in the mobile communications industry longer than I would like to admit. In the early 1970s, I started my career as a radio engineer for Motorola. At that time, Motorola designed and manufactured low-, mid- and high-tier private land mobile radios. Motorola had few competitors for the mid- and high-tier product lines (50- to 100-W radios). However, in the low tier, less than 25-W radio category, there were numerous contenders, mostly from European manufacturers with a 'Nordic Mobile Telephone' heritage.

But times were changing. In the late 1970s, the American public got their first taste of mobile communications when Citizen Band (CB) radio became popular ('10–4, good buddy'). It was an unlicensed, short-range, 'party-line' experience. Those skilled in the art knew that something better was needed. And the American communications industry responded. The Federal Communications Commission and major industry players, like AT&T and Motorola, specified America's first public mobile radio telephone system, AMPS (Advanced Mobile Telephone System). By the mid-1980s, AMPS was a proven technology and cellular subscriber growth was constantly exceeding forecasts.

By the early 1990s, cellular technology had become so popular that the first-generation analog systems could not keep up with the demand. New second-generation digital systems were developed to address the capacity shortfall. In the United States, three digital technologies were standardized and deployed: IS-136 (a TDMA technology utilizing the AMPS 30-kHz structure), IS-95 (a 1.25-MHz CDMA carrier scheme) and GSM (the European 200-kHz TDMA standard). This multi-standard wireless environment provided a unique proving ground for the three technologies. While IS-136 and IS-95 engaged in 'standards wars,' GSM gained a foothold in America. At the same time, GSM was achieving global acceptance because it offered a rich selection of capabilities and features that provided real incremental revenues for operators. As more and more countries adopted the technology, GSM experienced tremendous economies of scale for everything from chipsets to handsets, infrastructure and applications.

While the industry continued to experience stellar growth, American manufacturer dominance was challenged by Nordic companies, especially for the GSM technology. They brought to the United States, innovative, competitively priced products, backed by talented communications professionals with years of experience in designing, manufacturing, engineering and installing cellular equipment and systems throughout the world.

By the late 1990s, the Internet was pervasive and the wireless industry looked to mobile data as the growth opportunity. Once again, the industry undertook the task of defining new wireless systems—this third generation, 3G, was to be based on packet data. Three new wireless standards emerged; CDMA2000 (evolution of IS-95), EDGE (evolution of

GSM for existing spectrum) and WCDMA (evolution of GSM for new spectrum using a 5-MHz WCDMA carrier).

The evolution of GSM to 3G is about gradually adding more functionality, possibilities and value to the existing GSM network and business. The evolution begins with an upgrade of the GSM network to 2.5G by introducing GPRS technology. GPRS provides GSM with a packet data air interface and an IP-based core network. EDGE is a further evolutionary step of GSM packet data. EDGE can handle about three times more data subscribers than GPRS, or triple the data rate for one end-user. EDGE can be achieved through a very fast and cost-effective implementation. The only requirement is to add EDGE-capable transceivers and software.

With the continuation of EDGE standardisation towards GERAN (GSM/EDGE Radio access network), EDGE will achieve a full alignment with WCDMA. The goal for EDGE is to boost system capacity, both for real-time and best-effort services, and to become perfectly competitive with other 3G technologies.

What emerges with these evolutionary steps from GSM to GPRS, EDGE and WCDMA is a seamless 3G UMTS (Universal Mobile Telecommunications System) Multi-Radio network, one that maximizes the investments in GSM and GPRS.

It stands to reason that both EDGE and WCDMA will be mainstream 3G UMTS products from Nordic companies. This book, written by engineers from one of these Nordic companies, is an authoritative treatise on GSM evolution to 3G. The book provides an in-depth performance analysis of current and future GSM speech and GPRS/EDGE packet data functionality. Furthermore, the concept of a 3G UMTS Multi-Radio network (GSM/EDGE/WCDMA) is presented in depth as the best solution for wireless operators to evolve their networks towards 3G.

Times change, but some things do not. Nordic companies have been at the forefront of wireless communications for more than a half of a century. They have earned their pre-eminent position in the industry. I encourage you to listen to what this book has to say.

Mike Bamburak
VP Technology Architecture & Standards
AT&T

Over the years, scientists and dreamers have revolutionised the way we work and live through great inventions. Almost as quickly as news of the inventions spread, soothsayers rose to highlight real or imagined barriers to the success, popularity or the long-term use of these products and services. As occurred with electricity, the automobile and the television, soothsayers often misread the long-term impact of these inventions, underestimating qualities that led to their long-term success and adoption by the masses. Ultimately, all three inventions had tremendous social and economic impact on global societies, and the soothsayers were proven to have undervalued the importance of these great inventions over time.

In a slightly different way, the future of EDGE has been understated, underestimated, and undervalued by the latest pundits. Over the last few years, several global wireless organizations, including the GSA, the UWCC and now 3G Americas have stood their ground as advocates for EDGE because of the merits of the technology and its value to operators and customers as a spectrally efficient and cost-effective solution for third-generation (3G) wireless services. 3G Americas is firm in their belief that a comparative review of how EDGE meets three key criteria, performance, cost and the ease of transformation to 3G, will show that EDGE is indeed a superior technology choice.

The Reality of EDGE

On October 30, 2001, Cingular Wireless with its vendor partners announced its commitment to become the first operator in the world to deploy EDGE at 850 and 1900 MHz. With over 22-million wireless customers, Cingular is a major player in the global wireless marketplace. The reasons cited by Cingular for its EDGE selection included capacity and spectral efficiency competitive with any other technology choice (performance), the ability to deploy 3G EDGE in existing spectrum including 850 MHz (cost), a total capital cost of about $18 to $19 per Point of Presence (POP) in TDMA markets with plenty of go-forward capacity (cost), ridiculously low cost to upgrade the GSM network (only 10 to 15 percent of the network's cost), the enormous economies of scale and scope offered by the family of GSM technologies, ensuring the availability of equipment and applications at the lowest possible cost, and a transition path through GSM and GPRS achievable seamlessly through the use of the GAIT terminal (GSM-TDMA phone) that will ease transformation and result in customer retention.

Similarly, almost a year before Cingular's announcement, AT&T Wireless Services announced its commitment to EDGE. As of November 2001, reported operator commitments to EDGE in the Americas encompassed hundreds of millions of potential customers. These commitments validate the future of this third-generation technology. Cingular's commitment to EDGE in the 850-MHz band sets the stage for an accelerated uptake by operators throughout the Western Hemisphere. Regional US operators and many Latin American operators will find the opportunity to deploy EDGE in 850 MHz especially appealing. Furthermore, these commitments increase the possibility that Europe will recognize that EDGE's capacity and cost qualities make it an important complementary technology to WCDMA. As spectrum shortages inevitability occur in Europe, EDGE will provide an excellent solution for GSM operators as a complement to their WCDMA networks.

Benefits of EDGE

EDGE will benefit operators and customers because it is a cost-effective solution for 3G services. Cost efficiency is enabled by the economies of scope and scale demonstrated by the GSM family of technologies, including both TDMA and GSM, which represented nearly 80% of the world's digital cellular subscribers in 2001. More than half a billion GSM phones existed by mid-year 2001, and within a mere four months that number had risen by 100 000 000 phones to 600 000 000. Bill Clift, Chief Technology Officer of Cingular, noted that the cost differential between GSM and CDMA devices gets fairly significant at $15 to $20 per handset times millions of handsets each year. The economies of scale played a key role in the Cingular decision.

Another major benefit of EDGE cited by operators is that it enables TDMA and GSM carriers to offer 3G services while still realizing lower costs due to higher spectral efficiency and higher data rates. With the implementation of Adaptive Multi-Rate

(AMR) Vocoders, and Frequency Hopping, GSM is competitive with CDMA on spectral efficiency, which translates into higher capacity and faster data rates. EDGE offers transmission speeds of 384 kbps—fast enough to support full motion video—and throughput capacity 3 to 4 times higher than GPRS. Thus, EDGE is fast, EDGE is efficient and EDGE performs.

Additionally, the opportunity for international roaming with the GSM family of technologies offers yet another major incentive for operators to provide their customers with seamless communications services. Since EDGE and WCDMA will operate on the same GPRS core network, the EDGE/WCDMA customer will be able to travel the world seamlessly, staying connected with one number and one device.

Conclusion

EDGE will contribute to a bright future for 3G services, a vision shared by major analysts and industry groups. The Strategist Group predicts that revenue from wireless data will reach $33.5 billion globally by 2004. Frost & Sullivan expects that the proportion of operator revenues derived from non-voice services will be in excess of 45% by 2006. A UMTS Forum survey has estimated that non-voice revenues may take over voice revenues by 2004, while simple voice revenues will remain a critical revenue component comprising 34% of annual revenues in 2010. The UMTS Forum study also predicts that 3G revenues of $37.4 billion in 2004 will increase to $107 billion by 2006. All in all, predictions may vary but the consensus is clear that results will be positive for third-generation services.

This work offers the reader more than an evolutionary technical strategy of GSM's transition to 3G. It also provides a set of benchmarks for a core evaluation of the merits of EDGE as a central component of the wireless industry's fulfilment of its promise for higher degrees of service and convenience. This process is already being established, as evidenced by the first live EDGE data call completed by Nokia and AT&T Wireless on November 1, 2001. The connection of an EDGE handset with a laptop to the Internet for web browsing and streaming applications marked the first successful completion using EDGE 8-PSK modulation in both directions in the air interface. Indeed, it is another sign that EDGE will flourish in this new billion-dollar marketplace as a leading 3G technology in the Americas owing to its performance, cost, interoperability and ease of transformation. EDGE will outlast the neigh-sayers and in the long term, EDGE will far exceed expectations. And just as electricity, the automobile and the television changed our lives, EDGE will change our lives by providing 3G services for the masses.

Chris Pearson
Executive Vice President
3G Americas

I am honored to have been asked to provide a foreword and a few thoughts for the second edition of this book, which, although I am sure has been useful to the tried and true GSM operators, vendors, and researchers worldwide, has been particularly invaluable to those operators, like ourselves at Cingular Wireless, who have been intimately involved in actually deploying the latest new GSM networks. At Cingular Wireless, we have been especially fortunate to be among those who have 'tasted' the first fruits of many of the cutting-edge GSM advances and topics covered in this book, such as AMR, EDGE, QoS dimensioning and so on, as we, along with our four BSS vendors—Nokia, Ericsson, Siemens and Nortel—have worked together to turn the new technology described in this book into reality over the last two years.

Since 2002 to the date of this writing, Cingular has overlaid more than 10 000 GSM/EDGE base stations on top of its existing IS-136 TDMA network across the United States. All GSM base stations are now operating with live commercial traffic on AMR, while some selected markets are launching EDGE. In addition, we have pioneered the introduction of GSM into the 850-MHz band, which represents a significant milestone for GSM in penetrating the original cellular band in the United States, which has been home to four other air interfaces since the first in-service analog cellular networks were deployed in the early 1980s. Although providing spectrum for GSM in these 850 markets originally appeared as insurmountable, it has been systematically overcome and mastered by transforming some of the network simulators and tools described here in subsequent chapters, which were originally reserved for only research activity, into tools that can assist in the evaluation and planning of the high spectral efficiency techniques such as the tight ($\leq$12) BCCH and small frequency hopping pools with the high capacity (fractional load) needed for narrowband GSM deployment.

Throughout this GSM overlay journey at Cingular, our engineers have found the material in this book invaluable, as, in comparison to many other theoretical books or papers that simply explain the standards, this book provides many examples of real-world field trials to back up the theory and, in some cases, illustrate what is practically achievable. With our successful deployment of AMR and EDGE, Cingular is pleased to continue this tradition and provide, in some small part, a portion of our field-trial results on several of these new GSM advances herein.

Nevertheless, what is continually exciting is that the technology evolution and service applicability of GSM/EDGE is far from over, whereas, although much of the global focus has been on 'UMTS' and its promises, the GSM community has been quietly advancing the state of the art through continually 'upgrading the engine' on a tried and true frame.

For instance, in technology advancement, several new receiver algorithms have been discovered for the GMSK modulation used in GSM AMR/GPRS, which exploit some novel properties of the modulation to effectively create diversity with a 'virtual second antenna', which may then be used for interference cancellation. These ideas have spawned a feasibility study in GSM standards (GERAN) called *Single Antenna Interference Cancellation* (SAIC), which calls for applying these results for improved receiver performance in GSM mobile stations. Preliminary simulation and field-trial results are shown here with promising results, which, if deployed in synchronized networks could push GSM voice capacity above currently stated CDMA/WCDMA capacities. In the future, it would not come as a surprise if the SAIC ideas are extended to other modulations (e.g. 8-PSK used in EDGE), and other air interfaces as well.

In applications and data services, it seems that all operators and vendors, even from competing technologies, are still in the initial stages of attempting to move from 'best-effort' delivered data to those services that require the illusive end-to-end quality of service. New applications such as push-to-talk over GPRS/EDGE are just beginning to be deployed now as the move toward increasingly full conversational services over packet data are pursued. Success in this realm is, without any doubt, going to require new dimensioning, features and planning techniques, which are also described in subsequent chapters. It is these techniques coupled with finding the right mix of services that all operators and vendors strive for to keep the profitability and revenue stream intact while providing the additional functionality to maintain and attract new customers.

There is no doubt about the fact that many things are yet to be discovered in technology, planning and optimization of voice and data services as we enter a new realm where the Internet and wireless converge and people become used to 'always-on' services. Understanding how to provide these effectively with GSM/EDGE in a truly 'field-real' manner is the focus of this book. I trust that you will find it as useful and relevant as we have at Cingular Wireless.

Mark Austin, Ph.D.
Director of Strategic Technology
Cingular Wireless

Introduction

GERAN: Evolution towards 3G/UMTS

The wireless market has experienced a phenomenal growth since the first second-generation (2G) digital cellular networks, based on global system for mobile communications (GSM) technology, were introduced in the early 1990s. Since then, GSM has become the dominant global 2G radio access standard. Almost 80% of today's new subscriptions take place in one of the more than 460 cellular networks that use GSM technology. This growth has taken place simultaneously with the large experienced expansion of access to the Internet and its related multimedia services.

Cellular operators now face the challenge to evolve their networks to efficiently support the forecasted demand of wireless Internet-based multimedia services. In order to do this, they need to perform a rapid technology evolution rolling out new third-generation (3G) radio access technologies, capable of delivering such services in a competitive and cost-effective way. There are multiple recognised 3G technologies, which can be summarised into two main 3G evolution paths. The first one to be conceived and developed, and the more widely supported, is UMTS multi-radio (universal mobile telecommunications system), and the second one is cdma2000.

The first part of this book will describe the evolution GSM experienced after 5 years of continuous standardisation effort, and how the result of such an effort has provided GSM networks with a smooth, competitive and cost-efficient evolution path towards UMTS. The first steps of this evolution took place in Rel'97 standards, when general packet radio system (GPRS) was introduced to efficiently deliver packet-based services over GSM networks. Later, adaptive multi-rate (AMR), included in Rel'98, increased the spectral efficiency and quality-of-speech services to remarkable limits. Enhanced data rates for global evolution (EDGE) was introduced in Rel'99 and introduced more efficient modulation, coding and retransmission schemes, boosting the performance of data services. Release 99 also included the support of the UMTS traffic classes; hence providing the means for EDGE to support from the very beginning the same set of 3G services that UMTS was designed for. This synergy is completed in Rel'5 with the development of GERAN, a new radio access network architecture, based on GSM/EDGE radio access technologies, which is fully harmonised with UTRAN (UMTS terrestrial radio access network) through a common connectivity to the UMTS core network, and therefore integrated in the UMTS frame. Today, UMTS standardisation, carried out by the 3rd Generation Partnership Project (3GPP) standardisation body, jointly develops the GERAN and UTRAN as

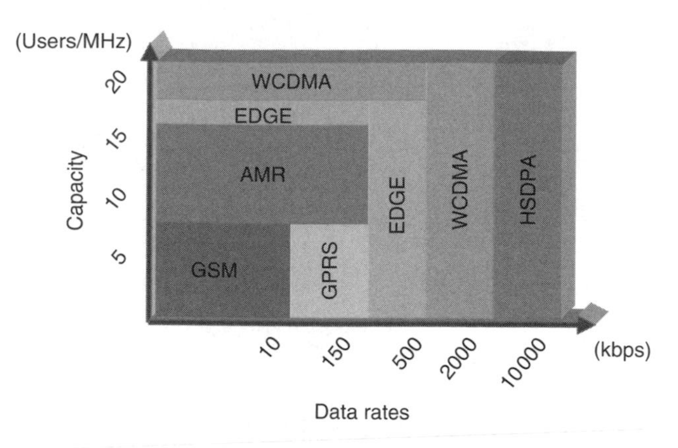

Figure 1. GSM evolution towards 3G/UMTS

part of the same concept, 3G UMTS multi-radio networks. This GSM evolution towards UMTS is illustrated in Figure 1.

Chapters 1 and 2 will provide a detailed description of the GSM evolution. Chapter 3 will provide an in-depth analysis of the GERAN quality-of-service (QoS) support in different releases. Finally, Chapter 4 will present the different standardised methods to support location services, their architecture and expected accuracy.

The second part of the book will thoroughly cover the performance efficiency of the technologies associated with the GSM evolution. Radio network performance is vital in order to maximise the return on investment from operators. Not only does it provide the ability to increase the capacity of the networks but also improves the QoS experienced by the end users. All the performance-related dimensions, such as link level voice and data performance, spectral efficiency, maximum and realistic data rates and service support capabilities are studied in depth in this part. Chapter 5 introduces the principles associated with GSM radio network performance analysis. All standardised functionality, such as AMR, SAIC, GPRS and EDGE, both for voice (Chapter 6) and data (Chapter 7) services, are carefully studied. Chapter 8 presents the relevant end-used performance aspects associated with the most relevant packet data services. Non-standardised solutions will be analysed as well in order to provide an overall perspective of the potential GSM/EDGE performance when advanced functionality such as dynamic channel allocation (Chapter 9) or link level enhancement (Chapter 11) techniques are introduced. Chapter 12 analyses the signalling channel's performance, since this may limit the overall network performance. A specific chapter on narrowband deployment scenarios (Chapter 10) will provide the guidelines to maximise the performance in these demanding radio conditions.

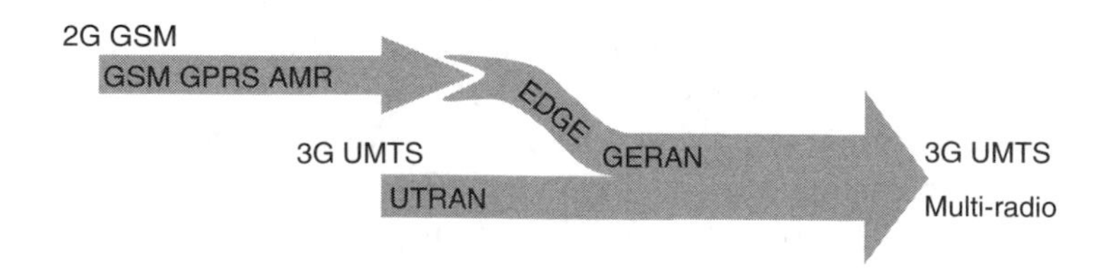

Figure 2. UMTS multi-radio technology building blocks

With the degree of complexity that the 3G evolution will bring, planning, optimising and operation of radio networks is a tough challenge that the operators have to tackle at a minimum cost. Chapter 13 presents new concepts based on automation and control engineering, designed to deal with all these issues, with functionality capable of fully automating the planning, optimisation and operation of radio networks, which not only decrease the operational expenditure (OPEX) of radio networks but maximise their performance as well.

Finally, the third part of the book introduces the principles of the main existing 3G radio technologies, benchmarking their technical capabilities and evolution paths. It covers in detail how GSM/EDGE and wideband code division multiple access' (WCDMA) frequency division duplex (FDD) can be efficiently and seamlessly integrated into the UMTS multi-radio network to ensure the achievement of a maximum performance at a minimum cost. Figure 2 illustrates the UMTS multi-radio technology building blocks.

The final chapter analyses the current global market dynamics and trends, in order to identify, from a global market perspective, the evolution paths available for different technologies.

For those readers interested in gaining a comprehensive insight into the contents of this book, Tartec (www.tartec.com) offers advanced training courses incorporating the structure and content of the book.

Abbreviations

1G	First Generation
2G	Second Generation
3G	Third Generation
4G	Fourth Generation
3GPP	3rd Generation partnership project
3GPP2	3rd Generation partnership project
8-PSK	Octagonal Phase Shift Keying
A	
A-GPS	Assisted GPS positioning system
AC	Admission Control
ACELP	Algebraic Code Excitation Linear Prediction
ACK	Acknowledge
ADPCM	Adaptive Differential Pulse Code Modulation
AFP	Automated Frequency Planning
AGCH	Access Grant Channel
AH	Antenna Hopping
ALI	Automatic Location Identification
AMH	Advanced Multilayer Handling
AMPS	American Mobile Phone System
AMR	Adaptive Multi-Rate codec
AoCC	Advice of Charge-Charging
AoCI	Advice of Charge-Information
AP	Access Point
APN	Access Point Name
API	Application Programming Interface
ARIB	Association of Radio Industries and Business in Japan
ARQ	Automatic Repeat Request
AT	Absolute Time
ATD	Absolute Time Difference
ATM	Asynchronous Transfer Mode
AuC	Authentication Center

B

BA	BCCH Allocation
BB	Baseband
BEP	Bit Error Probability
BER	Bit Error Rate
BCCH	Broadcast Control Channel
BDP	Bandwidth Delay Product
BGIWP	Barring of GPRS Interworking Profile(s)
BIM	Background Interference Matrix
BLER	Block Error Rate
BoD	Bandwidth on Demand
BS	Base station
BSIC	Base Station Identification Code
BSS	Base Station Subsystem
BSSAP	BSS Application Part
BSSGP	Base Station Subsystem GPRS protocol
BSC	Base Station Controller
BTS	Base Transceiver Station

C

CAMEL	Customised Application for Mobile Network Enhanced Logic
CC	Call Control
CCITT	Comité Consultatif International Télégraphique et Téléphonique
CCCH	Common Control Channel
CDMA	Code Division Multiple Access
CEPT	Conference Europeenne des Postes et Telecommunications
CFU	Call Forwarding Unconditional
CFNRc	Call Forwarding on Mobile Subscriber Not Reachable
CGW	Cell Gateway
CI	Cell Identity
C/I	Carrier/Interference
CIR	Carrier/Interference
CLNS	Connectionless Network Service
CM	Communication Management
CN	Core Network
CONS	Connection-Oriented Network Service
CRC	Cyclic Redundancy check
CRS	Cell Resource Server
CS	Circuit Switched
CS	Coding Scheme
CWTS	China Wireless Telecommunication Standard group in China

D

D-AMPS	Digital AMPS
DCA	Dynamic Channel Allocation

DCR	Dropped Call Rate
DCCH	Dedicated Control Channel
DFCA	Dynamic Frequency and Channel Allocation
DL	Down Link
DNS	Domain Name Server
DPCCH	Dedicated Physical Control Channel
DPDCH	Dedicated Physical Data Channel
DR	Directed Retry
DRX	Discontinuous Reception
DS-CDMA	Direct Sequence-Code Division Multiple Access
DTAP	Direct Transfer Application Process
DTM	Dual Transfer Mode
DTX	Discontinuous Transmission

E

E112	Emergency Service 911 in Europe
E911	Emergency Service 911 in US
ECSD	Enhanced Circuit Switched Data
EDGE	Enhanced Data Rates for Global Evolution
EFL	Effective Frequency Load
EFR	Enhanced Full Rate (speech coded)
EGPRS	Enhanced General Packet Radio System
EIR	Equipment Identity Register
E-OTD	Enhanced Observed Time Difference
ETSI	European Telecommunications Standards Institute

F

FACCH	Fast Associated Control Channel
FCCH	Frequency Correction Channel
FDD	Frequency Division Duplex
FER	Frame Error Rate
FH	Frequency Hopping
FLO	Flexible Layer One
FR	Full Rate
FTP	File Transfer Protocol

G

GAIT	GSM/ANSI Interoperability Team
GERAN	GSM EDGE Radio Access Network
GGSN	Gateway GPRS Support Node
GMLC	Gateway Mobile Location Center
GMSK	Gaussian Minimum Shift Keying
GPRS	General Packet Radio System
GPS	Global Positioning System
GRE	Generic Routing Encapsulation
GSM	Global System for Mobile Communications

GSN	GPRS Support Node
GTD	Geometric Time Difference
GTP	GPRS Tunnelling Protocol

H

HC	Handover Control
HLR	Home Location Register
HO	Handover
HPLMN	Home Public Land Mobile Network
HR	Half Rate
HSCSD	High Speed Circuit Switched Data
HSDPA	High Speed Downlink Packet Access
HSN	Hopping Sequence Number
HTML	HyperText Mark-up language
HTTP	HyperText Transfer Protocol
HW	Hardware

I

IETF	Internet Engineering Task Force
IFH	Intelligent Frequency Hopping
IM	Interference Matrix
IMEI	International Mobile Equipment Identity
IMGI	International Mobile Group Identity
IMSI	International Mobile Subscriber Identity
IP	Internet Protocol
IPv4	Internet Protocol version 4
IPv6	Internet Protocol version 6
IR	Incremental Redundancy
IS-54	First generation TDMA Radio Interface standard
IS-95	cdmaOne, one of 2nd generation systems, mainly in Americas and Korea
IS-136	US-TDMA, one of 2nd generation systems, mainly in Americas
ISDN	Integrated Services Digital Network
ISP	Internet Service Provider
ITU	International Telecommunication Union
ITU-T	Telecommunication standardization sector of ITU
IUO	Intelligent Underlay Overlay
IWMSC	Inter Working MSC

L

L2TP	layer Two Tunneling Protocol
LA	Link Adaptation
LAPD	Link Access Protocol for the D channel
LAPDm	Link Access Protocol for the Dm channel
LC	Load Control

LCS	Location Services
LIF	Location Interoperability Forum
LLC	Logical Link Control
LMU	Location Measurement Unit
M	
MA	Mobile Allocation
MAC	Medium Access Control
MAIO	Mobile Allocation Index Offset
MAP	Mobile Application Part
MCPA	Multi Carrier Power Amplifier
MCS	Modulation and Coding Scheme
ME	Mobile Equipment
MEHO	Mobile Evaluated Handover
MEO	Medium Earth Orbit
MHA	Mast Head Amplifier
MM	Mobility Management
MMS	Multimedia Messaging Service
MoU	Memorandum of Understanding
MPLS	Multi-protocol Label Switching
MRP	Multiple Reuse Pattern
MS	Mobile Station
MSC	Mobile Services Switching Centre
MTU	Maximum Transfer Unit
N	
NACC	Network Assisted Cell Change
NACK	Negative Acknowledgement
NC	Network Control
NCC	Network (PLMN) Colour Code
NCCR	Network Controlled Cell Re-selection
NMS	Network Management Subsystem
NSAPI	Network layer Service Access Point Identifier
NSS	Network and Switching Subsystem
O	
OMC	Operations and Maintenance Center
OSI	Open System Interconnection
OSS	Operation Sub-System
OTD	Observed Time Difference
P	
PACCH	Packet Associated Control Channel
PAGCH	Packet Access Grant Channel
PBCCH	Packet Broadcast Control Channel
PC	Power Control

PCCCH	Packet Common Control Channel
PCHM	Packet Channel Manager
PCU	Packet Control Unit
PCH	Paging Channel
PDCP	Packet Data Convergence Protocol
PDP	Policy Decision Point
PDP	Packet Data Protocol
PDU	Protocol Data Unit
PDTCH	Packet Data Traffic Channel
PEP	Policy Enforcement Point
PEP	Performance Enhancement Proxy
PFC	Packet Flow Context
PM	Policy Management
PNCH	Packet Data Notification Channel
PoC	Push to Talk over Cellular
PRACH	Packet Random Access Channel
PRM	Packet Resource Manager
PS	Packet Switched/Packet Scheduler
PSI	Packet System Information
PSPDN	Packet-Switched Public Data Network
PSTN	Public-Switched Telephone Network
PTM	Point-To-Multipoint
PTM-G	PTM Group call
PTM-M	PTM Multicast
PTM-SC	PTM Service Center
PTP	Point-To-Point

Q

QoS	Quality of Service
QR	Quarter Rate

R

RA	Routing Area
RAB	Radio Access Bearer
RACH	Random Access Channel
RANAP	Radio Access Network Application Part
RAT	Radio Access Technology
RB	Radio Bearer
RED	Random Early Detection
Rel'97	ETSI release 97
Rel'98	ETSI release 98
Rel'99	ETSI release 99
Rel'4	3GPP release 4
Rel'5	3GPP release 5
Rel'6	3GPP release 6
RLC	Radio Link Control

RM	Resource Manager
RNAS	Radio Access Network Access Server
RNC	Radio Network Controller
RR	Radio Resource
RRBP	Relative Reserved Block Period
RRLP	Radio Resource LCS Protocl
RRM	Radio Resource Management
RSC	Recursive Systematic Convolutional (coding)
RT	Real Time
RTCP	Real-time Control Protocol
RTD	Real-time Difference
RTO	Retransmission Timeout
RTP	Real-time Transport Protocol
RTSP	Real-time Streaming Protocol
RTT	Round Trip Time
RXLEV	Received Signal Level
RXQUAL	Received Signal Quality

S

SACCH	Slow Associated Control Channel
SAI	Service Area Identifier
SAIC	Single Antenna Interference Cancellation
SAPI	Service Access Point Indicator
SCH	Synchronisation Channel
SDCCH	Stand alone Dedicated Control Channel
SDP	Session Description Protocol
SE	Spectral Efficiency
SGSN	Serving GPRS Support Node
SI	System Information
SIM	GSM Subscriber Identity Module
SIP	Session Initiation Protocol
SM-SC	Short Message Service Center
SMSCB	Short Message Service Cell Broadcast
SMS-GMSC	Short Message Service Gateway MSC
SMS-IWMSC	Short Message Service Inter Working MSC
SMS-MO/PP	Mobile Originated Short Messages
SMS-MT/PP	Mobile Terminated Short Messages
SMS-PP	Point-to-Point Short Message Service
SMG	Special Mobile Group
SMLC	Serving Mobile Location Center
SMS	Short Message Service
SMSS	Switching and Management Sub-System
SMTP	Simple Mail Transfer Protocol
SNDCP	Sub Network Dependent Convergence Protocol
SNMP	Simple Network Management Protocol
SPSCH	Shared Physical Sub Channel

SS	Supplementary Services
SS7	Signalling System number 7

T

T1P1	North American Telecommunications Industry Association
T1P1.5	Sub Committee of T1P1
TA	Timing Advance
TAI	Timing Advance Index
TBF	Temporary Block Flow
TCH	Traffic Channel
TCP	Transport Control Protocol
TDD	Time Division Duplex
TDMA	Time Division Multiple Access
TF	Transport Format
TFC	Transport Format Combination
TFCI	Transport Format Combination Indicator
TFCS	Transport Format Combination Set
TFI	Temporary Flow Identity
TIA/EIA	Telecommunications Industry Alliance/Electronic Industries Alliance
TLLI	Temporary Logical Link Identity
TMSI	Temporary Mobile Subscriber Identity
TOA	Time of Arrival
TRAU	Transcoder/Rate Adapter Unit
TRHO	Traffic Reason Handover
TRX	Transceiver
TSL	Timeslot
TTA	Telecommunication Technology Association in Korea
TTC	Telecommunications Technology Committee in Japan
TU3	Typical Urban 3 km/h
TU50	Typical Urban 50 km/h

U

UDP	User Datagram Protocol
UE	User Equipment
UEP	Unequal Error Protection
UL	Uplink
UMTS	Universal Mobile Telecommunication services
URL	Uniform Resource Locator
USF	Uplink State Flag
USIM	Universal Subscriber Identity Module
UTD	Uniform Theory of Diffraction
UTRAN	UMTS Terrestrial Radio Access Network

V

VAD	Voice Activation Detector
VPN	Virtual Private Network

VoIP	Voice over IP
VLR	Visiting Location Register
VPLMN	Visited Public Land Mobile Network

W

WAP	Wireless Application Protocol
WCDMA	Wideband CDMA
WLAN	Wireless LAN
WML	WAP Mark-up Language
WSP	Wireless Session Protocol
WTP	Wireless Transport Protocol
WWW	World Wide Web

X

X.25	An ITU-T Protocol for Packet Switched Network
XHMTL	eXtensible HyperText Mark-up language

Part 1

GERAN Evolution

The standardisation of global system for mobile telecommunications (GSM) has been an ongoing process during the last decade, involving more than 100 different wireless operators and manufacturers around the world. The first phases of GSM standards were carried out by the European Telecommunications Standard Institute (ETSI), but today, the 3rd Generation Partnership Project (3GPP) jointly standardises GSM EDGE radio access network (GERAN) and UMTS terrestrial radio access network (UTRAN) as part of the UMTS multi-radio networks concept.

The first part of this book focuses on the standardised technical evolution of GERAN providing GSM networks with a smooth, competitive and cost-efficient evolution path towards UMTS.

Chapter 1 provides an insight about the status of the standardisation up to Rel'4. It covers circuit-switched evolution, including high-speed circuit-switched data (HSCSD) and adaptive multi-rate (AMR) and packet-switched general packet radio system (GPRS) definition. Chapter 2 focuses on further evolution, covering Rel'5 and Rel'6 new GERAN architecture and beyond. Chapter 3 provides an overview about the quality-of-service (QoS) architecture of GSM/EDGE for different releases and how the UMTS QoS has been efficiently adopted by GERAN. Finally, Chapter 4 covers the location service's methods and architectures.

1

GSM/EDGE Standards Evolution (up to Rel'4)

Markus Hakaste, Eero Nikula and Shkumbin Hamiti

The standardisation work of GSM-based systems has its roots in the 1980s, when a standardisation body 'Groupe Special Mobile' (GSM)[1] was created within the Conference Europeenne des Postes et Telecommunications (CEPT), whose task was to develop a unique digital radio communication system for Europe, at 900 MHz.

Since the early days of GSM development, the system has experienced extensive modifications in several steps to fulfil the increasing demand from the operators and cellular users. The main part of the basic GSM system development during the last decade until spring 2000 has been conducted in European Telecommunications Standards Institute (ETSI) Special Mobile Group (SMG) and its technical sub-committees, as well as in T1P1, which had the responsibility of the PCS1900 MHz specifications in the United States.

Currently, the further evolution of the GSM-based systems is handled under the 3rd Generation Partnership Project (3GPP), which is a joint effort of several standardisation organisations around the world to define a global third-generation UMTS (universal mobile communication system) cellular system. The main components of this system are the UMTS terrestrial radio access network (UTRAN) based on wideband code division multiple access (WCDMA) radio technology and GSM/EDGE radio access network (GERAN) based on global system for mobile communications (GSM)/enhanced data rates for global evolution (EDGE) radio technology.

In the following sections, an introduction of the evolution steps of the GSM specifications is presented.

[1] The original GSM acronym is 'Groupe Special Mobile'. This changed afterwards to Global System for Mobile communication, which is the current official acronym.

GSM, GPRS and EDGE Performance 2nd Ed. Edited by T. Halonen, J. Romero and J. Melero
 ISBN: 0-470-86694-2

1.1 Standardisation of GSM—Phased Approach

The need for continuous development of the GSM specifications was anticipated at the beginning of the specification work, which was, as a consequence, split into two phases. This phased approach was defined to make sure that specifications supported a consistent set of features and services for multi-vendor operation of the GSM products, both on the terminal and the network sides.

The GSM Phase 1 work included most common services to enable as fast as possible deployment of GSM in operating networks still providing a clear technological advance compared with existing analogue networks. The main features in Phase 1 include support for basic telephony, emergency calls, 300 to 9600 kbps data service, ciphering and authentication as well as supplementary services like call forwarding/barring. Also, short message service (SMS) was also included at this early phase of the specification work, although its commercial success came much later. From a radio performance point of view, frequency hopping, power control and discontinuous transmission were terminal mandatory from GSM Phase 1.

While the GSM Phase 1 networks were being built, GSM Phase 2 was being specified in ETSI SMG. The GSM Phase 2 specifications were frozen in October 1995, and they included a mechanism for cross-phase compatibility and error handling to enable evolution of the specifications. Many technical improvements were also introduced and included several new supplementary services like line identification services, call waiting, call hold, advice of charge and multi-party call. In the speech area, a half-rate channel mode codec was introduced to complement the already specified speech codec for GSM full-rate channel mode. In the data side, the main improvement included group 3 fax service support.

The first two phases of GSM provided a solid basis for the GSM system evolution towards the third-generation (3G) system requirements and items, which were, at the beginning of the work, better known as Phase 2+ items. In the core network (CN) side, the evolution led to the introduction of general packet radio system (GPRS) network architecture, especially designed for Internet connectivity. Similarly, the radio access network experienced significant enhancements in both packet and circuit modes to provide higher bit rates and better network capacity, in both packet and circuit-switched modes. From the users' perspective, in addition to better speech quality, this evolution consisted of significant enhancements to the capabilities to provide speech and data services. The following list gives an overview of the different work item categories:

- new bearer services and data-related improvements like high-speed circuit-switched (multislot) data (HSCSD), 14.4-kbps (single-slot) data, general packet radio service (GPRS) and enhanced data rates for global evolution (EDGE);
- speech-related items like enhanced full-rate (EFR) speech codec, adaptive multi-rate codec (AMR) with both narrow and wideband options, and tandem free operation (TFO) of speech codecs;
- mobile station (MS) positioning–related items like cell identity and timing advance, uplink time of arrival (TOA) and enhanced observed time difference (E-OTD) methods based on measurements within the cellular network, and assisted GPS method based on the GPS (global positioning system) technology;

- frequency band–related items like GSM400, 700, and 850 MHz, unrestricted GSM multi-band operation, release independent support of frequency bands;
- messaging-related items like SMS concatenation, extension to alphabet, SMS inter-working extensions, forwarding of SMSs and particularly multimedia messaging (MMS);
- new supplementary services like call deflection, calling-name presentation, explicit call transfer, user-to-user signalling, completion of calls to busy subscriber and new barring services;
- billing-related work items like payphone services, provision for hot billing and support for home area priority;
- service platforms like subscriber identity module (SIM) application toolkit and customised application for mobile network enhanced logic (CAMEL).

Although the phasing is used also when developing 3G standards, grouping into 'Releases' (later in this book referred to as Rel') is a more practical tool, currently used in the standardisation to manage changes between different specification versions and introduction of new features. New releases are normally produced every year, and they contain a set of all specifications having added functionality introduced to it compared with previous releases as a result of ongoing standardisation work [1].

1.1.1 GSM/TDMA Convergence through EDGE

Standardisation of EDGE was the main cornerstone for integrating the two major time division multiple access (TDMA) standards, GSM specified in ETSI and IS-136 specified in Telecommunications Industry Alliance/Electronic Industries Alliance (TIA/EIA) on the same evolution path. At the moment, new frequency bands are not available in the United States, so an evolutionary approach to the new services on existing frequency bands is important for the current US operators.

TIA/EIA-136 is a result of the evolution of the US cellular networks. While GSM was digital from the beginning, TIA/EIA-136 has its roots in analogue advanced mobile phone service (AMPS EIA-553), which has been digitised in stages. The first step was the introduction of digital traffic channels (TCHs) in the IS-54 standard version. The IS-136 releases introduced digital signalling channels, improved voice quality with an enhanced voice codec and standardised dual-band operation. Like in Europe, the next wave in complementing the capabilities of TIA/EIA-136 in the United States was targeted in finding suitable 3G high data rate solution.

The start of EDGE development goes back to 1997, when ETSI conducted a feasibility study on improved data rates for GSM evolution. At the same time, the Universal Wireless Communication Consortium's (UWCC) Global TDMA Forum prepared input for TIA's ITU IMT-2000 programme. In 1998, key technical parameters of these two developments were harmonised, forming a basis for a converged TDMA standard using 200-kHz carrier and octagonal phase shift keying (8-PSK) modulation–based radio interface.

In Europe, the work continued under ETSI SMG2's EDGE work items and the first-phase EDGE standard, including enhanced circuit-switched data (ECSD) based on HSCSD, and enhanced general packet radio service (EGPRS) based on GPRS, was finalised in spring 2000. In addition, the specifications supported an optional arrangement called *EDGE Compact*, which enabled the deployment of EDGE in as narrow as 1-MHz frequency bands.

In the United States, the work continued in TIA TR45 as further development of EDGE as part of the UWC-136 IMT-2000 proposal. In November 1999, the UWC-136 proposal was approved as a radio interface specification for IMT-2000. Currently, EDGE is part of the GERAN development that is carried out concurrently in 3GPP and UWCC to ensure high synergy between GSM and TDMA systems in the future.

The latest developments in the US market from the two largest IS-136 operators, AT&T and Cingular, define the mainstream evolution path likely to be followed by most IS-136 operators. IS-136 networks will fully migrate to GSM/EDGE/WCDMA UMTS technology making use, in some cases, of the GAIT (GSM/ANSI-136 Interoperability Team) functionality to ensure a smooth migration [2]. As a result of this, EDGE Compact is not likely to ever be implemented, since the evolution timing and the roaming requirements do not make this option interesting for operators.

1.1.2 GERAN Standardisation in 3GPP

In summer 2000, the specification work of GSM radio was moved from ETSI SMG2 to 3GPP, which is a project responsible of the UMTS standards based on the evolved GSM core network. This meant that Rel'99 was the last release of the GSM/EDGE standard that was specified in ETSI SMG, and all specifications related to GSM radio access network were moved under 3GPP responsibility with new specification numbers.

The moving of the specifications was motivated by both the need for improved working procedures and technical development closer to 3GPP. The core network part of the GSM specifications was transferred to 3GPP by the time the project work was initiated, and the arrangement of standardising the GSM network and radio access parts in two different standardisation bodies made it too cumbersome to work effectively. More importantly, activities aimed at closer integration of GSM/EDGE radio and WCDMA technologies had led to a decision to adopt the 3GPP-specific Iu interface for GERAN. Thereby, this integration achieves a true multi-radio UMTS standard, which is made up of two radio technologies, WCDMA and GSM/EDGE, that can be effectively and seamlessly integrated in order to maximise the efficiency and the quality of service provided to the end users with the introduction of the coming 3G multimedia services.

1.1.2.1 3GPP Organisation

The 3GPP is a joint effort of several standardisation organisations around the world targeting to produce specifications for a global 3G cellular system. These so-called organisational partners (OPs) have the authority to develop standards in their region, and they currently include ETSI in Europe, Association of Radio Industries and Business (ARIB) in Japan, Standardisation Committee T1—Telecommunications (T1) in the United States, Telecommunication Technology Association (TTA) in Korea, China Wireless Telecommunication Standard group (CWTS) in China and Telecommunications Technology Committee (TTC) in Japan.

3GPP also have market representation partners (MRP), whose role is to make sure that the standardisation is in line with the market requirements. In practice, the MRPs work together with OPs in the Project Coordination Group (PCG) to guide the overall direction of the 3GPP work. Potential future partners can also follow the work as observers.

More information and up-to-date links about the 3GPP-related activities and organisation can be found in http://www.3gpp.org/.

Most of the technical work in 3GPP is carried out by individual members, typically 3G operators and vendors, in technical specification groups (TSGs) and their working groups (WGs). The basic 3GPP work organisation is depicted in Figure 1.1.

Currently there are five TSGs in 3GPP:

- *TSG SA*—Service and System Aspects, taking care of the overall system architecture and service capabilities, and cross TSG coordination on these areas. Issues related to security, speech and multimedia codecs, network management and charging are also handled in SA.
- *TSG CN*—Core Network, taking care of the UMTS/GSM core network protocols. This includes the handling of layer 3 call control, mobility management and session management protocols between the mobile terminal and CN, as well as inter-working with external networks.
- *TSG T*—Terminals, taking care of the aspects related to the terminal interfaces. This includes the responsibility of issues like universal subscriber identity module (USIM), terminal equipment (TE) performance, service capability protocols, end-to-end service inter-working and messaging.
- *TSG RAN*—Radio Access Network, taking care of the aspects related to UMTS terrestrial radio access network. This includes the handling of specifications for the WCDMA-based radio interface towards the WCDMA terminal, RAN-specific interfaces Iub and Iur, and CN connectivity over Iu interface.
- *TSG GERAN*—GSM/EDGE Radio Access Network, taking care of the aspects related to the GSM/EDGE-based radio interface including testing for GERAN terminals,

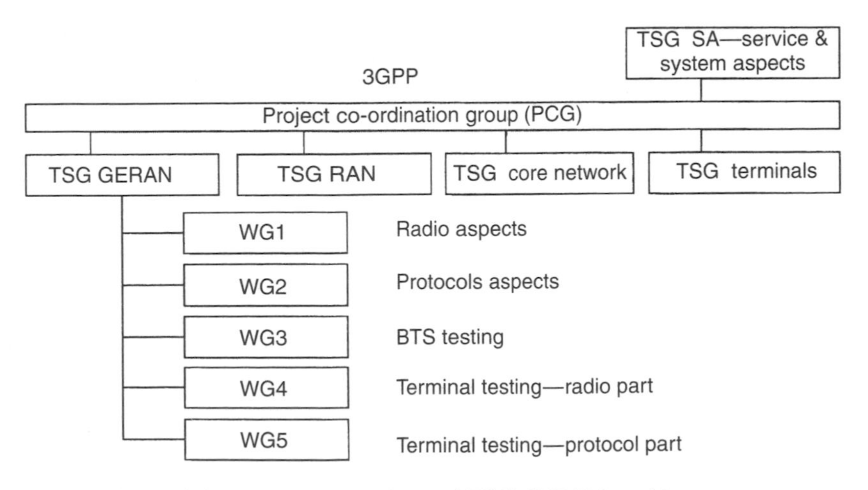

Figure 1.1. 3GPP organisation and TSG GERAN working groups

GERAN internal interfaces as well as external interfaces related to the connections towards the legacy GSM CN.

TSGs have the mandate to approve and modify technical specifications in their area. Although TSGs are reasonably independent in terms of making decisions, liaising between groups is intended to ensure the technical consistency and timely finalisation of the specifications for different releases. For TSG GERAN, this includes close cooperation, particularly with TSG SA on issues related to 3G bearer capabilities, architecture and security, and with TSG RAN on issues related to the common interfaces (Iu, Iur) and protocols (Packet Data Convergence Protocol (PDCP)) under TSG RAN responsibility, as well as overall optimisation of GSM/EDGE—WCDMA inter-operation.

While the TSGs carry the main responsibility for approval of the documents and overall work coordination, the detailed level technical work takes place in WGs. In TSG GERAN, there are currently five WGs:

- *WG1*—Radio Aspects, specifically having the responsibility over the specifications related to GSM/EDGE layer 1, radio frequency (RF), radio performance and internal Ater specification between Channel Codec Unit (CCU) and Transcoder and Rate Adapter Unit (TRAU).
- *WG2*—Protocols Aspects, specifically covering issues related to GERAN layer 2, like radio link control (RLC)/medium access control (MAC) and three radio resource (RR) specifications, A and Gb interfaces towards 2G CN and internal Abis specification between base station controller (BSC) and base transceiver station (BTS).
- *WG3*—Base Station Testing and O&M, taking care of all aspects related to conformance testing of GERAN BTSs, as well as GERAN-specific operation and maintenance issues for all GERAN nodes.
- *WG4*—Terminal Testing (Radio part), responsible for creating the specifications for the conformance testing of GERAN terminals in the area of radio interface layer 1 and RLC/MAC.
- *WG5*—Terminal Testing (Protocol part), responsible for creating the specifications for the conformance testing of GERAN terminals in the area of protocols above RLC/MAC.

1.2 Circuit-switched Services in GSM

The main drivers for the further development of the speech services and codecs are the user demand for better speech quality and the capacity gains enabling cost savings for network operators. Enhanced full-rate codec was the first significant improvement in speech quality when it was introduced in ETSI specifications in 1996 as well as in IS-641 standards in the United States during the same year. While the old GSM half-rate (HR) and full-rate (FR) codecs lacked the quality of the wireline telephony (32-kbps adaptive differential pulse code modulation (ADPCM)), EFR provided equivalent quality as a normal fixed connection even in typical error conditions. The EFR codec was jointly developed by Nokia and University of Sherbrooke and it has been widely used in operating GSM networks worldwide.

In spite of the commercial success of EFR, the codec left room for further improvements. In particular, the performance of the codec in severe radio channel error conditions could have been better. In addition to this, GSM half-rate codec was not able to provide adequate speech quality, so it was decided to continue the codec specification work with a new codec generation [3].

1.2.1 Adaptive Multi-rate Codec (AMR)

In October 1997, a new programme was initiated in ETSI, aimed at the development and standardisation of AMR for GSM system. Before the standardisation was officially started, a one-year feasibility study was carried out to validate the AMR concept. The main target of the work was to develop a codec that provided a significant improvement in error robustness, as well as capacity, over EFR.

The actual AMR codec standardisation was carried out as a competitive selection process consisting of several phases. In February 1999, ETSI approved the AMR codec standard, which was based on the codec developed in collaboration between Ericsson, Nokia and Siemens. Two months later, 3GPP adopted the AMR codec as the mandatory speech codec for the 3G WCDMA system. Some parts of the AMR codec work, such as voice activity detection (VAD) and optimised channel coding were finalised and included in the standard later, in June 1999.

The AMR codec contains a set of fixed-rate speech and channel codecs, fast in-band signalling and link adaptation. The AMR codec operates both in the full-rate (22.8 kbps) and half-rate (11.4 kbps) GSM channel modes. An important part of AMR is its ability to adapt to radio channel and traffic load conditions and select the optimum channel mode (HR or FR) and codec mode (bit rate trade-off between speech and channel coding) to deliver the best possible combination of speech quality and system capacity.

Even though the AMR codec functionality is not mandatory in Rel'98-compliant terminals, the large benefits this functionality will bring will probably drive the demand for the support of AMR codec in all terminals.

1.2.1.1 Speech and Channel Coding

The AMR speech codec utilises the algebraic code excitation linear prediction (ACELP) algorithm employed also in GSM EFR and D-AMPS EFR codecs. The codec is actually a combination of eight speech codecs with bit rates of 12.2, 10.2, 7.95, 7.4, 6.7, 5.9, 5.15 and 4.75 kbps. Table 2.1 in Chapter 2 summarises AMR speech codecs supported in the different releases. AMR narrowband (AMR NB) with Gaussian minimum shift keying (GMSK) modulation is supported already in Rel'98. Each codec mode provides a different distribution of the available bit rate between speech and channel coding. All of the codecs are defined for the full-rate channel mode, while the six lowest ones are defined also for the half-rate channel mode.

Channel coding performs error correction and bad-frame detection. The error correction in all the codec modes is based on the recursive systematic convolutional (RSC) coding with puncturing to obtain the required bit rates. Each codec mode utilises a 6-bit cyclic redundancy check (CRC) for detecting bad frames. In order to maximise the commonality with the existing GSM system, all channels use polynomials used for the previous GSM traffic channels.

1.2.1.2 In-band Signalling and Link Adaptation

In the basic AMR codec operation, shown in Figure 1.2, both the mobile station (MS) and the base transceiver station (BTS) perform channel quality estimation of the received signal. On the basis of the channel quality measurements, a codec mode command (downlink to the MS) or codec mode request (uplink to the BTS) is sent over the radio interface in in-band messages. The receiving end uses this information to choose the best codec mode for the prevailing channel condition. A codec mode indicator is also sent over the radio to indicate the current mode of operation in the sending side. The basic principle for the codec mode selection is that the mode chosen in the uplink may be different from the one used in the downlink direction, but the channel mode (HR or FR) must be the same.

The benefit of the in-band method is that it does not require a separate signalling channel for the message transfer. By sending the messages and the indicators together with the speech payload, the link adaptation operation can also be made faster, leading to improvements in the system performance.

The network controls the uplink and the downlink codec modes and channel modes. The MS must obey the codec mode command from the network, while the network may use any complementing information, in addition to codec mode request, to determine the downlink codec mode. The MS must implement all the codec modes. However, the network can support any combination of them, on the basis of the choice of the operator.

AMR also contains voice activity detection and discontinuous transmission (VAD/DTX). These are used to switch off the encoding and transmission during periods of silence, thereby reducing radio interference and extending the battery lifetime.

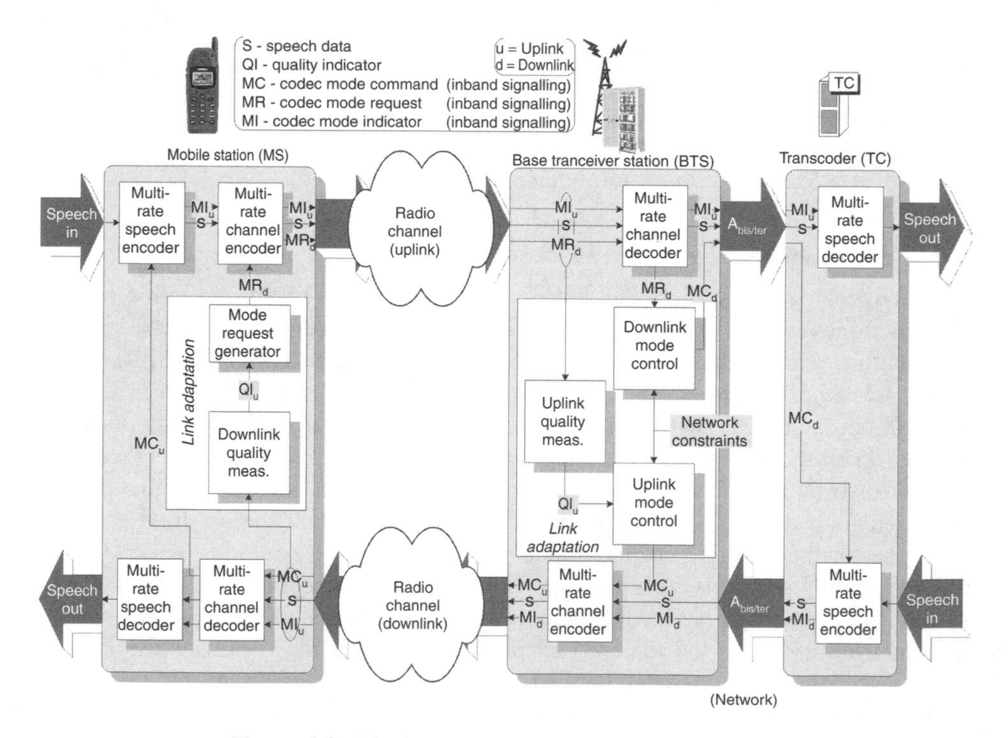

Figure 1.2. Block diagram of the AMR codec system

1.2.2 High Speech Circuit-switched Data (HSCSD)

The first-phase GSM specifications provided only basic transmission capabilities for the support of data services, with the maximum data rate in these early networks being limited to 9.6 kbps on one timeslot. HSCSD specified in Rel'96 was the first GSM Phase 2+ work item that clearly increased the achievable data rates in the GSM system. The maximum radio interface bit rate of an HSCSD configuration with 14.4-kbps channel coding is 115.2 kbps, i.e. up to eight times the bit rate on the single-slot full-rate traffic channel (TCH/F). In practice, the maximum data rate is limited to 64 kbps owing to core network and A-interface limitations.

The main benefit of the HSCSD feature compared to other data enhancements introduced later is that it is an inexpensive way to implement higher data rates in GSM networks owing to relatively small incremental modifications needed for the network equipment. Terminals, however, need to be upgraded to support multislot capabilities. The basic HSCSD terminals with relatively simple implementation (in size and cost compared to a single-slot terminal) available in the market today can receive up to four and transmit up to two timeslots and thus support data rates above 50 kbps. Figure 1.3 depicts the HSCSD network architecture based on the concept of multiple independent TCH/F channels.

In HSCSD, a new functionality is introduced at the network and terminals to provide the functions of combining and splitting the user data into separate n data streams, which will then be transferred via n channels at the radio interface, where $n = 1, 2, 3, \ldots, 8$. Once split, the data streams shall be carried by the n full-rate traffic channels, called *HSCSD channels*, as if they were independent of each other, for the purpose of data relay and radio interface L1 error control, until the point in the network where they are combined. However, logically the n full-rate traffic channels at the radio interface belong to the same HSCSD configuration, and therefore they shall be controlled as one radio link by the network for the purpose of cellular operations, e.g. handover.

For both transparent and non-transparent HSCSD connections, the call can be established with any number of TCH/F from one up to the maximum number of TCH/F, i.e. the minimum channel requirement is always one TCH/F.

If the wanted air interface user rate requirement cannot be met using a symmetric configuration, an asymmetric configuration can be chosen. The network shall, in this case, give priority to fulfilling the air interface user rate requirement in the downlink direction.

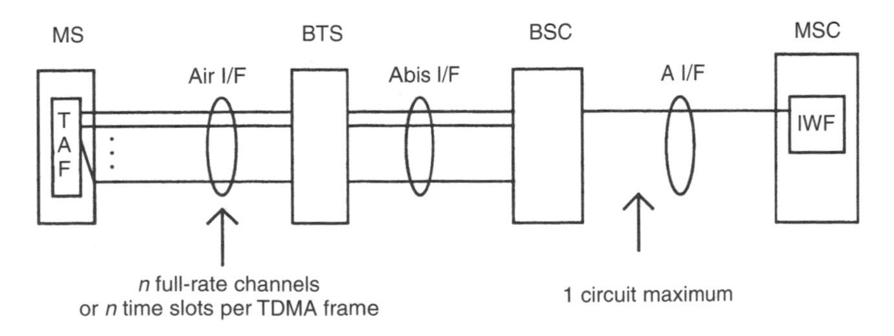

Figure 1.3. Network architecture for supporting HSCSD

The MS may request a service level upgrading or downgrading during the call, if so negotiated in the beginning of the call. This modification of channel requirements and/or wanted air interface user rate is applicable to non-transparent HSCSD connections only.

1.2.2.1 Radio Interface

Two types of HSCSD configurations exist: symmetric and asymmetric. For both types of configurations, the channels may be allocated on either consecutive or non-consecutive timeslots, taking into account the restrictions defined by the mobile station's multislot classes, described in detail in [2]. An example of the HSCSD operation with two consecutive timeslots is shown in Figure 1.4.

A symmetric HSCSD configuration consists of a co-allocated bi-directional TCH/F channel. An asymmetric HSCSD configuration consists of a co-allocated uni-directional or bi-directional TCH/F channel. A bi-directional channel is a channel on which the data are transferred in both uplink and downlink directions. On uni-directional channels for HSCSD, the data are transferred in downlink direction only. The same frequency-hopping sequence and training sequence is used for all the channels in the HSCSD configuration.

In symmetric HSCSD configuration, individual signal level and quality reporting for each HSCSD channel is applied. For an asymmetric HSCSD configuration, individual signal level and quality reporting is used for those channels. The quality measurements reported on the main channel are based on the worst quality measured among the main and the uni-directional downlink timeslots used. In both symmetric and asymmetric HSCSD configuration, the neighbouring cell measurement reports are copied on every uplink channel used. See [4] for more detail on signal level and quality reporting.

Transparent Data Transmission

In transparent data transmission, the V.110 data frames on the HSCSD channels carry data sub-stream numbers to retain the order of transmission over GSM, between the split/combine functions. Between these functions, channel internal multiframing is also used in order to increase the tolerance against inter-channel transmission delays. Depending on the location of the access point to external networks, the split/combine functionality is located at the base station sub-system (BSS) or in the IWF on the network side, and at the MS.

Non-transparent Data Transmission

Non-transparent mode of HSCSD is realised by modifying the radio link protocol (RLP) and L2R functions to support multiple parallel TCH/Fs instead of only one TCH/F

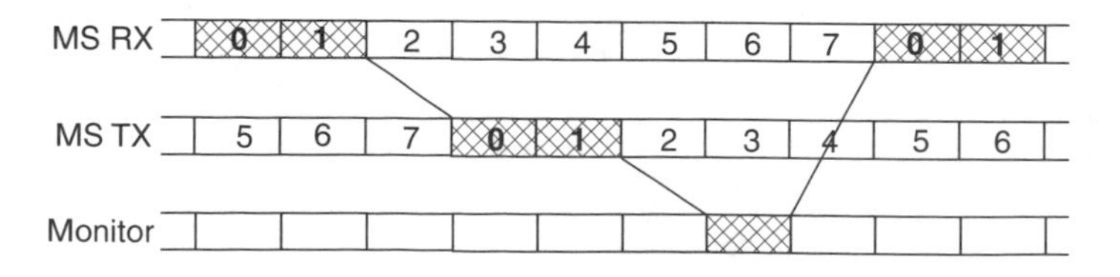

Figure 1.4. Double-slot operation in the air interface

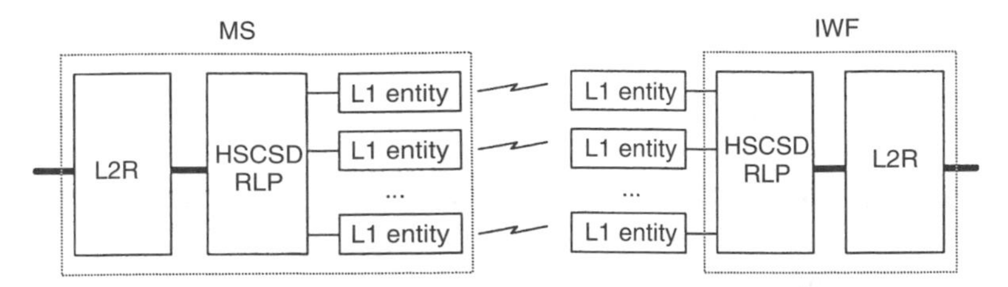

Figure 1.5. The HSCSD concept in non-transparent mode

(Figure 1.5). In addition, the RLP frame numbering is increased to accommodate the enlarged data-transmission rate.

1.3 Location Services

Different positioning methods found the basis for providing location services (LCS) in the GSM system. In general, these services can be implemented by using vendor-/operator-specific or standardised solutions. Both of these methods and their combination are likely to be used when LCS enter the operating networks.

The vendor-/operator-specific solutions utilise the information already available in the cellular network or measured by proprietary units installed for this purpose. The user location is thus estimated by combining and processing the available information with the help of tools and algorithms implemented for this purpose. In the standardised solution, the basic difference compared to the specific one is that the functionalities and capabilities required to retrieve the information for positioning are integrated into the system itself. The integrated solution affects the specifications and thus standardisation effort is needed, but benefit of this approach is that it enables multi-vendor support and provides a complete support for user privacy. In this chapter, a short overview of the history and drivers behind the standardisation of the location services is given. Chapter 4 describes in more detail the positioning methods, architecture and main technical principles behind them.

1.3.1 LCS Standardisation Process

The LCS standardisation began in ETSI SMG in 1995 after European operators asked for commercial services and separately in T1P1 in the United States after Federal Communications Commission (FCC) established the E911 requirements for the E911 service. The E911 requirement stated that by October 1, 2001, US operators must be able to locate the subscriber calling the emergency number 911.

However, several operators have applied and received waivers for the E911 requirements, and in spite of the different drivers behind the work in the United States and Europe, T1P1.5 (T1P1 is a subgroup of T1, a committee of TIA, the North American Telecommunications Industry Association) was given the responsibility to lead the overall standardisation activities in 1997. Four location methods were proposed and taken to the standardisation process:

- cell identity and timing advance;
- uplink time of arrival (TOA);

- enhanced observed time difference (E-OTD) methods based on measurements within the cellular network; as well as
- assisted GPS method based on the GPS technology.

In spring 1999, all four methods were included in the GSM specifications Rel'98 and Rel'99, including changes to GSM architecture and introduction of several new network elements for LCS purposes. Later in 2001, TOA method was excluded from the Rel'4 specifications because no commercial support for the scheme was foreseen because of the complexity in implementing the method in operating networks.

The first specification versions (Rel'98, Rel'99 and Rel'4) include the LCS support for circuit-switched connections over the A-interface towards the 2G CN only. In Rel'5, the focus is then moved towards providing the same LCS support in puncturing scheme (PS) domain connections over Gb and Iu interfaces. The main reasons for this gradual introduction of different modes were not only the regulatory requirements that, in practice, were applicable for coding scheme (CS) speech connections only but also the delays in the introduction of GPRS and its interfaces. When the penetration of GPRS users and terminals increases, LCS is expected to be more and more an attractive increment to the services provided to the GPRS subscribers in 'always connected' mode.

1.4 General Packet Radio System (GPRS)

1.4.1 Introduction of GPRS (Rel'97)

Soon after the first GSM networks became operational in the early 1990s and the use of the GSM data services started, it became evident that the circuit-switched bearer services were not particularly well suited for certain types of applications with a bursty nature. The circuit-switched connection has a long access time to the network, and the call charging is based on the connection time. In packet-switched networks, the connections do not reserve resources permanently, but make use of the common pool, which is highly efficient, in particular, for applications with a bursty nature. The GPRS system will have a very short access time to the network and the call charging could solely be based on an amount of transmitted data.

The GPRS system brings the packet-switched bearer services to the existing GSM system. In the GPRS system, a user can access the public data networks directly using their standard protocol addresses (IP, X.25), which can be activated when the MS is attached to the GPRS network. The GPRS MS can use between one and eight channels over the air interface depending on the MS capabilities, and those channels are dynamically allocated to an MS when there are packets to be sent or received. In the GPRS network, uplink and downlink channels are reserved separately, making it possible to have MSs with various uplink and downlink capabilities. The resource allocation in the GPRS network is dynamic and dependent on demand and resource availability. Packets can also be sent on idle time between speech calls. With the GPRS system, it is possible to communicate point-to-point (PTP) or point-to-multipoint (PTM); it also supports the SMS and anonymous access to the network. The theoretical maximum throughput in the GPRS system is 160 kbps per MS using all eight channels without error correction.

1.4.2 GPRS Network Architecture

In Figure 1.6, a functional view of the GPRS network is displayed. GPRS brings a few new network elements to the GSM network. The most important ones are the serving GPRS support node (SGSN) and the gateway GPRS support node (GGSN). Another important new element is the point-to-multipoint service centre (PTM-SC), which is dedicated to the PTM services in the GPRS network. Another new network element is the border gateway (BG), which is mainly needed for security reasons and is situated on the connection to the inter-PLMN backbone network. The inter-PLMN and intra-PLMN backbone networks are also new elements, both Internet protocol–based (IP-based) networks. In addition, there will be a few new gateways in the GPRS system like the charging gateway and the legal intcrception gateway.

While the current GSM system was originally designed with an emphasis on voice sessions, the main objective of the GPRS is to offer an access to standard data networks such as transport control protocol (TCP)/Internet protocol (IP) and X.25. These other networks consider GPRS just as a normal sub-network (Figure 1.7). A GGSN in the GPRS network behaves as a router and hides the GPRS-specific features from the external data network.

The mobile user can have either a static or a dynamic data network address, and the data traffic will always use the gateway indicated by this address (Figure 1.8). However, the home network operator could force all the traffic to use a home GGSN, for example, for security reasons.

A static address is permanently allocated for one subscriber. As it will point to a gateway of the home network, the data packets will always be routed through the home network. Figure 1.8 illustrates the case where a user is in his home network (case 1) and in a visited network (case 2).

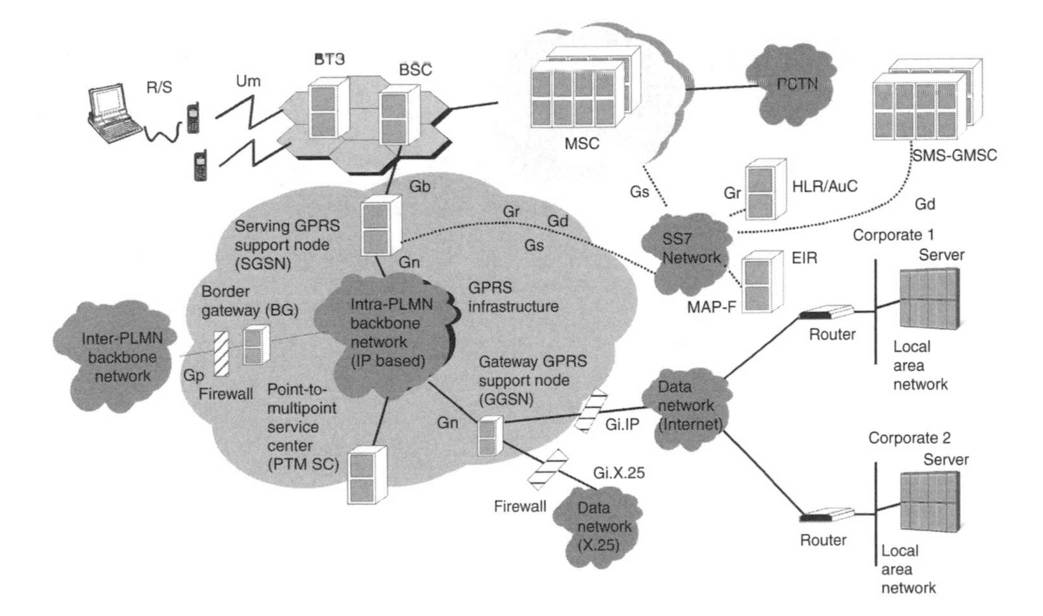

Figure 1.6. Functional view of GPRS

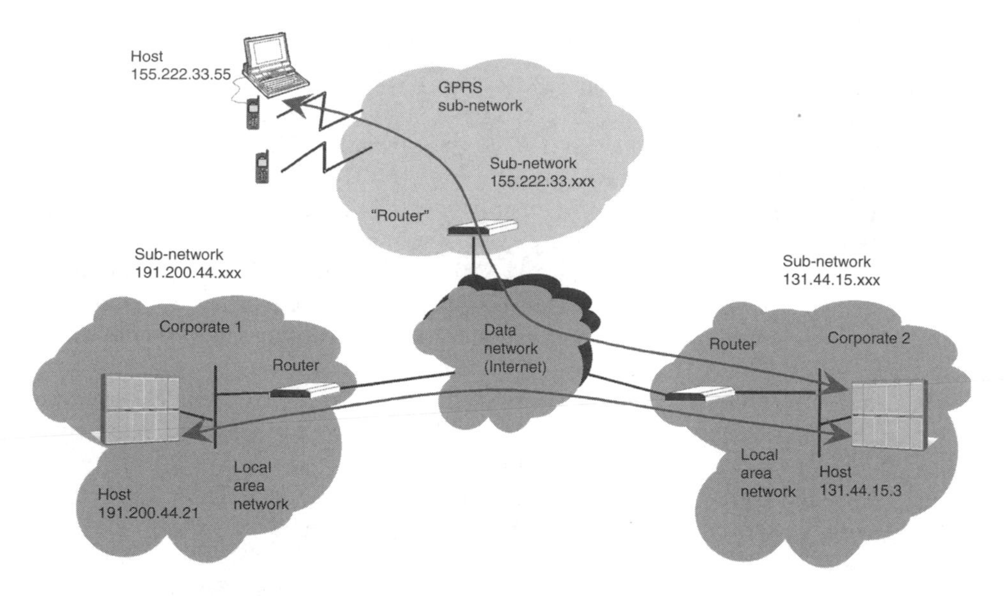

Figure 1.7. GPRS network seen by another data network

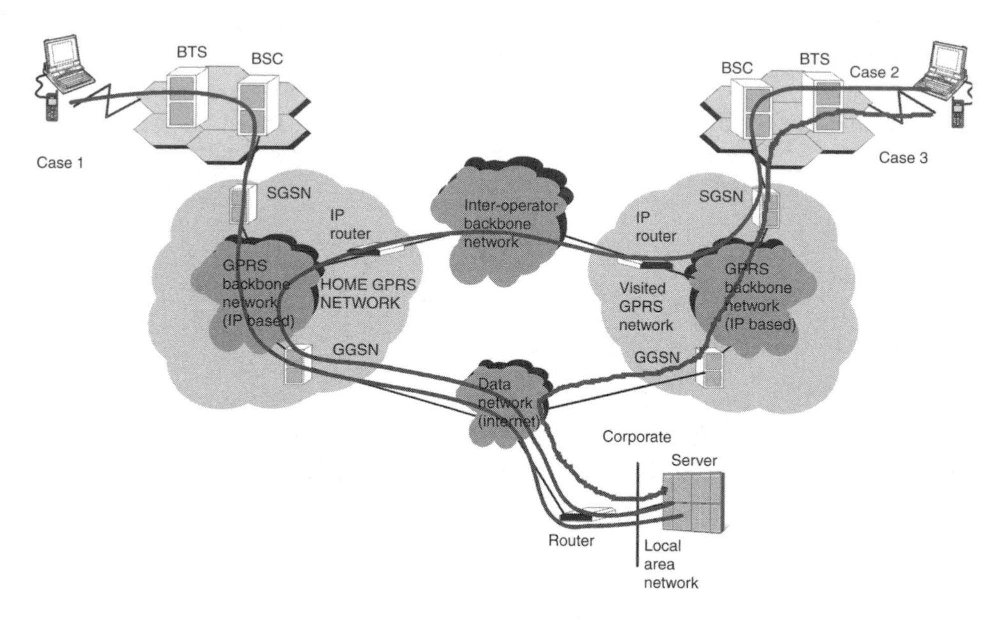

Figure 1.8. Data transfer

Each network can also have a pool of available addresses that can be allocated dynamically to the users by the GGSN. This decreases the number of addresses needed by an operator. A dynamic address is allocated to a user only for the time of a connection. To avoid routing the packet through the home network, the dynamic address can be allocated by a GGSN from the visited network (case 3 in Figure 1.8). It can also be allocated by a GGSN from the home network (case 2) to apply, for example, some specific screening.

The incoming and outgoing PTP traffic can be charged on the basis of the quantity of data transmitted, the protocol used, the length of connection, etc.

1.4.2.1 GPRS Mobiles

A GPRS MS can operate in one of three modes of operation (Table 1.1):

- *Class A mode of operation.* The MS is attached to both GPRS and other GSM services. The mobile user can make and/or receive calls on the two services simultaneously, subject to the quality of service (QoS) requirements, e.g. having a normal GSM voice call and receiving GPRS data packets at the same time.
- *Class B mode of operation.* The MS is attached to both GPRS and other GSM services, but the MS can only operate one set of services at a time. The MS in idle mode (and packet idle mode) is required to monitor paging channels (PCHs) for both circuit-switched and packet-switched services. However, the practical behaviour of the MS depends on the mode of network operation. For example, one mode of network operation is defined so that when an MS is engaged in packet data transfer, it will receive paging messages via the packet data channel (PDCH) without degradation of the packet data transfer.
- *Class C mode of operation.* The MS can only be attached to either the GSM network or the GPRS network. The selection is done manually and there are no simultaneous operations.

In GSM Phase 2, the MS uses one channel for the uplink traffic and one channel for the downlink traffic (Figure 1.9, '1-slot'). In GPRS, it is possible to have a multiple slot MS, e.g. a 2-slot MS with two uplink channels and two downlink channels (Figure 1.9, '2 slot'). The MS can have a different (asymmetric) uplink and downlink capability.

When an MS supports the use of multiple timeslots, it belongs to a multislot class as defined in [5]. Multislot class is defined as a combination of several parameters:

- The maximum number of received timeslots that the MS can use per TDMA frame.
- The maximum number of transmit timeslots that the MS can use per TDMA frame.
- Total number of uplink and downlink timeslots that can actually be used by the MS per TDMA frame.
- The time needed for the MS to perform adjacent cell signal level measurement and get ready to transmit (T_{ta}).
- The time needed for the MS to get ready to transmit (T_{tb}).

Table 1.1. GPRS mobile classes

Class A mode of operation	Class B mode of operation	Class C mode of operation
Full simultaneous use of packet and circuit mode connections	No simultaneous traffic but automatic sequential service	Alternate use or a GPRS-only MS

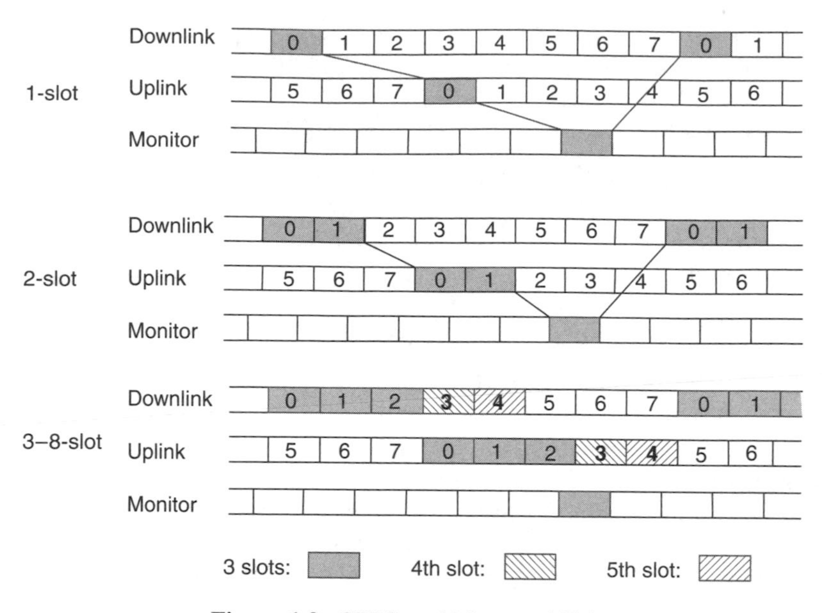

Figure 1.9. GPRS multislot capabilities

- The time needed for the MS to perform adjacent cell signal level measurement and get ready to receive (T_{ra}).
- The time needed for the MS to get ready to receive (T_{rb}).
- Capability to receive and transmit at the same time. There are two types of MS:
  - Type 1: The MS is not required to transmit and receive at the same time.
  - Type 2: The MS is required to transmit and receive at the same time.

A combination of these factors has led to a definition of 29 MS classes (in Rel'97 of GSM specifications). Most of the GPRS terminals (if not all) are type 1 owing to the simpler radio design. Table 1.2 shows the multislot classes. A complete multislot classes table (including Type 2 terminals) can be found in Annex B of [5].

1.4.2.2 Base Station Sub-system (BSS)

The BSS is upgraded with new GPRS protocols for the Gb interface (between the BSS and SGSN) and enhanced layer 2 (RLC/MAC see the section entitled *Medium access control and radio link control layer*) protocols for the air interface. The Gb interface connects the BSS and the SGSN, allowing the exchange of signalling information and user data. Both air and Gb interfaces allow many users to be multiplexed over the same physical resources. The BSS allocates resources to a user upon activity (when data are sent or received) and the data are reallocated immediately thereafter. The Gb interface link layer is based on frame relay, which is used for signalling and data transmission. The base station sub-system GPRS protocol (BSSGP) provides the radio-related QoS and routing information that is required to transmit user data between a BSS and an SGSN.

Table 1.2. Type 1 multislot classes

	Maximum number of slots			Minimum number of slots			
Multislot class	Rx	Tx	Sum	T_{ta}	T_{tb}	T_{ra}	T_{rb}
1	1	1	2	3	2	4	2
2	2	1	3	3	2	3	1
3	2	2	3	3	2	3	1
4	3	1	4	3	1	3	1
5	2	2	4	3	1	3	1
6	3	2	4	3	1	3	1
7	3	3	4	3	1	3	1
8	4	1	5	3	1	2	1
9	3	2	5	3	1	2	1
10	4	2	5	3	1	2	1
11	4	3	5	3	1	2	1
12	4	4	5	2	1	2	1

1.4.2.3 Mobile Services Switching Centre/Home Location Register (MSC/HLR)

The HLR is upgraded and contains GPRS subscription data. The HLR is accessible from the SGSN via the Gr interface and from the GGSN via the optional Gc interface. For roaming MSs, HLR is in a different public land mobile network (PLMN) than the current SGSN. All MSs use their HLR in home public land mobile network (HPLMN). The HLR is enhanced with GPRS subscriber information.

The MSC/visiting location register (VLR) can be enhanced for more efficient coordination of GPRS and non-GPRS services and functionality by implementing the Gs interface, which uses the BSSAP+ procedures, a subset of BSS application part (BSSAP) procedures. Paging for circuit-switched calls can be performed more efficiently via the SGSN. This is the case as well for the combined GPRS and non GPRS location updates. The Gs interface also allows easy Class B mode of operation for MS design. The SMS-GMSC and SMS-IWMSC functions are enhanced for SMS over GPRS. Also, the temporary mobile subscriber identity (TMSI) allocation procedure is modified.

1.4.2.4 SGSN

The SGSN is a main component of the GPRS network, which handles, e.g. the mobility management and authentication and has register function. The SGSN is connected to the BSC and is the service access point to the GPRS network for the GPRS MS. The SGSN handles the protocol conversion from the IP used in the backbone network to the sub-network-dependent convergence protocol (SNDCP) and logical link control (LLC) protocols used between the SGSN and the MS. These protocols handle compression and ciphering. The SGSN also handles the authentication of GPRS mobiles, and when the authentication is successful, the SGSN handles the registration of an MS to the GPRS network and takes care of its mobility management. When the MS wants to send (or receive) data to (from) external networks, the SGSN relays the data between the SGSN and relevant GGSN (and vice versa).

1.4.2.5 GGSN

The GGSN is connected to the external networks like the Internet and the X.25. From the external networks' point of view, the GGSN is a router to a sub-network, because the GGSN 'hides' the GPRS infrastructure from the external networks. When the GGSN receives data addressed to a specific user, it checks if the address is active. If it is, the GGSN forwards the data to the SGSN serving the MS, but if the address is inactive, the data are discarded. The mobile-originated packets are routed to the right network by the GGSN. The GGSN tracks the MS with a precision of SGSN.

1.4.3 GPRS Interfaces and Reference Points

The GPRS system introduces new so-called G-interfaces to the GSM network architecture. It is important to understand the function of every interface and reference point because this gives an insight to the GPRS system and consequent evolution. Figure 1.10 gives a logical architecture description with the interfaces and reference points of the GSM network with GPRS.

Connections of the GPRS system to the network and switching sub-system (NSS) part of the GSM network are implemented through signalling system number 7 (SS7) network (Gc, Gd, Gf, Gr, Gs), while the other interfaces and reference points are implemented through the intra-PLMN backbone network (Gn), the inter-PLMN backbone network (Gp) or the external networks (Gi). The different interfaces that the GPRS system uses are as follows:

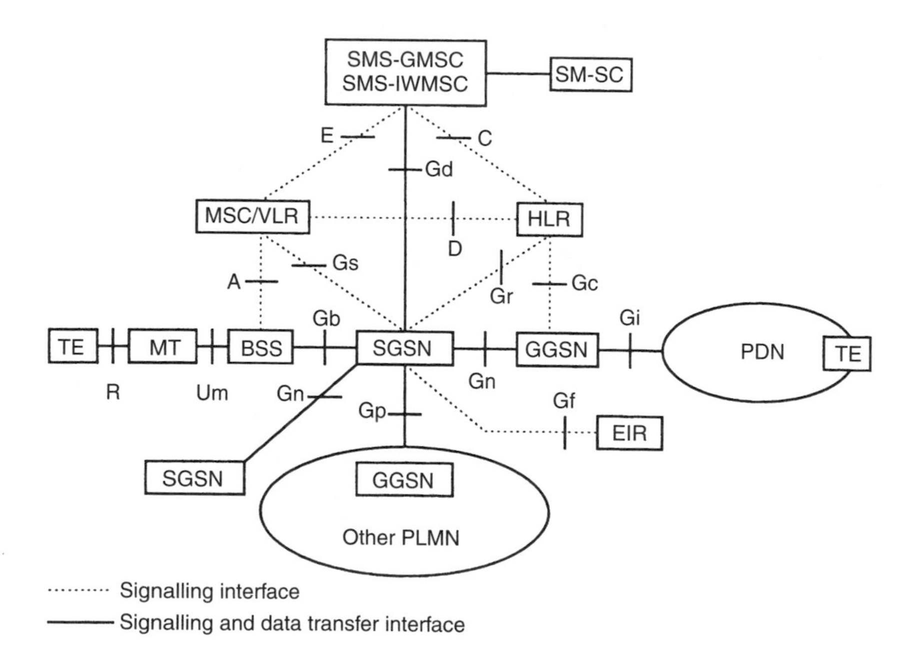

Figure 1.10. Logical architecture

- Gb between an SGSN and a BSS. The Gb interface is the carrier of the GPRS traffic and signalling between the GSM radio network (BSS) and the GPRS part. Frame relay–based network services (NSs) provide flow control for this interface.
- Gc between the GGSN and the HLR. The GGSN may request location information for network-requested context activation only via this optional interface. The standard also defines the use of a proxy GSN, which is used as a GPRS tunnelling protocol (GTP) to mobile application part (MAP) protocol converter, thus avoiding implementing MAP in GGSN.
- Gd between the SMS-GMSC and an SGSN, and between SMS-IWMSC and an SGSN. The Gd interface allows more efficient use of the SMS services.
- Gf between an SGSN and the Equipment Identity Register (EIR). The Gf gives the SGSN access to equipment information. In the EIR, the MSs are divided into three lists: black list for stolen mobiles, grey list for mobiles under observation and white list for other mobiles.
- Gn between two GSNs within the same PLMN. The Gn provides a data and signalling interface in the intra-PLMN backbone. The GTP is used in the Gn (and in the Gp) interface over the IP-based backbone network.
- Gp between two GSNs in various PLMNs. The Gp interface provides the same functionality as the Gn interface, but it also provides, with the BG and the Firewall, all the functions needed in the inter-PLMN networking, i.e. security, routing, etc.
- Gr between an SGSN and the HLR. The Gr gives the SGSN access to subscriber information in the HLR, which can be located in a different PLMN than the SGSN.
- Gs between an SGSN and an MSC. The SGSN can send location data to the MSC or receive paging requests from the MSC via this optional interface. The Gs interface will greatly improve effective use of the radio and network resources in the combined GSM/GPRS network. This interface uses BSSAP+ protocol.
- Um between a MS and the GPRS fixed network part. The Um is the access interface for the MS to the GPRS network. The MS has a radio interface to the BTS, which is the same interface used by the existing GSM network with some GPRS-specific changes.

There are two different reference points in the GPRS network. The Gi is GPRS-specific, but the R is common with the circuit-switched GSM network. The two reference points in the GPRS are as follows:

- Gi between a GGSN and an external network. The GPRS network is connected to external data networks via this interface. The GPRS system will support a variety of data networks and that is why the Gi is not a standard interface, but merely a reference point.
- R between terminal equipment and mobile termination. This reference point connects terminal equipment to mobile termination, thus allowing, e.g. a laptop-PC to transmit data over the GSM phone. The physical R interface follows, e.g. the ITU-T V.24/V.28 or the PCMCIA PC-Card standards.

1.4.4 GPRS Protocol Architecture

The GPRS system introduces a whole new set of protocols for the GSM Phase 2+ network. The inter-working between the new network elements is done with new GPRS-specific protocols. However, there are a number of existing protocols used at the lower layers of the protocol stacks, namely, TCP/user datagram protocol (UDP) IP. Figure 1.11 shows the transmission plane used in the GPRS system.

1.4.4.1 Physical Layer

The physical layer has been separated into two distinct sub-layers, the physical RF layer and the physical link layer.

The *physical RF layer* performs the modulation of the physical waveforms based on the sequence of bits received from the physical link layer. The physical RF layer also demodulates received waveforms into a sequence of bits that are transferred to the physical link layer for interpretation. The GSM physical RF layer is defined in GSM 05 series specifications, which define the following among other things:

- The carrier frequency characteristics and GSM radio channel structures [5].
- The modulation of the transmitted waveforms and the raw data rates of GSM channels [6]. Note that in the GPRS Rel'97 radio interface, only the original GSM modulation method, GMSK modulation, with the carrier bit rate of 270.833 kbps is defined. For one single timeslot, this corresponds to the gross data rate of 22.8 kbps.
- The transmitter and receiver characteristics and performance requirements [7].
- The *physical link layer* provides services for information transfer over a physical channel between the MS and the network. These functions include data unit framing, data coding and the detection and correction of physical medium transmission errors. The physical link layer operates above the physical RF layer to provide a physical channel between the MS and the network. The purpose of the physical link layer is to

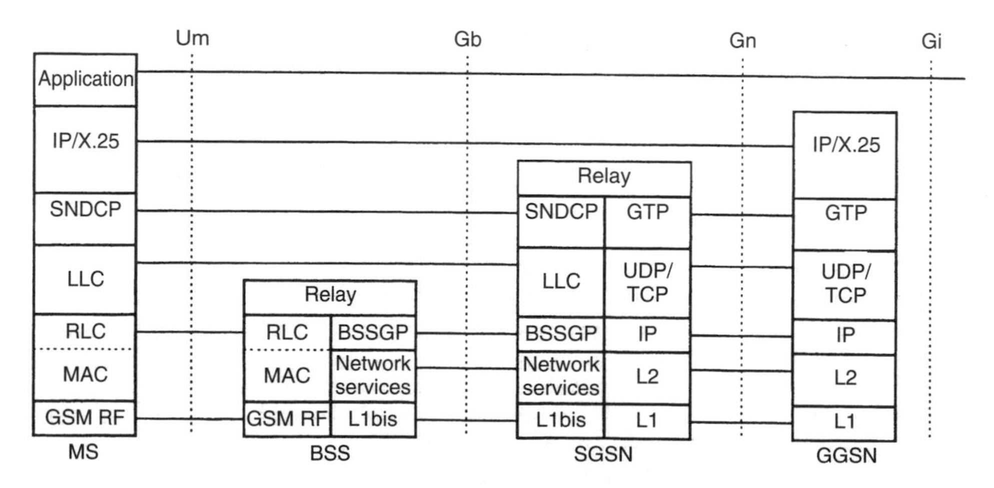

Figure 1.11. Transmission plane

convey information across the GSM radio interface, including RLC/MAC information. The physical link layer supports multiple MSs sharing a single physical channel and provides communication between MSs and the network. In addition, it provides the services necessary to maintain communications capability over the physical radio channel between the network and MSs. Radio sub-system link control procedures are specified in [4].

The physical link layer is responsible for the following:

- Forward error correction (FEC) coding, allowing the detection and correction of transmitted code words and the indication of uncorrectable code words. The coding schemes are described in the section entitled *Physical link layer*.
- Rectangular interleaving of one radio block over four bursts in consecutive TDMA frames, as specified in [8].
- Procedures for detecting physical link congestion.

The physical link layer control functions include

- synchronisation procedures, including means for determining and adjusting the MS timing advance to correct for variances in propagation delay [9],
- monitoring and evaluation procedures for radio link signal quality.
- cell selection and re-selection procedures.
- transmitter power control procedures.
- battery power saving procedures, e.g. discontinuous reception (DRX) procedures.

For more details, see the section entitled *Physical link layer*.

1.4.4.2 RLC/MAC

The RLC/MAC layer provides services for information transfer over the physical layer of the GPRS radio interface. These functions include backward error correction procedures enabled by the selective re-transmission of erroneous blocks. The RLC function offers a reliable radio link to the upper layers. The MAC function handles the channel allocation and the multiplexing, i.e. the use of physical layer functions. The RLC/MAC layer together form the open system interconnection (OSI) Layer 2 protocol for the Um interface and uses the services of the physical link layer (see the section entitled *Medium Access Control and Radio Link Control Layer*).

1.4.4.3 LLC

The logical link control layer offers a secure and reliable logical link between the MS and the SGSN to upper layers, and is independent of the lower layers. The LLC layer has two transfer modes—the acknowledged and unacknowledged. The LLC conveys signalling, SMS and SNDCP packets. See [10] for more information.

1.4.4.4 SNDCP

The sub-network-dependent convergence protocol (SNDCP) is a mapping and compression function between the network layer and lower layers. It also performs segmentation, reassembling and multiplexing. The SNDCP protocol is specified in [11].

Signalling has various sets of protocols, which are used with the existing GSM network elements. Internal signalling in the GPRS system is handled by protocols, which carry both data and signalling (LLC, GTP and BSSGP). The GPRS control plane is shown in Figure 1.12.

Figure 1.13 describes how network protocols are multiplexed. NSAPI is the network layer service access point identifier, which is used to identify the packet data protocol (PDP) context at SNDCP level. Service access point identifier (SAPI) is used to identify the points where LLC provides service to higher layers. SAPIs have different priorities. TLLI is the temporary logical link identity, which unambiguously identifies the logical link between the MS and the SGSN. The IP or the X.25 is the packet protocol offered to the subscriber by the GPRS system.

1.4.4.5 BSSGP

The primary functions of the base station sub-system GPRS protocol (BSSGP) (see [12]) include, in the downlink, the provision by an SGSN to a BSS of radio-related information used by the RLC/MAC function; in the uplink, the provision by a BSS to an SGSN of radio-related information derived from the RLC/MAC function; and the provision of functionality to enable two physically distinct nodes, an SGSN and a BSS, to operate node management control functions. See [12] for more details. The underlying network service is responsible for the transport of BSSGP packet data units (PDUs) between a BSS and an SGSN [12].

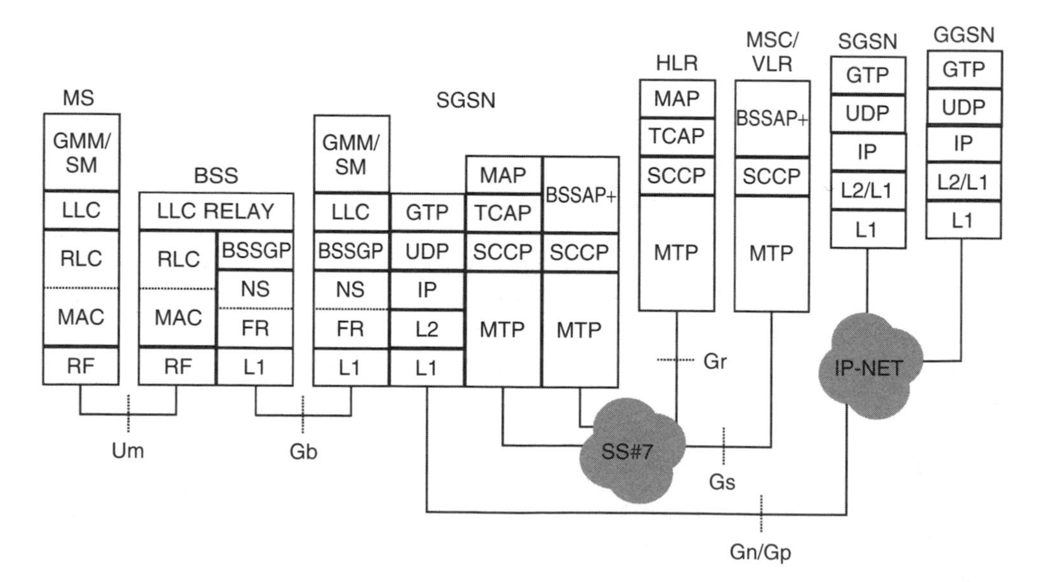

Figure 1.12. GPRS control plane

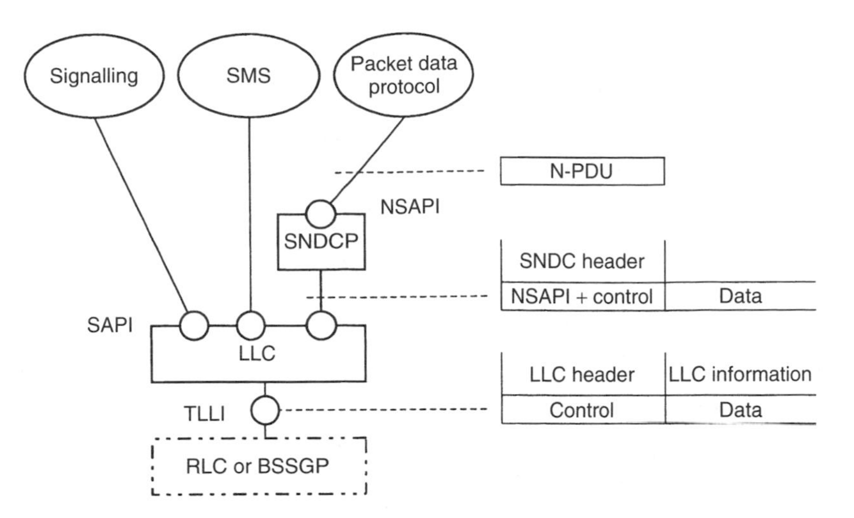

Figure 1.13. Multiplexing of network protocols

1.4.4.6 GTP

The GPRS tunnelling protocol (GTP) is used to tunnel data and signalling between the GSNs. The GTP can have proprietary extensions to allow proprietary features. The relay function relays PDP (packet data protocol) PDUs between the Gb and the Gn interfaces. For details see [13].

1.4.5 Mobility Management

The mobility management in the GPRS network is handled almost the same way as in the existing GSM system. One or more cells form a routing area, which is a subset of one location area. Every routing area is served by one SGSN. The tracking of the location of an MS depends on the mobility management state. When an MS is in a STANDBY state, the location of the MS is known on a routing area level. When the MS is in a READY state, the location of the MS is known on a cell level. Figure 1.14 shows the different mobility management states and transitions between them.

1.4.5.1 Mobility Management States

The GPRS has three various mobility management states. The IDLE state is used when the subscriber (MS) is passive (not GPRS attached). The STANDBY state is used when the subscriber has ended an active phase. An MS is in an active phase (READY state) when it is transmitting or has just been transmitting. The change between the states happens upon activity or when a timer expires. A description of the various states is given below.

IDLE State

The subscriber is not reachable by the GPRS network. The MS is only capable of receiving PTM-M data. The network elements hold no valid context for the subscriber and the subscriber is not attached to the mobility management. In order to change state, the MS has to perform a GPRS attach procedure.

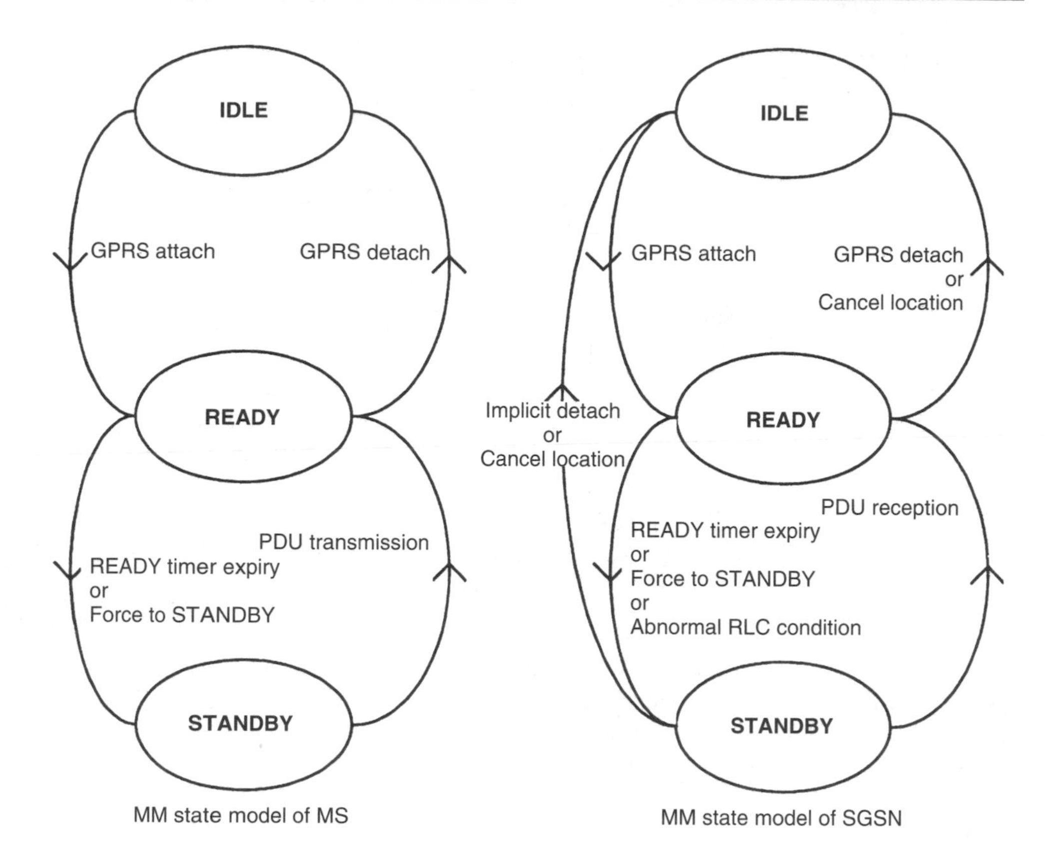

Figure 1.14. Mobility management state

STANDBY State

The subscriber is attached to the mobility management and the location of an MS is known on a routing area level. The MS is capable of receiving PTM data and pages for PTP data. The network holds a valid mobility management context for the subscriber. If the MS sends data, the MS moves to READY state. The MS or the network can initiate the GPRS detach procedure to move to IDLE state. After expiry of the MS reachable timer, the network can detach the MS. The MS can use the discontinuous reception (DRX) to save the battery.

READY State

The subscriber is attached to the mobility management and the location of an MS is known on a cell level. The MS is capable of receiving PTM and PTP data. The SGSN can send data to the MS without paging at any time and the MS can send data to the SGSN at any time. The network holds a valid mobility management context for the subscriber. If the READY timer expires, the MS moves to STANDBY state. If the MS performs a GPRS detach procedure, the MS moves to IDLE state and the mobility management context is removed. An MS in READY state does not necessarily have radio resources reserved. The MS can use the DRX to save the battery.

1.4.5.2 GPRS Attach and Detach

GPRS attach and GPRS detach are mobility management functions to establish and to terminate a connection to the GPRS network. The SGSN receives the requests and processes them. With the GPRS attach the mobile moves to READY state and the mobility management context is established, the MS is authenticated, the ciphering key is generated, a ciphered link established and the MS is allocated a temporary logical link identity. The SGSN fetches the subscriber information from the HLR. After a GPRS attach, the SGSN tracks the location of the MS. The MS can send and receive SMS, but no other data. To transfer other data, it has to first activate a PDP context.

When the subscriber wants to terminate a connection to the GPRS network, the GPRS detach is used. The GPRS detach moves the MS to IDLE state and the mobility management context is removed. The MS can also be detached from the GPRS implicitly when the mobile reachable timer expires. The GPRS detach is normally generated by the MS, but can also be generated by the network.

1.4.6 PDP Context Functions and Addresses

The packet data protocol (PDP) context functions are network level functions, which are used to bind an MS to various PDP addresses and afterwards used to unbind the MS from these addresses. The PDP context can also be modified. When an MS is attached to the network, it has to activate all the addresses it wants to use for data traffic with the external networks. Various PDP contexts have to be activated because these include address, QoS attributes, etc. After the subscriber has finished the use of activated addresses, these have to be deactivated. Also, change of address specific attributes have to be performed sometimes. The MS can use these functions when in STANDBY or READY state. The GGSN can also use these functions, but the SGSN is responsible for performing these functions. The MS can select a particular GGSN for access to certain services and also activate a PDP context anonymously, without using any identification.

1.4.6.1 Dynamic and Static PDP Address

The subscriber can use various kinds of PDP addresses. A static PDP address is permanently assigned to an MS by the HPLMN. An HPLMN dynamic PDP address is assigned to an MS by the HPLMN when an MS performs a PDP context activation. A visited public land mobile network (VPLMN) dynamic PDP address is assigned to an MS by the VPLMN when an MS performs a PDP context activation. It is indicated in the subscription if the MS can have a dynamic address or not.

1.4.6.2 PDP Context Activation

The PDP context activation can be done by the MS or by the network. The activate PDP context request is sent to the SGSN when a certain address requires activation. As an option, the request can also be made by the GGSN, if packets are received for an address without active context and the MS is GPRS attached. The request contains parameters for the context, like the TLLI, the protocol type, the address type, QoS, requested GGSN, etc.

1.4.6.3 PDP Context Modification

The PDP context can be modified by the SGSN with the modify PDP context. Only the parameters' QoS negotiated and radio priority can be changed. The SGSN sends the request to the MS, which either accepts the new QoS by sending a modify PDP context accept to the SGSN or does not accept it by sending a deactivate PDP context request for that context. In GPRS Phase 2 also, MS and GGSN can modify PDP context.

1.4.6.4 PDP Context Deactivation

The PDP context can be deactivated by the MS or by the network. Every address can be deactivated separately, but when the GPRS detach is performed, the network will automatically remove all the PDP contexts.

1.4.7 Security

The GPRS system uses GSM Phase 2–based security. These security functions include the authentication of the subscriber, the user identity confidentiality and the ciphering of the data traffic between the MS and the SGSN. The authentication of the subscriber is done the same way by the SGSN in the GPRS system as by the MSC/VLR in the Phase 2 GSM network. The TLLI is used to keep the subscriber identity confidential. The correspondence between the international mobile subscriber identity (IMSI) and the TLLI is only known by the MS and the SGSN. The ciphering function used between the MS and the SGSN is not the same as that used in the GSM Phase 2, but an optimised one for the packet-switched traffic. Security of the backbone is provided by using a private network, thus avoiding the possibility that an external *hacker* can address it. Each operator must guarantee its physical security.

1.4.8 Location Management

The location management procedures are used to handle the changing of a cell and/or a routing area, and the periodic routing area updates. If an MS stays a long time at the same place, the network has to receive an indication that the MS is still reachable. This is the reason why periodic routing area updates are made. All the MSs attached to the GPRS will perform a periodic routing area update. Only the Class B mode of operation mobiles, which are engaged in a circuit-switched communication, cannot. The MS performs a cell update when it changes cell within a routing area in READY mode. When the MS changes cell between the different routing areas, it performs a routing area update. There are two types of routing area updates, the intra-SGSN routing area update and the inter-SGSN routing area update. An SGSN can manage many routing areas and if the new routing area belongs to the management of a new SGSN, the inter-SGSN routing area update is used. If the new routing area belongs to the management of the same SGSN than the old one, the intra-SGSN routing area update is used. The old SGSN forwards user packets to the new SGSN, until it receives a cancel location from the HLR.

1.4.9 GPRS Radio Interface

This section gives an overview of the GPRS Rel'97 radio interface.

1.4.9.1 GPRS Packet Data Logical Channels

This section describes the logical channels of the GPRS radio interface. The reason for defining a new set of control channels for the GPRS packet data, which is partly parallel to that of the circuit-switched control channels, is to be able to flexibly allocate more signalling capacity for the packet data traffic without sacrificing the quality of the speech traffic.

The packet data logical channels are mapped onto the physical channels that are dedicated to packet data. The physical channel dedicated to packet data traffic is called a *packet data channel* (PDCH).

The packet data logical channels can be grouped into different categories as shown in Figure 1.15. These categories (packet common control channels, packet broadcast control channels, packet traffic channels and packet dedicated control channels) are introduced in more detail in the following list [14].

Packet Common Control Channel (PCCCH)

PCCCH comprises logical channels for common control signalling used for packet data:

- *Packet random access channel (PRACH)—uplink only.* PRACH is used by the MS to initiate uplink transfer to send data or signalling information.
- *Packet paging channel (PPCH)—downlink only.* PPCH is used to page an MS prior to downlink packet transfer.
- *Packet access grant channel (PAGCH)—downlink only.* PAGCH is used in the packet transfer establishment phase to send resource assignment to an MS prior to packet transfer.
- *Packet notification channel (PNCH)—downlink only.* PNCH is used to send a PTM-M (point to multipoint—multicast) notification to a group of MSs prior to a PTM-M packet transfer. (Note that the PTM-M service is not specified in GPRS Rel'97.)

Packet Broadcast Control Channel (PBCCH)—Downlink only

PBCCH broadcasts packet data specific system information. If PBCCH is not allocated, the packet data specific system information is broadcast on BCCH.

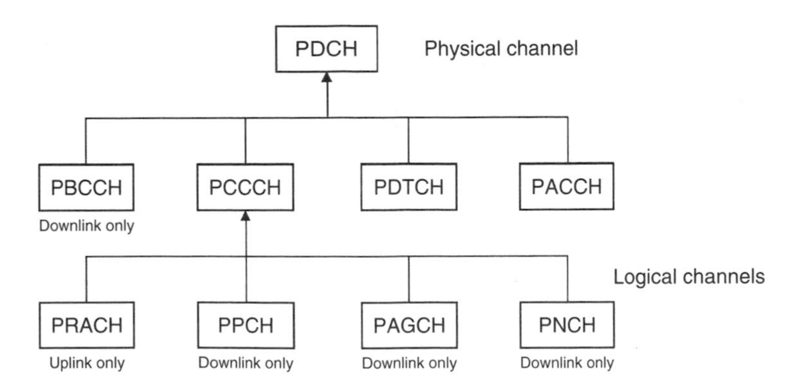

Figure 1.15. GPRS logical channels

Packet Traffic Channels

- *Packet data traffic channel (PDTCH).* PDTCH is a channel allocated for data transfer. It is temporarily dedicated to one MS. In the multislot operation, one MS may use multiple PDTCHs in parallel for individual packet transfer.

All packet data traffic channels are uni-directional, either uplink (PDTCH/U), for a mobile-originated packet transfer, or downlink (PDTCH/D), for a mobile terminated packet transfer.

Packet Dedicated Control Channels

- *Packet associated control channel (PACCH).* PACCH conveys signalling information related to a given MS. The signalling information includes, e.g. acknowledgements and power control information. PACCH also carries resource assignment and reassignment messages comprising the assignment of a capacity for PDTCH(s) and for further occurrences of PACCH. The PACCH shares resources with PDTCHs, which are currently assigned to one MS. Additionally, an MS that is currently involved in packet transfer can be paged for circuit-switched services on PACCH.
- *Packet timing advance control channel, uplink (PTCCH/U).* PTCCH/U is used to transmit random access bursts to allow estimation of the timing advance for one MS in packet transfer mode.
- *Packet timing advance control channel, downlink (PTCCH/D).* PTCCH/D is used to transmit timing advance information updates to several MSs. One PTCCH/D is paired with several PTCCH/Us.

1.4.9.2 Mapping of Packet Data Logical Channels into Physical Channels

Different packet data logical channels can occur on the same physical channel (i.e. PDCH). The sharing of the physical channel is based on blocks of four consecutive bursts, except for PTCCH. The mapping in frequency of PDCH onto the physical channel is defined in [5].

On PRACH and PTCCH/U, access bursts are used. On all other packet data logical channels, radio blocks comprising four normal bursts are used. The only exceptions are some messages on uplink PACCH that comprise four consecutive access bursts (to increase robustness). The following list shows some more details about how the packet data logical channels are mapped onto physical channels.

Packet Common Control Channel (PCCCH)

At a given time, the logical channels of the PCCCH are mapped on different physical resources than the logical channels of the CCCH. The PCCCH does not have to be allocated permanently in the cell. Whenever the PCCCH is not allocated, the CCCH shall be used to initiate a packet transfer. The PCCCH, when it exists, is mapped on one or several physical channels according to a 52-multiframe. In that case, the PCCCH, PBCCH and PDTCH share the same physical channels (PDCHs). The existence and location of the PCCCH shall be broadcast on the cell.

- *Packet random access channel (PRACH).* The PRACH is mapped on one or several physical channels. The physical channels on which the PRACH is mapped are derived by the MS from information broadcast on the PBCCH or BCCH.

  PRACH is determined by the uplink state flag (USF) being marked as free, which is broadcast continuously on the corresponding downlink. Additionally, a pre-defined fixed part of the multiframe structure for PDCH can be used as PRACH only and the information about the mapping on the physical channel is broadcast on PBCCH. During those time periods, an MS does not have to monitor the USF that is simultaneously broadcast on the downlink.
- *Packet paging channel (PPCH).* The PPCH is mapped on one or several physical channels. The exact mapping on each physical channel follows a pre-defined rule, as it is done for the paging channel (PCH).

  The physical channel on which the PPCH is mapped, as well as the rule that is followed on the physical channels, are derived by the MS from information broadcast on the PBCCH.
- *Packet access grant channel (PAGCH).* The PAGCH is mapped on one or several physical channels. The exact mapping on each physical channel follows a pre-defined rule.

  The physical channels on which the PAGCH is mapped, as well as the rule that is followed on the physical channels, are derived by the MS from information broadcast on the PBCCH.
- *Packet notification channel (PNCH).* The PNCH is mapped on one or several blocks on PCCCH. The exact mapping follows a pre-defined rule. The mapping is derived by the MS from information broadcast on the PBCCH.

Packet Broadcast Control Channel (PBCCH)

The PBCCH shall be mapped on one or several physical channels. The exact mapping on each physical channel follows a pre-defined rule, as it is done for the BCCH.

The existence of the PCCCH and, consequently, the existence of the PBCCH, is indicated on the BCCH.

Packet Timing Advance Control Channel (PTCCH)

Two defined frames of multiframe are used to carry PTCCH. The exact mapping of PTCCH/U sub-channels and PTCCH/D are defined in [5].

On PTCCH/U, access bursts are used. On PTCCH/D, four normal bursts comprising a radio block are used.

Packet Traffic Channels

- *Packet data traffic channel (PDTCH).* One PDTCH is mapped on to one physical channel. Up to eight PDTCHs, with different timeslots but with the same frequency parameters, may be allocated to one MS at the same time.

- *Packet associated control channel (PACCH).* PACCH is dynamically allocated on a block basis on the same physical channel that is carrying PDTCHs. However, one block PACCH allocation is used on the physical channel that is carrying only PCCCH when the MS is polled to acknowledge the initial assignment message.

  PACCH is of a bi-directional nature, i.e. it can dynamically be allocated both on the uplink and on the downlink, regardless of whether the corresponding PDTCH assignment is for uplink or downlink.

Different packet data logical channels can be multiplexed in one direction (either on the downlink or uplink) on the same physical channel (i.e. PDCH). The reader is advised to see details in [5]. The type of message that is indicated in the radio block header allows differentiation between the logical channels. Additionally, the MS identity allows differentiation between PDTCHs and PACCHs assigned to different MSs.

1.4.9.3 Radio Interface (Um)

Radio Resource Management Principles

A cell supporting GPRS may allocate resources on one or several physical channels in order to support the GPRS traffic [14]. Those physical channels (i.e. PDCHs), shared by the GPRS MSs, are taken from the common pool of physical channels available in the cell. The allocation of physical channels to circuit-switched services and GPRS can vary dynamically. Common control signalling required by GPRS in the initial phase of the packet transfer is conveyed on PCCCH (when allocated), or on CCCH. This allows capacity to be allocated specifically to GPRS in the cell only when a packet is to be transferred.

At least one PDCH, acting as a primary PDCH, accommodates packet common control channels that carry all the necessary control signalling for initiating packet transfer (i.e. PCCCH), whenever that signalling is not carried by the existing CCCH, as well as user data and dedicated signalling (i.e. PDTCH and PACCH). Other PDCHs, acting as secondary PDCHs, are used for user data transfer and for dedicated signalling.

The GPRS does not require permanently allocated PDCHs. The allocation of capacity for GPRS can be based on the needs for actual packet transfers ('capacity on demand' principle). The operator can as well decide to dedicate, permanently or temporarily, some physical resources (i.e. PDCHs) for GPRS traffic.

When the PDCHs are congested owing to the GPRS traffic load and more resources are available in the cell, the network can allocate more physical channels as PDCHs. However, the existence of PDCH(s) does not imply the existence of PCCCH. When no PCCCH is allocated in a cell, all GPRS attached MSs camp on the CCCH.

In response to a packet channel request sent on CCCH from the MS that wants to transmit GPRS packets, the network can assign resources on PDCH(s) for the uplink transfer. After the transfer, the MS returns to CCCH.

When PCCCH is allocated in a cell, all GPRS attached MSs camp on it. PCCCH can be allocated either as the result of the increased demand for packet data transfers or whenever there are enough available physical channels in a cell (to increase the QoS). The information about PCCCH is broadcast on BCCH. When the PCCCH capacity is inadequate, it is possible to allocate additional PCCCH resources on one or several PDCHs. If the network releases the last PCCCH, the MS performs cell re-selection.

The number of allocated PDCHs in a cell can be increased or decreased according to demand.

Multiframe Structure for PDCH

The mapping in time of the logical channels is defined by a multiframe structure. The multiframe structure for PDCH consists of 52 TDMA frames, divided into 12 blocks (of 4 frames), 2 idle frames and 2 frames used for the PTCCH according to Figure 1.16.

The mapping of logical channels onto the radio blocks is defined in this section by means of the ordered list of blocks (B0, B6, B3, B9, B1, B7, B4, B10, B2, B8, B5, B11).

The PDCH that contains PCCCH (if any) is indicated on BCCH. That PDCH is the only one that contains PBCCH blocks. On the downlink of this PDCH, the first block (B0) in the ordered list of blocks is used as PBCCH. If required, up to three more blocks on the same PDCH can be used as additional PBCCH. Any additional PDCH containing PCCCH is indicated on PBCCH.

On any PDCH with PCCCH (with or without PBCCH), up to the next 12 blocks in the ordered list of blocks are used for PAGCH, PNCH, PDTCH or PACCH in the downlink. The remaining blocks in the ordered list are used for PPCH, PAGCH, PNCH, PDTCH or PACCH in the downlink. In all cases, the actual usage of the blocks is indicated by the message type. On an uplink PDCH that contains PCCCH, all blocks in the multiframe can be used as PRACH, PDTCH or PACCH. Optionally, the first blocks in the ordered list of blocks can only be used as PRACH. The MS may choose to either ignore the USF (consider it as FREE) or use the USF to determine the PRACH in the same way as for the other blocks.

The mapping of channels on multiframes is controlled by several parameters broadcast on PBCCH. On a PDCH that does not contain PCCCH, all blocks can be used as PDTCH or PACCH. The actual usage is indicated by the message type. Two frames are used for PTCCH (see [5]) and the two idle frames as well as the PTCCH frames can be used by the MS for signal measurements and base station identification code (BSIC) identification.

An MS attached to GPRS shall not be required to monitor BCCH if a PBCCH exists. All system information relevant for GPRS and some information relevant for circuit-switched services (e.g. the access classes) shall in this case be broadcast on PBCCH. In order to facilitate the MS operation, the network is required to transmit certain types of packet system information (PSI) messages in specific multiframes and specific PBCCH blocks within the multiframes. The exact scheduling is defined in [5]. If no PCCCH is allocated, the MS camps on the CCCH and receives all system information on BCCH. In this case, any necessary GPRS-specific system information has to be broadcast on BCCH.

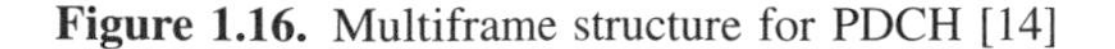

Figure 1.16. Multiframe structure for PDCH [14]

Radio Resource Operating Modes

Radio resource (RR) management procedures are characterised by two different RR operating modes [14], packet idle mode and packet transfer mode.

In *packet idle mode,* no temporary block flow (TBF) exists. Upper layers can require the transfer of a LLC PDU, which, implicitly, may trigger the establishment of TBF and transition to packet transfer mode.

In packet idle mode, the MS listens to the PBCCH and to the paging sub-channel for the paging group the MS belongs to. If PCCCH is not present in the cell, the MS listens to the BCCH and to the relevant paging sub-channels.

In *packet transfer mode*, the mobile station is allocated radio resource, providing a TBF on one or more physical channels. Continuous transfer of one or more LLC PDUs is possible. Concurrent TBFs may be established in opposite directions. Transfer of LLC PDUs in RLC acknowledged or RLC unacknowledged mode is provided.

When selecting a new cell, the MS leaves the packet transfer mode, enters the packet idle mode where it switches to the new cell, reads the system information and may then resume to packet transfer mode in the new cell.

The mobility management states are defined in [15]. Table 1.3 provides the correspondence between radio resource states and mobility management states.

Each state is protected by a timer. The timers run in the MS and the network. Packet transfer mode is guarded by RLC protocol timers.

The GPRS RR procedures and RR operating modes are defined in detail in [16].

Physical Link Layer

Different radio block structures are defined for the data transfer and for the control message transfer. The radio block consists of MAC header, RLC data block or RLC/MAC control block (Figure 1.17). It is always carried by four normal bursts. For detailed definition of radio block structure, see [17].

The MAC header (8 bits) contains control fields, which are different for uplink and downlink directions. The RLC header is of variable length and contains control fields, which are as well different for uplink and downlink directions. The RLC data field contains octets from one or more LLC packet data units (PDUs). The block check sequence (BCS) is used for error detection. The RLC/MAC control message field contains one RLC/MAC control message.

Channel Coding

Four coding schemes, CS-1 to CS-4, are defined for the packet data traffic channels. For all packet control channels (PCCHs) other than packet random access channel (PRACH)

Table 1.3. Correspondence between RR operating modes and MM states [14]

Radio resource BSS	Packet transfer mode	Measurement report reception	No state	No state
Radio resource MS	Packet transfer mode	Packet idle mode		Packet idle mode
Mobility management NSS and MS	Ready			Standby

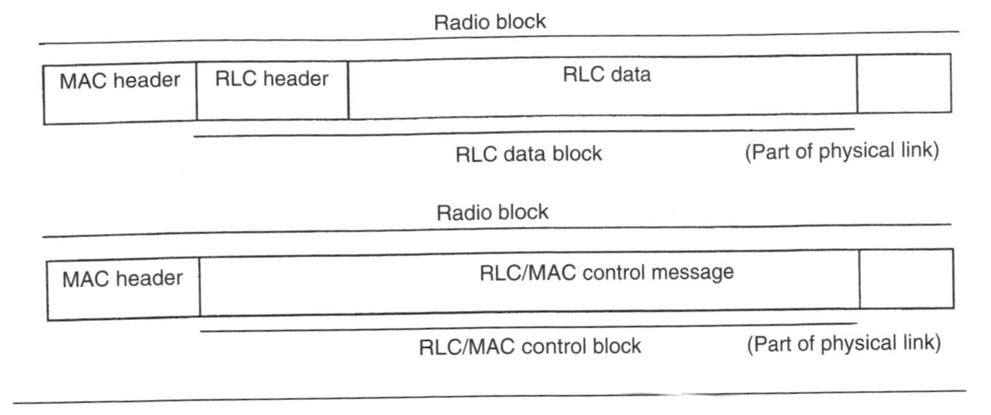

Figure 1.17. Radio block structures [14]

and packet timing advance control channel on uplink (PTCCH/U), coding scheme CS-1 is always used. For access bursts on PRACH, two coding schemes are specified. All coding schemes are mandatory for the MS. Only CS-1 is mandatory for the network.

Channel Coding for PDTCH

Four different coding schemes, CS-1 to CS-4, are defined for the radio blocks carrying RLC data blocks. The block structures of the coding schemes are shown in Figures 1.18 and 1.19.

The first step of the coding procedure is to add a BCS for error detection. For the CS-1 to CS-3, the second step consists of pre-coding USF (except for CS-1), adding four tail bits (TBs) and a half-rate convolutional coding for error correction that is punctured to give the desired coding rate. For the CS-4, there is no coding for error correction. The details of the codes are shown in Table 1.4, including the length of each field, the number of coded bits (after adding tail bits and convolutional coding), the number of punctured bits and the data rate (including the RLC header and RLC information).

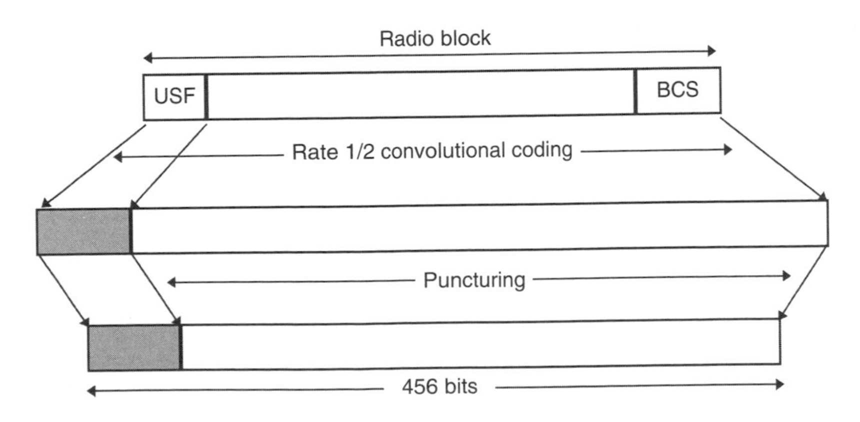

Figure 1.18. Radio block structure for CS-1 to CS-3

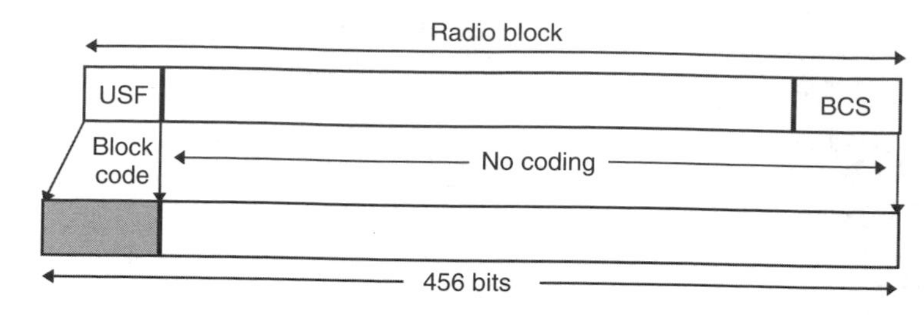

Figure 1.19. Radio block structure for CS-4 [14]

Table 1.4. Coding parameters for the coding schemes

Scheme	Code rate	USF	Pre-coded USF	Radio block excl. USF and BCS	BCS	Tail	Data rate (kbps)	Data rate kbps excl. RLC/MAC headers
CS-1	1/2	3	3	181	40	4	9.05	8
CS-2	~2/3	3	6	268	16	4	13.4	12
CS-3	~3/4	3	6	312	16	4	15.6	14.4
CS-4	1	3	12	428	16	–	21.4	20

The CS-1 is the same coding scheme as specified for the slow associated control channel (SACCH) in GSM 05.03. It consists of a half-rate convolutional code for FEC and a 40-bit FIRE code for the BCS. The CS-2 and CS-3 are punctured versions of the same half-rate convolutional code as CS-1. The CS-4 has no FEC. The CS-2 to CS-4 use the same 16-bit CRC for the BCS. The CRC is calculated over the whole uncoded RLC data block including the MAC header.

The USF has eight states, which are represented by a binary 3-bit field in the MAC header. For the CS-1, the whole radio block is convolutionally coded and the USF needs to be decoded as part of the data. All other coding schemes generate the same 12-bit code for the USF. The USF can be decoded either as a block code or as part of the data.

In order to simplify the decoding, the stealing bits (SB) (1-bit flag in either side of the training sequence code, as defined in [8]) of the block are used to indicate the actual coding scheme.

Channel Coding for PACCH, PBCCH, PAGCH, PPCH, PNCH and PTCCH

The channel coding for the PACCH, PBCCH, PAGCH, PPCH, PNCH and the downlink PTCCH is the same as the coding scheme CS-1. The coding scheme used for the uplink PTCCH is the same as for the PRACH.

Channel Coding for the PRACH

Two types of packet access bursts may be transmitted on the PRACH: an 8 information bits access burst or an 11 information bits access burst called the *extended packet access burst*. The reason for two different packet access bursts is that the 8-bit access burst cannot carry the additional information (USF) in the dynamic allocation mode. The mobile supports both access bursts. The channel coding for both burst formats is indicated as follows.

- *Coding of the 8 data bit packet access burst.* The channel coding used for the burst carrying the 8 data bit packet access uplink message is identical to the coding of the access burst as defined for the random access channel in [8].
- *Coding of the 11 data bit packet access burst.* The channel coding for the 11-bit access burst is the punctured version of the same coding as used for the 8-bit access burst.

Cell Re-selection

Initial PLMN selection and cell selection are done when the mobile station power is switched on. In GPRS packet, idle *cell re-selection* is performed autonomously by the MS. In GPRS packet transfer mode, the cell re-selection mechanism depends on the network operational mode. There are three network control modes: NC0, NC1 and NC2. In NC0, the MS performs the cell re-selection autonomously as in idle mode. In NC1, the MS also performs the cell re-selection autonomously but, in addition, it sends measurement reports to the network periodically. The measurement reports are analogous to the ones sent every 0.48 s in voice and CS data connections. In NC2, the MS sends periodical measurement reports and, additionally, the network sends cell re-selection commands to the MS, so the network has control of the cell re-selection process.

When the MS is attached to the circuit-switched core network (to MSC) and at the same time it is in packet transfer mode, the cell is determined by the network according to the handover procedures (handovers have precedence over GPRS cell re-selection).

In GPRS, two additional cell re-selection criteria, C31 and C32, are provided as a complement to the other GSM cell re-selection criteria. Cell re-selection criteria are specified in [4].

Also, a network controlled cell re-selection can be used in GPRS. The network can order mobile stations to send measurement reports to the network and to suspend its normal cell re-selection, and instead to accept decisions from the network. This applies to both packet idle mode and packet transfer mode. The MS measures the received RF signal strength on the BCCH frequencies of the serving cell and the neighbour cells as indicated in the BA-GPRS list and calculates the received level average (RLA) for each frequency, as defined in [4]. In addition, the MS verifies the BSIC of the cells. Only channels with the same BSIC as broadcast together with BA-GPRS on PBCCH are considered for re-selection.

The PBCCH broadcasts GPRS-specific cell re-selection parameters for serving and neighbour cells, including the BA-GPRS list. A BA-GPRS identifies the neighbour cells, including BSIC that need to be considered for the GPRS cell re-selection.

Timing Advance

The timing advance procedure is used to derive the correct value for timing advance that the MS has to use for the uplink transmission of radio blocks. The timing advance procedure comprises two parts:

- initial timing advance estimation
- continuous timing advance update.

The *initial timing advance estimation* is based on the single access burst carrying the packet channel request. The packet uplink assignment or packet downlink assignment then carries the estimated timing advance value to the MS. The MS uses this value for the uplink transmissions until the continuous timing advance update provides a new value (a few special cases for the initial timing advance are explained in [14]).

In the packet transfer mode, the MS uses the *continuous timing advance update* procedure. The continuous timing advance update procedure is carried on the PTCCH allocated to the MS. For the uplink packet transfer, within the packet uplink assignment, the MS is assigned the timing advance index (TAI) and the PTCCH. For the downlink packet transfer, within the packet downlink assignment, the MS is assigned the TAI and the PTCCH. The TAI specifies the PTCCH sub-channel used by the MS. On the uplink, the MS sends the assigned PTCCH access burst, which is used by the network to derive the timing advance.

The network analyses the received access burst and determines new timing advance values for all MSs performing the continuous timing advance update procedure on that PDCH. The new timing advance values are sent via a downlink signalling message (TA message) on PTCCH/D. The network can send timing advance information also in packet timing advance/power control and packet uplink Ack/Nack messages on PACCH. The mapping of the uplink access bursts and downlink TA messages on groups of eight 52-multiframes is shown in [14].

The BTS updates the timing advance values in the next TA message following the access burst.

Power Control Procedure

Power control is used in order to improve the spectrum efficiency and to reduce the power consumption in the MS. For the uplink, the MS follows a flexible power control algorithm, which the network can optimise through a set of parameters. It can be used for both open-loop and closed-loop power control. For the downlink, the power control is performed in the BTS. There is no need to specify the actual algorithms, but information about the downlink performance is needed. Therefore, the MSs transfer channel quality reports to the BTS. For the detailed specification of the power control see [4].

The MS shall calculate the RF output power value, P_{CH}, to be used on each individual uplink PDCH assigned to the MS [14]:

$$P_{CH} = \min((\Gamma_0 - \Gamma_{CH} - \alpha^*(C + 48), \mathrm{PMAX})$$

where

Γ_{CH} = an MS and channel-specific power control parameter. It is sent to the MS in any resource assigning message. The network can, at any time during a packet transfer, send new Γ_{CH} values to the MS on the downlink PACCH

Γ_0 = frequency band–dependent constant

$\alpha \in [0, 1]$ = a system parameter. Its default value is broadcast on the PBCCH. MS and channel-specific values can be sent to the MS together with Γ_{CH}

C = received signal level at the MS (see [14] for the derivation of this value)

$PMAX$ = maximum allowed output power in the cell

All power values are expressed in dBm.

P_{CH} is not used to determine the output power when accessing the cell on PRACH or RACH, in which case PMAX is used.

The BTS uses constant power on those PDCH radio blocks that contain PBCCH or that may contain PPCH. This power may be lower than the output power used on BCCH. The difference is broadcast on PBCCH. On the other PDCH radio blocks, downlink power control can be used. Thus, a procedure can be implemented in the network to control the power of the downlink transmission on the basis of the channel quality reports. The network has to ensure that the output power is sufficient for the MS for which the RLC block is intended as well as the MS(s) for which the USF is intended, and that for each MS in packet transfer mode, at least one downlink RLC block per multiframe is transmitted with an output power that is sufficient for that MS, on a block monitored by that MS.

The MS has to periodically monitor the downlink Rx signal level and quality from its serving cell.

In order to derive the value of C, the MS periodically measures the received signal strength. In the packet idle mode, the MS measures the signal strength of the PCCCH or, if PCCCH does not exist, the BCCH. In the packet transfer mode, the MS measures the signal strength on BCCH. The same measurements as for cell re-selection are used. Alternatively, if indicated by a broadcast parameter, the MS measures the signal strength on one of the PDCHs where the MS receives PACCH. This method is suitable in the case in which BCCH is in another frequency band than the one used by PDCHs. It requires that constant output power is used on all downlink PDCH blocks.

MS Measurements

The MS measures the signal strength of each radio block monitored by the MS. The C value is achieved by filtering the signal strength with a running average filter. The filtering is usually continuous between the packet modes. The different filter parameters for the packet modes are broadcast on PBCCH or, if PBCCH does not exist, on BCCH. The variance of the received signal level within each block is also calculated. The filtered value SIGN_VAR is included in the channel quality report.

The channel quality is measured as the interference signal level during the idle frames of the multiframe, when the serving cell is not transmitting. In the packet transfer mode, the MS measures the interference signal strength of all eight channels (slots) on the same carrier as the assigned PDCHs. In the packet idle mode, the MS measures the interference signal strength on certain channels that are indicated on the PBCCH or, if PBCCH does not exist, on BCCH. Some of the idle frames and PTCCH frames are used for the channel quality measurements, while the others are required for BSIC identification and the timing advance procedure (see [4] for details).

The MS may not be capable of measuring all eight channels when allocated some configurations of channels. The MS has to measure as many channels as its allocation allows, taking into account its multislot capability. The interference, Γ_{CH}, is derived by filtering the measured interference in a running average filter.

In packet transfer mode, the MS transfers the eight Γ_{CH} values and the RXQUAL, SIGN_VAR and C values to the network in the channel quality report included in the PACKET DOWNLINK ACK/NACK message.

The BSS has to monitor the uplink Rx signal level and quality on each uplink PDCH, active as well as inactive. The BSS also has to measure the Rx signal level and the quality of a specific MS packet transfer.

Scheduling the MS activities during the PTCCH and idle frames. The MS uses the PTCCH and idle frames of the PDCH multiframe for the following tasks:

- Neighbour cell–level measurement and BSIC identification for cell re-selection
- Continuous timing advance procedures
- Interference measurements.

The scheduling of these tasks is not specified in detail. During the frames when the MS receives TA messages, it can also make interference measurements. Additionally, during the frames when the MS transmits access bursts, it may also be possible to make measurements on some channels.

The MS tries to schedule the BSIC identification as efficiently as possible, using the remaining PTCCH frames and the idle frames and also considering the requirements for interference measurements.

Discontinuous reception (DRX). The MS has to support discontinuous reception (DRX; sleep mode) in packet idle mode. The DRX is independent from mobility management (MM) states, ready and standby. The negotiation of the DRX parameters is per MS. The DRX parameters and the overview of DRX operation are explained in [14].

Medium Access Control and Radio Link Control Layer

The medium access control and radio link control layer operates above the physical link layer in the protocol architecture. The MAC function defines the procedures that enable multiple MSs to share a common transmission medium, which may consist of several physical channels. The RLC function defines the procedures for a selective re-transmission of unsuccessfully delivered RLC data blocks. MAC/RLC is a complex protocol, and for details the reader is referred to [17]. This section gives only an introduction of the main functions of the RLC/MAC layer.

The basic principle of data transfer is illustrated in Figure 1.20. The network protocol data units (N-PDU), which correspond to full IP packets, are compressed and segmented into the sub-network protocol data units (SN-PDU) by the sub-network-dependent convergence protocol (SNDCP). IP-packet compression is optional. The SN-PDUs are encapsulated into one or several LLC frames. The size of the data part of the LLC frames is a parameter between 140 and 1520 bytes. LLC frames are segmented into RLC data blocks. At the RLC/MAC layer, a selective automatic repeat request (ARQ) protocol (including block numbering) between the MS and the network provides re-transmission of erroneous RLC data blocks. When a complete LLC frame is successfully transferred across the RLC layer, it is forwarded to the LLC layer.

There are two important concepts that constitute the core of RLC/MAC operation: temporary block flow and temporary flow identifier.

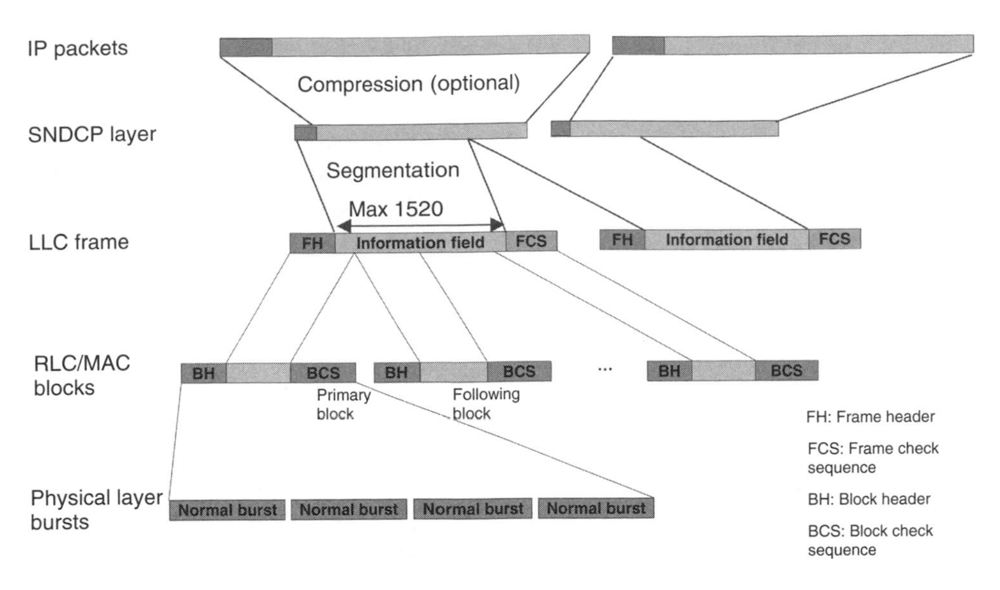

Figure 1.20. Transmission and reception data flow

Temporary Block Flow

A temporary block flow (TBF) [14] is a temporal connection between the MS and the network to support the uni-directional transfer of LLC PDUs on packet data physical channels. A TBF can use radio resources on one or more PDCHs and comprises a number of RLC/MAC blocks carrying one or more LLC PDUs. A TBF is temporary and is maintained only for the duration of the data transfer.

Temporary Flow Identity

Each TBF is assigned a temporary flow identity (TFI) [14] by the network. The assigned TFI is unique among concurrent TBFs in each direction and is used instead of the MS identity in the RLC/MAC layer. The same TFI value may be used concurrently for TBFs in opposite directions. The TFI is assigned in a resource assignment message that precedes the transfer of LLC frames belonging to one TBF to/from the MS. The same TFI is included in every RLC header belonging to a particular TBF, as well as in the control messages associated with the LLC frame transfer (e.g. acknowledgements) in order to address the peer RLC entities.

Uplink Radio Blocks Multiplexing

In uplink, the packet-switched users are multiplexed in the same PDCH by the MAC layer. There are three access modes supported by RLC/MAC protocol: dynamic allocation, extended dynamic allocation and fixed allocation.

Dynamic allocation. In order to make the multiplexing of mobile stations on the same PDCH possible, the uplink state flag (USF) is introduced. During the establishment of the uplink TBF, a USF is assigned to each mobile, which will be used by the network to indicate which terminal is allowed to transmit in the following uplink radio block.

With this method, each downlink block has to be decoded by all the MSs assigned to the PDCH in order to obtain the USF and the TFI, identifying the owner of the data or control information. In addition, even if no downlink information is available, a dummy radio block must be sent just to transmit the USF when required.

The USF is a 3-bit flag, so each PDCH can manage up to eight MSs, except on the PCCH, where the value '111' (USF = Free) indicates that the corresponding uplink radio block is reserved for the packet random access channel (PRACH) [14]. There are only 16 TA indexes for the continuous timing advance procedure to keep the MS synchronised with the TDMA frame at the BTS. If the maximum number of TBF multiplexed in UL is 7, the maximum number of TBF in DL is limited to 9 in the extreme case that the 16 TBF (7 in UL and 9 in DL) multiplexed on the same PDCH belongs to different MSs.

A special channel coding is used for the USF added to the different coding schemes that makes it quite reliable (even more than CS-1) and prevents uplink radio block losses. For multislot classes, the mobile must decode the USF in all PDCHs separately.

Whenever the MS detects an assigned USF value on an assigned PDCH, the MS shall transmit either a single RLC/MAC block or a sequence of four RLC/MAC blocks on the same PDCH. The number of RLC/MAC blocks to transmit is controlled by the USF_GRANULARITY parameter characterising the uplink TBF [14].

The fact that the MS has to receive a USF on each timeslot (TSL) introduces limitations on the maximum number of TSLs supported in the UL direction for certain MS classes. In most practical cases, with the simplest (E)GPRS radio implementation, the maximum number of TSLs in UL is limited to two. This can be seen in Figure 1.21 with a 3 + 2 MS allocation. Note that this limitation is very dependent on MS class.

Extended dynamic allocation. This procedure allows a higher number of TSLs in UL and eliminates the need of receiving USF on each TSL (which may create too many dummy blocks in DL). In this case, when a USF is received in a particular TSL, the MS is allowed to transmit on the same TSL and all the subsequent TSLs. Let us consider the allocation in Figure 1.22 where a 2 + 3 MS allocation is considered. When a USF is received on TSL 1, the MS will transmit on TSL 1 and 2. If simple dynamic allocation would be used, the MS would transmit only on TSL 1.

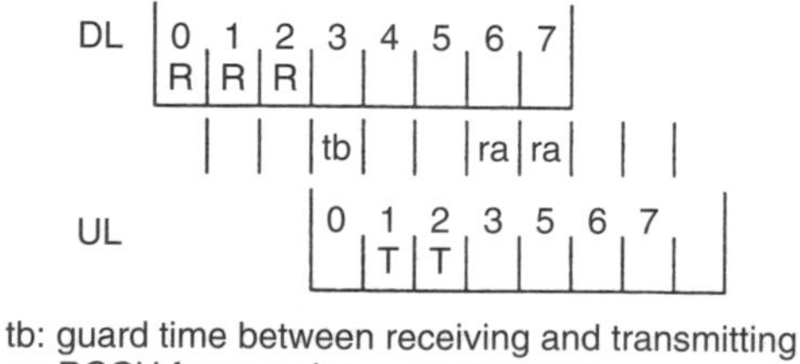

tb: guard time between receiving and transmitting
ra: BCCH frequencies measurements
R: MS receiving
T: MS transmitting

Figure 1.21. Dynamic UL allocation example

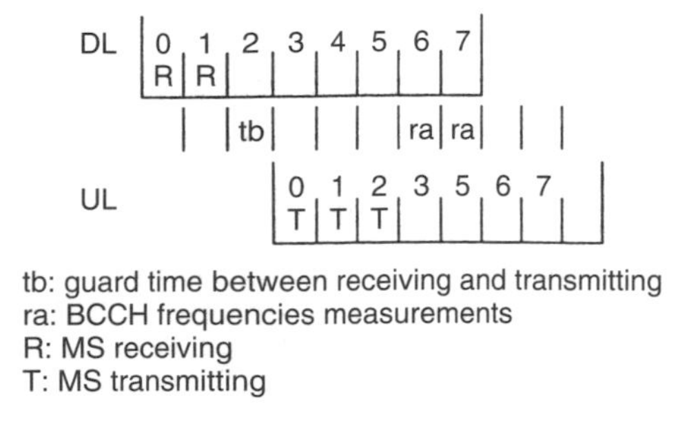

Figure 1.22. Extended dynamic UL allocation example

Fixed allocation. Using this procedure, the network communicates with the MS and UL transmission turn pattern by using a specific control message. In this case, USF is not utilised and, therefore, DL communication is point-to-point. This latter characteristic makes it more suitable for link optimisation techniques like DL power control and smart antennas. Once an MS has requested a resource assignment for the uplink, the network responds with a bitmap (ALLOCATION_BITMAP), indicating timeslots and radio blocks where the MS is allowed to transmit after a frame number contained in the parameter *TBF starting time*.

Time is slotted in *allocation periods* with a configurable duration that will be the basic time unit to assign resources for an uplink TBF. If more than one allocation is needed for the MS initial request, the network will send new bitmaps in packet ACK/NACK messages at the end of the current allocation period.

Modes for RLC/MAC Operation

In addition, there are two distinctive modes of operation of RLC/MAC layer:

- unacknowledged operation
- acknowledged operation.

Acknowledged mode for RLC/MAC operation. The transfer of RLC data blocks in the acknowledged RLC/MAC mode is controlled by a selective ARQ mechanism coupled with the numbering of the RLC data blocks within one temporary block flow. The sending side (the MS or the network) transmits blocks within a window and the receiving side sends packet uplink Ack/Nack or packet downlink Ack/Nack message when needed. Every such message acknowledges all correctly received RLC data blocks up to an indicated block sequence number (BSN). Additionally, the bitmap is used to selectively request erroneously received RLC data blocks for re-transmission.

Unacknowledged mode for RLC/MAC operation. The transfer of RLC data blocks in the unacknowledged RLC/MAC mode is controlled by the numbering of the RLC data blocks within one TBF and does not include any re-transmissions. The receiving side extracts user data from the received RLC data blocks and attempts to preserve the user information length by replacing missing RLC data blocks by dummy information bits.

Mobile-originated Packet Transfer

Uplink access. An MS initiates a packet transfer by making a packet channel request on PRACH or RACH, depending on the availability of PCCHs in the network. The network responds on PAGCH (if PCCHs are supported by the network) or access grant channel (AGCH), respectively. It is possible to use one- or two-phase packet access methods (see Figure 1.23).

In the one-phase access, the network will respond to the packet channel request by sending a packet uplink assignment message and reserving the resources on PDCH(s) for uplink transfer of a number of radio blocks. The reservation is done according to the requested resources indicated in the packet channel request. On RACH, there are only two cause values available for denoting GPRS, which can be used to request limited resources or two-phase access. On PRACH, the packet channel request may contain more adequate information about the requested resources, and, consequently, uplink resources on one or more PDCHs can be assigned by using the packet uplink assignment message.

In the two-phase access, the packet channel request is responded to with the packet uplink assignment, which reserves the uplink resources for transmitting the packet resource request. A two-phase access can be initiated by the network or an MS. The network can order the MS to send a packet resource request message by setting a parameter in packet uplink assignment message. The MS can require two-phase access in the packet channel request message. In this case, the network may order the MS to send a packet resource request or continue with a one-phase access procedure.

The packet resource request message carries the complete description of the requested resources for the uplink transfer. The MS can indicate the preferred medium access method to be used during the TBF. The network responds with the packet uplink assignment, reserving resources for the uplink transfer and defining the actual parameters for data transfer (e.g. medium access mode). If there is no response to the packet channel request within the pre-defined time period, the MS makes a retry after a random time.

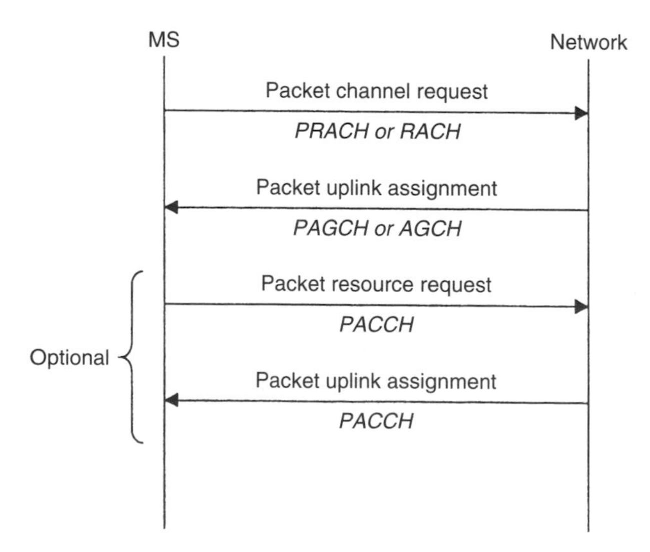

Figure 1.23. Access and allocation for the one or two-phase packet access, uplink packet transfer

On PRACH, a two-step approach is used and includes a long-term and a short-term estimation of the persistence [14]. The optimal persistence of the mobile stations is calculated at the network side.

The actual persistence values depend on

- the priority of the packet to be transmitted;
- the amount of traffic within higher priority classes;
- the amount of traffic within its own priority class.

Occasionally, more packet channel requests can be received than can be served. To handle this, a packet queuing notification is transmitted to the sender of the packet channel request. The notification includes information that the packet channel request message is correctly received and that the packet uplink assignment may be transmitted later. If the timing advance information becomes inaccurate for an MS, the network can send packet polling request to trigger the MS to send four random access bursts. This can be used to estimate the new timing advance before issuing the packet uplink assignment [14].

In case of dynamic or extended dynamic allocation, the packet uplink assignment message includes the list of PDCHs and the corresponding USF value per PDCH. A unique TFI is allocated and is thereafter included in each RLC data and control block related to that temporary block flow. The MS monitors the USFs on the allocated PDCHs and transmits radio blocks when the USF value indicates to do so. In case of fixed allocation, the packet uplink assignment message is used to communicate a detailed fixed uplink resource allocation to the MS. The fixed allocation consists of a start frame, slot assignment and block assignment bitmap representing the assigned blocks per timeslot [14]. The MS will wait until the start frame and then transmit radio blocks on those blocks indicated in the block assignment bitmap. The fixed allocation does not include the USF and the MS is free to transmit on the uplink without monitoring the downlink for the USF. If the current allocation is not sufficient, the MS may request additional resources in one of the assigned uplink blocks. A unique TFI is allocated and is thereafter included in each RLC data and control block related to that temporary block flow. Because each radio block includes an identifier (TFI), all received radio blocks are correctly associated with a particular LLC frame and a particular MS.

Contention Resolution

Contention resolution is an important part of RLC/MAC protocol operation, especially because one channel allocation can be used to transfer a number of LLC frames. There are two basic access possibilities, one-phase and two-phase access as seen in Figure 1.23. The two-phase access eliminates the possibility that two MSs can perceive the same channel allocation as their own. Basically the second phase of access will uniquely identify the MS by TLLI and that TLLI will be included in packet uplink assignment, therefore ruling out the possibility for a mistake.

Mobile Terminated Packet Transfer

Packet paging. The network initiates a packet transfer to an MS that is in the standby state by sending one or more packet paging request messages on the downlink (PPCH or PCH).

The MS responds to one packet paging request message by initiating a mobile-originated packet transfer. This mobile-originated packet transfer allows the MS to send a packet paging response message containing an arbitrary LLC frame. The message sequence described in Figure 1.24 is conveyed either on PCCCH or on CCCH. After the packet paging response is sent by the MS and received by the network, the mobility management state of the MS is ready.

The network can then assign some radio resources to the MS and perform the downlink data transfer.

Downlink packet transfer. The transmission of a packet to an MS in the ready state is initiated by the network using a packet downlink assignment message. In case there is an uplink packet transfer in progress, the packet downlink assignment message is transmitted on PACCH. Otherwise, it is transmitted in the PCCCH, in case there is one allocated in the cell. If that is not the case, it will be transmitted in the CCCH. The packet downlink assignment message includes the list of PDCH(s) that will be used for downlink transfer.

The network sends the RLC/MAC blocks belonging to one TBF on downlink on the assigned downlink channels. Multiplexing the RLC/MAC blocks addressed for different MSs on the same PDCH downlink is enabled with the TFI identifier, included in each RLC/MAC block [14].

The sending of the packet downlink Ack/Nack message is obtained through the periodical network-initiated polling of the MS. The MS sends the packet downlink Ack/Nack message in a reserved radio block, which is allocated together with polling.

Release of the resources. The release of the resources is initiated by the network by terminating the downlink transfer and polling the MS for a final packet downlink Ack/Nack message.

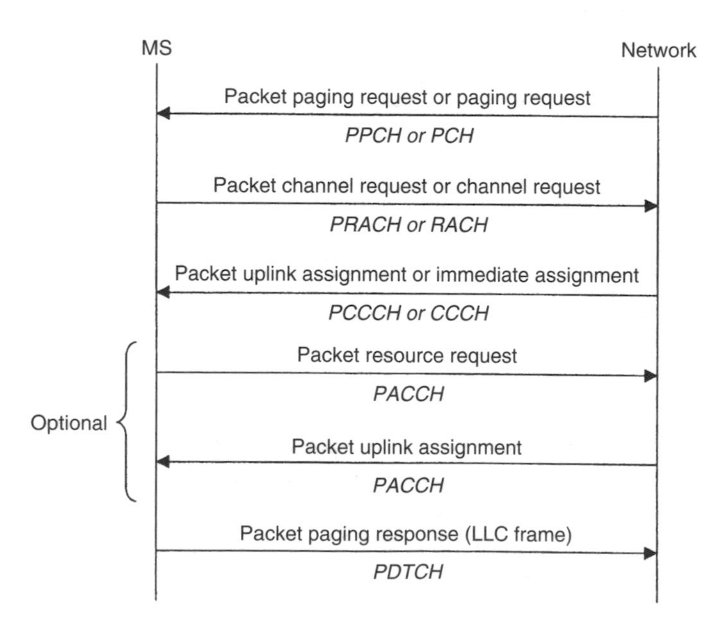

Figure 1.24. Paging message sequence for paging, downlink packet transfer

Simultaneous uplink and downlink packet transfer. During the ongoing uplink TBF, the MS continuously monitors one downlink PDCH for possible occurrences of packet downlink assignment or packet timeslot reconfigure messages on PACCH. The MS is therefore reachable for downlink packet transfers that can then be conveyed simultaneously on the PDCH(s) that respects the MS multislot capability.

If the MS wants to send packets to the network during the ongoing downlink TBF, it can be indicated in the acknowledgement that is sent from the MS. By doing so, no explicit packet channel requests have to be sent to the network. Furthermore, the network already has the knowledge of which PDCH(s) that particular MS is currently using so that the uplink resources can be assigned on the PDCH(s) that respect the MS multislot capability.

1.5 EDGE Rel'99

Enhanced data rates for GSM evolution (EDGE) is a major enhancement to the GSM data rates. GSM networks have already offered advanced data services, like circuit-switched 9.6-kbps data service and SMS, for some time. High-speed circuit-switched data (HSCSD), with multislot capability and the simultaneous introduction of 14.4-kbps per timeslot data, and GPRS are both major improvements, increasing the available data rates from 9.6 kbps up to 64 kbps (HSCSD) and 160 kbps (GPRS).

EDGE is specified in a way that will enhance the throughput per timeslot for both HSCSD and GPRS. The enhancement of HSCSD is called *ECSD* (enhanced circuit-switched data), whereas the enhancement of GPRS is called *EGPRS* (enhanced general packet radio service). In ECSD, the maximum data rate will not increase from 64 kbps because of the restrictions in the A-interface, but the data rate per timeslot will triple. Similarly, in EGPRS, the data rate per timeslot will triple and the peak throughput, with all eight timeslots in the radio interface, will reach 473 kbps.

1.5.1 8-PSK Modulation in GSM/EDGE Standard

The 'enhancement' behind tripling the data rates is the introduction of the 8-PSK (octagonal phase shift keying) modulation in addition to the existing Gaussian minimum shift keying. An 8-PSK signal is able to carry 3 bits per modulated symbol over the radio path, while a GMSK signal carries only 1 bit per symbol (Table 1.5). The carrier symbol rate (270.833 kbps) of standard GSM is kept the same for 8-PSK, and the same pulse shape as used in GMSK is applied to 8-PSK. The increase in data throughput does not come for free, the price being paid in the decreased sensitivity of the 8-PSK signal. This affects, e.g. the radio network planning, and the highest data rates can only be provided with limited coverage. The GMSK spectrum mask was the starting point for the spectrum mask of the 8-PSK signal, but along the standardisation process, the 8-PSK spectrum mask was relaxed with a few dB in the 400 kHz offset from the centre frequency [18]. This was found to be a good compromise between the linearity requirements of the 8-PSK signal and the overall radio network performance.

It was understood already at the selection phase of the new modulation method that a linear modulation would have quite different characteristics than the GMSK as a constant envelope modulation. A practical problem is how to define the receiver performance requirements for the physical channels with 8-PSK modulation, in particular, how to incorporate the phenomena resulting from the non-idealities in the transmitter and receiver. This

Table 1.5. Key physical layer parameters for GSM/EDGE

	8-PSK	GMSK
Symbol rate	270.833 kbps	270.833 kbps
Number of bits/symbol	3 bits/symbol	1 bit/symbol
Payload/burst	342 bits	114 bits
Gross rate/timeslot	68.4 kbps	22.8 kbps

problem is solved in the specification by using the concept of error vector of magnitude (EVM). The EVM effectively measures how far an observed signal sample is from the ideal signal constellation point (Figure 1.25). Thus the EVM takes into account the deviation in both I- and Q-axes, not explicitly distinguishing whether the signal impairment is due to the phase noise or the amplitude distortion. For the 8-PSK signal, [18] sets specific EVM percentage requirements for the transmitter and receiver performance.

A further peculiarity, introduced by bringing the second modulation method, 8-PSK, to GSM standard, is the need in downlink to blindly recognise the transmitted modulation method in the mobile station receiver. This is due to the characteristics of the EGPRS link quality control (LQC), where the used modulation and coding scheme (MCS) is adjusted according to the channel conditions to the most suitable one, and in downlink, no prior information is sent to the receiver but the receiver is able to find out the used MCS on the basis of (1) the blind modulation identification and (2) the decoding of the RLC/MAC header field that contains indication of the coding and puncturing scheme. The modulation identification is based on the different phase rotation characteristics in the GMSK and 8-PSK training sequences. In the GMSK training sequence, the symbol-by-symbol phase rotation is p/2, whereas in the 8-PSK training sequence, the rotation is 3p/8. Otherwise, the set of 8-PSK training sequences has identical information content (the same 26-bit sequence) as the GMSK training sequences.

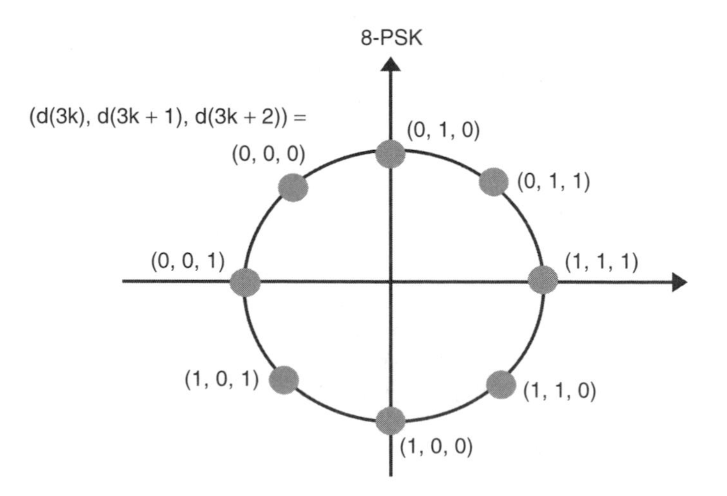

Figure 1.25. 8-PSK signal constellation

1.5.2 Enhanced General Packet Radio Service (EGPRS)

EGPRS is built on top of GPRS, which is the packet-switched data service of GSM. Examples of typical high bit rate packet services include fast file transfer, Internet service access, web browsing and remote email.

EGPRS has a major impact on the RF and the physical layer of the radio interface as well as on RLC/MAC protocol, but the changes to other protocols and protocol layers are minor. The EDGE RF specification and the definition of the burst structures are common to both EGPRS and ECSD. One large conceptual modification in EGPRS, compared with GPRS, is the link quality control which, in EGPRS, also supports incremental redundancy (type II hybrid ARQ), in addition to GPRS-type link adaptation mode (type I hybrid ARQ). The LQC includes nine different modulation and coding schemes (MCS-1–MCS-9) as well as related signalling and other procedures for link adaptation (switching between different MCSs).

Table 1.6 shows the EGPRS modulation and coding schemes (MCS) and their data throughputs. New GMSK coding schemes (MCS-1–MCS-4), different from GPRS GMSK coding schemes (CS-1–CS-4), are needed because of the incremental redundancy support. Figure 1.26 illustrates the coding and puncturing principle of the header and payload part of an MCS-9 radio block. In EGPRS radio blocks, the stealing bits indicate the header format. The RLC/MAC header is strongly coded (close to rate 1/3 code in 8-PSK modes and close to rate 1/2 code in GMSK modes) to allow reliable decoding of the header in incremental redundancy (IR) operation. There is also a separate header check sequence (HCS) for the error detection in the header part. The RLC data payload is encoded along with the final block indicator (FBI) and extension bit (E) fields. Also, the block check sequence (BCS) and tail bits (TB) are added before encoding. In the illustrated case (MCS-9), two RLC data blocks are mapped on one radio block. Note that in the case of MCS-8 and MCS-9, an RLC data block is interleaved over two bursts only. The RLC/MAC header is though interleaved over the whole radio block (i.e. four bursts).

EGPRS modulation and coding schemes are organised in families according to their RLC data block sizes. For example, the MCS-9 carries 1184 payload bits in two RLC data blocks (592 payload bits in each of them) during one EGPRS radio block of four consequent bursts. The MCS-6 carries 592 payload bits in one RLC data block and the MCS-3 carries 296 payload bits within the radio block. The number of payload bits in

Table 1.6. EGPRS modulation and coding schemes

Modulation and coding scheme	Code rate	Modulation	Data rate/ timeslot (kbps)	Family
MCS-9	1.0	8-PSK	59.2	A
MCS-8	0.92		54.4	A
MCS-7	0.76		44.8	B
MCS-6	0.49		29.6	A
MCS-5	0.37		22.4	B
MCS-4	1.0	GMSK	17.6	C
MCS-3	0.80		14.8	A
MCS-2	0.66		11.2	B
MCS-1	0.53		8.8	C

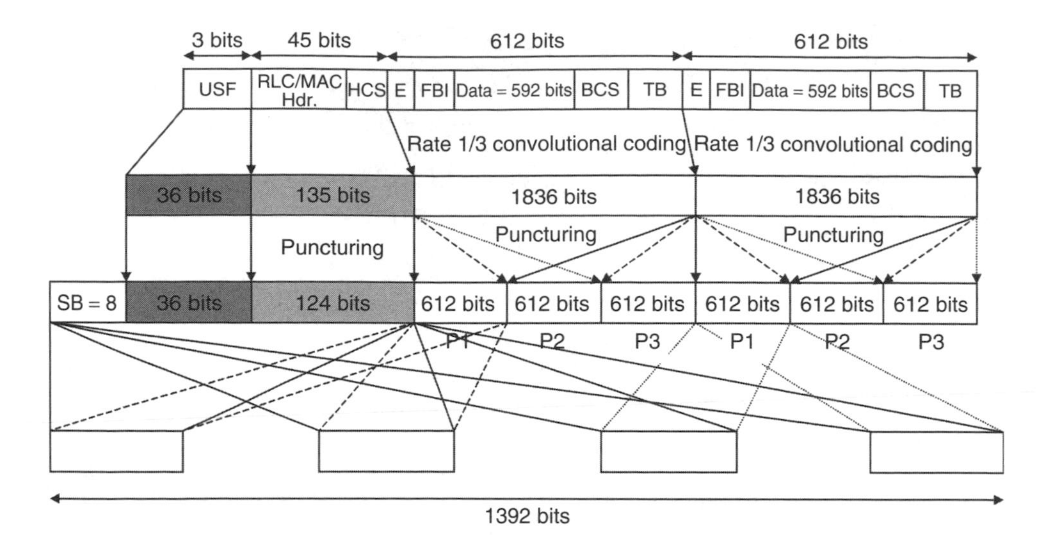

Figure 1.26. EGPRS coding and puncturing example (MCS-9: uncoded 8-PSK, two RLC blocks per 20 ms) [14]

MCS-6 and MCS-3 being sub-multiples of the higher coding schemes in the same family (family A) enables effective re-transmission of the negatively acknowledged RLC data blocks with the lower coding schemes within the family, if needed. A typical case where this functionality is useful is the sudden change in the channel conditions.

The transfer of RLC data blocks in the acknowledged RLC/MAC mode can be controlled by a selective type I ARQ mechanism or by type II hybrid ARQ (IR) mechanism, coupled with the numbering of the RLC data blocks within one temporary block flow. The sending side (the MS or the network) transmits blocks within a window and the receiving side sends packet uplink Ack/Nack or packet downlink Ack/Nack message when needed.

The initial MCS for an RLC block is selected according to the link quality. For the re-transmissions, the same or another MCS from the same family of MCSs can be selected. For example, if MCS-7 is selected for the first transmission of an RLC block, any MCS of the family B can be used for the re-transmissions. The selection of MCS is controlled by the network.

In the EGPRS type II hybrid ARQ scheme, the information is first sent with one of the initial code rates (i.e. rate 1/3 encoded data are punctured with PS 1 of the selected MCS). If the RLC data block is received in error, additional coded bits (i.e. the output of the rate 1/3 encoded data that is punctured with PS 2 of the prevailing MCS) are sent and decoded together with the already received code words until decoding succeeds. If all the code words (different punctured versions of the encoded data block) have been sent, the first code word (which is punctured with PS 1) is sent. Alternatively, it is possible to use incremental redundancy modes called *MCS-5–7* and *MCS-6–9*, in which the initial transmissions are sent with either MCS-5 or MCS-6 (respectively) and the re-transmissions are sent with MCS-7 or MCS-9 (respectively). In the EGPRS type I ARQ, the operation is similar to that of the EGPRS type II hybrid ARQ, except that the decoding of an RLC data block is solely based on the prevailing transmission (i.e. erroneous blocks are not stored). The motivation for supporting the incremental redundancy mode is the improved data throughput and the robustness for varying channel conditions and measurement errors. Incremental redundancy is defined to be mandatory for the MS and optional for the network.

Another RLC/MAC layer modification in Rel'99 due to EGPRS is the increased RLC window size. In GPRS with four-burst radio blocks (20 ms), the RLC window size of 64 is defined. With the higher coding schemes of EGPRS, MCS-7 to MCS-9, there are two RLC blocks per 20 ms radio block, which makes the RLC window size 64 too small and RLC protocol subject to stalling. Therefore, the EGPRS RLC window size was increased to 128, and a compression method was defined for the acknowledgement bitmap of the RLC blocks (see details in GSM 04.60).

1.5.2.1 Bit Error Probability (BEP) Measurements

In Rel'99, a generic enhancement to the channel quality measurements was also introduced. Earlier, the measure to be used for the channel quality indication was the RX_QUAL, which is an estimate of the pseudo bit error rate, i.e. computed by comparing the received bit sequence (before decoding) with the encoded version of the decoded bit sequence and counting the number of bit errors. It was shown in [19] that the alternative measure—bit error probability (BEP)—better reflects the channel quality in varying channel conditions—including frequency hopping and varying mobile speed. BEP is estimated burst by burst, for example, from the soft output of the receiver. The detailed realisation of the measurement itself is left for the equipment vendors, whereas the method of reporting the BEP measurements and the accuracy requirements are standardised. The measurements to be reported by the MS for the network are MEAN_BEP and CV_BEP, where the MEAN_BEP is the average of the block wise mean BEP during the reporting period and the CV_BEP is the average of the blockwise coefficient of variation (CV) (CV = Std(BEP)/Mean(BEP)) during the reporting period.

1.5.2.2 Adjustable Filtering Length for the EGPRS Channel Quality Measurements

A further standardised means of optimising the EGPRS performance is the filtering length of the LQC measurements, which can be varied [20]. The network and the MS can negotiate with each other—using the radio link control signalling—the optimum length of the filter that is used to smooth the BEP measurements that the MS sends to the network. For example, typically the filter length should be shorter for the MS with higher speed, so that the received channel quality information at the network would be as correct as possible. On the other hand, for slowly moving mobiles, the longer averaging of the measurements improves the accuracy.

1.5.3 Enhanced Circuit-switched Data (ECSD)

ECSD uses current HSCSD as a basis. The user data rates are not increased compared to HSCSD (up to 64 kbps), but these rates can be achieved with smaller numbers of timeslots and simpler MS implementation. Data rates to be provided with ECSD, although limited to 64 kbps, are still sufficient for providing various transparent and non-transparent services. ECSD enables inter-working with audio modems at higher data rates than in current GSM

Table 1.7. ECSD data rates

Coding scheme	Code rate	Modulation	Gross rate (kbps)	Radio interface rate (kbps)	User rates (kbps)
TCH/28.8 (NT/T)	0.419	8-PSK	69.2	29.0	28.8
TCH/32.0 (T)	0.462	8-PSK	69.2	32.0	32.0
TCH/43.2 (NT)	0.629	8-PSK	69.2	43.5	43.2

networks, inter-working with ISDN at various data rates and various video-based services ranging from still image transfer to videoconferencing services.

Higher data rates are defined for both transparent and non-transparent services. Radio interface rates (and user rates per timeslot) for ECSD are shown in Table 1.7. In the definition of the ECSD, only a few new traffic channels are defined. The control channels—both the common, dedicated and associated control channels—are the same as for the other circuit-switched services (like, e.g. for the 9.6 kbps circuit-switched data, HSCSD) except the fast associated control channel for the ECSD (FACCH/E), which uses GMSK modulation and is interleaved rectangularly on the four consequent TDMA frames. Further, the detection of the FACCH/E frame is based on the similar principles as the identification of the 8-PSK and GMSK modulated data blocks in EGPRS, i.e. the modulation method is blindly detected in the receiver, and if a block of four GMSK bursts is received in the middle of the 8-PSK data frames, the receiver knows that an FACCH/E frame was received.

The 28.8-kbps data rate is available for transparent and non-transparent services for both single and multislot configurations. The 32.0-kbps service is available only in multislot configuration (two-slot) and can be used for offering 64-kbps transparent service, while 43.2 kbps is available for non-transparent services only.

The ECSD architecture is largely based on HSCSD transmission and signalling. This ensures a minimum impact on existing specifications. The basic principle is to use the same transcoder and rate adaptation unit (TRAU) frame formats and multiple 16-kbps sub-channels in the network side. For example, 28.8-kbps service provided with one radio interface timeslot is supported with two 16-kbps sub-channels and 14.4-kbps TRAU frames on the network side.

Figure 1.27 shows the network architecture for providing (a) 57.6-kbps non-transparent service and (b) 56-/64-kbps transparent service.

ECSD 57.6-kbps non-transparent service is provided using two timeslots of 28.8 kbps as shown in Figure 1.27(a). The same frame formats as in HSCSD are used end to end between the MS/TAF and MSC/IWF. The BTS is responsible for the de-multiplexing of data between two radio timeslots and four Abis sub-channels.

The network architecture for supporting bit transparent 56-/64-kbps service is shown in Figure 1.27(b). It is provided using two air interface timeslots of 32 kbps. The rate adaptation functions are located in the BTS.

Higher data rates can be offered with limited coverage and, therefore, the link adaptation mechanism becomes essential. Switching between 8-PSK- and GMSK-based channels is done with the standard intra-cell handover procedure. Signalling for link adaptation is based on the existing HSCSD signalling mechanism.

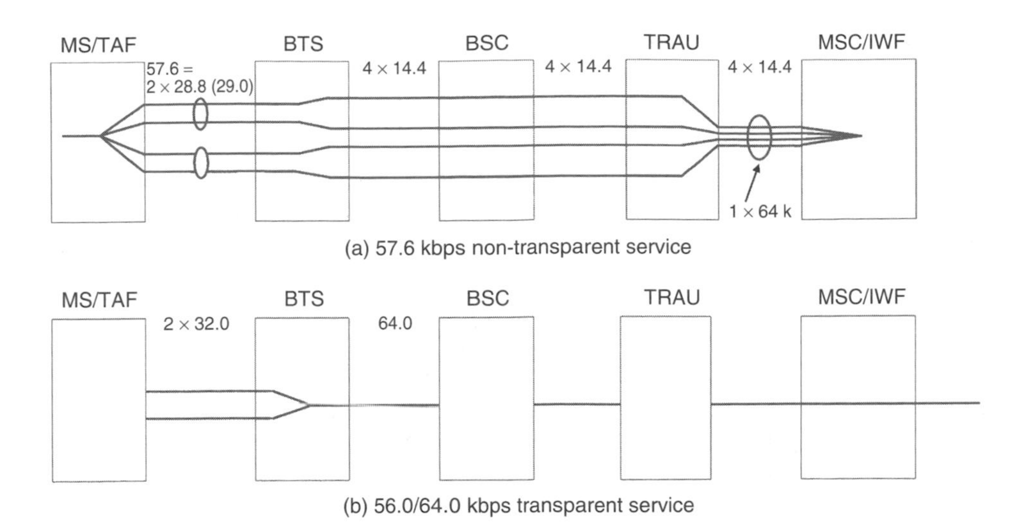

Figure 1.27. Network architecture

1.5.4 Class A Dual Transfer Mode (DTM)

With a Class A mobile station, a subscriber can simultaneously connect to the circuit-switched core network (to mobile switching centre (MSC)) and to the packet-switched core network (to SGSN), through A and Gb interfaces respectively. Until recently, the implementation of a Class A MS has been considered impractical and costly—owing to the uncoordination of the radio resources for the CS and PS connections, which implies duplicating many key parts of an MS like the transceiver. To enable practical low cost implementation of the Class A mobile, the Class A dual transfer mode (DTM) was defined. This functionality was deemed necessary in the Rel'99 specification—at the same schedule with the first UTRAN standard release (Rel'99) and with the inter-working between GSM and UMTS (e.g. inter-system handovers). This package of specifications ensures that an operator can provide the same type of service—e.g. simultaneous voice and data—in both radio access networks, GSM and UTRAN.

The DTM concept includes the following:

- *The single timeslot method.* TCH/H + PDTCH/H. In this method, AMR (adaptive multi-rate speech codec) voice frames are carried in the half-rate circuit-switched channel (TCH/H) and the user data in the half-rate packet-switched channel (PDTCH/H). The significance of the single timeslot method is that it enables implementation of the Class A DTM feature even in single timeslot MSs. Another advantage is that in case of a handover, the availability of the radio resources in the target cell can be easily guaranteed.
- *The multiple timeslot method.* TCH + one or more adjacent PDTCH. In this method, one full timeslot is reserved for the AMR voice frames and the adjacent timeslots are used for the user data transfer.

A further significant difference between the single and multiple timeslot methods is that in the former one, the PDTCH/H resource is always dedicated for the DTM user (dedicated mode of (E)GPRS MAC protocol), whereas in the latter one, the PDTCH resources can be shared (shared mode of (E)GPRS MAC protocol).

More details of Class A DTM can be found, e.g. in [21, 22].

1.5.5 EDGE Compact

Compact is a particular mode—or variant—of EGPRS, which is designed for the deployment in the narrow frequency band allocations—the minimum requirement being the deployment in less than 1 MHz of spectrum. It is known that the capacity (spectral efficiency) of the GSM network is limited by the BCCH reuse, and as available bandwidth decreases, its relative impact will be higher. This is due to the nature of the broadcast control channel concept in the GSM standard, which requires continuous transmission with constant transmit power in the beacon frequency (BCCH frequency). This causes the BCCH frequency reuse to be relatively large to ensure reliable reception of the system information in the broadcast and common control channels.

The key component of the compact mode is the new control channel structure, which is based on the discontinuous time rotating BCCH, which enables significantly lower reuse factors for the BCCH channel and thus operation in the narrow spectrum allocation. The characteristics of the compact mode are the following:

- It is a stand-alone high-speed packet data system (EGPRS variant with alternative control channel structure).
- It can be deployed in only 600 kHz of spectrum (+ guardband) by using three carriers in a 1/3 reuse pattern.
- It requires inter–base station time synchronisation.
- It uses discontinuous transmission on the control carriers and a new logical control channel combination (compact packet broadcast control channel (CPBCCH) based on a standard 52-multiframe). It uses timeslot mapping of control channels in a rotating fashion, which makes neighbour channel measurements more feasible during traffic mode.

More details on compact mode can be found in [23].

1.5.6 GPRS and EGPRS Enhancements in Rel'4

In GERAN Rel'4, a few new features are added to the GPRS and EGPRS protocols and signalling to improve the efficiency of the data transfer with certain typical traffic characteristics and the seamlessness of the service in the cell change. These features, the delayed temporary block flow release and the network-assisted cell change (NACC), are introduced briefly in the following sections.

1.5.6.1 Extended UL TBF

The early GPRS protocols are inefficient when dealing with the bursty TCP/IP traffic. The GPRS radio interface protocols are initially designed to free the unused radio resources as soon as possible. With bursty IP traffic, this may lead to the frequent set-up and release of the radio resources (more specifically TBFs), which results in an increased signalling load and inefficiency in the data transfer, as the release and set-up of the TBFs takes some time, and radio interface resources. Unnecessary TBF set-ups and releases can be avoided by delaying the release of the TBF. In DL direction during the inactive period, the connection is maintained by periodically sending dummy LLC frames in the downlink [20]. This DL enhancement was already introduced in Rel'97 specifications. Its counterpart in UL direction is called *Extended UL TBF* as was specified in Rel'4. In extended UL TBF, the UL TBF may be maintained during inactivity periods (the MS does not have more RLC information to send), and the network determines the release of the UL TBF.

1.5.6.2 Network-assisted Cell Change

The GPRS cell change was not initially designed for the services that would require seamless cell change operation, such as real-time services. Later on, these requirements gained more importance and a new type of cell change, network-assisted cell change (NACC), was defined. In the NACC, the network can send neighbour cell system information to an MS. The MS can then use this system information when making the initial access to a new cell after the cell change. This way, the MS does not need to spend some time in the new cell receiving the system information, and the typical break in the ongoing service decreases from seconds to a few hundreds of milliseconds [24].

References

[1] Kurronen K., *Evolution of GSM Towards the Third Generation Requirements*, Asian Institute of Technology (AIT) and Nokia, Hanoi, Vietnam, 25–26, 1997.

[2] Pirhonen R., Rautava T., Penttinen J., 'TDMA Convergence for Packet Data Services', *IEEE Personal Commun.*, **6**(3), 1999, 68–73.

[3] Järvinen K., 'Standardisation of the Adaptive Multi-rate Codec', *Proc. X European Signal Processing Conference (EUSIPCO)*, Tampere, Finland, 4–8, 2000.

[4] 3GPP TS 05.08 V8.6.0 (2000-09), 3rd Generation Partnership Project; Technical Specification Group GERAN; Digital Cellular Telecommunications System (Phase 2+); Radio Subsystem Link Control (Release 1999).

[5] 3GPP TS 05.02 V8.10.0 (2001-08), 3rd Generation Partnership Project; Technical Specification Group GSM/EDGE Radio Access Network; Digital Cellular Telecommunications System (Phase 2+); Multiplexing and Multiple Access on the Radio Path (Release 1999).

[6] 3GPP TS 05.04, Modulation, 3rd Generation Partnership Project; Technical Specification Group GERAN.

[7] GSM 05.05 Version 8.8.0 Release 1999: Digital Cellular Telecommunications System (Phase 2+); Radio Transmission and Reception.

[8] GSM 05.03 Version 8.11.0 Release 1999: Digital Cellular Telecommunications System (Phase 2+); Channel Coding.

[9] 3GPP TS 05.10, Radio Subsystem Synchronisation, 3rd Generation Partnership Project; Technical Specification Group GERAN.

[10] 3GPP TS 04.64, LLC Specification GPRS, 3rd Generation Partnership Project; Technical Specification Group GERAN.

[11] 3GPP TS 04.65, GPRS Subnetwork Dependent Convergence Protocol (SNDCP), 3rd Generation Partnership Project; Technical Specification Group GERAN.
[12] 3GPP TS 48.018, General Packet Radio Service (GPRS); Base Station System (BSS)—Serving GPRS Support Node (SGSN); BSS GPRS Protocol, 3rd Generation Partnership Project; Technical Specification Group GERAN.
[13] 3GPP TS 09.60, GPRS Tunnelling Protocol (GTP) across the Gn and Gp Interface, 3rd Generation Partnership Project; Technical Specification Group GERAN.
[14] 3GPP TS 03.64, General Packet Radio Service (GPRS); Overall Description of the GPRS Radio Interface; Stage 2, 3rd Generation Partnership Project; Technical Specification Group GERAN.
[15] 3GPP TS 03.60, GPRS Stage 2, 3rd Generation Partnership Project; Technical Specification Group GERAN.
[16] 3GPP TS 04.07, Mobile Radio Interface Signalling Layer 3—General Aspects, 3rd Generation Partnership Project; Technical Specification Group GERAN.
[17] GSM 04.60 V8.1.0 (1999-11), European Standard (Telecommunications Series), Digital Cellular Telecommunications System (Phase 2+); General Packet Radio Service (GPRS); Mobile Station (MS)–Base Station System (BSS) Interface; Radio Link Control/Medium Access Control (RLC/MAC) Protocol (GSM 04.60 Version 8.1.0 Release 1999).
[18] 3GPP TS 45.005, Radio Transmission and Reception, 3rd Generation Partnership Project; Technical Specification Group GERAN.
[19] Tdoc SMG2, 464/99, *Link Quality Control Measurements for EGPRS*, Dublin, Ireland, April 12–16, 1999.
[20] 3GPP TS 44.060, General Packet Radio Service (GPRS); Mobile Station (MS)–Base Station System (BSS) Interface; Radio Link Control/Medium Access Control (RLC/MAC) Protocol, Radio Resource Control Protocol, 3rd Generation Partnership Project; Technical Specification Group GERAN.
[21] 3GPP TS 43.055, Dual Transfer Mode, 3rd Generation Partnership Project; Technical Specification Group GERAN.
[22] Pecen M., Howell A., 'Simultaneous Voice and Data Operation for GPRS/EDGE: Class A Dual Transfer Mode', *IEEE Personal Commun.*, **8**, 2001, 14–29.
[23] ETSI SMG2 Tdoc 1549/99, Concept Proposal for EGPRS-136, Revision 1.6.
[24] 3GPP TSG GERAN #3 GP-010361 Network Assisted Cell Change; Concept Document.

2

Evolution of GERAN Standardisation (Rel'5, Rel'6 and beyond)

Eero Nikula, Shkumbin Hamiti, Markus Hakaste and Benoist Sebire

After the introduction of the Gb interface in Rel'97, all the consequent global system for mobile communications (GSM) standard releases (Rel'98, Rel'99, Rel'4) have been based on the assumption that there are two separate core network (CN) interfaces in GSM/GPRS (global system for mobile communications/General Packet Radio Service)—namely, the A interface between the base station subsystem (BSS) and the mobile switching centre (MSC) for the circuit-switched (CS) services, and the Gb interface between the BSS and the serving GPRS support node (SGSN) for the packet-switched (PS) services. At the same time, the universal mobile telecommunications service (UMTS) Rel'99 has adopted much of the functionalities of the GSM/GPRS core network, modified the functional split between the radio access network (RAN) and the core network and evolved the CN further. In the UMTS CN, the interface between the RAN and the CN is called Iu—with two component interfaces, the Iu-cs for the CS services and the Iu-ps for the PS services.

Other key points in the introduction to the GSM/EDGE Rel'5 development are the service continuity request of the operators who are deploying both GSM/EDGE and wideband code division multiple access (WCDMA) networks, and the dual mode nature of the handsets supporting these two different radio access technologies. It would be clearly beneficial for the whole GSM/WCDMA community—operators, service providers, subscribers, equipment manufacturers to name a few—if the similar set of services could be provided through both radio access technologies, GSM/EDGE and WCDMA. Against this assumption, there are two main alternatives for the GSM/EDGE to evolve, either to develop further the functionality and the capabilities of the A and Gb interfaces to match the same requirements as those set for the UMTS or to adopt the UMTS Iu interface

GSM, GPRS and EDGE Performance 2nd Ed. Edited by T. Halonen, J. Romero and J. Melero
 ISBN: 0-470-86694-2

and modify the GSM/EDGE BSS functionality accordingly. With the adoption of the UMTS Iu interface and the UMTS quality-of-service (QoS) architecture, GSM/EDGE radio access network (GERAN) and UMTS terrestrial radio access network (UTRAN) can be efficiently integrated under a single UMTS multi-radio network.

In the 3rd Generation Partnership Project (3GPP) GERAN Rel'5, the adoption of the Iu interface was selected as a way to go. Although this selection probably causes larger changes to the standard in the short term, it is a more future-proof solution—allowing the same services and CN solutions to be applied to both GERAN and UTRAN fluently. Section 2.1 presents an overview of the GERAN Rel'5 features with Iu alignment as a major topic. Section 2.2 explains the basics of the new architecture for GERANs that are currently being specified in 3GPP Rel'5, Section 2.3 gives an overview of Rel'6 features.

2.1 GERAN Rel'5 Features

In the 3GPP Rel'5 overall, the largest new functionality is the Internet multimedia subsystem (IMS) domain of the core network (Figure 2.1). The IMS provides a set of new services that are not available in the CS and PS domains of the Rel'4 CN. Some examples of the new IMS services are presence, multimedia and mixed media interactive communication [1].

IMS can be considered as a new architecture that provides a platform to launch a flexible set of new Internet-based multimedia services. This will have a major relevance to the

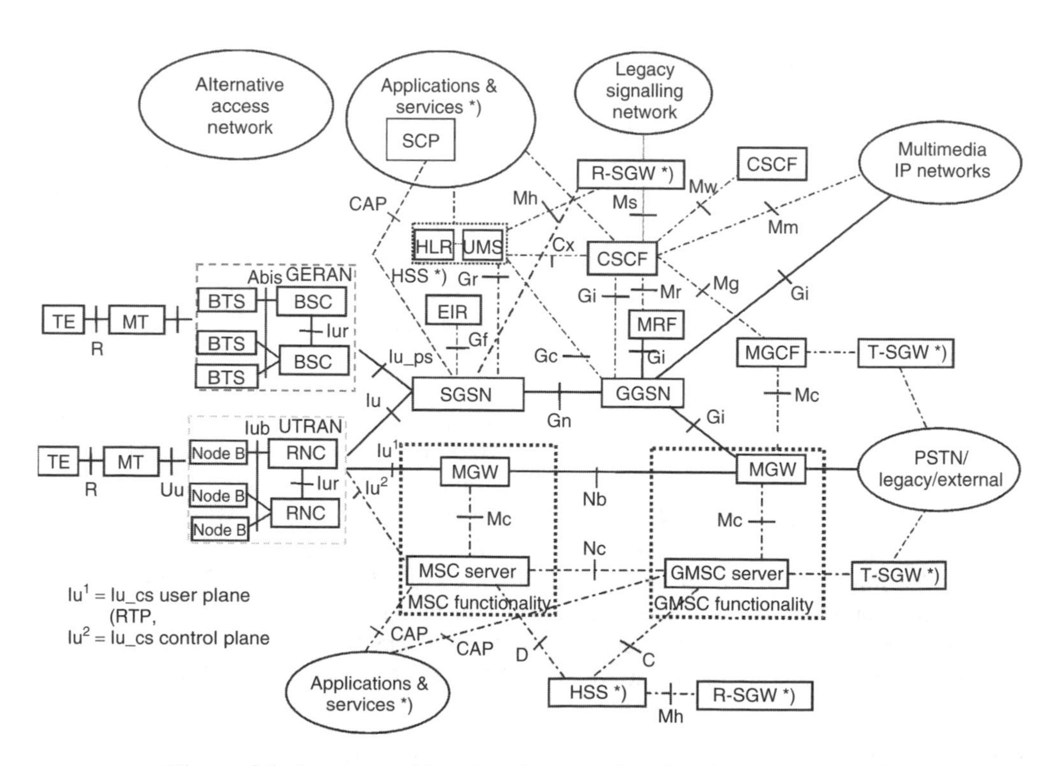

Figure 2.1. Internet multimedia subsystem domain of the core network

3rd-generation (3G) service provision and the subsequent operators' revenue generation. With the support of the Iu interface, the GERAN network can be directly connected to IMS, seamlessly delivering these services.

From the GERAN perspective, the support for the IMS services implies the following enhancements:

- Introduction of the Iu interface.
- Definition of the header adaptation mechanism for the real-time protocol (RTP)/user datagram protocol (UDP)/Internet protocol (IP) traffic. The mechanism can be either header compression or header removal or both.

The Iu interface for GERAN and the support for the header adaptation are standardised in Rel'5. The other major GERAN enhancements in Rel'5 are the following:

- wideband adaptive multi-rate (AMR) speech for enhanced speech quality
- half-rate octagonal-phase shift keying (8-PSK) speech for improved speech capacity
- fast power control for speech
- location service enhancements for Gb and Iu interfaces
- inter-BSC and base station controller (BSC)/radio network controller (RNC) network-assisted cell change (NACC).

The following subsections highlight the main characteristics of the new Rel'5 features; refer to the other chapters of this book for more details and given references for the more detailed descriptions.

2.1.1 Iu Interface for GERAN and the New Functional Split

Adopting the Iu interface and aligning the GERAN architecture with the UTRAN architecture is the major change to the GSM/EDGE Rel'5 standard. Not too many modifications are needed for the Iu-cs and Iu-ps interfaces themselves—they can be more or less used as such as they are in the existing specifications—but the majority of the modifications lies on the higher layers of the radio interface protocols, especially in the radio resource control (RRC) protocol. The RRC protocol happens to be the central point where the requirements from the two different worlds—namely, the legacy GSM/EDGE physical layer (PHY) and radio link control (RLC)/medium access control (MAC), and the expectations from the UMTS core network—conflict with each other.

Figure 2.2 shows the GERAN Rel'5 architecture. The legacy interfaces of GSM/EDGE—A and Gb—are intentionally included in the figure as they continue to coexist in the GERAN standard, and for many years to come they will coexist in the live networks in the field.

Generally speaking, the GERAN Iu alignment work includes the following:

- Adoption of the Iu interface. Both Iu-cs and Iu-ps interfaces are supported.
- Definition of the Iur-g interface between two GERAN networks. The Iur-g is a subset of the UTRAN Iur interface—enhanced with GERAN-specific requirements.

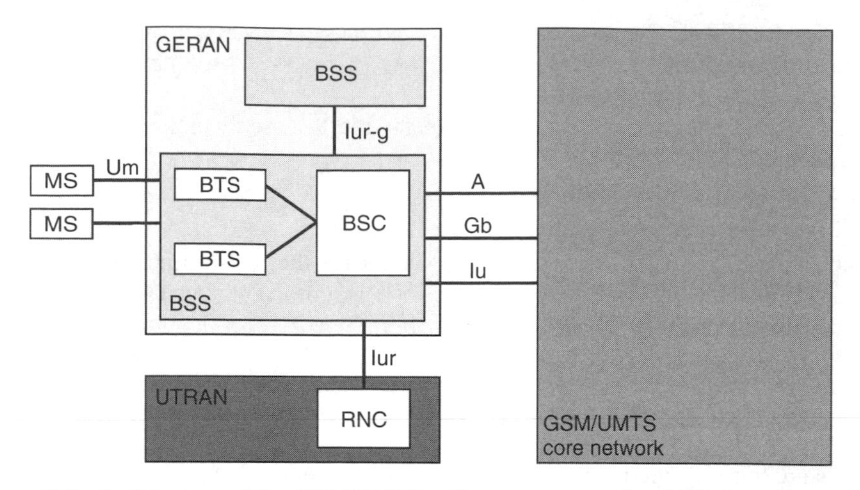

Figure 2.2. Architecture of the GSM/EDGE radio access network with Iu interface support

- Aligning the GERAN user plane and control plane protocol stacks with UTRAN. This implies that in the user plane, the subnetwork dependent convergence protocol (SNDCP) and the logical link control (LLC) protocols are replaced with the packet data convergence protocol (PDCP). In the control plane, the radio-related functions will be moved from the CN to the RAN and defined as part of the new RRC protocol.
- Adoption of the PDCP protocol from UTRAN. No major changes are needed to the UTRAN PDCP protocol for supporting GERAN, too.
- Definition of the GERAN RRC protocol. This constitutes the major part of the needed changes to align the GERAN architecture with UTRAN. The GSM radio resource (RR) protocol is a large one, and the new functional split between the GERAN and CN results in a large amount of modifications.
- The RLC and MAC protocols are not much affected by the new functional split as they are below the PDCP and RRC protocols in the user and control plane stacks, respectively. However, some enhancements are still needed, such as the following:
- Dedicated MAC mode (dedicated channels in general) for the new user plane protocol stack.
- Replacement of the link access protocol for the Dm channel (LAPDm) protocol with the RLC and MAC functions.
- Multiple simultaneous temporary block flows (TBFs) for a mobile station. This can partly be seen as a part of the functionality alignment with the UTRAN as multiple TBFs enable the support of the multiple parallel traffic flows with different QoS requirements.

Section 2.2 gives a deeper description of the new GERAN architecture and protocols. In the following paragraphs, the other new GERAN Rel'5 features—independent of the Iu alignment—are shortly visited.

2.1.2 Header Adaptation of the IP Data Streams

The header adaptation function is needed for the efficient transfer of the IP data streams—whether real time (e.g. RTP/UDP/IP) or non-real time (e.g. TCP/IP)—over the radio interface. The header adaptation is a function in the non-transparent mode of the PDCP protocol.

In the header compression, the transport and network level headers (e.g. RTP/UDP/IP) are compressed in such a way that the decompressed headers are semantically identical to the original uncompressed headers. Header compression is suited for standard Internet applications that are not designed to work only with GERAN and especially for multimedia applications, and therefore the scheme will be used with generic real-time multimedia bearers.

The header compression algorithms are standardised in the Internet Engineering Task Force (IETF) Robust Header Compression (ROHC) Working Group (WG) but the application of the IETF ROHC scheme to the 3GPP radio access technologies is jointly defined by the IETF and 3GPP.

2.1.3 Speech Capacity and Quality Enhancements

GERAN Rel'5 contains a number of new features that are aimed at the capacity increase and the quality improvement of the speech services.

2.1.3.1 Wideband AMR Speech

Wideband AMR codec is a new speech codec in the 3GPP specifications. Earlier, GSM has had the full rate (FR), enhanced full rate (EFR) and adaptive multi-rate (AMR) speech codecs. AMR codec was standardised in GSM Rel'98 and also adopted in the UTRAN Rel'99 as a mandatory speech codec. The AMR Rel'98 codec can be referred to also as an *AMR-NB* (AMR narrowband)—to distinguish it from the new AMR-WB (AMR wideband) codec.

The significance of the wideband AMR lies in the speech coding bandwidth. The bandwidth for the AMR-WB varies between 50 and 7000 Hz, whereas the bandwidth of the narrowband speech (e.g. AMR-NB) varies between 150 and 3500 Hz. This difference in the speech coding bandwidth relates directly to the speech-quality difference. The AMR-WB speech increases the naturalness and the intelligibility of the speech, and makes the AMR-WB codec suitable for applications with high-quality audio parts [2, 3].

The application of the AMR-WB codec in the radio interface is based on the similar principles as defined for the AMR-NB. The AMR-WB has nine different codec rates, ranging from 6.60 to 23.85 kbps. In the GSM/EDGE single timeslot configuration, all these codec rates can be supported in the 8-PSK speech channels, and the codec rates from 6.60 to 19.85 kbps can be supported in the Gaussian minimum shift keying (GMSK) speech channels. The link adaptation mechanism is similar to the one defined for the AMR-NB and is based on the inband signalling bits that are transmitted along the speech frames. The major difference in the deployment scenarios is that for preserving the superior quality of the wideband speech, no transcoders are allowed in the transmission path of the speech from one terminal to another.

The ITU-T codec subgroup has selected the 3GPP AMR-WB codec as its recommended wideband speech codec. This implies that the AMR-WB codec is the first truly global speech codec standard.

2.1.3.2 Fast Power Control for Speech

The basic power control (PC) of the GSM/EDGE operates on the 480-ms basis using the regular slow associated control channel (SACCH) frames for the reporting of the channel quality measurements and sending the PC commands. In Rel'99, faster PC with 20-ms basis was defined for the (enhanced circuit-switched data (ECSD)). The ECSD fast power control uses inband signalling for carrying the measurements and the PC commands and thus it is specific to the ECSD channels. In GERAN Rel'5, a third power control (PC) mechanism, called *enhanced power control* (EPC), is defined. The EPC uses a modified SACCH channel for carrying the measurements and the PC commands. The modified SACCH channel allows these messages to be carried on the single SACCH burst, resulting as a 1 per 120-ms frequency [4].

2.1.3.3 Half-rate 8-PSK Speech

As in the GSM/EDGE Rel'99, the 8-PSK modulation was applied only to enhance the capabilities of the data traffic channels (TCHs), (enhanced general packet radio service (EGPRS) and enhanced circuit-switched data ECSD); in Rel'5 the 8-PSK modulation is applied also to the speech services. Earlier in this chapter, the wideband AMR was already mentioned as one application area of the 8-PSK speech. In addition to that, the AMR-NB (Rel'98 AMR codec) can also be enhanced with the use of the 8-PSK modulation. After extensive studies, it was decided to standardise only the 8-PSK half-rate channels for the AMR-NB codec [5].

As the full-rate and half-rate GMSK channels are already in the standard, the significance of the half-rate 8-PSK speech channels is due to the better link level performance of the half-rate 8-PSK channels, compared with the half-rate GMSK channels. The improvement is largest for the highest codec rates that can be supported by the GMSK channels (7.95 and 7.4 kbps), as there is more room for the channel coding with 8-PSK channels. Further, the two highest modes of the AMR-NB (12.2 and 10.2 kbps) can be supported with half-rate 8-PSK channels but not half-rate GMSK channels.

Table 2.1 summarises all speech codecs supported in the GERAN specifications and their basic characteristics.

2.1.4 Location Service Enhancements for Gb and Iu Interfaces

The support for positioning a mobile station (MS), referred to in the standard as *location services* (LCS), is already defined in Rel'98 of the GSM standard. Three positioning mechanisms are supported for LCS: timing advance (TA), enhanced observed time difference (E-OTD) and global positioning system (GPS). Rel'98 defines the LCS only for the CS domain. In Rel'4, the first step was taken to define the LCS service support also for the PS domain. This solution assumes that in the case of the positioning request, the mobile station will suspend its GPRS connection, perform the needed location procedures using the existing LCS solution in the CS domain and resume the GPRS connection.

In Rel'5, the LCS service for the PS domain develops in two parallel tracks: one solution for the A/Gb mode and another solution for the Iu mode. In both these modes, the

Table 2.1. Speech codecs supported in GERAN

Speech codec		Bits per speech frame	Average code rate			
			GMSK channel		8-PSK channel	
			FR	HR	FR	HR
GSM FR		260	0.57	X	x	x
GSM HR		112	x	0.49	x	x
EFR		244	0.54	X	x	x
AMR-NB	12.2 kbps	244	0.54	X	x	0.36
	10.2 kbps	204	0.45	X	x	0.30
	7.95 kbps	159	0.35	0.70	x	0.23
	7.4 kbps	148	0.32	0.65	x	0.22
	6.7 kbps	134	0.29	0.59	x	0.20
	5.9 kbps	118	0.26	0.52	x	0.17
	5.15 kbps	103	0.23	0.45	x	0.15
	4.75 kbps	95	0.21	0.42	x	0.14
AMR-WB	23.85 kbps	477	x	X	0.35	0.70
	23.05 kbps	461	x	X	0.34	0.67
	19.85 kbps	397	0.87	X	0.29	0.58
	18.25 kbps	365	0.80	X	0.27	0.53
	15.85 kbps	317	0.70	X	0.23	0.46
	14.25 kbps	285	0.63	X	0.21	0.42
	12.65 kbps	253	0.55	X	0.18	0.37
	8.85 kbps	177	0.39	x	0.13	0.26
	6.60 kbps	132	0.29	x	0.10	0.19

same positioning mechanisms (TA, E-OTD and GPS) as in earlier releases are supported. The differences lie on the mode-specific protocol and signalling solutions that are tailored separately for the A/Gb and Iu modes. The enhancement compared with the Rel'4 solution for the PS domain LCS service is that it is not needed to suspend and resume the packet data connection for determining the location of the mobile but the location procedures are performed parallel to the connection.

Readers with more detailed interest on the LCS specifications are recommended to take a look at [6].

2.1.5 Inter-BSC and BSC/RNC NACC (Network-assisted Cell Change)

In Rel'4 of the GERAN specifications, the NACC procedures cover only cell re-selection to other GSM/EDGE cells within the same BSC where the BSC has system information available for the target cell. This limits the value of the NACC as inter-BSC cell changes and cell changes between GSM/EDGE and UTRAN cells in some network configurations are of frequent occurrence.

The inter-BSC and BSC-RNC NACC is an extension of the Rel'4 NACC feature. For the fast re-selection of cells, the MS needs to have knowledge of certain system information for the target cell before performing the cell change. If the cells belong to different BSCs/RNCs, the required system information has to be transported between BSCs/RNCs when updated to be available for the MSs when re-selecting cells [7].

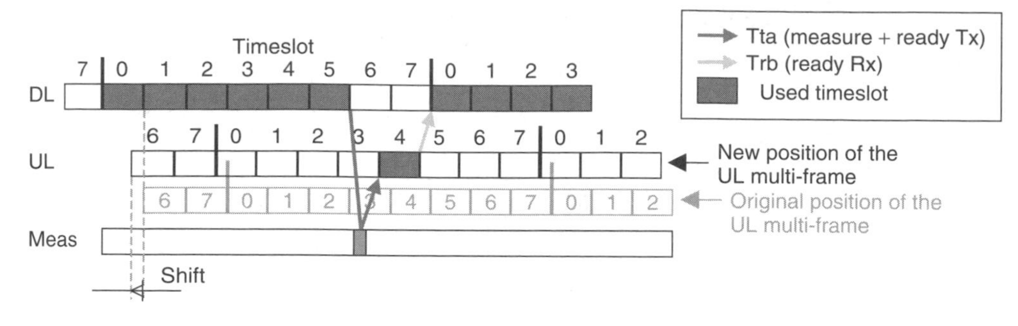

Figure 2.3. High multi-slot class concept for 6 + 1 allocations

2.1.6 High Multi-slot Classes for Type 1 Mobiles

There are two main types of MSs in the GERAN standard:

- *type 1 mobiles.* Not required to transmit and receive at the same time;
- *type 2 mobiles.* Required to be able to transmit and receive at the same time.

Owing to the need to perform neighbour cell measurements between the transmit (Tx) and receive (Rx) timeslots, the highest type 1 multi-slot class is in practice number 12 (4 Rx timeslots, 4 Tx timeslots, sum of Rx and Tx timeslots active at one time = 5). To overcome this restriction in the total number of active timeslots (five), the uplink (UL) multi-frame of those base transceiver stations (BTSs) in the network that would support high multi-slot classes is shifted (see Figure 2.3). This shift is transparent for any MS (legacy MS or new MS supporting high multi-slot classes) because this shift is just a TA minimum value controlled by the network. If the UL multi-frame is shifted 100 μs, then there is 100 μs available for the MS to switch from one frequency to another, and another 100-μs window for making the actual neighbour cell measurement, even for the 6 Rx timeslot + 1 Tx timeslot configuration.

The UL multi-frame shift (100 μs) restricts the maximum cell size to 18 km. However, this restriction should not become a problem as this solution can be deployed in selected areas, on a cell-per-cell basis, depending on the size of the cells, the expected data traffic and BTS capabilities [8].

2.2 GERAN Architecture

2.2.1 General

This section explains the basics of new architecture for GERAN that are currently being specified in 3GPP Rel'5. Description in this chapter assumes that the reader is familiar with basic protocols and interfaces used in pre-Rel'5 GSM/GPRS. In addition, the UMTS architecture as specified in [9] would provide useful insights for fully understanding the concepts presented here.

The evolution of GSM standards has been described in Chapter 1 and the beginning of Chapter 2 as well. Here, let us briefly recall the goals of GERAN Rel'5 specification. In Rel'5, the functional split between the GERAN and the core network (CN) has been harmonised with the functionality split that has been specified between UTRAN and the CN. As a consequence, we find that this harmonisation has three important impacts in the overall system:

- It enables GERAN to connect to the same 3G CN as UTRAN creating first steps towards efficient resource optimisations in multi-radio networks.
- It enables GERAN to provide the same set of services as UTRAN, making the radio technology invisible to the end user, while allowing operators to efficiently manage the available spectrum.
- Existing GERAN radio protocols need to undergo significant modifications, and this will obviously increase the complexity of radio interface protocols.

In the following sections, we can see details of the architecture. For more detailed information, the reader is advised to check the 3GPP specifications mentioned in the bibliography.

2.2.2 Architecture and Interfaces

The basic principles that are used for the design of GERAN architecture for Rel'5 are as follows:

- Separation of the radio-related and non-radio-related functionalities between the CN and the RAN. This principle is important in order to enable development of the platform that would allow for provisioning of services independent of the access type. For example, an operator could run one CN and offer services to the end users by utilising different radio access technologies, e.g. wideband code division multiple access (WCDMA), GSM/EDGE radio access networks (GERANs) or wireless local area network (LAN).
- Support of services for pre-Rel'5 terminals must be ensured. Obviously, it is important to evolve the GSM/EDGE radio access in a backwards compatible way.
- Maximise commonalities between GERAN and UTRAN but maintain the backward compatibility for GERAN. The key alignment point between the GERAN and UTRAN is the common interface towards the 3G CN (Iu interface). Consequently, GERAN should provide the same set of services as UTRAN, e.g. support the same radio access bearers (RABs) and QoS mechanisms as in UTRAN.
- Standardisation of the GERAN Rel'5 should support a true multi-vendor environment.

In accordance with those principles, the GERAN architecture for Rel'5 has been designed in 3GPP and is shown in Figure 2.2.

The following sections describe the details of the interfaces that are part of GERAN Rel'5 architecture.

2.2.2.1 Legacy Interfaces

The protocol model for legacy interfaces is based on the Rel'4 of GSM/EDGE and consequently the functionality split between the radio access and the CN is unchanged. The two interfaces towards CN are the following:

Gb interface. The Gb interface is used in GSM/GPRS to connect the GSM–serving GPRS support node (SGSN) and base station subsystem (BSS). This interface is needed in GERAN for supporting pre-Rel'5 terminals. There are two peer protocols that have been identified across Gb interface:

- The base station subsystem GPRS protocol (BSSGP) (see [10]). The primary functions of the BSSGP include: in the downlink, the provision by an SGSN to a BSS of radio-related information used by the RLC/MAC function; in the uplink, the provision by a BSS to an SGSN of radio-related information derived from the RLC/MAC function; and the provision of functionality to enable two physically distinct nodes, an SGSN and a BSS, to operate node-management control functions. See [10] for more details.
- The underlying network service (NS). The NS is responsible for the transport of BSSGP protocol data units (PDUs) between a BSS and an SGSN (see [10]).

A interface. Traditional interface in GSM that is used to connect BSS and MSC will be supported by GERAN. This interface is needed to support pre-Rel'5 terminals, as well as to provide new enhanced services such as, e.g. wideband telephony (with wideband adaptive multi-rate—WB-AMR—speech codec) and support for 8-PSK modulated speech channels. There are a number of available references that describe the protocol and procedures over the A interface; therefore, these descriptions are omitted from this chapter.

2.2.2.2 New Interfaces

The general protocol model for new interfaces is based on the protocol model used in UMTS and is shown in Figure 2.4. The layers and planes are logically independent of each other, which offers additional flexibility. For example, if needed, transport network protocol layers, may be changed in the future with virtually no impact on the radio network layer.

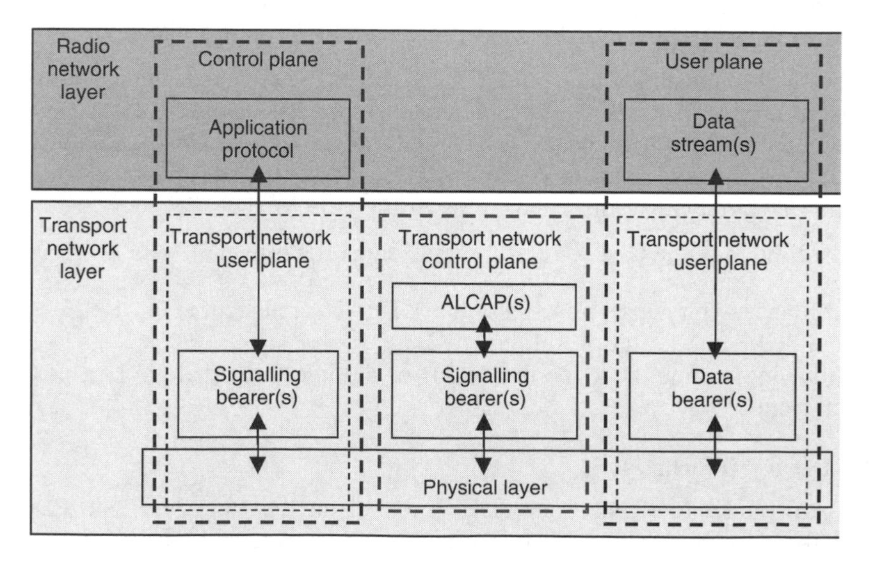

Figure 2.4. General UMTS protocol model

GERAN Rel'5 will connect to the UMTS CN using Iu interface. The Iu interface is specified at the boundary between the CN and the RAN. The Iu interface is used in Rel'99 to connect UTRAN and CN, and the basic idea is to adopt this interface for connecting GERAN to the UMTS CN. There are two flavours of Iu interface:

- *Iu-cs interface.* The interface between the MSC and its base station subsystem (BSS) and/or radio network signalling (RNS) is specified in the 25.41x-series of UMTS technical specifications and is called Iu-cs. The Iu-cs interface is used to carry information concerning BSS management, call handling and mobility management (MM).
- *Iu-ps interface.* The BSS-UMTS-SGSN interface is called Iu-ps interface and is used to carry information concerning packet data transmission and mobility management. The Iu-ps interface is defined in the 25.41x-series of UMTS technical specifications.

The protocols over Iu interface are divided into two structures:

- *User plane protocols.* These are the protocols implementing the actual radio access bearer service, i.e. carrying user data through the access stratum.
- *Control plane protocols.* These are the protocols for controlling the radio access bearers and the connection between theuser equipment (UE) and the network from different aspects (including requesting the service, controlling different transmission resources, handover and streamlining etc.). Also, a mechanism for transparent transfer of non-access stratum (NAS) messages is included.

The protocol stacks across Iu-cs and Iu-ps interfaces as specified in 3GPP TSs are shown in Figures 2.4 and 2.5. From these figures, it is seen that on the radio network layer there is no difference between Iu-cs and Iu-ps interfaces.

The radio network signalling (control plane) over Iu consists of the radio access network application part (RANAP). The RANAP protocol consists of mechanisms to handle all procedures between the CN and RAN. It is also capable of conveying messages transparently between the CN and the MS without interpretation or processing by the RAN. Over the Iu interface, the RANAP protocol is, e.g. used for provisioning of a set of general RAN procedures from the CN such as paging—notification. Further, RANAP provides procedures for the transfer of transparent non-access signalling, management of RABs as well as procedures related to the relocation function. The RANAP protocol is defined in [12].

The radio network layer UP consists of the Iu UP protocol. The Iu UP protocol is used to convey user data associated with RABs. The philosophy behind the design of the Iu UP protocol is to keep the radio network layer independent of the CN domain (circuit-switched or packet-switched) and to have limited or no dependency with the transport network layer. Meeting this objective provides the flexibility to evolve services regardless of the CN domain and to migrate services across CN domains. The Iu UP protocol is therefore defined with modes of operation that can be activated on a RAB basis rather than on a CN domain basis or (tele) service basis. The Iu UP mode of operation determines if and which set of features shall be provided to meet, e.g. the RAB QoS requirements. The Iu UP protocol is defined in [13].

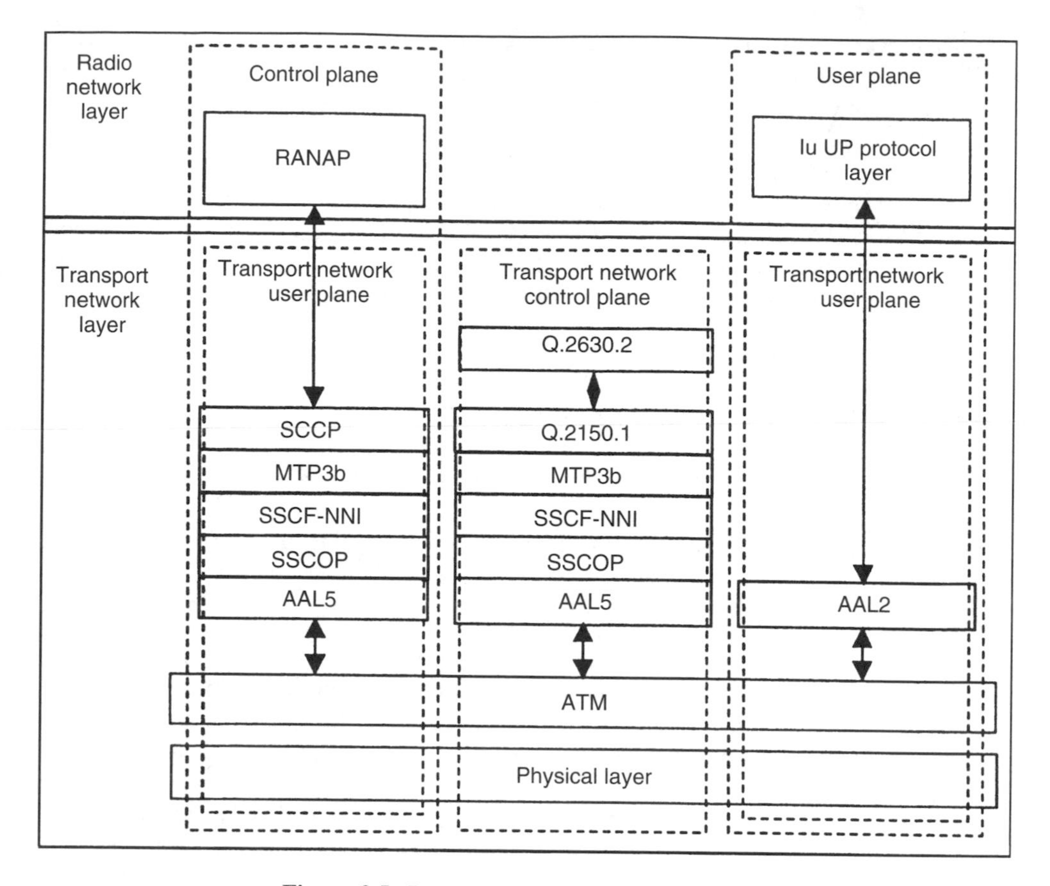

Figure 2.5. Protocol stack across Iu-cs interface

The difference between Iu-cs and Iu-ps in terms of protocol stack is visible in the transport network layer. In case of Iu-cs, the choice of signalling bearer for RANAP over the Iu interface is based on broadband signalling system No. 7, consisting of well-standardised components. In case of Iu-ps, the standard allows operators to choose one out of two standardised protocols to suit transport of Signalling Connection Control Protocol (SCCP) messages.

So far, we have seen that GERAN of Rel'5 could have four interfaces towards the CN. This could lead to quite a large number of possible combinations of interfaces towards the CN, thus increasing dramatically the complexity of both RAN and the MS; some simplifications are therefore necessary. The standard specifies two modes of operations for the MS:

- *A/Gb mode*, e.g. for pre-Rel'5 terminals or for Rel'5 terminals when connected to a GERAN with no Iu interface towards the CN;
- *Iu mode* (i.e. Iu-cs and Iu-ps), e.g. for Rel'5 terminals when connected to a GERAN with Iu interfaces towards the CN.

The MS shall operate only on one of the modes. This in practice means that if the MS is operating in A/Gb mode, then it is utilising services from the legacy GSM CN, while services from the UMTS CN are available when the MS is operating in Iu mode.

2.2.3 Radio Access Network Interfaces

2.2.3.1 BSS–MS Interface (Um)

Adoption of the Iu interface in GERAN Rel'5 requires significant modification on the radio interface protocol stack. These modifications are required primarily because of the significant difference in functionality split between GERAN and CNs depending on the interface that is used: A/Gb or Iu. This difference, together with the necessity to fulfil the backwards compatibility principle mentioned in the beginning of this chapter, leads to a rather complex protocol constellation over the air interface. Overall description of GERAN Rel'5 is given in [14] and the reader is referred to this specification for more details. As an illustration, Figures 2.6 and 2.7 show the user plane protocol stack for CS and PS domain, respectively [14].

As can be seen from Figure 2.7, the protocol stack over the air interface (Um) consists of common protocol layers (common to Gb and Iu-ps) as well as protocol layers specific

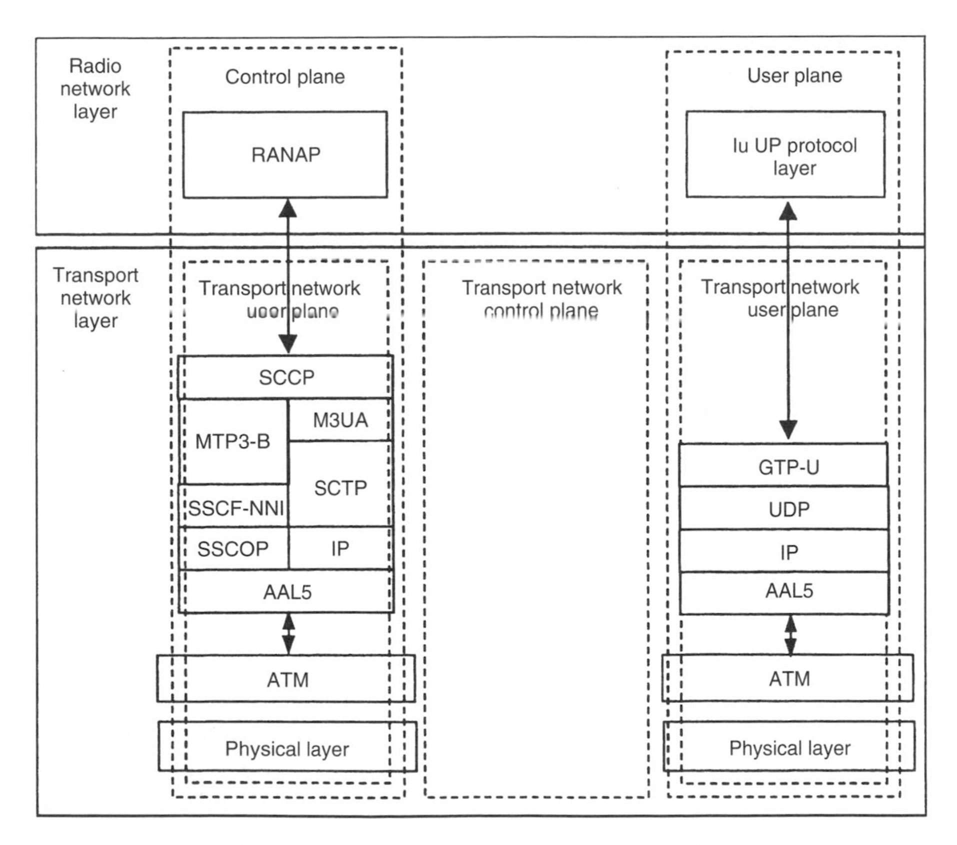

Figure 2.6. Protocol stack across Iu-ps interface

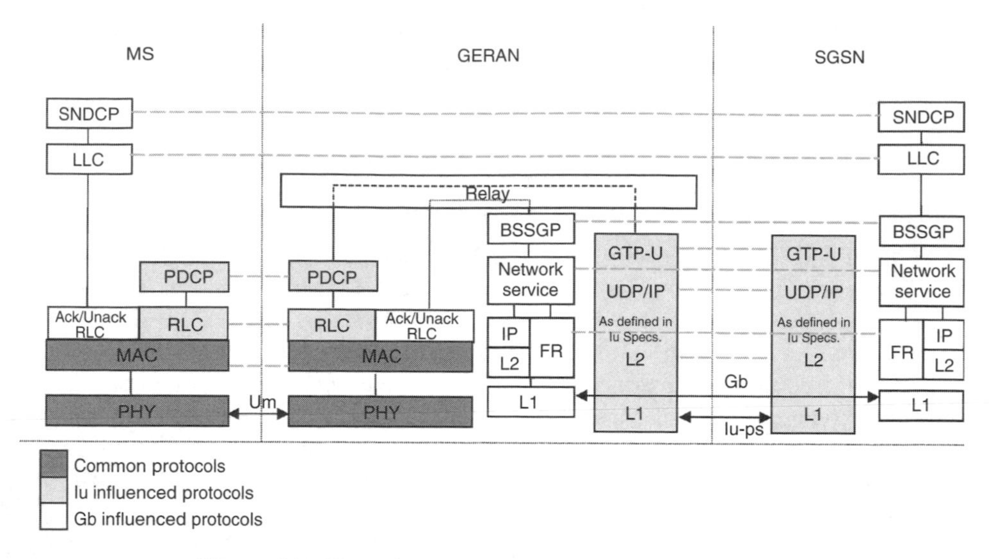

Figure 2.7. User plane protocol stack for PS domain [14]

to each interface. Similar figures for the control plane of the PS domain, as well as the user and control plane of the CS domain are available in [14]. The radio interface protocol architecture when GERAN connects to A or Gb is the same as defined in earlier releases. The sections below describe the protocol architecture specific to the Iu interface support.

Radio Interface Protocol Architecture

When designing protocols across the Um interface the usual layering principle has been followed:

- the physical layer (L1)
- the data link layer (L2) and
- the network layer (L3).

Further, the principle of user and control plane separation is applied. The radio interface protocol architecture is shown in Figure 2.8.

The RLC, MAC protocol and physical layer carry data for both control and UP. Layer 2 consists of RLC and MAC protocols as well as PDCP. The PDCP protocol is used only on the UP.

Layer 3 of the control plane is split into two sublayers, RRC and duplication avoidance functionality. The RRC protocol utilises services offered by RLC/MAC for transfer of data, except for data related to the operation on the broadcast control channel (BCCH) and common control channel (CCCH), where the data link layer is used. The duplication avoidance functionality provides the access stratum services to higher layers such as mobility management (MM) and call control (CC).

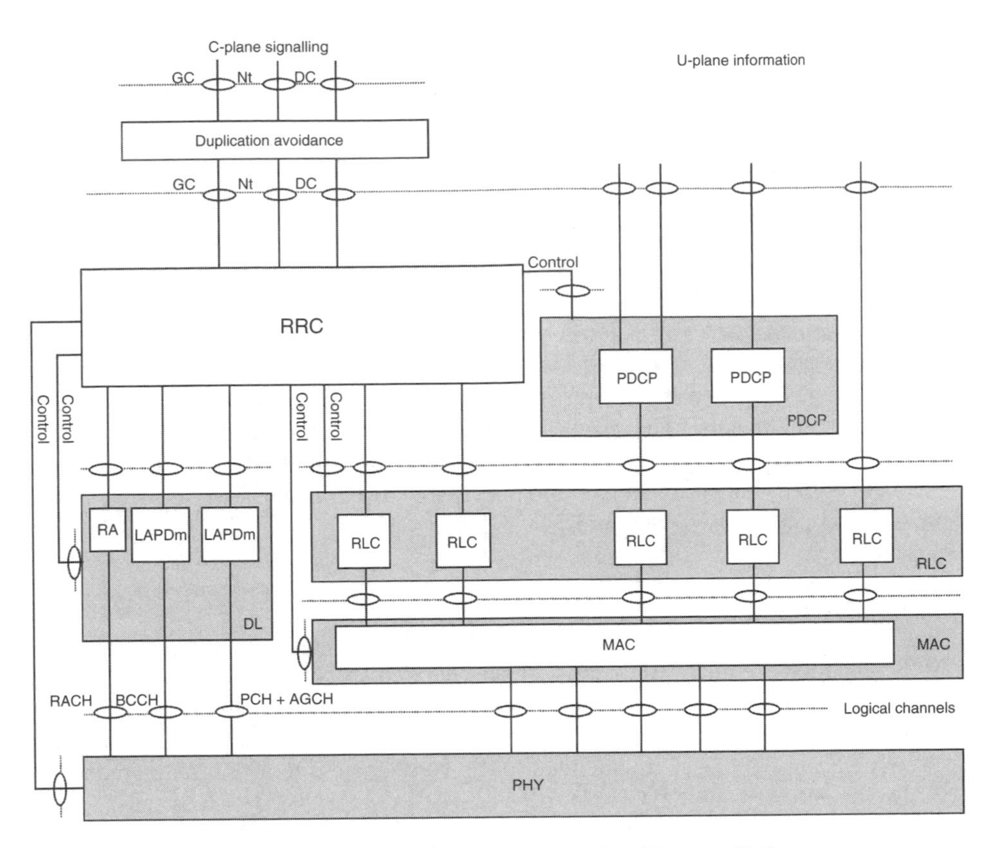

Figure 2.8. Radio interface protocol architecture [14]

Physical Layer (PHY) Protocol

The PHY of GERAN interfaces the MAC and the RRC protocols. Within a carrier, a series of timeslots of consecutive time division multiple access (TDMA) frames forms a basic physical channel. *Basic physical channels* are organised into multi-frame. Three types of multi-frame are defined:

- 26-multi-frame formed from 26 TDMA frames (120 ms) and is used for traffic and associated control channels.
- 52-multi-frame formed from 52 TDMA frames (240 ms) and used for traffic, associated control and common control channels.
- 51-multi-frame comprising 51 TDMA frames (235.4 ms). This multi-frame supports broadcast and common control. It always carries synchronisation and frequency correction bursts. It also supports short message service (SMS).

A *basic physical subchannel* (BPSCH) is defined as a basic physical channel or a part of a basic physical channel. Some BPSCHSs are reserved by the network for control and broadcast (51-multi-frame), others are assigned to dedicated connections with MSs called *dedicated BPSCH* (DBPSCH), or are assigned to a shared usage between several MSs and called *shared BPSCH* (SBPSCH).

A DBPSCH always follows the 26-multi-frame structure. It can be either full rate (DBPSCH/F) or half rate (DBPSCH/H). Among the 26 timeslots of a DBPSCH/F, one is always reserved for control and one is always idle. That leaves 24 timeslots available for traffic every 120 ms. Each timeslot is thus equivalent to 5 ms of traffic on DBPSCH/F. Similarly, each timeslot is equivalent to 10 ms of traffic on DBPSCH/H.

A SBPSCH always follows the 52-multi-frame structure. It can be either full rate (SBPSCH/F) or half rate (SBPSCH/H). A SBPSCH/H is available in MAC dual transfer mode (DTM) state only. Among the 52 timeslots of a SBPSCH/F, two are always reserved for TA procedures and two are always idle. That leaves 48 timeslots available for traffic every 240 ms. As on DBPSCH, each timeslot is thus equivalent to 5 ms of traffic on SBPSCH/F. Similarly, each timeslot is equivalent to 10 ms of traffic on SBPSCH/H.

The PHY offers logical channels and the associated transmission services. Logical channels are divided in two categories:

- traffic channels
- control channels.

Traffic channels are intended to carry either encoded speech or user data while control channels carry signalling or synchronisation data. Logical channels are multiplexed either in a fixed pre-defined manner or dynamically by the MAC on physical subchannels. Traffic channels of type TCH are intended to carry either speech or user data on dedicated BPSCHs (for one user only). TCH can be either full rate (TCH/F) or half rate (TCH/H), and is GMSK modulated. Traffic channels of type O-TCH (octal traffic channel) are intended to carry speech on dedicated BPSCHs (for one user only). O-TCH can be either full rate (O-TCH/F) or half rate (O-TCH/H) and is 8-PSK modulated. Traffic channels of type E-TCH (enhanced traffic channel) are intended to carry user data on dedicated BPSCHs (for one user only). E-TCH is full rate and always 8-PSK modulated only. PDTCHs are intended to carry user data on either SBPSCH or DBPSCH. The PDTCH is temporarily dedicated to one MS but one MS may use multiple PDTCHs in parallel. PDTCH allows several MSs to be multiplexed on the same SBPSCH. PDTCH also allows several traffic classes from the same MS to be multiplexed on the same PSCH (shared or dedicated). PDTCHs can be either full rate (PDTCH/F) or half rate (PDTCH/H), and can be either GMSK or 8-PSK modulated.

Control channels are intended to carry signalling or synchronisation data. Three categories of control channel are defined: broadcast, common and dedicated control channels.

Broadcast channels consist of frequency correction channels (FCCH), which carry information for frequency correction of the MS, synchronisation channels (SCH), which carry information for frame synchronisation of the MS and identification of a base transceiver station, broadcast control channel (BCCH), used for broadcasting general information on a base transceiver station (BTS) and packet broadcast control channel (PBCCH) used for broadcasting parameters valuable to the MS to access the network for packet transmission operation.

Common control type channels, known when combined as a common control channel (CCCH), consist of paging channel (PCH), used to page mobiles, random access channel

(RACH), used to request allocation of a dedicated control channel, access grant channel (AGCH), used to allocate a basic physical subchannel (PSCH) and notification channel, used to notify MSs of voice group and voice broadcast calls. Packet common control channels (PCCCH) are used to support packet-switched (PS) operations and consist of packet paging channel (PPCH), packet random access channel (PRACH), packet access grant channel (PAGCH) and packet notification channel (PNCH). PCCCH is optional for the network, and if a PCCCH is not allocated, the information for PS operation is transmitted on the CCCH. If a PCCCH is allocated, it may transmit information for CS operation.

There are a number of dedicated control channels to support DBPSCH. They are all derived from three types of control channel:

- Slow associated control channel (SACCH) is used mainly for the transmission of the radio measurement data needed for radio resource management algorithms. In the uplink, the MS sends measurements reports to the BTS, while in the downlink, the BTS sends commands to the MS. SACCH is also used for SMS transfer during a call.
- Fast associated control channel (FACCH), which is always associated with one TCH, and occurs on DBPSCH only.
- Stand-alone dedicated control channel (SDCCH).

Similarly, control channels for support of SBPSCH are the following:

- *The packet associated control channel (PACCH).* The PACCH is bi-directional. For description purposes PACCH/U is used for the uplink and PACCH/D for the downlink.
- *Packet timing advance control channel uplink (PTCCH/U).* Used to transmit random access bursts to allow estimation of the timing advance for one MS in packet transfer mode.
- *Packet timing advance control channel downlink (PTCCH/D).* Used to transmit tim ing advance updates for several MS. One PTCCH/D is paired with several PTCCH/Us.

For more information on the PHY see 3GPP TS 45.x series.

Medium Access Control (MAC) Protocol

The MAC sublayer allows the transmission over the PHY of upper layer PDUs in dedicated or shared mode. A MAC mode is associated with a BPSCH for use by one or more MSs (dedicated or shared mode, respectively). The MAC layer handles the access to and multiplexing on to the PSCHs of MSs and traffic flows.

MAC protocol for Rel'5 is heavily based on the MAC protocol of Rel'4 as specified in [15]. Main modifications in Rel'5 are related to the support of multiple TBFs, ciphering, support for radio bearers and support for RRC signalling. The functions of MAC include [14] the following:

- Configuring the mapping between logical channels and basic physical subchannels. The MAC is responsible for configuring the mapping of logical channel(s) on to the appropriate basic physical subchannel(s).

- Defining logical channels to be used for each radio bearer service.
- Assignment, reconfiguration and release of shared radio resources for a temporary block flow (TBF). The MAC layer may handle the assignment of radio resources needed for a TBF including needs from both the control and user plane. The MAC layer may reconfigure radio resources of a TBF.
- MS measurement reporting and control of the reporting. The MAC layer is responsible for sending information that controls the MS measurement reporting when using PBCCH or PACCH channels. The MAC layer also performs the reporting of the measurements from the MS to the network using PACCH.
- Broadcasting/listening of/to PBCCH and PCCCH. The MAC layer broadcasts/listens (to) the PBCCH of the serving cell for the sending/decoding of packet system information messages. The MAC layer listens to the BCCH of neighbouring cells for neighbour cell measurements. The MAC layer also sends paging information on the PCCCH or monitors the paging occasions according to the discontinuous reception (DRX) cycle.
- Timing advance control. The MAC layer controls the operation of timing advance on shared basic physical subchannels.
- Ciphering. The MAC is responsible for ciphering user data blocks when RLC operates in a transparent mode.
- Identification of different traffic flows of one or more MSs on the basic physical subchannels. Inband identification is needed to address a flow to an MS in the downlink or identify a flow from a MS in the uplink.
- Multiplexing/de-multiplexing of higher layer PDUs. This may include priority handling between data flows of one or more MSs, e.g. by attributes of radio bearer services.
- Multiplexing/de-multiplexing user and control plane data to/from the PHY for PDTCHs. The MAC layer is responsible for multiplexing/de-multiplexing RLC data blocks carried on PDTCH and RLC/MAC control blocks carried on PACCH.
- Scheduling of RLC/MAC data and control PDUs delivered to the physical layer on shared basic physical subchannels. This includes uplink state flag (USF) and relative reserved block period (RRBP) field monitoring for uplink transfer and sharing radio resources on the downlink.
- Splitting/recombining. This includes splitting/recombining of the RLC/MAC PDU flow belonging to one or more TBF(s) on to/from several shared logical channels. This function does not apply to RLC/MAC control blocks.

Figure 2.9 describes the MAC state model for GERAN in Iu mode. The model shows the MAC state for the MAC control entity of an MS and not for individual radio bearers (see [14]).

There are four states in the MAC state machine. The MAC control entity of an MS is in MAC-Idle state when there are no dedicated or shared basic physical subchannels. RRC can be in RRC-Cell_Shared state, RRC-GRA_abbPCH state or RRC-Idle mode (see

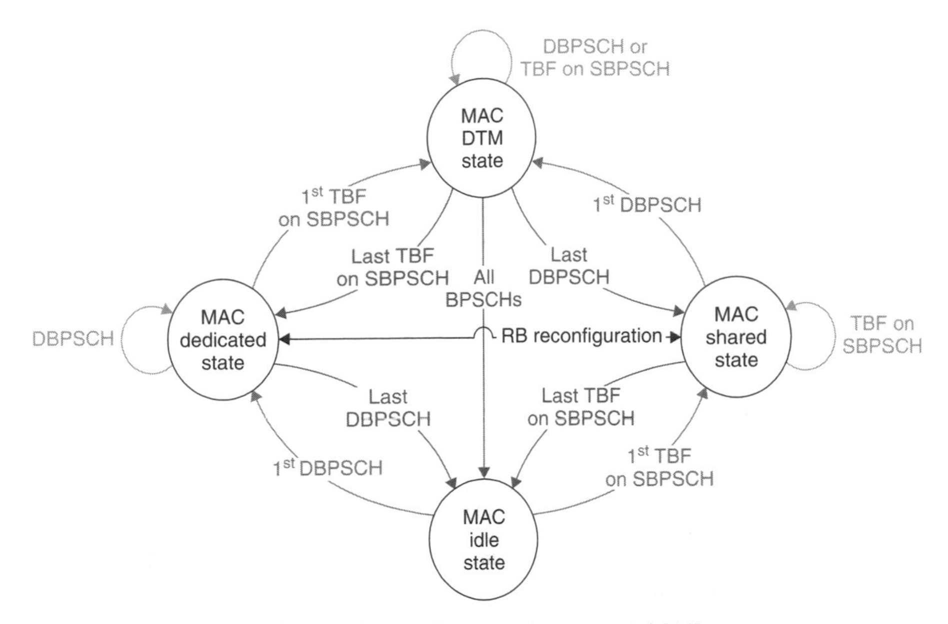

Figure 2.9. MAC protocol state model [14]

sections below). In this state, the MS camps on the PCCCH or on the CCCH. The MAC control entity is in MAC-Shared state when at least one TBF is ongoing but no dedicated BPSCHs have been allocated. The MAC control entity of an MS is in MAC-DTM state when it has one or more dedicated BPSCH(s)and one or more shared BPSCH(s)allocated. The MAC control entity will be in MAC-DEDICATED state when a dedicated BPSCH is used and no shared BPSCH is established. In MAC-Dedicated state MAC has no control functionality.

Radio Link Control (RLC) Protocol

The RLC protocol allows for the data transfer in transparent, acknowledged or unacknowledged modes, and, in addition, notifies unrecoverable errors to the upper layer.

When in transparent mode, RLC has no functionality and does not alter the data units of the upper layer.

In non-transparent mode, the RLC is responsible for ciphering RLC PDUs in order to prevent any unauthorised acquisition of data. RLC non-transparent mode is built based on the (E)GPRS if the RAB is mapped onto a packet data traffic channel (PDTCH). Alternatively, the RAB may be mapped onto an (E-)TCH (enhanced traffic channel). Signalling is performed on a PACCH when PDTCH is used, and on fast associated control channel (FACCH) and slow associated control channel (SACCH) when (E-)TCH is used. The mapping of an RAB onto the proper traffic channel (TCH) is dependent on the MAC mode in use and the targeted traffic class, as described in Chapter 3. In acknowledged mode, backward error correction (BEC) procedures are provided that allow error-free transmission of RLC PDUs through two possible selective automatic repeat request (ARQ) mechanisms: type I ARQ and hybrid type II ARQ (incremental redundancy). In unacknowledged mode, no BEC procedure is available.

For more details, see [15].

Packet Data Convergence Protocol (PDCP)

The PDCP layer is operational only on the UP of the radio interface protocol stack. This protocol layer is common to UTRAN and GERAN and is specified in [16]. The PDCP allows for the transfer of user data using services provided by the RLC, and header adaptation of the redundant network PDU control information (IP headers) in order to make the transport over the radio interface spectrum efficient. One important function of PDCP is header compression functionality, which consists of compressing transport and network level headers (e.g. RTP/UDP/IP) so that decompressed headers are semantically identical to the original uncompressed headers. Header compression is suited for generic Internet applications (i.e. not specific to GERAN), especially multimedia applications.

Radio Resource Control (RRC)

The RRC layer handles the control plane signalling of Layer 3 between the MSs and GERAN. The RRC performs the following functions (see [14]):

- *Broadcast of information provided by the non-access stratum (core network).* The RRC layer performs system information broadcasting from the network to all MSs. The system information is normally repeated on a regular basis. The RRC layer performs the scheduling, segmentation and repetition when such broadcasting is carried on a broadcast control channel (BCCH). This function supports broadcast of higher layer (above RRC) information. This information may be cell-specific or not. As an example, RRC may broadcast CN location service area information related to some specific cells.
- *Broadcast of information related to the access stratum.* The RRC layer performs system information broadcasting from the network to all MSs. The system information is normally repeated on a regular basis. The RRC layer performs the scheduling, segmentation and repetition when such broadcasting is carried on BCCH. This function supports broadcast of typically cell-specific information.
- *Establishment, re-establishment, maintenance and release of an RRC connection between the MS and GERAN.* The establishment of an RRC connection is initiated by a request from higher layers at the MS side to establish the first signalling connection for the MS. The establishment of an RRC connection includes an optional cell reselection, an admission control, and a layer 2 signalling link establishment. The release of an RRC connection can be initiated by a request from higher layers to release the last signalling connection for the MS or by the RRC layer itself in case of RRC connection failure. In case of connection loss, the MS requests re-establishment of the RRC connection. In case of RRC connection failure, RRC releases resources associated with the RRC connection.
- *Establishment, reconfiguration and release of radio bearers.* The RRC layer can, on request from higher layers, perform the establishment, reconfiguration and release of radio bearers in the user plane. A number of radio bearers can be established to an MS at the same time. At establishment and reconfiguration, the RRC layer performs admission control and selects parameters describing the radio bearer processing in layer 2 and layer 1 on the basis of information from higher layers.

- *Assignment, reconfiguration and release of radio resources for the RRC connection.* Depending on the RRC and MAC states, the RRC layer may handle the assignment of radio resources needed for the RRC connection including needs from both the control and user plane. The RRC layer may reconfigure radio resources during an established RRC connection. RRC signals to the MS to indicate resource allocations for purposes of inter-system handovers.
- *RRC connection mobility functions.* The RRC layer performs evaluation, decision and execution related to RRC connection mobility during an established RRC connection, such as handover, preparation of handover to UTRAN or other systems, cell re-selection and cell/GERAN registration area (GRA) update procedures, based on, e.g. measurements done by the MS.
- *Release of signalling connections.* The RRC layer provides the necessary functions for the GERAN or the MS to request the release of a signalling connection.
- *Paging/notification.* The RRC layer can broadcast paging information from the network to selected MSs on CCCH. Higher layers on the network side can request paging and notification. The RRC layer can also initiate paging during an established RRC connection.
- *Listening to the BCCH and common control channels (CCCH).* The RRC layer listens to the BCCH of the serving cell for the decoding of system information messages and to the BCCH of neighbouring cells for neighbour cell measurements; the RRC layer also monitors the paging occasions according to the DRX cycle and receives paging information on the CCCH.
- *Routing of higher layer PDUs.* This function performs at the MS side routing of higher layer PDUs to the correct higher layer entity, at the GERAN side to the correct RANAP entity.
- *Control of requested QoS.* This function shall ensure that the QoS requested for the radio bearers can be met. This includes the allocation of a sufficient number of radio resources.
- *MS measurement reporting and control of the reporting.* The measurements performed by the MS are controlled by the RRC layer including both GSM/EDGE air interface and other systems. The RRC layer is responsible for sending information that controls the MS measurement reporting when using BCCH or SACCH channels. The RRC layer also performs the reporting of the measurements from the MS to the network using SACCH.
- *Power control.* The RRC layer controls parameters for normal power control, enhanced power control and fast power control.
- *Control of ciphering.* The RRC layer provides procedures for setting of ciphering (on/off) between the MS and GERAN.
- *Integrity protection.* This function controls integrity protection and performs integrity protection of those RRC messages that are considered sensitive and/or contain sensitive information.

- *Support for location services.* Signalling between MS and GERAN to support positioning of an MS.
- *Timing advance control.* The RRC controls the operation of timing advance on dedicated basic physical subchannels.

The RRC protocol for GERAN Iu mode is designed by combining necessary procedures from the GSM RR sublayer specified in [17] and UTRAN RRC protocol specified in [18]. The basic idea of combining those two protocols is to align the GERAN Iu mode with UTRAN to a maximum extent while maintaining backwards compatibility by keeping critical procedures common to A/Gb and Iu modes. Further, the RRC protocol for Iu mode requires capabilities that are in line with UMTS bearer concept and should provide smooth interaction with Iu signalling, without causing modification to the Iu interface. New procedures that are created for GERAN Iu mode are based on the UTRAN RRC protocol and include the following procedures: RRC connection management procedures, RRC connection mobility procedures, radio bearer control procedures, signalling flow procedures, security mode control and delivery of non-access stratum messages. Other necessary procedures are modification of existing GSM RR procedures, e.g. system information, handover procedure, paging etc.

The result, on the protocol state machine level, is shown in Figure 2.10.

The state machine on the right applies to the GERAN A/Gb mode (denoted as RR modes), while the state machine on the left is applicable to the GERAN Iu mode. Let us look into the details of the Iu mode state machine.

Two modes of operation are defined for the MS, *RRC-idle mode* and *RRC-connected mode*. After power on, the MS stays in RRC-idle mode until it transmits a request to establish an *RRC connection*. In RRC-idle mode the MS is identified by non-access stratum identities such as international mobile subscriber identity (IMSI), temporary mobile subscriber identity (TMSI) and packet TMSI (P-TMSI). While in RRC-idle mode the MS has no relation to GERAN, only to the CN. For example, the MS has already attached to the network, performed necessary procedures, and released the RRC connection, so

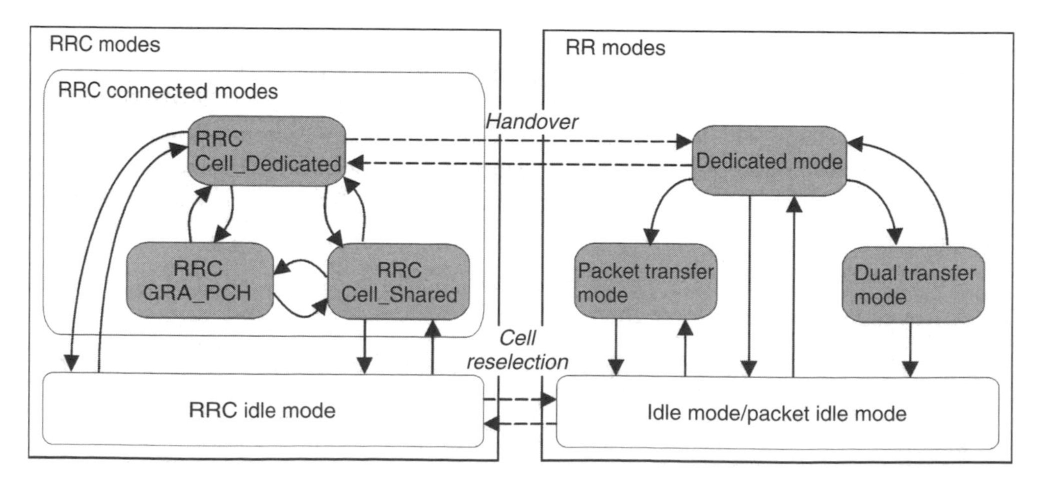

Figure 2.10. RRC and RR protocol state machine [14]

the MS is already known to the CN. For data transfer, a signalling connection has to be established. The RRC-connected mode is entered when the RRC connection is established at which point the MS is assigned a GERAN radio network temporary identity (G-RNTI) to be used as MS identity. The MS is identified within a GERAN with the G-RNTI. The MS leaves the RRC-connected mode and returns to RRC-idle mode when the RRC connection is released or at RRC connection failure.

RRC connection is established by using RRC connection management procedures that are specified on the basis of the corresponding UTRAN procedure described in [18]. After a successful establishment of the RRC connection the MS enters the RRC-connected mode and will be in one of the states shown in Figure 2.10. Three states are defined in RRC-connected mode: RRC-Cell_Shared, RRC-Cell_Dedicated and RRC-GRA_abbPCH. In RRC-Cell_Shared state, no dedicated BPSCH is allocated to the MS and the position of the MS is known by GERAN on the cell level. In RRC-Cell_Dedicated state, the MS is assigned one or more dedicated BPSCHs in the uplink and downlink, which it can use anytime. Furthermore, the MS may be assigned one or more shared BPSCHs. The position of the MS is known by GERAN on a cell level. In RRC-GRA_abbPCH state, no BPSCH is allocated to the MS. The location of the MS is known at the GRA level. Introduction of GRA is believed to decrease the MS power consumption and signalling load within GERAN. The GRA concept has been adopted for GERAN in Iu mode. The behaviour of the MS while in the RRC-GRA_abbPCH state is limited to monitoring the paging occasions according to the DRX cycle and receiving paging information on the (P)CCCH. The MS will listen to the (P)BCCH control channel of the serving cell for the decoding of system information messages and initiate a GRA updating procedure upon GRA change. If the network wants to initiate any activity, it shall make a paging request on the (P)CCCH logical channel within the GRA where the MS is. GRA updating is initiated by the MS, which, upon the detection of the new GRA, sends the network the registration area update information to the new cell. Any activity causes a transition to either the RRC-Cell_Shared or the RRC-Cell_Dedicated state, depending on the activity.

Here it is important to note one essential element in the radio interface protocol stack. Owing to legacy reasons, the RR management tasks are split between RRC layer and RLC/MAC layer. There is a complex interaction between those two layers when it comes to the RR assignment, maintenance and release. This interaction is shown in Table 2.2, but as a simplification one could note the main rule that the RRC layer is responsible for assignment, maintenance and release of *dedicated* channels, while the MAC layer is responsible for those of the *shared* channels. It is worth noting that even if the MAC layer is responsible for the management of the shared channel, the RRC layer will ensure the QoS for a particular radio bearer and therefore configure the MAC layer correspondingly.

2.2.3.2 BSS–BSS and BSS–RNS Interface (Iur-g)

The interface between GERAN BSSs or between GERAN BSS and UTRAN RNS is called the *Iur-g interface*. The support of the Iur-g interface in GERAN allows for registration areas to span across several BSS areas. Similarly, an interface between GERAN BSS and UTRAN RNS would allow for registration areas that are made up of both GERAN and UTRAN cells, typically in the same geographical location. The benefits of introducing registration areas are visible in the fact that it will reduce the amount of signalling in the network since the MS would make fewer updates. In case of registration areas that consist

Table 2.2. Interaction between RRC and MAC layers

Currently allocated channel(s)		Allocation of new resources		Current control plane states	
TCH or PDTCH on DBPSCH	PDTCH on SBPSCH	TCH or PDTCH on DBPSCH	PDTCH on SBPSCH	MAC state	RRC state
1 (or more)	—	RRC	RRC	Dedicated	Cell_Dedicated
1 (or more)	1 (or more)		MAC	DTM	
	1 (or more)			Shared	Cell_Shared
—	—	NA		Idle	
—	—		NA		GRA_PCH
—	—	NA			RRC-idle mode

of both UTRAN and GERAN cells, the introduction of Iur-g will reduce the amount of signalling on the radio interface.

Procedures over the inter BSS Iur-g and BSS–RNS Iur-g are exactly the same. The difference is in the parameters that are used within specific messages on that interface.

The Iur-g interface has been designed according to the following principles:

- This interface shall be open.
- It is assumed that the Iur-g is based on the existing Iur specification (3GPP TS 25.42x series) and uses a subset of the messages and procedures defined for the radio network subsystem application part (RNSAP). This interface shall support the exchange of signalling information between a BSS and an RNC. The Iur-g does not support user data.
- From a logical standpoint, this interface is a point-to-point interface between two BSSs or between one BSS and one RNC within a public land mobile network (PLMN). From a physical point of view, the interface could share Iu or other transmission resources.
- The Iur-g interface is optional as the Iur in UTRAN today. The presence of the Iur-g is transparent to the UE/MS: the GERAN specifications shall ensure that all mobiles function correctly, irrespective of the presence or absence of the Iur-g interface. This 'transparency principle' can be used to allow infrastructure manufacturers to implement this interface independently of other features. This implies that the UE/MS should behave like a UTRAN UE in UTRAN and a GERAN MS in GERAN, and UTRAN legacy mobiles will also be supported.

Following the principles outlined above, the protocol stack across the Iur-g interface has been designed on the basis of the UTRAN Iur interface and is shown in Figure 2.11.

When comparing this with the protocol stack specified across the Iur interface in UTRAN, we find that in Iur-g there is no UP. One reason for this choice is that procedures that are required on the Iur-g interface are fulfilled using the subset of RNSAP basic mobility procedures and RNSAP global procedures and are run over the control plane only. The list of procedures that need support over Iur-g are related to the MS operation on RRC-GRA_abbPCH state. Those are paging, cell update and GRA update procedures.

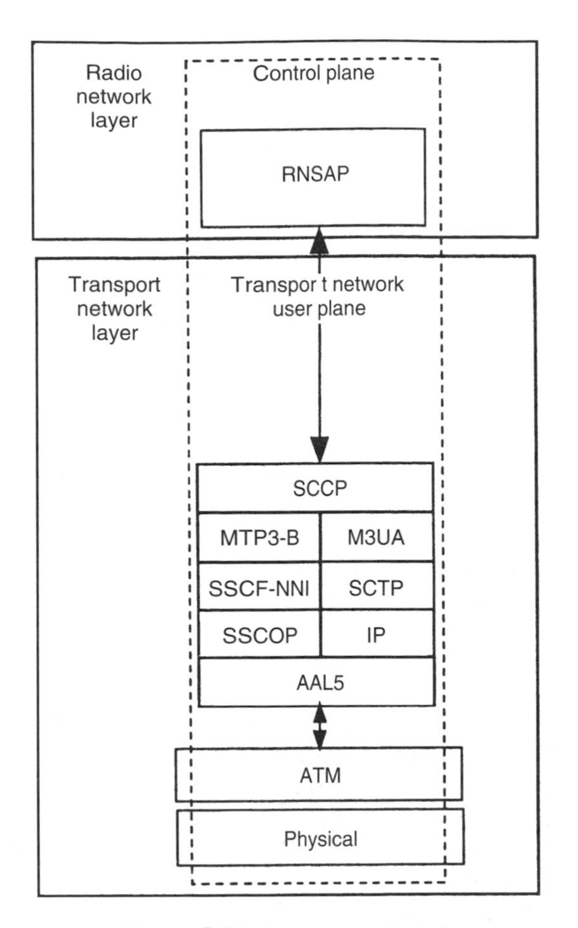

Figure 2.11. Iur-g protocol stack

The list of elementary procedures in RNSAP that support those are uplink signalling transfer, downlink signalling transfer, relocation commit, paging and error indication.

2.3 GERAN Rel'6 Features

Overall, 3GPP Rel'5 featured the emergence of the IMS architecture and services. The 3GPP Rel' 6 continues to develop the IMS framework in the UMTS CN and at the same time it makes the necessary optimisation and adaptation in the RANs—GERAN and UTRAN. One significant area to complement in the IMS specifications is the efficient provision of the real-time IMS services—including inter-working between the IMS and Public-Switched Telephone Network (PSTN) networks. From GERAN perspective, the requirement for the efficient support of IMS real-time services has led to the definition of Flexible Layer One (FLO) in Rel'6.

- Other significant work areas in GERAN Rel'include
- Single Antenna Interference Cancellation (SAIC)

- Multimedia Broadcast Multicast Service (MBMS) in GERAN
- Enhancement of Streaming QoS class services in GERAN

The above-mentioned GERAN Rel'6 work areas will be briefly introduced in the following sections.

2.3.1 Flexible Layer One

In GSM/EDGE, the provision of real-time services have been standardised in previous 3GPP releases. For each new service, optimised radio bearers were defined after careful optimisations with specific channel coding [24] and performance requirements [25]. A good example is AMR, which was introduced in Rel'98. Taking, for instance, AMR at 10.2 kbps, 3GPP specs clearly defines the channel coding:

- Split the 204-bit input block into 65 class 1a bits and 139 class 1b bits
- Attach a 6-bit cyclic redundancy check CRC to the class 1a bits
- Apply a recursive systematic convolutional coding of rate 1/3
- Apply a specific puncturing scheme
- Finally, diagonally interleave the remaining coded bits over 8 bursts.

The result is a very optimised but unfortunately not flexible service support. A radio bearer that is optimised for AMR cannot be used for anything else and, in general, a channel coding that is optimised for one real-time service cannot be used for another one. Consequently, for each new real-time service that is to be provided efficiently, new coding schemes need to be standardised, implemented and tested. With more and more services to be supported, it quickly becomes a burden for implementation as memory consumption and testing time increase. For example in Rel'5, AMR-WB and 8-PSK half-rate channels for AMR-NB took 18 months to be standardised and 58 pages of new coding schemes had to be introduced in [24] alone. Besides, with the IMS, the need for new real-time services is increasing and the pace at which they need to be introduced is becoming faster. To ensure the success of these new services, an optimised support is needed. The problem is that detailed characteristics of the IMS services are still unknown and it will be very difficult to know them long enough in advance to standardise and optimise a channel coding. Hence the need for a new approach in defining optimised radio bearers for the support of real-time services in GERAN. Hence the need for a flexible layer one.

2.3.1.1 Flexible Layer One Principles

Before going into the details of the FLO architecture, it is important to understand a few general high-level principles introduced by FLO. These principles are also described in [19]–[21]. In brief, rather than having fixed coding schemes in specifications, FLO provides a framework that allows the coding scheme to be specified and optimised at call setup. With FLO, the Layer 1 offers transport channels (TrCHs) to Layer 2. Each transport channel can carry one data flow with a certain QoS and several transport channels can be multiplexed in the very same radio packet[1]. Transport blocks (TB) are exchanged

---

[1] A radio packet is the block of coded bits to be sent before interleaving. One radio packet is transmitted for every TTI.

between Layer 1 and Layer 2 on transport channels: in the downlink, transport blocks are sent from Layer 2 to Layer 1, and in the uplink, transport blocks are sent from Layer 1 to Layer 2. The rate at which transport blocks are exchanged is called the *transport time interval* (TTI). For every TTI, one transport block is exchanged on every *active* transport channel. A transport channel is *inactive* when there is no transport block to be exchanged. Layer 3 configures the transport channels at call setup.

The configuration of a transport channel is called the Transport Format (TF). A transport format includes parameters such as input block size and CRC size. The transport formats are given by Layer 3 to fulfil the QoS requirements of the data flow to be carried on the transport channel. One transport channel can have several transport formats. Every TTI, for each transport block, Layer 2 selects the appropriate transport format on the corresponding active transport channel (link adaptation).

Only a limited number of combinations of the transport formats of different TrCHs are allowed. For instance, if 8 transport channels had 8 different transport formats each, not all 64 possible combinations would be allowed. A valid combination is called a *Transport Format Combination* (TFC) and the set of valid combinations is the *Transport Format Combination Set* (TFCS). Successive radio packets can very well use different TFCs with Layer 2 selecting the appropriate TFC (link adaptation).

In order to tell the receiver which TFC is used, a Transport Format Combination Indicator (TFCI) indicates the TFC of the radio packet. The TFCI is a layer 1 header that can be compared with the stealing bits.

Table 2.3 summarises the changes introduced by FLO.

2.3.1.2 Flexible Layer One Architecture

The FLO architecture is a simplified version of the UTRAN one (see [26]). The two radio access technologies being different and the focus being made on real-time services, quite a few simplifications were possible. Also, three important principles were used when designing the architecture:

- Keep things simple;
- Avoid introducing too many options;
- Reuse existing GSM/EDGE elements as much as possible.

Table 2.3. Changes introduced by FLO

GSM/EDGE	FLO
Logical channels	Transport channels (TrCH)
Speech frames (speech) RLC blocks (EGPRS) data frames (ECSD)	Transport blocks (TBs)
Coding scheme	Transport format combination (TFC)
Coding scheme is fixed for each logical channel	TFs are configured for each TrCH
Only one logical channel per radio block	A TFC can combine several TrCHs in the same radio packet
Coding scheme is given by the stealing bits	TFC is given by the TFCI

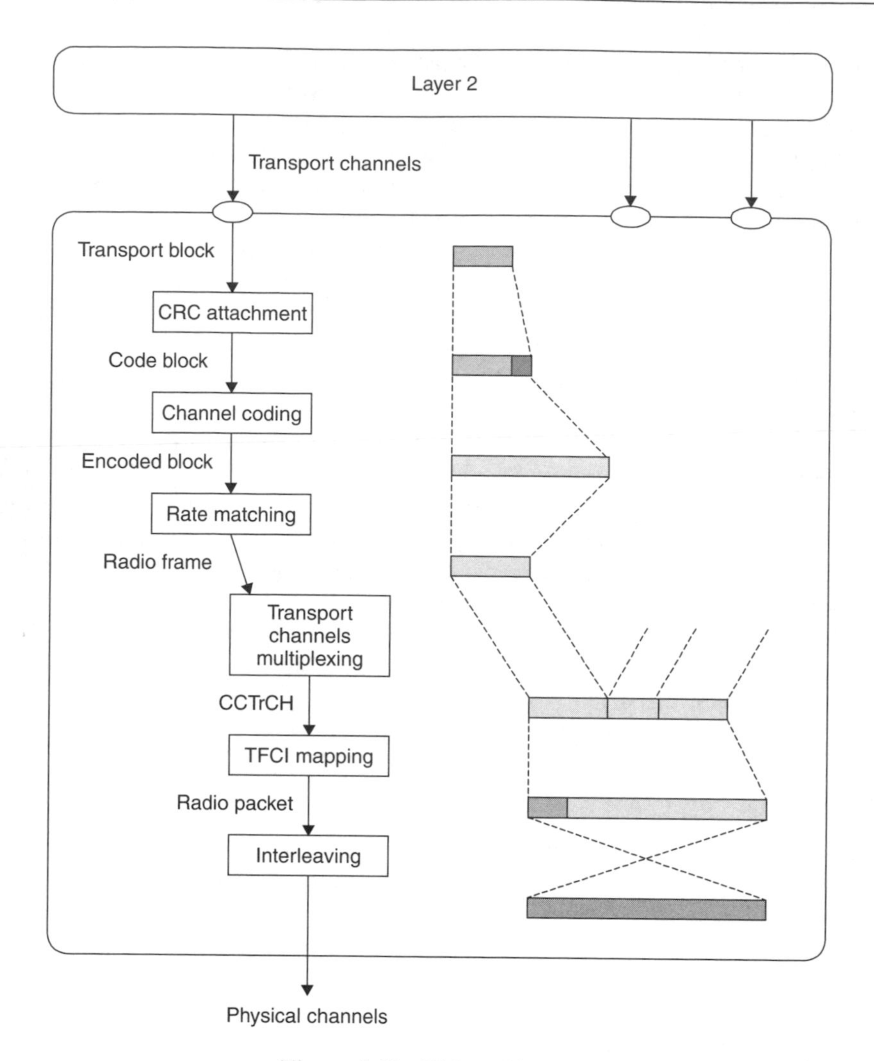

Figure 2.12. FLO architecture

The result is a six-building-blocks architecture depicted in Figure 2.12. For every TTI, transport blocks coming from active transport channels are first individually processed in CRC attachment, channel coding and rate matching, and then processed together in transport channel multiplexing, TFCI mapping and interleaving. After interleaving, bursts of coded bits are ready to be sent over the radio medium.

CRC Attachment

Error detection is optionally ensured on each transport block through a cyclic redundancy check. The size of the CRC to be used is fixed on each transport channel and configured by Layer 3 (0, 6, 12 or 18 bits). Code blocks are output from the CRC attachment and the entire transport block is used to calculate the parity bits.

Channel Coding

Forward error correction is provided on code blocks with the same 1/3 non-recursive non-systematic convolutional code of constraint length 7 as in EGPRS. Encoded blocks are output from the Channel Coding.

Rate Matching

In rate matching, coded bits of encoded blocks are repeated or punctured when needed. If an encoded block is too large, bits are punctured and the effective coding rate increases. Conversely, if an encoded block is too small, bits are repeated and the effective coding-rate decreases. The Layer 3 assigns the rate matching attribute (RMA) for each TrCH. They define priorities between the coded bits of different TrCHs. The higher the RMA is, the more important the coded bits are. After rate matching, the encoded blocks are called *radio frames*.

Transport Channel Multiplexing

This building block simply concatenates the radio frames from active transport channels into a Coded Composite TrCHs (CCTrCH).

TFCI

To indicate which TFC is being used, a TFCI is attached at the beginning of the CCTrCH producing a radio packet.

Interleaving

Radio packets are finally interleaved over GMSK or 8-PSK bursts. The interleaving is flexible: it can be 20-ms block rectangular or 40-ms block diagonal, it can be on full-rate or half-rate channels.

2.3.1.3 Configuration Example

In order to make things clear, an example of how FLO can be configured to support an AMR call is provided.

Number of Transport Channels

First of all, three transport channels are required: one for the class 1a bits (TrCH A), one for the class 1b bits (TrCH B) and another one for the associated signalling (TrCH C). Why different transport channels? Because of different QoS requirements: the class 1a bits need to be protected by a short CRC, the class 1b bits do not require any CRC and the associated signalling always require a long CRC (see [24]). Furthermore, class 1a bits typically require stronger error protection than class 1b bits. To avoid another transport channel and increase the overhead, the two inband bits are transmitted together with the class 1a bits.

Transport Block Sizes

The transport block size is fixed for the associated signalling: 184 bits always (see [24] and [15]). For the class 1a and 1b, the transport block sizes depend on the AMR codec mode to be transmitted. In this example, we consider four codec modes:

- 4.75 kbps 41 bits for TrCH A (including 2 inband bits) and 56 bits for TrCH B;
- 5.90 kbps 57 bits for TrCH A (including 2 inband bits) and 63 bits for TrCH B;
- 7.4 kbps 63 bits for TrCH A (including 2 inband bits) and 87 bits for TrCH B;
- 12.2 kbps 83 bits for TrCH A (including 2 inband bits) and 163 bits for TrCH B.

Additionally, AMR SID frames need to be sent: 41 bits on TrCH A.

CRC

For the CRC, the three transport channels are configured as follows:

- 6-bit CRC for TrCH A (as for existing TCH/AFS);
- 0-bit CRC for TrCH B (as for existing TCH/AFS);
- 18-bit for TrCH C (as for existing SACCH/TP).

Channel Coding

The channel coding block applies the same coding for all three transport channels, the same 1/3 convolutional code as in EGPRS.

Rate Matching Attributes

Concerning the rate matching attributes, they need to be selected to match the existing coding rate (TCH/AFS in [24]). As a result, the RMA for the class 1a bits is 6 and the one for class 1b bits is 5. The RMA for the associated signalling does not matter since it is always transmitted alone (when TrCH C is active, it is the only one).

TFCI

The TFCI is chosen to be 3 bits allowing the same configuration as for AMR in GSM/EDGE: 4 modes, 1 silence descriptor frame (SID) frame and associated signalling (FACCH).

Interleaving

The interleaving is made block diagonal over 8 bursts as for TCH/AFS.

Transport Format Combination Set

Figure 2.13 shows the resulting TFCS. It contains 6 transport format combinations:

- One for the associate signalling (TFC1) where only TrCH C is active;
- Four for four AMR modes where both TrCH A and B are active (TFC2-5);
- One for the SID frame where only TrCH A is active (TFC6).

Layer 3 configures these transport format combinations and it is Layer 2 that performs link adaptation by selecting the appropriate ones for every radio packet, at every TTI.

2.3.2 Single Antenna Interference Cancellation (SAIC)

Another significant development in GERAN Rel'6 is the definition and introduction of SAIC. Joint detection interference suppression techniques for TDMA-based systems such

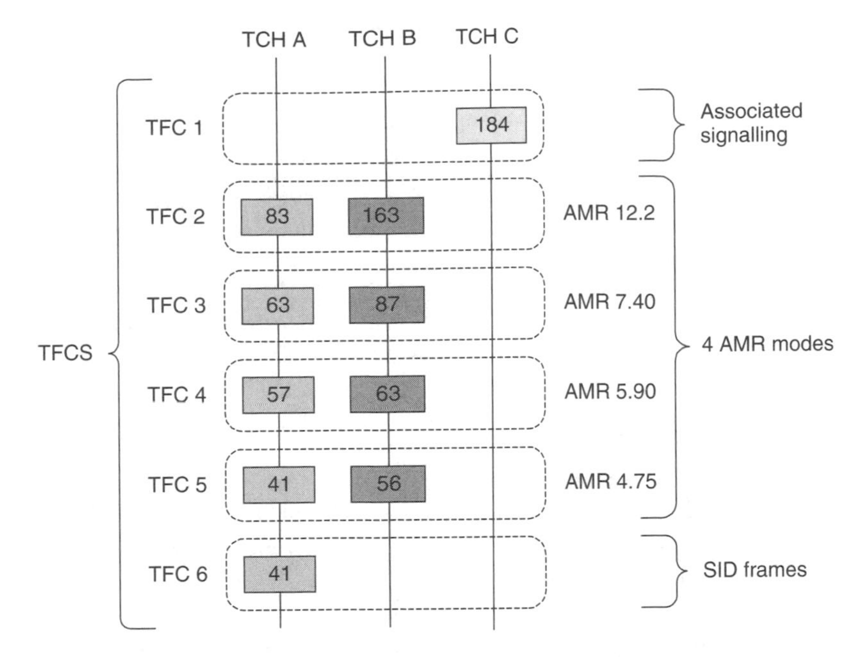

Figure 2.13. TFCS for AMR example

as GSM/EDGE were studied already in mid 1990's by a number of researchers [22, 23]. Significant co-channel interference performance improvement potential was identified already then but the requirement for relatively loose frequency reuse and for RAN synchronisation kept the practical applications on hold until these days.

The drivers for the newly born interest on interference suppression in GSM/EDGE are the new GSM/EDGE networks in the United States, which require tighter frequency reuse factors than in earlier GSM networks, and finding of blind interference suppression methods that work also without RAN synchronisation. Increased processing power in mobile station is a further driver, even though relatively low-complexity algorithms for blind interference suppression can be found.

The main idea in SAIC is to cancel or suppress the co-channel interference—coming from other cells transmitting in the same frequency according to the frequency reuse plan. Better co-channel interference performance by a mobile station can be translated to spectral efficiency gain in the radio network level once the radio network capacity is interference limited and the number of SAIC capable MSs is significant. In 3GPP GERAN standard, SAIC will mainly be translated to tighter C/I performance requirements, but also a signalling mechanism for MS to indicate its SAIC capability to network may be standardised.

SAIC is described in more detail and its performance characterisation is presented in Section 6.8.

2.3.3 *Multimedia Broadcast Multicast Service (MBMS) in GERAN*

Broadcast and Multicast are methods for transmitting data from applications running on a single node (e.g. server) to a number of nodes in the network. This concept is not new in cellular networks. During past 3GPP Releases there have been two methods to achieve this:

- A cell broadcast service (CBS) allowing for low bit-rate data to be transmitted to all subscribers in a set of given cells over a shared broadcast channel. This service offers a message-based service (see 3GPP TS 23.041 and 3GPP TS 25.324).
- An IP-Multicast service allowing for mobile subscribers to receive multicast traffic (see 3GPP TS 23.060).

IP-Multicast service as specified in 3GPP TS 23.060 does not allow for multiple subscribers to share radio or CN resources and as such does not offer any advantages as far as resource utilization within the PLMN and over the RAN. On the other hand CBS is designed for message-based service and offers only limited capabilities in terms of bit rates.

In order to support provisioning of future multimedia services to a large number of subscribers at the same time, and still maintain the efficiency of the network, 3GPP has decided to work on a solution called *Multimedia Broadcast Multicast Service*. For example, an SGSN will send data once to a BSS regardless of the number of Base Stations and MSs that wish to receive it. The benefit of multicast and broadcast on the air interface is that many users can receive the same data on a common channel, thus not clogging up the air interface with multiple transmissions of the same data. 3GPP has already produced Stage 1 document (3GPP TS 22.146), and currently is working on a Stage 2 and Stage 3 specifications.

2.3.4 Enhancement of Streaming QoS Class Services in GERAN A/Gb Mode

Chapter 1 outlined a brief history of how GPRS was developed from Rel'97 onwards. Development of new services and requirement to provide those services over existing networks has led to a continuous evolution of system features. GPRS has been designed having non-real-time services in mind, but during recent years the need for real-time services, especially services of streaming nature, has arisen. There has been a number of features that have improved the capabilities of GPRS to support services with more stringent delay requirements. Those features enabled operators to introduce services using streaming QoS class.

During Rel'6, 3GPP TSG GERAN has decided to work on improving further the performance of streaming over GERAN A/Gb mode. The work is ongoing, and one area of particular interest is finding ways to reduce packet loss during cell change.

References

[1] 3GPP TR 22.941, IP Based Multimedia Services Framework.

[2] Rotola-Pukkila J., Vainio J., Mikkola H., Järvinen K., Bessette B., Lefebvre R., Salami R., Jelinek M., 'AMR Wideband Codec—Leap in Mobile Communication Voice Quality', *Eurospeech 2001*, Aalborg, Denmark, September 3–7, 2001.

[3] Bessette B., Lefebvre R., Salami R., Jelinek M., Rotola-Pukkila J., Vainio J., Mikkola H., Järvinen K., 'Techniques for High-Quality ACELP Coding of Wideband Speech', *Eurospeech 2001*, Aalborg, Denmark, September 3–7, 2001.

[4] 3GPP TS 45.008, Radio Subsystem Link Control, 3rd Generation Partnership Project; Technical Specification Group GERAN.
[5] Contribution to 3GPP TSG GERAN AdHoc, Sophia Antipolis, France: On GERAN Speech Capacity with Different Quality Criteria, February 12–16, 2001.
[6] 3GPP TS 43.059, Functional Stage 2 Description of Location Services in GERAN (Rel'5).
[7] 3GPP TR R3-012181, Study on Network Assisted Inter BSC/RNC Cell Change for REL'5.
[8] 3GPP Tdoc GP-011755, High Multislot Classes for Type 1 Mobiles, August 27–31, 2001.
[9] 3GPP TS 23.110, UMTS Access Stratum; Services and Functions, 3rd Generation Partnership Project; Technical Specification Group GERAN.
[10] 3GPP TS 48.018, General Packet Radio Service (GPRS); Base Station System (BSS)–Serving GPRS Support Node (SGSN); BSS GPRS Protocol, 3rd Generation Partnership Project; Technical Specification Group GERAN.
[11] 3GPP TS 48.016, General Packet Radio Service (GPRS); Base Station System (BSS)–Serving GPRS Support Node (SGSN) Interface; Network Service, 3rd Generation Partnership Project; Technical Specification Group GERAN.
[12] 3GPP TS 25.413, UTRAN Iu Interface RANAP Signalling, 3rd Generation Partnership Project; Technical Specification Group GERAN.
[13] 3GPP TS 25.415, UTRAN Iu Interface User Plane Protocols, Radio Resource Control Protocol, 3rd Generation Partnership Project; Technical Specification Group GERAN.
[14] 3GPP TS 43.051 v5.3.0, GSM/EDGE Radio Access Network (GERAN) Overall Description; Stage 2, 3rd Generation Partnership Project; Technical Specification Group GERAN.
[15] 3GPP TS 44.060, General Packet Radio Service (GPRS); Mobile Station (MS) Base Station System (BSS) Interface; Radio Link Control/Medium Access Control (RLC/MAC) Protocol, Radio Resource Control Protocol, 3rd Generation Partnership Project; Technical Specification Group GERAN.
[16] 3GPP TS 25.323, Description of the Packet Data Convergence Protocol (PDCP), Radio Resource Control Protocol, 3rd Generation Partnership Project; Technical Specification Group GERAN.
[17] 3GPP TS 44.018, Mobile Radio Interface Layer 3 Specification, Radio Resource Control Protocol, 3rd Generation Partnership Project; Technical Specification Group GERAN.
[18] 3GPP TS 25.331, Radio Resource Control (RRC) Protocol Specification, 3rd Generation Partnership Project; Technical Specification Group GERAN.
[19] Sébire B., Bysted T., Pedersen K., 'Flexible Layer One for the GSM/EDGE Radio Access Network (GERAN)', *ICT2003*, Tahiti, French Polynesia, February 23–March 1, 2003.
[20] Sébire B., Bysted T., Pedersen K., 'IP Multimedia Services Improvements in the GSM/EDGE Radio Access Network', *VTC2003*, Jeju, South Korea, April 22–25, 2003.
[21] 3GPP TR 45.902, Flexible Layer One.

[22] Giridhar K., Chari S., Shynk J. J., Gooch R. P., Artman D. J., 'Joint Estimation Algorithms for Co-Channel Signal Demodulation', *ICC '93*, Geneva, 1993, pp. 1497–1501.
[23] Ranta P. A., Hottinen A., Honkasalo Z. C., 'Co-Channel Interference Cancelling Receiver for TDMA Mobile Systems', *ICC '95*, Seattle, USA, June 18–22, 1995, pp. 17–21.
[24] 3GPP TS 45.003
[25] 3GPP TS 45.005
[26] 3GPP TS 25.212

3

GERAN QoS Evolution Towards UMTS

Erkka Ala-Tauriala, Renaud Cuny, Gerardo Gómez and Héctor Montes

For voice calls and short message service (SMS), the same mobile network elements have offered both the bearer capabilities as well as the actual mass-market end-user services. Wireless Application Protocol (WAP) browsing started a new era in this respect. For WAP browsing, the traditional mobile network elements offer just the data connectivity bearer for the IP packets, while the actual end-user service is provided by an operator or Internet-hosted server using IP packets as the unifying protocol layer between the server and the mobile terminal.

In the beginning of WAP browsing, this data connectivity bearer was provided with circuit-switched GSM data connections, later on with High Speed Circuit-Switched Data (HSCSD) and more recently by using general packet radio system (GPRS) technology over GSM, EDGE or universal mobile telecommunications system (UMTS) radio access network. When GPRS technology is being used, this data bearer connectivity is provided via Gi Access Points (APs). The relationship between APs and associated Access Point Names (APNs) is explained later on in this chapter.

This same data connectivity bearer for the IP packets can be used, in the same way as for WAP, for any higher-layer application protocol that can operate on top of an IP stack. Different protocols are used for different applications and end-user services. These higher-layer protocols can be used for client-to-server type of applications like WEB and WAP browsing, Multimedia Messaging Service (MMS), multimedia streaming and e-mail, as well as for new emerging peer-to-peer IP-based applications that are using Session Initiation Protocol (SIP).

Circuit-switched GSM data and HSCSD connections treat all the IP packets equally, regardless of the end user, service or the higher-level application protocol. For GPRS-based connections, 3GPP has a standardised architecture and solution for providing data connectivity bearer level QoS differentiation on the end user, user group, service or

GSM, GPRS and EDGE Performance 2nd Ed. Edited by T. Halonen, J. Romero and J. Melero
 ISBN: 0-470-86694-2

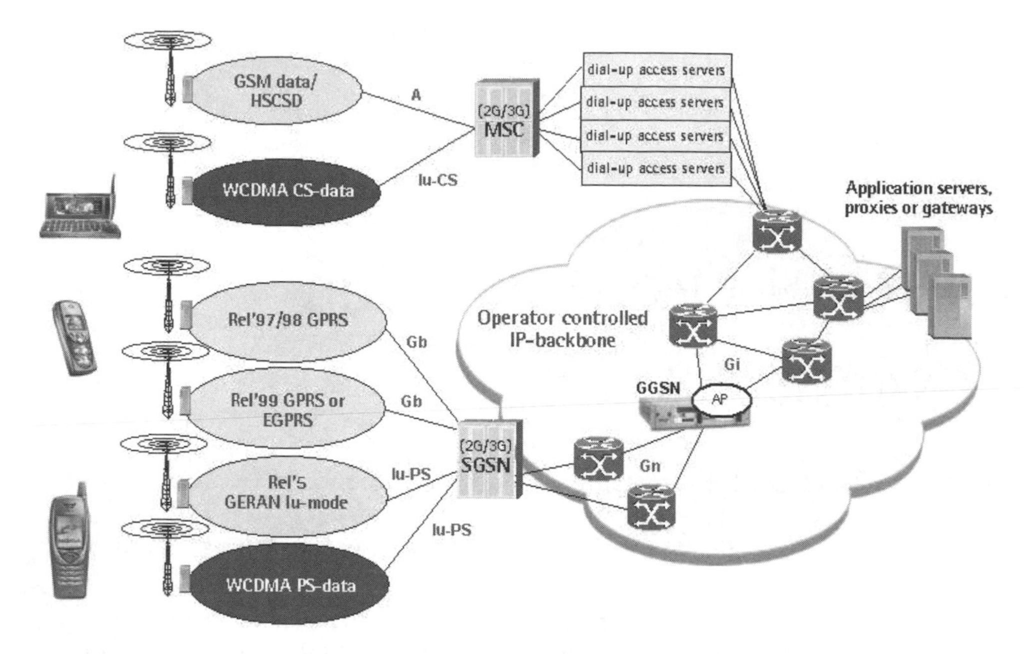

Figure 3.1. Different end-to-end data connectivity bearer options discussed in this chapter

higher-level application protocol basis. These standardised mechanisms are also covered in this chapter in more detail.

Figure 3.1 illustrates the different end-to-end data connectivity bearer chains that are examined in this chapter.

IP packets and higher-layer application protocols on top of IP can use all these different radio access data bearer variants for the communication between the clients and servers. However, the difference between these variants is on the provided data speeds as well as on the supported data connectivity bearer level QoS differentiation capabilities.

On the GPRS-based connections, Rel'97/98 GPRS, Rel'99 GPRS or EGPRS and Rel'5 GSM/EDGE radio access network (GERAN) Iu mode are referring to the different GERAN standardisation baselines. Wideband code division multiple access (WCDMA) PS data refers to UMTS terrestrial radio access network (UTRAN). The differences in the QoS capability along different standard releases are discussed in more detail towards the end of this chapter after the introduction of the 3GPP QoS architecture going across the different standard baselines.

3.1 Mobile Network as a Data Transport Media for IP-based Services

In the fixed Internet backbone network, lower-layer protocols like Ethernet, Asynchronous Transfer Mode (ATM), Frame Relay or point-to-point protocol (PPP) are typically used over Synchronous Digital Hierarchy (SDH) or Synchronous Digital NETwork (SONET) infrastructure. These are referred to as layer $1 + 2$ protocols, carrying IP as the layer 3 protocol, while transmission control protocol (TCP) and user datagram protocol (UDP) are the most commonly used layer 4 protocols. Different end-user application-related higher-layer protocols run on top of TCP/IP or UDP/IP.

Examples of such applications and higher-layer protocols are web/wap browsing, which use Hyper Text Transfer Protocol (HTTP) or Wireless Application Protocol (WAP); Multimedia Streaming, which makes use of Real-time Streaming Protocol (RTSP) for signaling and Real-time Transfer Protocol (RTP) for conveying media data; e-mail, which employs several protocols such as Simple Mail Transfer Protocol (SMTP), and a new generation of peer-to-peer services based on Session Initiation Protocol (SIP). This layered protocol architecture is illustrated in Figure 3.2. Examples from the end-user applications and related higher-layer protocols are also shown in the following paragraphs.

The GPRS protocol architecture introduced in Chapter 1 of this book is a particular layer $1 + 2$ protocol implementation that is used for carrying the IP packets in the GSM/EDGE access network. For the application layer IP packets, the whole mobile network is a transparent data pipe. In other words, the information on the application level IP-packet header is not used for routing the traffic between the mobile client and the AP in the gateway GPRS support node (GGSN).

Either TCP connections are established or UDP packets are carried between the client and a server or proxy. The higher-level protocols like HTTP or RTP use either TCP or UDP as the lower-layer carrier. This protocol stack structure is illustrated in Figure 3.3.

This illustrates the protocol stack handling from the standard point of view. In practice, handling of TCP, UDP or higher-layer application protocols may also be embedded to some mobile network elements like serving GPRS support nodes (SGSNs) or GGSNs due to, for example, IP protocol optimisation or charging reasons. Such specific solutions are not standard related and are not within the scope of this book.

At the PDP-context activation, the mobile terminal is being allocated an IP address, known also as the *Packet Data Protocol* (*PDP*) address. The PDP-context activation is done to some specific access point name (APN). The used APN can be defined in the terminal settings or alternatively so-called default or Wild card APN can be used. One APN can point to one or several physical APs located at GGSNs. The mapping from an APN to a GGSN's physical IP address in the Gn-interface between SGSN and GGSN is done with a Domain Name Server (DNS). Figure 3.4 shows the relationship between the terminal settings, APNs, DNS, APs as well as example DNS and terminal settings [1].

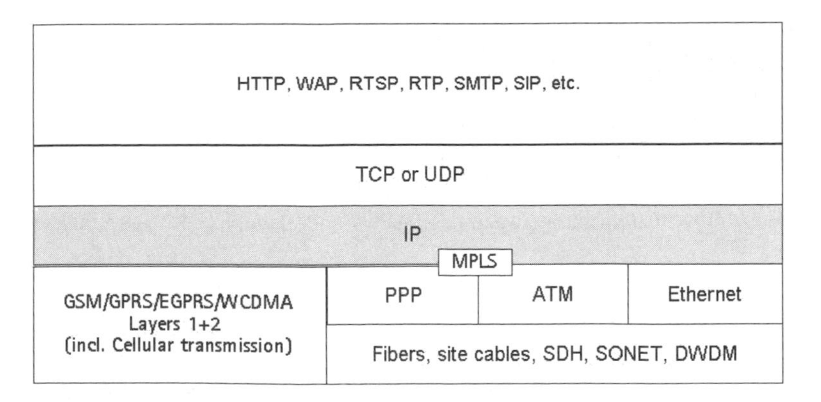

Figure 3.2. Layered general protocol architecture for IP-based applications

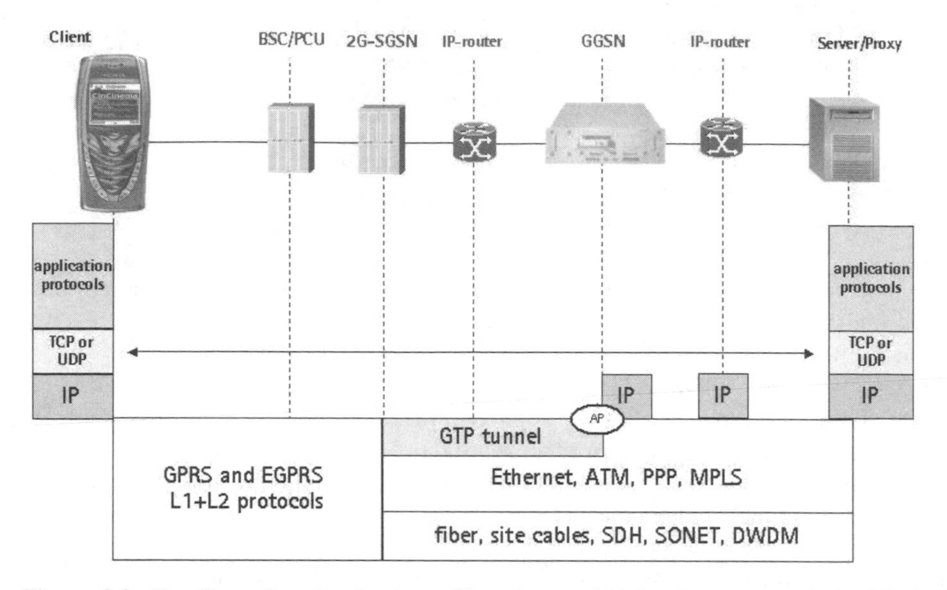

Figure 3.3. Handling of application layer IP packets and higher-layer protocols in GPRS

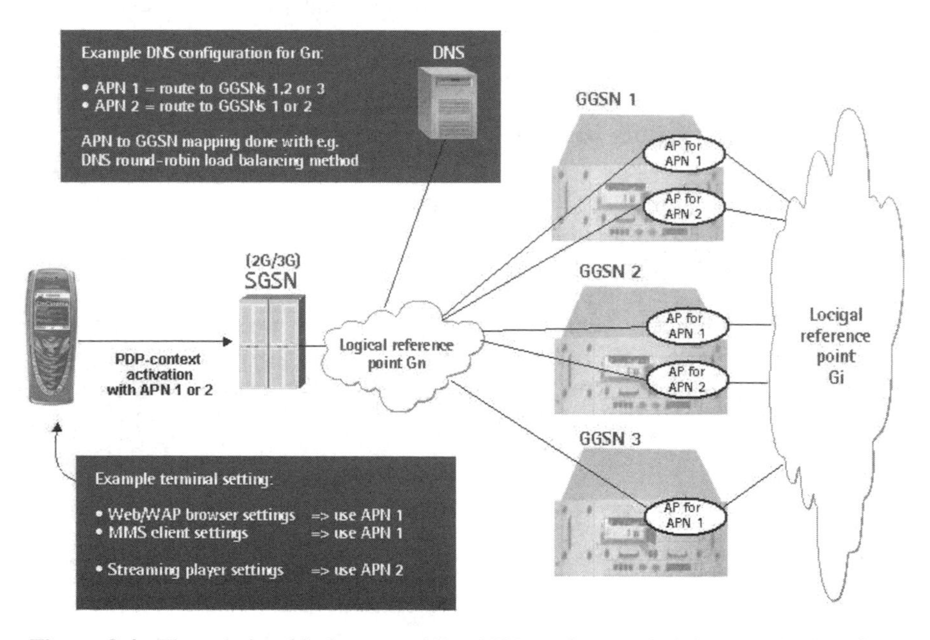

Figure 3.4. The relationship between APs, APNs and example DNS+terminal settings

The routing of the application layer IP packets in the GPRS architecture is based on the allocated AP for the uplink direction and on both the allocated GGSN AP and PDP addresses for the downlink direction. In the Gi-reference point of the GGSN, an AP is linked either to specific physical or virtual interface. This can be, for example, a specified physical Ethernet, a SDH/SONET interface or a virtual tunnel using Generic

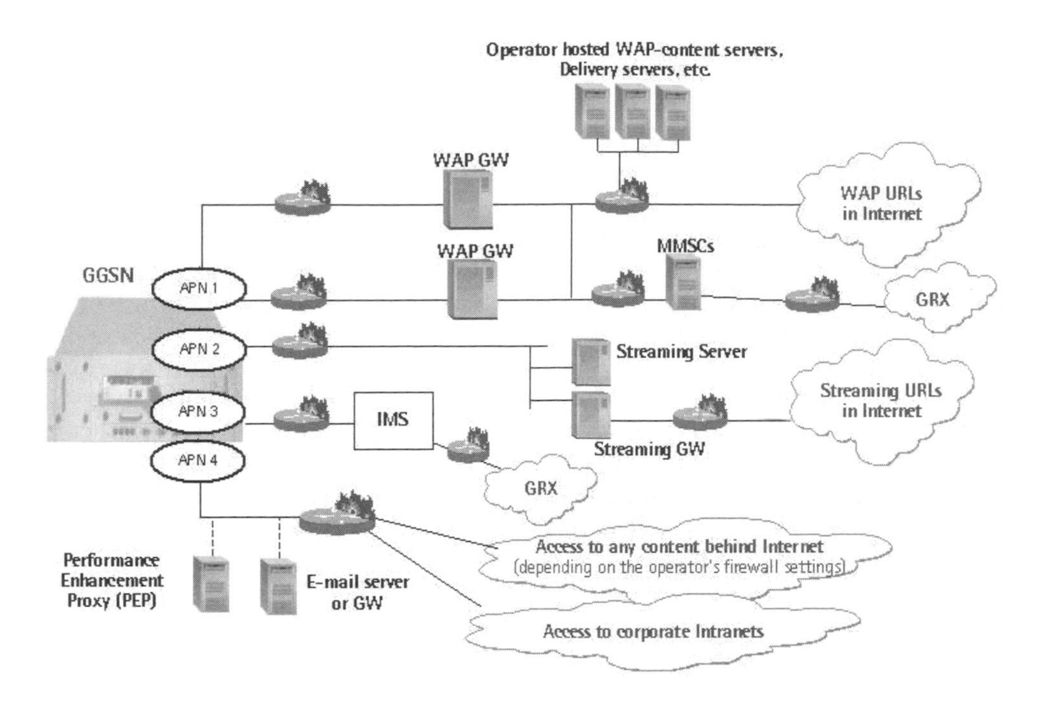

Figure 3.5. Example of APN, firewall, server and proxy arrangements in Gi

Routing Encapsulation (GRE) [2], Layer Two Tunneling Protocol (L2TP) [3], Virtual Local Area Network (VLAN) [4], Multi-protocol Label Switching (MPLS) [5] or IPSec protocol [6]. This physical interface or virtual tunnel can in turn point to some specific firewall, gateway, proxy, corporate Intranet or directly to the public Internet.

The PDP-context QoS between the terminal and GGSN is provisioned on the APN basis and the arrangements behind all the APs in different GGSNs linking to the same APN are typically equal. As a result, in the rest of this chapter, architecture and arrangements per APN are being discussed.

An example of Gi APN, firewall, server and proxy arrangement is illustrated in Figure 3.5. The used APN arrangements are typically influenced, for example, by charging, security or QoS. In the figure, GRX refers to an inter-operator GPRS Roaming Exchange, MMSC refers to Multimedia Messaging Service Center and IMS to IP-Multimedia Subsystem. Performance Enhancement Proxies (PEP) are introduced in more detail in Chapter 8. This may vary in different networks and depends on the individual operator preferences as well as on the used mobile network equipment capabilities. The relationship between the APNs and 3GPP QoS architecture is discussed later on in this chapter.

3.2 Example of IP-based Applications Using Mobile Network as Data Bearer

In the Gi-reference point of the GGSN, an APN is linked either to specific physical or virtual interface. Depending on the used IP-based applications, different higher-level protocols are used between the mobile and different Gateways (GWs), proxies or servers located behind one or several APNs. In Section 3.2.1, the basic architecture for the standard-based WAP browsing, MMS, audio/video streaming and 3GPP Rel'5 standardised IMS services are introduced.

3.2.1 WAP Browsing

In the fixed Internet, WEB browsing became very popular during the 1990s. The traditional WEB browsing is based on the retrieval of WEB pages from a server to a client. The WEB pages are built with Hyper Text Mark-up Language (HTML), and they have been retrieved from a client to a server by using HTTP over one or more TCP/IP connections.

Initial WAP Forum standards were defined to optimise browsing services for mobile handset screens and wireless links. These standards defined a new page format for mobile handset screens known as WAP Mark-up Language (WML). For retrieving WML pages from a client to a server, a new wireless optimised protocol stack, WAP, was defined between the mobiles and WAP Gateways (WAP GW), while HTTP was specified to be used between the WAP GW and the actual server hosting the WML pages. During the year, several WAP versions have been issued.

In the WAP 1.x specifications, the usage of a UDP/IP-based WAP protocol stack was defined between the mobile-embedded browser client and WAP GW. In the mobile browser settings, the IP address of the used WAP GW was defined, and the mobile station sent all the WAP-browsing data to the IP address of this gateway. In addition to the WAP GW IP address, either the dial-up access server telephony number or the correct GPRS APN had to be configured to the mobile browser settings.

WAP 2.0 specifications included two important additions. First, WAP Forum and W3C WEB-browsing standardisation body agreed to harmonize HTML and WML so that same pages can be accessed with both fixed and mobile browsers. This evolution of HTML and WML is known as eXtensible HyperText Mark-up Language (XHTML). In addition, WAP 2.0 specification includes the option to use either UDP/IP-based WAP stack or alternatively TCP/IP-based HTTP stack for the mobile embedded XHTML browsing connections [7].

3.2.2 Multimedia Messaging Service (MMS)

Multimedia messaging standards did not specify any new IP-based protocols. Instead, it standardised the format of the messages (for text, sound and pictures) as well as the MMS usage over already existing protocols like WAP. As a result, the MMS traffic also typically uses either WAP or XHTML connections for message retrieval and delivery. In the case of mobile terminated messages, the network typically sends one or several SMS (via a connection from the MMSC to an SMS Center, SMSC) to the mobile terminal, to notify on the arrival of an MMS, triggering a PDP-context activation and/or WAP or XHTML connection set-up from a mobile to an MMSC. An example arrangement is illustrated in the Figure 3.6 [8].

The MMS client settings in the terminal should include the correct WAP GW IP address as well as the GPRS APN providing access to the MMSCs.

3.2.3 Audio/Video Streaming

Audio and video clips can be transferred to mobile stations (MS) by using, for example, MMS. However, with MMS, the clip is played only when it has been completely delivered

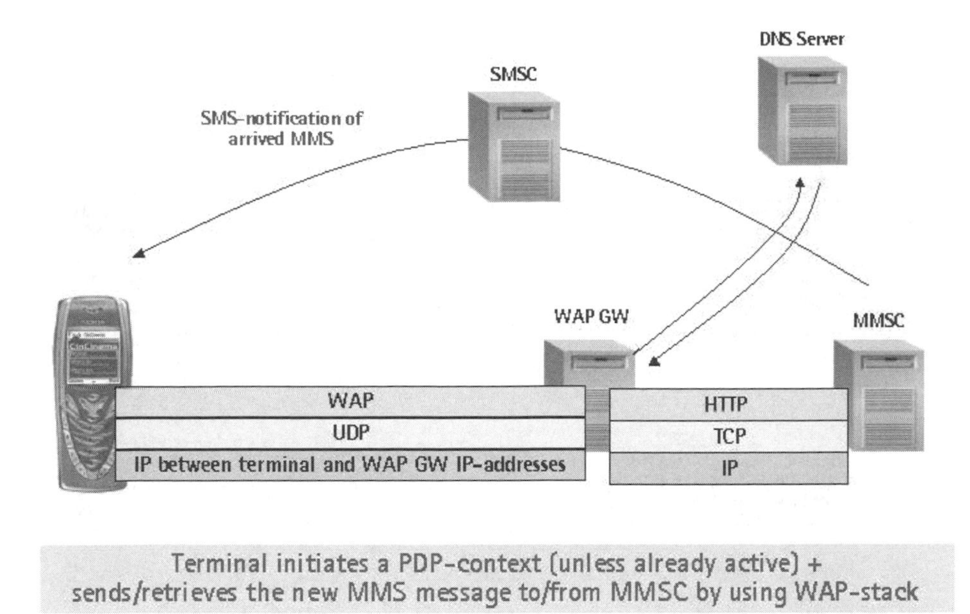

Figure 3.6. Example arrangements for MMS connections using WAP 1.x

to the MS. When audio or video is streamed, a small buffer space is created on the user's terminal, and data starts downloading into it. As soon as the buffer is full (usually just a matter of seconds), the file starts to play. As the file plays, it uses up information in the buffer, but while it is playing, more data is being downloaded. As long as the data can be downloaded as fast as it is used up in playback, the file will play smoothly.

Compared with, for example, MMS technology and so-called local playing, streaming technology allows longer-lasting clips and/or higher audio or video quality to be used with a same terminal memory size. On the other hand, from the radio network QoS mechanism perspective, streaming technology is much more demanding than, for example, usage of MMS and local playing for the audio/video clip downloading.

Although there are a lot of proprietary solutions in the market, streaming technology typically uses Real-time Streaming Protocol. This protocol is used to negotiate the session set-up, pausing it as well as to tear down streaming sessions [9]. RTSP uses TCP/IP as the lower-layer bearer. Once the 'PLAY' command has been given over the RTSP protocol, the actual media streams are transferred with RTP, and the quality of the transmission is controlled with Real-time Control Protocol (RTCP). Both of these typically use UDP/IP as the lower-layer bearer [10]. The usage of these protocols is illustrated in Figure 3.7. In addition, there is an example on the PDP context and GERAN radio bearer (RB) QoS set-up for RTSP, RTP and RTCP streaming protocol connections at the end of this chapter.

For the mobile environment, 3GPP Rel'4 includes standardised protocol and codec definitions between the terminal clients and streaming server and/or GWs [11]. However, proprietary vendor-specific streaming solutions may be used over mobile infrastructure as long as the client and server software comes from the same manufacturer.

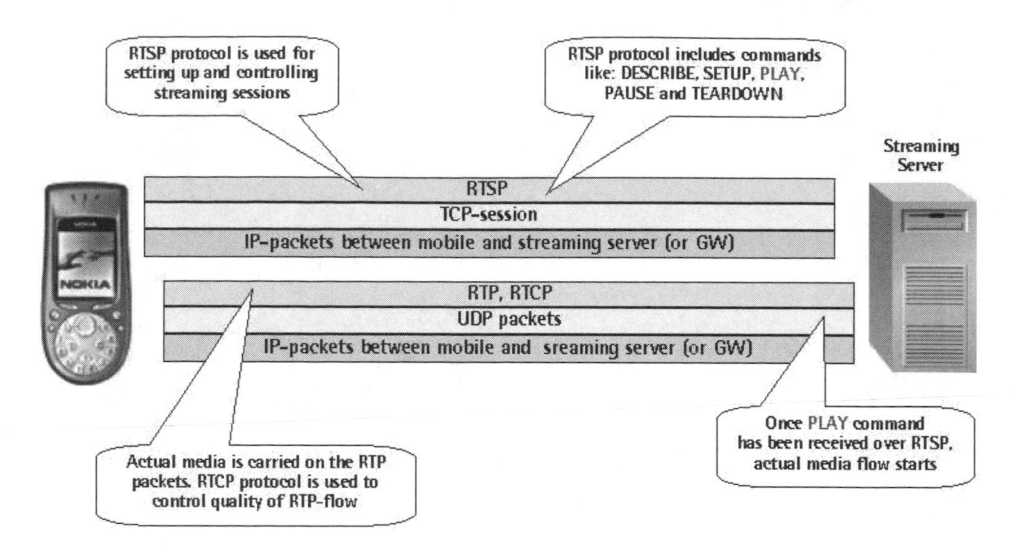

Figure 3.7. The usage of IP-based protocols for streaming audio/video

3.2.4 IMS Services

3GPP Rel'5 core network standards define the usage of IMS for setting up the so-called *peer-to-peer multimedia sessions* between two or more mobile stations [12]. These sessions are set up by using SIP between the mobiles and IMS, while the actual sessions, for example, gaming, voice or video over IP data can be routed directly from mobile to mobile without any server between them. Between the mobiles, various protocols may be used depending on the used IMS applications. For example, HTTP or RTP and RTCP may be used for these connections. The basic concept of the IMS is illustrated in Figure 3.8.

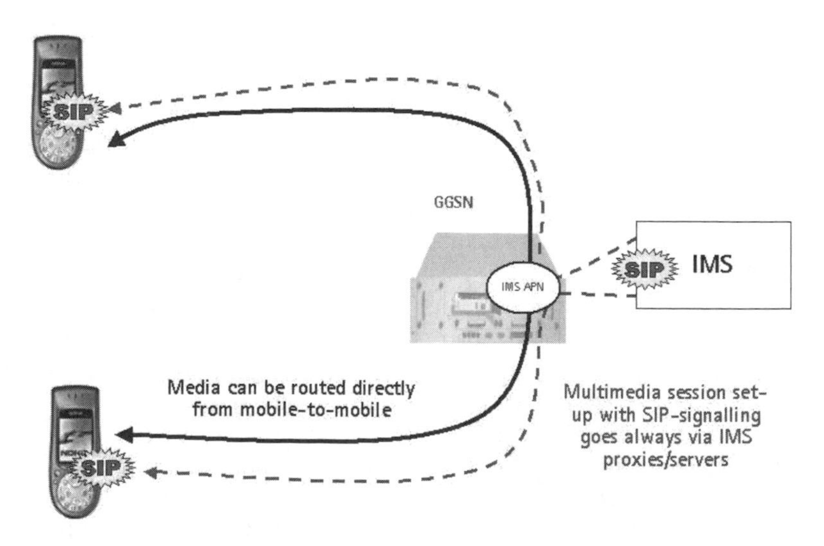

Figure 3.8. The basic concept of IMS for peer-to-peer multimedia sessions

3GPP standardised bearer QoS authorisation for the IMS services is discussed in more detail later on in this chapter.

3.3 End-to-end QoS in the 3GPP QoS Architecture

The evolution of the QoS management architecture for mobile networks began with Rel'97/98, where GPRS was introduced. The main feature of Rel'97 was the introduction of the concept of GPRS bearer service (BS), which was the base of the GPRS QoS management framework. The GPRS bearer service is based on the concept of PDP context, which is, in essence, a logical connection set up between mobile station (MS) and gateway GPRS support node (GGSN) to carry all the IP traffic to and from MS.

Although the PDP context was introduced by the GPRS architecture, the 3GPP standards have further developed the concept to support a variety of new requirements. The first GPRS standard (Rel'97/98) already allows one mobile station (MS) to have multiple simultaneous PDP contexts, but they all need a dedicated PDP address. This is enhanced in Rel'99, as it enables the usage of several PDP contexts per PDP address, each having a QoS profile of its own. The first PDP context opened for a PDP address is called the *primary or default PDP context*. The subsequent contexts opened for the same PDP address are *secondary* contexts. However, the usage of secondary PDP contexts requires that they be connected to the same APN as the primary PDP context.

The basis of this architecture is that QoS bearer differentiation is done on the PDP-context basis. Thus, all the applications of a single user sharing the same PDP context have the same QoS attributes. For differentiated QoS treatment, parallel primary and/or secondary PDP contexts need to be activated for the same MS. In addition to the MS capabilities, this depends on whether the different services are available behind the same or different APNs. This is illustrated in Figure 3.9.

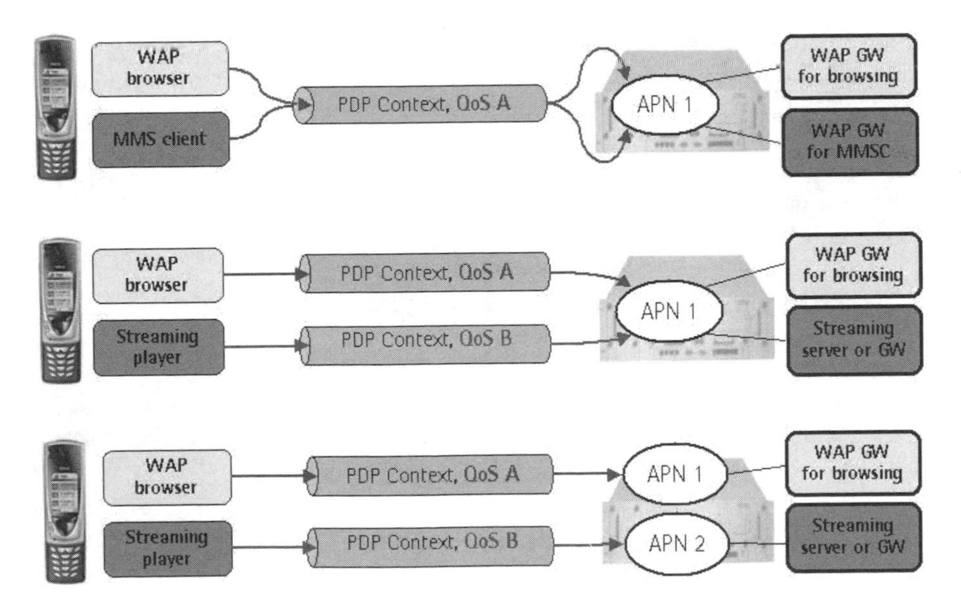

Figure 3.9. PDP-context QoS differentiation examples

Depending on the mobile station implementation, QoS is driven either by the subscriber and APN-specific QoS profiles in the Home Location Register (HLR), by the mobile station requesting explicit attribute values or both [13]. This is explained in more detail in the following paragraphs.

This difference of the GERAN standard baselines and associated QoS architecture is illustrated in Figure 3.10.

On the basis of the Rel'97 GPRS Gb bearer service (Gb BS), the 2G-SGSN can map priority information to each LLC frame based on the priority of the associated PDP context.

Rel'99 includes some features allowing a better control of the QoS provision in the radio domain, like sending the QoS requirements from the 2G-SGSN to the BSS by means of the *BSS packet flow context* (PFC). A given BSS PFC can be shared by one or more activated PDP contexts with identical or similar negotiated QoS requirements. The data transfer related to PDP contexts that share the same PFC constitute one aggregated packet flow, whose QoS requirements are called *aggregate BSS QoS profile*. The aggregate BSS QoS profile defines the QoS that must be provided by the BSS for a given packet flow between the MS and the SGSN, that is, for the Um and Gb interfaces combined. Rel'99 GERAN specification allows the support of all the other standardised QoS-parameters except Conversational traffic class.

In addition, Rel'99 standardised the usage of IETF Differentiated Services (Diff-Serv) [14, 15] for the user plane IP-packet prioritisation as a part of the core network bearer service (CN BS).

The standardisation of GERAN Iu mode was completed in the 3GPP Rel'5. With GERAN Iu mode, GERAN and UTRAN QoS management frameworks are fully harmonised, including support for Conversational traffic class, since GERAN adopts the UMTS architecture with the 3G-SGSN being connected to both types of radio access networks (RANs) through the Iu interface. The Iu interface towards the packet-switched (PS) domain of the CN is called Iu PS and uses Iu bearer service (Iu BS).

In Figure 3.11, GPRS BS refers to a PDP context with Rel'97/98 QoS attributes, while the UMTS BS refers to a PDP context with Rel'99 QoS attributes, respectively. The IP bearer service (IP BS) in Rel'5 standardises the usage of Policy Decision Function (PDF) for IMS service-related QoS authorisation. The usage of PDF is also discussed in more detail later on in this chapter.

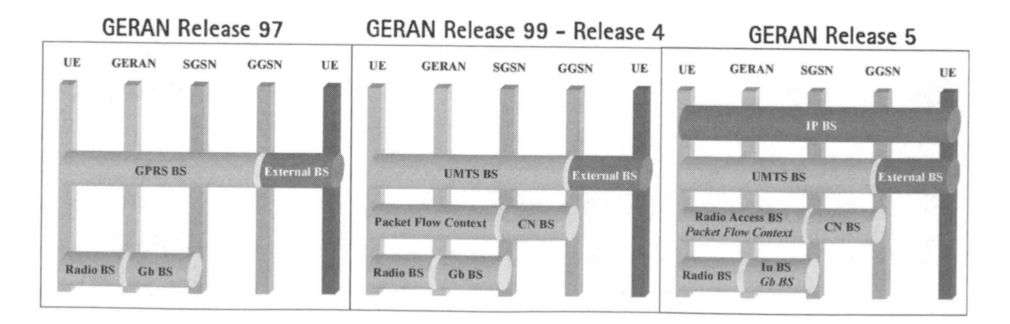

Figure 3.10. Evolution of QoS architecture in GERAN-based mobile networks

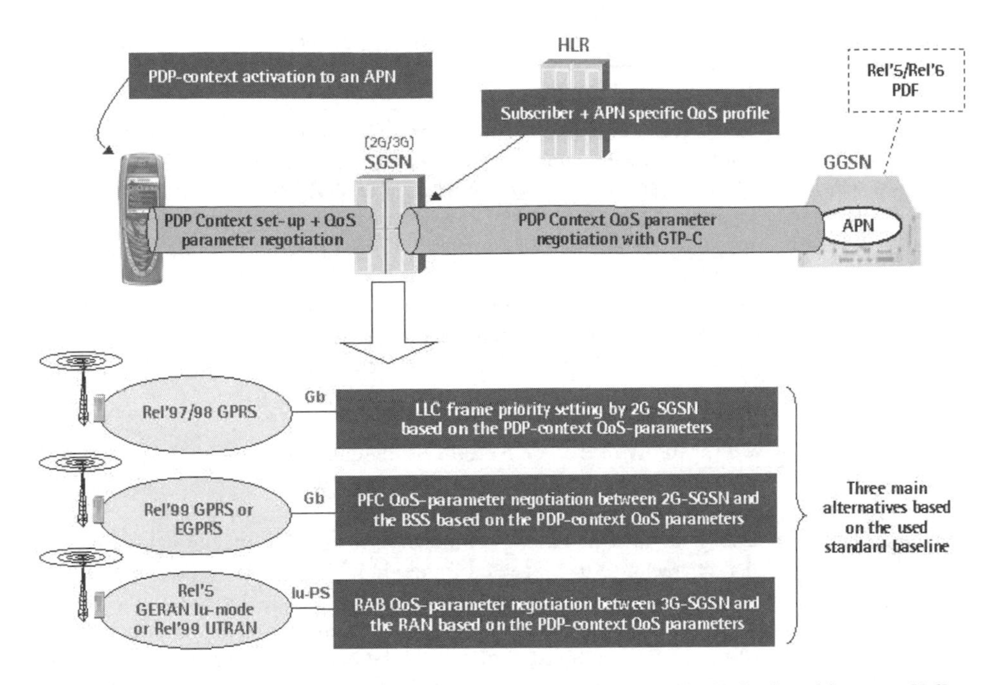

Figure 3.11. End-to-end QoS negotiation alternatives in the 3GPP QoS architecture [16]

The PDP-context activation signalling and QoS attribute negotiation takes place between the mobile station and the SGSN. SGSN gets from the HLR, subscriber and APN-specific QoS profiles via a standardised interface using Mobile Application Part (MAP) protocol. SGSN takes care of the logical link control (LLC) frame–priority mapping and/or BSS Packet Flow Context (BSS PFC) set-up with BSS GPRS Protocol (BSSGP) towards GERAN and of the radio bearer set up with radio access network application part (RANAP) protocol towards UTRAN or GERAN supporting Rel'5 Iu mode. In addition, SGSN performs the GTP-tunnel set-up towards the GGSN with GTP-C protocol. These higher-level QoS negotiation alternatives are shown in Figure 3.11.

3.4 PDP-context QoS Parameter Negotiation

In the PDP-context negotiation, the QoS attributes are negotiated between the terminal and the network.

QoS requirements expressed through the QoS profile in Rel'97/98 are defined in terms of the following attributes: precedence, delay, reliability and throughput classes (see Table 3.1).

The service precedence indicates the relative importance of maintaining the service commitments under abnormal conditions (e.g. which packets are discarded in the event of problems such as limited resources or network congestion). The delay class determines the per-packet GPRS network transit delay. The combination of the transmission modes of the different network protocol layers supports the reliability class performance requirements. User data throughput is specified in terms of a set of throughput classes that characterise the expected bandwidth required for such traffic.

Table 3.1. QoS attributes in Rel'97/98

QoS attribute	Description
Precedence class	There are three different service precedence levels, which indicate the priority of maintaining the service. High precedence level guarantees service ahead of all other precedence levels. Normal precedence level guarantees service ahead of low priority users. Low precedence level receives service after the high and normal priority commitments have been fulfilled.
Delay class	Delay attribute refers to end-to-end transfer delay through the GPRS system. There are three predictive delay classes and one best-effort class.
Reliability class	There are five different reliability classes. Data reliability is defined in terms of residual error rates for the following cases: • Probability of data loss • Probability of data delivered out of sequence • Probability of duplicate data delivery and • Probability of corrupted data.
Throughput class	User data throughput is specified in terms of a set of throughput classes that characterise the expected bandwidth required for a PDP context. The throughput is defined by both peak and mean classes: • The *peak throughput* specifies the maximum rate at which data are expected to be transferred across the network for an individual PDP context. There is no guarantee that this peak rate can be achieved or sustained for any time period; this depends upon the MS capability and available radio resources. The peak throughput is independent of the delay class that determines the per-packet GPRS network transit delay. • The *mean throughput* specifies the average rate at which data are expected to be transferred across the GPRS network during the remaining lifetime of an activated PDP context. A best-effort mean throughput class may be negotiated, and means that throughput shall be made available to the MS on a per need and availability basis.

Rel'99 introduced the UMTS QoS traffic classes and attributes (described in Table 3.2), and GPRS QoS at session management level integrated with them.

The harmonisation of GERAN and UTRAN QoS management frameworks in the Rel'99 implies that the same QoS parameters are also employed in lower-level protocols. However, as the Rel'97/98 MS need to work in the Rel'99 standard-based network, mapping from the Rel'97/98 attributes to the Rel'99 attributes has also been standardised (see Table 3.3).

When defining the UMTS QoS classes, the restrictions and limitations of the air interface have been taken into account. It is not reasonable to define the same mechanisms used in fixed networks due to different error characteristics of the air interface. The QoS mechanisms provided in the cellular network have to be robust and capable of providing reasonable QoS resolution.

Rel'99 defines four different QoS classes (or traffic classes):

- conversational class
- streaming class

Table 3.2. QoS attributes in Rel'99

QoS attribute	Description
Traffic class	Type of application for which the radio access bearer (RAB) service is optimised ('conversational', 'streaming', 'interactive' or 'background').
Delivery order	Indicates whether the bearer shall provide in-sequence service data unit SDU delivery or not.
Maximum SDU size (octets)	Defines the maximum allowed SDU size.
SDU format information (bits)	List of possible exact sizes of SDUs. If unequal error protection shall be used by a RAB service, SDU format information defines the exact subflow format of the SDU payload.
Delivery of erroneous SDUs	Indicates whether SDUs with detected errors shall be delivered or not. In case of unequal error protection, the attribute is set per subflow.
Residual BER	Specifies the undetected bit error ratio for each subflow in the delivered SDUs. For equal error protection, only one value is needed. If no error detection is requested for a subflow, residual bit error ratio indicates the bit error ratio in that subflow of the delivered SDUs.
SDU error ratio	Defines the fraction of SDUs lost or detected as erroneous. SDU error ratio is defined only for conforming traffic. In case of unequal error protection, SDU error ratio is set per subflow and represents the error ratio in each subflow. SDU error ratio is only set for subflows for which error detection is requested.
Transfer delay (ms)	Indicates maximum delay for 95th percentile of the distribution of delay for all delivered SDUs during the lifetime of a bearer service, where delay for an SDU is defined as the time from a request to transfer an SDU at one service access point SAP to its delivery at the other SAP.
Maximum bit rate (kbps)	Maximum number of bits delivered at a SAP time within a period of time divided by the duration of the period.
Guaranteed bit rate (kbps)	Guaranteed number of bits delivered at a SAP within a period of time (provided that there are data to deliver), divided by the duration of the period.
Traffic handling priority	Specifies the relative importance for handling of all SDUs belonging to the RAB compared with the SDUs of other bearers.
Allocation/retention priority	Specifies the relative importance compared with other RABs for allocation and retention of the radio access bearer. The allocation/retention priority attribute is a subscription parameter, which is not negotiated from the mobile terminal.

- interactive class
- background class.

The main differentiating factor between these classes is how delay sensitive the traffic is; conversational class is meant for traffic that is very delay sensitive while background class is the most delay-insensitive traffic class.

Table 3.3. Rel'97/98 QoS attribute mapping to Rel'99 QoS attributes

Rel'97/98 QoS attribute	Rel'99 QoS attribute
Precedence class	Allocation/retention priority
Delay class	Interactive and background traffic classes and interactive class traffic handling priorities
Reliability class	Specified combinations of residual BER, SDU error ratio and delivery of erroneous SDUs
Throughput class	Maximum bit rate

Conversational and streaming classes are mainly intended to be used to carry real-time traffic flows. Conversational real-time services, like voice over IP, video over IP or real-time gaming, are the most delay-sensitive applications, and those data streams should be carried in conversational class. Streaming class is also meant for delay-sensitive application audio or video streaming, but unlike conversational applications, they can compensate some delay variation with client-embedded buffering mechanisms.

Interactive and background classes are mainly meant to be used by applications like WEB or WAP browsing, MMS or e-mail. Owing to looser delay requirements, compared to conversational class, both provide better error rate by means of retransmission schemes. The main difference between interactive and background classes is that interactive class is the classical data communication scheme that, on an overall level, is characterised by the request response pattern of the end user, for example, WEB or WAP browsing, while background class is the scheme that, on an overall level, is characterised by the fact that the destination is not expecting the data within a certain time. The scheme is thus more or less delivery time insensitive, for example, background e-mail or file downloading. Responsiveness of the interactive applications is ensured by separating interactive and background applications. Traffic in the interactive class has higher priority in scheduling than background class traffic, so background applications use transmission resources only when interactive applications do not need them. This is very important in the wireless environment where the radio interface bandwidth is small compared with fixed networks.

The standardised HLR capabilities can be used for creating differentiated subscriber and APN-specific QoS profiles, for example, for 'VIP', 'Budget' and 'Machine' users. Figure 3.12 illustrates HLR connection toward SGSN, with parallel PDP contexts for the application having different QoS requirements as well as subscriber differentiation possibilities via the HLR QoS profiles in a Rel'99 environment. The placement of the different application servers behind different APNs illustrates only an example and the actual real-life implementation may vary operator by operator. The figure is applicable also to the Rel'97/98 environment, except that with Rel'97/98 Streaming class PDP contexts or secondary PDP contexts are not supported (but parallel primary PDP contexts may still be used).

The following is specified on the PDP-context QoS attribute negotiation:

- For each PDP-context Rel'97/98 or Rel'99 QoS attribute, the MS can request either explicit standard defined values or alternatively just subscribed values from the network.

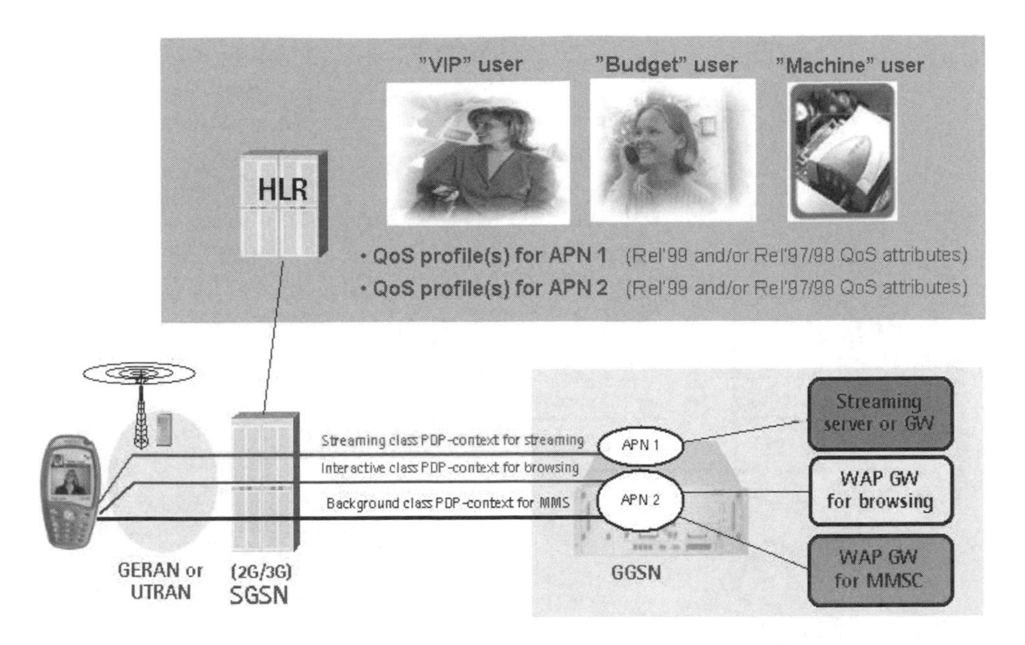

Figure 3.12. Example application and end-user differentiation with Rel'99 QoS framework

- On the basis of an approved 3GPP Rel'99 Change Requests (CR) from June 2002 [17, 18], in case the MS requests subscribed values from the network for the Rel'99 traffic class attribute and/or Rel'99 guaranteed bit rate attribute, only 'interactive' or 'background' values are allowed for the traffic class parameter and no guaranteed bit rate is allocated by the network.
- In case the MS requests for any parameter subscribed values in the PDP-context activation, the values are obtained from the HLR. In the HLR, different QoS parameters may be configured for each subscriber and APN combination.
- In case the MS requests explicit numerical values, then the same HLR profile works as the upper limit given for each subscriber and APN combination.
- The same HLR QoS profile may be also used for the differentiated subscriber QoS treatment as illustrated earlier in this chapter with 'VIP', 'Budget' and 'Machine' users.

3.4.1 QoS Authorisation for IMS and Non-IMS Services with PDF

Before Rel'5, the individual subscriber's QoS authorisation is completely based on the subscriber and APN-specific QoS profile stored in the HLR. In the 3GPP R'5 core network standards, new IMS has been standardised in order to enable IP-based mobile-to-mobile peer-to-peer connectivity for applications like gaming, messaging, voice over IP or video over IP.

As a part of the 3GPP Rel'5 IMS-standardisation, the Policy Decision Function (PDF), formerly known as Policy Control Function (PCF), has been standardised between the

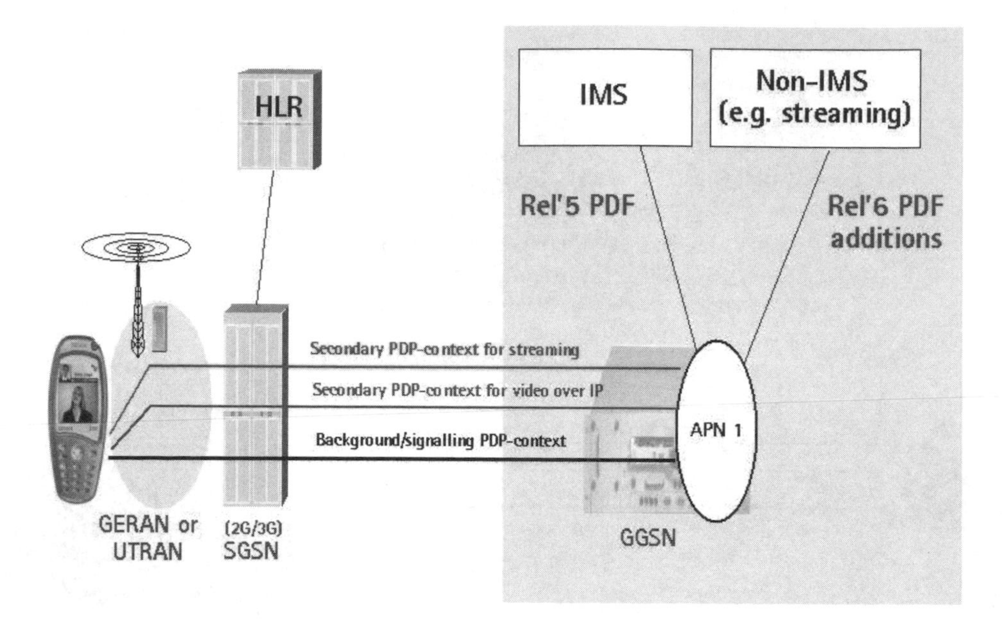

Figure 3.13. IMS and non-IMS service's bearer QoS authorisation with Rel'5 and Rel'6 PDF

GGSN and IMS [18]. The role of PDF is to allow the mobile operator to authorise and, in this way, to control the QoS of the IMS-related PDP contexts.

The IMS sends an authorisation token to the MS in the SIP-signalling as a part of the IMS session set-up using, for example, an interactive class primary PDP context. Now, as a part of the IMS service like video over IP, when the MS initiates a secondary PDP context for the real-time media component, it will send this authorisation token as a part of the secondary PDP-context activation to the network. The GGSN uses this token as well as the information received from the IMS via the PDF to either accept or downgrade the QoS attributes of the MS-initiated secondary PDP context.

In the 3GPP Rel'6 standardisation, there have been initiatives to extend the PDF to also include the QoS authorisation and control of PDP contexts used for selected non-IMS services like streaming [10] or Multimedia Broadcast Multicast Services (MBMS). Both Rel'5 and Rel'6 PDFs are illustrated in Figure 3.13.

It should be noted that these concepts are also planned to work on the earlier standard versions of GERAN or UTRAN radio access networks. Thus, the usage of 3GPP Rel'5 or Rel'6 PDF for QoS authorisation should be possible, for example, with Rel'99-compliant GERAN or UTRAN BSCs/RNCs and base stations. However, the used mobile stations would need to support the Rel'5 and/or Rel'6 PDF for the QoS authorisation to work properly.

3.5 Negotiated PDP-context QoS Enforcement in GERAN (and UTRAN)

In the GERAN and UTRAN radio domains, the QoS mechanisms can be divided into control and user plane functions. During the RAB or PFC establishment procedure, some control plane functions take care of the admission control of new connections, the selection of appropriate protocol transmission modes and error protection algorithms as well as the allocation of radio resources. Furthermore, the control plane preserves the negotiated QoS while the service is ongoing. The RAN should be able to react against overloaded situations, as well as identify potential downgrade in the QoS provisioning.

The goal in a QoS-aware environment is to allow predictable service delivery to certain types of traffic, regardless of what other traffic is flowing through the network at any given time. It aims to create a multi-service RAN where traditional bursty traffic may share the same resources with traffic with more rigorous latency, jitter, bit rate and/or packet loss requirements. The functions that take care of the QoS provisioning once the bearer is established belong to the user plane. An appropriate traffic conditioning and scheduling allows the differentiation between links against time-variation radio conditions.

Figure 3.14 shows the major radio-related functional blocks involved in the QoS provisioning. Radio configuration and performance-indicator operating parameters are monitored and provisioned through the management interfaces. Monitored parameters include statistics regarding traffic carried at various QoS levels. These statistics may be important for accounting purposes and/or for tracking compliance to service level specifications (SLSs) negotiated with customers. An SLS is presumed to have been negotiated between the end user and the provider that specifies the handling of the end-user's traffic, that is, a set of parameters, which together define the QoS offered to a traffic stream by the provider's network.

All the functionalities described in the following sections are supported in Rel'99, Rel'4 and Rel'5, while Rel'97/R98 is out of the scope of the following sections. Note that RAB parameters are introduced in GERAN Rel'5 for the Iu interface, though the scope of the functionalities is also applicable to Rel'99 and Rel'4 by considering the PFC attributes instead of RAB attributes.

3.5.1 Control Plane QoS Mechanisms

Control plane QoS mechanisms are responsible for two main tasks: connection admission control and establishment, and QoS preserving mechanisms.

The first one is related to the RAB/PFC and RB establishment procedure once the SGSN has validated the service requested by the user equipment and has translated the PDP-context QoS profile into RAB parameters.

The second task takes place once the bearers are established; additional mechanisms should try to preserve the negotiated QoS. For this purpose, load control function is responsible for controlling overload situations.

3.5.1.1 Admission Control

The purpose of the admission control function is to accept or deny new users in the radio access network. In this manner, the admission control tries to avoid overload situations basing its decisions on interference and resource availability measurements (Figure 3.15). It is performed at initial MS access (i.e. RAB assignment), handover and RAB reconfiguration (Rel'5) or BSS PFC creation and modification (Rel'99). These cases may give different answers depending on user priority and load situation.

Admission control can take into account radio and transport resources, radio link control (RLC) buffer size and base station physical resources. When performing admission

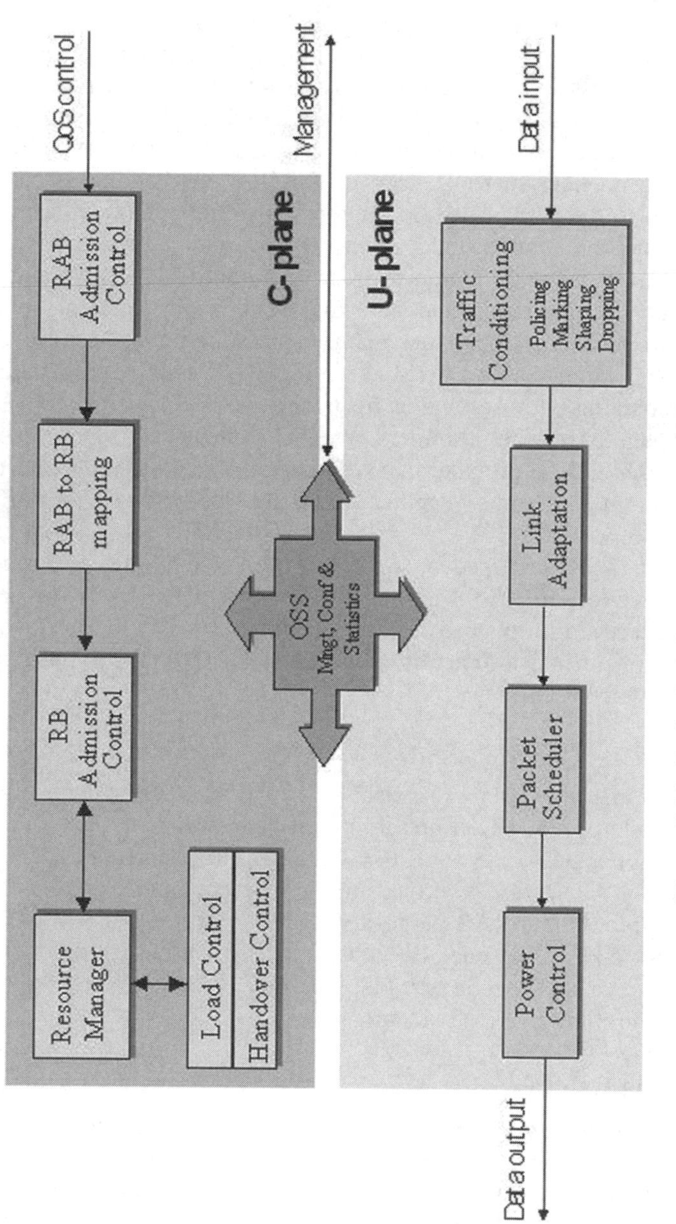

Figure 3.14. Radio QoS functional blocks coordination

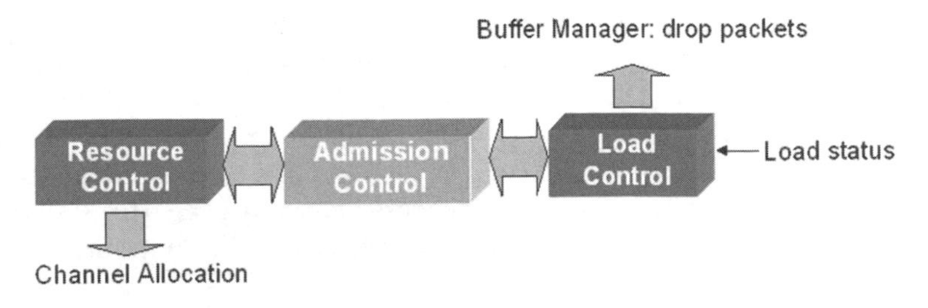

Figure 3.15. Admission control in GERAN

decision, this function requests resources from other entities and based on this information, different decisions can be made, such as accept/reject the connection, queue the request or perform a directed retry (see Section 6.3). The resource control performed by the admission control also supports the service retention through the *allocation/retention priority* attribute. This feature is service independent, providing users differentiation depending on their subscription profiles. High priority users are able to pre-empt lower priority users when admission control is performed and there is no available capacity. Furthermore, once they are admitted, this feature allows high priority users to retain their connections against lower priority users in case of overload.

In *circuit-switched* domain, the admission control decision is normally based on the availability of time slots in the target cell. However, it is also possible that new calls are blocked even though there are time slots available to support it (this is called *soft blocking*). Soft blocking can be based on interference levels or quality measurements in surrounding cells. In case of Dynamic Frequency and Channel Allocation (see Chapter 9), the admission control based on interference levels can be done quite accurately.

In *packet-switched* domain, admission control manages non-real-time and real-time connection requests differently. In case of non-real-time services such as interactive and background traffic classes, the experienced throughput decreases gradually when the number of users increase, up to the blocking situation. By limiting the non-real-time load, it is possible to provide a better service to already admitted users.

Admission control evaluates whether there is enough capacity to provide the requested guaranteed bit rate for services that require guaranteed bit rate over shared channels (e.g. streaming traffic class). Since radio link conditions are unknown at connection set-up phase and it can change when the user is moving, a statistical-based admission control is required [19]. The admission control can be performed with different levels of risk. This is done in terms of different blocking and user satisfaction probability, depending on the user subscription profile.

3.5.1.2 RAB/PFC to RB Mapping

Different services must be conveyed over appropriate radio bearers that enable QoS requirements to be satisfied. To achieve it, this function takes care of selecting appropriate conditions for the transmission over a chosen logical channel (e.g. traffic channel (TCH), packet data traffic channel (PDTCH)) such as transmission modes (acknowledge, non-acknowledge), modulation and coding schemes (MCSs), error protection type (EEP, unequal error protection UEP), etc. For instance, the use of retransmissions causes additional delay, but at the same time provides much better reliability, which is needed for many data applications. The main input to select these conditions is the *traffic class* attribute though other attributes from the QoS profile such as bit rate, reliability and delay requirements should be taken into account in this procedure. Several RB parameter combinations are possible to configure the corresponding bearer for different traffic classes. A description of the different protocols of the GERAN architecture can be found in Chapter 2.

Conversational services are the most delay-sensitive applications, hence retransmissions are not normally performed. Conversational packet-switched connections are not possible over the Gb interface. Packet-switched connections over the Iu-PS interface are likely to be fulfilled with non-transparent packet data convergence protocol (PDCP) mode, which performs IP/UDP/RTP header adaptation. There is no optimised voice bearer in physical layer (with optimised error protection schemes) for voice over IP (VoIP) in GERAN Rel'5/Rel'6, only generic conversation bearer is supported.

Streaming class applications can be supported with acknowledged or non-acknowledged mode. High reliability requirements can impose the use of retransmissions though this implies an increase of the delay/jitter. Therefore, a possible solution to minimise the transmission delay can be to limit the number of retransmissions by selecting more robust MCS. By minimising the number of retransmissions and with proper scheduler algorithm design, it is possible to fulfil delay requirements. However, limiting the number of retransmissions by selecting a more protective MCS has an impact on the throughput performance. It should be noted that incremental redundancy (IR) typically requires high retransmission rates and therefore limiting the number of retransmissions will impact on IR performance. In non-acknowledge mode, the link adaptation algorithm should select the highest MCS that matches the reliability QoS requirements (depending on the radio link conditions) and maximises the throughput. Though the use of shared medium access control (MAC) mode achieves a better resource utilisation, dedicated MAC may be used for certain services and/or users in order to provide a better quality in terms of delay, throughput and SDU error ratio.

Interactive and *background* services require the use of the acknowledged RLC mode in order to provide high reliable transport capability at the expense of introducing a variable throughput and delay. Retransmissions over the air provide much better reliability, needed for this kind of services. Since acknowledge mode is used, the link adaptation algorithm must select dynamically the MCS that achieves the maximum throughput per timeslot. These services can use dedicated or shared MAC mode though it is recommended to select shared MAC mode in order to minimise resource utilisation.

Figures 3.16 and 3.17 optimise the different network protocol transmission modes for each traffic class in GERAN Rel'99/Rel'4 (Gb interface) and Rel'5 (Iu-PS interface).

3.5.1.3 Resource Manager

The optimisation of the radio resource management is a key issue in GERAN. An efficient resource management over the radio domain is one of the most important and effective QoS mechanisms. The resource manager has the main task of determining the needed resources to fulfil the QoS requirements of a new connection and verifying whether there

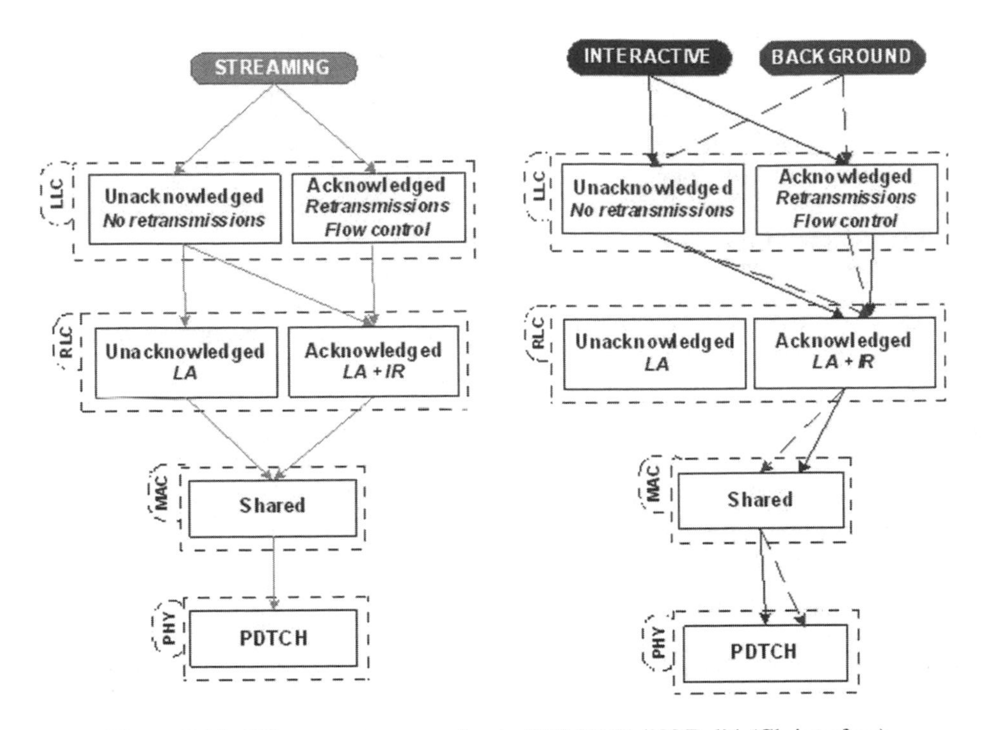

Figure 3.16. RB parameters mapping in GERAN Rel'99/Rel'4 (Gb interface)

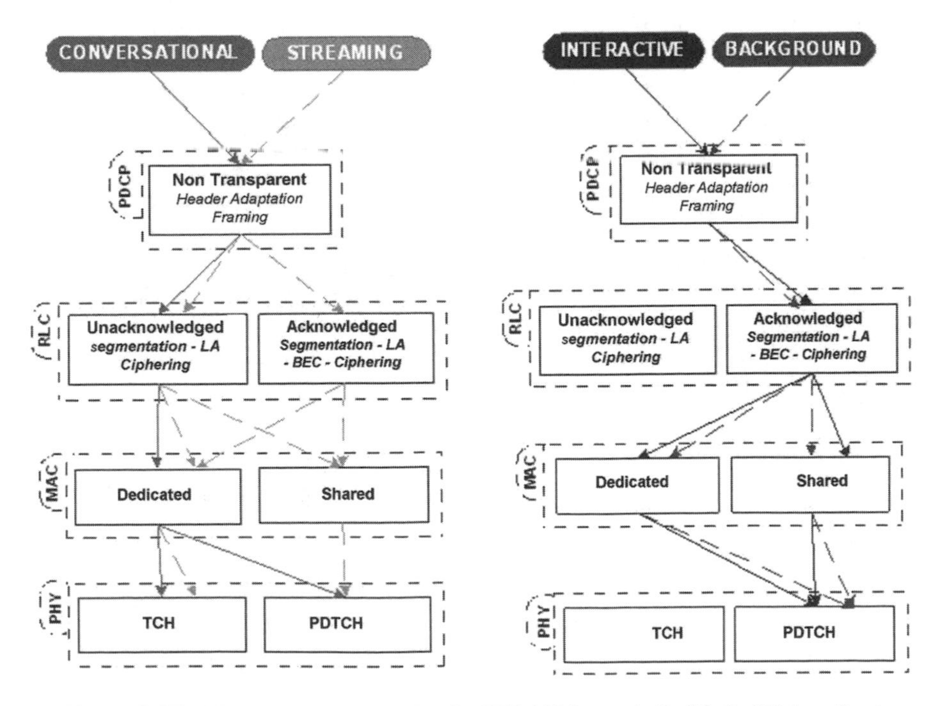

Figure 3.17. RB parameters mapping in GERAN Iu mode Rel'5 (Iu-PS interface)

are enough available resources in the cell to support these requirements. Resource management is different for real-time and non-real-time services. Non-real-time services do not require any guaranteed bit rate so they can be allocated in any available channel. The main criterion used in channel allocation for these services is to allocate channels that give the maximum capacity for the temporary block flow (TBF) within the limits defined by the MS capability. However, real-time service allocation over dedicated channels consists of determining the availability of resources and reserving them on the basis of guaranteed throughput requirements and mobile MS capability. In case of allocation over shared channels, additional information about the quality of the radio channels (throughput per timeslot) is needed in order to allocate necessary resources to guarantee QoS requirements [20].

The different options to allocate resources for real-time services are to dedicate the channel for the real-time connection, to share the channel between a real-time connection and one or several non-real-time connections or to share the channel between more than one real-time connection, independently of the type of service [21].

3.5.1.4 Load and Handover Control

Handovers can be triggered by different reasons. They are performed when the MS leaves the dominance area of the actual serving cell. In other cases, they are performed when the current call is experiencing bad quality. A third kind of handover is referred to as the traffic handover when the current cell is congested, whereas neighbour cells are not.

The aim of the load control is to control the load in overload situations, but it also has the ability to control it even before the overload situation happens, that is, preventive load control. Whenever the network detects an overload situation in a cell, load control can trigger load reason handovers to adjacent cells and select which connections to move. Different types of services and user subscription profiles may be taken into account in this selection so that services and users with the lowest priorities are the most susceptible to be moved.

3.5.2 User Plane QoS Mechanisms

User plane QoS mechanisms take care of the QoS provisioning once the bearer is established. One of the tasks of the user plane in the radio domain is to provide service differentiation by means of a proper traffic conditioning and scheduling of the radio blocks similar to those performed in the IP transport domain following the DiffServ framework proposed in the IETF [14, 15].

Furthermore, other tasks of the user plane include being in charge of dynamically adapting the link to compensate fluctuations of the radio conditions as well as controlling output power.

3.5.2.1 Link Adaptation

Wireless links are affected by radio link conditions. This implies that the performance in terms of throughput, delay and block error ratio is continuously varying. These variations in the QoS provisioning must be compensated as much as possible by selecting proper modulation and coding schemes. This mechanism is known as *link adaptation* (see Chapter 7 for more information).

More specifically, the link adaptation function is able to select dynamically the channel coding, channel mode (full-rate (FR)/half-rate (HR)), interleaving depth and modulation, depending on the service requirements, system load, channel state and interference. The link adaptation algorithm may need cooperation with resource allocation to meet the negotiated RB for real-time connections.

The throughput depends on the link quality in terms of carrier/interference (C/I) values. Thus, the more protective modulation and coding scheme, the less throughput for a given C/I condition. The performance of the ideal link adaptation corresponds to the envelope of all MCS curves, since its aim is to maximise the throughput.

The link adaptation algorithm in non-acknowledged mode should select the highest MCS that matches the reliability QoS requirements, depending on the radio link conditions. Since retransmissions are not performed in this mode, reliability requirements must be achieved using a proper coding scheme. However, in acknowledged mode, the target is normally to achieve the maximum throughput/timeslot, that is, depending on the radio link conditions, the link adaptation algorithm should select the MCS that maximise the throughput/timeslot. Reliability QoS requirements in acknowledged mode can be achieved by means of retransmissions. Delay requirements impose limits to the throughput maximisation criteria.

3.5.2.2 Traffic Conditioning

Any particular packet will experience different effects when it is sent via the radio interface. Traffic conditioner provides conformance between the negotiated QoS for a service and the data unit traffic. This function is performed by different mechanisms:

- Policing function is the process of comparing the arrival of data packets against a QoS profile, and based on that, forwarding, marking, shaping or dropping them so as to make the output stream conformant to the profile.
- Marker is responsible for tagging the packets exceeding the guaranteed bit rate in order to treat them accordingly.
- Shaper is the procedure of delaying (queuing) packets within a traffic stream to cause it to conform to some defined traffic profile.
- Dropper selectively discards packets that arrive at its input, based on a discarding algorithm.

Depending on the type of service, packet conditioner should work in a different way. For instance, since non-real-time services have no stringent delay requirements, packets exceeding the maximum bit rate can be shaped until they can be forwarded. However, real-time services are delay sensitive, hence these packets must be treated differently: packets between the guaranteed and the maximum bit rate can be marked so that the packet scheduler handles them accordingly as non-guaranteed traffic while packets above the maximum bit rate should be shaped.

3.5.2.3 Packet Scheduler

Packet scheduler (see Chapter 7) is the core function in the QoS provisioning. Since the radio domain is most probably the bottleneck of the whole network, an efficient treatment of the radio blocks based on priorities is a key issue. This function takes care of scheduling radio resources for different RABs/PFCs for both uplink and downlink directions. On the basis of the radio interface interference information, capacity requests from the MS (uplink) and received data in the buffers (downlink), packet scheduler will allocate radio resources for the RAB/PFC. This algorithm should take into account the QoS profile in order to prioritise between services and user subscription profiles (through the allocation/retention priority attribute).

Conversational services should be handled with dedicated radio resources due to their stringent delay requirements, although background traffic from the same user could be multiplexed on to the same radio channel. Therefore, an absolute priority packet scheduling policy could allow background traffic to be sent only during silent periods of conversational speech traffic.

Streaming services requires QoS guarantees in terms of throughput. Some streaming traffic has a bursty nature and, because of this, shared channels are more adequate for this kind of traffic than dedicated channels. Hence, packet scheduler for shared channels must ensure certain bandwidth for streaming connections [19], providing a higher priority to the streaming traffic up to the guaranteed bit rate against other non-real-time connections. These priorities may depend on several parameters such as traffic class, QoS requirements, or some internal operator parameters. This separation between packets up to and above the guaranteed bit rate may be managed by a marking function, which differentiates precisely between both types of packets. Another option is the use of dedicated channels that have better link performance.

An example of packet scheduler for shared channels is depicted in Figure 3.18 [20]. In this scheme, packet scheduling ensures certain bandwidth for streaming connections using a Deficit Round Robin (DRR) approach, which is responsible for dynamically assigning the allocated bandwidth for each flow. The deficit value is used for tracking the context lead/lack compared to the guaranteed bit rate. As long as the deficit value is above zero, the scheduler shares the capacity to that context. For streaming services, a different priority is allocated when the deficit is above/below zero, corresponding to guaranteed/non-guaranteed bit rate, in a way that the streaming traffic up to the guaranteed bit rate is dealt with the highest priority, whereas streaming traffic exceeding the guaranteed bit rate up to maximum bit rate is managed as non-real time traffic.

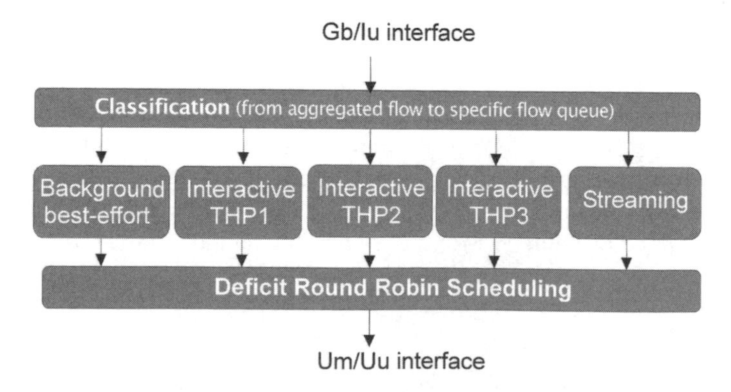

Figure 3.18. Example of packet scheduler's scheme

The operator could control the packet scheduling not only on the basis of specific parameters of the scheduling algorithm but also by means of the allocation/retention priority attribute (e.g. 'VIP', 'Budget' and 'Machine'). In that way, both service and user differentiation are performed.

3.5.2.4 Power Control

The objective of the power control function is the reduction of interference levels while QoS of the connections is maintained, which causes an increase of capacity and an improvement of the spectral efficiency. Additionally, power control in uplink is used to reduce the power consumption of the MS.

Power control tries to keep the received level and the received quality in a range of good values (defined by an upper threshold and lower threshold) by changing the transmitter power 'on the other side' of the radio connection.

Power control could be used in order to distinguish between different types of users (e.g. 'VIP' or 'Budget') and between different kinds of services. Hence, with the use of different maximum base transceiver station (BTS) power values, it should be possible to prioritise between services, for example, the use of the P0 parameter to decrease interference level to speech calls.

3.6 End-to-end QoS Management

From the mobile operator point of view, end-to-end management of the QoS is of key interest.

Policy management is a possible and optional way to solve the problem of managing complex networks. Basically, it offers a centralised QoS control point, which administrates the network in order to achieve a consistent service behaviour for the need of a specific QoS. That means that all QoS-related functions distributed along the network could be managed from a central point of administration. Policy management is based on rules containing conditions and actions.

In order to apply a proper policy management to the radio domain, different radio QoS functions must be modelled abstracting the implementation details from the parameters of interest for configuration and management. Once the manageable parameters are identified, the operator is able to build policy rules in the form 'if condition, then action'. The semantics of a policy rule are such that if the set of conditions evaluates to TRUE, then the set of actions is executed. The scope of these rules can vary: global policies have a global effect on the end result of applying the given policy, that is, the end result of such policy relies on it being executed globally across the entire path of network service; domain policies are meaningful within one domain of QoS mechanisms, for example, radio domain; local policies are only meaningful within one network element. From the rule management viewpoint, it is very essential that both the global and domain policies are administrated by a central controlling point. Local policies can reside within the control of local element, so long as such local policies have no effect on the QoS service globally.

In the mobile networks, the QoS management includes, independent of the standards baseline, provisioning of HLR QoS profiles, APN/service access arrangements and policy-based or static configuration of the QoS algorithms in the network elements. In this process, the operator's QoS personnel also need to have information on the network planning and dimensioning, targeted service and/or application-specific Key Performance Indicators (KPIs) and forecasted traffic mix. Also, the personnel need to have access to the network and application performance KPI and measurement data in order to verify the effects of the provisioned QoS settings. This framework is illustrated in Figure 3.19.

3.6.1 Example of Service Activation Procedure

An example of streaming service activation is described in this section [10]. Figure 3.20 illustrates the different steps that form an example 3GPP QoS-aware streaming service activation. Also, in this example, at least Rel'99-compliant GERAN QoS implementation is assumed.

Firstly, the user initiates the streaming client application, which requests a primary PDP context towards an APN, providing streaming content. The QoS requirements of the application signalling in the MS are mapped on QoS attributes. Since the primary PDP context is used for RTSP signalling, it requires high reliability. Thereby, a QoS profile with interactive traffic class, high priority and low error rate is appropriate for this kind of traffic.

After the SGSN has validated the service for that user by querying theHLR, local admission control is performed (e.g. based on the state of the buffers, the CPU load, etc.). Then, the SGSN maps the QoS attributes on to RAB PFC QoS attributes and triggers a RAB PFC assignment procedure in the RAN BSS (note that the concept of RAB attributes are introduced for the Iu interface in Rel'5 GERAN Iu mode, while a similar procedure can be performed for the BSS PFC in the Gb interface for previous releases). In the

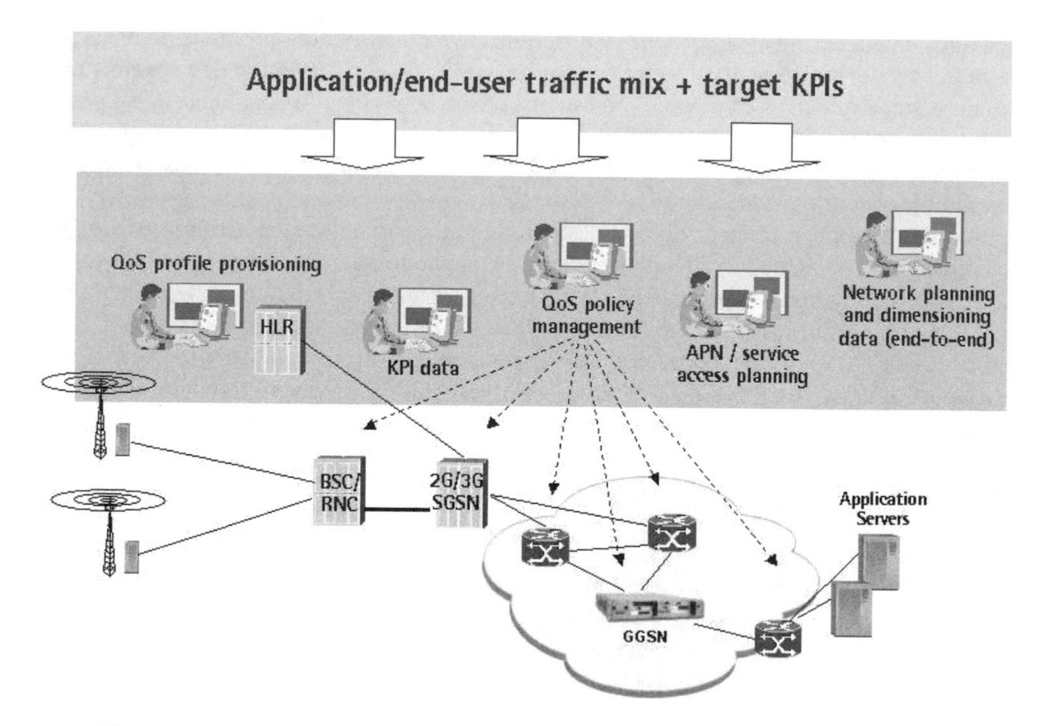

Figure 3.19. Working field of the operating QoS personnel in a 3GPP environment

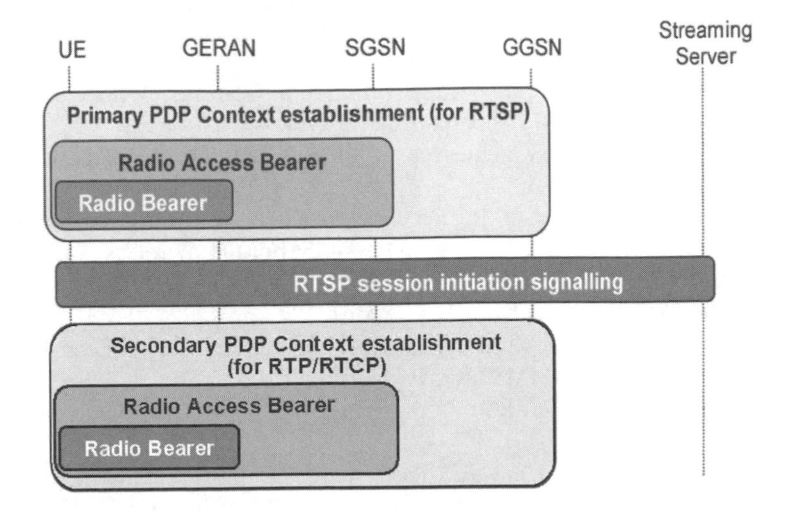

Figure 3.20. Streaming service activation procedure example

RAN BSS, admission control is mainly based on the availability of radio resources. Once accepted, the new RAB PFC request, RAB PFC attributes are mapped on to proper RB parameters used in the physical and link layers (e.g. retransmission requirements, etc.). An RB according to these parameters is established.

After RTSP session initiation signalling, the MS triggers secondary PDP-context activation procedure towards the same APN, for real-time protocol (RTP) traffic and for the real-time control protocol (RTCP) traffic.

The MS converts user data application requirements for RTP and RTCP traffic into QoS profile of the secondary PDP context. The SGSN validates the service request by checking from the HLR QoS profile if the requested QoS level can be provided to this subscriber. In addition, the SGSN performs admission control based on the available bandwidth for the new PDP context. In the case of 3GPP Rel'6-compliant streaming client, Rel'6 PDF could potentially also be used to validate the requested QoS parameters from a streaming server or proxy, supporting this option.

In case the flow is accepted, resource reservation is performed by decreasing the bandwidth quota by the guaranteed bit rate of the new PDP context. The QoS profile is mapped on to RAB/PFC QoS attributes, which are sent to the GERAN. The GERAN performs an admission control and then does the mapping between RAB/PFC and RB parameters. Both local admission control and resource reservation are performed in the GGSN in the same way as in the SGSN before the PDP context is admitted.

References

[1] 3GPP 23.060, Release 1999: General Packet Radio Service (GPRS), Service Description, Stage 2.

[2] RFC-2784, Farinacci D., Li T., Hanks S., Meyer D., Traina P., Generic Routing Encapsulation (GRE).

[3] RFC-2661, Townsley W., Valencia A., Rubens A., Pall G., Zorn G., Palter B., Layer Two Tunneling Protocol (L2TP).
[4] *IEEE Standards for Local and Metropolitan Area Networks: Virtual Bridged Local Area Networks*, IEEE-SA Standards Board, 1998.
[5] RFC-3031, Rosen E., Viswanathan A., Callon R., Multiprotocol Label Switching Architecture.
[6] RFC-2401, Kent S., Atkinson R., Security Architecture for the Internet Protocol.
[7] WAP Architecture, Version 12-July-2001, WAP Forum, Wireless Application Protocol Architecture Specification, WAP-210-WAPArch-20010712.
[8] WAP MMS Architecture Overview, version 25-April-2001, WAP Forum, Wireless Application Protocol Multimedia Messaging Service Architecture Overview Specification, WAP-205-MMSArchOverview-20010425-a.
[9] RFC-2326, Shulzrinne, H., Rao A., Lanphier, R., Real Time Streaming Protocol (RTSP).
[10] Montes H., Gómez G., Cuny R., Paris J. F., 'Deployment of IP Multimedia Streaming Services in Third Generation Mobile Networks', *IEEE Wireless Commun. Magazine*, 2002, 84–92.
[11] 3GPP 26.234, Release 4: Transparent End-to-End Packet Switched Streaming Services (PSS); Protocols and Codecs.
[12] 3GPP 22.941, Release 5: IP Based Multimedia Services Framework, Stage 0.
[13] 3GPP 23.107, Release 1999: Quality of Service (QoS) Concept and Architecture.
[14] RFC-3246, Davie B., Charny A., Bennett J., Benson K., Le Bouded, J., Courtney W., Davari S., Firoiu V., Stiliadis D., An Expedited Forwarding PHB (Per-Hop Behaviour).
[15] RFC-2597, Heinanen J., Baker F., Weiss W., Wroclawski J., Assured Forwarding PHB Group.
[16] 3GPP 23.207, Release 5: End-to-End Quality of Service (QoS) Concept and Architecture.
[17] 3GPP 23.107, CR 112, 3GPP TSG-SA2 Meeting #25 Naantali, Finland, June 24–28 2002.
[18] 3GPP 23.060, CR 397, 3GPP TSG-SA2 Meeting #25 Naantali, Finland, June 24–28 2002.
[19] Montes H., Gómez G., Fernández D., 'An End-to-End QoS Framework for Multimedia Streaming Services in 3G Networks', *13th IEEE International Symposium on Personal, Indoor and Radio Mobile Communications PIMRC'02*, Lisbon, Portugal, September 2002.
[20] Montes H., Fernández D., 'An Enhanced Quality Of Service Method for Guaranteed Bit rate Services Over Shared Channels in (E)GPRS Systems', *52nd IEEE VTC Spring'02*, Alabama (EEUU), May 2002.
[21] Montes H., Gómez G., Paris J. F., 'An End-to-End QoS Management Framework for Streaming Services in 3G Mobile Networks', *8th IEEE and IASTED International Joint Conference on Wireless and Optical Communications WOC'02*, Banff, Alberta, Canada, July 2002.

4

Mobile Station Location

Mikko Weckström, Maurizio Spirito and Ville Ruutu

The need to determine the mobile terminal position has been growing continuously during the last few years. Requirements for emergency services such as the North American E911 are one of the key drivers for the development of accurate position technologies such as the enhanced observed time difference (E-OTD) and time of arrival (TOA). In Europe, the key drivers are the commercial services, even though the European emergency services (E112) will utilise the same mobile location technologies.

Location system (LCS) functionality is going to enable a new set of services for cellular users in the coming years. With general packet radio system (GPRS), it is economically feasible for users to be permanently connected to Internet-based services. With LCS, the available services can have new location-related information providing new revenue generation opportunities for wireless operators.

For the network infrastructure, this means new challenges as well. In order to be able to locate a number of mobile terminals continuously, some new components and signalling protocols are required. Since the network has not been designed for accurate positioning, there has to be some new ways to combine the radio measurement data or to carry out some new measurements. In the future, GSM/ EDGE radio access network (GERAN) and UMTS terrestrial radio access network (UTRAN) systems will offer the LCS services a variety of technologies such as E-OTD, and assisted GPS positioning system (A-GPS). The full capability is expected to be available by the end of 2004. Current GSM networks offer the basic capabilities mostly via cell identity–based (CI-based) methods. Already there are some commercial CI-based services available. Cellular network operators have been deploying the emergency positioning service in the United States, which is the main driver for E-OTD and TOA owing to the Federal Communications Commission (FCC) regulations.

In the following sections, the different technologies used for mobile positioning will be presented, with some analysis on the achieved accuracy of the different methods.

Portions reprinted, with permission from Spirito M. A., Pöykkö S., Knuuttila O., 'Experimental Performance of Methods to Estimate the Location of Legacy Handsets in GSM', *Proceedings of IEEE Vehicular Technology Conference—VTC 2001*, 2001.

GSM, GPRS and EDGE Performance 2nd Ed. Edited by T. Halonen, J. Romero and J. Melero
 ISBN: 0-470-86694-2

More detailed information can be found from some of the sources listed in the references section [1, 2].

4.1 Applications

There are a large number of applications that the location functionality enables, and the methods described in this section are just a subset of the potentially available ones. Applications are divided into three different service levels as well. These levels are as described in the Location Interoperability Forum (LIF): basic service level, enhanced service level and extended service level. Figure 4.1 illustrates these levels.

The service levels are related mainly to accuracy. The method for the basic service level is usually CI-based but some other methods can be used as well. The other service levels use more accurate methods. The enhanced service level uses the E-OTD and the extended service level the A-GPS. But more important is that these service levels enable the support of different applications. Some of these applications can be found in Figure 4.2.

As accuracy increases, more applications are available, but with a higher cost of the LCS solution. These applications also have evolution paths from the network sub-subsystem–based (NSS-based) architecture towards base station subsystem (BSS) and

Service level	Accuracy/architecture	Application
Basic service level	400–1500 m/NSS	Friend finder, location sensitive billing
Enhanced service level	40–150 m/NSS/BSS	Emergency, tracking finding
Extended service level	< 50 m/NSS/BSS	Accurate navigation, emergency, safety

Figure 4.1. LCS service levels

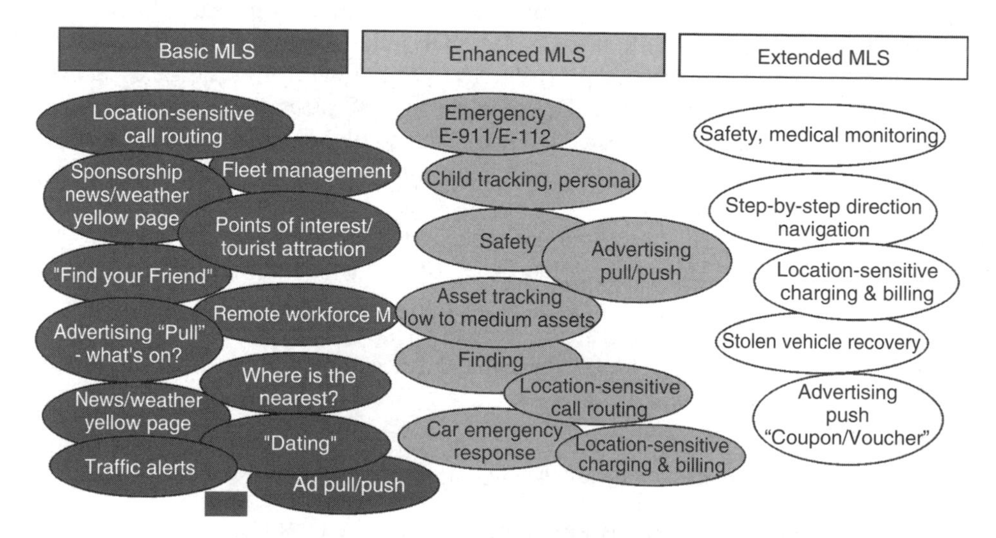

Figure 4.2. Segmentation examples of mobile location services

mobile station–based (MS-based) solutions that will ensure larger location capacity in the network. All of these applications will connect to the system via standard interfaces, such as LIF application programming interface (API). The API allows different application providers to easily connect to the system and offer their applications.

4.2 Location Architectures

This section will introduce the relevant LCS architectures. The first methods implemented are usually those enabling the basic service level. In this case, the architecture is usually NSS-based (included only in GSM standards Rel'98 and Rel'99) and hence offers moderate capacity with low cost. The NSS architecture relies on standard interfaces for collecting the information required to calculate the position estimate. The NSS architecture can use information such as CI or timing advance (TA) and, in some cases, measurements of the signal strength received by the handset Received Signal Level (RXLEVs). Basically, more information allows better accuracy. The basic NSS architecture is described in Figure 4.3, where the serving mobile location centre (SMLC) and the gateway mobile location centre (GMLC) are connected to the mobile services switching centre (MSC).

The location interfaces are the Ls for the SMLC and the Lg for the GMLC. These interfaces are based on the standards and hence should be the same for all cellular network suppliers. This should help in the case of multi-vendor networks where the cellular network consists of equipment from many different suppliers.

In Figure 4.3, the first element that is specific for positioning is the location measurement unit (LMU). For E-OTD location method, the LMU is measuring the timings of the base transceiver stations (BTSs) and reporting these values to the SMLC to make it possible to calculate the position estimate together with the timing reports sent by the mobile. In the case of the TOA method, LMUs receive transmissions from the handset. The SMLC is then sending the estimated geographical coordinates of the handset to the GMLC that is sending the information to and from the application. The Ls and Lb interfaces are between SMLC and base station controller (BSC) or mobile services switching centre MSC, and the Lg interface is between the GMLC and MSC. These are

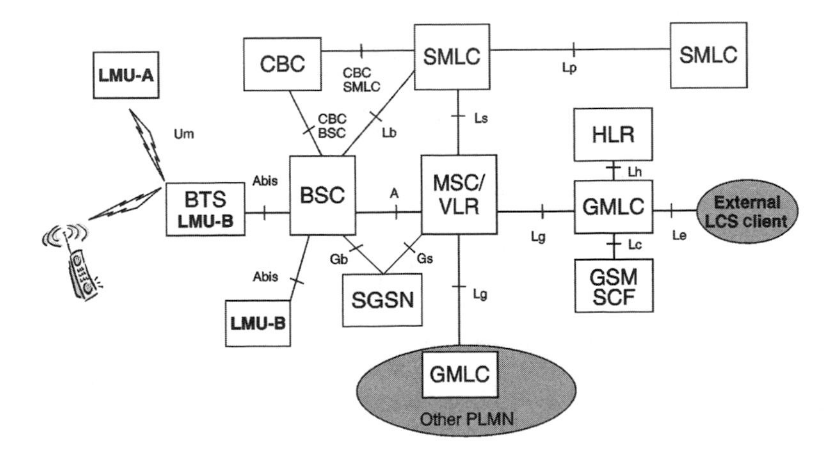

Figure 4.3. Generic LCS architecture

the LCS-specific signalling system number 7 (SS7) interfaces. The Lb and Ls have in essence similar stack structure, but the information content is slightly different. In the case of the Lb, the network database is smaller, as the SMLC is serving only one BSC, and therefore it enables greater location calculation capacity in the SMLC. The next step in architecture evolution is the BSS-based architecture that enables the system to use the radio interface information more efficiently and therefore does not have to use the signalling necessarily for the measurement information transport towards the MSC. The BSS architecture utilises the measurement data that are specific for enhanced methods like E-OTD more efficiently. Figure 4.4 displays the basic BSS-based architecture that is the only architecture option in GSM standards Rel'4 and later.

LCS support for General Packet Radio Service (GPRS) is not included in GSM Rel'98 and Rel'99, but later releases have increasing support.

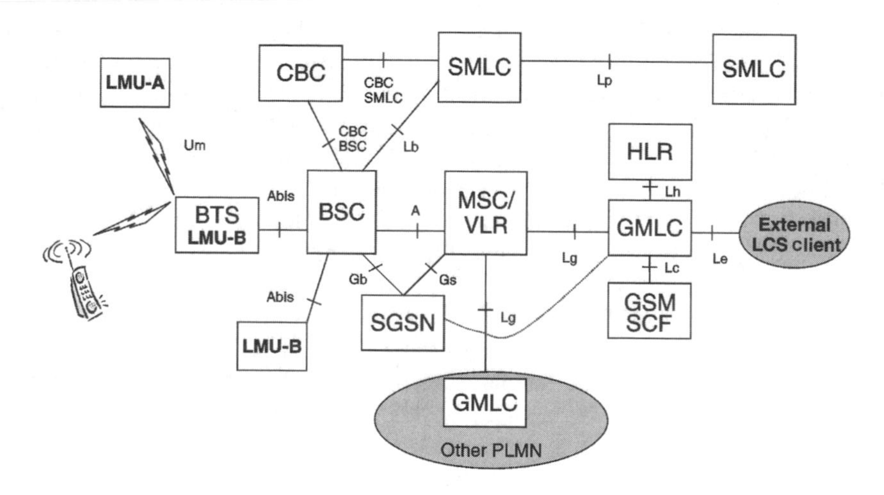

Figure 4.4. BSS LCS architecture

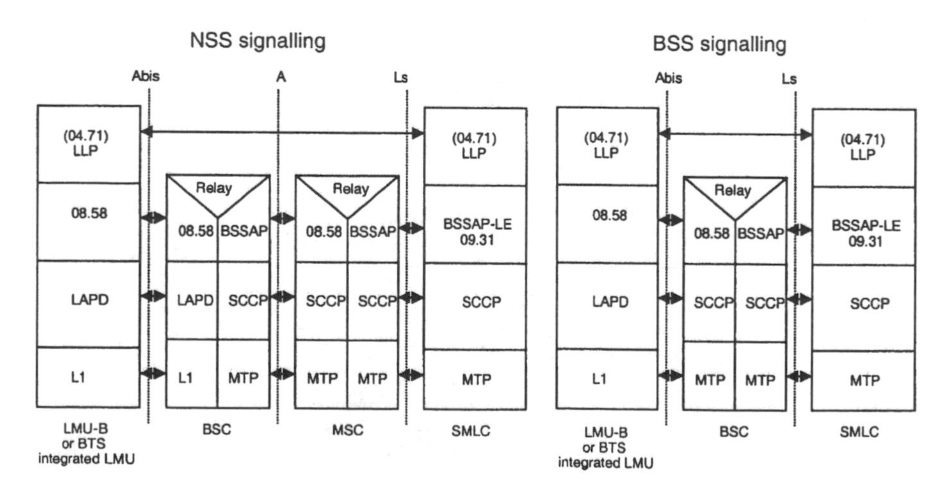

Figure 4.5. Basic signalling structures of NSS and BSS architectures

The third architecture is MS based. It uses the assistance data that is generated in the cellular network as well as the MS capability of performing LCS measurements and determining its own location. In this architecture, the SMLC calculates the assistance data that is sent to the mobile via standard messages. The calculation of the location is done in the mobile terminal. Mobile-based E-OTD and A-GPS both belong to this category.

Figure 4.5 illustrates the basic signalling protocols for the NSS- and BSS-based architecture.

4.3 Location Methods

There are many location techniques devised for application in mobile communications networks. Most of them estimate the MS coordinates by processing with *multilateration* techniques absolute distances and/or relative distances between the MS and multiple base stations. A general analysis and comparison of such *multilateration* techniques can be found in [3, 4]. In this section, the structure of LIF is followed going through first the basic service level and then enhanced service level. Finally, the A-GPS method for extended service level will be presented. There are MS-based and MS-assisted methods, so when describing the appropriate method the assistance order is also described. Figure 4.6 illustrates the position accuracy related to these different methods.

4.3.1 Basic Service Level

Timely deployment of LCS for all MSs, including the legacy ones, requires that mobility management information, already available in cellular networks, should be used for implementing location technologies capable of offering basic service level accuracy. Such technologies are of crucial importance from a commercial point of view. In fact, they allow operators and service providers to start offering location-based services to all customers with minimal additional costs, while waiting for location technologies capable of supporting enhanced and extended service levels to be available commercially. From a technical point of view, even when more accurate technologies will be fully available, basic service level location methods will still be needed; for instance, when they are accurate enough

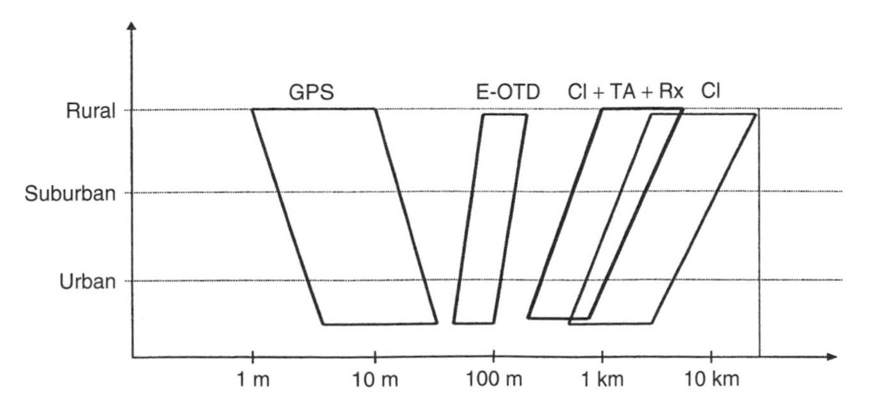

Figure 4.6. Accuracy of LCS methods

to achieve the requested location accuracy or when more accurate techniques are not supported in the whole network. Besides, basic service level location technologies may serve as *back-up methods*, since the enhanced and extended service level methods can fail if, for instance, an insufficient number of base stations/satellites is available for measurements.

Measurements available for the implementation of basic service level location methods in a GSM network are CI, TA, and strength of signals received by the handset (RXLEVs). Along with the estimated coordinates, all methods provide a definition of a *confidence region*, which defines a geographical region where the actual location of the target terminal can be found with a given probability (confidence coefficient).

4.3.1.1 Cell Identity (CI) Location Method

In cellular networks, base stations serve only a limited geographical area, thus the identity of the base station serving an MS provides the simplest location information. In GSM, the serving base station identity is available with the parameter known as CI.

One way to determine a CI-based location estimate is to assume that the mobile is at the serving cell's antenna coordinates. That is the most likely estimate when the serving cell is omni-directional. For a sector cell, a more accurate CI-based location can be estimated at the centre point of the serving sector. To allow this calculation, however, orientation and basic radiation properties of the serving base station antenna, and cell size are needed in addition to the serving cell's antenna coordinates. The basic principle of the CI location method is represented in Figure 4.7.

4.3.1.2 Cell Identity and Timing Advance (CI + TA) Location Method

In addition to CI, also TA for the serving base station is available in GSM. The TA represents the round trip delay between the mobile and the serving BTS expressed as an integer multiple of the GSM bit period. The TA is used in the GSM system to avoid overlapping of bursts transmitted by different users in the same cell at the serving base station site [5]. According to the GSM specifications, the TA is an integer number between 0 and 63. A simple model for a TA measurement can be expressed as TA $=$ round $\{d_{\mathrm{BTS}}/(cT_{\mathrm{b}}/2)\}$,

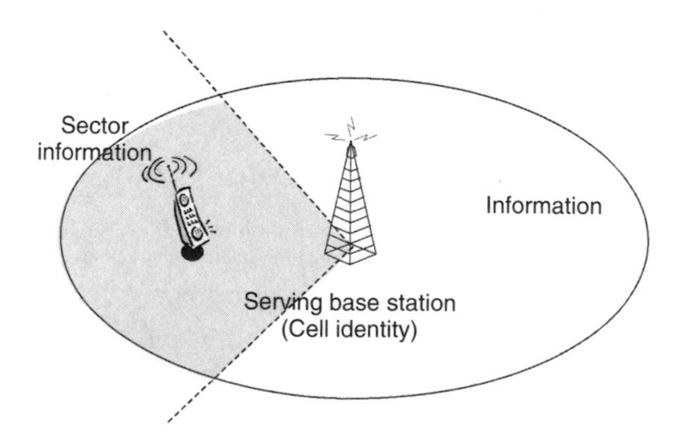

Figure 4.7. Cell identity (CI) location method. (Reproduced from Spirito M. A., Pöykkö S., Knuuttila O., 'Experimental Performance of Methods to Estimate the Location of Legacy Handsets in GSM', *Proceedings of IEEE Vehicular Technology Conference—VTC 2001*, 2001 by permission of IEEE.)

where $T_b = 3.69$ µs is the GSM bit period, c the speed of radio waves, and d_{BTS} the distance between handset and serving base station estimated by the system. In the CI + TA location method an estimate of the distance between handset and serving base station, d_{TA}, is obtained from a TA made available by the network as follows:

$$d_{TA} = \begin{cases} \frac{1}{4}(cT_b/2) & \text{if TA} = 0 \\ \text{TA}\ (cT_b/2) & \text{if TA} > 0 \end{cases}$$

The TA information as such cannot be used to improve the location accuracy when the serving cell is omni-directional; in fact, in such circumstances and in the absence of any additional information, the most likely handset location is at the serving cell's antenna coordinates. On the other hand, if the cell is a sector cell, the handset's location can be estimated on a circular arc having radius equal to the distance estimated from TA (d_{TA}), centre at the serving cell's antenna coordinates. The serving cell's antenna azimuth and radiation pattern determine the orientation and angular width of the arc, respectively. The radial thickness of the arc is determined by the TA measurement error margin, which is comparable to the TA resolution of half the bit period ($cT_b/2 \approx 550$ m).

The performance of the CI + TA location method can be eventually improved by additionally taking into account the reliability of d_{TA} as an estimate of the distance between terminal and serving base station. The estimate d_{TA} is in fact characterized by an error that makes it differ from the actual distance between MS and serving BTS. The error is determined by interference, Non–Line-Of-Sight and multi-path propagation over the radio channel; other impairments affecting the algorithms are used to estimate the d_{BTS}; quantization error is introduced when rounding the estimated round trip propagation delay to the nearest multiple of the bit period. The CI + TA location principle is represented in Figure 4.8.

4.3.1.3 Cell Identity, Timing Advance and RXLEVs (CI + TA + RXLEV) Location Method

The CI + TA + RXLEV method estimates the handset's coordinates by combining CI, TA (used to estimate the distance between handset and serving base station) and RXLEV

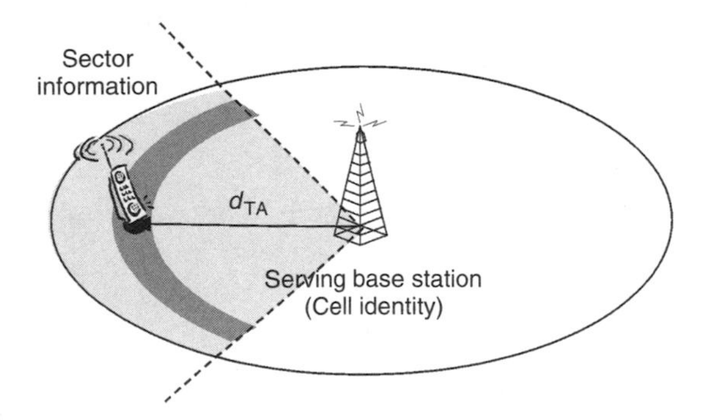

Figure 4.8. Cell identity and timing advance (CI + TA) location method. (Reproduced from Spirito M. A., Pöykkö S., Knuuttila O., 'Experimental Performance of Methods to Estimate the Location of Legacy Handsets in GSM', *Proceedings of IEEE Vehicular Technology Conference—VTC 2001*, 2001 by permission of IEEE.)

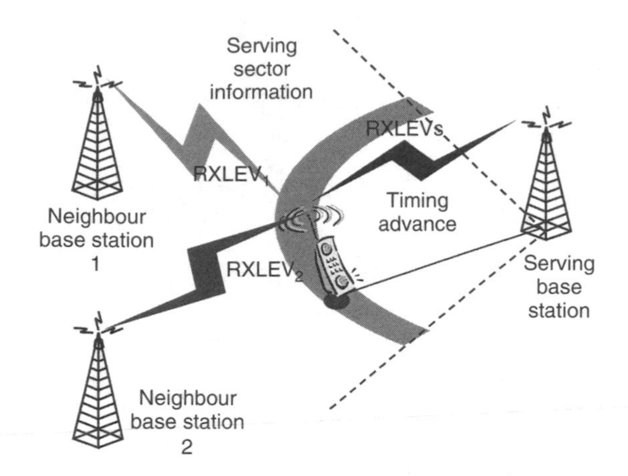

Figure 4.9. Cell identity, timing advance and RXLEVs (CI + TA + RXLEV) location method

information. RXLEVs are measurements of the strength (i.e. power) of signals received by the mobile terminal from the serving cell and from up to six strongest neighbour cells, introduced in the GSM system to support mobility management procedures.

The level of a signal received by a mobile terminal, or more precisely the attenuation that signal has experienced, depends also on the reciprocal position of the terminal and the base station from which the signal was transmitted. By using suitable propagation models, it is thus possible, for instance, to estimate the distance between the handset and each transmitting base station. Each estimated distance defines a circle on which the handset may be located. When at least three distinct circles are available, the handset location can be estimated at their intersection. The CI + TA + RXLEV location principle is represented in Figure 4.9.

4.3.2 Enhanced Service Level

Enhanced service level location methods offer better accuracy compared with those belonging to the basic service level category. However, the cost of better performance is a more complex system requiring additional equipment.

4.3.2.1 Enhanced Observed Time Difference

Like other enhanced service level location methods described in this chapter, E-OTD relies on timing measurements of cellular radio signals in the GSM network. In E-OTD the MS measures the time difference between the receptions of bursts transmitted from the reference base station and a neighbour base station, i.e. the observed time difference $\text{OTD} = t_{r\times N} - t_{r\times R}$, when the MS receives signals at the moments $t_{r\times R}$ and $t_{r\times N}$ from the reference base station and a neighbour base station respectively (Figure 4.10). Normally the reference base station is the serving base station of the MS, since it is expected to have the best radio link. The OTD measurement is repeated between the reference base station and other neighbour base stations in order to collect a set of OTD measurements based on which the location can be estimated.

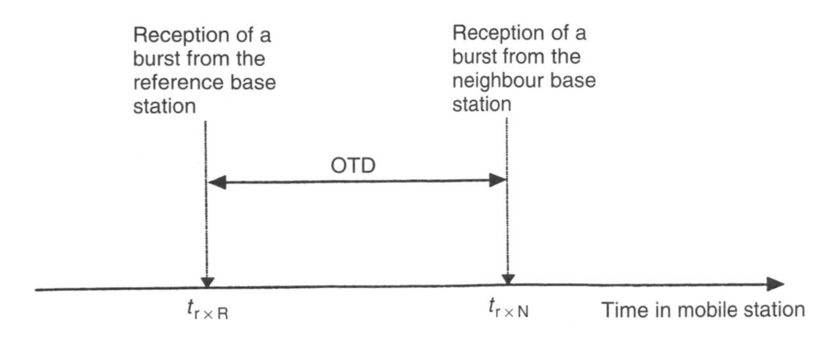

Figure 4.10. Basic E-OTD measurement by a mobile station

The reason this method is called *enhanced OTD* is that GSM specifications have included the OTD measurement feature already before any location support. However, this is intended for pseudo-synchronous handovers, and has a low measurement resolution that is inadequate for location purposes [6].

The basic equation for E-OTD location is OTD = RTD + GTD. OTD is composed of two parts. The first one is called the real-time difference (RTD). It is the synchronization difference between the reference and neighbour base stations, i.e. the relative difference in transmission times of their bursts. If the reference base station transmits its burst at the moment $t_{t\times R}$, and the neighbour base station at the moment $t_{t\times N}$, then RTD $= t_{t\times N} - t_{t\times R}$. If the base stations are synchronised and transmit at the same time, RTD is zero. The second part of OTD is the geometric time difference (GTD). It describes the contribution of different propagation times (different distances) between the MS and two base stations to OTD.

GTD is the important variable, since it includes actual information about location: GTD $= [d_{mN} - d_{mR}]/c$ (Figure 4.11). Here d_{mN} is the distance between the MS and the neighbour base station, d_{mR} is the distance between the MS and the reference base station, and c is the speed of radio waves.

With the E-OTD method, the MS measures OTD, and assuming that RTD is known, it is possible to determine GTD. Geometrically, GTD defines a set of points in the plane that have the property that the distance difference from such a point to the neighbour and reference base stations, respectively, is constant and equal to GTD $\times\, c$. But this is nothing but a definition of a hyperbola having its foci at the positions of the two base stations. Thus a GTD value limits the possible location for the MS observing a constant OTD value between the two base stations to be on a hyperbola.

When at least two hyperbolas are obtained, the location estimate can be found at their intersection. In some cases, two hyperbolas can have two intersections. Then a unique solution requires one additional hyperbola, or other additional information (e.g. coverage area of the reference cell) is needed to select one of the intersections. This is demonstrated in the Figure 4.12. If only hyperbolas 1 and 2 are available, there are two possible intersections, A and B. Adding hyperbola 3 to the picture it becomes evident that the intersection B is the correct answer. If hyperbola 3 is missing, then information about the coverage area of the reference base station (grey area) can be used to select intersection B.

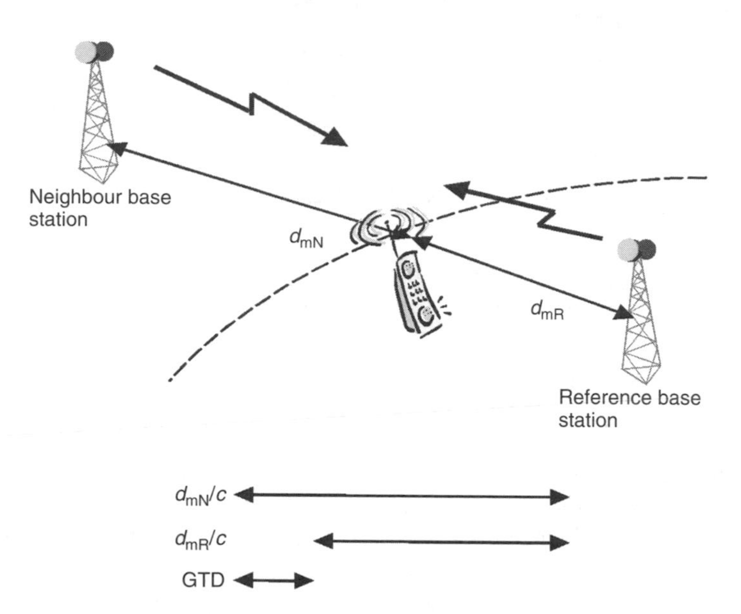

Figure 4.11. Geometrical time difference

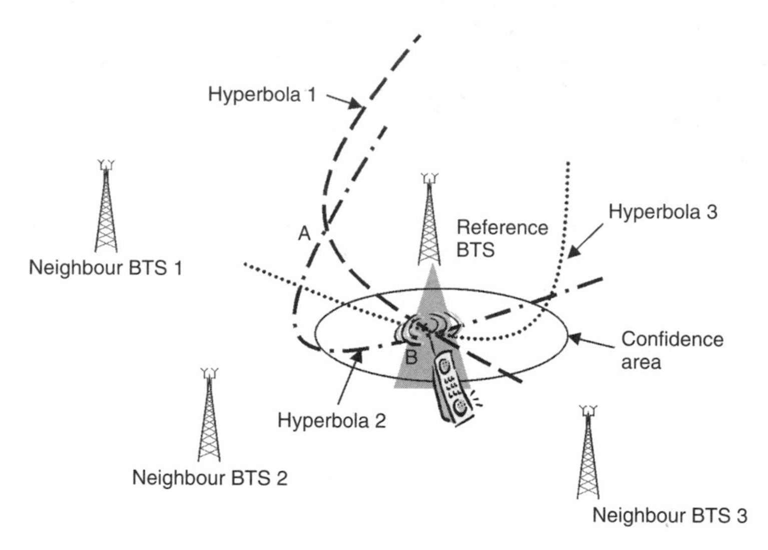

Figure 4.12. Location estimation based on hyperbolas

The MS reports together with OTD measurements also the quality figures related to each measurement using the standardised radio resource LCS protocol (RRLP). These quality figures are the number of single OTD measurements for a certain base station pair, and their standard deviation. These can be used in location calculation to estimate the reliability of each hyperbola and to form a confidence area. Typically, the confidence area for E-OTD is an ellipse that describes an area inside which the MS is located with a certain probability (e.g. 90%).

E-OTD requires that RTD values are known. One possibility is that the whole network is synchronised so that RTD values are always zero. A more practical approach is to measure RTDs. According to GSM standards, RTD values can be measured using LMUs. RTD values can be calculated from the equation: RTD = OTD – GTD, where OTD is measured by the LMU and GTD can be calculated from the known coordinates of the LMU and base stations.

According to GSM specifications, an LMU can also attach to the measurement report the Absolute Time (AT) timestamp for the last measured burst from the reference base station. When this is done, the RTD values can be called absolute time difference (ATD) values. AT timestamp is given using the global positioning system (GPS) as a reference clock, i.e. an LMU needs to have a GPS clock available. At the moment only GPS is mentioned in the standards as a candidate for the reference clock. AT timestamp allows comparison of measurements from different LMUs directly if all LMUs have the GPS clock as a common reference clock. This allows, depending on implementation, easier calculation procedure and, e.g. for A-GPS location method, delivery of GPS time to the MS. In addition, using AT allows LMUs to use only internal antenna that measures signals only from the BTS sectors of the site where the LMU is situated. Such installation approach requires an LMU at every BTS site, but avoids using external receiver antennas for LMUs that usually require more laborious and expensive installation work.

The above-mentioned approach of defining hyperbolas for location estimation from OTD measurements is called *hyperbolic E-OTD*. Another possibility mentioned in GSM specifications is the so-called *circular E-OTD*. The basic equation for each base station is $d_{bM} - d_{bL} = c^*(t_{r\times M} - t_{r\times L} + t_{offset})$. Here d_{bM} and d_{bL} are the distances from a base station to the MS and the LMU respectively. $t_{r\times M}$ is the reception time of a signal from the base station at the MS, and $t_{r\times L}$ is the reception time of a signal from the base station at the LMU. $t_{r\times M}$ and $t_{r\times M}$ are measured against the internal clocks of the MS and the LMU, and there normally exists a time-offset t_{offset} between these clocks. As earlier, c denotes the speed of radio waves. The MS and the LMU need to measure at least three base stations so that it is possible to solve the MS location and the unknown clock offset t_{offset}. This approach is called *circular*, since after solving a group of at least three equations, the distances between the MS and the base stations is known allowing circles around the base stations to be defined. Then the location estimate of the mobile can be found at the intersection of the circles. In practice, circular E-OTD uses the same OTD measurement reported by the MSs and LMUs as the hyperbolic E-OTD, and it is possible to obtain from the group of circular equations the hyperbolic equations.

There are two variants of the E-OTD method defined in GSM specifications: MS-assisted E-OTD and MS-based E-OTD. In the case of MS-assisted E-OTD, the SMLC sends a request for OTD measurements to the MS. The request can include a list of base stations that the mobile should measure. The MS can also be told so-called expected OTD values, which are estimates of OTD values relative to the reference base station. The intention is to help the MS to perform measurements by giving the MS some idea when signals from neighbour base stations should arrive. Once the MS has performed OTD measurements, it reports them to the SMLC, which calculates the location estimate.

In the MS-based E-OTD, the mobile performs the location estimation itself using base station coordinates and RTD values received from the network. The MS can get this information from the E-OTD request, by specifically asking for the information, or from broadcast sources.

E-OTD requires the MS to support the needed measurement functionality. In practice, this can be done in modern GSM MSs with only software changes. That is why E-OTD is seen as a good candidate for mass location purposes since it is easy to include in every MS. Since each mobile performs the necessary measurements itself, location capacity is not limited by the measurements as in the case of methods relying on network-based measurement equipment. E-OTD measurements can be performed relatively fast, e.g. 1- or 2-s measurement normally should be sufficient.

E-OTD is relatively accurate. Simulations performed in T1P1 (GSM 05.50) show that E-OTD accuracy varies much depending on environment, number of hyperbolas, etc. For example, in an urban area, 67% accuracy is approximately 200 m with two hyperbolas, and approximately 100 m in a suburban area. Some field tests indicate 67% accuracy up to 50 to 150 m, depending on the environment. Some field test results are introduced later in Section 4.4.2.1 entitled *E-OTD field test* in this chapter.

Location accuracy for E-OTD depends on several issues. The quality of measurements in the MS and LMUs is naturally one of the factors. The physical environment is another. Multi-path propagation and especially non-line-of-sight conditions can decrease accuracy. The number of different base station sites that the MS can receive determines the number of hyperbolas available. More hyperbolas mean better accuracy. Poor geometry of base stations relative to the MS can degrade accuracy. For example, if all base stations happen to be more or less aligned with the MS (e.g. on a straight highway), the accuracy transversal to the line formed by base stations and the MS is poor. Insufficient number of LMUs might mean that there is no full RTD information that can make the accuracy worse. It is also extremely important to have correct and updated network information available for good accuracy. Inaccuracies in base station coordinates directly cause inaccuracy, and incorrect base station identity and channel information can lead to wrong identification of base stations.

4.3.2.2 Time of Arrival/Time Difference of Arrival

In GSM standards Rel'98 and Rel'99, TOA location method was introduced. In this method, random access bursts in the random access channel from the MSs is measured

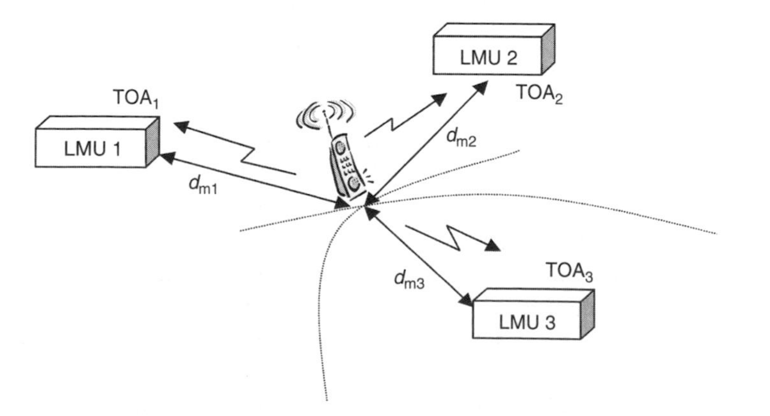

Figure 4.13. Time of arrival method

by LMUs in different positions in the network (Figure 4.13). The mobile is made to perform an asynchronous handover in order to make it send usual random access bursts. Thus, if the mobile is in idle state, a signalling channel is opened to it first. In Rel'4, TOA method was removed. However, a simplified version called *Uplink Time Difference Of Arrival* (U-TDOA) has been proposed to be included again in Rel'6. In this variant, no handovers are used. Instead, usual transmissions from a mobile are measured. In both TOA and U-TDOA, the difference in TOA values measured by LMUs 1 and 2 in two different positions determines a hyperbola: $c^*(\mathrm{TOA}_1 - \mathrm{TOA}_2) = d_{\mathrm{m1}} - d_{\mathrm{m2}}$, where c is the speed of radio waves, and d_{m1} and d_{m2} denote the distance from the MS to LMUs 1 and 2 respectively. Then, location estimation proceeds the same way as described in the case of E-OTD.

The accuracy of TOA/U-TDOA is comparable to that of E-OTD [7]. TOA/U-TDOA can be implemented so that no changes in the mobile are required. However, TOA/U-TDOA requires LMUs capable of TOA measurements and in addition also a common clock in LMUs (in practice GPS clock) so that the TOA measurements are comparable with each other. Coordination between base stations is needed in order to get the LMUs to measure the MS when it transmits signals. This means that extra signalling load is generated. Also, the capacity is limited since several LMUs are needed to locate one MS at a time.

Certain US operators have announced the adoption of U-TDOA for E911 purposes. TOA/U-TDOA offers the possibility to locate even existing mobiles, which means that accuracy requirements by FCC for it are not so tight as for methods requiring changes for mobiles. For commercial applications, limitations of TOA/U-TDOA, like cost of sensitive LMUs and capacity issues, may limit its attractiveness.

4.3.3 Extended Service Level

Methods belonging to the extended service level offer the best performance. This is achieved with both extra equipment in the network and additional hardware in the MS.

4.3.3.1 Assisted GPS

Global positioning system (GPS) is a satellite-based, widely used location system. It relies on a fleet of medium earth orbit (MEO) satellites that transmit two L-band spread spectrum signals. A GPS receiver measures relative times of arrival of signals sent simultaneously from at least three (2D location) or four (3D location) satellites. On the basis of these measurements, the distances between the satellites and the receiver can be estimated. Each distance defines a sphere around the position of a satellite, and by finding their intersection the location is estimated.

A straightforward way to use GPS for mobile location is to integrate a GPS receiver into an MS. However, this approach has its limitations, since originally GPS was not planned for use indoors or in urban environments. GPS signals are weak, and a normal stand-alone GPS receiver needs line of sight to satellites. Integration of a GPS receiver in a handset also increases the manufacturing price and power consumption. Antenna integration can also be challenging. The time delay before the location estimate is obtained can sometimes be long, depending on the GPS receiver implementation details, and history of usage (i.e. whether the GPS receiver has been on earlier).

Wireless networks can deliver assistance information to the mobile and its GPS receiver in order to improve its performance and to reduce some of the above-mentioned problems. Assistance data can consist of the following information:

- Reference time
- Reference location
- Differential GPS (DGPS) corrections
- Navigation model (contains satellite ephemeris and clock corrections)
- Ionospheric model
- Universal Time Co-ordinates
- UTC model
- Almanac
- Acquisition assistance
- Real-time integrity.

Assistance data speeds up acquisition, decreases power consumption, increases sensitivity and expands coverage area (e.g. urban environment).

As mentioned earlier, LMUs can be used to help generate assistance data for GPS, namely reference time. The network knows, on the basis of LMU measurements, the correspondence between the GPS time and GSM timing, e.g. what is the exact GPS time when a certain GSM frame of a certain base station is starting. The MS can then detect the arrival of a frame from the serving base station, and estimate what is the corresponding GPS time. Together with reference position this helps to improve sensitivity, since the GPS receiver has a good understanding of timing of signals that it needs to receive.

When the GSM network informs the MS about the data that the GPS satellites are sending (e.g. almanac, ephemeris data, clock corrections, ionospheric and UTC model), the GPS receiver does not need to decode the information itself. This helps under poor signal conditions. Since the data rate of GPS signals (50 bit/s) is low compared to that of GSM, GPS satellite data transfer within assistance data also allows faster location determination and decreases power consumption in the MS.

Acquisition assistance data essentially describes the Doppler shift and code phase of the signal from a certain satellite at the reference time and position. As the name indicates, this part of assistance data is intended to help the GPS receiver with signal acquisition.

Real-time integrity describes the real-time status of the GSP satellite constellation. It lists those GPS satellites that the MS should not use for location estimation since they are estimated not to function properly.

In GERAN and UTRAN standards, three basic variants of GPS are supported. The first one is simple stand-alone GPS, where the MS includes a GPS receiver that operates independently from the cellular network. The two other variants are two versions of A-GPS, where the network participates in GPS location (Figure 4.14). In MS-based network assisted GPS, the MS receives assistance data from the network, but calculates the GPS location estimate itself. In MS-assisted network-based GPS, the MS performs actual measurements and reports them to the network that calculates the location estimate.

GPS has good accuracy, up to 2 m with differential corrections. The removal of intentional accuracy degradation features, such as the selective availability (SA), from GPS signals allows basic GPS achieving accuracy up to 10–20 m. Thus, the accuracy of

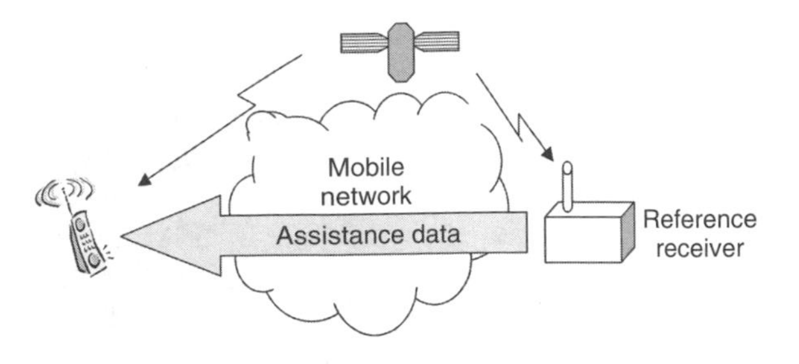

Figure 4.14. Assisted GPS

A-GPS is not usually the challenge. Instead, availability indoors and in other demanding environments is a more relevant concern.

At the moment, US GSM operators have not announced the use of A-GPS to fulfil E911 emergency call requirements (although A-GPS is used for E911 with some other cellular technologies). Still, it is expected that, once A-GPS capable MS penetration grows, A-GPS will eventually play an important role also in this field.

4.4 LCS Performance

4.4.1 Basic Service Level Performance

4.4.1.1 Location System Implementation

A simplified version of the logical architecture of a location system incorporating the basic service location methods described in Section 4.2 is represented in Figure 4.15. Besides the 'classical' GSM network elements (BTS, BSC, MSC, visiting location register (VLR)

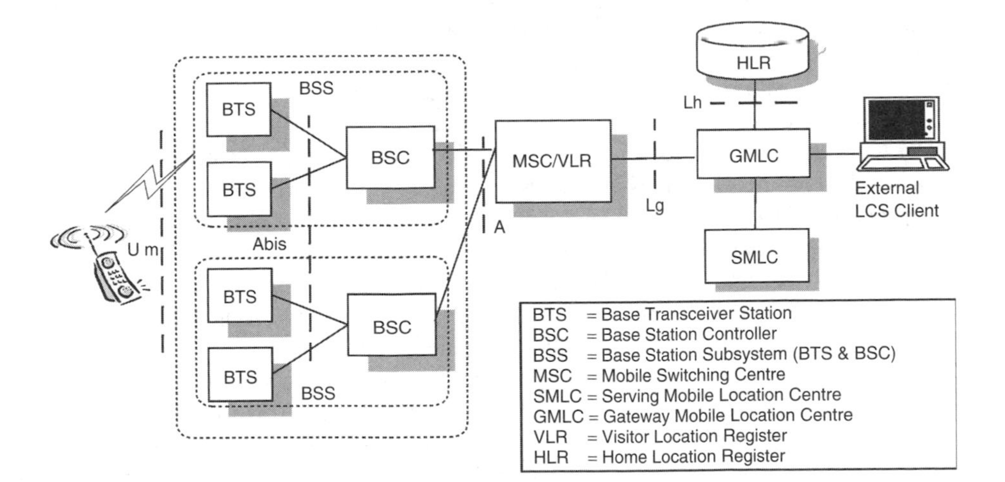

Figure 4.15. Architecture of a location system implementation. (Reproduced from Spirito M. A., Pöykkö S., Knuuttila O., 'Experimental Performance of Methods to Estimate the Location of Legacy Handsets in GSM', *Proceedings of IEEE Vehicular Technology Conference—VTC 2001*, 2001 by permission of IEEE.)

and home location register (HLR)), new elements introduced in the GSM LCS specifications Rel'98 and Rel'99 are shown, namely SMLC and GMLC. The SMLC is the network element, or corresponding functionality, performing the actual location calculation. The GMLC acts as interface between the GSM network and external LCS clients.

To perform a location calculation, CI, TA and RXLEVs must first be made available in the network side and then delivered to the SMLC. In GSM, the TA is available in the network only for a mobile terminal in *dedicated mode* (i.e. a mobile communicating with the serving base station using a dedicated channel). A GSM phone measures RXLEVs both in dedicated mode and in *idle mode* (i.e. the phone is turned on but no two-way communication takes place) however, it reports RXLEVs to the fixed part of the network only in dedicated mode. In order to obtain TA and RXLEVs from a handset in idle mode, and thus to be able to locate it, the MSC can establish a dedicated connection unnoticeable to the user. In this way, the TA can be determined and the mobile terminal is forced to report RXLEVs to the network, thus location calculation is enabled.

The forced measuring of TA and reporting of RXLEVs for GSM handsets in idle mode, as described above, is in line with subscriber privacy mechanisms defined in GSM LCS specifications Rel'98. The target handset may be positioned only if allowed in the handset subscription profile, unless required by local regulatory requirements. The LCS client access to location information is restricted unless otherwise stated in the target terminal subscription profile; thus, the subscriber has full control of delivering his/her location information. For emergency services, the target terminal making an emergency call can be positioned regardless of the privacy attribute value of the GSM subscriber associated with the target terminal making the call. The GMLC authenticates all applications and encrypts the traffic towards location-based applications.

4.4.1.2 Experimental Performance

This section presents experimental performance of CI, CI + TA and CI + TA + RXLEV location methods incorporated in the system shown in Figure 4.15. A more extended analysis of the data can be found in [8]. CI, TA and RXLEV measurements were collected from Network 11, in a city centre and a suburban area in early 2001. Measurements in the 900 MHz band were collected driving a certain number of test routes. Inside a vehicle, a laptop, connected to a GPS receiver and a GSM phone, was periodically storing in log files GPS reference locations as well as *measurement reports* and time stamps. One *measurement report* includes the data a mobile terminal is forced to report to the network in the system implementation of Figure 4.16; namely, CI, TA, and one RXLEV for the serving cell and one RXLEV for each of the strongest neighbour cells (up to six). CI, TA and RXLEV data were gathered in 76 009 (102 269) locations in an urban (suburban) area from 44 (46) serving cells and 76 (83) neighbour cells.

The analysis of the data collected during the test routes shows that in the suburban area half of the TA values were between 1 and 2 while in the urban area the TA values were mainly 0 and 1. In the urban area, TA values up to four were measured, but TA ≥ 3 were measured only at very few locations. In roughly 55% of all measurements the TA was zero. The number of neighbours in the urban area in over 80% of the cases was at least four; the mean RXLEV values were between −60 dBm for the first neighbour cell and −80 dBm for the sixth neighbour cell. In the suburban area, TA values up to four were measured, the number of neighbours was as high as in the urban area but, due to a

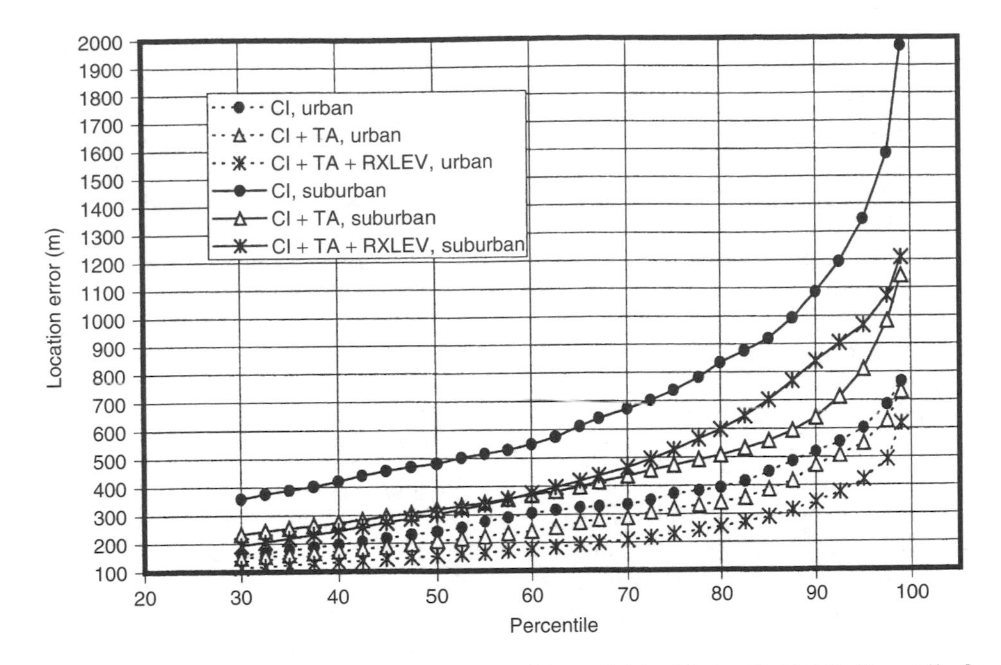

Figure 4.16. Location error statistics. (Reproduced from Spirito M. A., Pöykkö S., Knuuttila O., 'Experimental Performance of Methods to Estimate the Location of Legacy Handsets in GSM', *Proceedings of IEEE Vehicular Technology Conference—VTC 2001*, 2001 by permission of IEEE.)

less dense network, the mean RXLEV values were between 5 and 10 dBm lower than in the urban area.

The measurements collected have been processed off-line with the CI, CI + TA and CI + TA + RXLEV location methods described in Section 4.3. Each method determines a location estimate by processing only one measurement report. The *location error*, defined as the distance between a GPS reference location and the corresponding location estimate, is used as a measure of location accuracy. Location error statistics are shown in Figure 4.16.

It can be seen that the CI + TA + RXLEV method outperforms the other two methods in the urban area, while in the suburban area it performs essentially as well as CI + TA. Both CI + TA and CI + TA + RXLEV methods are significantly more accurate than the CI method in both environments for all percentiles. Therefore, CI + TA and CI + TA + RXLEV methods reduce not only the average location error but also the maximum error. For example, when comparing CI and CI + TA + RXLEV methods, the 99% percentile reduces from 2 to 1.2 km in the suburban area, and from 770 to 620 m in the urban area. Controlling the maximum location error is crucial; in fact, larger than expected location errors may negatively affect the customer's perception of service reliability and can decrease the customer's interest for LCS.

The absolute values of the location errors depend strongly on the network as well as the environment where the measurements are performed. The results shown in Figure 4.16 can be used to estimate the accuracy of the methods in different conditions if the network dependence is somehow reduced. This can be done by properly scaling the location error.

A good candidate for the scaling factor is the cell size, which is, however, not a clearly defined quantity. Since in the CI method the handset location estimate is always at the base station coordinates, a statistically meaningful definition for the average cell size (i.e. cell radius) in each environment is the 95% percentile of the CI location error. In the following, the accuracy performance is further analysed in terms of *relative location error*, defined as the location error scaled to the cell size defined above (e.g. 603 m in urban and 1345 m in suburban area).

It can be seen from the results in Figure 4.17 that for a given percentile level, the CI + TA relative accuracy is better in the suburban than in the urban area, while the CI + TA + RXLEV relative accuracy is almost the same in both environments. This clearly indicates that the location accuracy obtained by the CI + TA + RXLEV method scales according to the cell size. In other words, in networks with smaller cells the absolute CI + TA + RXLEV accuracy would most likely improve as compared to the accuracy shown in Figure 4.16. The scalability of the CI + TA + RXLEV accuracy with the cell size is very relevant from the commercial application viewpoint. This is because the location accuracy requirements in densely populated areas, where cells are small, are high due to the proximity of the points of interest (e.g. restaurants). In contrast, in less densely populated areas, where cells are large, the points of interest (e.g. gas stations) are more separated and the accuracy requirements can be relaxed.

The relative accuracy of CI, CI + TA and CI + TA + RXLEV methods for different values of measured TA is compared in Figure 4.18 (for illustration purposes, the 67% percentile is shown). The CI location error increases for increasing TA values. This is due to the fact that the TA gives an indication of the distance between handset and serving base station but the CI location estimate is always at the serving base station site, irrespective of the measured TA. Similarly, the CI + TA location accuracy degrades as TA increases (the only exception being for TA = 4, which is, however, not statistically meaningful). In fact, the geographical region in which a mobile terminal can measure a

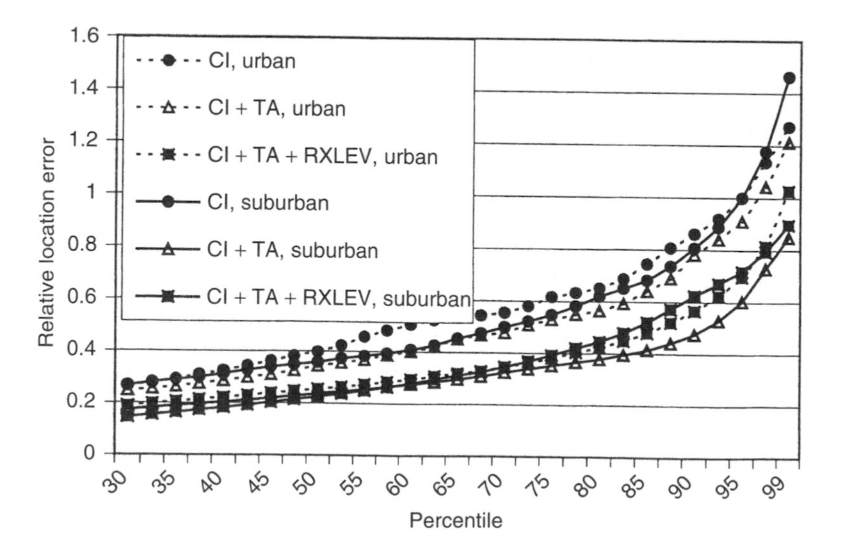

Figure 4.17. Relative location error statistics

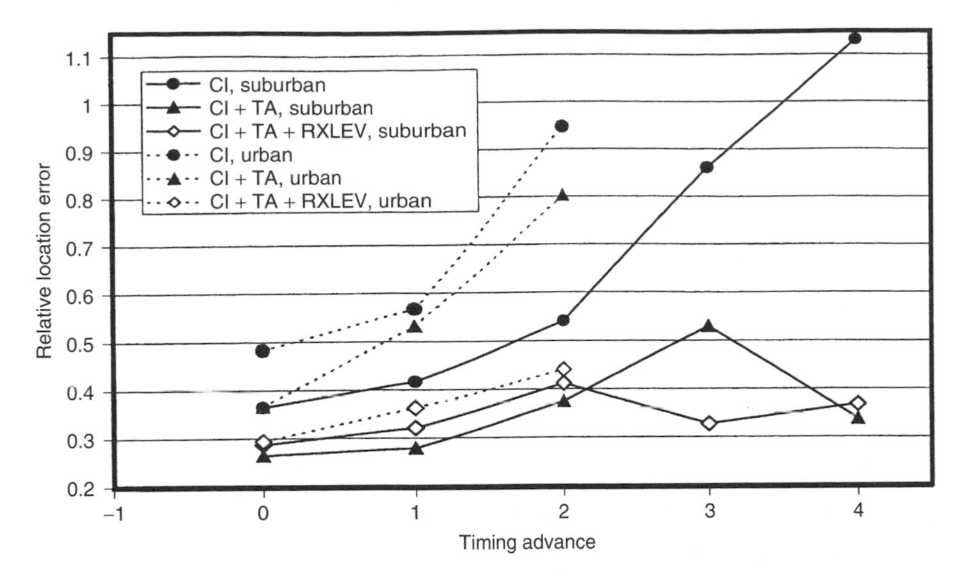

Figure 4.18. 67% percentiles of the relative location error as a function of measured TA. (Reproduced from Spirito M. A., Pöykkö S., Knuuttila O., 'Experimental Performance of Methods to Estimate the Location of Legacy Handsets in GSM', *Proceedings of IEEE Vehicular Technology Conference—VTC 2001*, 2001 by permission of IEEE.)

certain TA, the TA value increases in size with the TA but the CI + TA location estimate is always placed along the serving cell's antenna azimuth, at a distance from the serving base station derived from the TA. The degradation of CI + TA + RXLEV accuracy for increasing TA values is much smaller as compared with the other two methods.

The accuracy results provided in this section show that irrespective of the environment, CI + TA and CI + TA + RXLEV methods significantly reduce both average and maximum errors obtained with the CI method. Additionally, the CI + TA + RXLEV location error turns out to scale along with the cell size making this method very suitable for services such as finding/guidance and notification, as it delivers highly accurate location estimates in densely populated areas where the demand for accurate location information is strong. Finally, the CI + TA + RXLEV accuracy proves not to deteriorate as much as the CI and CI + TA accuracy, when the distance between the MS and the serving base station (or equivalently the TA) increases. A more extensive analysis of the experimental results [6] also proves that, especially in the urban environment, an increase in the number of measured neighbours leads to an improvement in the CI + TA + RXLEV location accuracy.

4.4.2 Enhanced Service Level Performance

4.4.2.1 E-OTD Field Test

In this section, some results are described from an overlay E-OTD test system in Network 7 (Figure 4.19). The system consists of a demonstrator serving mobile location centre (SMLC-D), a modem gateway, a demonstrator mobile station (MS-D) and several demonstrator location measurement units (LMU-D).

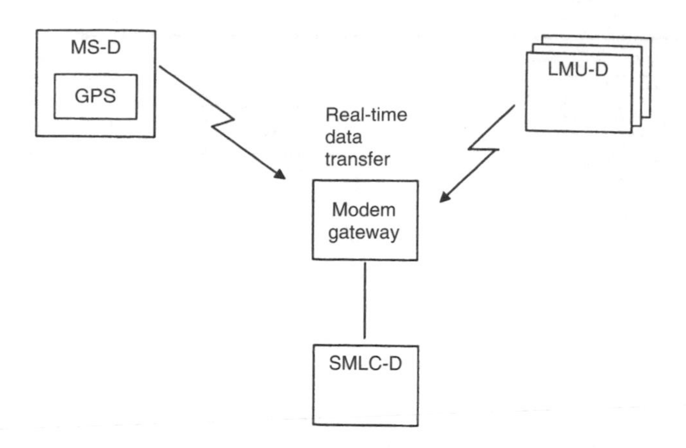

Figure 4.19. E-OTD field test system

Normal GSM data calls are used to deliver information between the SMLC-D and the LMU-Ds and MS-D via the modem gateway. LMU-D, MS-D and SMLC-D correspond functionally to the standardised solution, but signalling solution is proprietary. The MS-D includes a GPS receiver that is used to define the real position.

In the typical measurement procedure, the SMLC-D sends measurement commands to LMU-Ds that start to send RTD measurements periodically back. Therefore, SMLC-D is aware of the RTD situation all the time. Then, SMLC-D sends an E-OTD measurement command to the MS-D to be located. The mobile performs E-OTD measurements for one location that takes 1 to 2 s and sends the measurement results back to the SMLC-D that calculates the E-OTD position estimate.

The test area was a suburban commercial and residential environment. The typical distance between cells in the area is 200 to 1000 m. The area has apartment and office buildings with typically six to eight storeys, but also smaller residential houses with one to two floors.

To give an example, a test run was done with seven LMU-Ds. The MS-D was moving on a 15-km route with velocity 0 to 50 km/h. The distance between the E-OTD location estimate and the real location, i.e. the location error, had the distribution as shown in Table 4.1.

Another example test run was done with six LMU-Ds. Now the MS-D was moving on a shorter 5-km route. The location errors were as in Table 4.2. Better accuracy is probably due to the environment that was mainly with low residential houses (1 to 2 floors).

In the third example test run, the MS-D was moving again on a shorter 5-km route (Table 4.3). Now the area was densely built-up mainly with six floor apartment buildings. The location errors indicate less accurate results than in the previous runs. This is due to the less favourable environment.

Table 4.1. E-OTD field test example (moving MS)

Number of measurements	Median accuracy	67% accuracy	90% accuracy
267	58 m	90 m	232 m

Table 4.2. E-OTD field test example (moving MS)

Number of measurements	Median accuracy	67% accuracy	90% accuracy
150	35 m	42 m	85 m

Table 4.3. E-OTD field test example (moving MS)

Number of measurements	Median accuracy	67% accuracy	90% accuracy
200	87 m	120 m	240 m

Table 4.4. E-OTD field test results (stationary MS)

Number of measurements	Median accuracy	67% accuracy	90% accuracy
45	28 m	31 m	33 m

Table 4.4 illustrates a stationary test in a single location (four LMU-Ds were on). In this case, the location errors are very consistent.

References

[1] Pent M., Spirito M. A., Turco E., 'Method for Positioning GSM Mobile Stations Using Absolute Time Delay Measurements', *IEE Electron. Lett.*, **33**(24), 1997, 2019, 2020.

[2] Spirito M. A., Guidarelli Mattioli A., 'Preliminary Experimental Results of a GSM Phones Positioning System Based on Timing Advance', *Proceedings of 1999 IEEE Vehicular Technology Conference VTC 1999 Fall*, Amsterdam, The Netherlands, September 1999, pp. 2072–2076.

[3] Spirito M. A., 'Accuracy of Hyperbolic Mobile Station Location in Cellular Networks', *IEE Electron. Lett.*, **37**(11), 2001, 708–710.

[4] Spirito M. A., 'On the Accuracy of Cellular Mobile Station Location Estimation', *IEEE Trans. Vehicular Technol.*, **50**(3), 2001, 674–685.

[5] ETSI TC-SMG, European Digital Telecommunications System (Phase 2); Radio Subsystem Synchronization (GSM 05.10), ETSI—European Telecommunications Standards Institute.

[6] Spirito M. A., 'Further Results on GSM Mobile Stations Location', *IEE Electron. Lett.*, **35**(11), 1999, 867–869.

[7] ETSI TC-SMG, European Digital Telecommunications System (Phase 2); Background for Radio Frequency (RF) Requirements (05.50), ETSI—European Telecommunications Standards Institute.

[8] Spirito M. A., Pöykkö S., Knuuttila O., 'Experimental Performance of Methods to Estimate the Location of Legacy Handsets in GSM', *Proceedings of IEEE Vehicular Technology Conference—VTC 2001*, 2001.

Part 2

GSM, GPRS and EDGE Performance

The second part of the book focuses on performance delivered by global system for mobile communications (GSM)/enhanced data rates for global evolution (EDGE) networks. The performance is measured in two different and complementary ways. First, field trial results help to get a realistic understanding of basic GSM performance capabilities. Second, simulation results envisage performance of features not yet available in real networks. In any case, the methodology used in the performance analysis is consistent throughout the book.

The basics of GSM/EDGE radio performance, the methodology followed to characterise network performance and measure spectral efficiency and the definition of the GSM basic set of functionality and its associated baseline performance are the topics covered in the first chapter of this part.

Chapter 6 presents GSM/adaptive multi-rate (AMR) and Single Antenna Interference Cancellation (SAIC) voice performance. It starts from basic enhanced full-rate (EFR) speech codec performance and later on analyses additional capacity enhancement techniques already available in many of the existing GSM networks. GSM system performance with AMR codec is thoroughly presented in this chapter. Finally, further evolution of the speech services in Rel'5 and Rel'6, including SAIC and Flexible Layer One (FLO), is addressed in its last sections.

General packet radio system (GPRS) and enhanced GPRS (EGPRS) packet-switched performance is studied in Chapter 7. The analysis covers link performance, radio resource management algorithms, system level performance and end-to-end performance considering different applications over (E)GPRS. This chapter gives indications of how (E)GPRS services can be introduced in an existing GSM and how mixed (E)GPRS and speech traffic can be planned and dimensioned.

Chapter 8 focuses on end-to-end service performance. It covers the main effects that have an impact on user-perceived quality, from physical layer up to application protocol dynamics. The service performance of the most relevant data services like web browsing, FTP download, gaming, push-to-talk application, and so on are studied following a methodological approach. Hints about how to improve service performance are also provided.

Dynamic frequency and channel allocation is an advanced radio resource management technique that pushes the limit of GSM network capacity. Chapter 9 describes this technique and provides accurate performance figures with EFR and AMR codecs.

GSM deployments in narrowband scenarios are studied in Chapter 10. Network migration to GSM with gradual frequencies re-farming leads typically to a narrowband available for GSM. Different BCCH planning strategies and specific radio resource management solutions are described in this chapter.

Chapter 11 focuses on the GSM EDGE radio access network (GERAN) link budget calculation and describes the different techniques that can be applied to enhance both uplink (UL) and downlink (DL) performance.

Chapter 12 tackles control channel performance in GSM/EDGE from two points of view: signalling channel's link level performance and capacity of the different channel configurations.

Finally, Chapter 13 covers GSM's innovative automation functionality to be applied in the planning, parameter optimisation and troubleshooting tasks.

5

Basics of GSM Radio Communication and Spectral Efficiency

Juan Melero, Jeroen Wigard, Timo Halonen and Javier Romero

The purpose of this chapter is to introduce the principles of GSM/EDGE radio performance, defining the tools and methodology utilised throughout the book to study the performance enhancements provided by the multiple voice and data functionality presented. All the analysis presented in different chapters is based on commercial network's trials and simulation results. Appendix E describes the simulation tool and configurations used for the simulation-based analysis and Appendix F presents the trial networks mentioned across the book as *Network #*.

Section 5.1 describes a GSM radio interface basic link performance including a description of the transmission and receiving chain and the basic GSM standardised functionality, such as frequency hopping (FH), power control and discontinuous transmission. In Section 5.2, the most relevant key performance indicators (KPIs) to monitor network performance are presented. Section 5.3 defines the spectral efficiency and the different existing ways to quantify it and Section 5.4 describes the effective frequency load (EFL) measurement methodology, utilised across the whole book to characterise networks performance and measure spectral efficiency from real networks trial data and simulation results. Finally, Section 5.5 defines the baseline performance of GSM systems analysing their achievable performance.

5.1 GSM Radio System Description

This section introduces very briefly the principles of GSM radio communication, its channel structure and basic functionality, such as frequency hopping, power control and discontinuous transmission. An introduction about their impact in link performance is

GSM, GPRS and EDGE Performance 2nd Ed. Edited by T. Halonen, J. Romero and J. Melero
 ISBN: 0-470-86694-2

included as well. It focuses on the basic GSM speech service since the equivalent (E)GPRS analysis is covered more extensively in Chapters 1 and 7. For a more detailed description of the subjects introduced in this section, the reader is recommended to consult [1, 2].

5.1.1 Basic Channel Structure

The GSM standard is based on a multi-carrier, time-division multiple access and frequency division duplex, MC/TDMA/FDD [1]. Table 5.1 shows the frequency bands that have been defined for GSM.

The carrier spacing is 200 kHz allowing, for example, for 124 and 374 radio frequency channels in the 900 and 1800 MHz bands respectively, thus leaving a guardband of 200 kHz at each end of the sub-bands. Each radio frequency is time divided into TDMA frames of 4.615 ms. Each TDMA frame is subdivided into eight full slots. Each of these slots can be assigned to a full-rate (FR) traffic channel (TCH), two half-rate (HR) TCHs or one of the control channels. A slot is equal to one timeslot (TSL) on one frequency. The time and frequency structure is displayed in Figure 5.1.

Table 5.1. GSM standardised frequency bands

GSM frequency band	Available frequencies	Where available
400 MHz	450.4–457.6 MHz paired with 460.4–467.6 MHz or 478.8–486 MHz paired with 488.8–496 MHz	Europe
800 MHz	824–849 MHz paired with 869–894 MHz	America
900 MHz[1]	880–915 MHz paired with 925–960 MHz	Europe, Asia Pacific, Africa
1800 MHz	1710–1785 MHz paired with 1805–1880 MHz	Europe, Asia Pacific, Africa
1900 MHz	1850–1910 MHz paired with 1930–1990 MHz	America

[1]Including the extended GSM (EGSM) band.

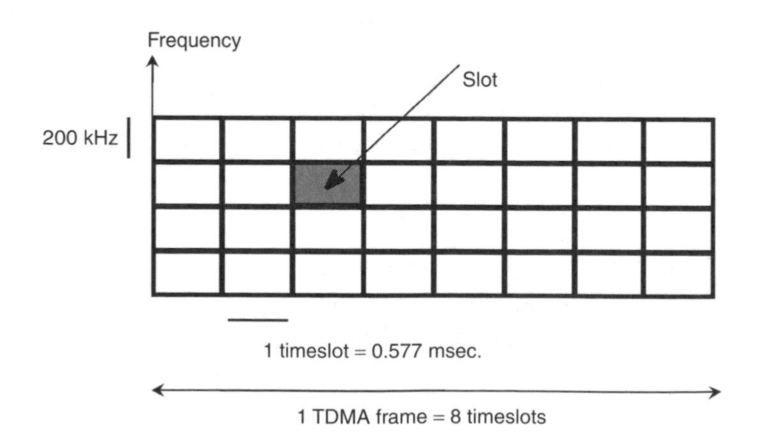

Figure 5.1. The multiple access scheme in GSM

The data transmitted in one slot is denoted as a burst. There are five different types of bursts: the normal burst, the access burst, the frequency correction burst, the synchronisation burst and the dummy burst [3]. The format and information of the individual bursts depends on the type of channel it belongs to.

Channels can be described at two levels: at the physical and the logical level. A physical channel corresponds to one timeslot on one carrier, while a logical channel refers to the specific type of information carried by the physical channel. Different kind of information is carried by different logical channels, which are then mapped or multiplexed on physical channels.

Logical channels can be divided into two groups: control channels and traffic channels. Traffic channels are used to carry user data, which can be either speech or data. A logical traffic channel for speech and circuit-switched data in a GSM is called a TCH, which can be either full rate (TCH/F) or half rate (TCH/H). A logical traffic channel for packet-switched (PS) data is referred to as *packet data traffic channel* (PDTCH).

The full-rate traffic channel (TCH/F) is a 13-kbps coded speech or data channel with a raw data rate of 9.6, 4.8 or 2.4 kbps. The half rate supports 7, 4.8 and 2.4 kbps [3]. The TCH channels support bi-directional transmission in order to carry speech and data communication.

A PDTCH/F corresponds to the resource allocated to a single mobile station on one physical channel for user data transmission. Because of the dynamic multiplexing on to the same physical channel of different logical channels, a PDTCH/F using Gaussian minimum shift keyingGMSK modulation carries information at an instantaneous bit rate ranging from 0 to 22.8 kbps. A PDTCH/F using 8-PSK modulation carries information (including stealing symbols) at an instantaneous bit rate ranging from 0 to 69.6 kbps. These bit rates will then be mapped into certain user payload depending on the channel coding scheme applied.

A PDTCH/H corresponds to the resource allocated to a single mobile station on half a physical channel for user data transmission. The maximum instantaneous bit rate for a PDTCH/H is half that for a PDTCH/F.

All packet data traffic channels are unidirectional, either uplink (PDTCH/U), for a mobile-originated packet transfer or downlink (PDTCH/D) for a mobile-terminated packet transfer.

Signalling and controlling information is carried on the other type of logical channels, the control channels. In GSM, there are common as well as dedicated control channels. These channels are described in Chapter 12.

The GSM specifications describe which physical channels to use for each logical channel [3]. Several combinations of the different channels are possible. Figure 5.2 shows the organisation of the TCH/F speech channel and the corresponding slow associated control channel (SACCH). This 26-frame multiframe only considers one timeslot per TDMA frame.

Of the 26 frames, 24 are used for traffic. One frame is used for the SACCH channel, while the last one is an idle frame. During this idle frame time interval period, a mobile can receive other control channels and measure the received signal level from neighbouring cells. The complete GSM frame, timeslot and burst structure is shown in Figure 5.3.

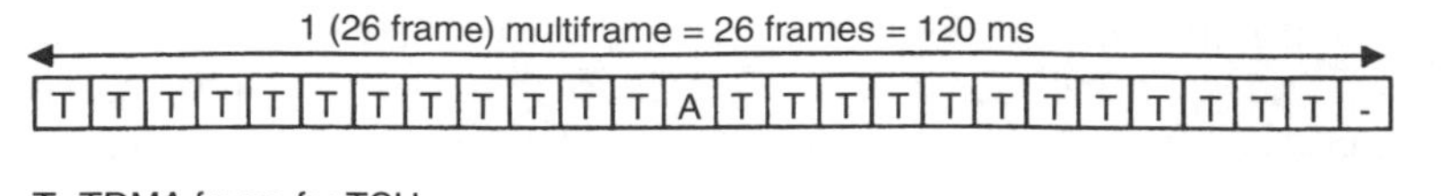

T: TDMA frame for TCH
A: SACCH frame for TCH
- : idle frame

Figure 5.2. Organisation of TCH and the corresponding control channels [3]

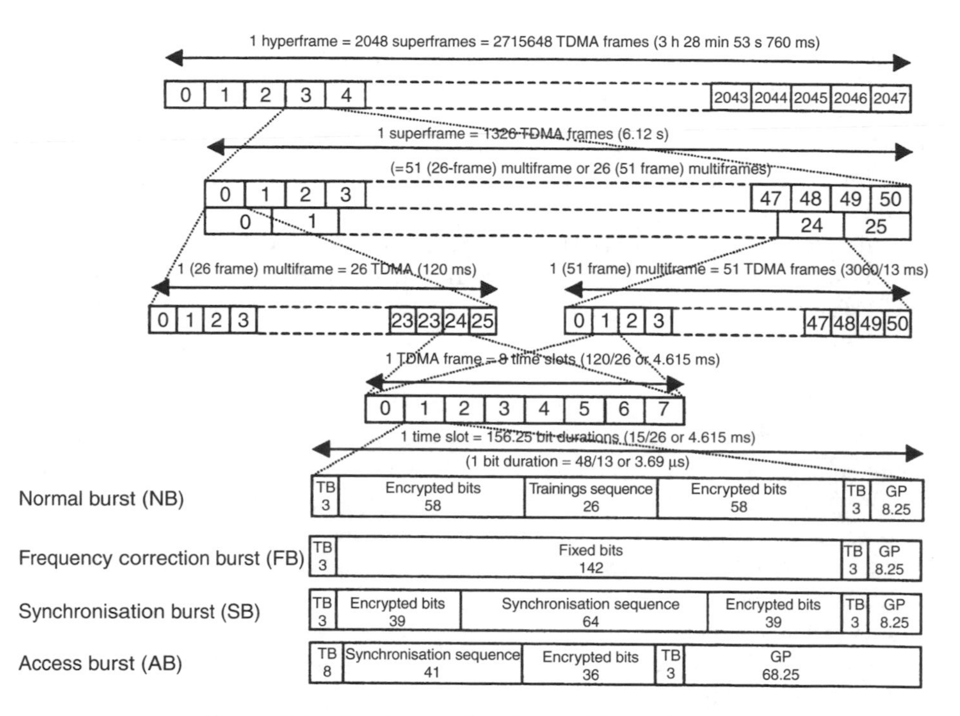

Figure 5.3. Timeframes, timeslots and bursts in GSM

5.1.2 Transmitting and Receiving Chain

Figure 5.4 shows the transmitting and receiving chain of a GSM receiver. Several successive operations have to be performed to convert a speech signal into a radio signal and back.

The following operations take place on the transmitting side:

- *Source coding.* Converts the analogue speech signal into a digital equivalent.
- *Channel coding.* Adds extra bits to the data flow. This way redundancy is introduced into the data flow, increasing its rate by adding information calculated from the source data, in order to allow detection or even correction of bit errors that might be introduced during transmission. This is described in more detail below.
- *Interleaving.* Consists of mixing up the bits of the coded data blocks. The goal is to have adjacent bits in the modulated signal spread out over several data blocks.

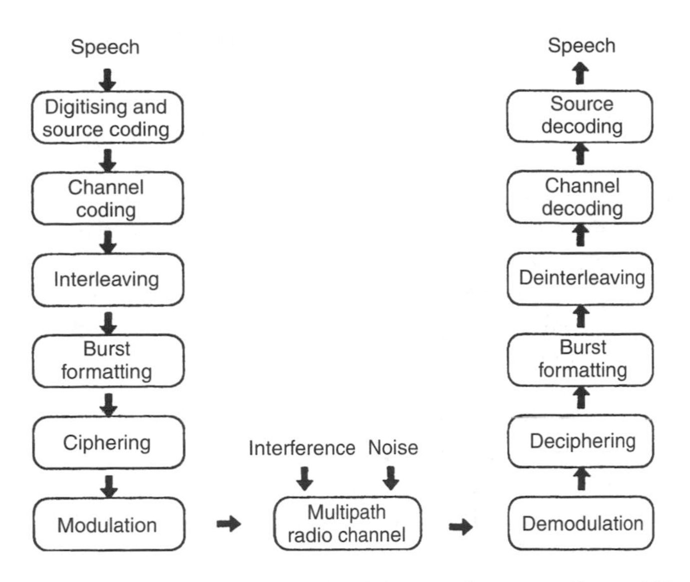

Figure 5.4. Flow diagram of the link operations according to [4]

The error probability of successive bits in the modulated stream is typically highly correlated, and the channel coding performance is better when errors are decorrelated. Therefore, interleaving improves the coding performance by decorrelating errors and their position in the coded blocks.

- *Ciphering*. Modifies the contents of these blocks through a secret code known only by the mobile station and the base station.
- *Burst formatting*. Adds synchronisation and equalisation information to the ciphered data. Part of this is the addition of a training sequence.
- *Modulation*. Transforms the binary signal into an analogue signal at the right frequency. Thereby the signal can be transmitted as radio waves.

The receiver side performs the reverse operations as follows:

- *Demodulation*. Transforms the radio signal received at the antenna into a binary signal. Today most demodulators also deliver an estimated probability of correctness for each bit. This extra information is referred to as soft decision or soft information.
- *Deciphering*. Modifies the bits by reversing the ciphering code.
- *Deinterleaving*. Puts the bits of the different bursts back in order to rebuild the original code words.
- *Channel decoding*. Tries to reconstruct the source information from the output of the demodulator, using the added coding bits to detect or correct possible errors, caused between the coding and the decoding.
- *Source decoding*. Converts the digitally decoded source information into an analogue signal to produce the speech.

In GSM the data bits are coded. The channel coding introduces redundancy into the data flow by increasing the bit rate. For the TCH/FS mode, there are three bit-coding classes. The class 1a bits have a 3-bit cyclic redundancy check (CRC) and all class 1 bits are encoded by a convolution code. The class 2 bits remain unprotected. Note that the coding is different in case of a PDTCH, which is used to carry packet switched data. This is described in Chapter 1.

The reordering and interleaving process mixes the encoded data block of 456 bits, and groups the bits into eight sub-blocks (half bursts). The eight sub-blocks are transmitted on eight successive bursts (interleaving depth equals 8). The channel coding can be seen in Figure 5.5, while the reordering and interleaving is displayed in Figure 5.6.

Because of multipath propagation, the erroneously received bits tend to have a bursty nature. The convolutional code gives the best performance for random positioned bit errors, and therefore reordering and interleaving is introduced in the GSM signal transmission flow. However, the reordering/interleaving only improves the coding performance if the eight successive bursts carrying the data information of one speech block are exposed to uncorrelated fading. This can be ensured either by spatial movement (high user speed) or by frequency hopping.

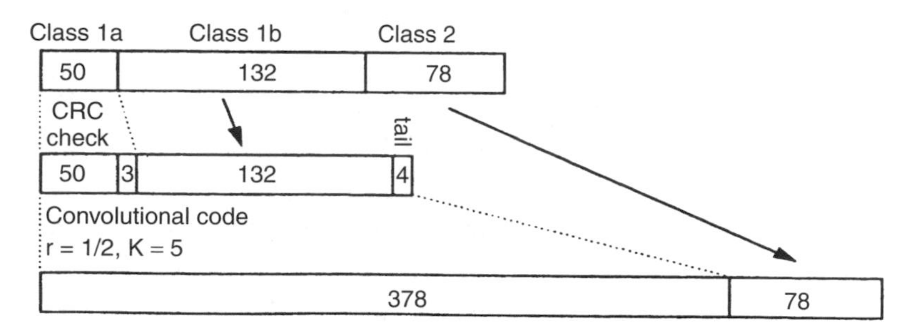

Figure 5.5. Channel coding of the TCH/FS

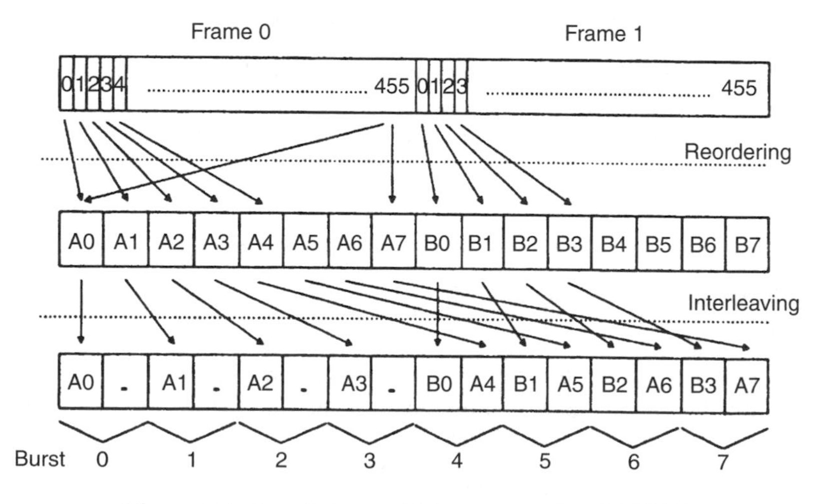

Figure 5.6. Reordering and interleaving of the TCH/FS

5.1.3 Propagation Effects

Radio propagation in mobile communications is complex. The transmitted signal from the base station may have been exposed to a combination of various propagation phenomena such as free space attenuation, diffraction, reflection and other propagation effects before the mobile station receives it. A statistical approach is therefore mostly used for characterisation of the mobile radio channel:

- Mean path loss based on some nth power law with range and an absolute power level at a given distance.
- Variations in the local mean path loss, normally modelled by a lognormal distribution.
- Fast fading superimposed on the slow variation of the local mean signal. Rayleigh or Rician distributions are often used to characterise the fast-fading profile.

5.1.3.1 Path Loss

A signal transmitted by a radio connection is attenuated along the propagation path. The power captured by a receiving antenna decreases with increasing distance to the transmitter. This is caused by the fact that the signal is transmitted not only in a straight line but as a sphere, whose radius is the separation between the transmitter and the receiver. There is less power on the same piece of surface of a larger sphere than on a smaller sphere. In cities, there are also losses in power due to reflection, diffraction around structures and refraction within them. The attenuation of received power with distance is called *path loss*.

Terrain features affect this propagation loss. Therefore, when the path loss has to be predicted, all the terrain features should be taken into account to get an accurate result.

Path loss is relatively easy to calculate when free space is assumed and when there is a direct path between the two antennas (a line-of-sight path). Then the path loss is given by [5]

$$L_{\mathrm{dB}} = 32 + 20 \log f + 20 \log d$$

where f is the frequency in MHz and d is the distance in km.

The equation above gives the best-case result for the path loss. In reality, the path loss will always be higher because of terrain features.

A simple model for the mean path loss as a function of the distance, including terrain features, is given by [6]

$$L_{\mathrm{dB}} = \gamma \cdot 10 \log d + k$$

where d is the distance between base station and mobile station, γ is the propagation path slope, and k is a constant.[1] For free space propagation, γ equals 2 (as in the equation above). For land mobile communications, the path loss slope γ is typically within the range of 3 to 5.

5.1.3.2 Fading

A signal nearly never arrives by only one path. Because of scattering from the terrain obstacles, there is more than one path by which the signal arrives at the destination.

[1] The constant k includes the dependency of mobile and base station height, frequency, land coverage, etc.

Radio signals therefore arrive at the mobile receiver from different directions, with different time delays. They combine at the receiver antenna to give a resultant signal, which depends crucially on the differences in path length that exist in the multipath field, since the various spectral components that comprise the transmitted signal can combine constructively or destructively. As the mobile station moves from one location to another, the phase relationship between the spectral components within the various incoming waves changes, so the resultant signal changes rapidly. It is worth noting that whenever relative motion exists, there is also a Doppler shift in frequency components within the received signal.

Especially in built-up areas this can cause problems because the antenna might be below the surrounding buildings, so that there is no line-of-sight (LOS) transmission path. This situation can be seen in Figure 5.7.

Fading can be classified in two parts: shadow fading (slow fading) and multipath fading (fast fading).

Shadow fading is caused by shadowing from the terrain obstacles and generates a variation in the received signal level local mean. It typically exhibits a statistic lognormal distribution, i.e. a normal pdf (probability density function) of dB values [7]:

$$p(m) = \frac{1}{\sqrt{2\pi\sigma^2}} \mathrm{e}^{-(m-\overline{m})^2/2\sigma^2}$$

where $\overline{m}$ is the mean value of the random variable m, and σ is the standard deviation—both expressed in dB. Typical measured standard deviation values for the lognormal fading are in the range 5 to 10 dB.

The multipath fading component is caused by the combination of the signal arriving via different paths. Measurements suggest that the received signal at each location is made up of a number of replicas of the signal with random amplitude and phase [8]. The amplitudes and phases are assumed to be statistically independent, with the phase of each replica distributed uniformly between 0 and 2π. The amplitudes are assumed to sum up to the mean value of the envelope. Figure 5.8 illustrates typical envelope (received signal strength) variations as an MS moves.

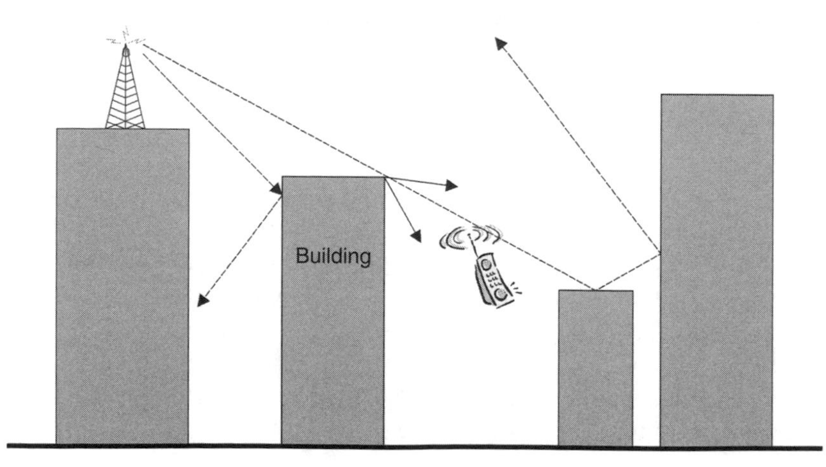

Figure 5.7. Radio propagation in urban areas

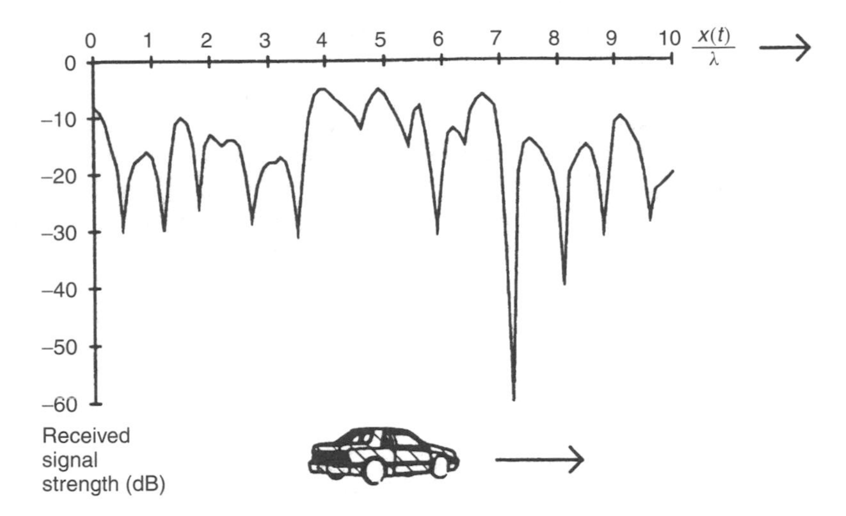

Figure 5.8. Example of signal envelope variations caused by multipath propagation

The pdf of fast fading in urban environments can mostly be described by a Rayleigh distribution

$$p(r) = \frac{r}{\sigma^2} e^{-r^2/2\sigma^2} \quad r \geq 0$$

where σ^2 is the mean local signal level.

In LOS conditions, a Rician distribution better characterises the fading profile.

5.1.4 Basic TCH Link Performance with Frequency Hopping

This section describes the basic link level performance of TCH logical channels. The PDTCH one, which is used for packet-switched traffic, is studied in detail in the first section of Chapter 7.

Specifications of the required performance of individual GSM mobile stations in terms of interference tolerance and sensitivity for certain *test conditions* are described in [9]. The required performance in the case of interference can be seen in Table 5.2 for frequency hopping (FH) and non-hopping (NH). It is shown for the typical urban (TU) [4] channel profile with two different speeds (3 and 50 km/h). The channel profile and speed combinations are denoted as TUx, where x is the speed. The factor α is a value that lies between 1 and 1.6, depending on the channel condition.

GSM specifications support a slow FH mode, which means that the frequency of the transmission is changed on a burst-by-burst basis. The uplink (UL) and downlink (DL)

Table 5.2. Reference interference performance at 9 dB C/I [GSM0505]

	TU3 NH	TU3 FH	TU50 NH	TU50 FH
Max. FER	$21\alpha\%$	$3\alpha\%$	$6\alpha\%$	$3\alpha\%$

Note: FER—Frame erasure rate.

frequencies are always duplex frequencies. Two different kinds of frequency hoppings are specified for GSM: random and sequential hopping.

Two parameters describe the hopping frequency used in each frame: the mobile allocation index offset (MAIO) and the hopping sequence number (HSN). The MAIO can take as many values as there are frequencies in the hopping list and is used to indicate the offset within the hopping list used to identify the frequency used. The hopping list is given by the mobile allocation (MA) list. The HSN can take 64 different values defining the hopping sequence used. If the HSN is set to zero, the hopping mode is sequential, while the frequencies vary pseudo-randomly when it differs from zero.

The FH algorithm is described in [3]. For a given set of parameters, the index to an absolute radio frequency channel number (ARFCN) within the mobile allocation is obtained with the following algorithm, where FN is the frame number and N is the number of frequencies:

```
if HSN = 0 (cyclic hopping) then:
MAI, integer (0.. N-1) : MAI = (FN + MAIO) modulo N
else:
MAI, integer (0.. N-1) : MAI = (S + MAIO) modulo N
```

where S is calculated from the HSN, time parameters, the number of frequencies and a predefined table, which can be found in [3].

Two sequences with the same HSN, but different MAIOs, will never overlap each other, i.e. the hopping sequences are orthogonal. Two channels having different HSN, but the same frequency list and the same timeslot, will interfere with each other only in $1/m$ of the bursts, where m is the number of frequencies in the hopping sequence. This way the interference can be averaged out over the network [2].

Some control channels have some restrictions with regards to FH. For example, the broadcast control channel (BCCH), which takes up one timeslot on the BCCH transceiver (TRX), cannot use FH. The reason for this is that the BCCH frequency has to be fixed in order for the mobiles to be able to find it.

The activation of FH in GSM improves link performance due to two main effects, frequency diversity and interference diversity.

5.1.4.1 Frequency Diversity

Frequency hopping increases the fading decorrelation among the bursts that compose each frame. This is due to the different fast-fading profiles of different frequencies. As the decorrelation increases, the efficiency of the channel coding will increase.

Figure 5.9 shows the gain from sequential and random hopping under *test conditions* [10]. The TU channel profile with 3 and 50 km/h has been used. The maximum hopping frequency diversity gain of 7 dB for slow moving mobiles is achieved with sequential hopping and eight frequencies (equal to the interleaving depth). This is also found in [4]. The gain of random hopping for TU3 would slowly continue increasing with larger number of frequencies getting closer to the maximum value of 7 dB. The largest relative gain increase is already achieved with a limited number of frequencies, i.e. four frequencies achieve over 75% of the gain of eight frequencies. It can also be seen that for the TU50 *test conditions*, the gain from FH is marginal. This is due to the fact that the

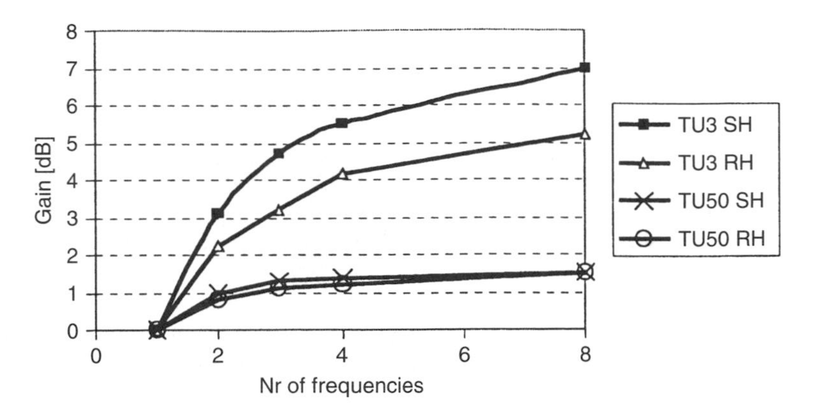

Figure 5.9. Gain at 2% FER from random and sequential hopping (RH and SH) for a TU3 and TU50 in test conditions

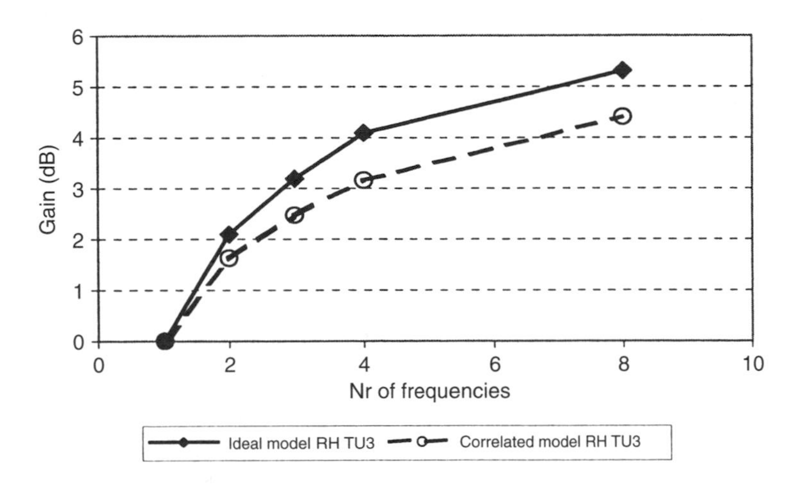

Figure 5.10. TU3 random hopping frequency diversity gain with correlated frequencies

higher speed of the mobiles already gives frequency diversity without FH. The bit error rate (BER) is the same for all hopping configurations, since the BER is calculated before the decoding with no gain from FH.

Random hopping gives approximately 1 to 2 dB lower gain than sequential hopping for a lower number of hopping frequencies [9]. However, random hopping still provides the best overall network performance once the improvement from interference diversity is taken into account [11] and therefore it is the most utilised hopping scheme in practical deployments.

The previous figure assumed that the different frequencies used had a fully uncorrelated fading profile. In practice, the hopping frequencies will have a certain correlation factor depending on the frequency separation between them and the channel profile in use [12]. Figure 5.10 represents the frequency diversity gain for TU3 and random FH when the frequencies in the hopping list are consecutive, i.e. the effective hopping bandwidth is limited to 200 kHz * (*number_of_hopping_frequencies*). For configurations between four and eight frequencies, there is roughly 1-dB degradation.

5.1.4.2 Interference Diversity

When using random hopping sequences in the network, or non-grouped frequency planning sequential hopping, the GSM network performance experiences an additional gain. This gain is referred to as *interference diversity gain*, and is the consequence of a more homogeneous distribution of the interference in the network.

In order to achieve the random interference effect, the sites in a certain area should be allocated different HSNs, which would change their instantaneous frequencies in use, independently of the rest of the sites. Therefore, the interference received by a certain mobile will not always be originated from the same source, as in the case of non-hopping or grouped sequential hopping, but it will change with every TDMA frame (4.615 ms).

Apart from the benefits of the reduced worst-case interference for the poor quality users, random hopping also improves the performance of the GSM channel coding and interleaving. In the GSM decoder, both the mean and the variance of the errors in the frame have an effect on the decoding success. Thus, it is beneficial to reduce the number of errors, as well as to make sure that these errors are spread along the frames. Random FH plays a role in both these key issues.

Two different components can be identified in the interference diversity gain:

- *Interference averaging effect.* This gain is related to the location change of the interference source, which distributes the overall interference in the network between all the mobiles. As illustrated in Figure 5.11, without random hopping, one mobile would suffer from heavy interference, while the others are in excellent conditions. By using random frequency hopping, the heavy interference is spread out over all the mobiles,

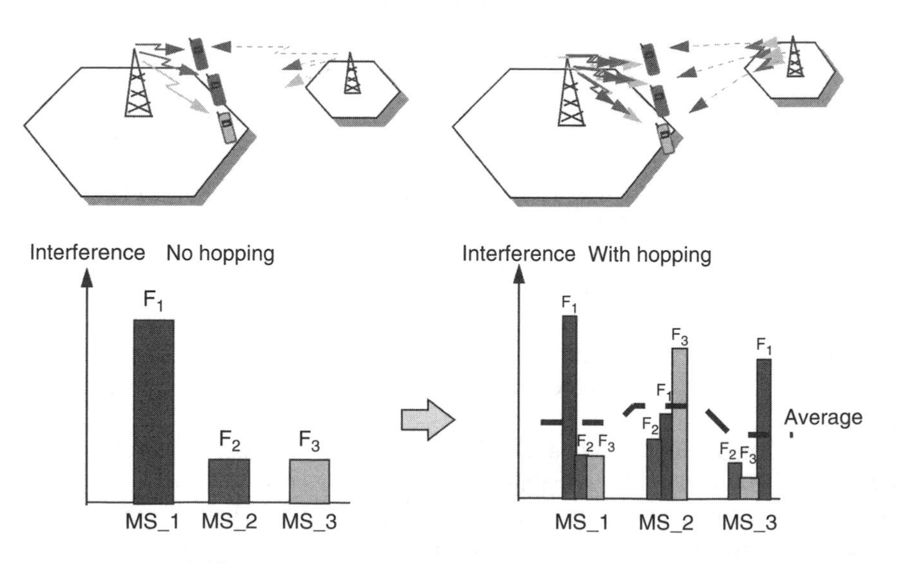

Figure 5.11. Interference averaging by random frequency hopping

and the result is a similar average interference for all of them. This interference-averaging effect does not degrade much the quality of the best performing mobiles while greatly improving the quality of those with high interference.

- *Fractional loading gain.* Fractional loading gain is obtained when not all frequencies in the hopping lists are constantly used by the TRXs of a sector, i.e. the frequencies are not 'fully loaded.' This can be achieved using random hopping with more frequencies included in the hopping list than effective TRXs in use. Because of this effect, the interference is spread across the available spectrum further randomising the probability of interference collision, hence coding and interleaving schemes become more efficient and there is an effective link performance gain. Figure 5.12 shows the fractional load gain of a TU3 random frequency hopping scenario with different frequency loads[2] and hopping list lengths.

5.1.5 Discontinuous Transmission (DTX)

Voice activity detection (VAD) is a feature in GSM, which can be used to omit transmission when a silent period is detected during a speech connection. This is called discontinuous transmission (DTX). The network operator controls the activation of DTX.

When an MS is not in DTX mode, it sends and receives 100 TDMA bursts per SACCH multiframe (0.48 s), which can be used for the estimation of the RXQUAL [13]. When a mobile station is in DTX mode, i.e. there is a silent period, only 12 TDMA bursts are sent in 480 ms. The burst transmission in DTX mode is shown in Figure 5.13.

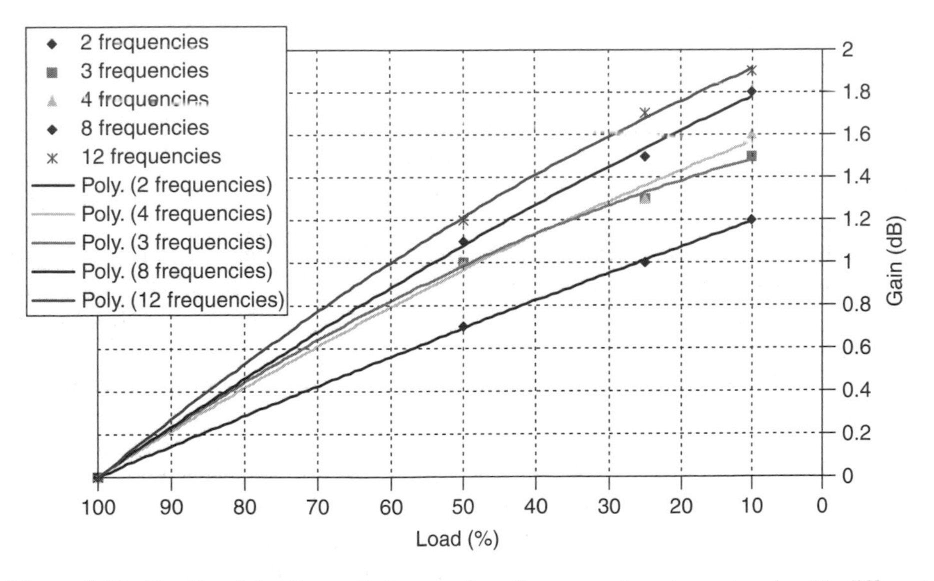

Figure 5.12. Fractional loading gain in a random frequency hopping network with different frequency hopping list lengths

[2] Section 5.3 introduces the concept and the numeric definition of Frequency Load.

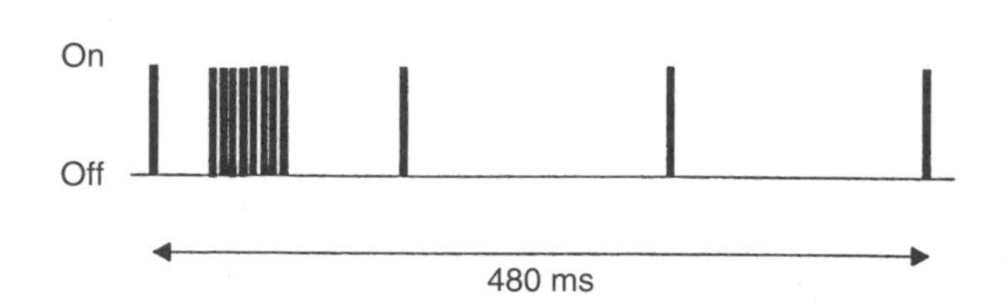

Figure 5.13. Schematic time–power diagram of a SACCH multiframe in DTX mode [7]

Speech detection is carried out at the transmitting end by the earlier-mentioned VAD, which distinguishes between speech superimposed on environmental noise and noise without speech being present. The output of the VAD is used to control the transmitter switch. If the VAD fails to detect every speech event, then the transmitted speech will be degraded due to clipping. On the other hand, if the VAD identifies noise as speech often, then the effectiveness of DTX is diminished. Both these result in degraded performance [14].

At the receiving end, the background acoustic noise abruptly disappears whenever the radio transmitter is switched off. Since switching can take place rapidly, it has been found that this noise switching can have a very negative effect on the end-user perception. In bad scenarios, the noise modulation greatly reduces the intelligibility of the speech. This problem can be overcome by generating at the receiver a synthetic signal known as 'comfort noise' whenever the transmitter is switched off. If the comfort noise characterisation is well matched to that of the transmitted noise, the gaps between talk spurts can be filled in such a way that the listener does not notice the switching during the conversation. Since the noise constantly changes, the comfort noise generator should be updated constantly [15].

From the 12 bursts, which are sent in DTX mode, 4 are the SACCH frame, which is being used for signalling. The eight others contain the silence descriptor frame (SID frame) refreshing the comfort noise characteristics.

DTX has an influence on the estimation of RXQUAL. RXQUAL is measured by the mobile station for the downlink and by the base station for the uplink. The estimated RXQUAL values can be averaged before they are used in the power control and the handover algorithm.

The estimation is done by evaluating the BER before decoding over a SACCH multiframe (0.48 s) and then maps the value over to an RXQUAL value. The way the BER is estimated is not specified; however, its accuracy has to fulfil the specifications requirements [10].

Figure 5.14 shows one way of estimating the BER. A is the frame, which has been transmitted through a radio channel and which has deinterleaved, but not yet decoded. If this frame is compared with the original encoded frame (B), we get the actual BER before decoding, called the *real BER*. The estimated BER is found by encoding the received frame after it has been decoded and then comparing it with A, i.e. comparing A and C. C is the estimate of B. If the decoder can correct all the errors in A, C will be the same as B, so the estimated BER is the same as the real BER. However, if errors are still there after the decoding, the estimated and real BER might differ.

Since an RXQUAL value is calculated for every SACCH multiframe, the BER is estimated over 100 TDMA bursts in the case of a non-DTX mode and 12 TDMA bursts if in a DTX mode. The first is called *RXQUAL-FULL* and the second *RXQUAL-SUB*.

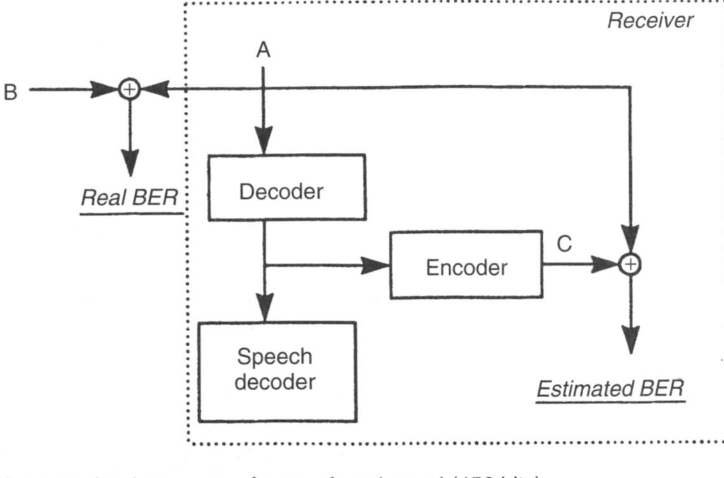

Figure 5.14. BER estimation before decoding method. (There are other methods to estimate the BER and RXQUAL)

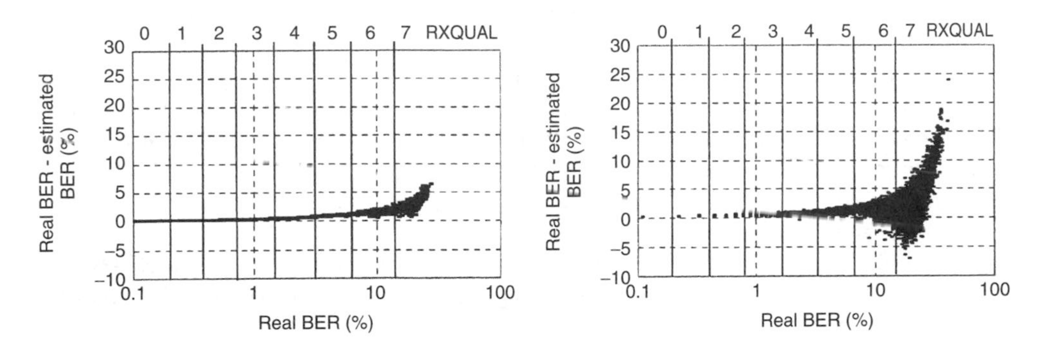

Figure 5.15. Error of estimated BER as a function of the real BER in non-DTX (a) and DTX (b) modes

Figure 5.15 shows the difference between the real BER and the estimated BER as a function of the real BER for both DTX mode and non-DTX modes. A GSM link simulator has been used to generate the results. Each of the operations in the GSM transmission path, including a fading radio channel and thermal noise (white Gaussian), is included. A typical urban channel (TU) [10] with a speed of 50 km/h has been used with random hopping over eight uncorrelated frequencies. Every point in the DTX mode case is averaged over 12 bursts, while 100 bursts are used in the non-DTX mode case.

It can be seen that the estimation error is highest for high *real BER* and for the DTX mode case compared to the non-DTX mode case, i.e. the variance in DTX mode is higher. The reason for this is the lower number of bursts, which are used for the estimation.

The estimation error is not symmetric. The average estimation error is positive. In the non-DTX mode there are no negative points at all, while in the DTX mode the error is clearly more positive than negative. This means that the estimated BER is lower than the real BER.

An explanation can be found in Figure 5.14. As long as all errors in *A* are corrected in the decoder, no estimation error will be made. However, if errors are still there after corrections in the decoder, this will introduce errors in *C*. These errors are not just errors on random places, since the error patterns in *B* and *C* are quite correlated. If the first 10 bits of *B* are corrupted and the decoder can only correct 5 of them, then it is very likely that there are some errors in the first 10 bits in *C*. The estimated BER will be lower, since some errors of *B* are also present in *C*, so they are not detected.

Sometimes, however, the error will be negative, especially for a high real BER. This is due to the fact that the error patterns become more random. No negative error can be seen in the case of the non-DTX mode, since in this case the averaging over 100 bursts removes this, while when averaging over 12 bursts, some negative values still exist. However, if the real RXQUAL is compared with the estimated RXQUAL (based on the estimated BER), then it is found that only a small percentage of the values differ. The difference is never more than 1 RXQUAL value. When compared with the specifications [10], this means that the requirements are 100% fulfilled.

5.1.6 Power Control

Power control is a feature that tries to reduce the interference. In the uplink it has the additional advantage of saving the battery power of the mobile station.

The values of the link quality (RXQUAL) and the link signal power (RXLEV) are used as input to the power control algorithm. Up and downlink power control act independently. The RXQUAL and RXLEV values can be weighted and averaged before being used in the power control algorithm. These averaged values are called *RXLEV-AV* and *RXQUAL-AV*. The power is regulated according to the following parameters:

- *U_RXQUAL_P*. If P × 1 out of N × 1 values (voting) of the averaged RXQUAL, (RXQUAL-AV) are lower (or better quality) than the *Upperthreshold RXQUAL-AV*, then the output power is decreased.
- *L_RXQUAL_P*. If P × 2 out of N × 2 values of the averaged RXQUAL, (RXQUAL-AV) are higher (or worse quality) than the *Lowerthreshold RXQUAL-AV*, then the output power is increased.
- *U_RXLEV_P*. If P × 3 out of N × 3 values of the averaged RXLEV, (RXLEV-AV) are greater than the *Upperthreshold RXLEV-AV*, then the output power is decreased.
- *L_RXLEV_P*. If P × 4 out N × 4 values of the averaged RXLEV, (RXLEV-AV) are lower than the *Lowerthreshold RXLEV-AV*, then the output power is increased.

If any one of the above conditions determines that the power should be increased, the power is indeed increased, even though one of the parameters may determine that the power should be decreased. This way good quality and signal level is always guaranteed.

A power increase step and a power decrease step can be specified, but if the level or the quality of a connection drops rapidly, for example, if an MS moves from outdoor to indoor, then it is possible to increase the power with one step to the desired level. The power control is only used once in a certain time interval, which can be specified, although the maximum performance is achieved at its maximum rate, once in each 480 ms period.

5.2 Cellular Network Key Performance Indicators (KPIs)

5.2.1 Speech KPIs

The performance of cellular radio networks and, in particular, of speech services can be measured using multiple different key performance indicators (KPIs). Furthermore, many of these KPIs can be collected either by drive tests or from network management system (NMS) statistics. Since NMS statistics represent the overall network performance, they are more representative than drive test results. Drive-test-based KPIs should only play a relevant role in monitoring the overall network performance when such KPIs are not available from NMS statistics. Additionally, statistics can represent busy hour or daily average performance. This book uses busy hour values in order to analyse the performance in high load conditions.

Traditionally, cellular operators have measured and benchmarked their network performance using mostly the BER and the dropped call rate (DCR) to quantify the speech quality and the rate of lost connections respectively. Additionally, call success rate (CSR) and handover success rate (HSR) have been used to measure the performance of the signalling channels associated with call originations and handovers. There are, however, a number of additional KPIs that can be used to measure the network performance, and in particular, the speech quality. Figure 5.16 illustrates the different KPIs and the location in the receiving chain where they are measured. These indicators are not only utilised to measure the network performance but can also be used by the network for multiple radio resource management functions, such as handovers, power control, adaptive multi-rate codec (AMR) link and channel mode adaptation, GPRS link adaptation, etc.

An important aspect to consider when measuring and benchmarking the overall network performance is the significant variance associated with the performance of different terminals. This is mainly due to differences in radiated power and sensitivity, which can potentially impact the link budget balance and therefore the experienced performance.

Figure 5.17 displays a typical performance distribution for different terminal models taken from the most commonly used ones in a commercial network (Network 14).

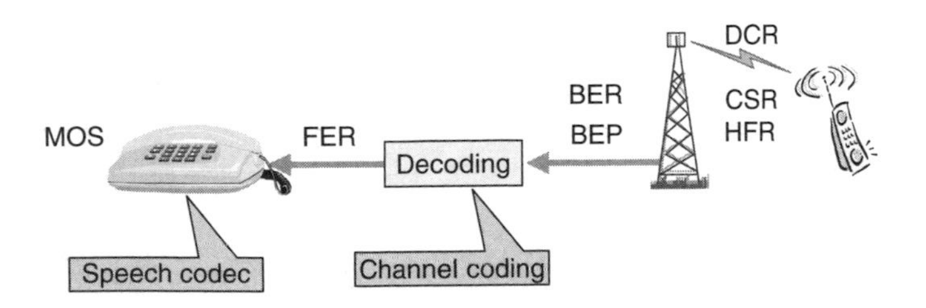

Figure 5.16. Main speech KPIs

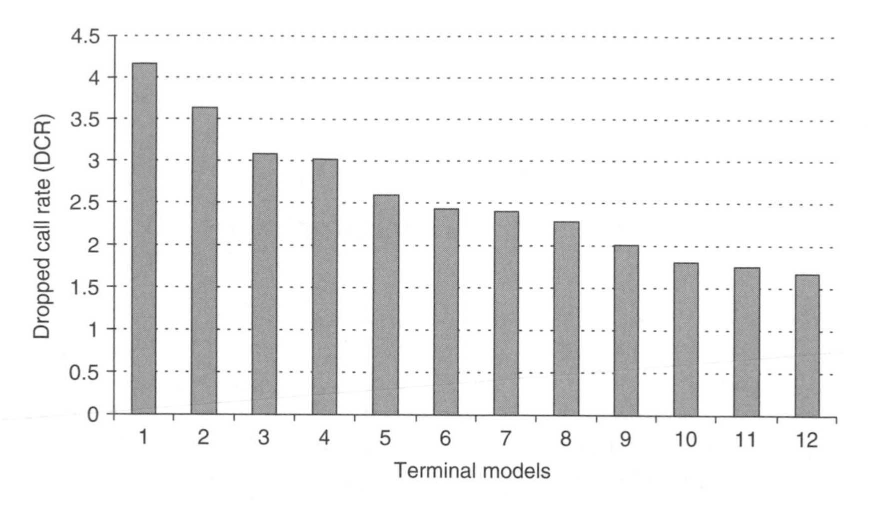

Figure 5.17. DCR Performance of different terminal models (Network 14)

5.2.1.1 Mean Opinion Score (MOS)

Speech quality can be quantified using mean opinion score (MOS). MOS values, however, can only be measured in a test laboratory environment. MOS values range from 1 (bad) to 5 (excellent). They are generated from the subjective voice quality perceived and reported by the group of listeners participating in the test. Additionally, listening tests performed with different conditions (like speech material, language used and listening conditions) could lead to different MOS results. Therefore, absolute MOS values between the different tests cannot be directly compared. A thorough description of this KPI and the methodology used to measure it can be found in [16].

There are, however, multiple tools in the market that try to measure and quantify the subjective speech quality of the link from data collected through drive tests using different algorithms.

5.2.1.2 Bit Error Rate (BER)

BER is a measurement of the raw bit error rate in reception before the decoding process takes place. Any factor that impacts the decoding performance, such as the use of FH or the robustness of the channel coding associated with a certain speech codec, will impact the correlation between BER and Frame Erasure Rate (FER), or the perceived end-user voice quality. Therefore, in order to properly measure the link quality using this KPI, this correlation will have to be considered for all the existing codecs and channel profiles. In GSM, BER is mapped into RXQUAL, the values of which range from 0 to 7 as displayed in Table 5.3.

5.2.1.3 Frame Erasure Rate (FER)

Frame erasure rate is a very powerful performance indicator since it is highly correlated with the final voice quality that the end user perceives. Different speech codecs will have, however, a slightly different FER to MOS correlation since the smaller the speech

Table 5.3. RXQUAL values and corresponding BER

RXQUAL	BER (%)	Assumed value (%)
0	<0.2	0.14
1	0.2–0.4	0.28
2	0.4–0.8	0.57
3	0.8–1.6	1.13
4	1.6–3.2	2.26
5	3.2–6.4	4.53
6	6.4–12.8	9.05
7	>12.8	18.1

codec bit rate is, the more sensitive it becomes to frame erasures. According to the studies presented in [17], the FER values in which different speech codecs will begin to experience MOS degradation, and the rate of such degradation is quite uniform. Therefore, FER can be efficiently used as a speech quality performance indicator.

The traditional problem associated with the usage of FER as a KPI in GSM systems has been the lack of FER-based NMS statistics. Therefore only drive test data have been available to measure the FER of the system. FER is, however, available to the network in the uplink (UL), and the downlink (DL)FER could be estimated from the available information in the network. Rel'99 specification includes the enhanced measurement report, which together with other enhancements brings the DL FER reporting. With this enhancement, terminals will report DL FER, thus permitting the network to generate DL FER statistics based on these reports. The FER performance data presented in this section have been collected through drive tests.

Finally, it is important to define how FER data should be analysed, both in terms of the averaging to apply to the raw data and the threshold values that define the occurrence of bad quality samples. The impact of frame erasures in the speech quality degradation will not only be related with the average FER value but with the frame erasure distribution as well. The FER associated with real connections has a bursty nature. Once the channel coding is not able to restore the original information, the FER degradation is easily very high. Figure 5.18 displays a typical FER profile with 2 s (four 480 ms multiframe periods) averaging samples. When a single FER degradation event occurs, the FER values are likely to be very high, and this can have a significant impact on the average FER per call.

Figure 5.19 displays the same data processed with different FER averaging windows, in terms of percentage of samples with FER over a threshold. Surprisingly, the higher the averaging window the higher the percentage of samples over the selected threshold. This is due to the previously displayed bursty nature of the FER. The use of high averaging windows may lead to erroneous conclusions about the associated speech quality. Therefore, in order to be able to determine the speech quality from the FER data in a better way, small averaging windows should be selected.

FER statistics should have a threshold defined, which determines the level of non-acceptable speech degradation. For the recommended small average windows, this threshold will be in the range of 2 to 8%. The percentile of FER samples above this threshold will represent the proportion of bad speech quality samples. Figure 5.20 displays the outage of 2 s samples with FER over different thresholds.

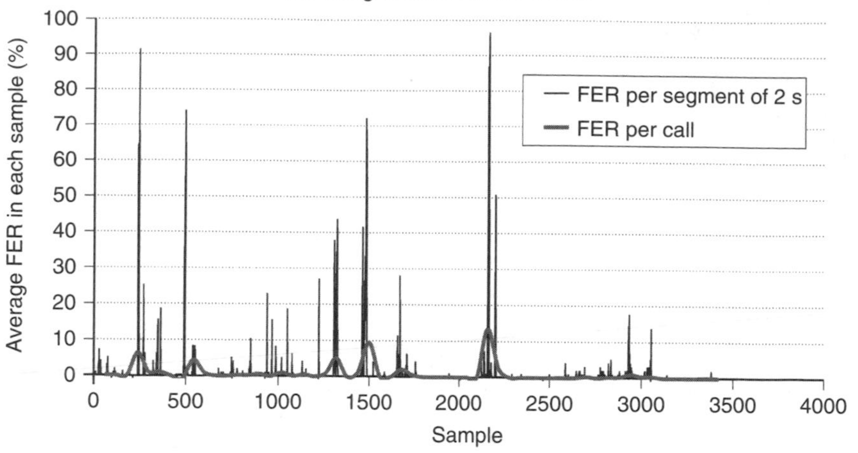

Figure 5.18. FER instantaneous distribution (Network 1)

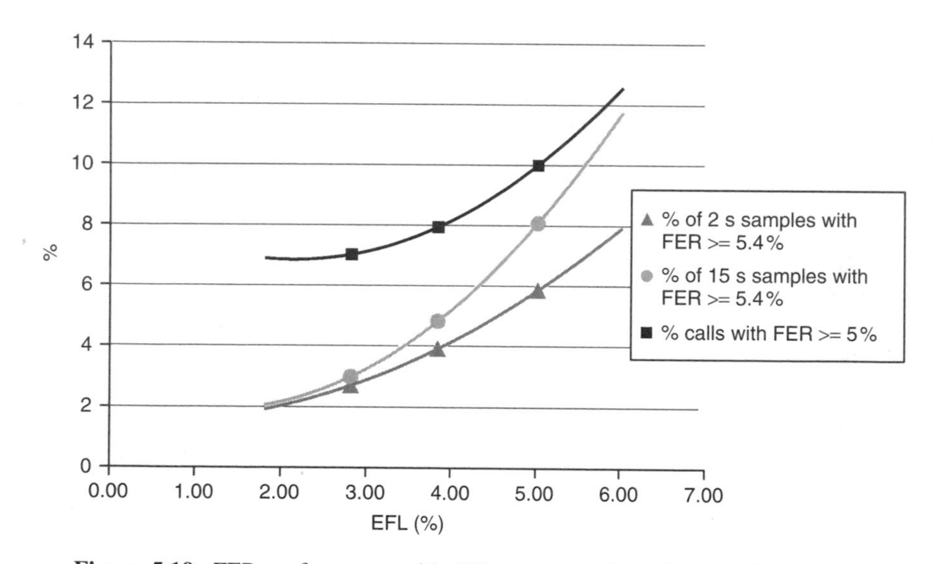

Figure 5.19. FER performance with different averaging windows (Network 1)

5.2.1.4 Bit Error Probability (BEP)

Rel'99 includes an additional signal quality indicator, the bit error probability. The terminals will report the mean BEP and its coefficient of variation (standard deviation/mean value). According to [13], the received signal quality for each channel shall be measured on a burst-by-burst basis by the mobile station and the base station subsystem (BSS) in a manner that can be related to the BEP for each burst before channel decoding using, for example, soft output from the receiver. The scale of the reported mean value will range from 1 to 32 and that of the coefficient of variation will range from 1 to 8.

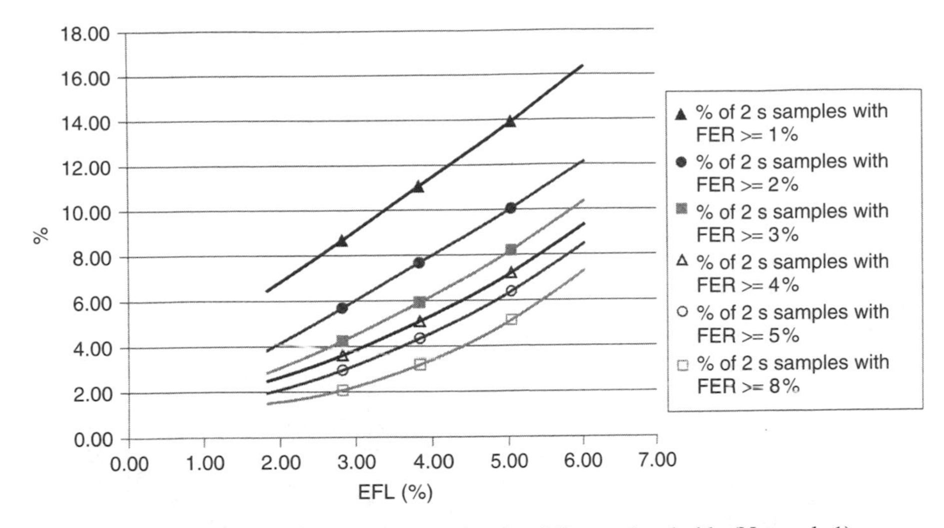

Figure 5.20. Outage of FER samples for different thresholds (Network 1)

BEP has a higher accuracy than RXQUAL and its coefficient of variation can be used to characterise the channel profile.

5.2.1.5 Dropped Call Rate (DCR)

DCR measures the percentage of connections lost and it has always been a very important KPI since it is generally considered that a dropped call has a very negative impact on the end-user-perceived quality of service (QoS). In GSM, the mechanism of one of the main triggering events of a speech dropped call relies on a counter (radio link timeout) that counts down or up depending on the successful reception of the SACCH frames. This mechanism was designed to drop connections that the link had degraded below a certain threshold. Therefore, although DCR is correlated with the end-user-perceived voice quality, it is rather an indication of the performance of the SACCH signalling link.

There are different ways to calculate DCR. It can be calculated as dropped calls per Erlang, dropped calls over originated calls and dropped calls over all calls handled by each cell, including incoming handovers. In this book, we will use dropped calls over originated calls only, since it is the most common way in which operators calculate DCR.

Figures 5.21 and 5.22 shows the DCR performance associated with multiple mature live networks measured from NMS statistics and drive tests.

Although both the drive test and the NMS statistics belong basically to the same set of networks, the average DCR across all the measured networks is 1.8% in the case of NMS statistics and 2.8% in the case of drive tests. This difference is mostly related to the drive test route selection, which is generally designed to cover all the potential problematic areas, and has a high handover per call ratio. Additionally, there is a high variance in the drive test results due to the small amount of events (number of calls) collected to achieve a statistically significant result. Figure 5.23 shows the DCR values collected from NMS and drive tests for the same network during multiple days. The average DCR from drive

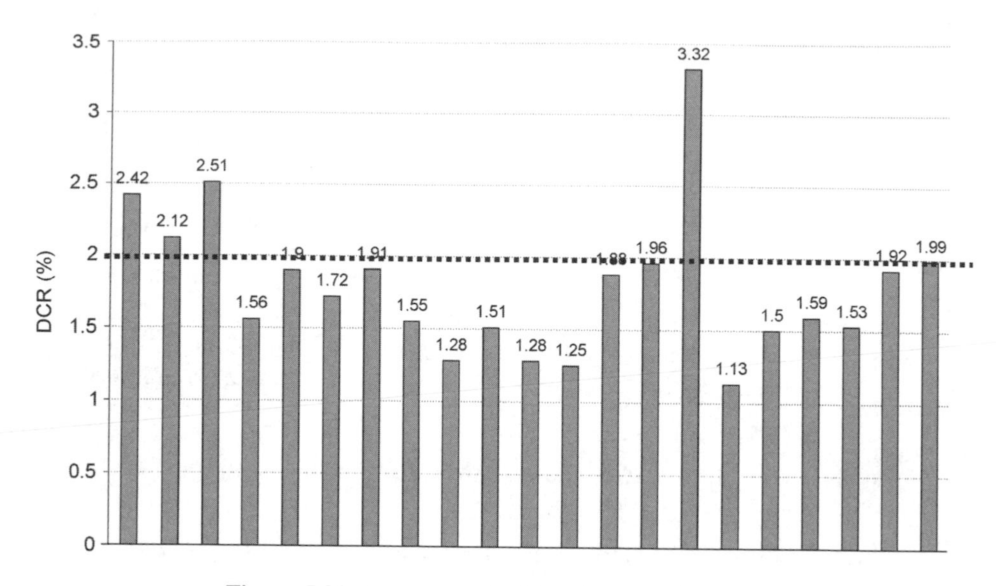

Figure 5.21. Real networks DCR from NMS statistics

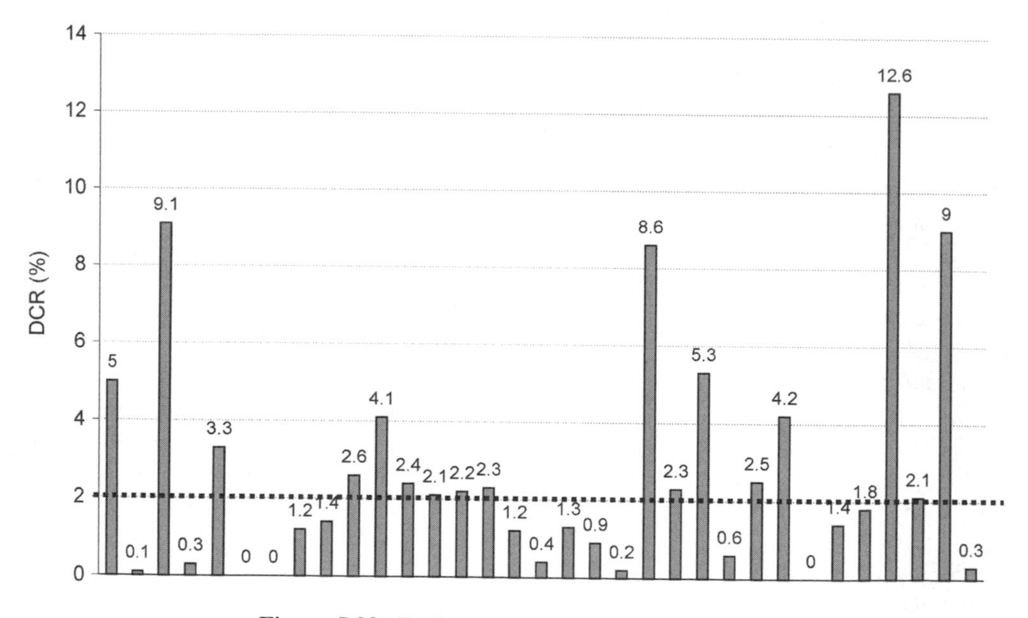

Figure 5.22. Real networks DCR from drive tests

tests was in this case 2.5 times higher than the one collected from NMS statistics and there is an obvious high variance in the drive test data, which may lead to erroneous conclusions.

In conclusion, the DCR should always be measured from NMS statistics, and well-performing networks should have a DCR ranging from 1 to 2% depending on the baseline performance (see Section 5.5), features in use and the network load.

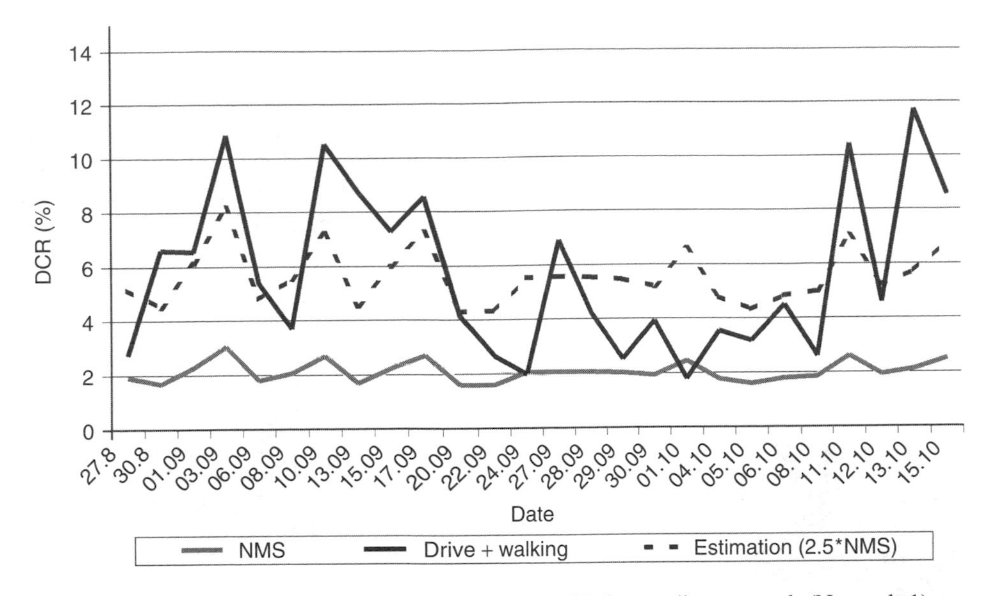

Figure 5.23. Drive tests and NMS statistics—DCR from a live network (Network 1)

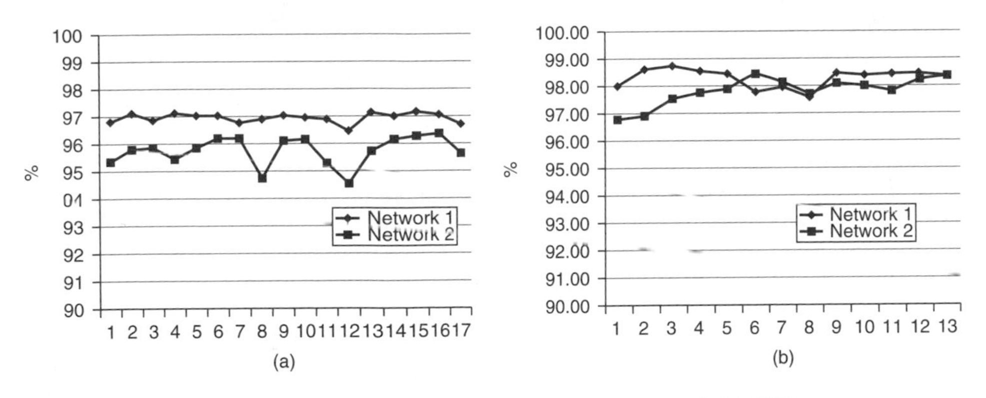

Figure 5.24. Live networks busy hour (a) CSR and (b) HSR

5.2.1.6 Call Success Rate (CSR) and Handover Success Rate (HSR)

Finally, there are other KPIs, such as call success rate (CSR) and handover success rate, which measure the success rates of the signalling process related to originations and handover, respectively. Figure 5.24 displays the values collected for different days and configurations from live networks for CSR and HSR. Well-performing networks should have both the CSR and the HSR above 95%.

5.2.1.7 KPIs Correlation

There is no bad performance indicator, since all of them point out different aspects of the network performance. The important factor is to be able to understand what each one represents and the correlation between the different relevant indicators. Additionally, such correlation may be dependent on different factors. For instance, BER can be used to measure speech quality but in order to analyse the voice quality from the BER data, the correlation between BER and speech quality is needed. This correlation is not constant, since it is dependent on the decoding gain, which is dependent on the channel coding and the channel profile in use. Figure 5.25 displays the significantly different correlation between BER and FER for hopping and non-hopping channels, measured in a live network.

Finally, Figure 5.26 displays the correlation of the most common KPIs for the same network and different load levels. The BER and FER data have been collected from drive

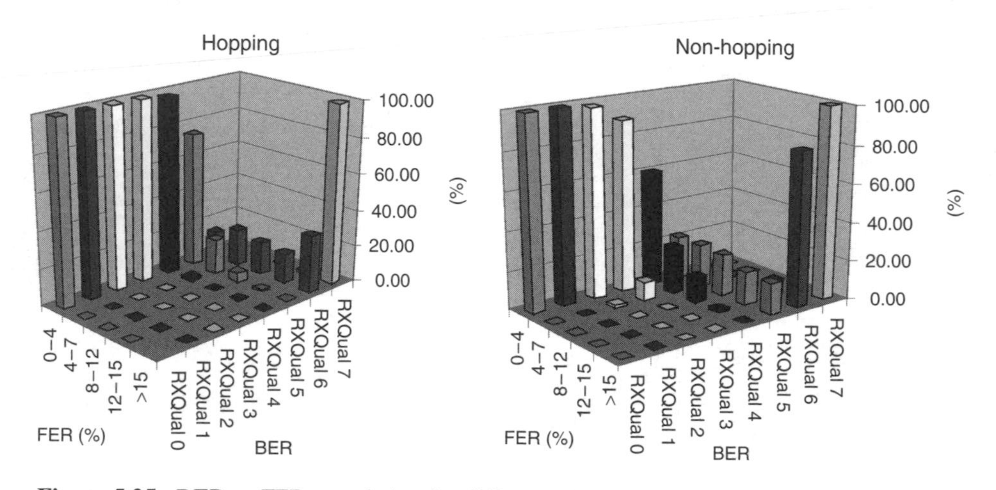

Figure 5.25. BER to FER correlation for different channel profiles (hopping/non-hopping) (Network 1)

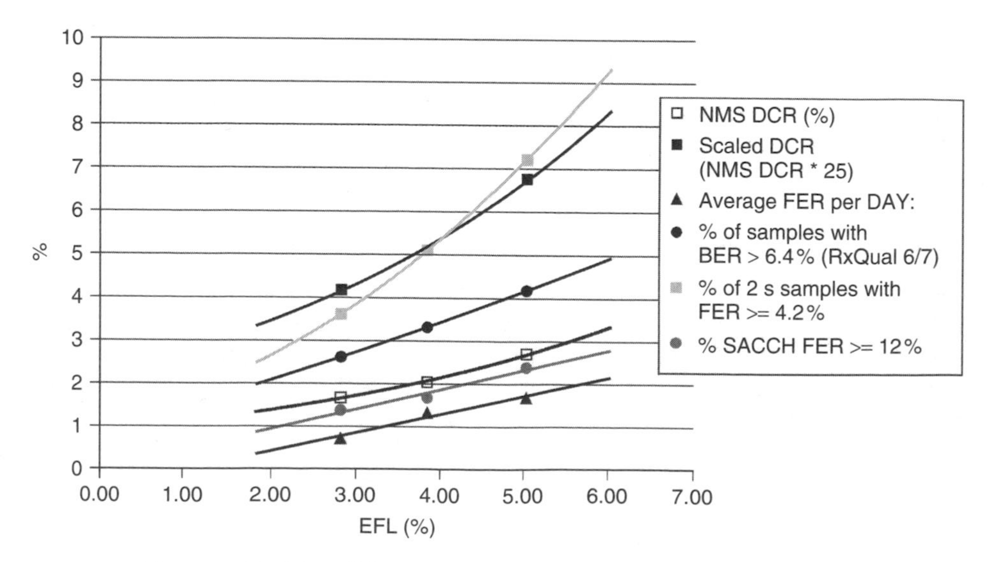

Figure 5.26. Main KPIs correlation from a live network (Network 1)

tests. The statistically significant NMS DCR statistics have been used. As presented in Figure 5.23, the DCR from the drive tests routes is higher than the overall network DCR, so the NMS DCR values need to be scaled accordingly in order to obtain the estimated DCR performance associated with the drive tests routes and perform the KPI correlation.

The percentage of 2 s samples with FER equal to or higher than 4.2% is highly correlated to the percentage of dropped calls. For the purpose of this book, 2% DCR and 2% outage of 2 s FER samples with FER equal to or higher than 4.2% (2% bad quality samples outage) have been selected in order to benchmark the network performance. From multiple analysis performed, BER and FER have shown a very linear degradation as the interference (load) increased. On the other hand, DCR tends to have a linear degradation up to a certain 'saturation point', when the degradation starts to become exponential. This is due to the nature of the radio link timeout and the other DCR triggering mechanisms.

5.2.2 *Data KPIs*

GSM/EDGE data services can be offered in circuit-switched (CS) mode with high speed circuit-switched data (HSCSD)/enhanced circuit-switched data (ECSD), and in packet-switched (PS) mode with (E)GPRS. This book will focus mostly on PS services delivered through the (E)GPRS PS core network. Measuring (E)GPRS performance is somehow different from measuring voice performance. The end-user perception will be determined in most data services by the application interfaces. Such interfaces will filter events such as dropped connections, which in the case of speech services are critical in the end-user quality perception, so end users only perceive delays and effective throughputs. There are many different data services with different performance requirements. Chapter 8 introduces a set of the most relevant packet data services, their associated performance indicators and requirements, and the end-user experience with the use of different radio technologies. This section concentrates on the introduction and analysis of network-performance-related KPIs, which are generic for any given service and identify how efficiently the network is delivering packet-switched traffic.

The performance requirements are agreed between the network and the mobile station at the service establishment phase. The radio resource management should then do as much as possible to maintain the negotiated quality. Data KPIs can be measured at different levels and in different places. The radio access network has control of the performance below logical link control (LLC) level (or packet data convergence protocol (PDCP) layer in GERAN Rel'5), but it cannot control what is happening on the core network side. The most important (E)GPRS radio link performance parameters in the radio access network are reliability, throughput and delay.

5.2.2.1 Reliability

Reliability describes the maximum probability of erroneous radio link control (RLC) delivery to the LLC layer. Reliability depends mainly on the operation mode of the RLC layer. In most of the cases, errors are not tolerated by data applications and retransmissions are needed. In order to avoid further delays, retransmissions are done preferably in the RLC layer by operating in acknowledge (ACK) mode. In RLC ACK, there is a small probability of undetected block errors due to the limited CRC field length (16 and 12 for GPRS and EGPRS respectively) included in each radio block. The residual BER in RLC

ACK mode is 10^{-5} and 2×10^{-4} for GPRS and EGPRS respectively. There are some delay-sensitive data services that operate in the UNACK mode and can handle a certain number of erroneous RLC blocks. The mode of operation is negotiated at the service establishment phase.

5.2.2.2 Throughput

Throughput is measured as the amount of data delivered to the LLC layer per unit of time, i.e. RLC payload. Throughput is only relevant when the amount of data that is being transmitted is large enough and can be easily measured by the network over the duration of a temporary block flow (TBF) as displayed in Figure 5.27. Each TBF has an associated measured throughput sample. Each network will have a different throughput probability distribution depending on the load and network configuration. Instead of using the whole probability distributions, it is more practical to compute the average and percentile throughput values. The typical variation with load of the average TBF throughput can be seen in Figure 5.28. Throughput percentile values characterise the tails of the throughput distribution. Tenth percentile (minimum throughput achieved by 90% of the TBFs) is considered in Chapter 7 as one of the main quality criterion. Real-time services with guaranteed throughputs require instantaneous throughput measurement at RLC level that can be done with a running average filter.

User throughput can also be measured at the application layer, for example, by measuring the downloading time of a large file using the file transfer protocol FTP. This end-to-end measurement should be done in one of the extremes of the communication (either at the server or at the mobile station). This measurement can be done with a simple application that monitors application layer throughputs. For real-time services, like streaming, instantaneous throughput can be measured with a running average filter at application level.

5.2.2.3 Delay

Delay is the time it takes for an LLC protocol data unit (PDU) to be transferred from the serving GPRS support node (SGSN) to the mobile station or from the mobile station to the SGSN. LLC level delay includes the required time to re-transmit erroneous RLC blocks. LLC delay probability distribution can be measured at the Packet Control Unit PCU. Delay is relevant for short packets transmission and for real-time data services. In the RLC ACK mode, the delay is very dependent on the radio access round trip time (RA RTT)

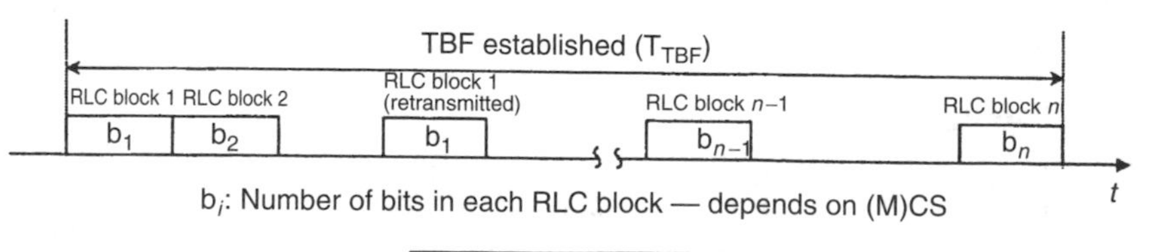

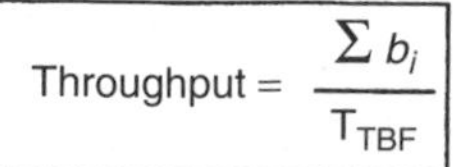

Figure 5.27. Temporary block flow throughput

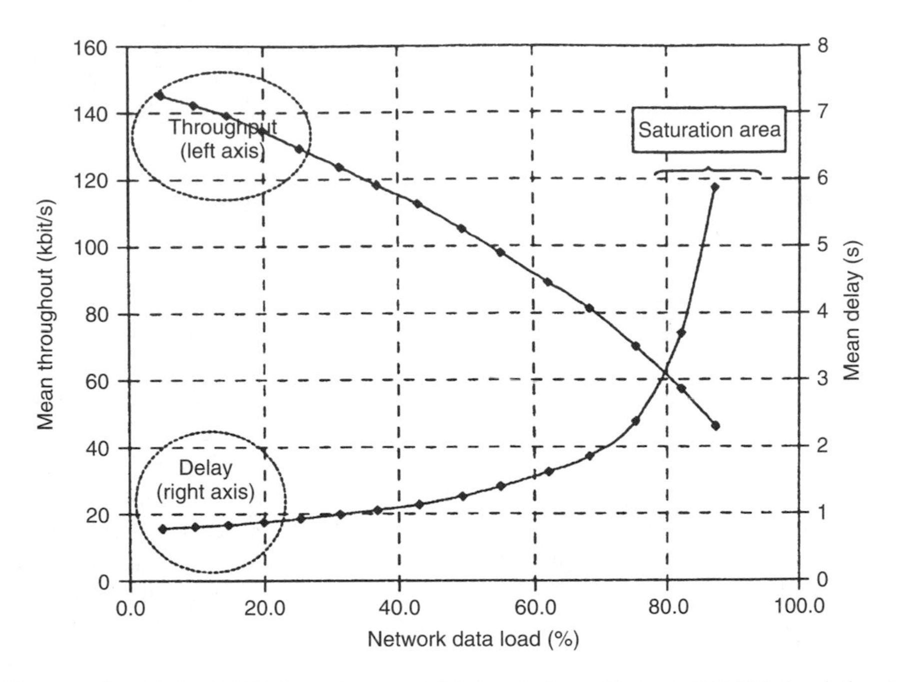

Figure 5.28. Typical TBF throughput and LLC variation with load (EGPRS simulations)

(see Section 7.6), and it may have a considerable impact on the application performance. RA RTT defines the total time since an RLC packet is transmitted until ACK/NACK information is received. The polling strategy (see Chapter 7) plays an important role in RA RTT. Its typical value is of the order of 250 ms including TBF establishment delay. With adequate network architecture optimisation, it is possible to minimise the RA RTT. As the data load increases, the LLC delay increases smoothly until the network is close to saturation point, where LLC delays are so large that it is not possible to offer any data service efficiently. The evolution of the delay with load is displayed in Figure 5.28. For real-time data services working in RLC NACK mode, the delay is minimised by allocating the maximum priority in the scheduler.

Delay can be measured at the application layer as well. A common way of measuring (E)GPRS network response time is by sending PING commands to a server. A heavily loaded network may have too long response times that may disrupt application and transport layer protocols.

There are some data performance indicators that are defined at network level and do not refer to the performance or individual users. Some of the most important are (E)GPRS load, throughput per cell, timeslot (TSL) capacity, percentage of satisfied users, TBF blocking and throughput reduction factor.

5.2.2.4 (E)GPRS Load

A useful way to measure data load is to measure the average amount of TSL utilised by (E)GPRS services, which can be expressed in terms of 'Data Erlangs' or 'TSL utilisation.' This measurement includes user data, retransmission and control information sent over

PDTCH channels. This definition is very useful in understanding how many TSL are, on average, fully utilised by (E)GPRS services. Both *Data Erlangs* and *TSL utilisation* can be easily calculated in real (E)GPRS network from the available counters. *Data Erlang* can be defined as:

$$\textit{Data Erlang} = (\text{Total number of transmitted radio blocks} \times 0.02\ \text{s})/$$
$$\times\ \text{Duration of busy hour(s)}$$

Note that 1 Erlang corresponds to a continuous transmission of 1 radio block every 20 ms.

A Data Erlang is made up of initial RLC block transmissions, successive retransmissions of previously sent RLC blocks and control RLC blocks. Retransmissions are done in the RLC acknowledgement mode. Either the network or the mobile station retransmits RLC blocks that have not been acknowledged. Retransmissions can be divided in two classes: regular retransmissions and pre-emptive retransmissions. Regular retransmissions are done when RLC blocks have been negatively acknowledged. Pre-emptive retransmissions are done for a radio block that has not been positively or negatively acknowledged (no information has been received since the radio block was sent). Those blocks are called *pending-ACK blocks*. Pre-emptive retransmissions of pending-ACK blocks are done when there is no other RLC block to transmit under the following two conditions. The first one is when the RLC protocol is stalled and the transmitting window cannot advance. The second one is when there is no more user data to transmit but acknowledgement information has not been received yet. Pre-emptive retransmissions are optional for both network and MS side. Although pre-emptive retransmissions are good from the RLC protocol point of view, they actually increase the Data Erlang in the network even though only a small fraction of the pre-emptive retransmissions are useful.

5.2.2.5 Throughput Per Cell/Sector

Throughput per cell is a measure of the amount of bits that can be transmitted in a cell. It is normally measured in kbps/cell units. (E)GPRS networks can easily calculate this indicator by measuring how much information has been transmitted during a certain period. Throughput per cell does not take into account retransmissions or control information below the LLC layer. Figure 5.29 shows an example of the variation of throughput per cell as a function of *Data Erlangs*.

5.2.2.6 TSL Utilisation

A measurement of how much hardware (HW) resources are being utilised by (E)GPRS service is given by the *TSL utilisation* concept. TSL utilisation is defined as

$$\textit{TSL utilisation (\%)} = \text{Data Erlang/available TSL for (E)GPRS.}$$

TSL utilisation takes into account the average number of TSLs that are available for (E)GPRS. It is a measure of how much the network is loaded with data services. Networks with *TSL utilisation* close to 100% are close to saturation and end-user performance is likely to be very poor. In Figure 5.28, the network data load is measured as a percentage of *TSL utilisation*.

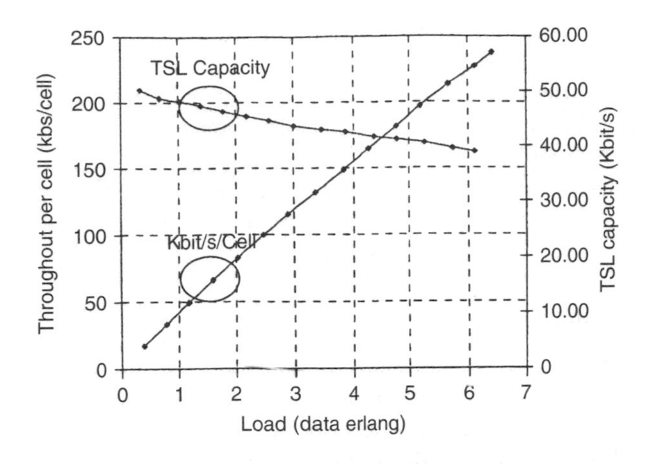

Figure 5.29. Kilobits per second/cell and TSL capacity as a function of Data Erlangs (EGPRS simulations)

5.2.2.7 TSL Capacity

TSL capacity is a measure of how much data the network is able to transmit with 1 Data Erlang. It is a measure of how efficiently HW resources are being utilised to transmit data. TSL capacity depends mainly on network interference levels. It may also depend on RLC protocol efficiency. It can be calculated as

TSL capacity = Throughput per cell/*Data Erlangs*.

TSL capacity depends on the RLC algorithm design (e.g. the use of pre-emptive transmission or not, polling strategy, etc.) and data traffic burstiness. An RLC efficiency factor can be introduced as

TSL capacity = RLC efficiency factor × Throughput per cell/Optimum Data Erlangs

where Optimum Data Erlangs is the Data Erlangs calculated excluding non-useful retransmission of pending ACK blocks. Therefore, it assumes optimum RLC protocol behaviour. The RLC efficiency factor tends to be lower with GPRS (mainly because of RLC window stalling, see Section 7.2.1) and in the case of bursty traffic (see Section 7.4.7). EGPRS with non-bursty traffic has an RLC efficiency factor close to 1 as it does not suffer that much from stalling.

A typical TSL capacity can be seen in Figure 5.29. The more loaded the network, the higher the interference levels and the lower the TSL capacity.

5.2.2.8 Satisfied Users

Even in best-effort services, it is necessary to introduce a network level indicator of end-to-end performance. The user-satisfaction criteria depend on the type of services. For best-effort data, a combination of minimum TBF throughput and maximum LLC delay can be considered. The network can produce statistics of TBF throughput and, possibly, LLC delays. The degree of user satisfaction is measured as a percentage of samples that fulfil the user-satisfaction criteria. Operators' experience and customer feedback will help to identify the percentage and the satisfaction criteria to be associated with a good performing network, in the same way that nowadays 1% DCR is considered to be good performance for speech services. In Chapter 7, different satisfaction criteria will be considered with 90% of good quality samples.

Satisfied users and throughput per cell are closely related in best effort networks. The more loaded the network the higher the throughput per cell but, on the other hand, a lower percentage of satisfied users can be achieved.

5.2.2.9 TBF Blocking

TBF blocking probability shows how close cells are to saturation. When the blocking limit is reached, packets have to be dropped in the core network. Blocking changes exponentially with network load, so once a network is experiencing blocking, small increases in the offered load will be translated into much higher blocking probabilities. With high TBF blocking, the performance collapses, and LLC delay increases exponentially, as displayed in Figure 5.30. Although there are seven TSLs available for (E)GPRS services in the network in this example, the maximum Data Erlangs will never reach seven. The EGPRS network should be dimensioned to ensure a low TBF blocking probability. This is a similar concept to speech blocking dimensioning, in which the number of TSLs is dimensioned to achieve a maximum blocking probability of 2% (typically).

5.2.2.10 Throughput Reduction Factor

Throughput reduction factor (in short, reduction factor) is a measure of how the TBF throughput is reduced when the same TSLs are shared by more than one user. By knowing the TSL capacity and the reduction factor, it is possible to compute average user throughput as

$$\text{TBF throughput} = \text{TSL capacity} \times \text{Number of allocated TSL} \times \text{Reduction factor.}$$

Low loaded cells have a reduction factor close to 1, meaning that very few TSLs are shared between the users. Ideally, reduction factor would not depend on the number

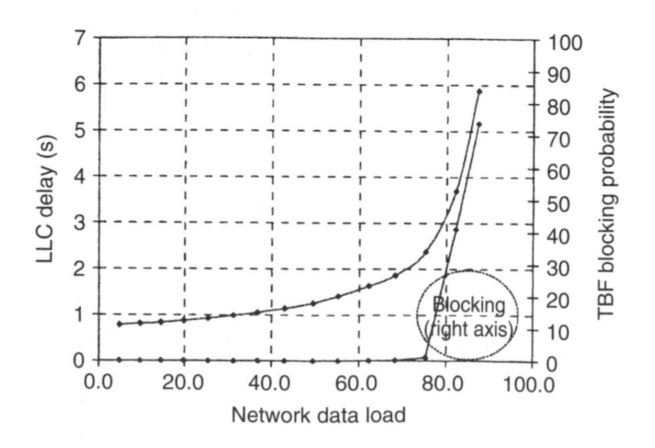

Figure 5.30. LLC delay degradation with TBF blocking (EGPRS simulations)

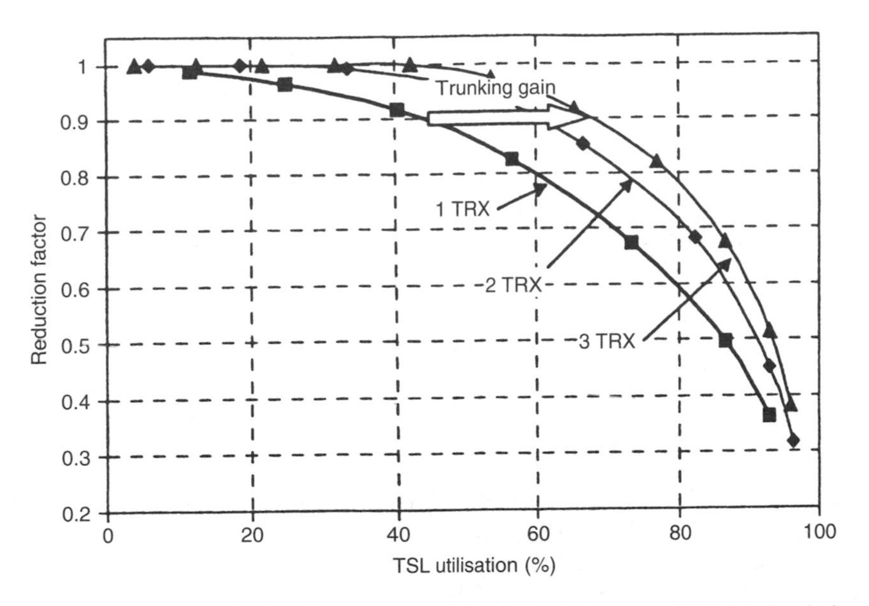

Figure 5.31. Typical reduction factor versus TSL utilisation curves (EGPRS simulations)

of allocated TSLs. In practice, with limited available TSLs and real channel allocation algorithms, there are some differences. In any case, the reduction factor concept is a simple and useful tool for dimensioning (E)GPRS networks.

Figure 5.31 shows reduction factors computed in an EGPRS network with eight TSLs available for EGPRS. Best-effort traffic is assumed. When the number of available TSLs increases, the available resources are used more efficiently and the same reduction factors can be achieved with higher *TSL utilisation*. This can be used to measure trunking efficiency gains. An in-depth (E)GPRS trunking performance analysis is performed in Chapter 7.

5.3 Spectral Efficiency

This section introduces the concept of spectral efficiency and the different ways used to quantify it. Cellular operators invest money to get frequency spectrum licences for a certain number of years. It is therefore very important to ensure maximum return on such investment. Spectral efficiency quantifies the amount of traffic a network can carry for a given spectrum. Furthermore, it is a measure of the radio performance efficiency, thus higher spectral efficiency will imply higher quality of service provided to the end users for a given traffic load.

Figure 5.32 shows the performance characterisation, in terms of load and quality, of two different networks, or a single network with a different set of functionality. Network B is more spectral efficient than Network A because, for a certain benchmarked quality, Network B can carry more traffic, and for a certain traffic load, Network B has a better performance. If certain functionality that increases the spectral efficiency is introduced in Network A, the quality will immediately improve and considering the previous quality as the benchmarking quality level, there will be additional traffic load that can be carried

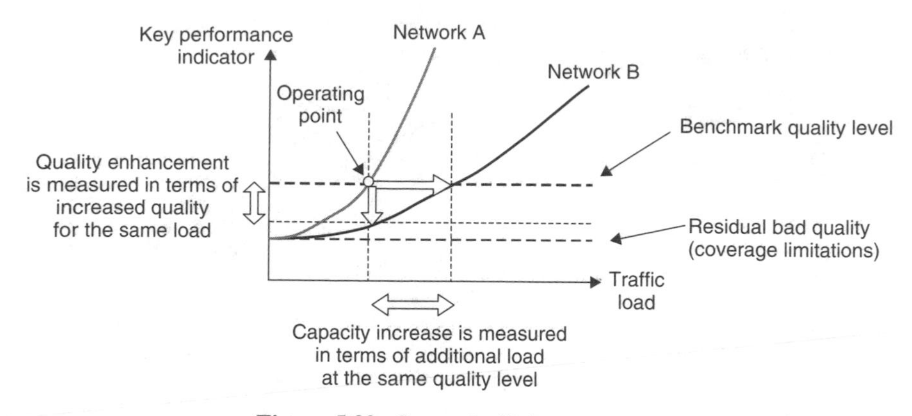

Figure 5.32. Spectral efficiency concept

by the network. The relative value of this additional traffic load is the capacity increase provided by the introduced functionality.

Even though the most traditional way to express spectral efficiency has been Erlangs per megahertz per square kilometre (Erl/MHz/km^2), there have been many other ways to quantify the same concept in the wireless industry. Some of the ways to measure it are

- effective reuse
- fractional load
- frequency load
- effective frequency load (EFL).

It is important at this point to highlight the terminology used across this book, since some commonly used terms have a different meaning in different regions worldwide:

- *Cell/sector*. Both terms represent the same concept. They refer to the logical entities that have a single BCCH associated with each one of them. Several cells/sectors can be grouped under the same site if they are fully synchronised.
- *Site*. A site refers to a single or multiple cells/sectors sharing the same physical location and being fully synchronised. The radio performance benefits from synchronised cells/sectors are better handover performance of synchronised handovers and the possibility of using MAIO management across the synchronised cells/sectors.

5.3.1 Effective Reuse

One of the common ways used to try to quantify the spectral efficiency is the effective reuse concept. Effective reuse describes how often the same frequency can be reused in the network. Basically, the capacity of the network can be estimated, if the potential effective reuse of a particular solution is known. It is calculated as

$$R_{\text{eff}} = \frac{N_{\text{freqsTOT}}}{N_{\text{TRXave}}}$$

where R_{eff} = effective reuse; N_{freqsTOT} = total number of used frequencies; N_{TRXave} = average number of TRXs per sector.

Effective reuse is a measure of the rate of the available spectrum by the number of TRXs per sector. However, this measure has some limitations:

- The deployed number of TRXs does not imply that the TRXs are fully loaded.
- The relationship between the number of TRXs and the maximum traffic carried will be conditioned by the blocking dimensioning criteria, which may vary from network to network.
- The number of Erlangs per TRX for the same blocking rate will increase as the TRX configuration increases.
- Therefore, effective reuse is not an accurate indicator to measure the spectral efficiency since the correlation between number of TRXs and carried traffic is not fixed.

5.3.2 Fractional Load

Another way to measure the efficiency of hopping networks, related to the effective reuse, is the fractional load. Fractional loading means that the cell has been allocated more frequencies than the number of deployed TRXs. This is only possible for RF hopping TRXs. The fractional load reflects the ability to load the hopping network with certain TRX configuration. It, however, has the same effective reuse-associated limitations. It can be calculated as

$$L_{\text{frac}} = \frac{N_{\text{TRX}}}{N_{\text{freqs/cell}}}$$

where

L_{frac} = fractional load
N_{TRX} = number of TRXs in a cell
$N_{\text{freqs/cell}}$ = number of frequencies allocated to a cell (frequency hopping list length).

5.3.3 Frequency Allocation Reuse

Frequency allocation reuse (FAR) indicates how closely the frequencies are actually reused in a network. It is calculated as

$$\text{FAR} = \frac{N_{\text{freqsTOT}}}{N_{\text{freqs/MA}}}$$

where

FAR = frequency allocation reuse
N_{freqsTOT} = total number of used frequencies
$N_{\text{freqs/MA}}$ = average number of frequencies in MA lists.

This term measures the ability of hopping networks to implement very tight reuses. If the network does not utilise fractional loading, the frequency allocation reuse is the same as effective reuse. FAR also gives a good indication about frequency allocation, but does not provide a good measure for spectral efficiency since it does not reflect how loaded the network is.

5.3.4 Frequency Load

The frequency load can be written as

$$L_{\text{freq}} = L_{\text{HW}} \cdot L_{\text{frac}}$$

where

L_{freq} = frequency load
L_{HW} = the busy hour TSL occupation
L_{frac} = fractional load

This indicator reflects how loaded a frequency of a sector is, i.e. how much traffic is carried by the available spectrum. Therefore, it is a good indication of the network spectral efficiency. However, it is dependent on the deployed reuse. In order to remove this limitation, the definition of a spectral efficiency indicator that can be used in any network configuration, the EFL measure, is introduced.

5.3.5 Effective Frequency Load

The effective frequency load (EFL) quantifies how loaded each frequency available in the system is [18]. It is independent of the frequency reuse and of the TRX configuration deployed. It can be calculated as

$$\text{EFL}(\%) = \frac{\text{Frequency load}}{\text{FAR}}$$

or

$$\text{EFL}(\%) = \frac{L_{\text{HW}}}{R_{\text{eff}}}$$

which can be translated to

$$\text{EFL}(\%) = \frac{\text{Erl}_{\text{BH}}}{\text{Tot\#freq}} \frac{1}{\text{Ave\#}\left(\dfrac{\text{TSL}}{\text{TRX}}\right)}$$

where

Erl_{BH} = average busy hour carried traffic (Erlangs per sector)
Tot# freq = total number of available frequencies in the system
Ave#(TSL/TRX) = average number of TSLs per TRX available for traffic usage

EFL can be directly translated into other spectral efficiency indicators. For example, it can be translated into Erl/MHz/sector, using the following formula:

$$\text{Spectral efficiency (Erl/MHz/sector)} \cong 40 \times \text{EFL}\ (\%)$$

where Ave#(TSL/TRX) has been considered to be 8.

EFL can be easily calculated from the network and it provides a direct measure of how loaded the frequency spectrum is. If one TRX was deployed in every sector for each available frequency and all the TRXs were fully loaded (8 Erlangs/TRX), the EFL would be 100% and the associated spectral efficiency (Erl/MHz sector) would be 40. In this book, EFL has been selected as a measure of traffic load and spectral efficiency. As presented in the next section, an EFL-based methodology has been developed to characterise real networks performance and quantify their spectral efficiency for any benchmarking quality.

5.3.5.1 FR and HR Mixed Traffic Scenario

$$\text{HR EFL} = \text{HR Erlangs}/(\text{Tot\#freq} \times 2 \times \text{aver\#(TSL/TRX)}).$$

Total EFL can be defined as

$$\text{Total EFL} = \text{FR EFL} + \text{HR EFL}.$$

The overall performance depends on the share of HR and FR users. Relative performance analysis with different features or different networks can be done only with the same traffic share.

5.3.5.2 EFL for Data Services

EFL describes how much the frequencies are loaded with (E)GPRS traffic. It can be calculated using the *Data Erlangs* KPI defined in the previous section. In the FH layer, the TSL capacity depends only on the EFL value, independently of the number of hopping frequencies or the amount of *Data Erlangs*. This principle is illustrated in Figure 5.33. The figure shows the evolution of TSL capacity as EFL increases. It can be noticed that networks with different number of hopping frequencies have the same performance.

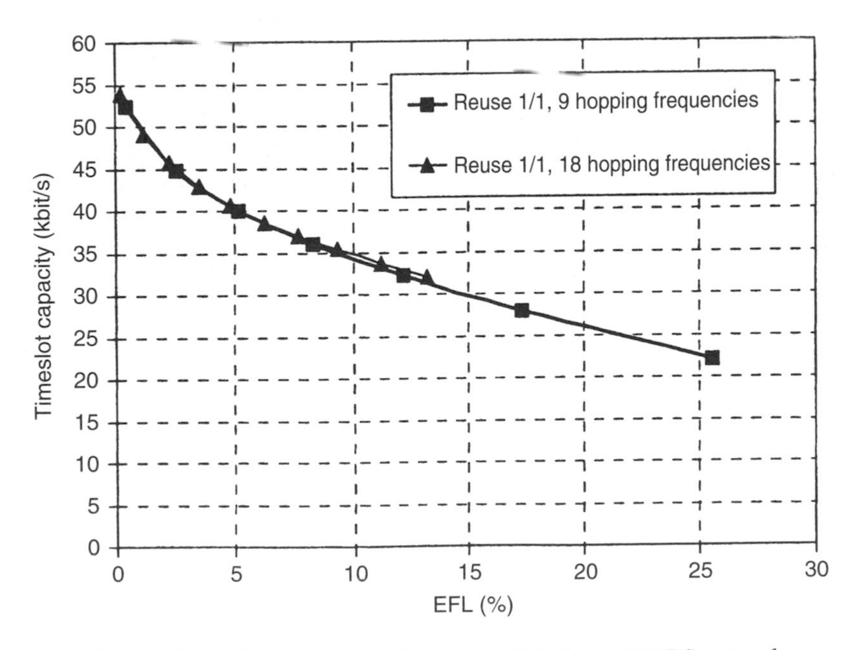

Figure 5.33. TSL capacity evolution with EFL in an EGPRS network

5.3.6 EFL for Mixed Voice and Data Services

When voice and data services are supported together in the hopping layer, it is possible to define a total Erlang and a total EFL:

$$\text{Total Erlangs} = \text{Voice Erlangs} + \text{Data Erlangs}$$

$$\text{Total EFL} = \text{Voice EFL} + \text{Data EFL}.$$

EFL analysis can be done considering Voice and/or Data KPIs versus Total EFL. In mixed scenarios, the service performance depends on the share of Data and Voice Erlangs.

5.4 EFL Trial Methodology

The spectral efficiency of deployed cellular networks is very hard to quantify. This is due to the fact that cellular networks have a pseudo-static operating point in terms of traffic load. The performance measurements collected will identify the performance of the network associated with the traffic load operating point. However, this information is not sufficient to quantify the spectral efficiency of the measured network, i.e. it is not possible to quantify the maximum capacity the network can support with a certain benchmark quality. In order to do so the network should be operating in maximum load conditions.

5.4.1 Network Performance Characterisation

The methodology presented in this section allows the performance characterisation of deployed cellular networks. Once this is done, it is possible to estimate the load associated with any benchmark quality. It is therefore possible to calculate the maximum network capacity for a given benchmark quality or how much the network quality will degrade as load increases. As described in the previous section, the EFL is directly proportional to the traffic and inversely proportional to the available bandwidth. In order to characterise the performance of a network, the quality associated with different EFL values needs to be identified. In real networks, the traffic load is fixed but the number of frequencies deployed can be modified in order to increase the EFL. During trials, the network performance can be characterised by modifying the number of frequencies used.

Figure 5.34 displays the characterisation of Network 1 carried out during extensive trials. Both the BCCH and the hopping layer have been characterised since each one of these layers displays a very differentiated performance. There is hardly any degradation on BCCH performance as load increases since the BCCH TRX is always transmitting full power. On the other hand, the hopping layer shows a clear degradation as load increases. The overall network performance will be the combination of the performance of both layers weighted by the traffic distribution. In the hopping layer, each point represents different hopping configurations, modifying the number of frequencies from 23 to 14. In the BCCH layer, the different traffic load operating points are achieved by means of different channel allocation priority strategies, which define the traffic distribution across the BCCH and hopping layers.

As shown in the above figure, even though Network 1 had a particular traffic load operating point, once the network performance is characterised, the performance associated with any traffic load and the maximum traffic load for certain quality requirements

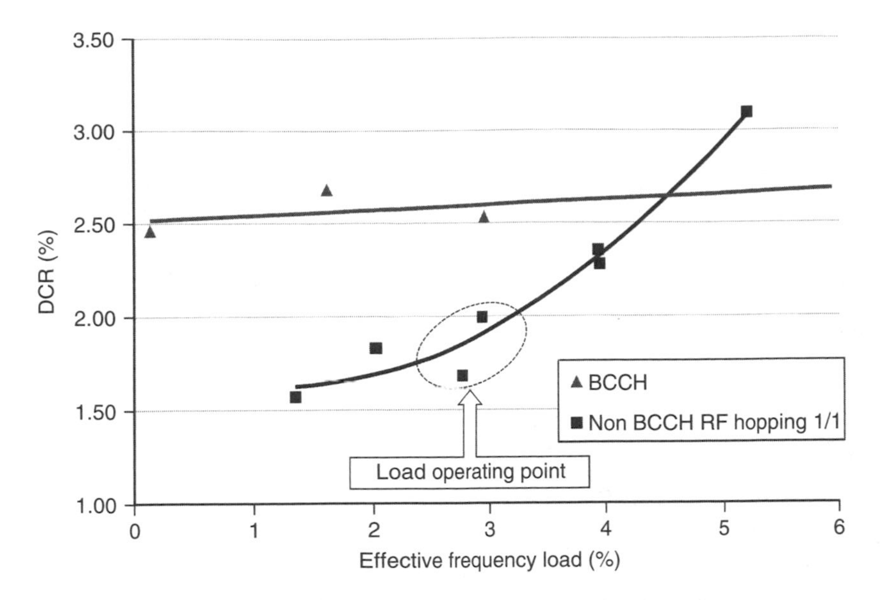

Figure 5.34. Characterisation of Network 1 performance

(spectral efficiency) can be calculated. This has obvious benefits in order to deploy the most adequate technology strategy according to the required quality and the forecasted traffic growth.

5.4.2 *Trial Area Definition*

The trial area has to be representative of the network to be characterised. In this book, results from multiple trials will be presented. In order to be representative of the networks to be characterised, most of them were performed in the high-traffic-level areas of the network and had an adequate size (30–40 sites).

The trial area was always divided into the main trial area and a boundary area. All the results presented and analysed correspond to the main trial area (15–20 sites). The boundary area (15–20 sites) implements the same functionality and configuration as the main trial area but with a slightly modified planning ensuring that the main trial area is isolated from the external network and serving as a transition buffer to the different planning of the external network. Figure 5.35 illustrates the trial area definition for Network 1.

Each network configuration trialled, using a different set of functionality, was tested for as long as required in order to collect all the information required to characterise the performance associated with such configuration. All available performance indicators were collected, processed and analysed, but the results presented in this book are usually provided in terms of dropped call rate and frame erasure rate.

The information presented from the trials, in terms of traffic load and quality, represents the average value of the network during the busy hour period. This average value, however, will have a distribution across the network that should be understood in order to properly analyse the results. Figures 5.36 and 5.37 represent the distribution of traffic load and dropped calls across the sectors of the Network 1 trial area.

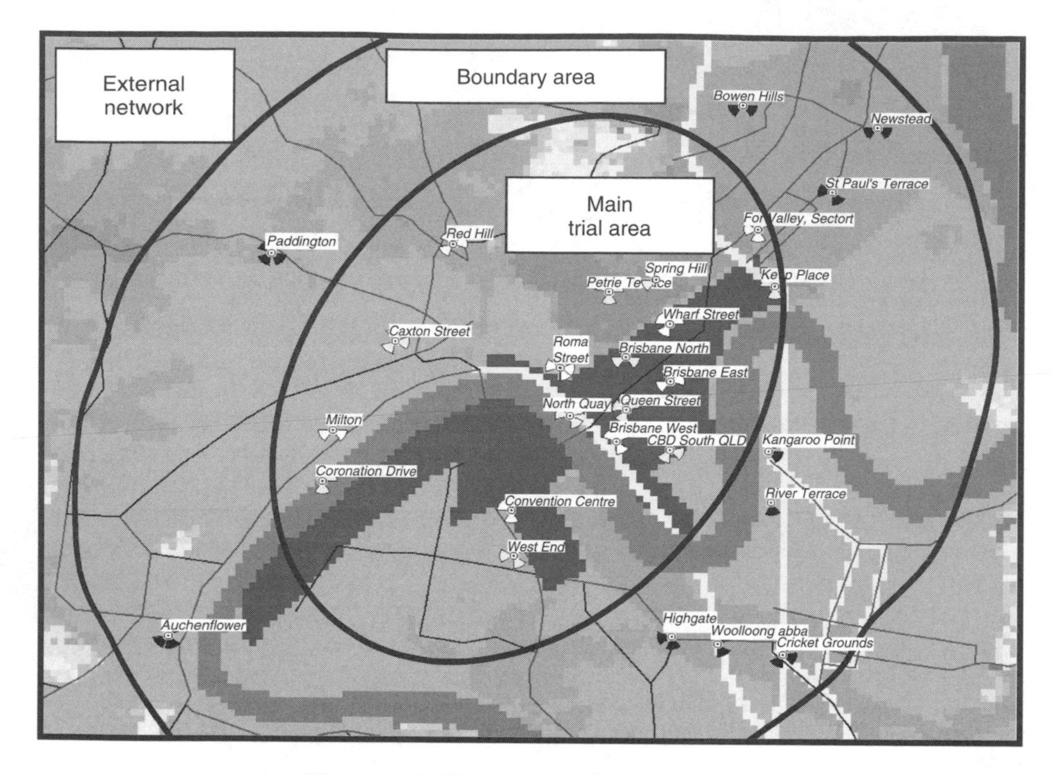

Figure 5.35. Network 1 trial area definition

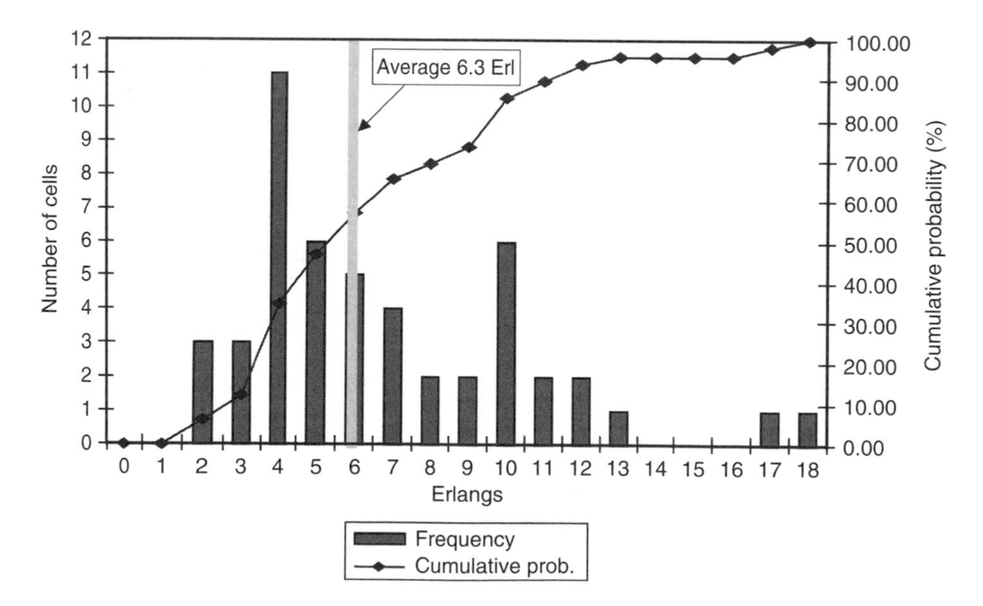

Figure 5.36. Traffic load distribution in Network 1 main trial area

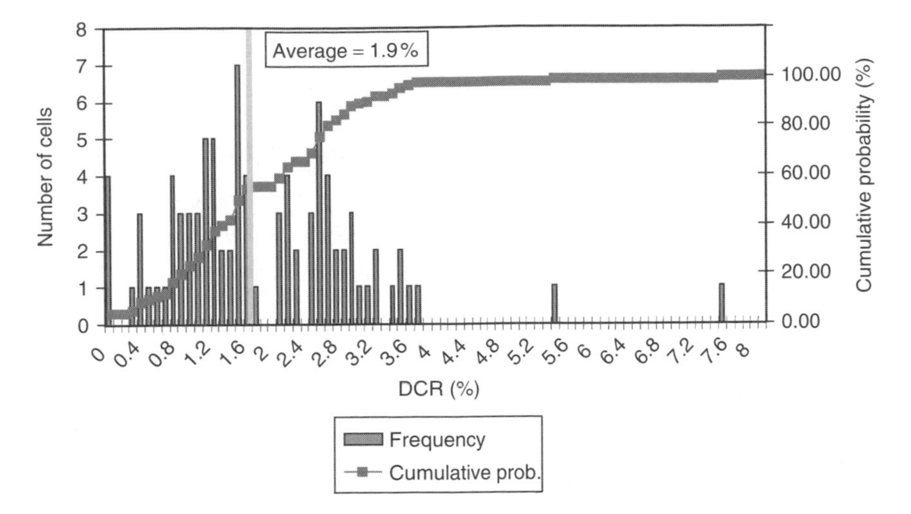

Figure 5.37. Dropped call rate distribution in Network 1 trial area

5.4.3 *Methodology Validation*

The presented methodology is based on the assumption that the performance of a network is fixed for a given EFL value. The high EFL-trialled values were obtained by decreasing the utilised spectrum. This means that doubling the load or halving the number of utilised frequencies should have the same impact on performance since they generate equivalent EFL values. In other words, the performance characterisation of the network should be independent of the available bandwidth.

Figure 5.38 presents the results of a simulation study that utilised different bandwidths for the hopping layer. It is clearly seen that the network performance characterisation is equivalent for all bandwidths with the exception of the 2-MHz case. This is due to the lower hopping gains associated with this case. However, as explained in Section 5.1, the

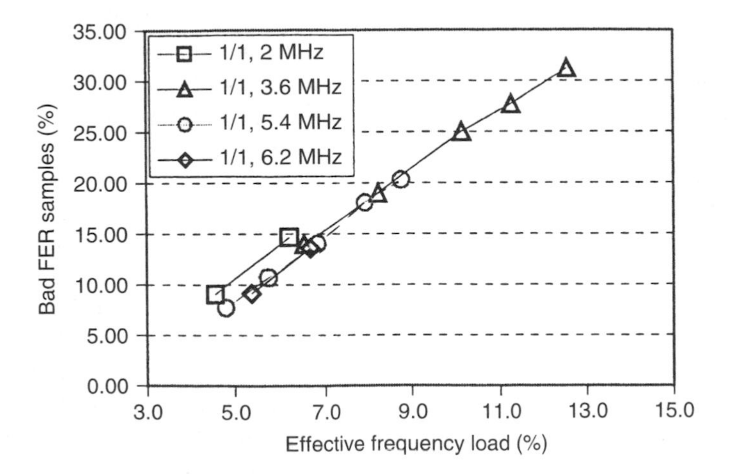

Figure 5.38. EFL network characterisation with different bandwidths based on simulations

additional gains from FH get marginal over a certain number of frequencies in the hopping list. Therefore, the methodology is accurate as long as the number of frequencies of the configurations utilised to characterise the network does not impact the performance of the basic functionalities deployed in the network. The deviation suffered from this effect will worsen the calculated performance as compared with the real performance.

5.5 Baseline Network Performance

The baseline performance of a network is defined in this book as its performance with the basic set of standard GSM functionality. Such functionality has been considered to be RF hopping with advanced reuse schemes, MAIO management, DL and UL power control and DL and UL discontinuous transmission (DTX). Chapter 6 will analyse the performance enhancements associated with this set of functionality. All the performance analysis carried out throughout this book is going to show the achievable performance with different features and functionality based on relative gains obtained from simulations and trials on top of the GSM achievable baseline performance, which is based on real networks performance and presented in this section.

The baseline performance of any cellular network, using any technology, is going to be heavily conditioned by the initial network planning, the traffic distribution and the complexity of the propagation environment. In terms of network planning, the most important factors are the antenna heights and locations. In order to ensure a satisfactory baseline performance, the antenna locations should be carefully selected. High antennas may provide low-cost coverage for initial deployment but will have a negative impact on interference-limited environments, limiting the achievable spectral efficiency. Antennas below roof levels, for which signal propagation is contained by the surrounding buildings, have proven to provide the best performance for interference-limited environments and hence highest spectral efficiency. Following this principle, microcellular networks, because of their characteristic antenna locations, will be able to reach a higher spectral efficiency through the implementation of tighter effective reuses. Different networks will have a very different performance depending on these factors.

The network performance characterisation is generally quite similar across different networks. As presented in the previous section, the generation of this characterisation requires measuring the performance associated with multiple EFL values. Figure 5.39 presents the hopping layer characterisation for two networks with exactly the same set of functionality deployed (RF hopping 1/1, MAIO management, UL PC and UL DTX). There is a significant baseline performance offset but both networks display a similar quality degradation slope as traffic load increases.

In order to estimate the achievable GSM baseline network performance, the measured characterisation of a live network with the defined basic set of GSM standard functionality will be used. Figure 5.40 displays the hopping layer performance associated with mature macrocellular networks using such defined basic set of GSM standard functionality. Network 1 performance has been fully characterised. From Networks 3, 4 and 5 only the performance information from the traffic load operating point is available. The dotted line represents the estimated characterisation for Network 5 according to the previous analysis. Considering 2% DCR as the benchmark quality point, the displayed estimation indicates that 12% EFL in the hopping layer is achievable for Network 5. Networks 3 and 4 measured performances are located in the area marked as *higher performance*, which indicates

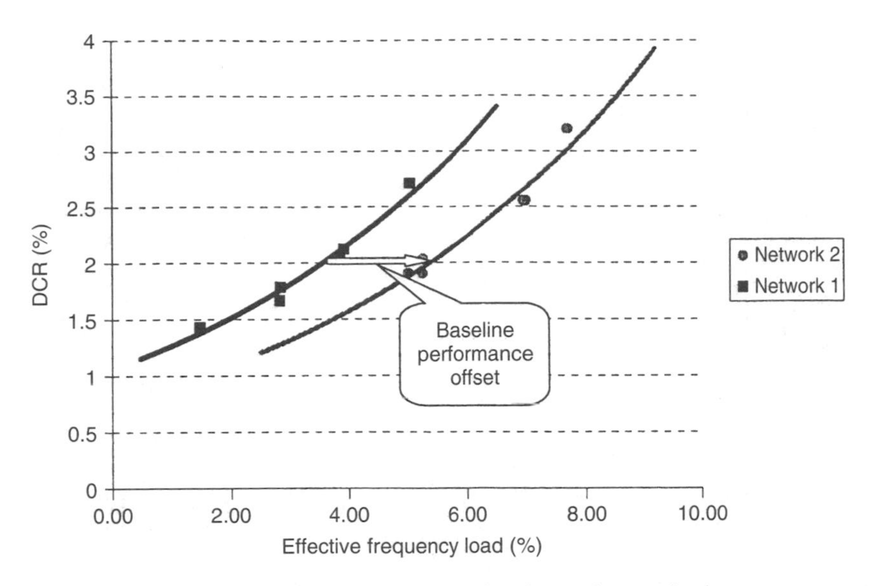

Figure 5.39. Live networks hopping layer characterisation

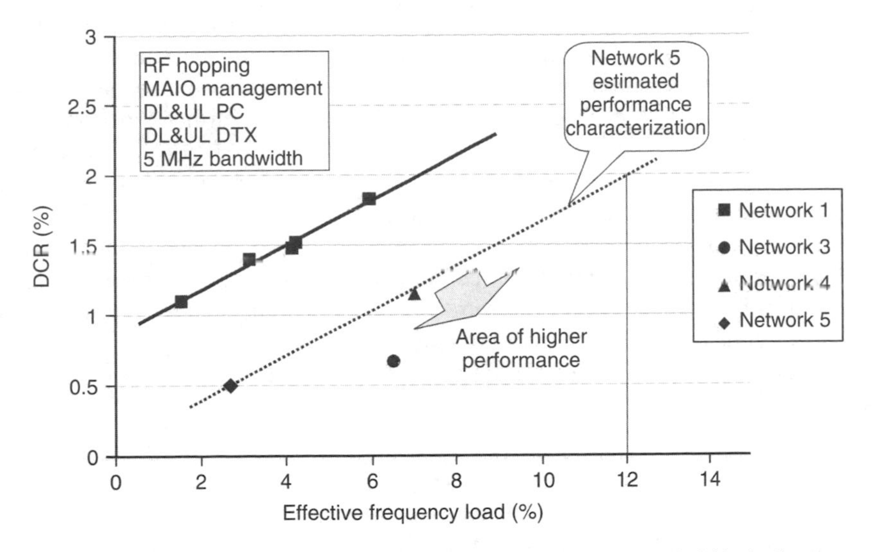

Figure 5.40. GSM baseline network performance for 5-MHz bandwidth deployments

that these networks should achieve as well the previously mentioned achievable baseline performance. This is the value used throughout the book as the achievable GSM baseline performance for 5-MHz bandwidth deployments. Such value can only be achieved when the baseline network planning is optimised accordingly, whereas typically, networks not fully optimised display a baseline performance of 6 to 8% EFL in the hopping layer.

In terms of BCCH achievable performance, the assumption throughout the book is that a BCCH reuse of 12 can be achieved guaranteeing adequate control channels performance.

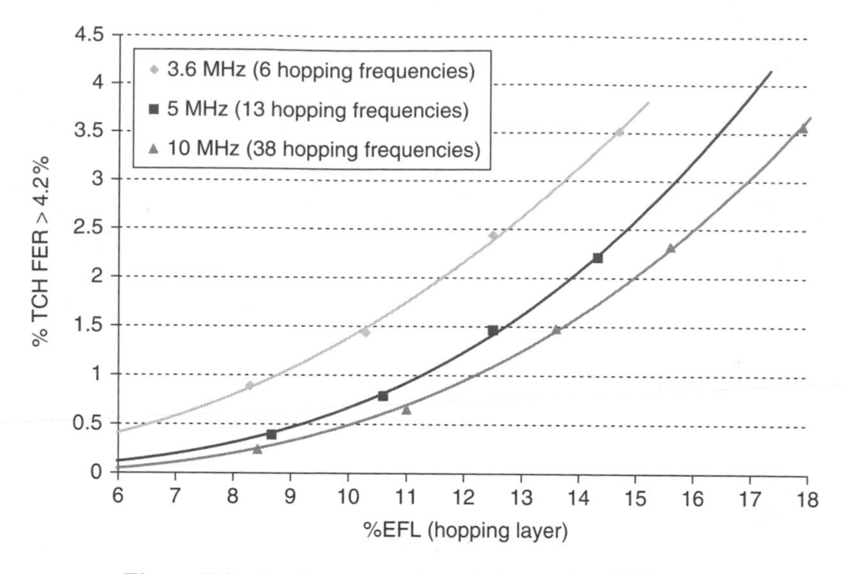

Figure 5.41. Performance gains relative to the 5 MHz case

With this reuse the speech traffic is likely to experience worse performance than the benchmarking quality (as displayed in Figure 5.34), but since this targeted quality is related to the overall performance, with the right traffic load and traffic distribution between BCCH and hopping TRXs, the targeted requirement will be fulfilled. Further functionality, such as the transmit diversity techniques presented in Chapter 11, AMR or the 'BCCH underlay' presented in Chapter 10 may enhance the BCCH performance, either improving the quality or allowing for even tighter reuses to be deployed.

The previous analysis was performed for the case of 5-MHz deployment; however, the available deployment bandwidth will condition the performance of the hopping layer. As presented in Section 5.1, the gains from FH, frequency and interference diversity are going to be dependent on the number of hopping frequencies in use. Figure 5.41 illustrates the hopping performance differences, relative to the 5-MHz case, for the 3.6 and 10 MHz bandwidth cases.

References

[1] Mouly M., Pautet M. B., The GSM System for Mobile Communications, 1992.

[2] Nielsen Thomas T., Wigard J, *Performance Enhancements in a Frequency Hopping GSM Network*, Kluwer Academic Publishers, Dordrecht, 2000.

[3] ETSI TC-SMG, GSM 05.02: European Digital Cellular Telecommunications System (Phase 2)—Multiplexing and Multiple Access on the Radio Path, September 1994.

[4] COST231, Digital Mobile Radio towards Future Generation Systems, COST 231 Final Report, 1999.

[5] Mehrotra A., *Cellular Radio Performance Engineering*, Artech House, London, 1994.

[6] European Radiocommunications Office (ERO), Final Document for the ERO Workshop on Traffic Loading and Efficient Spectrum Use, 1998.
[7] Mogensen P., Jensen C., Andersen J. B., 1800 MHz Mobile Netplanning based on 900 MHz Measurements, COST 231 TD(91)-008, Firenze, January 1991.
[8] Madsen B., Data Receiver for the Pan European Mobile Telephone System (GSM), Dancall Radio A/S, September 1990.
[9] Mogensen P., Wigard J., Frederiksen F., Performance of Slow Frequency Hopping in GSM (Link-level). In COST 231, September 1995.
[10] ETSI TC-SMG, GSM 05.05: European Digital Cellular Telecommunication System (Phase 2)—Radio Transmission and Reception, March 1995.
[11] Wigard J., Mogensen P., Johansen J., Vejlgaard B., 'Capacity of a GSM Network with Fractional Loading and Random Frequency Hopping', *IEEE Proc. PIMRC*, Taipei, 1996, pp. 723–727.
[12] Klingenbrunn T, Mogensen P., 'Modelling Frequency Correlation of Fast Fading in Frequency Hopping GSM Link Simulations', *IEEE Vehicular Technology Conference*, 1999, pp. 2398–2402.
[13] ETSI TC-SMG, GSM 05.08: European Digital Cellular Telecommunication System (Phase 2)—Radio Sub-system Link Control, March 1996.
[14] ETSI Section 6.12, Comfort Noise Aspect for Full Rate Speech Traffic Channels, January 1993.
[15] Mehrotra A., *GSM System Engineering*, Artech House, Boston, 1997.
[16] ITU-T P.800, Telecommunication Standardization Sector of ITU (08/96), Series P: Telephone Transmission Quality, Methods for Objective and Subjective Assessment of Quality.
[17] ETSI TC-SMG, GSM 06.75: Performance Characterization of the GSM Adaptive Multi-Rate (AMR) Speech Codec.
[18] Wigard J, Mogensen P, Michaelsen P. H, Melero J, Halonen T, 'Comparison of Networks with Different Frequency Reuses, DTX and Power Control Using the Effective Frequency Load Approach', *VTC2000*, Tokyo, Japan, 2000.

6

GSM/AMR and SAIC Voice Performance

Juan Melero, Ruben Cruz, Timo Halonen, Jari Hulkkonen,
Jeroen Wigard, Angel-Luis Rivada, Martti Moisio, Tommy Bysted,
Mark Austin, Laurie Bigler, Ayman Mostafa and Rich Kobylinski and
Benoist Sebire

Despite the expected growth of data services, conventional speech will be the dominant revenue generation service for many years. As voice traffic continues growing, many networks are facing capacity challenges, which may be addressed with the right technology deployment. This chapter presents the global system for mobile communications (GSM) voice performance, highlighting the spectral efficiency enhancements achievable with different functionality. Section 6.1 introduces the performance improvements associated with basic baseline GSM speech performance functionality (as defined in Chapter 5), which includes frequency hopping (FH), power control (PC) and discontinuous transmission (DTX). In Section 6.2, frequency reuse partitioning methods are presented. Section 6.3 presents trunking gain GSM functionality such as traffic reason handover (TRHO) or Directed retry (DR). Half rate is introduced in Section 6.4. Rel'98 adaptive multi-rate (AMR) is thoroughly analysed in Section 6.5 and Source Adaptation is presented in Section 6.6. Section 6.7 introduces Rel'5 EDGE AMR enhancements and finally, SAIC functionality performance gains are presented in Section 6.8. The performance results described in this chapter include comprehensive simulations and field trials with real networks around the world.

6.1 Basic GSM Performance

This section will describe the network level performance enhancements associated with the basic set of standard GSM functionality defined in the previous chapter: FH, PC and DTX. Section 5.5 in Chapter 5 introduced the GSM baseline network performance based on the deployment of this basic set.

GSM, GPRS and EDGE Performance 2nd Ed. Edited by T. Halonen, J. Romero and J. Melero
 ISBN: 0-470-86694-2

6.1.1 Frequency Hopping

The principles of operation of frequency hopping (FH) and the associated link level gains were introduced in Chapter 5. This section will focus on hopping deployment strategies and overall network performance. FH can be deployed using two differentiated strategies: Baseband (BB) hopping and radio frequency (RF) hopping. Both kinds of hopping are illustrated in Figure 6.1.

In BB hopping, the number of hopping frequencies is limited to the number of deployed transceivers (TRXs), since the hopping is executed by the call being switched across the TRXs, whose frequencies are fixed. The broadcast control channel (BCCH) can be included in the hopping sequence.[1]

When RF hopping is used, the call stays on one TRX, but the frequency of the TRX changes for every frame. The number of hopping frequencies is only limited by the number of available frequencies, thus providing greater deployment flexibility.[2] In practice, the BCCH TRXs have fixed frequencies since the BCCH has to be transmitted all the time. These frequencies are generally not included in the hopping sequence since the overall performance improves when the hopping and non-hopping bands are not mixed. This means that when RF hopping is deployed, there are usually two effective layers per sector, the BCCH TRX (non-hopping) and the non-BCCH TRX (hopping).

When the TRX configuration is very large across the whole network (over 64 TRXs per sector), BB hopping will perform similarly to RF hopping. Additionally, the benefits from BB hopping are applicable to the BCCH TRX as well, and for large configurations the combining loss is potentially lower since BB hopping can use remote tune combiners (RTCs) while RF hopping requires wideband (WB) combining. However, in most practical deployments, RF hopping will outperform BB hopping due to its flexibility and robustness since its performance is not conditioned by the number of TRXs deployed or the accuracy of the frequency planning. The additional frequency diversity gains from RF hopping and

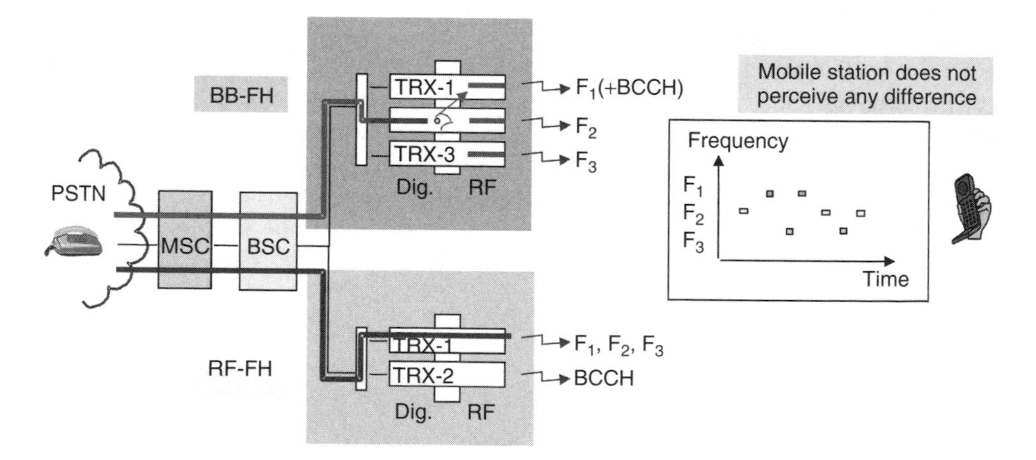

Figure 6.1. Baseband and RF hopping

[1] Some terminals experience problems with BB hopping when BCCH is included in the hopping list and DL power control is used because of the wide range of signal fluctuation.

[2] GSM specifications limit the maximum number of hopping frequencies per hopping list to 64.

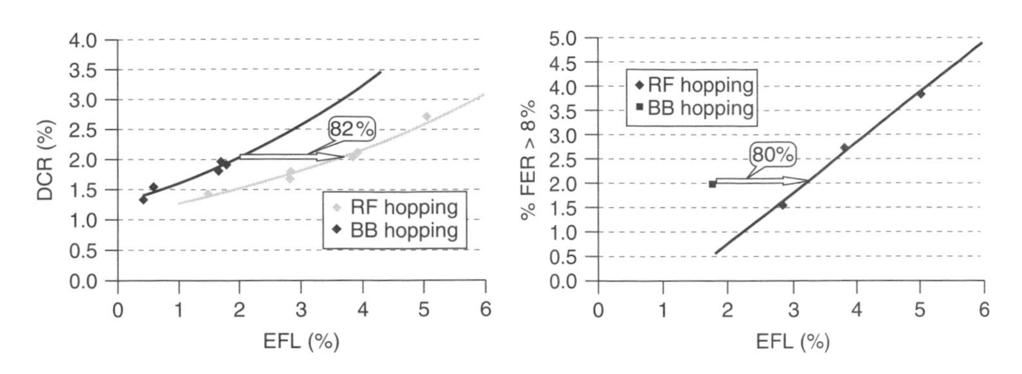

Figure 6.2. Network 1 RF hopping versus BB hopping DCR and FER-trialled performance

the use of cross polarisation antennas and air combining will compensate for the potential additional combining losses. All the studies throughout this book have considered RF hopping deployments, although the performance of BB hopping would be equivalent under the previously stated conditions.

Figure 6.2 displays the performance of the RF and BB hopping layers for Network 1 in terms of dropped call rate (DCR) and frame erasure rate (FER). Since the average number of TRXs per sector was 2.5, BB hopping gains were, in this case, not comparable to those of RF hopping.

With a hopping list of n frequencies, $64 \times n$ different hopping sequences can be built. These are mainly defined by two parameters, as was explained in Chapter 5. Firstly, the mobile allocation index offset (MAIO), which may take as many values as the number of frequencies in the hopping list. Secondly, the hopping sequence number (HSN), which may take up to 64 different values. Two channels bearing the same HSN but different MAIOs never use the same frequency on the same burst, while two channels using different HSNs only interfere with each other $(1/n)$th of the time.

One of the main hopping deployment criteria is the frequency reuse selection. The different frequency reuses are characterised with a reuse factor. The reuse factor indicates the sector cluster size in which each frequency is used only once. The reuse factor is typically denoted as x/y, where x is the site reuse factor and y the sector reuse factor. This means that a hopping reuse factor of 3/9, corresponds to a cellular network with nine different hopping lists which are used every three sites. FH started to be used in the mid-1990s. Initially, conservative reuses, such as 3/9, were used. However, 3/9 minimum effective reuse was limited to nine. Later on, 1/3 proved to be feasible and finally 1/1 was considered by many experts as the optimum RF hopping frequency reuse. Figure 6.3 illustrates the hopping *standard* reuses of 3/9, 1/3 and 1/1.

As the frequency reuse gets tighter, the interference distribution worsens, but, on the other hand, the associated FH gains increase. Figures 6.4 and 6.5 display the performance of hopping reuses 1/1 and 3/9. The results are based on simulations using Network 1 propagation environment and traffic distribution. Reuse 3/9 shows better carrier/interference (C/I) distribution and therefore bit error rate (BER) performance. However, reuse 1/1 has a better overall frame error rate (FER) performance due to the higher frequency and interference diversity gains. As a conclusion, the most efficient hopping reuse will be dependent on the generated C/I distribution and the associated link level gains.

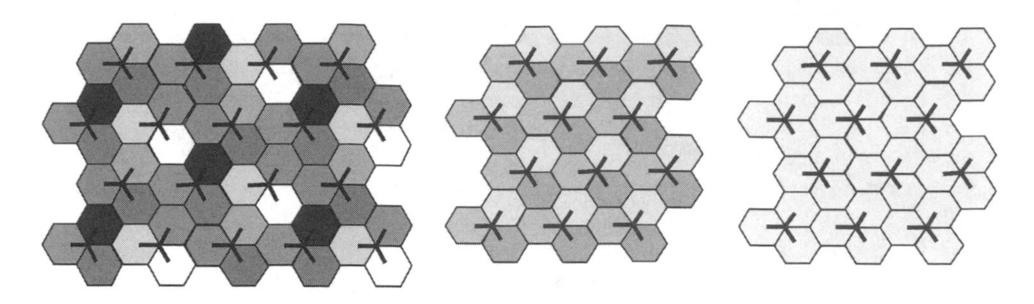

Figure 6.3. Hopping standard reuses: 3/9, 1/3 and 1/1

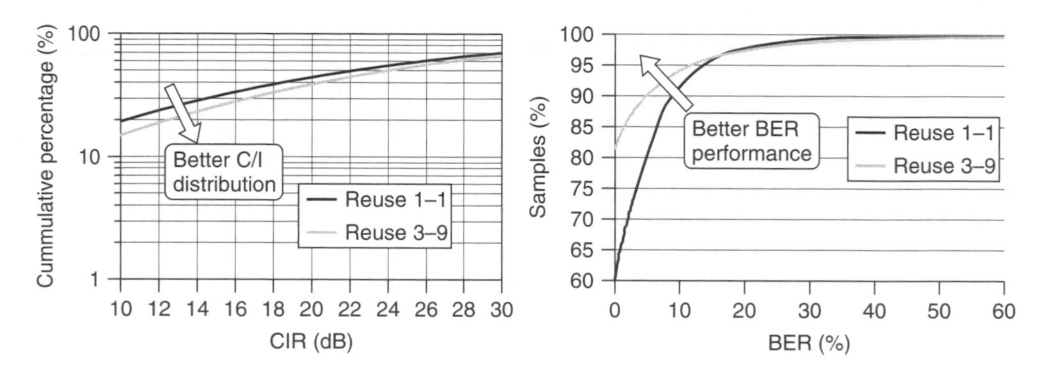

Figure 6.4. C/I and BER Network 1 simulation performance

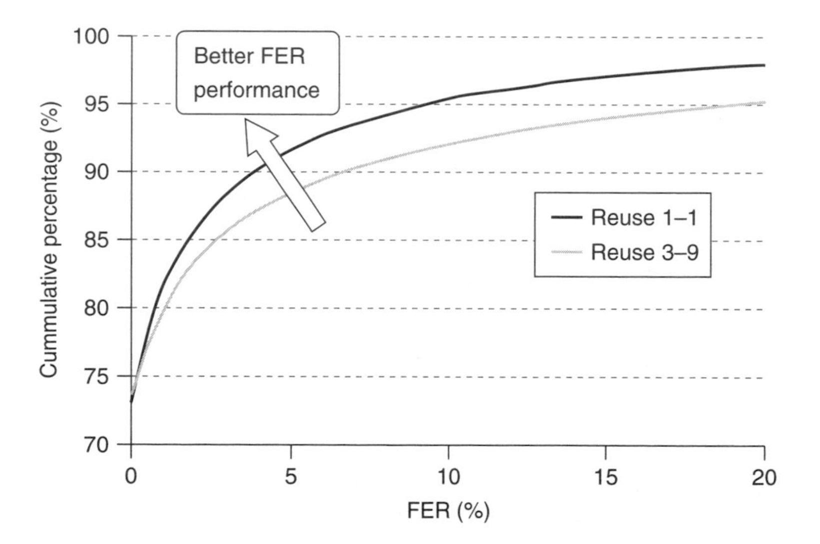

Figure 6.5. FER Network 1 simulation performance

Reuse 1/3 has proven to have equivalent or even slightly higher performance in deployment scenarios with regular site distribution and antenna orientation. However, most practical networks are quite irregular both in terms of site distribution and antenna orientation. Reuse 1/1 has the advantage of better adaptability to any network topology and provides higher hopping gains, especially in the case of limited number of available hopping frequencies. Finally, the MAIO management functionality can be used in the 1/1 deployment to ensure that there is no interference between sectors of the same site.

The MAIO management functionality provides the ability to minimise the interference by controlling the MAIOs utilised by different TRXs within a synchronised site containing multiple sectors. There are as many possible MAIOs as number of frequencies in the hopping list. Therefore, the number of TRXs that can be controlled within a site without repetition of any MAIO value is limited by the number of hopping frequencies. As the effective reuses decrease, adjacent and co-channel interference (CCI) will begin to be unavoidable within the sites. When this intra-site inter-sector interference is present, the use of different antenna beamwidth will impact the final performance. Appendix A analyses the MAIO management limitations, the best planning strategies and the impact of different antenna beamwidths.

Figure 6.6 displays the trial performance of RF hopping 1/3 and 1/1 when deployed in Network 2. In this typical network topology, both reuses had an equivalent performance.

The logic behind 1/1 being the best performing hopping reuse relies on the higher hopping gains for the higher number of frequencies. As presented in Chapter 5, most of the hopping gains are achieved with a certain number of hopping frequencies and further gains associated with additional frequencies are not so high. On the other hand, reuse 1/1 has the worst possible interference distribution. In order to get most hopping gains, to improve the interference distribution and to keep the high flexibility and adaptability associated to 1/1 reuse, the innovative *advanced* reuse schemes (n/n) were devised. The

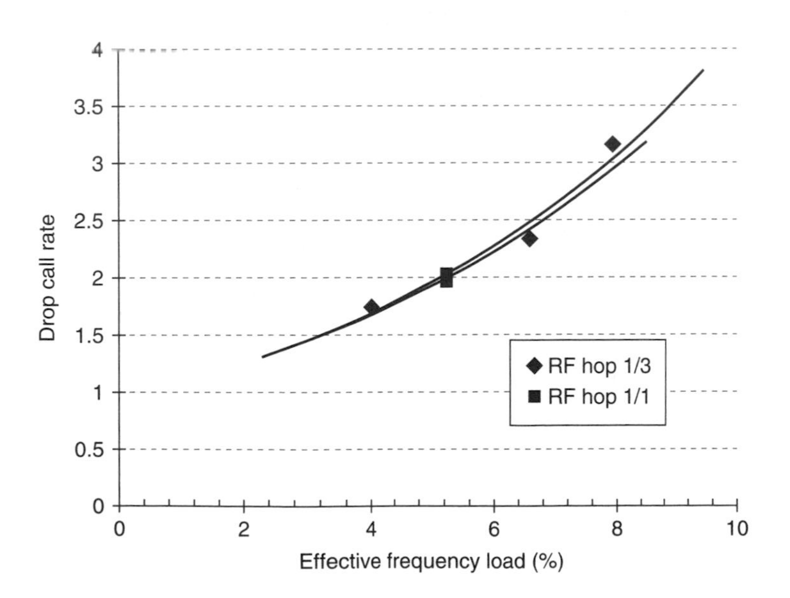

Figure 6.6. Network 2 RF hopping 1/3 and 1/1 trialled performance

idea is to introduce a site reuse in the system but keep the 1/1 scheme within each site. The end result is an improved C/I distribution while maintaining most of the link level gains. Reuse 1/1 would belong to this category with a site reuse of 1. Figure 6.7 illustrates some examples of *advanced* reuse schemes.

The optimum FH *advanced* reuse scheme will be dependent on the number of available frequencies. For example, if 18 hopping frequencies are available, and assuming that 6 frequencies provide most of the hopping gains, the optimum scheme would be 3/3. On the other hand, the MAIO management limitations presented in Appendix A are determined by the number of hopping frequencies within the site and as n (reuse factor) increases the efficiency of MAIO management will decrease. Therefore, the optimum *advanced* reuse scheme will be determined by both the total number of available hopping frequencies and the existing TRX configuration.

Figure 6.8 shows the performance achieved by 1/1 and 2/2 reuses both in terms of DCR and FER in Network 1 trials. The measured capacity increase ranged between 14 and 18%. In this scenario, with 23 hopping frequencies available, reuse 3/3 is expected to outperform both 1/1 and 2/2.

As a conclusion, the best performing hopping schemes are the RF hopping *advanced* reuse schemes. Additionally, these schemes do not require frequency planning or optimisation in order to deliver their maximum associated performance. Other hopping deployment strategies can yield equivalent results, but have intensive planning and optimisation

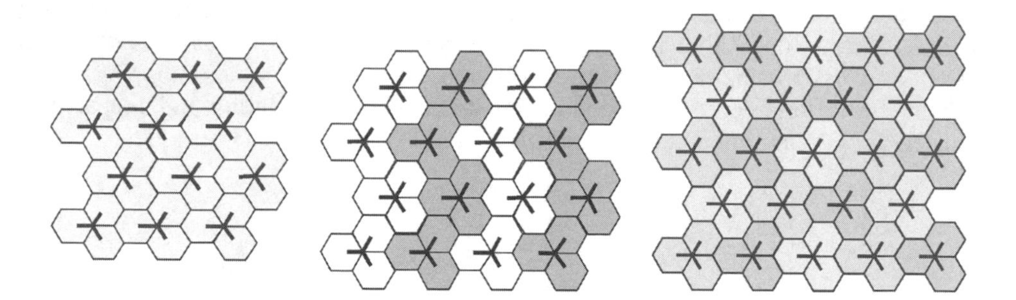

Figure 6.7. *Advanced* reuse schemes. 1/1, 2/2 and 3/3

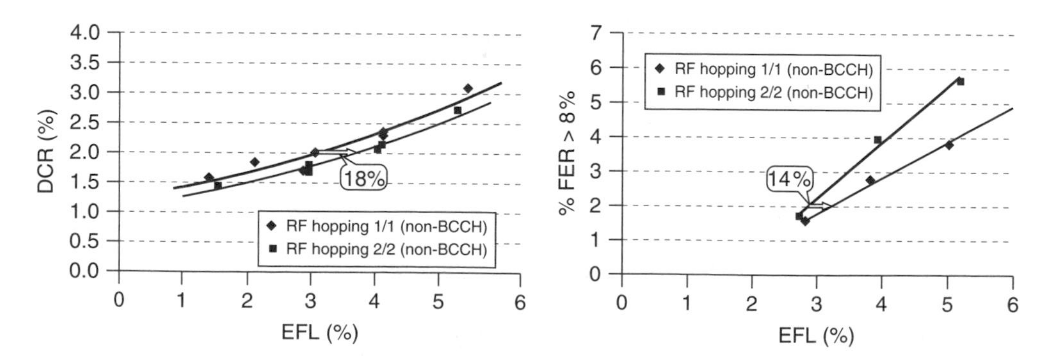

Figure 6.8. Network 1 RF hopping 1/1 versus 2/2-trialled DCR and FER performance

requirements in order to achieve their expected performance and are not so flexible in order to be upgraded or re-planned. Examples of these other strategies could be BB hopping with large TRX configurations or heuristic planning,[3] which require complex planning tools and are very hard to optimise. These deployment strategies can make use of the automated planning functionality described in Chapter 13 in order to achieve high performance without intensive planning and optimisation effort.

6.1.2 Power Control

Power control (PC) is an optional feature in GSM that reduces the interference in the system by reducing both the base station and mobile station transmitted power. Additionally, it contributes to extend the terminal's battery life. The PC algorithm is described in more detail in Chapter 5. PC can be efficiently combined with FH.

The control rate of PC is 0.48 s (once per slow associated control channel (SACCH) multiframe). For relatively slow moving mobiles, the interference reduction gain from PC can be substantial, whereas for fast moving mobiles it may not be possible to efficiently track the signal variation. Hence, conservative power settings must be used, which limits the interference reduction gain. The average PC gain in carrier interference (C/I), measured by network simulations including FH and for a mobile speed of 50 km/h, has been found to be approximately 1.5 to 2.5 dB at a 10% outage level [1].

Figure 6.9 displays the impact of downlink power control (DL PC) in the distribution of received signal level. The 10-dB offset shows the efficiency of DL PC. Much higher dynamic ranges (range of power reduction) do not provide additional gains and, in some cases, can cause some problems since the DL PC speed may not be fast enough in some cases to follow the signal fading variations.

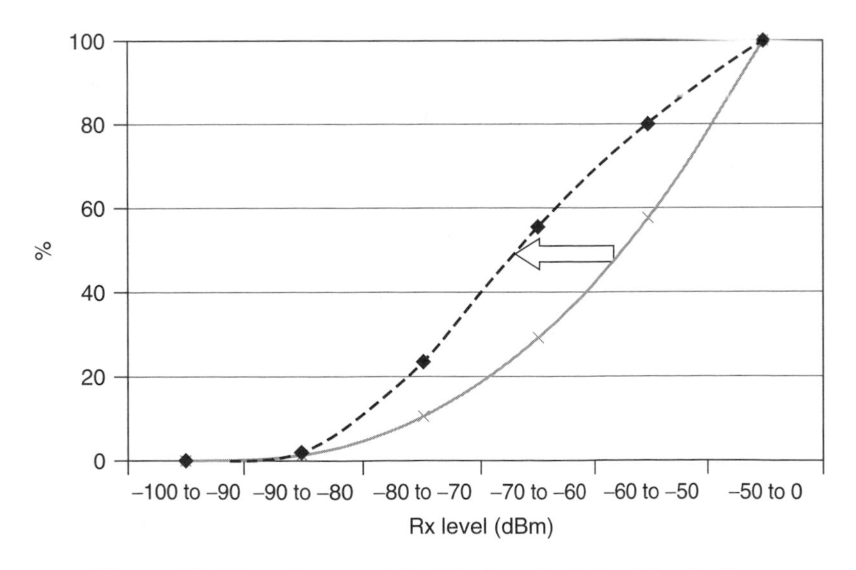

Figure 6.9. Network 1 DL PC trialled received signal level offset

[3] Irregular hopping lists in order to optimise the C/I distribution.

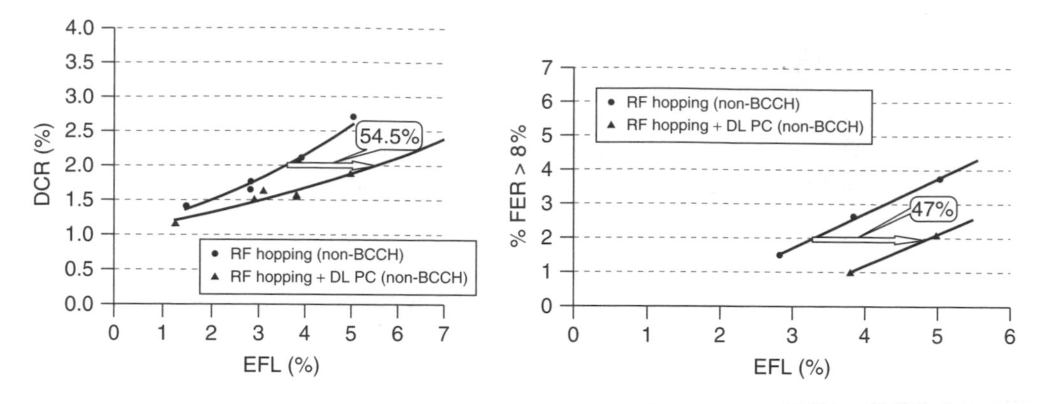

Figure 6.10. Network 1 DL PC DCR and FER performance gains

The power control settings have an important effect on the DL PC performance. From multiple analysis carried out, the DL PC performance reaches its optimum value when the following criteria are fulfilled:

- fast averaging and voting parameters are set in order to define a fast algorithm;
- the power step should be smaller (around 2 dB) when decreasing the power and higher (around 4 dB) when increasing it;
- the dynamic range should be large enough (16–20 dB) but the largest value is not recommended.

The network level performance enhancements achieved with the introduction of DL PC has been measured in real networks. Figure 6.10 displays the measured gain from DL PC both in terms of DCR and FER in Network 1. A 50% capacity increase over the hopping network can be generally considered when DL PC is activated in the network.

6.1.3 Discontinuous Transmission

An effective way to decrease the interference and thus increase the network performance is to switch the transmitter off when speech is not present. This is known as *discontinuous transmission* (DTX). In a normal conversation, each person talks, on average, less than 50% of the time. DTX functionality is described in detail in Chapter 5.

Results from simulations have shown a linear relationship between the DTX factor and the C/I distribution improvement when combined with random FH [1]. This implies that a DTX factor of 0.5 gives an interference reduction of approximately 3 dB [2]. When the SACCH and noise description frame are considered, the reduction is down to 2.5 dB. The gain from DTX is not dependent on the network load and location of the mobile stations, as is the case for PC. Uplink DTX has, as UL PC, the second aim to extend battery power in the mobile station.

When network operators turn on DTX in their network, they should carefully analyse the BER (RXQUAL) statistics since the BER estimation accuracy is affected by the introduction of DTX, as presented in Chapter 5. The BER is estimated over 100 time

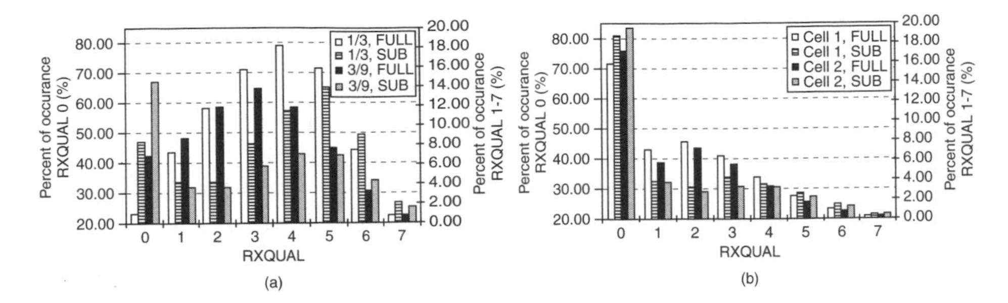

Figure 6.11. Simulations (a) and Network 8 (b) RXQUAL-FULL and RXQUAL-SUB distribution

division multiple access (TDMA) bursts in the non-DTX mode and over 12 TDMA bursts in the DTX mode. The first is called *RXQUAL-FULL* and the second *RXQUAL-SUB*.

Network simulations showed that the higher variance causes the RXQUAL-SUB values to be more spread when compared with the RXQUAL-FULL values. While the network actually gets better with DTX, this cannot directly be extracted from the RXQUAL network statistics since they are worse with DTX [3]. Figure 6.11 shows the distribution of (a) RXQUAL-FULL and (b) RXQUAL-SUB values from simulations (1/3 and a 3/9 reuse high loaded hopping network) and real network measurements. The amount of RXQUAL values equal to 6 and 7 is higher when using the SUB values. This is caused by the high variance of the SUB values. In [4], simulation results can be found for different frequency reuses with and without PC and DTX. This effect will impact not only the generated statistics but also the overall network performance as well since multiple radio resource management (RRM) algorithms, such as handover and PC, rely on accurate BER information.

The impact on network performance when DTX is introduced has been analysed both with simulations and trials. Owing to the different issues mentioned, such as definition of realistic voice activity factor, inaccuracies in the RXQUAL reporting and additional interference from SACCH and noise description frame, the realistic gain from DTX is hard to quantify via simulations and will be limited in practical deployments. Figure 6.12

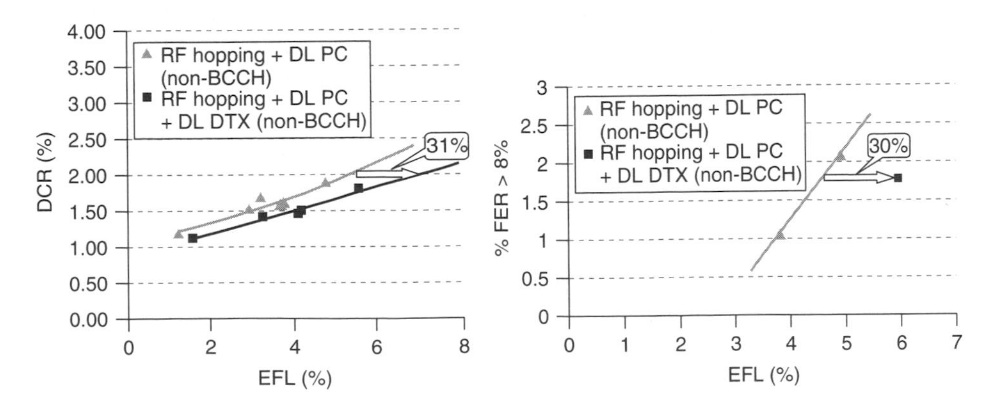

Figure 6.12. Network 1 DL DTX DCR and FER performance gains

displays the Network 1 measured performance enhancements, both in terms of DCR and FER, with the introduction of DL DTX in the hopping network. A 30% capacity increase can be considered when DL DTX is activated in a hopping network using PC.

The conclusion of this section is that the use of RF hopping, deployed with the *advanced* reuse schemes and making use of the MAIO management functionality yields the best hopping performance without demanding planning requirements. Additionally, DL PC and DTX work efficiently in hopping networks and combined together double the previous hopping network capacity. The results from this section are considered in this book to define the GSM baseline configuration for speech services providing the baseline performance. Further sections of this chapter will analyse the additional enhancements other functionality will provide on top of the defined GSM baseline performance and, unless otherwise stated, this baseline configuration will always be in use in the simulations and trial results presented.

6.2 Reuse Partitioning

The basic concept of reuse partitioning is to decrease the effective reuse utilising channel allocation schemes able to assign connections to different layers using different reuses. Therefore, the available spectrum of a radio network is divided into different bands, which constitute different layers with different reuses. In this book, the layer with looser reuse that ensures full coverage is called *overlay* and the different layers with tighter reuses, which provide higher effective capacity, are called *underlay* layers. Mobile stations close to the base station can use the underlay network, while mobile stations on the edge of the cell usually utilise the overlay layer. The dynamic allocation of mobile stations to the underlay layers provides an effective capacity increase, since these layers are deployed with tighter reuses.

There are many different practical implementations for reuse partitioning, such as concentric cells,[4] multiple reuse pattern (MRP),[5] intelligent underlay–overlay (IUO)[6] and cell tiering.[7] All these are based on two or more layers deployed with different reuses, and channel allocation strategies that allocate the calls into the appropriate layer according to the propagation conditions. However, each approach has its own special characteristics. For instance, IUO channel allocation is based on the evaluation of the C/I experienced by the mobile station, while concentric cells utilise inner and outer zones where the evaluation of the inner zone is based on measured power and timing advance. More insight into some of the different reuse partitioning methods, and especially about the combination of FH and frequency partitioning, is given in [2].

Many different studies from frequency reuse partitioning have been carried out with an estimated capacity gain typically of 20 to 50% depending on system parameters such as available bandwidth, configuration used, etc. This section does not focus on basic reuse partitioning, but rather analyses its combination with FH since the gains of both functionality can be combined together.

---

[4] Motorola proprietary reuse partitioning functionality.
[5] Ericsson proprietary reuse partitioning functionality.
[6] Nokia proprietary reuse partitioning functionality.
[7] Nortel proprietary reuse partitioning functionality.

6.2.1 Basic Operation

Frequency reuse partitioning techniques started to be deployed in the mid-1990s. At that time some terminals did not properly support FH, especially when it was combined with PC and DTX. When FH terminal support started to be largely required by operators, the terminal manufacturers made sure these problems were fixed and FH started to be extensively deployed. It was therefore important to combine both FH and frequency reuse partitioning in order to achieve the best possible performance.

Basic reuse partitioning implements a multi-layer network structure, where one layer provides seamless coverage and the others high capacity through the implementation of aggressive frequency reuses.

Figure 6.13 displays the basic principle of frequency reuse partitioning operation. The overlay layer provides continuous coverage area using conventional frequency reuses that ensures an adequate C/I distribution. Overlay frequencies are intended to serve mobile stations mainly at cell boundary areas and other locations where the link quality (measured in terms of C/I ratio or BER) is potentially worse. The underlay layer uses tight reuses in order to provide the extended capacity. Its effective coverage area is therefore limited to locations close to the base transceiver station (BTS), where the interference level is acceptable.

The channel allocation, distributing the traffic across the layers, can be performed at the call set-up phase or later on during the call by means of handover procedures. Different implementations base the channel allocation on different quality indicators. The best possible quality indicator of the connection is C/I. Therefore, IUO, which makes use of such an indicator, is considered to be the best reuse partitioning implementation and will be the one further analysed across this section.

In IUO implementation, the C/I ratio is calculated by comparing the downlink signal level of the serving cell and the strongest neighbouring cells, which use the same underlay frequencies. These are reported back from the terminals to the network, and the C/I ratio is computed in the network according to

$$\frac{\mathrm{C}}{\mathrm{I}} = \frac{P_{\mathrm{own\_cell}}}{\sum_{n=1}^{6} P_{i\_\mathrm{BCCH}}}$$

where $P_{\mathrm{own\_cell}}$ is the serving cell measured power, corrected with the PC information, and P_{i_BCCH} is the power measured on the BCCH of interfering cell i.

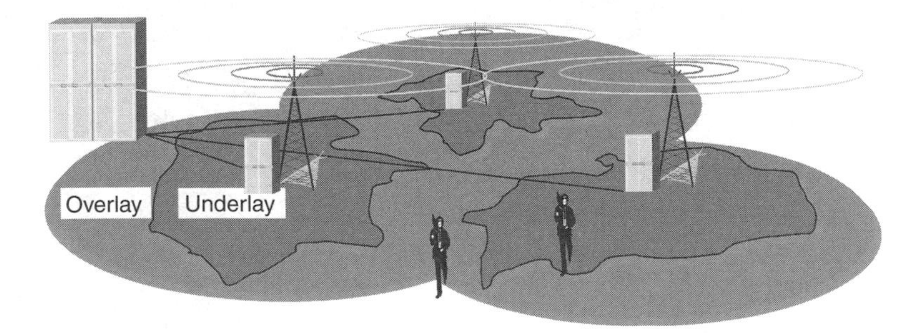

Figure 6.13. Basic principle of reuse partitioning

6.2.2 *Reuse Partitioning and Frequency Hopping*

Both FH and reuse partitioning reduce the effective frequency reuse pattern and thereby enable capacity gains in cellular mobile systems. The gains of hopping are associated with a better link level performance due to the frequency and interference diversity. On the other hand, the gains from reuse partitioning are related to an efficient channel allocation that ensures that calls are handled by the most efficient layer/reuse. These gains can be combined since they have a different nature. In the IUO implementation, the combination of frequency hopping (FH) and reuse partitioning is defined as *intelligent frequency hopping* (IFH).[8]

Figure 6.14 displays the IFH concept and how the frequency band is divided into different layers/sub-bands. IFH supports BB and RF hopping. As with conventional FH, in BB hopping mode, the BCCH frequency is included in the hopping sequence of the overlay layer, but not in RF hopping mode.

6.2.2.1 Basic Performance

The performance of IFH is going to be presented on the basis of extensive trial analysis. Figure 6.15 illustrates Network 2 IFH performance. Network 2 traffic configuration was quite uniform with four TRXs per cell across the whole trial area. There were two TRXs in both the overlay and the underlay layers. DL PC and DTX were not in use in this case.

These results suggest that IFH provides an additional capacity increase of almost 60%. However, the gains from IFH are tightly related with the gains of PC since both functionalities tend to narrow down the C/I distribution, effectively increasing the system capacity. Therefore, it is important to study the performance of IFH together with DL PC in order to deduce its realistic additional gains on top of the previously defined GSM baseline performance.

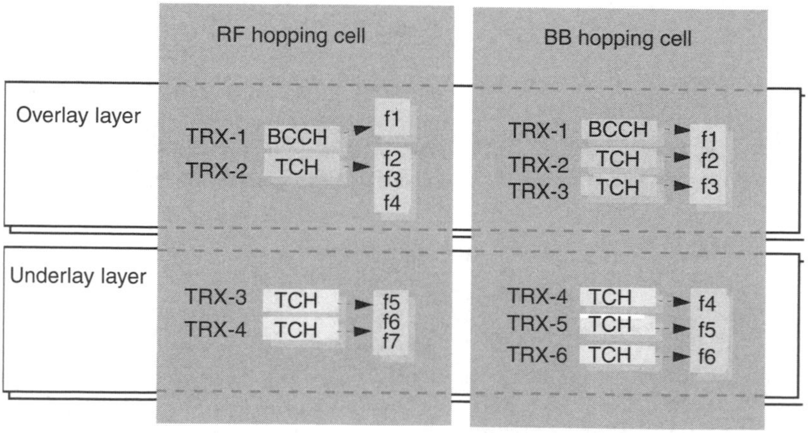

Figure 6.14. IFH layers with RF and BB hopping

[8] Nokia proprietary functionality.

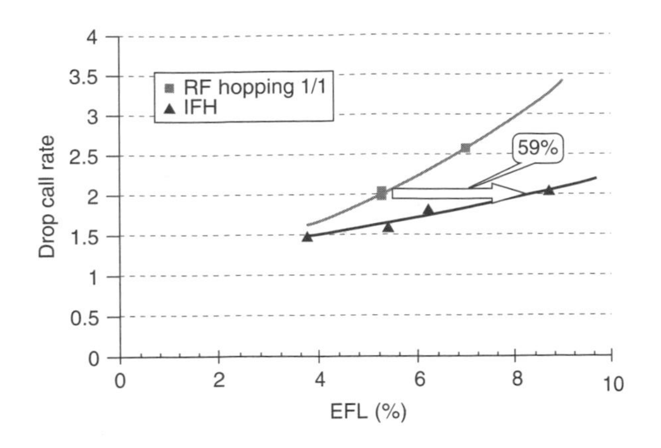

Figure 6.15. Network 2 IFH performance increase

6.2.2.2 IFH with Power Control Performance

Simulation analysis showed the gains from DL PC and IFH were not cumulative, so a careful analysis is required to find out the additional gain of IFH over the defined GSM baseline performance. The simulation analysis suggests this gain is roughly 25%. Figure 6.16 shows the results collected in Network 5 trials. Its configuration is equivalent to Network 2. The measured gain was between 18 and 20%. However, in order to achieve this gain, the PC and IFH parameters have to be carefully planned for the algorithms not to get in conflict.

As a conclusion, additional gains of the order of 20% on top of the GSM baseline performance are expected with the introduction of IFH. However, this introduction requires careful planning and parameter optimisation and the gains are dependent on network traffic load and distribution. Therefore, unlike FH, PC and DTX, the use of reuse partitioning is only recommended when the capacity requirements demand additional spectral efficiency from the network.

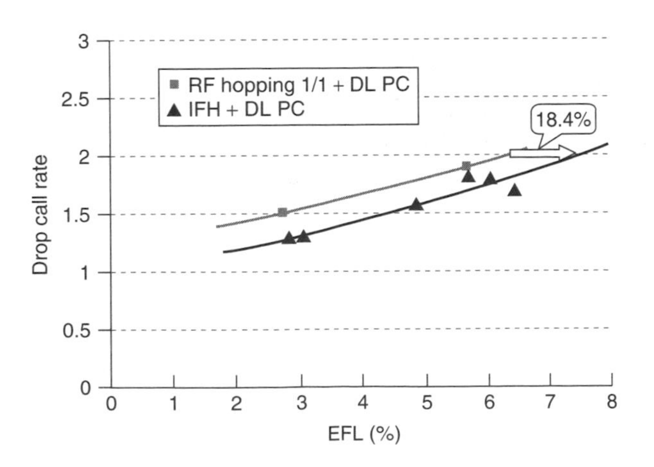

Figure 6.16. Network 5 IFH and DL PC performance

6.3 Trunking Gain Functionality

This section describes various solutions that increase the radio network trunking efficiency through a more efficient use of the available hardware resources. This is possible with an effective share of the resources between different logical cells. Sharing the load of the network between different layers allows a better interference distribution to be achieved as well. This study focuses on the trunking gains obtained with DR and traffic reason handover functionality.

6.3.1 Directed Retry (DR)

Directed retry (DR) is designed to improve the trunking efficiency of the system thus reducing the blocking experienced by the end users. As its name points out, this functionality re-directs the traffic to neighbouring cells in case of congestion in call set-up (mobile originated or mobile terminated). The cell selection is based on downlink signal level. The performance of this functionality is very dependent on the existing overlapping between cells since it is required that at least one neighbouring cell has sufficient signal level for the terminal to be re-directed. The higher the overlapping, the higher the trunking efficiency gain. If the overlapping between two cells were 100%, the trunking efficiency of both cells would be ideally equivalent to that of a single logical cell with the resources of both cells. Figure 6.17 displays the gain of DR with different overlapping levels based on static simulations. The gain ranges from 2 to 4%. In order to achieve substantial gains, the required overlapping level is higher than the typical one of single-layer macrocellular networks.

6.3.2 Traffic Reason Handover (TRHO)

Traffic reason handover (TRHO), sometimes referred to as *load sharing* [5], is a solution that optimises the usage of hardware resources both in single-layer and multi-layer networks. The main difference with DR is that in the case of TRHO, when a certain load

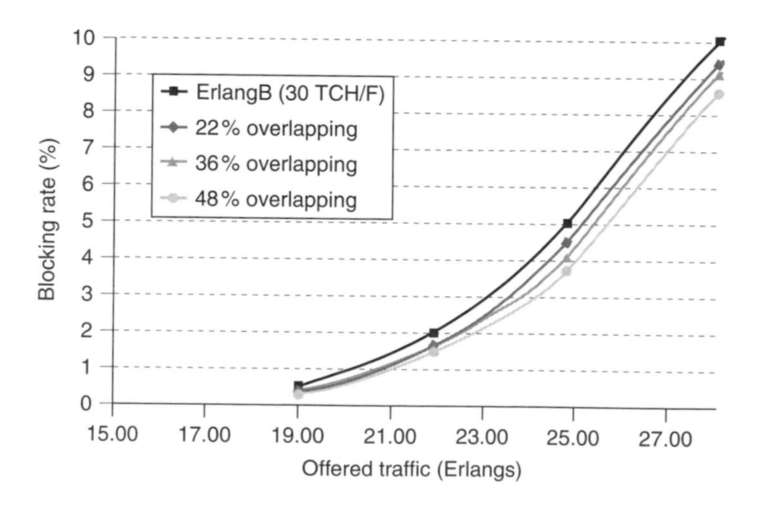

Figure 6.17. Performance of directed retry measured from static simulations. Four TRXs per cell and different overlapping level configurations

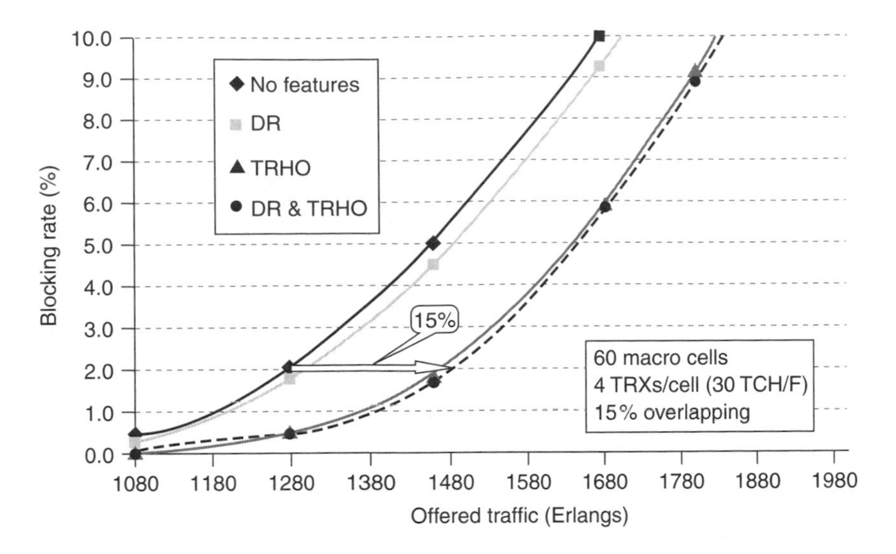

Figure 6.18. TRHO and DR performance from static simulations

is reached, all the terminals connected to the serving cell can be subject to reallocation to neighbouring cells. This makes this functionality much more efficient than DR since there is a much higher probability of finding a suitable connection to be reallocated.

There are several different implementations that may impact the end performance. Some basic implementations are based on releasing the load of the serving cell by forcing a certain number of connections to 'blindly' be handed over to neighbouring cells. Some others change the power budget handover margins directing the mobile stations in the cell border to less loaded neighbouring cells. Finally, some advanced solutions can monitor the load in the possible target cells to ensure successful handovers and system stability.

The performance of TRHO functionality has been verified extensively with simulations and real network trials. Figure 6.18 shows static simulation results using advanced implementation of the functionality. The network configuration was based on 36 cells with 4 TRXs per cell and 15% overlapping with neighbouring cells, which is considered here as a typical macrocellular scenario. For a 2% blocking rate, TRHO increases 15% the capacity of each cell through an overall trunking efficiency gain. Additionally, DR does not provide substantial additional gain for large TRX configurations, although its use is recommended since both TRHO and DR are complementary functionalities.

6.4 Performance of GSM HR Speech Channels

As described in Chapter 5, a logical traffic channel for speech is called *TCH*, which can be either in full-rate (TCH/F) or half-rate (TCH/H) channel mode. When TCH/H is in use, one timeslot may be shared by two connections thus doubling the number of connections that can potentially be handled by a TRX and, at the same time, the interference generated in the system would be halved for the same number of connections. TCH/H channels are interleaved in 4 bursts instead of 8 as the TCH/F, but the distance between consecutive bursts is 16 burst periods instead of 8.

In the first GSM specifications, two speech codecs were defined, the full-rate (FR) codec, which was used by the full-rate channel mode, and the half-rate (HR) codec, which was used by the HR channel mode. Later on, the enhanced full-rate (EFR) codec, used as well by the FR channel mode, was introduced to enhance the speech quality. Finally, AMR was specified introducing a new family of speech codecs, most of which could be used dynamically by either channel mode type depending on the channel conditions.

In terms of link performance, the channel coding used with the original HR codec provides equivalent FER robustness to one of the EFR codecs. This is displayed in Figure 6.19(a). However, when the quality of the voice (mean opinion score (MOS)) is taken into account, there is a large difference between EFR and HR. This is displayed in Figure 6.19(b). Considering the error-free HR MOS performance as the benchmark MOS value, the EFR codec is roughly 3 dB more robust, that is, for the same MOS value, the EFR codec can support double the level of interference. On the other hand, as mentioned before, HR connections generate half the amount of interference (3 dB better theoretical performance).

From these considerations it is expected that HR has a network performance similar to that of EFR, in terms of speech quality outage. Figure 6.20 illustrates this, displaying network-level simulations comparing the performance of EFR and HR. The quality benchmark criterion, bad quality samples outage, must be selected to reflect the lower performance of HR in terms of MOS. For instance, the selected outage of FER samples is that which degrades the speech quality below the error-free HR MOS. In the case of EFR, the considered FER threshold is around 4%, whereas in the case of HR, this threshold is down to 1% approximately. With these criteria both HR and EFR have a similar performance in terms of spectral efficiency. The reader should notice that the network performances shown here depends on the selected criteria, i.e. other sensible outages could have been chosen, and the performances of EFR and HR could have changed consequently, although the overall conclusion of similar performance would still be valid. These results do not consider speech quality or HW utilisation and cost. Obviously, the use of HR has an associated degradation in speech quality since the HR speech codec has a worse error-free MOS, that is, for the good-quality FER samples, the speech quality of EFR would be

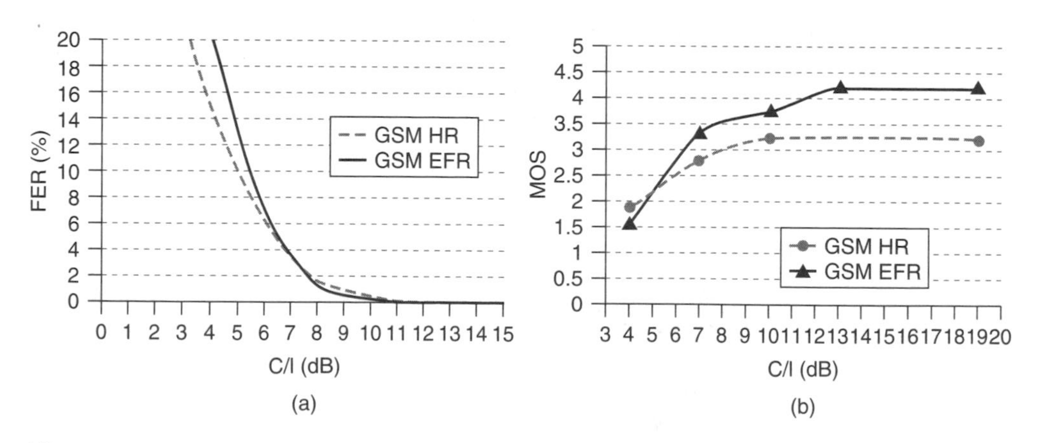

Figure 6.19. Link level performance in terms of (a) FER and (b) MOS of FR and HR codecs in TU3 for ideal frequency hopping [6]

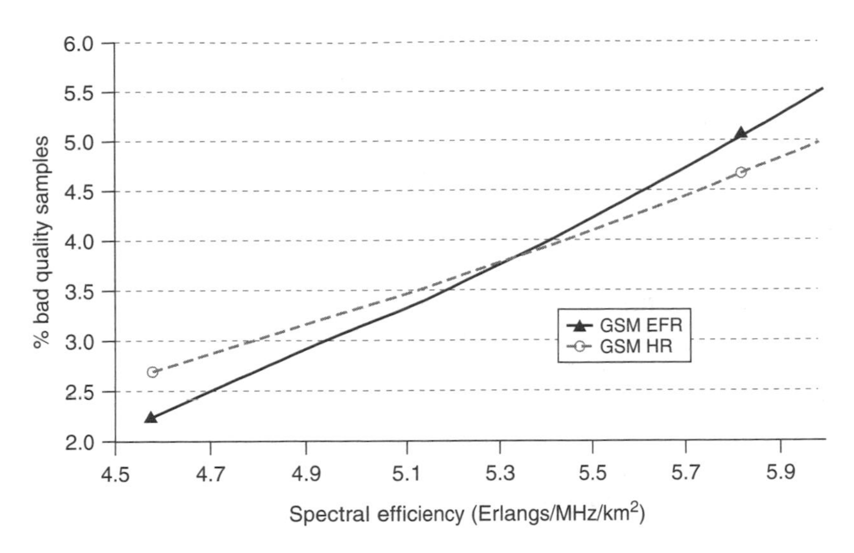

Figure 6.20. HR versus EFR network simulations (hopping layer)

higher than that of HR. On the other hand, the HW utilisation improves so the number of TRXs required for the same traffic is lower. Appendix B analyses the HR-associated HW utilisation gains, providing guidelines for HR planning and dimensioning.

The main reason why the original HR codec has not been widely deployed has been the market perception of its speech quality degradation. However, some operators have successfully deployed HR as a blocking relief solution only utilised during peak traffic periods. The introduction of AMR will undoubtedly boost the use of HR channel mode since the speech codecs used in this mode are a subset of the ones used in the FR channel mode; so, with an efficient mode adaptation algorithm there should not be substantial MOS degradation when the HR channel mode is used.

6.5 Adaptive Multi-rate (AMR)

6.5.1 Introduction

The principles of adaptive multi-rate (AMR) were described in Chapter 1. AMR is an efficient quality and capacity improvement functionality, which basically consists of the introduction of an extremely efficient and adaptable speech codec, the AMR codec. The AMR codec consists of a set of codec modes with different speech and channel coding. The aim of the AMR codec is to adapt to the local radio channel and traffic conditions by selecting the most appropriate channel and codec modes. Codec mode adaptation for AMR is based on received channel quality estimation [7]. The link level performance of each AMR codec, measured in terms of TCH FER, is presented in this section.

Increased robustness to channel errors can be utilised to increase coverage and system capacity. First, link performance is discussed; then system performance is examined by using dynamic GSM system simulations and experimental data. Mixed AMR/EFR traffic scenarios with different penetration levels are studied. Finally, the performance of AMR HR channel mode utilisation and its HW utilisation gains are carefully analysed.

6.5.2 GSM AMR Link Level Performance

With AMR, there are two channel modes, FR and HR, and each of these use a number of codec modes that use a particular speech codec. Figure 6.21 includes the link level performance of all AMR FR codec modes: 12.2, 10.2, 7.95, 7.4, 6.7, 5.9, 5.15 and 4.75 kbps. Channel conditions in these simulations are typical urban 3 km/h (TU3) and ideal FH.

As displayed in Figure 6.21, robust AMR FR codec modes are able to maintain low TCH FER with very low C/I value. With AMR FR 4.75 codec mode, TCH FER remains under 1% down to 3-dB C/I, whereas 8.5 dB is needed for the same performance with EFR codec. Therefore, the gain of AMR codec compared with EFR is around 5.5 dB at 1% FER. However, the gain decreases 1 dB when FH is not in use as presented in Figure 6.22 where the channel conditions were TU3 without FH.

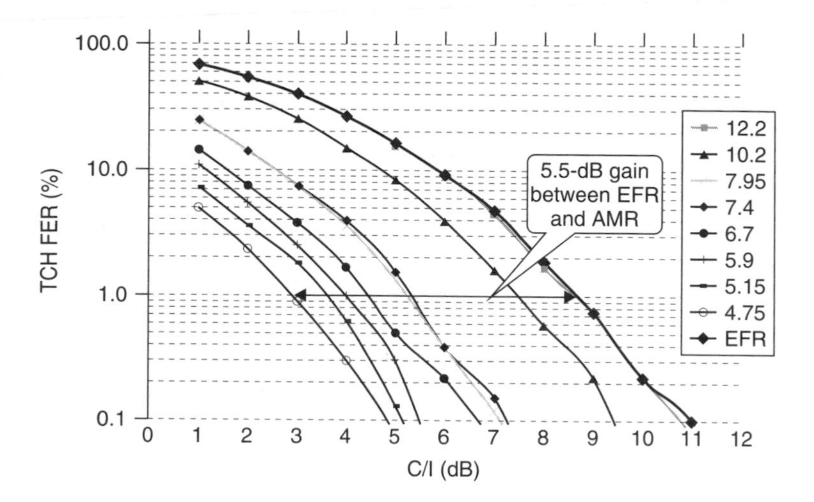

Figure 6.21. AMR full-rate link level TCH FER results (TU3, iFH)

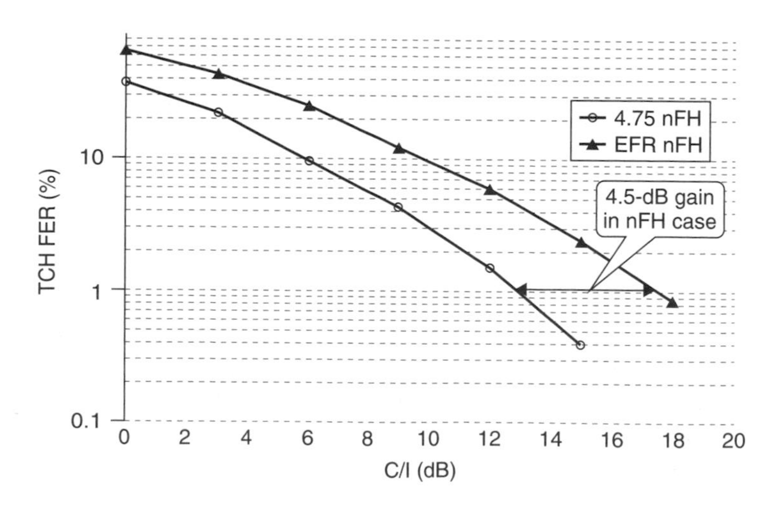

Figure 6.22. AMR full-rate link level TCH FER results (TU3 without FH)

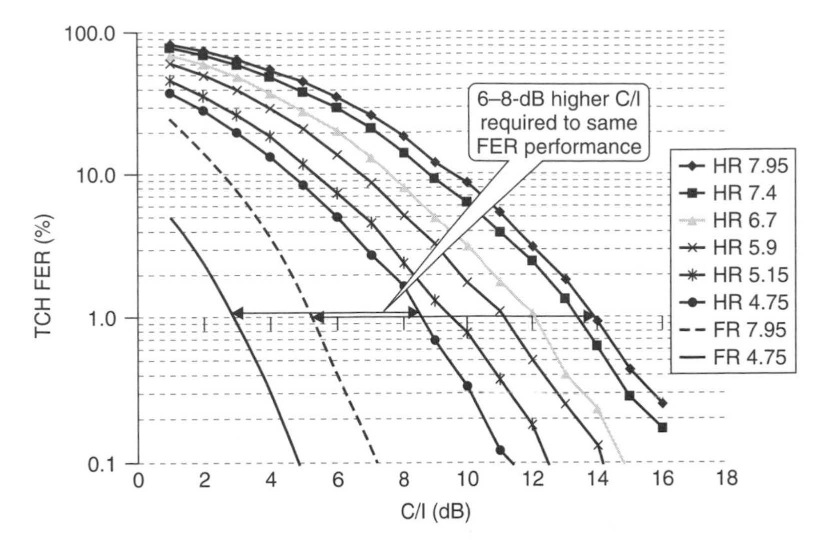

Figure 6.23. AMR half-rate link level TCH FER results (TU3, iFH)

Figure 6.23 shows the link performance of all AMR HR codec modes: 7.95, 7.4, 6.7, 5.9, 5.15 and 4.75 kbps. In the case of AMR HR codec modes, although the same speech codecs as in FR mode are in use, higher C/I values are required for the same performance. For example, there is around 8 dB difference between FR 7.95 kbps and HR 7.95 kbps. This is due to the different channel coding used. The gross bit rate of FR channel mode codecs is 22.8 kbps, whereas it is only 11.4 kbps for the HR channel mode codecs [6]. Therefore, for the same speech coding, HR channel mode has considerably less bits for channel coding and therefore it has higher C/I requirements. Finally, the HR 4.75 codec mode has a similar link performance to the FR 12.2 codec mode. All these points need to be taken into account when the channel mode adaptation strategy between full- and half-rate channel modes is defined.

The link performance enhancements of AMR codec improve as well the effective link budget and therefore coverage of speech services, as presented in Chapter 11.

6.5.2.1 Speech Quality with AMR Codec

With AMR, there are multiple speech codecs dynamically used. Some of them tend to have higher speech quality (stronger speech coding), others tend to be more robust (stronger channel coding). Figure 6.24 shows the MOS performance of the AMR FR codec modes. These curves illustrate how each codec mode achieves the best overall speech quality in certain C/I regions. The codec mode adaptation algorithm should ensure that the MOS performance of the overall AMR codec is equivalent to the envelope displayed in the figure. This envelope displays the performance of the AMR codec in the case of ideal codec mode adaptation, that is, the best performing codec mode is selected at each C/I point. In the high C/I area, EFR and AMR codecs obtain about the same MOS. However, with less than 13-dB C/I, the EFR codec requires approximately 5 dB higher C/I to obtain the same MOS than the AMR codec. Figure 6.25 shows equivalent MOS results for AMR HR channel mode. For the same MOS values, HR codecs have higher C/I requirements.

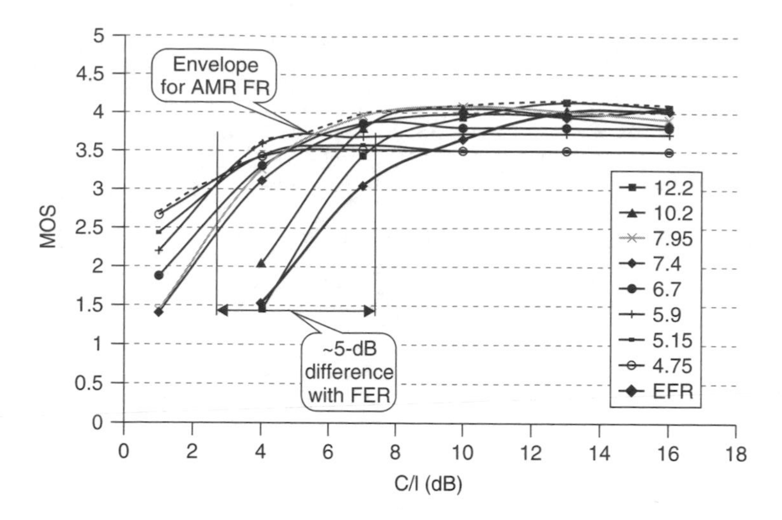

Figure 6.24. EFR and AMR full-rate MOS results from [6]

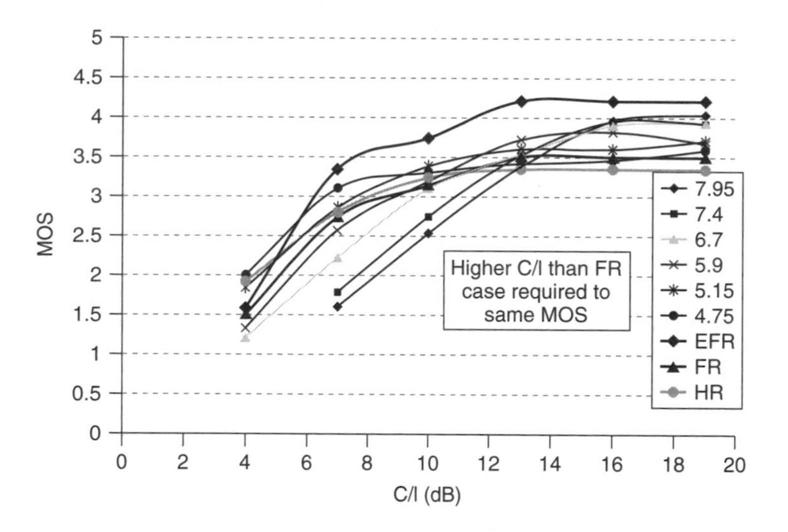

Figure 6.25. AMR half-rate, EFR, FR and HR MOS results [6]

6.5.2.2 Link Performance Measurements

Extensive laboratory and field validation of the simulated results presented previously has been conducted. The results for a TU50 non-hopping channel profile are shown in the following figures. Figure 6.26 shows the FER performance of AMR FR codec modes 12.2, 5.90 and 4.75, together with link adaptation (using codecs 12.2, 7.40, 5.90 and 4.75; 11, 7 and 4 dB as thresholds). Around 6-dB gain at 1% FER is found for the link adaptation,[9] which confirms the link gains predicted in the simulations.

[9] As per Figures 6.21 and 6.24, AMR FR 12.2 is used as an approximation of EFR in this subsection. The assumption is pessimistic in terms of speech quality.

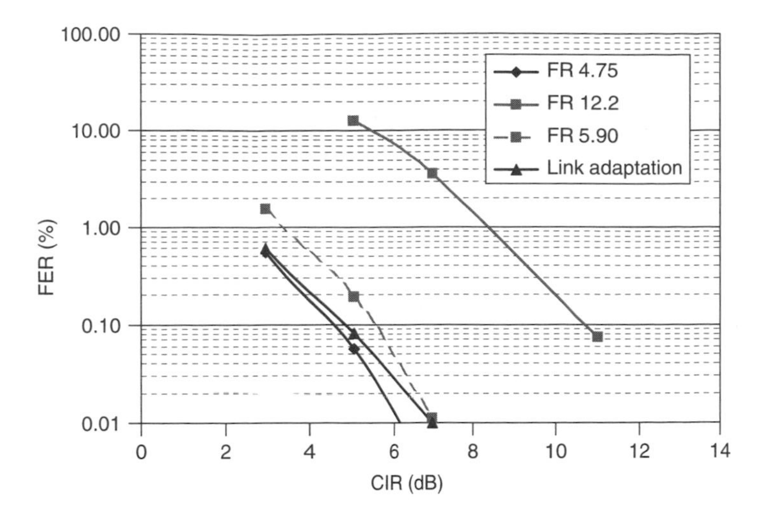

Figure 6.26. Experimental AMR full-rate link level FER results (DL TU50 no hopping, laboratory)

Also, speech quality performance was evaluated in laboratory environment. For that purpose, MOS was estimated by means of Perceptual Evaluation of Speech Quality (PESQ), an objective measurement recommended by the ITU [8]. The results are shown in Figure 6.27, for AMR FR codec modes 12.2 and 4.75, together with link adaptation with 11, 7 and 4 dB as thresholds. It is seen that AMR FR 12.2 quality falls below 3 in the PESQ scale for C/I values below 7 dB, whereas AMR FR 4.75 can be used until C/I values of 2 dB before its quality drops to the same level. As expected, the link adaptation plot is an envelope of the two: at high C/I value it sticks to the AMR FR 12.2 performance, providing its good error-free speech quality, but it behaves as the AMR FR 4.75 in low C/I. The resulting gain is also around 5 dB for a PESQ value of 3.

Finally, the link level capacity gains provided by AMR have been assessed through field trials in Network 14. At present, there is not enough AMR-capable mobile station (MS) penetration to allow the use of the effective frequency load (EFL) methodology presented in Chapter 5. Instead, link level performance must be calculated. Figure 6.28 shows the FER performance of EFR and AMR FR 5.90. In the drive test used to gather these results, the MS camped all the time in the BCCH layer. The plot was built by correlating a C/I estimation with FER on a SACCH period basis (i.e. every 480 ms); then, the average FER was computed for each integer C/I value. Both C/I and FER were measured by the receiver. The mobile speed varied between 3 and 50 km/h, and since hopping was not in use, the full potential gain could not be achieved. Nevertheless, as much as 4-dB gain was found at 1% FER level, in line with the link level simulation results. These results confirm that the performance figures presented in this chapter are achievable in live networks.

6.5.3 GSM AMR System Level Performance

This section introduces different AMR implementation strategies and the associated performance gains. The performance evaluations are based on dynamic simulation results

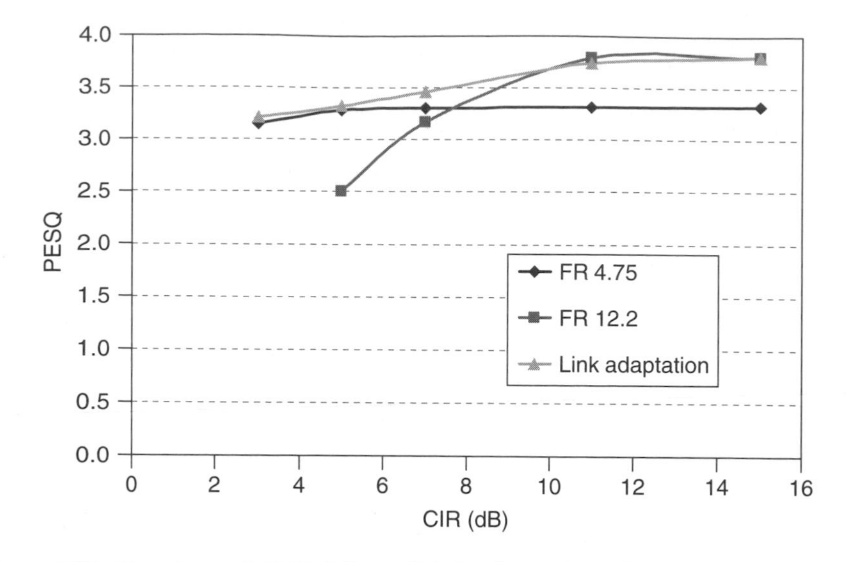

Figure 6.27. Experimental AMR full-rate link level speech quality results (DL TU50 no hopping, laboratory)

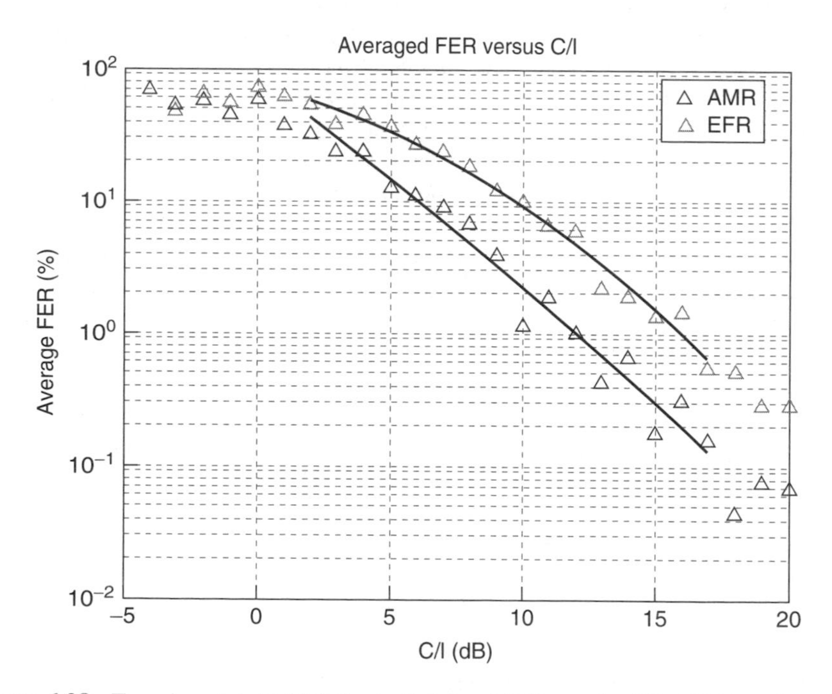

Figure 6.28. Experimental AMR full-rate link level FER results (DL TU50 no hopping, real network)

and in most cases refer to the performance of RF hopping configurations. Also, field trial results are provided to back up simulations. The BCCH performance is evaluated separately.

In the mixed traffic case (EFR and AMR capable terminals), two deployment strategies are studied. The first one deploys AMR in a single hopping layer and achieves its gains with the use of more aggressive PC settings for AMR mobiles, thus decreasing the interference level in the network. Owing to better error correction capability against channel errors, lower C/I targets can be set for AMR mobiles; hence tighter PC thresholds can be used. Another deployment strategy would be to make use of the reuse partitioning techniques to distribute the traffic across the layers according to the terminals' capabilities.

The use of HR channel mode is potentially an efficient way to increase capacity and HW utilisation in the network. However, when using HR codec modes, the connections require higher C/I in order to achieve the same quality as AMR FR, as displayed in Figure 6.23. Therefore, in order to determine the performance and resource utilisation capabilities of HR channel mode, a careful analysis is needed.

6.5.3.1 Codec Mode Adaptation

The codec mode adaptation for AMR is based on received channel quality estimation. The GSM specifications [7] include a reference adaptation algorithm based on C/I. In this method, the link quality is estimated by taking burst level C/I samples. Then, samples are processed using non-adaptive Finite Impulse Response (FIR) filters of order 100 for FR and 50 for HR channels. The described C/I-based quality evaluation method was used in the codec mode adaptation simulations in this study. The target of these simulations was to demonstrate the effect of codec mode adaptation in the system level quality and to find out optimum thresholds for subsequent performance simulations. Table 6.1 defines the C/I thresholds used in the codec mode adaptation simulations.

The GSM specifications [7] define that a set of up to four AMR codec modes, which is selected at call set up, is used during the duration of the call (the codec set can be updated at handover or during the call). The set of codec modes presented in Table 6.2 was included in the following performance evaluation simulations. In order to utilise the total dynamic range that AMR codec offers, the lowest 4.75 kbps and the highest 12.2 kbps modes were chosen in this set.

Figure 6.29 displays the TCH FER as a function of the C/I threshold sets included in Table 6.1. The relative usage of each codec mode is shown as well. The tighter the thresholds, more robust codec modes are used and lower FER can be achieved. On the other hand, tighter thresholds and robust codec modes may impact negatively the MOS performance. The 12.2 codec mode is clearly the most used mode. However, its usage drops rapidly when thresholds are tightened. With *set4* thresholds, for example, mode

Table 6.1. Simulated C/I threshold sets to demonstrate the effect of codec mode adaptation in FR channel

	set1	set2	set3	set4	set5
12.2–7.4 threshold	9 dB	11 dB	13 dB	15 dB	17 dB
7.4–5.9 threshold	5 dB	7 dB	9 dB	11 dB	13 dB
5.9–4.75 threshold	2 dB	4 dB	6 dB	8 dB	10 dB

Table 6.2. Defined mode sets for the performance simulations

AMR full-rate set (kbps)	AMR half-rate set (kbps)
12.2	
7.4	7.4
5.9	5.9
4.75	4.75

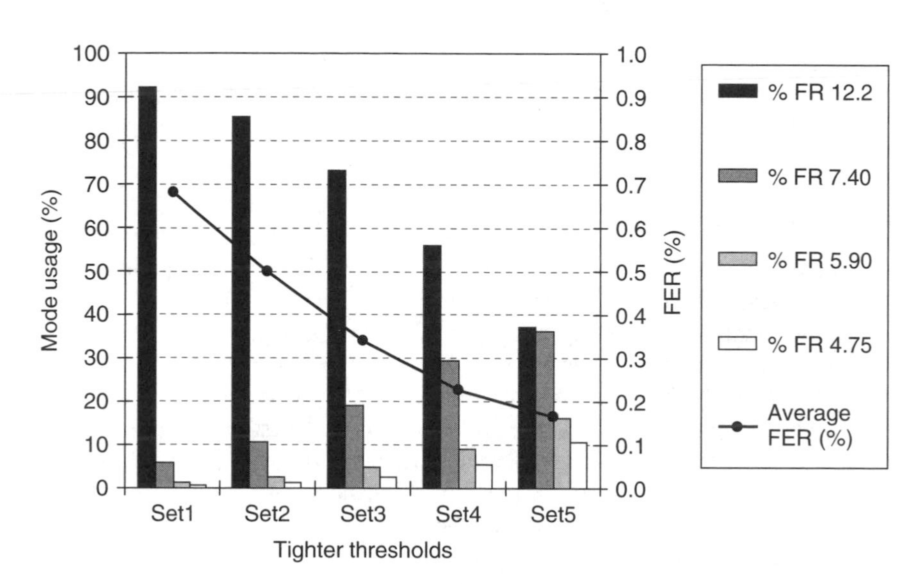

Figure 6.29. Average TCH FER and mode usage for different settings

12.2 is used less than 50% of the time and the usage of the mode 4.75 has increased over 10%. In order to find out the threshold settings with best speech quality performance, an MOS evaluation is needed.

Figure 6.30 displays a MOS performance analysis for the different threshold settings. The metrics used are the percentage of bad quality connections; a connection is considered as bad quality if its average MOS is below 3.2. In the system simulator, the MOS of the ongoing connections can be estimated by mapping the measured TCH FER values to corresponding MOS values. MOS values presented in [6] were used in this mapping. It can be seen that the best performing setting is *set3*, even though its average FER is higher than *set4* and *set5*. As explained in this section, the selection of the optimum thresholds will ensure that the AMR codec MOS performance follows the envelope of the ideal codec mode adaptation, that is, the best performing codec is always used. In the following simulations, the codec mode usage distributions are not presented, but this example gives the typical proportions likely to be seen in different AMR configurations.

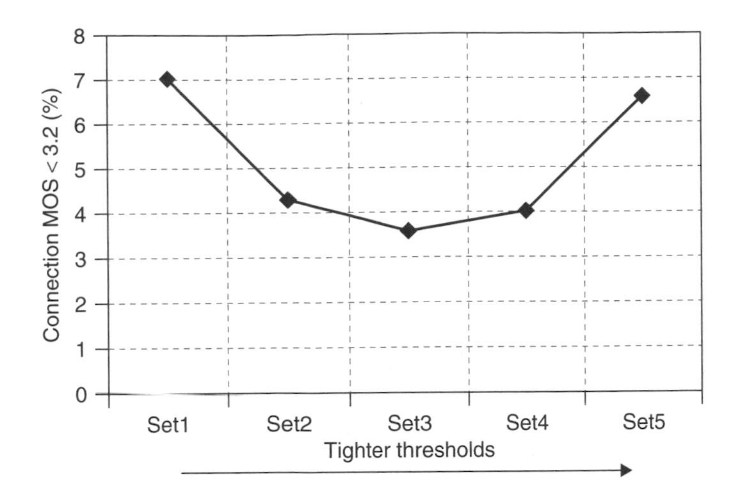

Figure 6.30. MOS performance as a function of threshold tightening

6.5.3.2 Single-layered Scenario Deployment

The previous section showed the speech quality increase accomplished with the AMR codec. There are several possibilities to turn this increased quality into additional capacity in the network. Naturally, if there were only AMR mobiles in the network, the network capacity could be easily increased by using tighter frequency reuses. However, most practical cases will have a mixture of AMR and non-AMR capable terminals in the network. In the mixed traffic scenario, one way to obtain the AMR gains would be through the adjustment of the PC settings so that AMR connections generate less interference. This can be effectively done defining different PC thresholds for AMR and non-AMR mobiles. The use of tighter PC thresholds for AMR mobiles will effectively lower the transmission powers used and therefore less interference will be generated. The following network simulations present the performance results using this technique.

Figure 6.31 displays the simulation results for a single-layered hopping network, with different AMR terminal penetration, and different PC thresholds for EFR and AMR traffic. Performance was measured in terms of FER. The number of bad FER samples using a 2 s averaging window was calculated. Samples with FER higher than 4.2% were considered bad quality samples. The use of AMR has a major impact on the system quality for the same traffic load. On the other hand, the use of AMR increases substantially the network capacity for the same quality criteria. Figure 6.32 illustrates the relative capacity gains for different AMR penetrations compared to the EFR traffic case. There is more than 50% capacity gain with 63% AMR penetration, and up to 150% capacity increase in the case of full AMR penetration.

Since different speech codecs have different MOS performance, it is important to analyse the MOS-based performance of AMR. Figure 6.33 shows that the AMR MOS-based gain is close as well to 150%. The calculation of MOS in network level simulators is estimative only, and assumptions must be made. From this analysis, it can be stated that MOS gain is in the same range as FER gain, as expected from the link level results

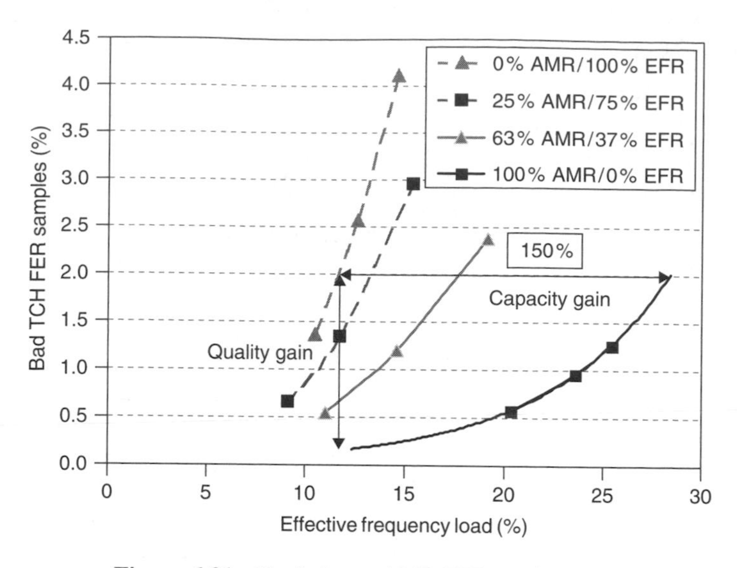

Figure 6.31. Single-layer AMR FER performance

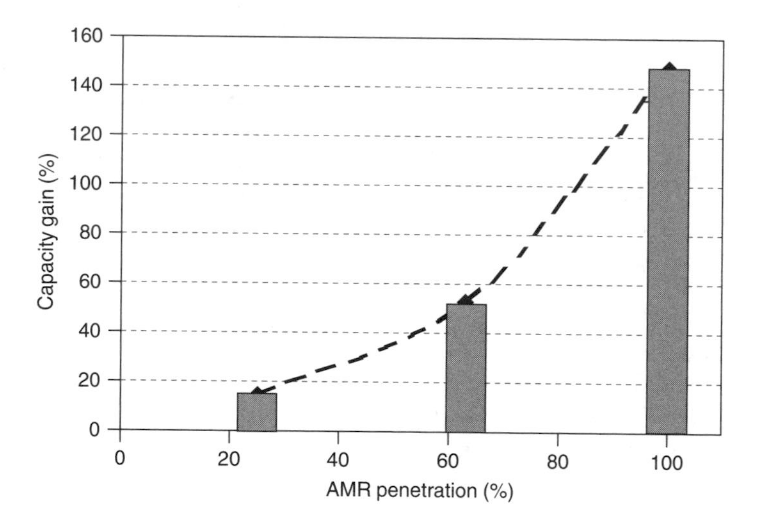

Figure 6.32. TCH FER-based AMR capacity gain

presented in the previous section, where similar gains were pointed out for FER- and MOS-based link performance.

These results assume a good performance link adaptation. Nevertheless, the core of the algorithm itself is not specified in the standards, and its implementation is challenging, mostly in terms of the C/I estimation and fading compensation [7]. Non-ideal link adaptation in the MS can impact the overall capacity gain in the DL direction. However, this is heavily MS-dependant, and for the reference case that has been studied here, it can be concluded that the gain of AMR with full terminal penetration on top of the defined GSM baseline performance is around 150%, both in terms of FER and MOS.

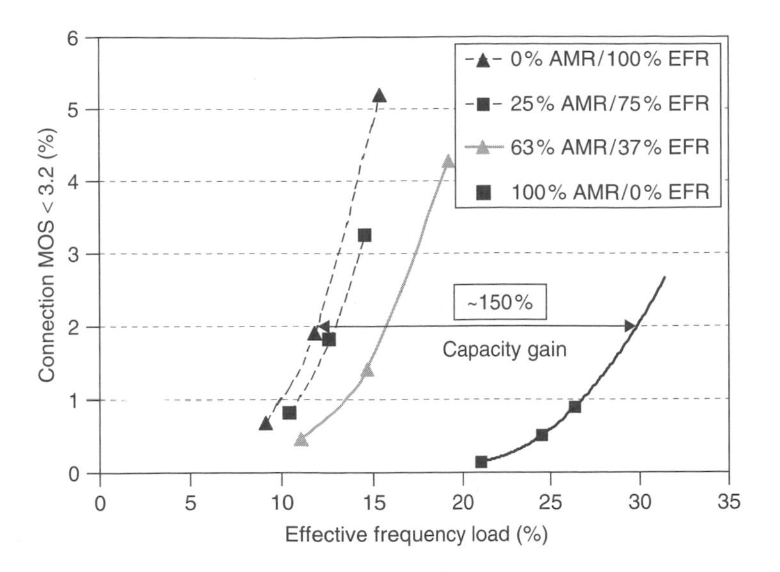

Figure 6.33. MOS-based AMR capacity gain

6.5.3.3 Multiple Layered Scenario Deployment

The use of the AMR functionality, together with reuse partitioning techniques, allows the network performance in the mixed traffic scenario to be further increased. AMR mobiles can be handed over to the underlay layer, which deploy tighter reuses, while the non-AMR mobiles would stay in the overlay layer where the looser reuses guarantee adequate performance. In Figure 6.34, the performance of the simulated multiple layer network is compared with the single-layer network. In the single-layer case, if the quality of the EFR connections is to be guaranteed, the potential AMR gains cannot be realised. In the multiple layer scenario, the AMR gains are realised and at the same time the quality of EFR connections guaranteed. This is translated to the slopes of the capacity increases displayed in the figure.

6.5.3.4 Half-rate Channel Mode

The MOS performance evaluation presented in this section demonstrated that AMR HR channel mode can be used without any noticeable speech quality degradation in high C/I conditions, since the HR channel mode uses the same set of speech codecs as the FR mode. In order to efficiently deploy HR, it is essential to have a quality-based channel mode adaptation algorithm that selects the adequate (good quality) to change from FR to HR. In the following simulations, the channel mode adaptation was based on the measured RXQUAL values. All incoming calls were first allocated to FR channel, and after a quality measurement period, it was decided whether the connection continues in FR mode or a channel mode adaptation is performed.

The effect of AMR HR utilisation on the FER-based performance is illustrated in Figure 6.35. Both AMR FR only and AMR FR and HR have similar performance. In other words, the use of HR neither improves nor degrades the overall network performance in terms of spectral efficiency. The gain from AMR HR channel mode comes from the higher

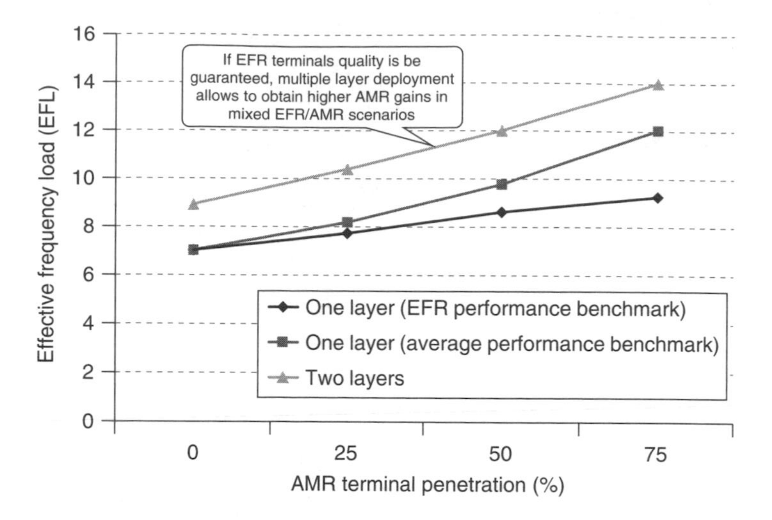

Figure 6.34. Single versus multiple layer AMR performance. Capacity increase relative to the EFR single-layer case

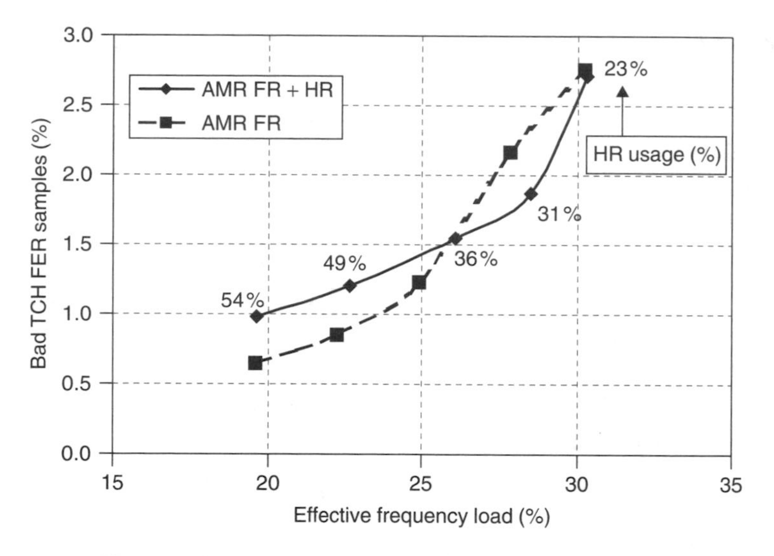

Figure 6.35. AMR half-rate channel mode performance

HW utilisation, which leads to a lower number of TRXs required to support the same traffic load. When the required number of timeslots for different traffic loads is evaluated, this number is reduced when HR channel mode is in use. This is illustrated in Figure 6.36. Notice that the half-rate utilisation, and therefore the HW efficiency, is higher at lower load operation points, as the better C/I distribution enables the HR usage. Appendix B presents in detail HR channel mode resource utilisation increase and dimensioning guidelines.

Since the 12.2 and 10.2 codec modes are not available when HR channel mode is in use, there will be some impact on speech quality when HR channel mode is introduced.

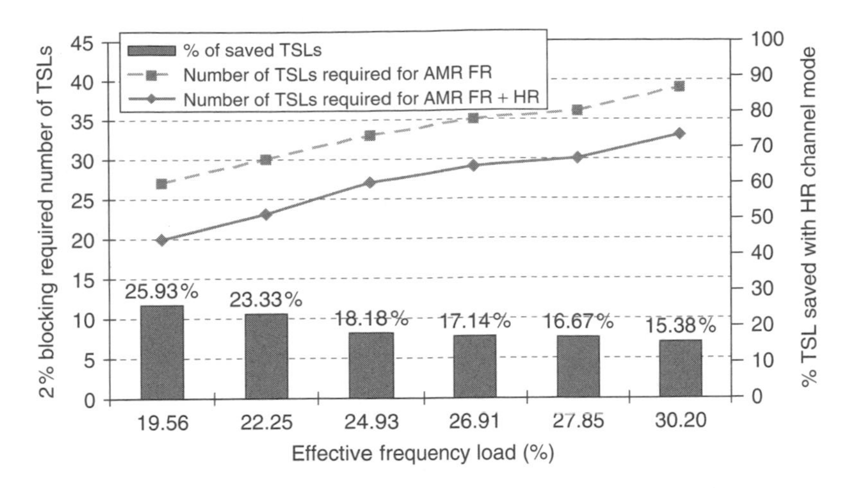

Figure 6.36. Half-rate gain in terms of timeslot requirements

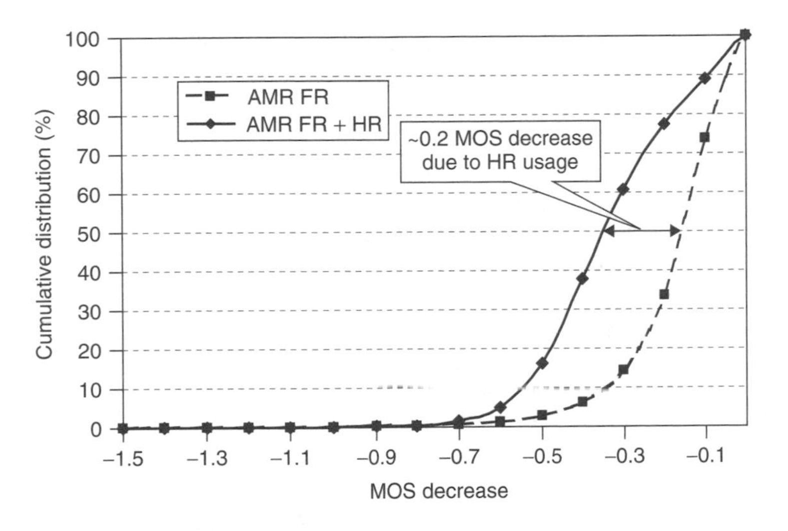

Figure 6.37. HR channel mode impact on MOS performance

Figure 6.37 presents the speech quality distribution of the connections as a function of MOS degradation, where the reference is EFR speech quality in error-free conditions. There is only an additional 0.2 MOS degradation when AMR HR channel mode is used which is due to the unavailability of the high AMR speech codecs (12.2 and 10.2) in the HR channel mode.

Finally, the use of AMR HR channel mode requires fewer TRXs for the same traffic load so the effective deployed reuse is looser. Looser reuses have some positive performance side effects such as higher MAIO management control, and higher performance gains in case dynamic frequency and channel allocation (DFCA) (Chapter 9) is introduced. Therefore, the introduction of HR channel mode reduces the TRX configuration and indirectly increases the performance.

6.5.3.5 AMR Codec Performance in BCCH Frequencies

As previously mentioned, AMR link level gain is slightly lower without FH. Figure 6.38 presents network simulation results for EFR and AMR codecs and different BCCH reuses. In these results, AMR BCCH reuses 9 and 12 outperform the EFR BCCH reuses 12 and 18, respectively. This means that BCCH reuses can be tightened in case of full or very high AMR penetration. The tightening of the BCCH reuse is of special importance in the narrowband (NB) deployment scenarios, as explained in Chapter 10. With the introduction of AMR, the signalling channels may become the limiting factor in terms of network performance. This subject is studied in Chapter 12.

6.6 Source Adaptation

6.6.1 Introduction

In the previous section, AMR link adaptation, using both codec and channel mode adaptation, was presented. Source adaptation is an innovative technique, which enables the use of new criteria for the AMR codec selection based on source signal instantaneous characteristics.

Source adaptation is linked to the voice activity detection process. When the voice activity has certain profiles, the speech quality is not degraded with the usage of lower AMR codecs; for instance, low energy voice sequences can be encoded with low-resolution speech codecs without any perceptual quality degradation. Therefore, with adequate codec adaptation algorithms, average bit rate during active speech can be reduced without reducing the speech quality. Figure 6.39 illustrates this concept, in which the speech codec is changed according to the detected voice activity without any perceived speech quality degradation as compared to the continuous use of the 12.2 kbps codec.

Two aspects have to be considered in order to integrate this concept with the existing AMR codec mode adaptation:

- The first issue is how to integrate this source-dependant codec decision algorithm with the codec mode adaptation. The quality of the local link must still be taken into

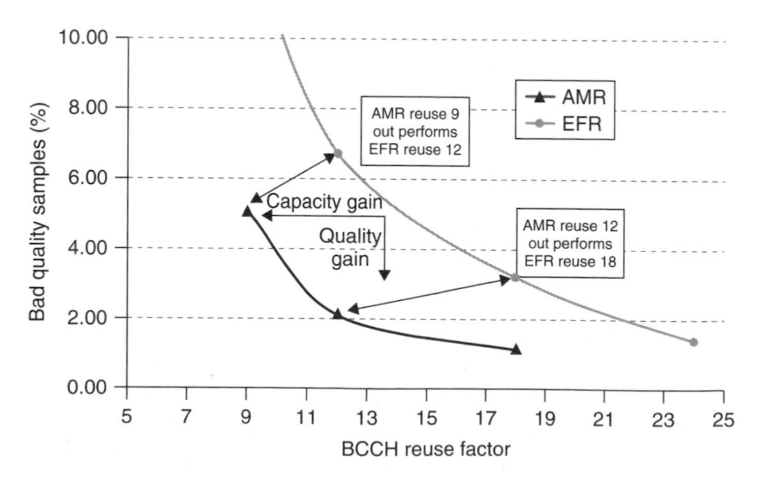

Figure 6.38. BCCH layer performance with AMR

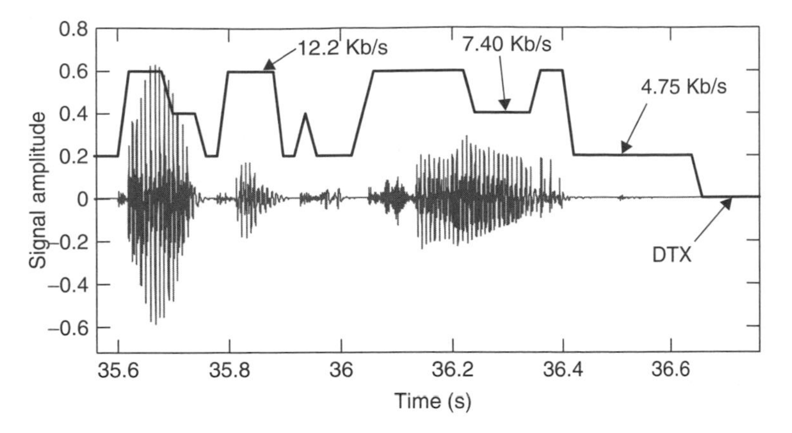

Figure 6.39. Source-adapted codec changes in a sample speech signal

account in the codec selection, that is, source dependency is added to the overall codec adaptation algorithm. One simple way to address this question is to choose the most robust codec between the two issued by both algorithms (link adaptation and source adaptation). For instance, if link conditions are good, but the speech characteristics allow reduction in bit rate, and hence the usage of a more robust codec, the latter should be used.

- The second question is how to turn the potential reduction of gross bit rate into network capacity gain. The way to achieve this is to utilise lower transmission power, generating less interference and therefore increasing network capacity. In order to achieve this, the source adaptation algorithm should be linked to a fast power control (FPC) functionality with the same cycle as AMR link adaptation, that is, 40 ms. This enables further reduction in transmitted power on top of the traditional PC algorithm.

It should be noted that DL source adaptation could be implemented as a proprietary feature in the network encoder side without any need of new signalling or impact in the remote decoder of the receiver; thus, there is no need for changes in the standards. Nevertheless, UL source adaptation does require changes in the standards, since the terminal is not allowed to override Codec Mode Commands (CMC, see [7]) sent from the network.

6.6.2 *System Level Performance*

The following simulation results present the achievable capacity gains with Source Adaptation on top of AMR. The Active Codec Set was chosen to include 3 codec modes, AMR FR 12.2, 7.40 and 4.75. Compared to the 4-codec ACS, studied in the previous section, the performance degradation is negligible. Such set was selected since standards state [7] that codec changes can be made only to neighbouring codecs. Thus, if codec is to be downgraded from 12.2 to 4.75 because of source adaptation, it may be convenient to reduce the number in intermediate steps.

Figure 6.40 shows the FER performance of basic AMR and AMR with source adaptation. In this case, for practical purposes, a sample is considered to be bad quality if

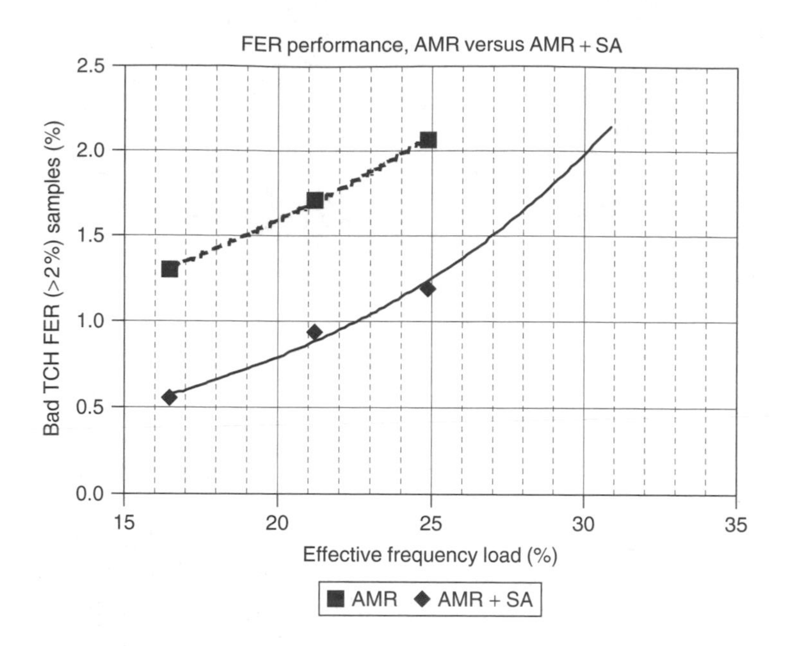

Figure 6.40. Performance of source adaptation compared to AMR

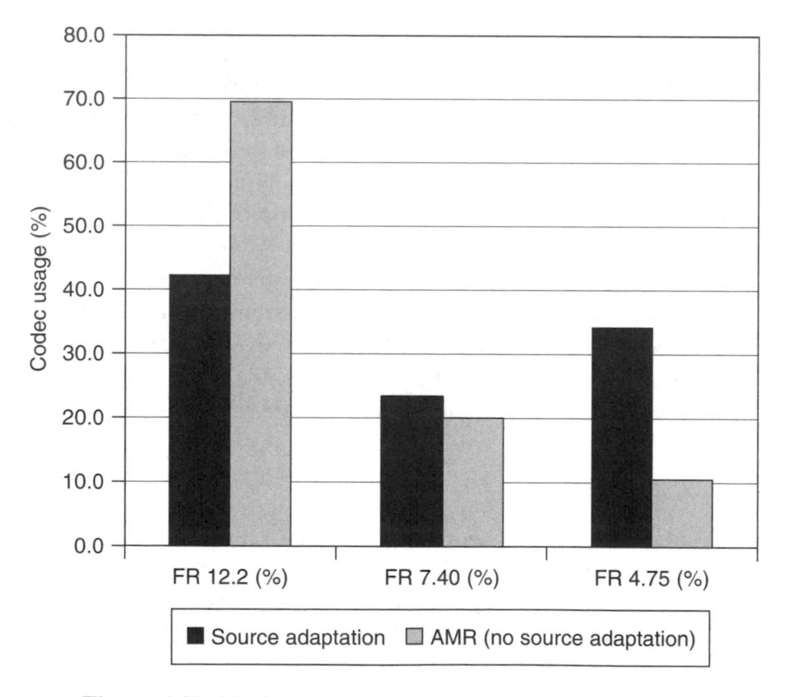

Figure 6.41. Mode usage with and without source adaptation

its FER is higher than 2%. From these simulations, a gain of 25% is achieved with the introduction of source adaptation.

As explained in the previous sections, this gain is basically due to a reduction in the transmitted power (FPC function), which is made possible because of the usage of more robust codecs. This effect is illustrated in Figure 6.41, where a comparison between codec distributions with and without source adaptation is shown. It can be seen that the usage of AMR FR 12.2 is reduced from 70% to almost 40%, whereas AMR FR 4.75 usage is increased from 10% to around 35%. As stated before, this bit-rate reduction does not impact perceived quality.

6.7 Rel'5 EDGE AMR Enhancements

6.7.1 Introduction

This section presents the performance analysis of the AMR enhancements included in Rel'5 and presented in Chapter 2.

GERAN Rel'5 specifies a new octagonal phase shift keying (8-PSK) modulated HR channel mode (O-TCH/AHS) for AMR NB speech. This new physical channel allows the use of all the available NB AMR speech codecs, with code rates ranging from 0.14 (AMR4.75) to 0.36 (AMR12.2).

Additionally, Rel'5 brings one additional method to increase the speech capacity of the network. A new *Enhanced Power Control* (EPC) method can be used for dedicated channels with both Gaussian minimum shift keying (GMSK) and 8-PSK modulation.[10] EPC steals bits from the SACCH L1 header to be used for measurement reports (uplink) and PC commands (downlink). This way the PC signalling interval can be reduced from SACCH block level (480 ms) down to SACCH burst level (120 ms). Thanks to optimised channel coding scheme (40-bit FIRE code replaced with 18-bit CRC code) of the new SACCH/TPH channel, the link level performance is actually improved by 0.6 dB (TU3 ideal FH at 10% BLER). See Reference [9] for further details.

This section also describes a major speech quality enhancement feature of Rel'5, the introduction of the WB AMR speech codecs. Link level results for WB-AMR in both GMSK FR and 8-PSK HR channels are presented.

6.7.2 EDGE NB-AMR Performance

Figure 6.42 displays the link level performance of the new 8-PSK Half-Rate channels, which is better than the performance of GMSK channels for all the AMR codec modes except for the most robust one (4.75 kbps). The higher the AMR codec mode is, the better the O-TCH/AHS channel mode performs compared to the TCH/AHS mode.

Listening tests (33 listeners, clean speech) were conducted in order to verify the subjective performance of the EDGE AMR HR channel mode compared to standard GSM AMR HR. Figure 6.43 presents the collected results. At each operating point (C/I), the AMR mode with highest MOS value is selected. The results show a clear benefit for the EDGE AMR HR over GSM AMR HR. Owing to the higher AMR modes in EDGE,

[10] Note that this is not the same as fast power control (FPC), which is specified for ECSD in Rel'99. ECSD FPC uses a 20-ms signalling period.

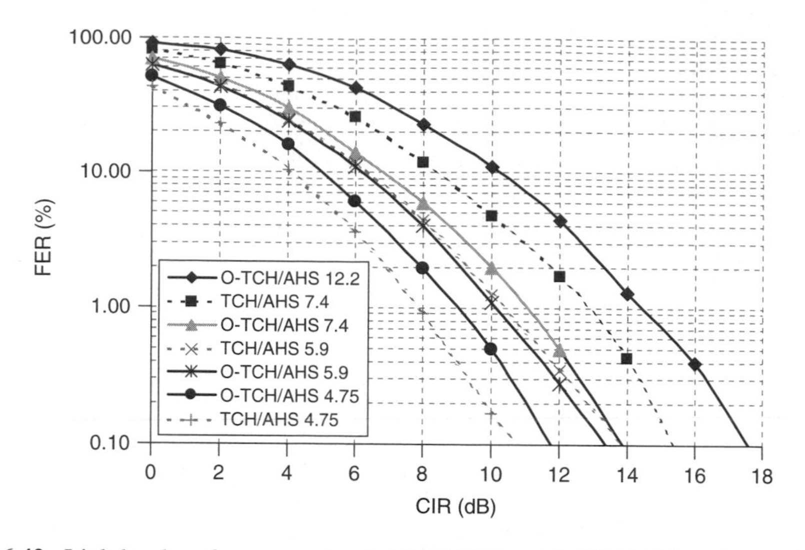

Figure 6.42. Link level performance of main TCH/AHS and O-TCH/AHS codec modes. TU3 iFH channels

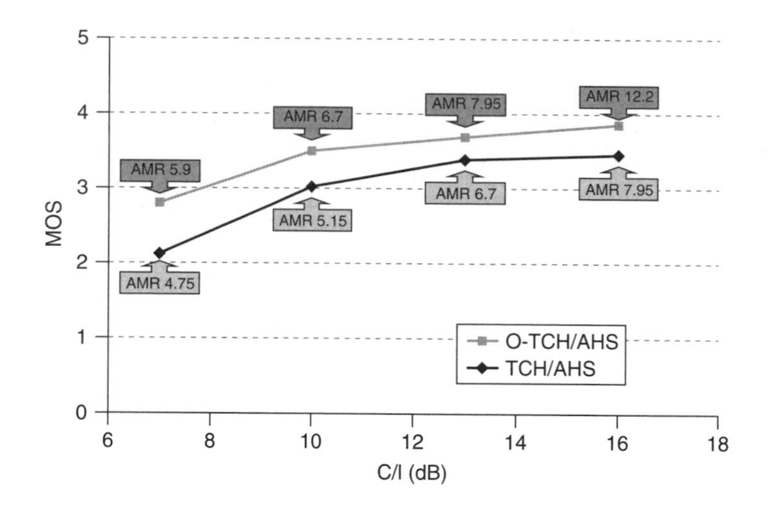

Figure 6.43. Comparison of MOS test results between GSM AMR and EDGE AMR half-rate. Best codec is selected for each operating point

the subjective performance gains would be even higher in typical practical conditions, where C/I is high and background noise is present.

From all these results, it can be concluded that EDGE AMR HR channel should be prioritised over GSM AMR HR channel, that is, when EDGE AMR HR is available, there is no need to do channel mode adaptation to GSM AMR HR. This is mainly because of the possibility to also use the two highest AMR codec modes (10.2 and 12.2).

Network simulations have confirmed the expected results from the above presented link level performance. Although no significant differences in terms of FER are observed when

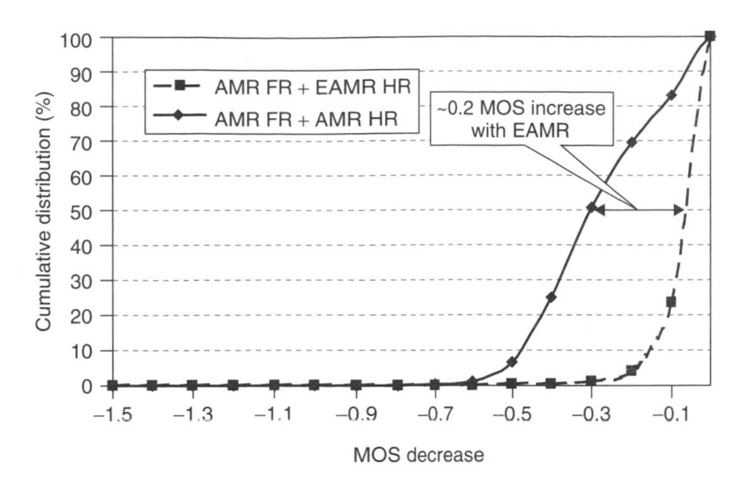

Figure 6.44. O-TCH/AHS channel mode impact on MOS performance

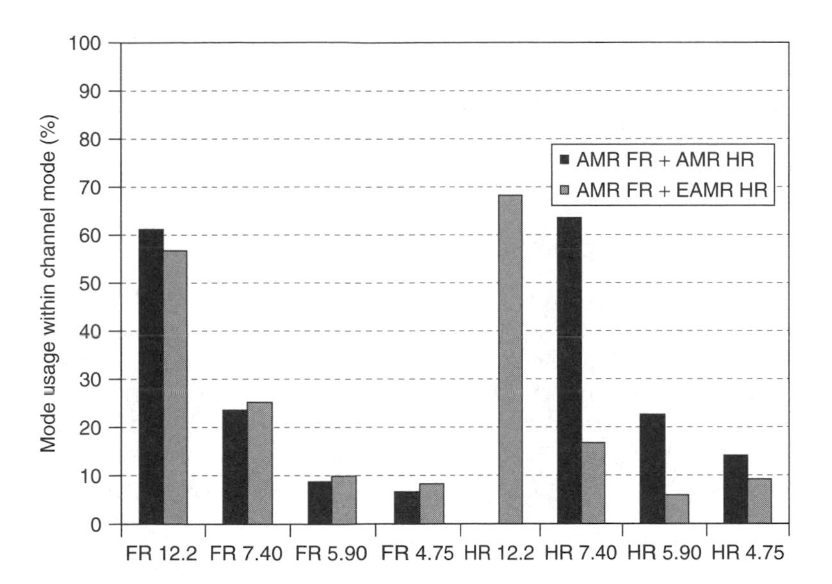

Figure 6.45. Mode usage when using O-TCH/AHS and TCH/AHS

using O-TCH/AHS, there is certainly an increase in the average speech quality perceived by the end-user, which is shown in Figure 6.44. In this figure, both plots correspond to the use of Codec and Channel Mode adaptation, but in 'AMR FR-HR' HR modulation is GMSK (as in Section 6.5.3.4), whereas in 'E-AMR' HR modulation is 8-PSK, adding the 12.2 kbps codec. The figure shows the quality distribution of the connections as a function of MOS degradation, in a way similar to Figure 6.37. The speech quality gain is roughly 0.2 on the MOS scale.

Figure 6.45 shows the compared codec mode usage histograms. FR codecs remain practically the same, but the presence of the additional 12.2 kbps codec in the O-TCH/AHS alters the HR distribution significantly. Almost 70% of the time an 8-PSK HR channel is used, the 12.2 kbps codec will be selected. It is the high usage of this codec that provides speech quality improvement. The final conclusion then is that although E-AMR NB does not provide significant capacity gains, it increases the average speech quality.

6.7.3 EPC Network Performance

The faster PC cycle of EPC can be transformed into system capacity gain. This gain depends heavily on the mobile station speed and shadow-fading component of the signal. An extensive set of network level simulations were performed in order to study the effect of EPC on system level. The simulation set-up was similar to the one used in the basic GSM HR simulations presented earlier. Only non-BCCH layer was studied with 1/1 RF hopping and 4.8 MHz of total spectrum. DTX was not used and bad quality call criteria was 1% FER averaged over the whole call. In the simulations, the EPC performance was compared to the standard PC performance, using the same RxQual/RxLev-based PC algorithm. Figure 6.46 shows the performance results at different speeds and shadow-fading profiles.

These results show that the gain from EPC is substantial for fast moving mobiles (around 20%) but quite small in the case of slow moving mobiles (the relative gain of EPC increases slightly also when the slow-fading variation is higher). This points out the sufficient speed of the GSM standard PC mechanism to follow shadow-fading variations for slow moving mobiles.

6.7.4 EDGE Wideband AMR Codecs

GERAN Rel'5 specifies new AMR WB speech codecs, which present a major improvement in speech quality. With the introduction of a new 8-PSK FR physical channel, WB AMR modes up to 23.85 kbps can be supported. The highest modes offer performance exceeding standard wireline quality. Owing to improved speech coding, WB codecs offer better quality than NB codecs even with similar codec bit rates. WB AMR codec modes in Rel'5 are defined for GMSK FR (TCH/WFS) and for 8-PSK FR (O-TCH/WFS) and HR (O-TCH/WHS). The usage of GMSK HR with WB codecs has not been standardised.

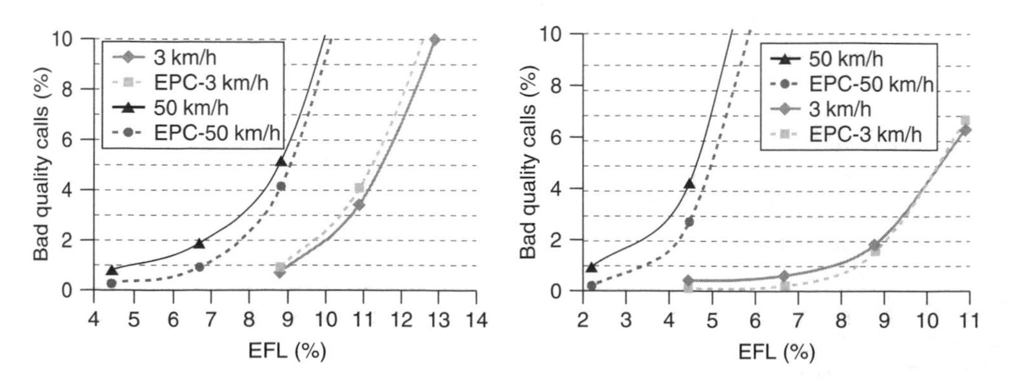

Figure 6.46. TCH-AHS EPC performance for slow fading standard deviation of 6 and 8 dB

Three codec modes are available in all channels: 12.65, 8.85 and 6.60 kbps. For 8-PSK FR channels (O-TCH/WFS), two additional WB-AMR codec modes can be used: 23.85 and 15.85, which offer very high quality speech in good channel conditions.

Figure 6.47 (a) shows the link level performance of WB AMR codecs on the GMSK FR (TCH/WFS) and 8-PSK HR (O-TCH/WHS) channels. The C/I requirements for 1% FER are between 5 and 9 dB for GMSK and about 4 dB higher with 8-PSK. Figure 6.47 (b) shows the curves for 8-PSK FR. The C/I requirement range for the three lowest modes is now between 4 and 6 dB.

Finally, Figure 6.48 summarises the C/I requirements of all the modes. It can be concluded that especially the 8-PSK FR channels seems to provide a very attractive option to provide truly high quality speech without compromising capacity.

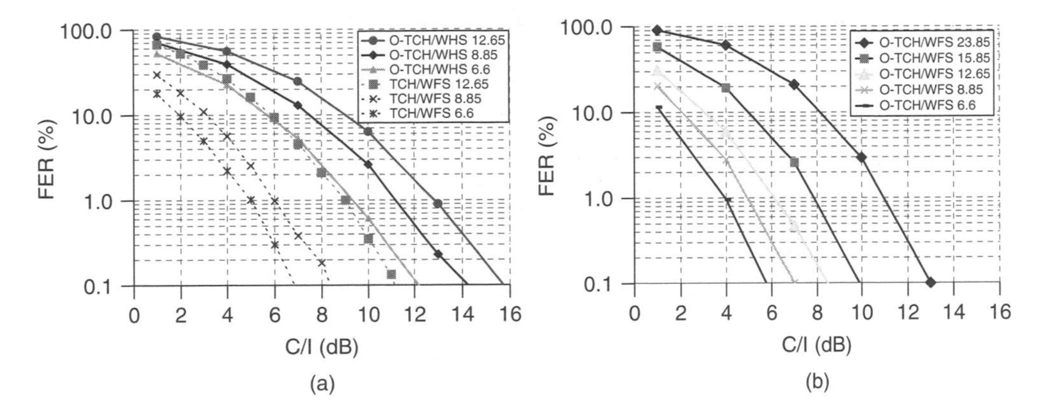

Figure 6.47. Link level performance of O-TCH/WHS, TCH/WFS and O-TCH/WFS AMR modes (TU3 iFH)

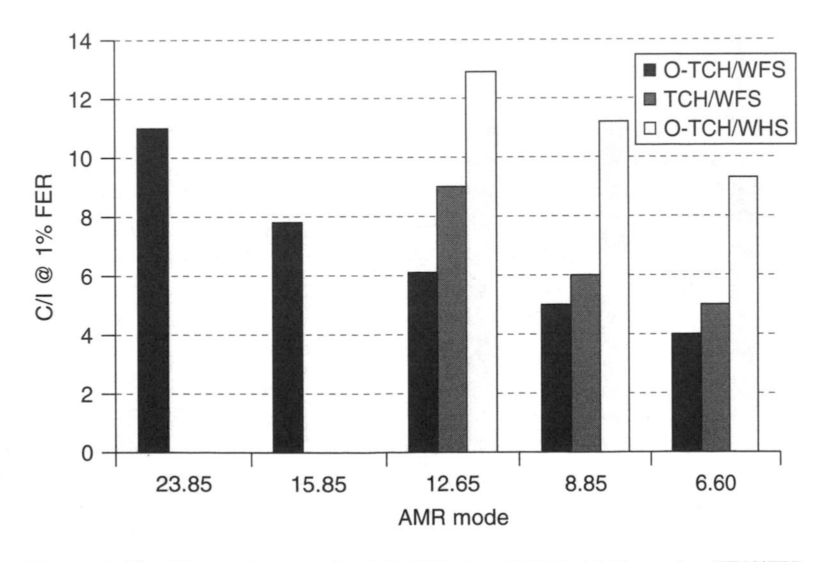

Figure 6.48. C/I requirement for 1% FER for all WB-AMR modes (TU3iFH)

6.8 Single Antenna Interference Cancellation (SAIC)

SAIC is a generic term comprising several different single antenna receiver techniques, which attempt to cancel interference through signal processing. It is mainly intended for usage in the downlink, since terminal size, cost and aesthetics requirements preclude the usage of multiple antenna techniques.

Within the 3GPP standardisation work carried out by Technical Specification Group (TSG) GERAN, an SAIC feasibility study is currently ongoing, investigating the expected performance of SAIC terminals in typical network scenarios and the need for signalling of the MS SAIC capability, for example, by using the MS classmark information.

The goal is to have SAIC standardised for Rel'6. However, owing to some operators' requests to have SAIC terminals earlier than Rel'6 natural timeframe, SAIC could become a release-independent functionality, with the first commercial terminals available in early 2004. This section attempts to explain the current status of the SAIC development, and analyse it in terms of link and network performance enhancements.

6.8.1 SAIC Techniques Overview

In GSM, one of the main factors limiting the spectral efficiency is the MS receiver performance when subject to co-channel interference (CCI). Conventional receivers have traditionally been optimised for Gaussian noise; nevertheless, as the nature of interference is partly known (e.g. modulation type and its subsequent characteristics, training sequence, etc), it is possible to optimise receivers to improve the performance in the presence of such interference. For mobiles having a single antenna, a number of techniques exist, all being designated as SAIC. Although SAIC techniques can be applied for both CCI and Adjacent Channel Interference (ACI), the main focus in this section will be SAIC for suppression of CCI.

In the literature, a number of SAIC techniques exist but basically they can be divided into two families:

- Joint Detection/Demodulation (JD)
- Blind Interference Cancellation (BIC).

The basic principle of Joint Detection [10, 11] is to do simultaneous detection of the desired and the interfering signal. In order to do this, JD methods rely on the identification of Training Sequence Code (TSC) of the interferer and its offset from the TSC of the desired signal. On this basis, a joint channel estimation is conducted using the 26 known bits from the interfering TSC and the corresponding bits from the desired signal. The resulting refined channel estimate takes into account the effect of the interfering burst, which improves the accuracy of the estimate and thereby decreases the BER. The two channel estimates are then used in a joint detector, which simultaneously detects the desired and the interfering signal.

JD works for both GMSK and 8-PSK modulations, although simulations have demonstrated that the best performance is achieved for a GMSK-modulated interferer. Because JD detects simultaneously the desired and the interfering signal, its performance is generally superior to other cancellation techniques, although at the cost of higher complexity.

Compared with JD, BIC techniques do not explicitly need a channel estimate of the interfering signal to perform detection of the desired signal. Several BIC methods exist, using different properties of the interfering signal; for example, for a GMSK-modulated interferer it is possible to use its constant envelope property or the fact that the same information is carried in the real and imaginary part of the signal [12, 13]. Basically, BIC techniques have a lower complexity than JD-based methods, making them more cost efficient and therefore attractive for first-generation SAIC terminals. The cost for this lower complexity is a reduction in the potential performance gain, compared with methods based on JD. The SAIC results presented in this section are based on two different low complexity BIC solutions: the first one is Mono Interference Cancellation (MIC) [13]; the other is called *GMSK-IC* (Interference Cancellation). Both the solutions have previously been used to generate input to the feasibility study ongoing in TSG GERAN [14, 15].

Identification of SAIC capability is another important issue currently being debated. The network is likely to require the knowledge of which mobiles are SAIC-capable, and potentially to what degree (e.g. GMSK only, GMSK and 8-PSK, etc). Such information may be beneficial for the network to control the C/I conditions different terminals are subject to: for instance, through tighter effective reuses for SAIC MS, or by means of SAIC-optimised Power Control, or even SAIC-tuned DFCA algorithms (see Chapter 9). Several possibilities exist: special signalling could be defined (one way to do this is to have explicit classmark information) or the network could be required to identify SAIC-capable terminals in an unassisted way.

6.8.2 SAIC Link Performance and Conditioning Factors

SAIC receivers are, as explained in the last section, optimised to extract additional information from the incoming signals to improve the detection of the desired signal. As a consequence, the performance of SAIC receivers can generally have large variations, depending on different parameters. The three most important parameters affecting SAIC receiver performance are the delay between desired and interfering signal, their respective TSC, and finally a new parameter named *Dominant to rest of Interference Ratio (DIR)*. Each algorithm behaves differently with respect to these variables, and this makes the characterisation of SAIC performance a challenging task.

Delays, especially between the main interferer and the desired signal, affect the performance of SAIC mobiles. Especially for implementations based on JD such delays can be critical, resulting in a significant drop in performance gains already at 5-symbols delay. BIC techniques are generally not so sensitive to delays and some algorithms can have a nearly constant gain with up to 20 symbols delay. This means that achievement of maximum SAIC benefits, especially in the case of JD, requires synchronised networks, which guarantee low delays between desired and interfering signals. In such a case, an inter-site distance of 3 km would imply an approximate maximum possible delay of 4 symbols,[11] and a median value much below that figure. Figure 6.49 [16] shows some laboratory results using a SAIC prototype, where the performance for different delays between desired signal and a single interferer has been tested. For this scenario, it can be seen that at large delays the link level gain is 4 dB at 1% FER, which is approximately half the gain that can be achieved when the interfering and desired signal are fully

[11] Bit length is about 1.1 km; assuming interference comes always from first tier, maximum distance between MS and interferer BTS can be 4.5 km at most.

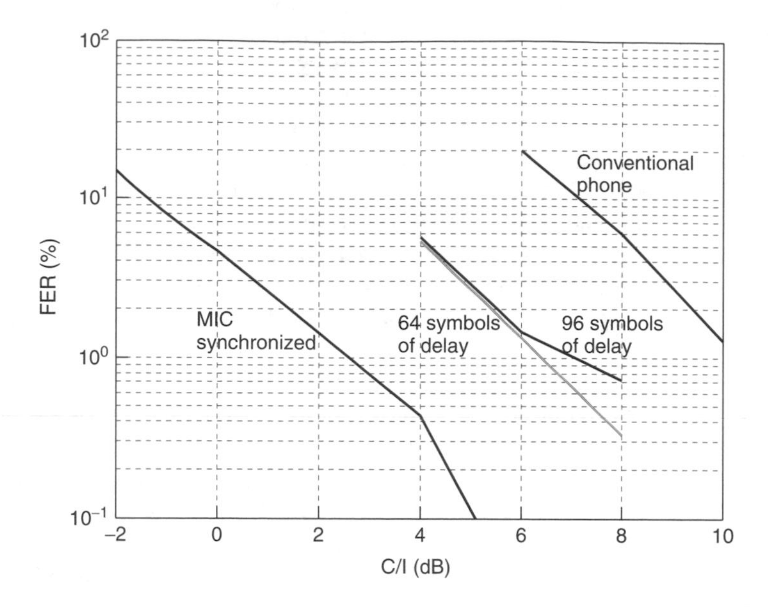

Figure 6.49. Performance of MIC prototype for several delays between desired signal and interferer. One single interference source was used (TCH FS, TU50 non-hopping)

synchronised. Thus, to obtain the highest possible gain for SAIC mobiles the network needs to be synchronised.

Another important aspect that needs to be taken into account is the influence of the TSCs in use. The TSC used in GSM were originally optimised to ensure good auto-correlation properties, whereas the cross-correlation properties were not considered. For unsynchronised networks, this is not a problem because of the delay between the signals from the different base stations. But when running a synchronised network the TSCs will arrive to the mobile approximately at the same time, and for some combinations of TSCs this may degrade the SAIC performance. The best way to cope with this problem is to either make proper TSC planning, thus reducing the likelihood of the worst TSC combinations, or to rely on DFCA algorithms to dynamically assign the most adequate TSC for each connection.

The SAIC performance demonstrated in Figure 6.49 was based on a scenario where only a single interfering signal was present. In practice, a mobile will receive interference from a number of BTSs at the same time and this will in general reduce the capability of SAIC mobiles to suppress interference. This can easiest be explained for JD-based algorithms, where simultaneous detection of the desired and a single interfering signal is done. When more than one interferer is present a basic JD receiver can only remove the strongest interferer i.e. the levels of the other interferers will limit the performance of the receiver.

The interference suppression capability of a SAIC receiver is normally linked to the power of the dominating interferer compared with the rest of interference, which can be described by the DIR factor:

$$DIR = \frac{I_{\mathrm{DOM}}}{\sum_{n=1}^{N} I_n - I_{\mathrm{DOM}} + N}$$

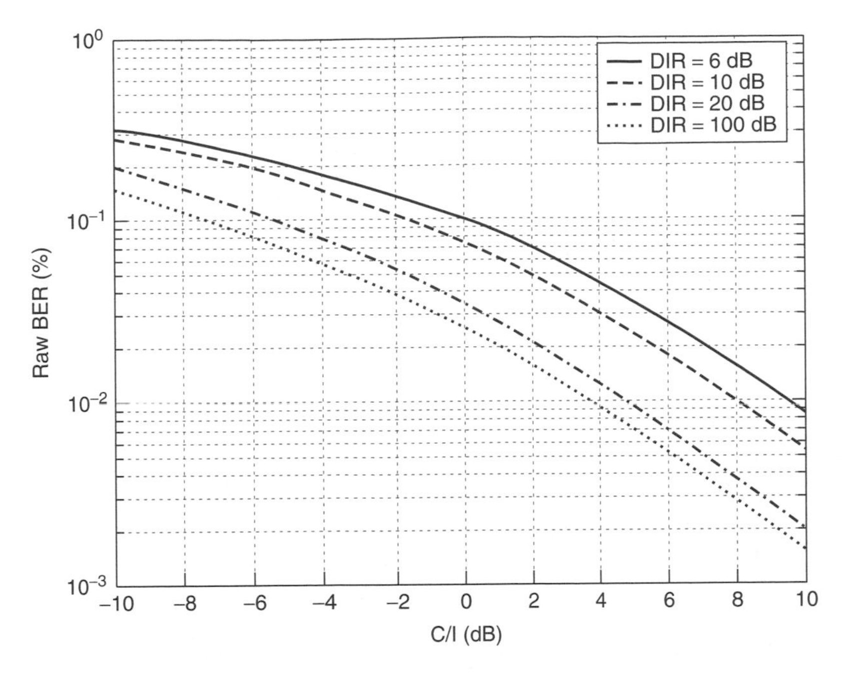

Figure 6.50. Link performance as a function of DIR

where I_{DOM} is the power of the dominant interferer, M is the total number of co- and adjacent channel interferers and N is the receiver noise. In Figure 6.50, the performance of an SAIC receiver is shown for various DIR.[12] In practice, the interferer scenario will be more complicated. As expected, the gains are higher when a single interferer is dominating but still some gain is possible for low DIR values.[13] Figure 6.51 shows a sample of a typical cumulative distribution of DIR taken from network simulations.

6.8.3 SAIC Network Performance

The dependency of SAIC performance with interference delay, TSC and DIR has important implications in the network performance. It is obvious from the discussion in the previous section that higher gains can be expected in synchronised networks having good TSC planning, especially for Joint Detection (which will lose, anyway, part of its potentiality even in the synchronised case, especially in large-cell areas). Nevertheless, the implications of DIR in performance deserve more attention. It has been identified through simulations that the higher the load level is, the lower the DIR values are, which, as shown in Figure 6.50, leads to a reduction in the potential link and network gains from SAIC. Figure 6.52 shows the simulated network performance of AMR FR 5.90 for a conventional and an SAIC receiver. The overall gain is close to 40%.[14] This is lower than the

[12] A simple model based on two interferers has been used for these simulations.

[13] In a real network, the DIR experienced by ongoing connections can vary between −5 dB to about 30 dB, and the gain provided by SAIC would change accordingly.

[14] Half Rate SAIC performance is likely to provide higher gains due to the more favorable associated DIR distribution.

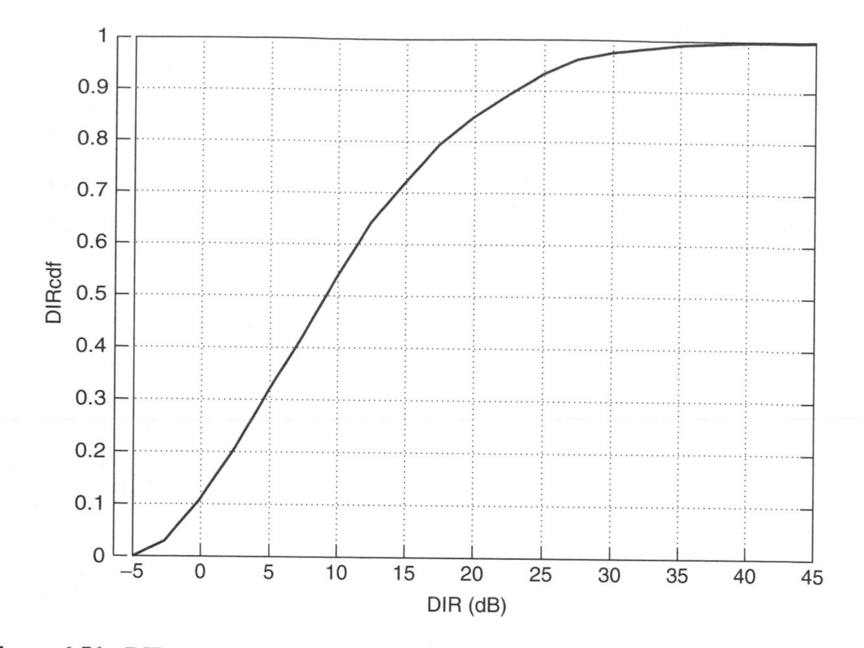

Figure 6.51. DIR cumulative distribution function for a heavily loaded hopping network

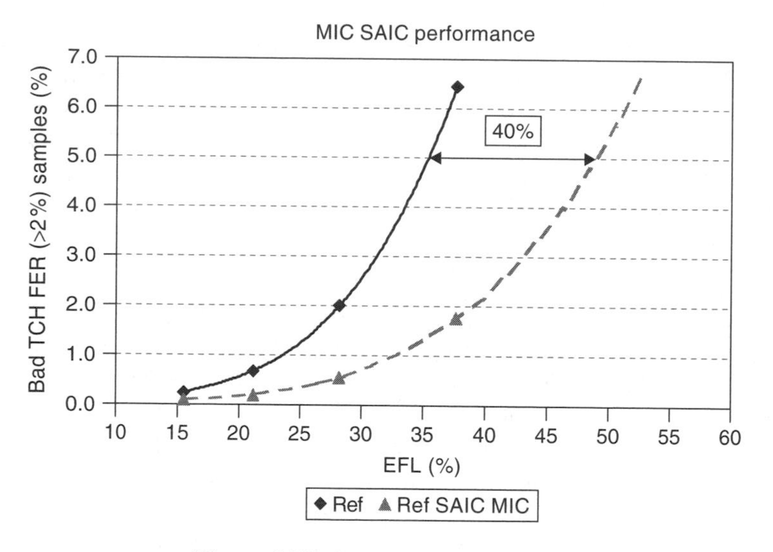

Figure 6.52. SAIC FER performance

gains theoretically expected from the link performance results displayed in Figure 6.49, which is a logic consequence of the realistic DIR conditions occurring in the network.

Also, PC actually reduces the benefits of SAIC, since it decreases DIR in the network. Finally, another important factor to be taken into account is ACI. Simulations have shown that performance degradation in the presence of ACI is higher for SAIC traffic than for

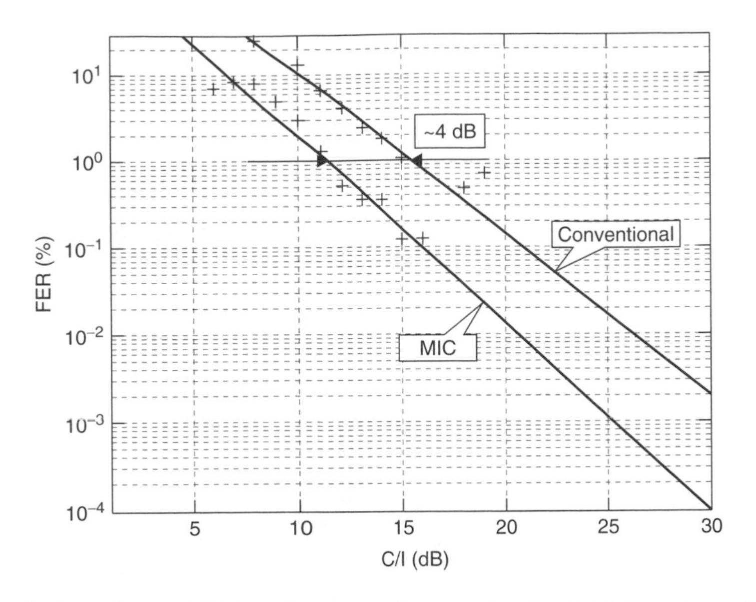

Figure 6.53. Experimental FER performance of conventional and MIC receivers (TCH FS, random hopping over 5 frequencies)

non-SAIC. ACI implications on SAIC are under study in standardisation, but it should be noted here that ACI is much dependant on network planning, for example, it is possible to avoid intra-site ACI if enough spectrum is available.

Nevertheless, trials conducted to assess SAIC network performance gains have demonstrated promising results. In an unsynchronised network, a C/I decrease at 10% outage of up to 2.7 dB was registered. In synchronised networks, the reported gain ranges from 2–3 dB to more than 4 dB, depending on the number of interferers. Figure 6.53 (Network 14) shows the measured link performance of a conventional receiver and an SAIC prototype operating in a synchronised test network. C/I estimates from the receiver were correlated with measured FER. The interference was artificially generated in five neighboring sectors. This explains why the gain is higher than predicted by simulations. This result is, nevertheless, indicative of the important benefits achievable through SAIC.

6.9 Flexible Layer One

Flexible Layer One (FLO) was presented and explained in Chapter 2. FLO defines a parameterised, easily configurable physical layer devised to optimally support any new Internet multimedia subsystem (IMS) and the support of FLO to AMR FR codec 12.2 will be studied here.

6.9.1 FLO for Circuit-switched Voice

Figure 6.54 shows how FLO Transport Channels (TrCHs) are configured to support AMR 12.2 codec. Two TrCHs, labeled A and B, are required, because of different quality requirements of class 1a and 1b bits. On TrCH A, the transport block is 83 bits long, whereas on TrCH B, it is 163 bits long. A 6-bit CRC is attached on TrCH A to make a

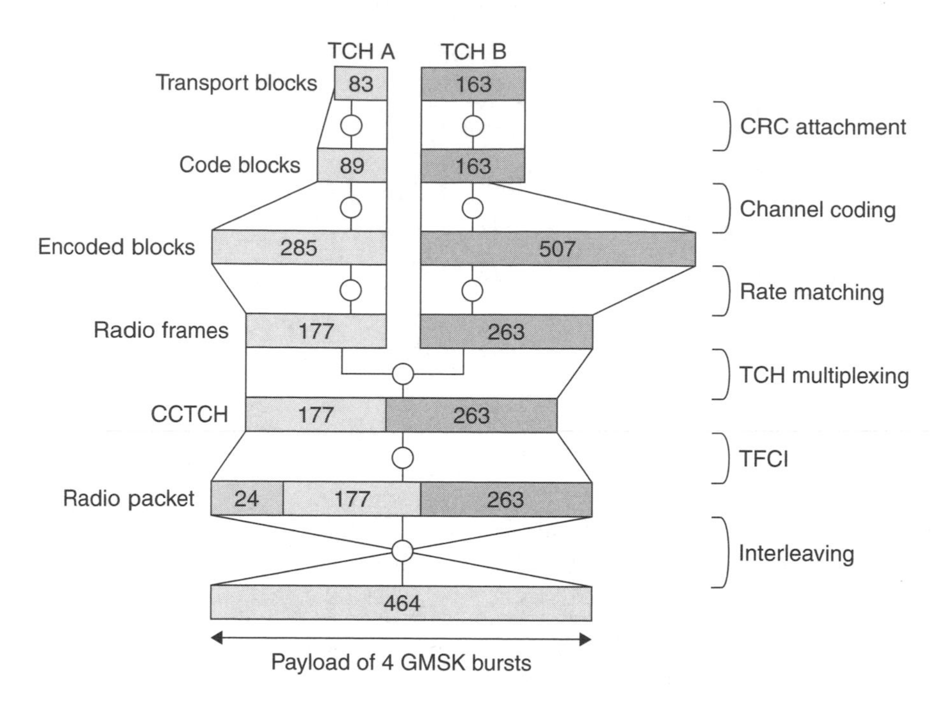

Figure 6.54. AMR FR 12.2 codec encapsulated in FLO bearers

code block of 89 bits. The two code blocks are then processed through channel coding, yielding 285 coded bits for TrCH A and 507 bits for TrCH B. After rate matching, 177 bits are left on TrCH A and 263 bits on TrCH B. Concatenation of the two radio frames results in a Codec Composite TrCH (CCTrCH) of 440 coded bits. The 24-bit Transport Format Combination Indication (TFCI) is then attached to the CCTrCH, creating a 464-bit radio packet. The radio packet is finally diagonally interleaved over eight bursts.

The requirement that the size of the radio packet is equal to the 464-bit payload available on GMSK FR channels is complied with. On the other hand, the coding rate for TrCH A is 0.503 and for TrCH B 0.620, which are close to the coding rates in AMR FR 12.2: 0.500 for class 1a bits and 0.5949 for class 1b bits. The small increase is mainly due to the overhead introduced by FLO (TFCI and tail bits for the second transport channel).

Figure 6.55 compares the link-level performance of AMR FR 12.2 with the AMR 12.2 over FLO in TU3 with ideal FH. As we can see, the performance difference is negligible. In both cases, 9 dB of C/I is required to reach 1% of FER. This demonstrates the ability of the FLO architecture to produce optimised coding schemes.

6.9.2 FLO for VoIP

In Voice over Internet Protocol (VoIP), AMR frames are encapsulated in real-time protocol (RTP) packets and carried over UDP/IP. This means that an RTP/UDP/IP header is to be transmitted together with the encoded speech. The header is compressed [17] to reduce the impact of the subsequent overhead, which is as a result of 51 bits in average. For the same speech bit rate, this overhead will obviously reduce the channel coding protection.

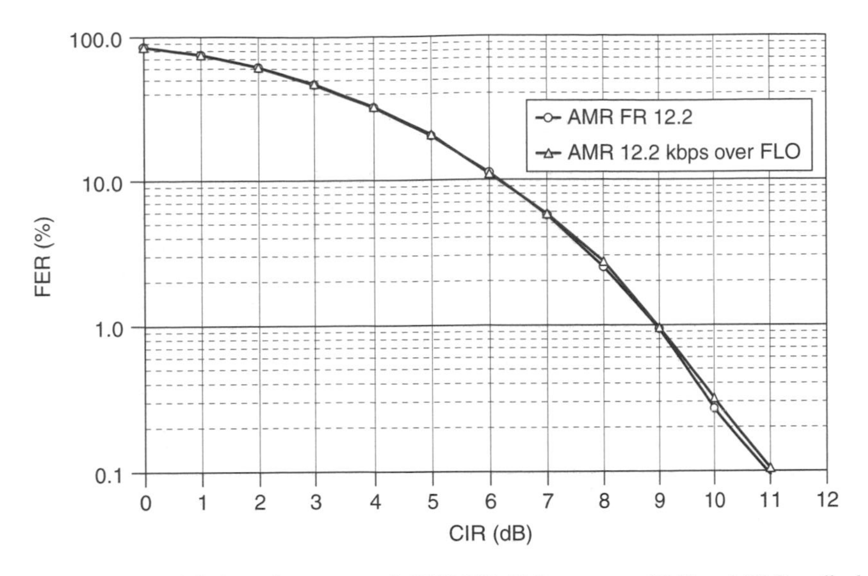

Figure 6.55. Compared link performance of AMR FR 12.2 over non-FLO and FLO radio bearers

In Rel'5, the only radio bearers that can be used for VoIP are based on the (E)GPRS coding schemes. Unfortunately, they were not designed for real-time unacknowledged dedicated services such as VoIP, but instead optimised for non-real-time acknowledged shared services:

- The interleaving depth is fixed to 20 ms, although a 40-ms or an 80-ms depth is allowed by the delay requirements of VoIP.
- The VoIP payload to be carried does not always exactly fit into one RLC packet. So, padding has often to be used. For instance, with VoIP AMR 10.2, MCS-3 has to be used with a padding of 40 bits.
- The RLC/MAC header contains some fields that are useless for unacknowledged dedicated services. For instance, the uplink state flag (USF) is not needed.

With FLO, all these problems are solved, thereby improving the performance:

- The interleaving scheme can be selected at call setup.
- The size of the transport blocks is negotiated and no padding is required.
- The RLC/MAC header is optimised for unacknowledged dedicated services [18].

With AMR FR 12.2 taken as an example, Figure 6.56 compares the link-level performance of VoIP in Rel'5 with the one of the same service in Rel'6 using FLO (VoIP with AMR 12.2 as codec mode is called *VoIP 12.2* here). In Rel'5, MCS3 (GMSK) or MCS5 (8-PSK) can be used for VoIP 12.2. With FLO, the channel coding is tailored at call setup during which some freedom is allowed. FLO offers several alternatives for the channel coding: the diagonal interleaving depth can be either 40 or 80 ms, and the modulation can be GMSK or 8-PSK. The performance for each of the four combinations is shown in Figure 6.56.

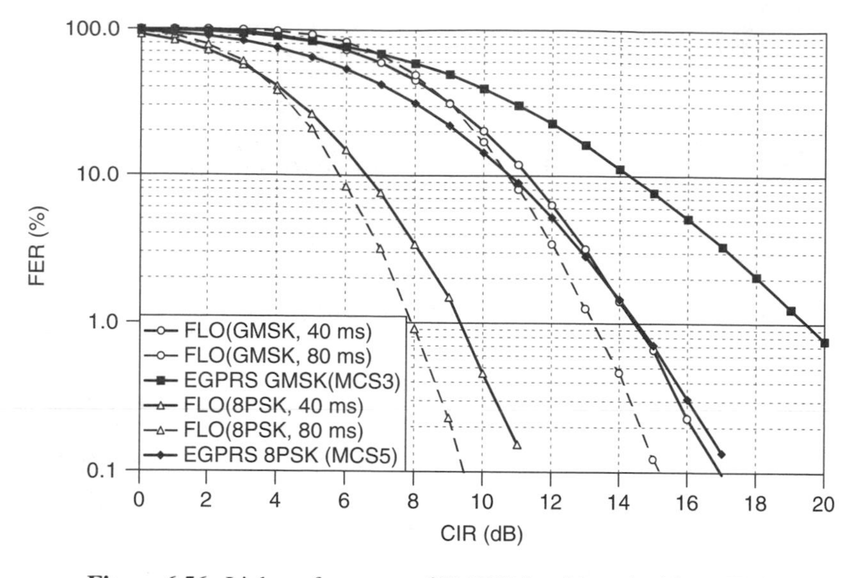

Figure 6.56. Link performance of VoIP12.2, with and without FLO

At 1% FER, the resulting gains range between 5 and 6.6 dB. For instance, with GMSK modulation and 80-ms diagonal interleaving, the link-level performance of VoIP 12.2 at 1% FER is improved by FLO by more than 6 dB compared to Rel'5 (non-FLO), where MCS3 is used.

Even with FLO, there is a significant loss in performance for VoIP because of overhead. Comparison of Figure 6.55 and Figure 6.56 shows a link degradation of more than 4 dB at 1% FER (80 ms interleaving, GMSK modulation).

It can be concluded that, for CS speech, FLO is capable of providing robustness similar to the one achievable with non-FLO physical layer. Also, it improves noticeably the performance for IMS real-time services, partially overcoming the limitations of non-FLO radio bearers.

References

[1] Johansen J., Vejlgaard B., *Capacity Analysis of a Frequency Hopping GSM System*, M.Sc.E.E. Thesis Report, Aalborg University, Aalborg, June 1995.

[2] Nielsen Thomas T., Wigard J., *Performance Enhancements in a Frequency Hopping GSM Network*, Kluwer Academic Publishers, Dordrecht, 2000.

[3] Wigard J., Nielsen Thomas T., Skjærris S., Mogensen P., 'On the Influence of Discontinuous Transmission in a PCS1900/GSM/DCS1800 Type of Network', *VTC'99*, Houston, 1999.

[4] Wigard J., Mogensen P., Michaelsen P. H., Melero J., Halonen T., 'Comparison of Networks with different Frequency Reuses, DTX and Power Control using the Effective Frequency Load Approach', *VTC2000*, Tokyo, Japan, 2000.

[5] Watanabe F., Buot T., Iwama T., Mizuno M., 'Load Sharing Sector Cell in Cellular Systems', *IEEE Proc. PIMRC'95*, 1995, pp. 547–551.

[6] 3GPP TR 26.975 v5.0.0 Release 5: Performance Characterization of AMR Speech Codec.
[7] 3GPP TS 45.009 v5.3.0 Release 5: Link Adaptation.
[8] ITU-T Recommendation P.862 (02/2001), Perceptual Evaluation of Speech Quality (PESQ), An Objective Method for End-to-End Speech Quality Assessment of Narrowband Telephone Networks and Speech Codecs.
[9] Bellier T., Moisio M., Sebire B., 'Speech Capacity Enhancements in the GSM/EDGE Radio Access Network (GERAN)', *ICT2002*, Beijing, China, June 23–26 2002.
[10] Giridhar K., Chari S., Shynk J. J., Gooch R. P., Artman D., 'Joint Estimation Algorithms for Cochannel Signal Demodulation', *IEEE Proc. Int. Conf. Communications*, Geneva, Switzerland, 1993, pp. 1497–1501.
[11] Ranta P. A., Hottinen A., Honkasalo Z. C., 'Co-channel Interference Cancelling Receiver for TDMA Mobile Systems', *Proc. IEEE Int. Conf. Communications*, Seattle, WA, 1995, pp. 17–21.
[12] P. Leung, R. Berangi, Indirect Cochannel Interference Cancellation", Wireless personal communication 2001.
[13] *Method for Interference Suppression for TDMA and/or FDMA Transmission*, US Patent Application Publication No. US 2002/0141437 A1, October 3 2002.
[14] 3GPP TSG GERAN#11 Tdoc GP-022892: Draft Feasibility Study on Single Antenna Interference Cancellation (SAIC) for GSM Networks (Release 6).
[15] 3GPP TSG GERAN#13 Tdoc GP-030276: SAIC Link Level Simulation Model.
[16] 3GPP TSG GERAN#11 Tdoc GP-022557: Laboratory & Field Testing of SAIC for GSM Networks.
[17] RFC 3095, RObust Header Compression (ROHC).
[18] 3GPP TR 45.902, Flexible Layer One.

7

GPRS and EGPRS Performance

Javier Romero, Julia Martinez, Sami Nikkarinen and Martti Moisio

This chapter covers performance analysis of general packet radio service (GPRS) and enhanced GPRS (EGPRS) as they are specified in the Rel'97 and Rel'99 specifications respectively. The analysis is carried out considering potential deployment scenarios and covers most of the practical issues related to packet-switched data dimensioning.

Rel'99 introduces the Enhanced Data rates for Global Evolution (EDGE) radio interface, which is the gateway to enhanced spectrum efficiency (SE) and higher data rates. GSM/EDGE radio access network is called GERAN in 3GPP, whereas GPRS system using EDGE radio interface is called EGPRS. Rel'99 and later GERAN releases rely on the EDGE radio interface. The transition to EDGE will be smooth for operators who have an already deployed GPRS as EGPRS is fully backwards compatible with GPRS and does not require core network upgrades. Most of them are introducing EGPRS-capable base stations gradually, minimizing network changes as much as possible.

There are many cases and many deployment scenarios to consider. Current operational GSM networks utilize different deployment and frequency planning strategies. Many operators have traditionally used baseband hopping, while fractional reuse networks (with radio frequency (RF) hopping) are now being widely adopted, especially in new network deployments. Most of the analysis of this chapter can be applied to any network.

Section 7.1 introduces the basic link performance, which is key to understanding system level performance. Achievable throughput depending on radio link conditions is studied under interference and noise-limited scenarios. The benefits of Link Adaptation (LA) and incremental redundancy (IR) are studied, both with and without frequency hopping.

Section 7.2 describes the main radio resource management (RRM) functionality required in (E)GPRS. Algorithm design and its impact on performance are discussed. Polling, acknowledgement strategy, LA, IR, channel allocation, scheduler and power control functionality are covered.

GPRS and EGPRS system level capacity and dimensioning guidelines are covered in Sections 7.3 and 7.4. Most of the analysis is performed with dynamic simulations with a sophisticated simulation tool described in Appendix E. Hopping and non-hopping scenarios are studied. Spectral efficiency with different quality criteria is discussed as well.

GSM, GPRS and EDGE Performance 2nd Ed. Edited by T. Halonen, J. Romero and J. Melero
 ISBN: 0-470-86694-2

The performance analysis covers topics from pure soft-capacity deployments to hardware (HW)-limited scenarios. Two different sets of coding schemes (CS) are considered for GPRS: CS1–2 and CS1–4. The impact of the different mobile station (MS) timeslot (TSL) capabilities is studied. Throughput distribution maps and the effect of different PS traffic patterns are included as well. Deployment recommendations are given on the basis of previous simulation analysis. Finally, this section describes the effect of different traffic patterns and how it affects PS traffic performance.

Section 7.5 studies system performance in mixed packet-switched and circuit-switched traffic cases. The interaction between CS and PS traffic is described. A procedure to compute the maximum data capacity while maintaining speech quality is introduced as well. Section 7.6 presents (E)GPRS performance estimation based on collected field trial results from a major European city. Section 7.7 briefly discusses application performance over (E)GPRS, which is covered in more detail in Chapter 8. Finally, Section 7.8 presents some GPRS and EGPRS laboratory and field trial measurements.

7.1 (E)GPRS Link Performance

In this section, the performance of one single radio link is studied under interference and noise-limited conditions. The performance is studied in terms of block error rate (BLER) and throughput per TSL for the different (modulation and) coding schemes, (M)CS. It should be noted that the core of EGPRS data capacity improvement is purely based on link level enhancements, and therefore, it is of utmost importance to understand the link level performance before studying system level capacity issues.

7.1.1 Introduction

Link level simulations are used to evaluate the performance of the physical layer. These link level results can be utilized as an input to further studies including network level simulations, see Appendix E. The link is a one-way connection between one transmitter and one receiver. A general simulation chain is shown in Figure 7.1.

The simulation chain assumes a certain modulation method (Gaussian minimum shift keying (GMSK) or octagonal phase shift keying (8-PSK)), channel coding, channel estimation and receiver algorithm. The input parameters are channel type (e.g. typical urban,

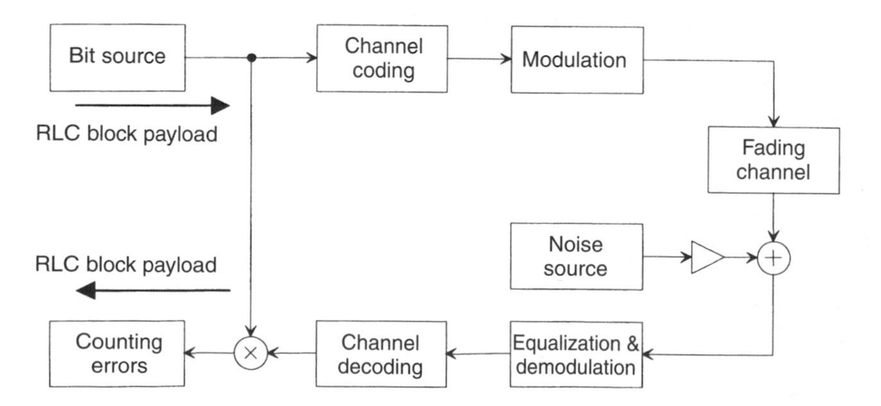

Figure 7.1. General structure of a link level simulation chain

rural area,...), mobile speed, frequency band, the use of frequency hopping (FH), average signal level relative to average interference level (Carrier-to-interference Ratio, CIR) or average energy per symbol relative to noise density (E_S/N_0). Output parameters are the channel bit error rate before decoding (raw BER), information bit error rate after decoding (BER), block error rate after decoding (BLER) and throughput.

EGPRS differentiates from GPRS by incorporating a number of improvements. EGPRS utilizes 8-PSK modulation (for coding schemes MCS-5 to MCS-9) to enable higher data rates. EGPRS applies incremental redundancy to help link adaptation and improve link performance. Radio block headers in EGPRS are coded robustly and separated to enable IR combining and other features. The radio block structure is quite different than that in GPRS. The GMSK coding schemes are different with EGPRS (MCS-1–4 instead of CS1–4). All GMSK modes of EGPRS have incremental redundancy.

In GSM specifications, the minimum base transceiver station (BTS) and MS performance for (E)GPRS is established in [1]. Minimum performance is determined in terms of receiver sensitivity and minimum-received CIR for a reference BLER value. In the following subsections, simulated link level results of (E)GPRS will be presented. Note that the results may change depending on the considered receiver and transmitter implementation.

This section presents (E)GPRS link level results for a typical urban (TU) channel model, described in GSM specifications in [1]. The simulated band is 900 MHz, but very similar results would be obtained with the 800-MHz band. For 1800 and 1900-MHz bands, the same results could be applied dividing the MS speed by 2 (as the fast-fading process is two times faster due to the halved wavelength). This means that the results for the TU3 (typical urban at 3 km/h) channel in the 900-MHz band are the same as the TU3 channel in the 800-MHz band; and also the same as the TU with an MS speed of 1.5 km/h for 1800 and 1900-MHz bands.

Simulation results are shown for interference and noise-limited scenarios. Interference-limited simulations correspond to MS testing setup in the interference-limited scenario described in GSM specifications [2].

7.1.2 (E)GPRS Peak Throughputs

(E)GPRS transmits segments of logical link control (LLC) blocks with a maximum bit rate that depends on the (M)CS utilized. The amount of LLC bits transmitted in a radio block is given by the amount of channel coding applied on each (M)CS. Table 7.1 summarizes the amount of LLC bits in a radio block of 20 ms and the peak throughput offered to the LLC layer for the different (M)CS.

The maximum EGPRS peak throughput for GMSK-modulated radio blocks is given by MCS-4, which has lower peak throughput than CS-4. EGPRS may have slightly lower throughput than GPRS CS-1–4 under very good radio link conditions if MCS-5–9 cannot be used. This may happen in UL direction with certain terminals that may not be capable of transmitting with 8-PSK modulations due to transmitter limitations.

7.1.3 RF Impairments

The reference clock signal in the MS and the BTS has a phase noise due to imperfections of frequency synthesizers. Phase noise adds to interference noise and may become the main noise source at high C/I values. 8-PSK modulation is more sensitive to phase noise

Table 7.1. (E)GPRS peak throughput values

MAC header RLC header (E)GPRS RLC data

LLC segment

	Coding scheme	Modulation	RLC blocks/radio blocks	FEC code rate	User bits/20 ms	Bit rate (bps)
GPRS	CS-1		1	0.45	160	8 000
	CS-2		1	0.65	240	12 000
	CS-3	GMSK	1	0.75	288	14 400
	CS-4		1	n/a	400	20 000
EGPRS	MCS-1		1	0.53	176	8 800
	MCS-2		1	0.66	224	11 200
	MCS-3		1	0.85	296	14 800
	MCS-4		1	1.00	352	17 600
	MCS-5	8-PSK	1	0.38	448	22 400
	MCS-6		1	0.49	592	29 600
	MCS-7		2	0.76	448 + 448	44 800
	MCS-8		2	0.92	544 + 544	54 400
	MCS-9		2	1.00	592 + 592	59 200

than GMSK due to the lower inter-symbol distance. Especially, low-protected MCSs are more affected because they are not able to correct errors introduced by phase notice.

When power amplifiers transmit at full power they may lose linearity, meaning that the transmitted signal is distorted. From the receiver point of view, signal distortion is another source of noise noticeable at high C/I values. GMSK modulation has quite constant amplitude and therefore is quite robust to power amplifiers' non-linearity. 8-PSK modulation has larger amplitude variations, meaning that power amplifier non-linearity will distort the signal more and introduce more noise. To avoid excessive signal distortion, 8-PSK is transmitted with 2 to 4 dB lower power compared with GMSK. Typically, 2-dB back-off is enough for the BTS, while 4 dB is typically required for the MS due to worse power amplifier non-linear behavior.

Both phase noise and power amplifier non-linearity are RF impairments that affect higher MCS, thus degrading the performance mainly at high C/I values. Figure 7.2 shows EGPRS BLER with and without RF impairments. MCS-9 is the coding scheme more affected by RF impairments, as it has no channel coding at all.

Simulation results in Figure 7.2 assumes typical phase noise spectrum for BTS and MS and memory-less power amplifier non-linearity model. All link level performance results in this section consider the effect of RF impairments.

7.1.4 Interference-limited Performance

The radio link is interference limited when the interference levels are well above the receiver sensitivity level. Most of the networks are interference limited as the same frequencies are reused in different cells, in order to efficiently utilize the available spectrum.

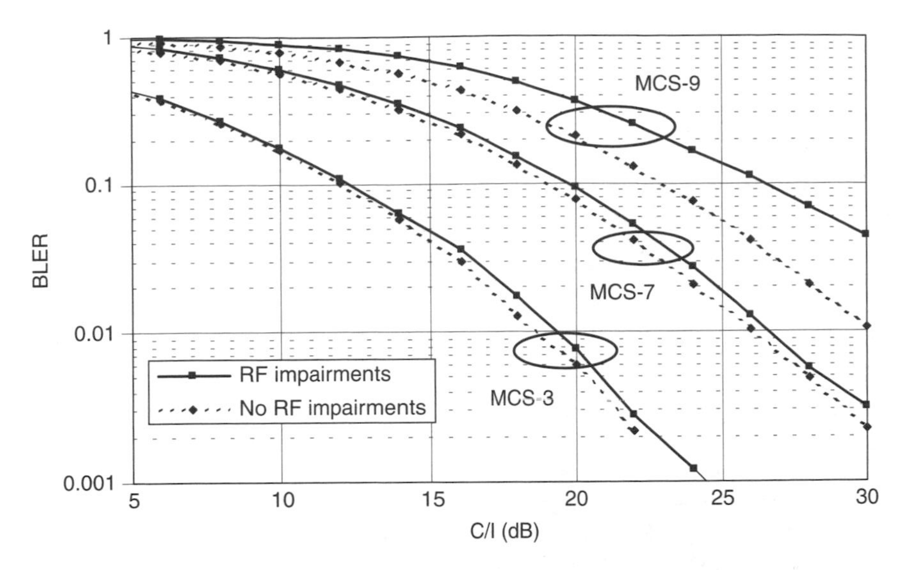

Figure 7.2. RF impairments' effect in link performance

An interference-limited network is usually considered to be capacity limited, meaning that the interference level is setting the limits of the network spectral efficiently. On the other hand, noise-limited networks are considered to be coverage limited, meaning that the network has cell range rather than spectrum efficiency limitations.

7.1.4.1 BLER Performance

Some of the received radio link control (RLC) blocks can be erroneous. Link level robustness depends on the coding rate and the modulation format. Different (M)CS achieve different BLER as can be seen in Figure 7.3. In general, the higher the (M)CS, the higher the BLER probability for the same CIR. The change from MCS-4 (no forward error correction (FEC) coding) to MCS-5 (maximum FEC coding but 8-PSK modulation) may break this rule.

Minimum BTS and MS performance in interference-limited scenarios have been included in GSM specifications. Minimum performance is specified as the minimum C/I required to achieve 10% BLER for different channel conditions, see Table 7.2. Minimum performance can be exceeded by both MS and BTS depending on the receiver implementation.

7.1.4.2 Acknowledged Mode Throughput

When error-free data delivery is necessary, the (E)GPRS network can retransmit erroneous radio blocks. There is a basic retransmission mechanism specified in Rel'97 and an advanced one (based on IR) included in Rel'99 for EGPRS. IR has been introduced in Chapter 1 and will be described in detail later on.

With the basic retransmission mechanism, the achieved throughput can be calculated as a function of the BLER:

$$\text{Throughput} = \text{Peak\_Throughput} \times (1 - \text{BLER})$$

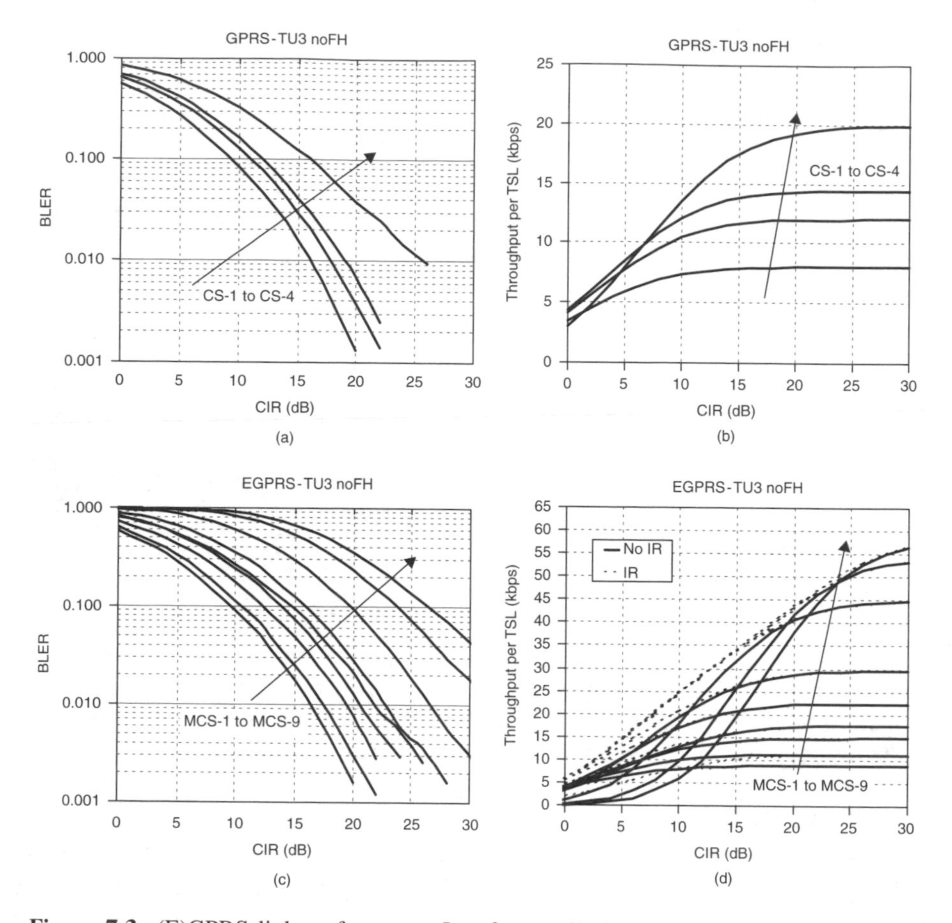

Figure 7.3. (E)GPRS link performance. Interference limited. No hopping. TU3 channel. 900-MHz band

where BLER is given in Figure 7.3. When IR is applied, the probability of erroneous blocks in successive retransmissions is much lower than that in the initial transmission. Therefore, in order to calculate the throughput per TSL, the BLER of the first and successive transmissions have to be considered. EGPRS throughput values with IR are shown in Figure 7.3. Retransmissions are done with the same MCS. No memory limitations have been considered, so IR is always possible. An IR memory capable of storing 20 000 soft values (bit values before decoding) per user gives performance close to the optimum. It is worthwhile mentioning the good performance of MCS-9 that in principle could always be used. In practice, the limited IR memory, the link performance under different propagation conditions and the possible delay requirements imposed by upper layers, requires the usage of the other MCS.

7.1.4.3 The Effect of Frequency Hopping

Frequency hopping helps to combat fast fading by randomizing C/I values of each burst. Randomization of C/I has the effect of averaging the C/I values over the duration of the

Table 7.2. Minimum C/I for BLER <10% in interference-limited scenarios. 900-MHz band. From GERAN specifications [1][a]

Type of channel	Propagation conditions TU3 (no FH)	TU3 (ideal FH)	TU50 (no FH)	TU50 (ideal FH)	RA250 (no FH)
PDTCH/CS-1	13 dB	9 dB	10 dB	9 dB	9 dB
PDTCH/CS-2	15 dB	13 dB	14 dB	13 dB	13 dB
PDTCH/CS-3	16 dB	15 dB	16 dB	15 dB	16 dB
PDTCH/CS-4	19 dB	23 dB	23 dB	23 dB	[b]
PDTCH/MCS-1	13	9	9	9	9
PDTCH/MCS-2	15	13	13	13	13
PDTCH/MCS-3	16	15	16	16	16
PDTCH/MCS-4	21	23	27	27	[b]
PDTCH/MCS-5	18	14.5	15.5	14.5	16
PDTCH/MCS-6	20	17	18	17.5	21
PDTCH/MCS-7	23.5	23.5	24	24.5	26.5[c]
PDTCH/MCS-8	28.5	29	30	30	[b]
PDTCH/MCS-9	30	32	33	35	[b]

[a] TUx = typical urban channel at x km/h; RA250x = rural area x km/h.
[b] Reference performance cannot be met.
[c] Performance is specified at 30% BLER.

radio blocks, so the probability of having consecutive bad C/I values is minimized. On the other hand, it increases the variance of the C/I values; therefore, it is more likely that some of the bursts experience bad C/I. The effect of a burst experiencing bad C/I can be compensated by channel coding and interleaving as it was explained in Chapter 5. Nevertheless, if radio blocks are not well protected (i.e. they don't have enough coding) the randomization of the C/I values has a negative effect, as the channel coding is not robust enough to correct the effect of some bursts having bad C/I. As a consequence, the use of frequency hopping for some of the coding schemes is not beneficial. In those cases, the frequency hopping gain turns to be negative.

Figure 7.4 shows the link performance difference at 10% BLER for some selected MCSs. MCS-1 and MCS-7 (coding rate of 0.5 and 0.76 respectively) benefit from frequency hopping (FH) having a direct link level gain of 4 and 1.8 dB. On the other hand, MCS-3 and MCS-9 (coding rate of 0.85 and 1) have a loss of 0.9 and 3.9 dB respectively. It should be noted that MCS-9 is interleaved only over two bursts, in order to avoid bad performance with FH.

Higher frequency hopping gains can be achieved with lower BLER values. Nevertheless, when LA is applied for throughput maximization, the normal operation point of the (M)CS never goes lower than 10% BLER. This is because LA will switch to a higher (M)CS when BLER is low.

When the users are moving faster (e.g. in a TU50 channel) fast-fading process tends to be more uncorrelated having a very similar effect of FH. Therefore, the gain or loss of FH is marginal.

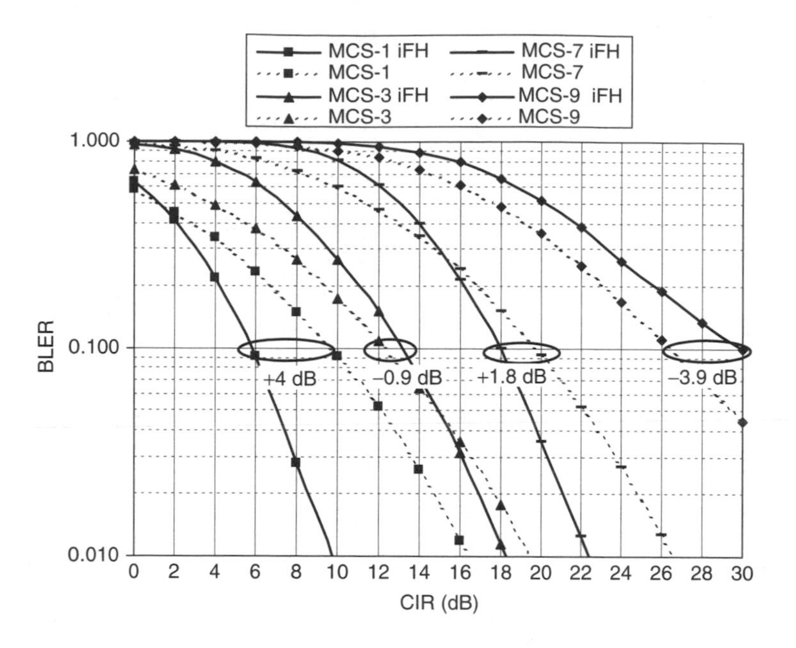

Figure 7.4. EGPRS link performance. With and without hopping. TU3 channel. 900-MHz band

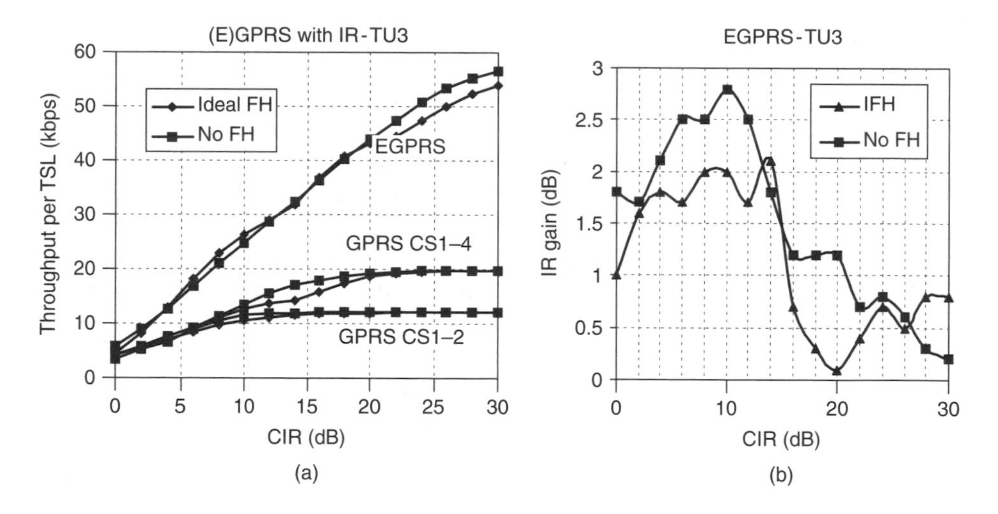

Figure 7.5. Throughput per TSL and IR gain versus CIR. TU3 channel. 900-MHz band

7.1.4.4 Ideal Link Adaptation

(E)GPRS network may automatically select the most appropriate (M)CS, depending on the radio link conditions, in order to maximize throughput. The link level performance will, therefore, follow the envelope of the throughput versus CIR curves of the different (M)CSs (Figure 7.3). Figure 7.5 represents the achieved link performance with LA, in an ideal situation with optimum CIR-switching points. For the same throughput, there is an appreciable link level gain with EGPRS compared with GPRS that varies from a couple of dBs for 5 kbps up to 15 dB for throughput values close to 20 kbps. The higher the throughput, the larger the gain. Frequency hopping results tend to be some dB worse, specially in the high-throughput area where high (M)CS are utilized. Figure 7.5 displays the gain of IR, which always improves link performance, in some cases over 3 dB (in no FH case). The envelope of throughput per TSL versus C/I depends on the MCS-switching points in case that IR is not used; but when IR is used, the envelope is smoother. IR gain changes according to this envelope so that IR gain is not constant but changes with C/I values. The IR gain is larger without frequency hopping in most of the cases. It should be noted that real RRM algorithms will have an impact on the represented link level results. In reality, LA is based on the radio link condition estimation reported by the MS or measured by the BTS, which is not perfect. Therefore, a realistic LA algorithm will make non-optimum (M)CS decisions. Nevertheless, the use of IR, which actually adapts the transmission to radio link conditions by increasing the coding rate on successive retransmissions, has been proved to be quite an efficient way to complement the LA algorithm. In effect, the use of IR reduces LA implementation requirement, and it is better to talk about LA & IR together as an efficient mechanism to adapt the transmission to changing radio link conditions. In conclusion, the combination of LA&IR algorithm allows the network to operate very close to optimum conditions.

7.1.4.5 Fast-moving Terminals

In case the MS is in a fast-moving vehicle (250–300 km/h) like high-speed trains, there is an impact in link performance. The phase of arrival of the signals changes very rapidly due to the Doppler effect so that it cannot be considered to be constant during the radio access burst duration. Phase errors may make the receiver to erroneously detect some bits independently of the received signal-to-noise ratio causing a BER floor (see Figure 7.6). The less protected (M)CSs cannot correct raw bit errors; therefore, they may be unusable in practice. GPRS maximum coding scheme usage is limited to CS-1–3, while CS-3 still suffers from high block error rate. 8-PSK modulated signals suffer even more from phase

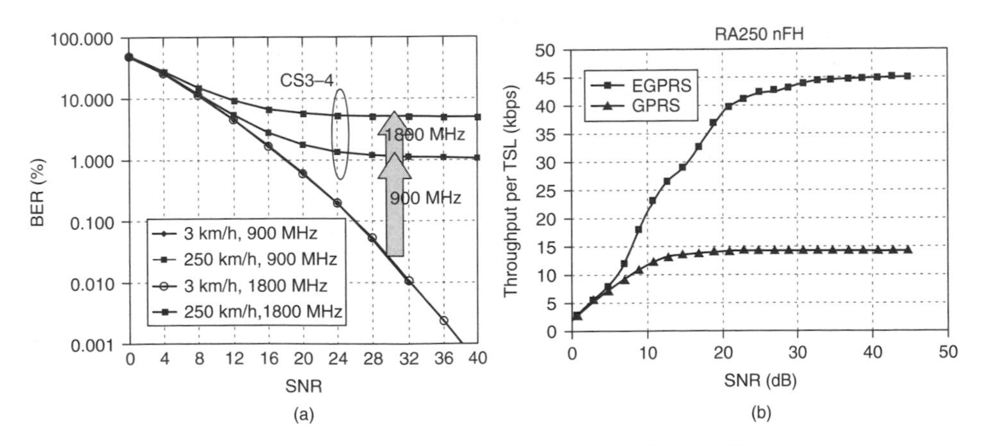

Figure 7.6. (a) Raw BER for CS1 to CS4. (b) GPRS CS-1–4 and EGPRS performance with fast-moving terminals (RA250 nFH channel)

noise, but more protected MCSs are able to correct the bit errors so that EGPRS provides higher throughput than GPRS (see Figure 7.6).

7.1.4.6 GPRS/EGPRS Throughput versus Cell Range

In order to calculate the coverage or throughput as a function of the distance for (E)GPRS, a noise-limited scenario is considered, with performance limited by thermal noise and transmitter output power. The link level results and the link budget calculation are used for this purpose.

Link budget calculation is used to map path loss to E_s/N_0 as described in Chapter 11. Using Figure 7.7, E_s/N_0 can be translated into throughput per TSL. Figure 7.8 (b) shows how EGPRS DL throughput changes with the path loss. E_s/N_0 is a function of the received signal power, symbol rate and noise figure. The received power depends, for example, on the transmitter output power, the transmitter and receiver antenna gains, cable losses and path loss. Calculations in Figure 7.8 are based on Table 11.3 calculations.

Path loss (L) is a function of the distance. The vehicular test model specified in [3] has been considered in this analysis. According to the reference, the path loss can be calculated according to:

$$L = 40(14 \times 10^{-3}\Delta h_b)\log_{10}(R) - 18\log_{10}(\Delta h_b) + 21\log_{10}(f) + 80 \text{ dB}$$

where R is the distance between base station and mobile station in kilometers, f is the carrier frequency in megahertz and Δh_b is the base station antenna height, in meters, measured from the average rooftop level.

Once the path loss as a function of the distance is known, E_s/N_0 versus cell range can be obtained. Finally, combining this function with the throughput per TSL obtained from link level simulations, the throughput per TSL for each coding scheme versus cell range can be obtained.

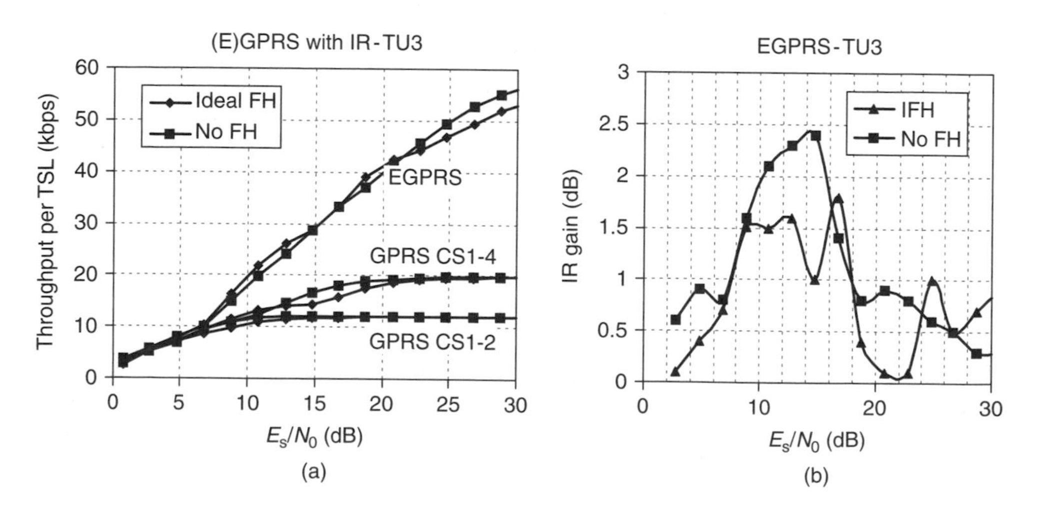

Figure 7.7. Throughput per TSL and IR gain versus E_s/N_0 for GPRS LA and EGPRS LA & IR. TU3 channel. 900-MHz band

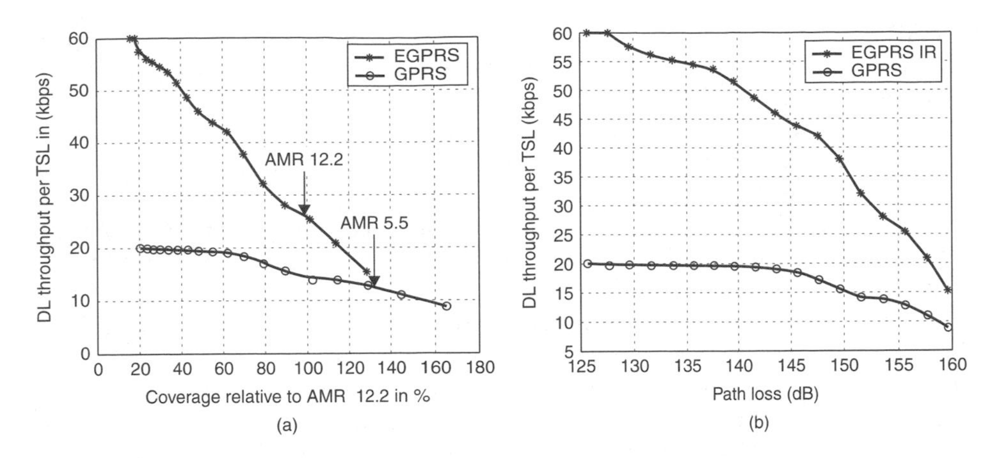

Figure 7.8. (a) GPRS CS-1–4 and EGPRS throughput versus coverage relative to AMR 12.2. 4-dB UL diversity gain assumed for AMR. (b) GPRS CS-1–4 and EGPRS throughput per TSL versus path loss with IR. Both graphs: TU3 channel, ideal link adaptation, 900-MHz band and ideal frequency hopping

On the basis of the link budget presented in Section 11.1, the path loss for adaptive multi-rate (AMR) 5.15 voice codec (see Chapter 6) in uplink is 163.2. Assuming the vehicular path loss model, the cell range, considering an antenna height of 15 m, is 13.36 km. In the same way, AMR 12.2 voice codec path loss is 158.4 and the corresponding cell range is 9.95 km. The downlink (DL) throughput per TSL versus relative ranges can be represented, taking AMR 12.2 range as a reference. In the case of AMR 5.15, the coverage relative to AMR 12.2 is calculated as $13.36 \times 100/9.95 = 134\%$.

Figure 7.8 (a) shows DL throughput per TSL versus relative cell range for EGPRS with IR and GPRS in a TU3 channel when using ideal LA. Link budget calculations are according to Table 11.3., Figure 7.8(b) shows DL throughput values versus path loss with the same assumptions.

7.2 (E)GPRS Radio Resource Management

The radio link control (RLC)/medium access control (MAC) layer involves a number of new procedures optimized for data transmission in packet-switched mode. A description of the new RLC/MAC mechanisms that have been introduced in the standards is presented in Chapter 1. The design of the algorithms that make use of the new mechanisms is not included in the (E)GPRS specifications. Nevertheless, it is very important to understand these algorithms and their overall performance impact.

7.2.1 Polling and Acknowledgement Strategy

Polling is a mechanism in which the packet control unit (PCU) asks the mobile station to send information about the DL conditions and about the successfully received RLC blocks. Polling is part of the (E)GPRS selective request ARQ mechanism, where non-correctly received RLC blocks are retransmitted. The PCU polls the mobile station by transmitting an RLC data block with a polling bit set to 1. This message is sent on the packet-associated control channel (PACCH) (Section 1.4.9.1). The mobile responds to the polling by transmitting a packet DL acknowledge (ACK)/negative acknowledge (NACK) message on the PACCH sub-channel scheduled by the PCU, as displayed in Figure 7.9. A packet DL ACK/NACK message contains radio link condition information and acknowledgement bitmaps for the RLC receiving window at the MS. Note that polling can be used in both RLC-acknowledged and unacknowledged RLC mode.

Polling can also be coordinated with the temporary block flow (TBF) dropping algorithm. By detecting excessive losses of DL ACK/NACK messages, it is possible to detect inoperative radio link conditions.

It is really important to find an optimal frequency of the polling mechanism. The polling algorithm may be able to adapt to different radio link conditions by polling more often when high BLER is detected. Additionally, it has to take into account the frequency at which RLC blocks are scheduled. If the mobile is not polled frequently, the RLC window could get stalled during the transmission. On the other hand, if the mobile station is polled very often, it will create a lot of signaling overhead in UL.

The transmitter has a window in order to address RLC blocks. In this window, the RLC blocks are classified as untransmitted, pending ack, acked and nacked. An RLC block is in untransmitted status if it has not been transmitted yet for the first time. Once it is transmitted, it changes to pending ack state. When a packet DL ACK/NACK message is received, the PCU updates the transmitting window as follows. If the block has been positively acknowledged, the block is set to acked. If the block has been negatively acknowledged, the status is set to nacked. If no acknowledgement information is received for a transmitted block, the status is still pending ack. Once the transmitting window is updated, the window is slid up to the first unacknowledged block, see steps 1 to 3 in Figure 7.10.

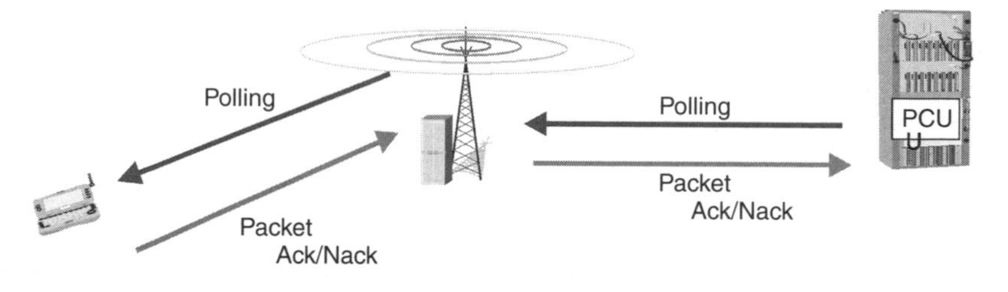

Figure 7.9. Polling mechanism

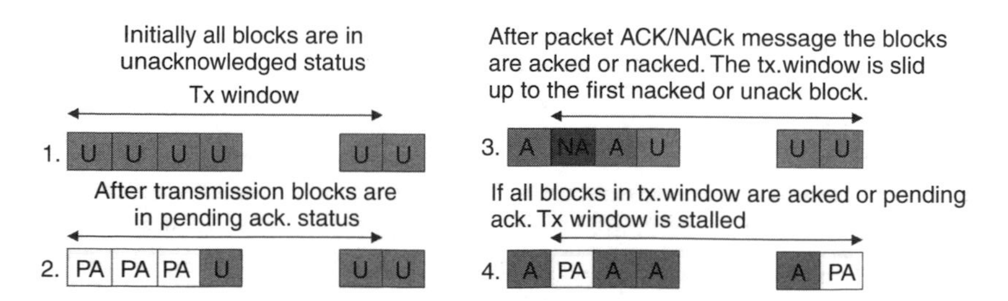

Figure 7.10. RLC window stalling

The polling mechanism is applicable only in the DL direction; however, a similar mechanism is used for UL. In the UL direction, the PCU may send packet UL ACK/NACK messages to the mobile in order to inform the MS about which blocks have been successfully received. When a packet UL ACK/NACK message is received, the mobile updates its transmit window. If the block is positively acknowledged, the status is set to *acked* and if the block is negatively acknowledged, the status is set to *nacked*. A block is not negatively acknowledged until a timer expires, in order to avoid retransmissions when the RLC block has already been retransmitted within the RLC acknowledgement delay. A counter is used instead of a timer in Rel'99 onward. After that, the transmit window is slid further up to the first unacknowledged RLC block. The algorithm that controls the sending of UL ACK/NACK messages is called UL acknowledgement strategy.

7.2.1.1 The Effect of RLC Window Stalling

Stalling may occur in UL and DL directions. When a nacked block is retransmitted, its status is changed from nacked to pending ack. Also when an untransmitted block is transmitted, its status is changed to pending ack. If there are no blocks whose status is either nacked or untransmitted within the RLC window, meaning that all the RLC blocks have been transmitted but no acknowledgement information has been received yet, the window is said to be stalled. In this situation, new incoming RLC blocks cannot be transmitted until new acknowledgements are received so that the RLC window can advance. In a stalling situation, the RLC layer may retransmit pending ack blocks one after another, starting from the beginning of the transmit window. Those are preemptive retransmissions that will help to solve the RLC window congestion.

Figure 7.10 graphically shows the stalling process. In DL, the size of the RLC window is 64 RLC blocks for GPRS and 1024 (max.) for EGPRS. The risk of stalling is reduced in EGPRS by means of two improvements. Firstly, the transmit window is much larger. Secondly, the retransmission mechanisms are more efficient.

7.2.2 Link Adaptation Algorithms for (E)GPRS

7.2.2.1 Introduction

In (E)GPRS, the information is transmitted in packets using different channel coding schemes. The purpose of coding is to improve the transmission quality when radio link conditions are bad, since the coding introduces some redundancy in the packets in order to detect and correct bit errors that occur during the transmission.

By increasing the amount of redundancy, the error correction can be improved, but at the same time the net bit rate is reduced. In order to maximize the throughput, robust coding schemes should be used in poor radio link conditions and coding schemes with less protection in good radio conditions.

Link adaptation (LA) algorithms are used to select the optimum (M)CS as a function of the radio link conditions. The selection of the coding scheme is based on channel quality estimates.

In a typical LA, the receiver measures the carrier-to-interference ratio or the raw bit error rate, and makes an estimate for the channel quality by averaging the measurements over some time interval [4]. The LA algorithm makes the decision about the coding scheme by comparing the estimated channel quality with certain threshold values in order to obtain maximum throughput. On the basis of the available channel quality estimates (CIR, BLER and bit error probability (BEP)), different LA algorithms are possible.

LA Algorithms for DL and UL

Radio link measurements are the same in RLC acknowledged and unacknowledged modes. The same packet DL ACK/NACK and packet UL ACK/NACK messages are sent in both cases.

In DL, the PCU polls the mobile, which responds by sending the GPRS or EGPRS packet DL ACK/NACK message. These messages contain measurements with information about the link quality. The GPRS report contains the available interference levels in serving cell, the RXQUAL (received signal quality), C (received signal level) and SIGN_VAR (average variance of received signal level). Similar information is included for EGPRS, apart from mean and co-variance averaged values of BEP, which are included instead of RXQUAL and SIGN_VAR values. On the basis of the information of radio link conditions, the network selects the optimal coding scheme to use in DL acknowledged and unacknowledged mode.

In the uplink, the network commands the mobile station with the packet UL ACK/NACK message the coding scheme has to use. The selection of the coding scheme is based on the same LA algorithm used in DL using measurements performed by the base station.

Coordination with Power Control Algorithm

Both LA and power control algorithms have link quality measurements as input. The quality thresholds of both algorithms should be coordinated. Power control should not interfere with LA in order to avoid a situation where a reduction in output power results in a change of coding scheme and a decrease in throughput. The LA algorithm usually has priority over the power control algorithm.

Quality-of-service Requirements Input to LA

The crosspoints or thresholds for LA can be determined on the basis of the user's quality-of-service (QoS) profile. The general rule in ACK mode is to select the (M)CS that maximizes throughput per TSL. Nevertheless, for certain users with very strict delay requirements, it could be better to use a lower coding scheme in order to avoid too many retransmissions that introduce more delay. If RLC unacknowledged mode is used, the crosspoints could be set to force the use of lower coding schemes to achieve the minimum reliability QoS requirements.

7.2.2.2 LA in GPRS

An ideal LA algorithm for GPRS (TU3 channel with ideal FH) would follow the envelope of the C/I versus throughput curves with different coding schemes, see Figure 7.11. In GPRS, the algorithms can be typically based on BLER and CIR estimation. Both measurements can be calculated from reported information from the MS or the BTS. If C/I is

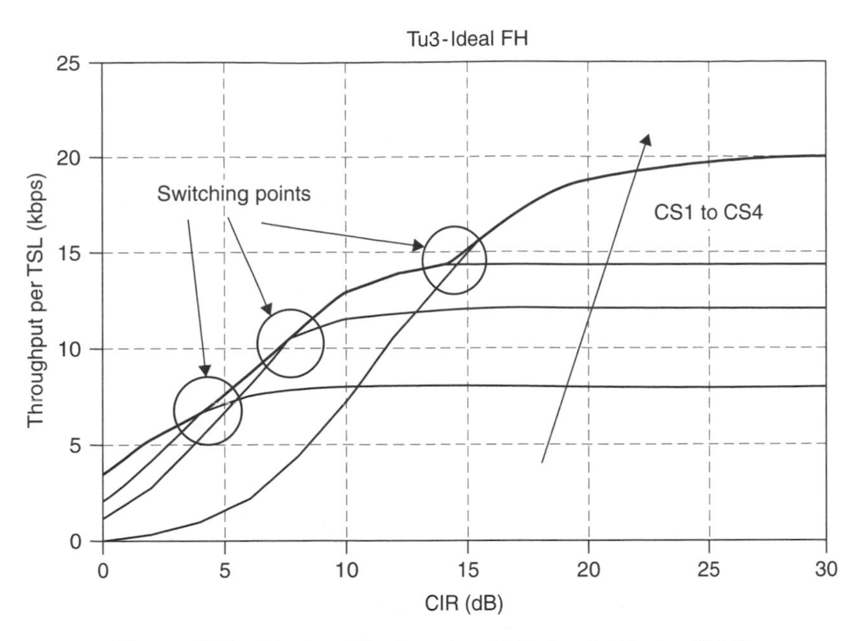

Figure 7.11. LA algorithm based on CIR thresholds for GPRS

used for LA, different thresholds have to be set up for the hopping and non-hopping cases. This is because the C/I versus throughput curve of each CS changes with the use of FH.

7.2.2.3 LA in EGPRS

As mentioned in previous sections, the retransmission mechanism is more efficient in EGPRS due to the use of incremental redundancy. LA should be coordinated with the IR mechanism for optimum performance.

Incremental Redundancy

IR is a physical layer performance enhancement for the acknowledged RLC mode of EGPRS. IR adjusts the code rate to actual channel conditions by incrementally transmitting redundant information until decoding is successful. The automatic repeat request (ARQ) protocol takes care of requesting and retransmitting incorrectly received blocks. IR improves the reception of retransmissions by combining the information from the original failed transmission with the new retransmitted block and, in this way, increases the probability of correct reception. Both the modulation and coding schemes and the puncturing schemes may be changed in subsequent retransmissions [5]. IR is an EGPRS-specific implementation of type II hybrid ARQ transmission scheme [6].

Figure 7.12 shows an example of IR operation with MCS-9. The original data are first convolutionally encoded with a coding rate of 1/3. Before the first transmission, the encoded data are punctured by removing two-thirds of the encoded bits. The punctured bits are transmitted over the radio interface. The effective channel-coding rate is 1/1, i.e. the same number of bits as the original data is transmitted. The receiver is able to decode and recover the original data if radio link conditions are good enough. Otherwise,

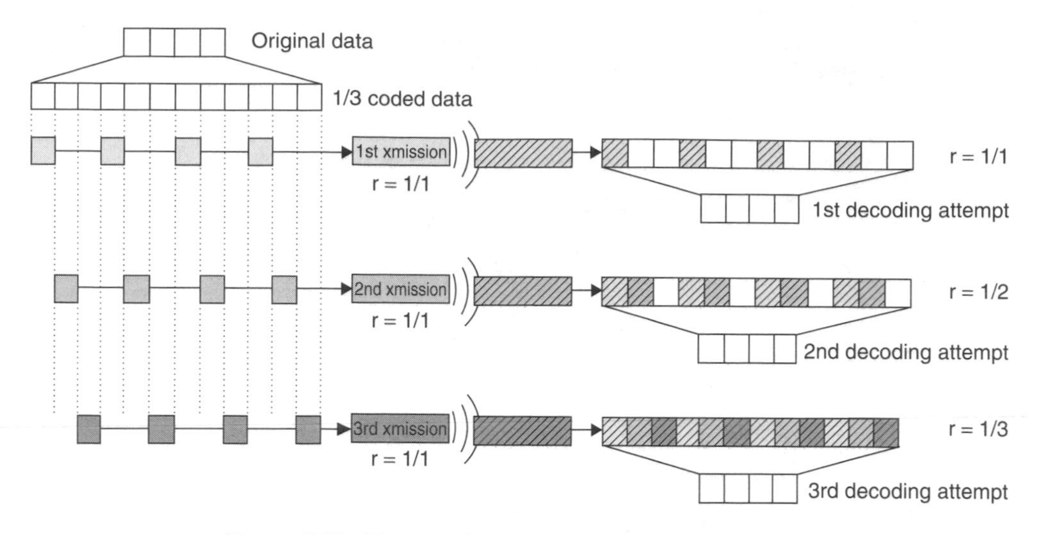

Figure 7.12. IR transmission and combining with MCS-9

the network will resend the same coded data but with a different puncturing scheme: different bits are now removed. The receiver keeps in its memory the information of the received bits before decoding (soft values). When the second transmission is received, the receiver has now two-thirds of the received encoded bits, so the effective coding rate (after combining with the first transmission) for the second transmission is 1/2. The probability of receiving correctly the original data is now higher than in the first transmission. If the block is still incorrect, the network will send again the same encoded data with a different puncturing scheme, containing the rest of the original encoded data. The effective coding rate for the third transmission is 1/3, and the probability of receiving correctly the data is much higher.

Without IR, the BLER of each transmission is potentially the same. However, with IR, the BLER of retransmission after combining is much lower than BLER of the initial transmission. Therefore, the final net throughput without IR is lower than the throughput with IR. For this reason, the LA crossing points or thresholds need to take into account the use of IR.

IR is mandatory for the MS but not for the network. If the MS runs out of memory, it cannot store incorrect received blocks to be combined with the new ones. This situation can be reported to the network, and the LA mechanism should act accordingly assuming that IR will not be done at the MS side.

EGPRS Retransmissions

EGPRS has nine different coding schemes (MCS-1–MCS-9). The coding schemes MCS-1–MCS-4 use GMSK modulation and MCS-5–MCS-9 use 8-PSK modulation. These coding schemes are grouped in three families as specified in Section 1.5.2. Family A consists of coding schemes MCS-1 and MCS-4, family B consists of MCS-2, MCS-5 and MCS-7 and family C consists of MCS-3, MCS-6, MCS-8 and MCS-9.

In GPRS, all the retransmissions of an RLC block have to be performed with the same coding scheme as the original transmission. This fact increases the risk of stalling in the transmit window, and may create additional delays. In EGPRS, retransmissions of an RLC block with different coding schemes are allowed. However, these retransmissions must be done using the MCS from the same family. MCS-6 and MCS-9 can be freely selected as they have the same RLC block size. One MCS-9 radio block contains two RLC blocks, each one of which could be sent separately in two MCS-6 radio blocks. The same principle applies to MCS-5 and MCS-7. In these cases, IR combining between two different MCSs is possible. Note that IR is always possible when retransmitting with the same MCS.

In order to allow retransmissions with other coding schemes, re-segmentation and padding can be used. However, as these procedures modify the data before the channel coding, IR combining is not possible in these cases. When using padding, dummy bits are added to an MCS-8 block to make it suitable for MCS-6, forming an MCS-6-padded block (see Figure 7.13). Padding is only possible with MCS-8. When using re-segmentation, an RLC block is split in half and separately transmitted, but must be acknowledged as one. The following re-segmentations are possible (see Figure 7.13): one MCS-4 block into two MCS-1 blocks, one MCS-5 block or one MCS-7 block into two MCS-2 blocks, one MCS-6 block or one MCS-9 block into two MCS-3 blocks and one MCS-6-padded block into two MCS-3-padded blocks.

Padding and re-segmentation are useful, in principle, when there is the need to retransmit the same radio block with lower MCS due to radio link condition changes. IR combining of some of the retransmitted radio blocks is normally enough to achieve optimum performance if the packet delay is not a critical factor. Therefore, padding and re-segmentation can be avoided in most of the situations.

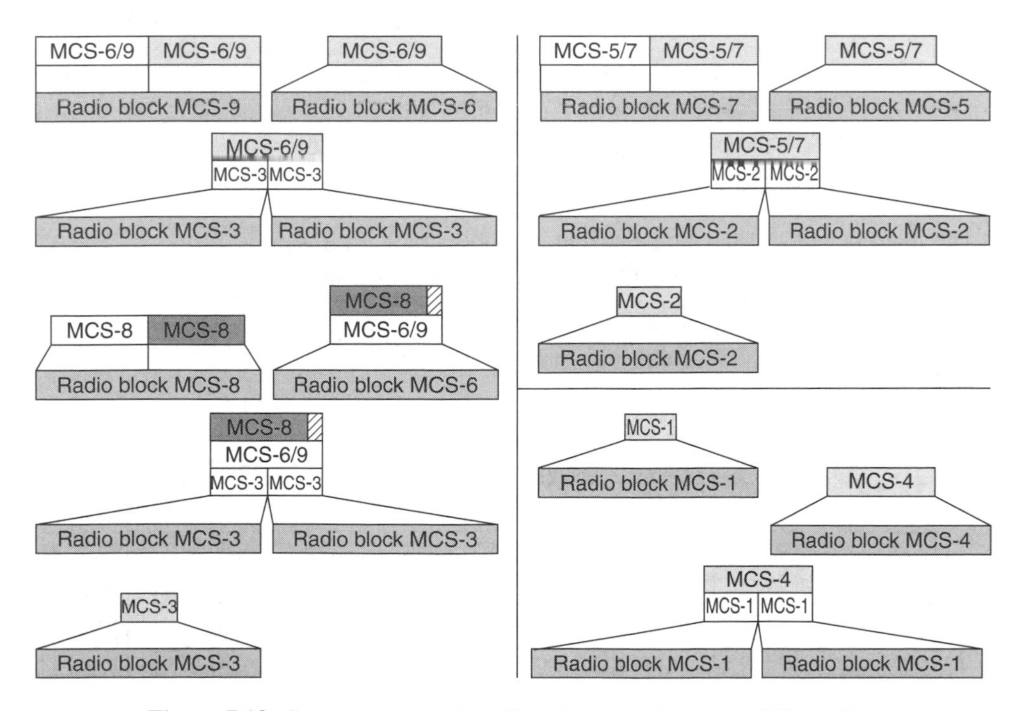

Figure 7.13. Segmentation and padding between the three MCS families

7.2.2.4 EGPRS Link Adaptation Implementation Example

Link adaptation can be based on BEP values estimated in the receiver. Mean (MEAN_BEP) and normalized covariance (CV_BEP) BEP values are reported by the MS as explained before. Look-up tables (CV_BEP, MEAN_BEP), like the one in Figure 7.14 are then used to decide the coding scheme used in the transmission of the radio blocks. The LA look-up table in the figure is optimized for throughput maximization without IR. In case IR is used, a different table should be used.

7.2.3 (E)GPRS Channel Allocation

Channel allocation algorithms are responsible for assigning channels to (E)GPRS MSs. There are two issues to consider in this process: the interaction with CS and interaction with other (E)GPRS users.

7.2.3.1 Interaction with CS Services

In GSM/EDGE networks, CS and (E)GPRS connections have to live together sharing the same resources (timeslots). Two different strategies are possible: sharing all the available resources between PS and CS or splitting these resources in two separate pools. In both cases, the channel allocation algorithm selects the resources to be used for each connection. A channel is considered to be a particular TSL using certain frequency parameters.

Two Separate Pools of Resources

This resource sharing method is based on splitting all the channels into two different pools, the circuit-switched pool and the packet-switched pool, see Figure 7.15. This means that voice and CS data connections are allocated to channels that belong to the CS pool, and (E)GPRS connections are allocated to channels within the PS pool, as far as there is capacity available in the pools. Upgrades and downgrades of the two pools are possible in order to adapt the size of the pools to traffic load conditions. A minimum number of channels can be reserved for each pool to ensure a certain minimum PS or CS capacity.

The main benefits of having two separate pools are as follows:

The PS channels always remain consecutive to each other, and thus the probability of allocating the maximum number of TSLs supported by the MS is maximized. This improves statistical multiplexing so that better utilization of available TSLs can be achieved.

	8-PSK MEAN BEP																															
8-PSK_CV_BEP	1	2	3	4	5	6	7	8	9	10	11	12	13	14	15	16	17	18	19	20	21	22	23	24	25	26	27	28	29	30	31	32
1	5	5	5	5	5	5	5	5	5	5	5	6	6	6	6	7	7	7	7	7	7	7	8	8	8	8	8	8	9	9	9	9
2	5	5	5	5	5	5	5	5	5	5	5	6	6	6	6	7	7	7	7	7	7	7	8	8	8	8	8	8	9	9	9	9
3	5	5	5	5	5	5	5	5	5	5	6	6	6	6	7	7	7	7	7	7	7	8	8	8	8	8	8	8	9	9	9	9
4	5	5	5	5	5	5	5	5	5	6	6	6	6	6	7	7	7	7	7	7	7	8	8	8	8	8	8	8	9	9	9	9
5	5	5	5	5	5	5	5	5	6	6	6	6	6	7	7	7	7	7	7	7	7	8	8	8	8	8	8	8	9	9	9	9
6	5	5	5	5	5	5	5	6	6	6	6	6	7	7	7	7	7	7	7	7	7	8	8	8	8	8	8	8	9	9	9	9
7	5	5	5	5	5	5	6	6	6	6	6	7	7	7	7	7	7	7	7	7	7	8	8	8	8	8	8	8	9	9	9	9
8	5	5	5	5	5	6	6	6	6	6	7	7	7	7	7	7	7	7	7	7	7	8	8	8	8	8	8	8	9	9	9	9

Figure 7.14. LA look-up table for 8-PSK modulation. From EGPRS specifications [7]

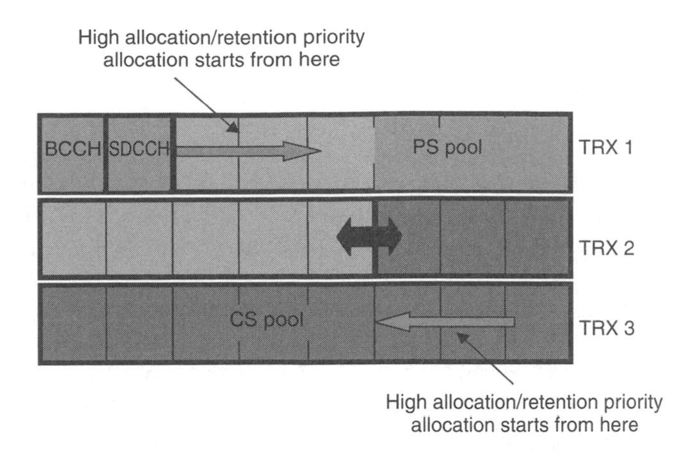

Figure 7.15. Two-pool channel allocation method

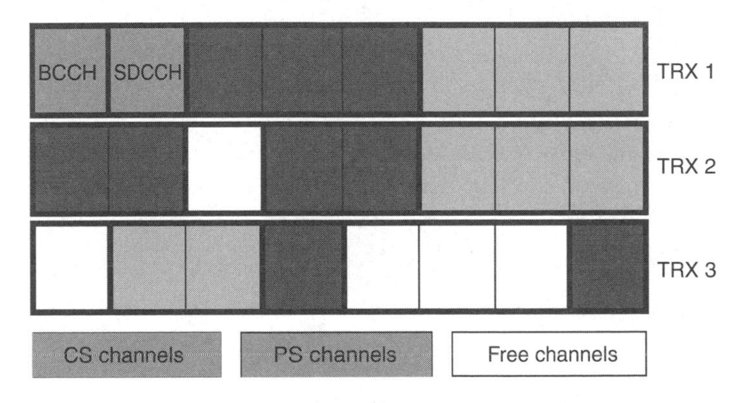

Figure 7.16. One-pool method

CS services need stable TSL allocation and have continuous traffic flow. But on the other hand, PS services need frequent TSL allocations with shorter duration, and the traffic tends to be bursty. The channel allocation algorithm can be optimized for the two kinds of traffic inside each pool, and the interaction between the two pools is minimized.

The main drawback is that there is the need to do continuous repacking of the connections towards the end of the pools, so that non-utilized channels are at the border of the two pools, and can be converted to PS or CS when required.

One Common Pool of Resources

This way of sharing the resources consists in forming a pool of channels in which CS and PS connections are allocated. Both types of connections have to compete for the use of the resources in the cell. The concept is depicted in Figure 7.16.

The main advantage of this method is that the pool of available channels is larger, so it gives more freedom for channel selection. There are also some drawbacks of this method. It is not possible to optimize channel allocation for PS and CS services separately. It may happen that the free resources available in the pool are fragmented into non-consecutive channels, and multi-slot PS and CS mobiles cannot use these TSLs efficiently.

7.2.3.2 Interaction with Other EGPRS Users

Once the potential (E)GPRS channels are identified, there is the need to select most suitable channels for the (E)GPRS connection. The simplest way to do that is to select consecutive channels that are less utilized. Considering only best-effort users, the channel utilization can be measured as the amount of users who are using the channel. If there are already N mobiles in one channel, a new mobile would get a portion of $1/(N+1)$ of that TSL capacity. The channels that give the maximum capacity, i.e. the channels with fewer allocated mobiles, will be selected for the new connection. If (E)GPRS connections have different priorities, the channel allocation may consider the different user priorities in order to calculate the utilized portion of channels. It is also possible to reserve absolute values of channel utilization for guaranteed throughput services over (E)GPRS.

7.2.4 (E)GPRS Scheduler

Once the resources, timeslots and frequencies have been assigned, the PS users can be multiplexed with other PS users sharing the same channels. Multiplexing is controlled by the scheduling mechanism in the UL and DL direction.

The scheduler assigns turns to each user depending on certain criteria. It is possible to see the scheduler as a queuing system for each TSL as shown in Figure 7.17. Scheduling algorithms for best effort and interactive traffic classes, and for guaranteeing throughput for streaming services will be presented next.

7.2.4.1 Scheduling Algorithms

Round Robin

The simplest scheduling algorithm is the round robin, in which each user is assigned a transmission turn with the same priority. When all the users have transmitted once, the

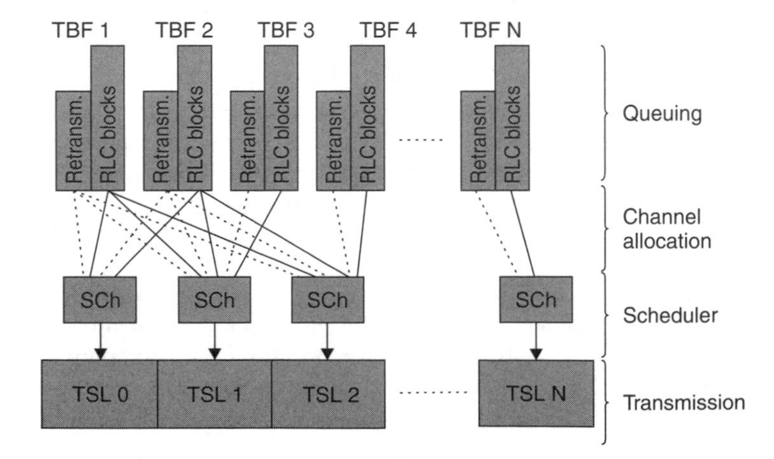

Figure 7.17. The scheduler assigns turns to the different users for each TSL

first user starts to transmit again. Round robin can be applied to each TSL. More elaborate scheduling mechanisms with similar effect to the previously described are also possible.

Weighted Round Robin

Weighted round robin does not share equally the scheduling turns between the users because the users have different priorities. The number of assigned turns is proportional to a weighting factor or priority. The weights assigned to each user are fixed. Weighted round robin is useful when (E)GPRS subscribers are divided in different classes (e.g. gold, silver and bronze), with different subscription characteristics and different service levels.

Scheduling Algorithms to Guarantee Throughput

Streaming services need guaranteed throughput through the air interface. There are two possibilities in order to guarantee throughput: dedicate timeslots (similar to CS) or share the timeslots with other services. When the timeslots are shared, the scheduler assigns transmission turns to guarantee a minimum throughput. The rest of the transmission turns can be used for other users or extra traffic over the guaranteed throughput. See Figure 7.18.

7.2.5 GPRS and EGPRS Multiplexing

Although EGPRS Rel'99 terminals are fully backwards compatible, there are some implementation issues to be considered since Rel'97 terminals are not capable of receiving any data modulated with 8-PSK. This may create problems when GPRS and EGPRS terminals are sharing TSLs and there is the need to assign the uplink state flag (USF) to a Rel'97 terminal. In order to be received by the GPRS terminal, the USF has to be sent with a GMSK-modulate radio block. Forcing GMSK modulation in DL will limit the MCS selection so it may have an impact on EGPRS downlink throughput. Forcing GMSK in DL would require re-segmentation of RLC blocks that need to be retransmitted and were originally transmitted with 8-PSK modulations. In practice, re-segmentation can be avoided by not sending USF to Rel'97 terminal when no GMSK-modulated block can be produced in the RLC layer. This later solution may have some impact on Rel'97 terminal UL throughput under large amount of EGPRS retransmissions in DL.

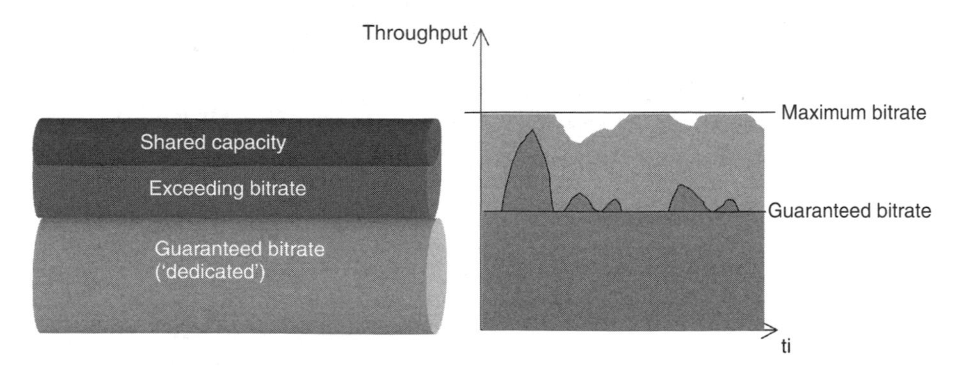

Figure 7.18. Use of shared channels for streaming services

Some of the following improvements can be implemented in the RRM in order to cope with the GPRS and EGPRS multiplexing problem:

- Modify the channel allocation so that it avoids sharing TSLs between GPRS and EGPRS.
- Synchronize DL and UL scheduler in a way that, whenever possible, USFs for Rel'97 GPRS terminals are sent over Rel'97 radio blocks.
- Use the USF granularity to minimize the number of USFs that need to be sent.

7.2.6 Power Control

Power control can also be used with packet-switched services. Power control for (E)GPRS is more challenging than for speech due to the bursty nature of packet-switched services. (E)GPRS power control can be applied to uplink and downlink.

7.2.6.1 Uplink Power Control

The purpose of UL power control is mainly to save battery from the mobile station. However, it helps also to reduce interference. When the radio conditions are good, the MS transmits with lower-output power. The algorithm is described in Chapter 1. It is performed autonomously by the mobile with the help of some parameters provided by the network. No interference is taken into account in the algorithm, only path loss is considered.

The operator can modify α and Γ_{CH} parameters described in the section entitled *Power control procedure* in 1.4. Γ_{CH} determines the minimum mobile output power, and α determines how much a change in the received level (C) is translated into a change of MS output power (it changes the slope). Figure 7.19 shows mobile station output power versus DL-received signal level for different values of α and Γ_{CH}. Note that by setting different parameter values, it is possible to control the minimum level when UL power

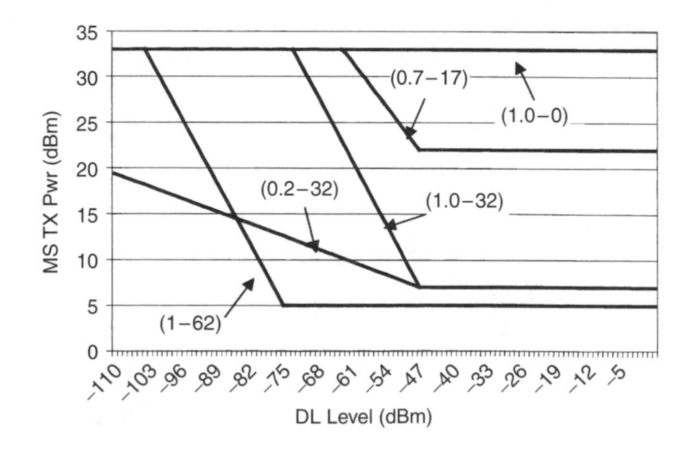

Figure 7.19. Mobile station output power (MS TX Pwr) versus DL-received signal level (DL level) for different (alpha–gamma) values

control starts to operate, as well as how aggressively the power control is applied when this minimum level is surpassed.

The mobile station shall use the same output power on all four bursts within one radio block. When accessing a cell on the PRACH or RACH (random access channels described in Section 1.4.9.1) and before receiving the first power control parameters during packet transfer, the mobile shall use the output power defined by PMAX that is the maximum MS transmitted power.

7.2.6.2 Downlink Power Control

The objective of downlink power control (DL PC) is to optimize the base station output power maintaining the same quality in the ongoing connections. The DL PC algorithm decreases the interference in the system. In a mixed scenario with packet- and circuit-switched services, DL PC results in a reduction of the interference generated from PS to the speech services.

Figure 7.20 shows how the DL PC algorithm is composed of two steps mainly. Firstly, the mobile performs quality measurements in downlink and secondly, depending on these measurements, the base station output power is updated. Depending on the kind of measurements and the criteria used to change the output power, several DL PC algorithms can be defined.

BTS transmitter power can be written as

$$\text{BTS PWR} = \text{BCCH PWR} - \text{P0} - \text{Delta}$$

where BCCH PWR is the BCCH channel transmission power, P0 is a constant value that introduces negative power offset and delta is a component that depends on radio link conditions. Different P0 can be assigned to individual users. When dynamic UL RLC block allocation is in use (see section entitled *Medium access control and radio link control layer* in Chapter 1), the maximum value of delta is limited to 10 dB. This is to prevent USF being lost by another MS sharing the same TSL with the user receiving the radio block. This is critical as the USF has one-to-many transmission nature and any user sharing the same TSL need to decode the USF in order to transmit in the UL direction. In case of static UL RLC block allocation, USFs are not needed, and therefore Delta values are not limited.

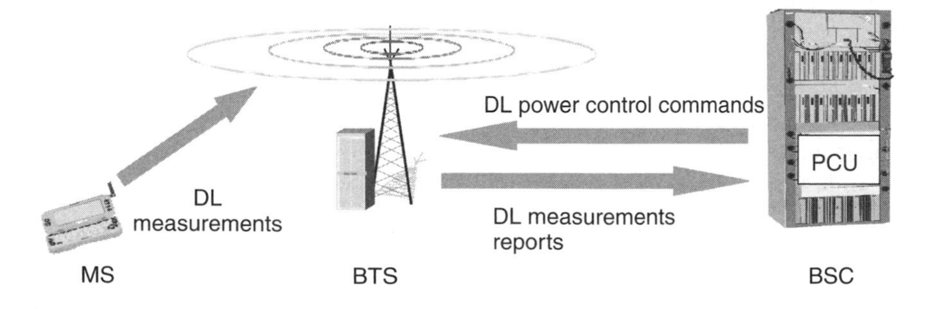

Figure 7.20. Downlink power operation

DL PC should act outside the LA work window in order to avoid a situation where a decrease in output power results in a change of coding scheme and a decrease in throughput. If only a subset of (M)CS is utilized, e.g. CS-1–2, the DL PC work window is larger and more gain can be achieved. The implementation of all modulation and coding schemes in EGPRS reduces significantly the DL PC work window, and the interference reduction is very limited unless a certain data capacity is sacrificed. In any case, DL PC is always convenient when mixing voice and data services on the same hopping layer, in order to reduce interference levels generated to speech services. P0 can be used for this purpose. A reduction of 3 dB in (E)GPRS transmitted power does not have a big impact on average data capacity, but it definitely has a major impact on speech quality, see Section 7.5.

7.3 GPRS System Capacity

7.3.1 Introduction

In this chapter, the performance of GPRS systems is evaluated by means of network level simulations. The focus is on downlink performance since in most applications and scenarios, the data flow is clearly asymmetrical, thus the data capacity bottleneck will be in DL. Typically, uplink traffic may consist mostly of transport control protocolTCP-acknowledgement messages, although occasionally certain UL capacity is required in order to send user data.

Performance indicators (throughput, delay) are measured between base station controller (BSC) (PCU) and MS, i.e. they are not end-to-end values. The analysis is focused on the characterization of data transport capabilities of the GPRS base station subsystem (BSS) according to Rel'97 specifications. End-to-end values depend heavily on issues like higher protocol layers (including the application level), GPRS core network elements and the quality of the fixed line connection in between the two entities. These issues are out of the scope of this analysis and will be handled separately in Section 7.7.

7.3.2 Modeling Issues and Performance Measures

A basic description of the simulator is described in Appendix E. In the following text, the GPRS-specific features and modeling are described. Figure 7.21 schematically describes the GPRS/EGPRS simulation model. All the necessary details affecting the analyzed performance values are taken into account in the simulations.

7.3.2.1 Radio Link Modeling

Only downlink user traffic is simulated. Downlink packet ACK/NACK messages are modeled on uplink. These blocks experience interference that correlates with the downlink CIR, according to

$$\text{CIR\_uplink} = \text{CIR\_downlink} + N(2, 4)$$

where all the units are expressed in dB, and $N(2, 4)$ is a random normally distributed value with mean of 2 and variance of 4. This relationship is based on separate simulations with both up- and downlink traffic assuming 90/10 asymmetry.

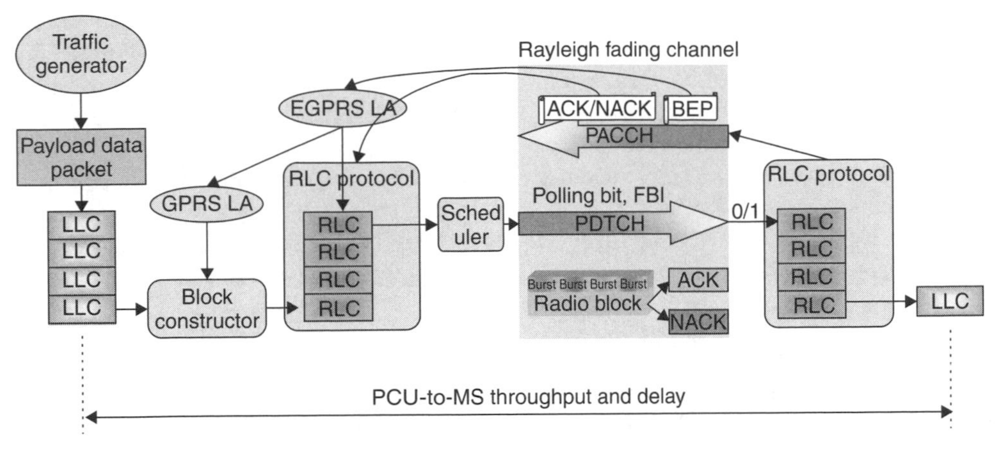

Figure 7.21. GPRS/EGPRS functionality in the simulator

Link level results use two-phase mapping tables according to principles in [8]. First, burst-level CIR values are mapped on to raw BER values (two tables are needed, one for GMSK and one for 8-PSK). Next, the arithmetic mean and standard deviation of raw BER samples over the interleaving period are calculated. These values are then mapped on to block error probability (BLEP) according to used coding scheme. Once the BLEP is calculated, a random experiment is done to decide whether the block had an error.

7.3.2.2 Traffic Model

All GPRS sessions arrive at the network according to Poisson process and end after every data block has been positively acknowledged. The traffic pattern generated for all GPRS users is FTP like traffic (file of 120-kB download) and follows the guidelines described [3]. Note that layers above LLC are not implemented in the simulator.

7.3.2.3 Radio Resource Management Model

The channel allocation algorithm selects the least-occupied consecutive slots from the same transceiver (TRX). If all the slots are full (maximum number of users in the same TSL is limited to 6), the TBF is blocked and a new TBF establishment procedure is started after 5 s. Round robin type of scheduling is applied, so each active TBF transmits a fixed amount of time after which the transmission turn is given to the next TBF in the queue. Scheduling is done on slot basis, i.e. every slot in each TRX has its own queue. Network polling is modeled in such a way that every MS has a polling counter, which is initially set to zero. Every time an RLC block is retransmitted as a result of negative acknowledgement, the polling counter is increased by 3 and in all other cases (normal transmission or retransmission due to stalling), the polling counter is increased by 1. When the value of the counter reaches 18, polling bit is set true in the next RLC block and the counter is reset to zero.

Cell reselection is triggered by power budget algorithm with penalty timer and hysteresis (see values in Table E.3 in Appendix E). If the MS is in packet-idle mode, it just attaches itself to the new cell. In packet transfer mode, current TBF is released and TBF establishment attempt is made to the new cell. If at least one channel could be allocated, MS can continue data transmission after 1 s (estimated processing and signaling delay). The unacknowledged LLC PDUs transmitted in the old cell are copied and retransmitted in the new cell. If channel allocation fails, the MS will experience a delay of 5 s after which it returns to the packet-idle mode.

Link adaptation is based on C/I and uses all four GPRS coding schemes. The algorithm exploits the minimum C/I values optimized for coding schemes 2 to 4. Frequency hopping is random radio frequency (synthesized) hopping. Mobile allocation index offset (MAIO) management (see Chapter 5) is used with 1/1 reuse scheme. Power control is not used.

7.3.2.4 MAC Layer Modeling

TBF establishment procedure (downlink) is triggered whenever new data arrives at the PCU addressing a certain MS. A downlink TBF establishment delay (defined in Table E.3, Appendix E) is taken into account before the actual data transmission can take place.

TBF blocking occurs when PCU is unable to allocate radio resources for the use of MS. A timer *T3190* is started and reestablishment of the TBF cannot be started until the timer expires (expiry limit of the timer *T3190* is presented in Table E.3).

TBF release procedure is entered whenever there are no LLC data in the PCU's buffer waiting for transmission. Since delayed TBF release procedure is used in DL, data transmission can proceed without any additional delays whenever new data are available. Before the actual release, all blocks inside the transmitting window must be correctly received and the expiry limit of the delayed TBF release timer must be reached. This timer is started whenever there is no packet Ack/Nack message with final block indicator (FBI) bit set to true is received by the PCU and it is reset if new data arrive for the specific MS. The expiry limit of the release timer is presented in Table E.3, Appendix E.

TBF dropping is based on lost pollings in the uplink direction. Whenever the network receives a valid RLC/MAC control message from the MS, it shall reset counter *N3105*. The network shall increment the counter for each radio block (allocated to that MS with the relative reserved block period (RRBP) field) for which no RLC/MAC message is received. If *N3105* reaches *N3105*max, the network releases the downlink TBF. *N3105* max is a parameter defined in Table E.3.

7.3.2.5 RLC/LLC Layer Modeling

All the relevant features of the RLC protocol are taken into account in the simulations. These include active transmitting and receiving windows, selective ARQ with block sequence numbers, final block indicators, polling bits and downlink packet-acknowledgement message with measurement reports. The RLC layer operates in acknowledged mode, LLC in unacknowledged mode.

7.3.2.6 Upper Layer Modeling

Layers above LLC (TCP/IP, RTP/IP,...) are not explicitly modeled in the analysis of this section. The effect of upper layers will be considered in Section 7.7.

7.3.2.7 Simulation Parameters

Some of the most important GPRS-specific simulation parameters and their default values are presented in Appendix E. Note that the general simulation parameters and network setup presented in Appendix E are also utilized in these simulations.

7.3.3 GPRS Performance in a Separate Non-hopping Band

In this section, performance analysis of the GPRS system in a separate non-hopping band is presented. The analysis is done with two different LA algorithms. The first algorithm uses only the two most robust coding schemes (CS-1 and CS-2) and the second algorithm uses all four GPRS coding schemes. Only CS-1 and CS-2 are used in most of the initial GPRS networks, but it does not achieve the full benefits of GPRS.

Performance of the GPRS system is analyzed with different reuse patterns using one TRX with eight TSLs available (except in the trunking gain analysis). All simulated users are best-effort GPRS users, i.e. only GPRS traffic is simulated and there is no prioritization between different users. The case with mixed CS and PS traffic in the same TRX traffic is analyzed separately in Section 7.5.

First, basic GPRS capacity and performance figures are presented starting with the behavior of user-specific mean TBF throughput and mean gross LLC frame delay curves. These are both good indicators of the quality experienced by an average GPRS user. Next, 10th percentile throughput and 90th percentile delay are shown, which are good network level QoS indicators. This basic performance analysis concludes with the spectrum efficiency values achieved with a GPRS system. This analysis includes both the effect of different reuse patterns and different user-specific QoS criteria.

After evaluating the general GPRS performance, the effects of two special cases related to GPRS deployment are discussed. First, GPRS performance on BCCH layer is studied. Next, TSL capacity analysis is presented. In the last part of this section, the trunking gain analysis shows how the traffic-handling capacity of each individual TRX increases when more TRXs are added to the cell.

7.3.3.1 Mean Throughput and Mean Delay

Figure 7.22 shows the evolution of mean throughput and delay of different reuse patterns and LA. Throughput is measured as the average throughput delivered to the LLC layer during the duration of the TBF. Delay refers to LLC frame delay: the time from when the LLC frame starts to be transmitted by the PCU until it is entirely received by the MS. As was stated in the previous section, RLC operates in ACK mode (retransmissions are possible). Throughput values range from 90 to about 20% of the maximum theoretical values of the highest coding schemes used in the LA algorithm.

Best throughput values are achieved with loose reuses, as expected. The performance levels experienced with reuse patterns 2/6, 3/9 and 4/12 are quite close to each other while 1/3 has clearly lower but still satisfactory performance. However, it should be noted that this comparison does not take into account the amount of spectrum needed for each reuse pattern.

As the load increases, there are two main reasons for throughput reduction: increasing interference levels and TSL sharing between users. Both effects will be analyzed later on by means of TSL capacity and throughput reduction concepts, respectively. For 1/3 reuse

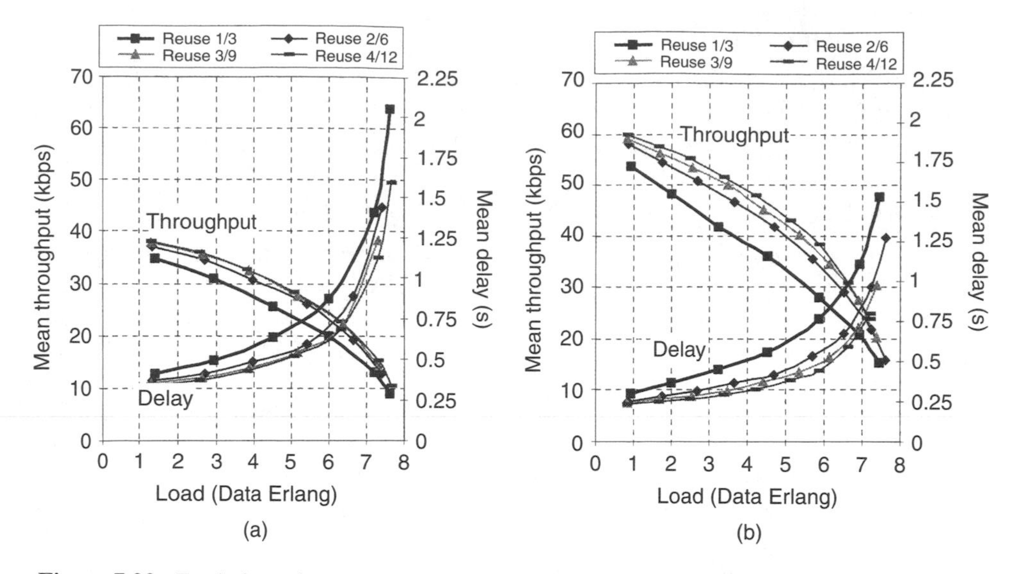

Figure 7.22. Evolution of mean throughput and mean delay in a non-hopping GPRS network with different reuse patterns. (a) GPRS with CS-1–2 and (b) GPRS with CS-1–4. Three TSL-capable MS. One TRX. Macrocell environment

and 6 Data Erlangs, 50% of the throughput reduction is due to interference and another 50% is due to TSL multiplexing. The same 6 Data Erlangs with 3/9 reuse has 15 and 85% throughput reduction due to interference and TSL sharing respectively.

As the load (Data Erlangs) gets close to the absolute maximum (8 Erlangs for 1 TRX), the dominant performance degradation effect in the TBF multiplexing over the same timeslots and performance is similar, regardless of the frequency reuse.

When comparing the performance of the used LA schemes, it is seen that in terms of mean throughput values, the additional coding schemes CS-3 and CS-4 bring around 50% higher throughputs. The gain is smaller when the reuse gets tighter. This is due to higher interference levels, which decrease the usefulness of less-robust coding schemes. Still, even with small reuse distance and high load, the users close to the base station can utilize the higher coding schemes effectively.

7.3.3.2 10th Percentile Throughput and 90th Percentile Delay

In the previous section, only the mean values were compared. Mean values are proportional to the overall spectrum efficiency of the whole network, but they do not tell about the distribution of quality among users. This section focuses on user quality at the lower bound of the distributions. Figure 7.23 shows the minimum-throughput values achieved by 90% of the users (10th percentile throughput), and maximum delays for 90% of the transmitted LLC frames (90th percentile delay).

Note also that the decrease in throughput is now more noticeable than in the mean-value curves. This is because interference levels are directly affecting low-throughput users. The average quality is not influenced that much due to the high quality experienced by the majority of the users, but these percentile figures are immediately affected.

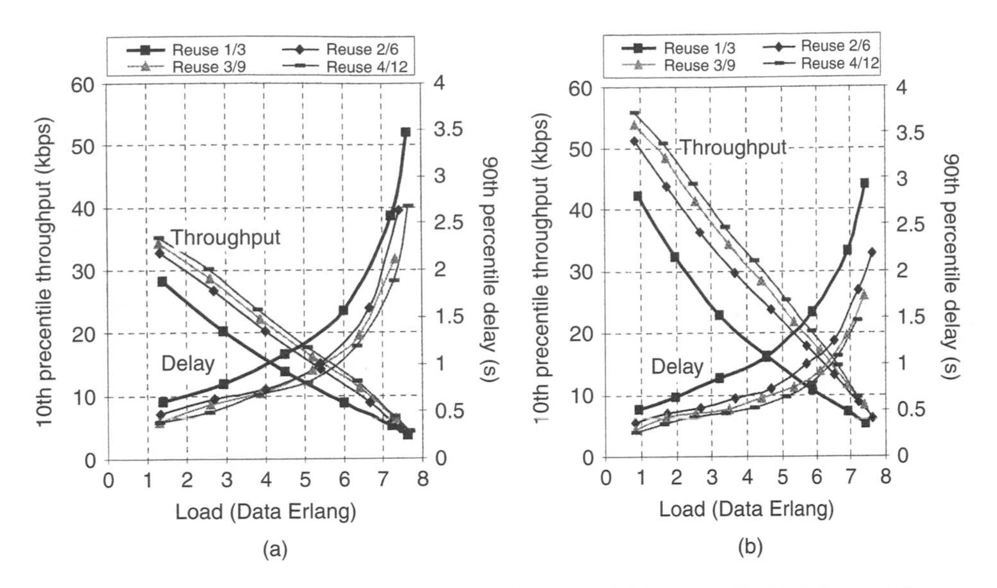

Figure 7.23. Evolution of 10th percentile throughput and 90th percentile LLC frame delay in a non-hopping GPRS network with different reuse patterns. (a) GPRS with CS-1–2 and (b) GPRS with CS-1–4. Three TSL-capable MS. One TRX. Macrocell environment

Minimum throughput values are up to around 50% lower than the average values. Note that a simple round robin scheduler is used, therefore scheduler-based throughput equalization techniques are not applied to users with bad throughputs. With 50% load (4 data Erlangs), 90% of the users achieve 23- and 33-kbps net throughput in reuse 4/12 with CS-1–CS-2 and CS-1–CS-4, respectively.

7.3.3.3 Spectrum Efficiency and TBF Blocking

The previous sections considered user-perceived throughput and delay. In this one, the spectrum efficiency of the network is shown. The GPRSsystem traffic is increased until it is fully loaded. At this saturation point, TBF blocking starts to grow exponentially according to queuing theory and additional increase in offered load does not increase the spectrum efficiency any more. Figure 7.24 shows the evolution of spectrum efficiency and TBF blocking as a function of network load.

The tighter the reuse is, the better spectrum efficiency can be achieved. But, pure load-based comparison is not always sensible, since it does not take any quality requirement into account. This spectrum efficiency and quality trade-off is tackled in a later section.

From Figure 7.24, it can also be seen that TBF blocking starts to take place with the load close to 8 Data Erlangs (absolute maximum limit). The increase of blocking is very rapid and the network will reach the state of congestion. Reaching this congestion point should be avoided when dimensioning of the GPRS network.

7.3.3.4 Performance in the BCCH Layer

Earlier sections only considered non-BCCH TRXs. In this section, the performance of GPRS deployment in the BCCH layer is analyzed. There are seven TSLs available for

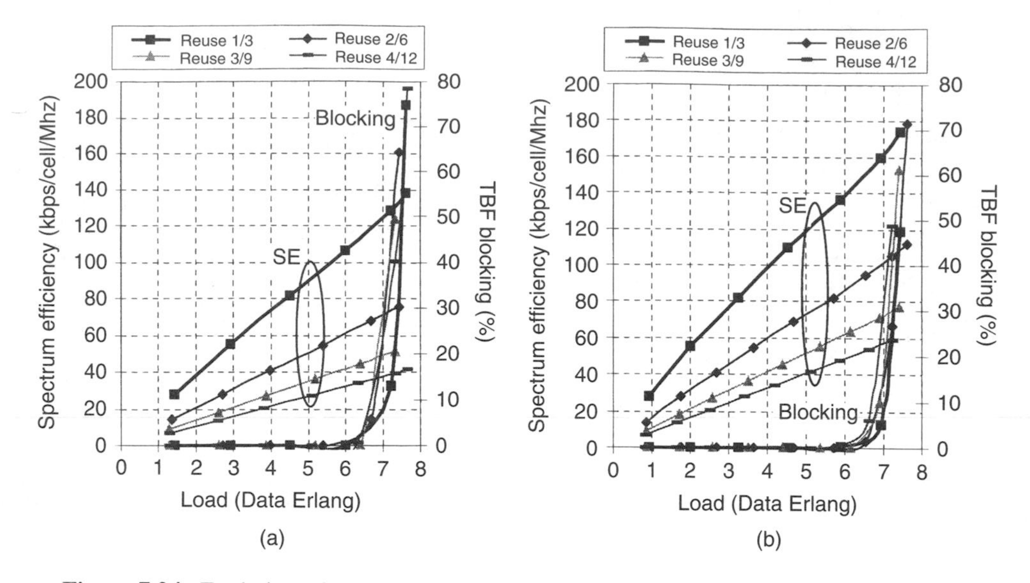

Figure 7.24. Evolution of spectrum efficiency and TBF blocking in a non-hopping GPRS network with different reuse patterns. (a) GPRS with CS-1–2 and (b) GPRS with CS-1–4. Three TSL-capable MS. One TRX. Macrocell environment

GPRS as TSL 0 is reserved for BCCH. The analysis is performed with two different reuse patterns (3/9 and 4/12) and the achieved results are compared against the results with the same reuse patterns in the non-BCCH layer (Figure 7.25). The non-BCCH layer has only seven TSLs available to allow a fair comparison with BCCH. The analysis is performed with the LA algorithm using all four available GPRS coding schemes. Note that reuse 3/9 is very aggressive for standard BCCH deployments. From the radio bearer–quality point of view, the BCCH layer differs from the non-BCCH layer considerably. The most important characteristic of the BCCH layer is that all the downlink bursts in the BCCH frequency must be transmitted constantly and with full power.

In the BCCH layer, the TSL capacity, see Section 5.2, is constant and independent of GPRS load. Therefore, the throughput variation is only due to user multiplexing over the same TSL. The difference in performance with the non-BCCH layer is due to the lower interference levels at low network loads in the non-BCCH layer. As the load increases and the network becomes fully loaded, the interference levels are similar and throughputs are equivalent.

7.3.3.5 TSL Capacity

TSL capacity analysis is useful for GPRS performance characterization and network dimensioning with different number of allocated TSLs and different number of available TSLs for GPRS (see Section entitled *TSL capacity* in 5.2.2.7). TSL capacity tells how much data the network can transmit with a fully utilized TSL. TSL capacity depends on interference levels and RLC protocol performance. It does not depend on HW dimensioning. Figure 7.26 shows TSL capacity for non-BCCH layer with different reuses. Figure 7.26 (a)shows TSL capacity, not taking into account preemptive retransmissions.

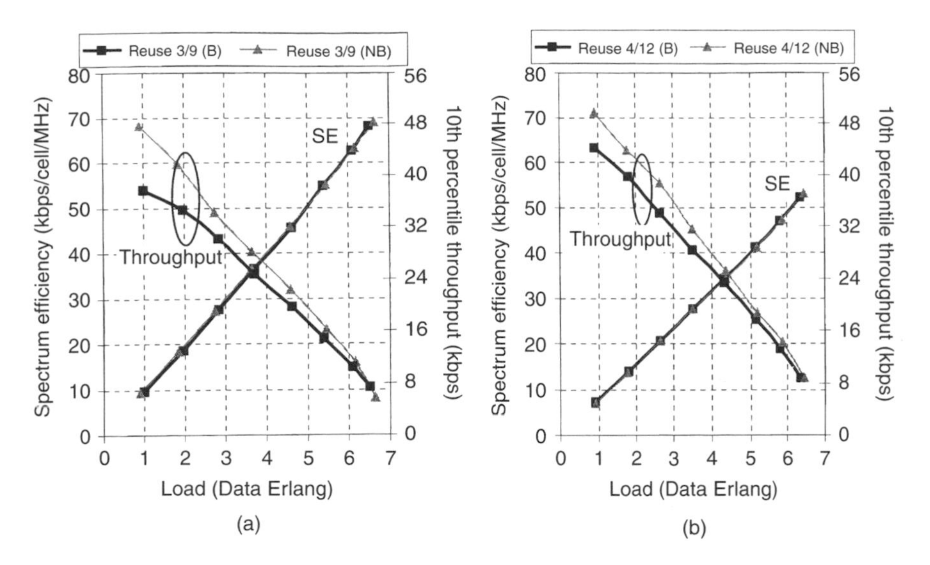

Figure 7.25. Evolution of spectrum efficiency and 10th percentile throughput in a non-hopping GPRS network with different reuse patterns in both BCCH ('B') and non-BCCH ('NB') layers using CS-1–4. (a) reuse 3/9 and (b) reuse 4/12. Three TSL-capable MS. Both graphs are with one TRX and seven TSLs available

Preemptive retransmissions are pending ACK RLC blocks that may be transmitted during TBF release phase and RLC protocol stalling situation. If these retransmissions are not accounted for, the TSL capacity depends only on interference levels. As the network load increases, the TSL capacity decreases because of higher interference levels. When preemptive retransmissions are considered, the TSL capacity curve changes, see Figure 7.26 (b). The effect is more evident for higher interference levels and low Data Erlang values. For the traffic model considered (FTP), the dominant effect causing the TSL capacity reduction is the RLC window stalling. When the network is heavy loaded (Data Erlangs approaching 8), the network multiplexes users on the same TSL, thus reducing the number of preemptive retransmissions as they have lower priority than any user data. Simulation assumes three TSL MS; however, if lower MS TSL capability were used, the probability of RLC protocol stalling would be lower (as throughput is lower). Figure 7.27 shows GPRS CS1–4 TSL capacity in the BCCH layer. If preemptive retransmissions are not considered, the TSL capacity is constant because of the continuous transmission in the DL direction. TSL capacity with GPRS CS1–2 is very close to 12 kbps for any BCCH reuse.

7.3.3.6 Trunking Efficiency Analysis

There is a trunking gain when more than one TRX is dedicated for GPRS. Figure 7.28 shows simulation results in a non-hopping GPRS network with 1/3 reuse pattern dedicating one, two or three TRXs in each sector exclusively for GPRS. LA with CS-1–4 is utilized in the analysis. The figure represents blocking probability and throughput reduction factors as a function of TSL utilization (see Sections 5.2.2.6, 5.2.2.10). It is a measure of how user throughput is affected by TSL sharing among several users. TSL utilization is a measure of how many of the available TSLs for (E)GPRS are being utilized, on average,

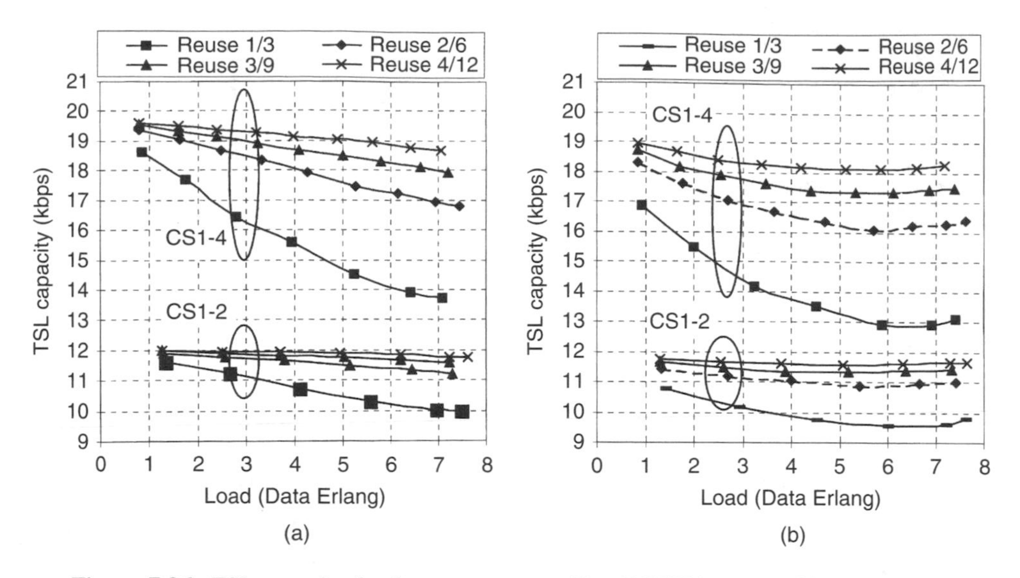

Figure 7.26. TSL capacity for frequency reuses. Non-BCCH layer. (a) No preemptive retransmissions and (b) with preemptive retransmissions. Three TSL-capable MS

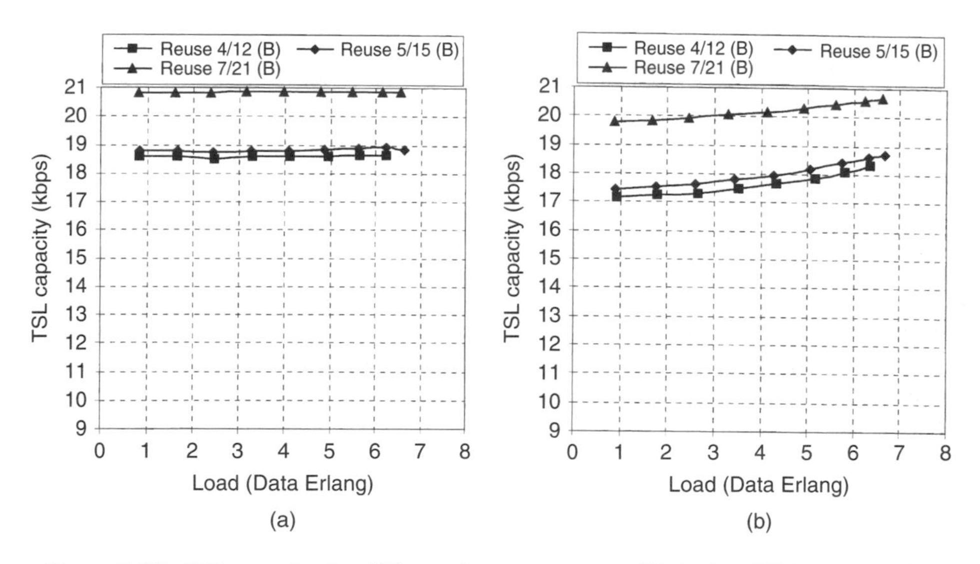

Figure 7.27. TSL capacity for different frequency reuses. CS-1–4. BCCH layer. (a) No preemptive retransmissions and (b) with preemptive retransmissions. Three TSL-capable MS

to carry data traffic. It varies between 0 (no data traffic) and 100% (full utilization of the available resources).

Figure 7.28 shows TBF blocking probability and SE for the same network with different number of TRXs. By increasing the number of TRXs, it is possible to reach higher TSL utilization for the same TBF blocking probability. This effect is equivalent to the one experienced by speech traffic. Basically, the existing hardware can be used more

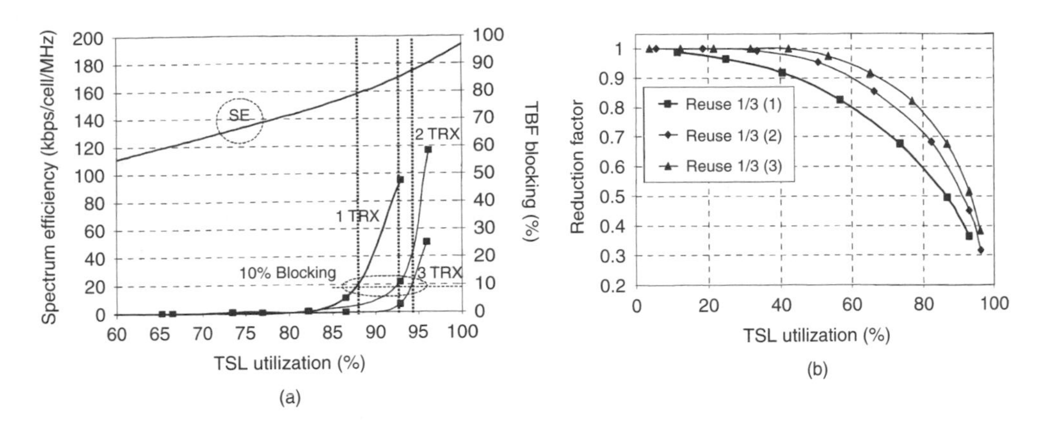

Figure 7.28. Evolution of spectrum efficiency and throughput reduction factors in a non-hopping macrocell GPRS network with reuse pattern of 1/3 and with different number of TRXs and CS-1–4. Three TSL MS. Three TSL-capable MS. One TRX

effectively when the number of available channels increases and the results show that GPRS service is no exception.

Throughput reduction factors are also affected by increasing the number of TRXs. By doing so, the throughput reduction factors increase and, therefore, it is possible to achieve the same average throughput with higher TSL utilization. This means that SE can be increased while maintaining the same average throughput. Table 7.3 summarizes SE values achieved with different reduction factors. Table 7.3 shows, for example, that for a reduction factor of 10% (this defines network quality) the SE is 50% higher with three TRXs compared to one TRX. This is purely a trunking efficiency gain.

Simulation results have shown that GPRS throughput reduction factors are the same for any frequency reuse including fractional reuses (e.g. 1/1 hopping). Only in the case where interference levels are very high, the reduction factor curve changes

Reduction factors and TBF blocking from Figure 7.28 will change with different number of allocated TSLs. Trunking efficiency depends on the MS TSL capabilities (Nu) and total number of TSLs available for GPRS (Ns). Doubling Ns has the same effect as dividing Nu by two, from the trunking efficiency point of view. In effect, trunking efficiency depends exclusively on the rate (N) of the two factors $N = \text{Ns/Nu}$. For example, one TRX with one TSL allocation has the same trunking efficiency as three TRXs with three TSL allocations. The latter assumes an ideal channel allocation algorithm that is able to share evenly the load over the available TSL. A properly designed channel allocation algorithm should get close to this situation.

Table 7.3. EGPRS maximum SE [kbps/Cell/MHz] with different number of TRXs. Three TSL mobile stations

Reduction factor	0.9	0.8	0.7	0.5
1 TRX	52.5	68.4	75.6	96.6
2 TRX	68.4	80.4	87.5	101.5
3 TRX	80.5	89.6	95.2	103.6

In order to improve trunking efficiency, one TSL allocation should be preferred, but the price paid is a lower user throughput. TBF blocking curves and reduction factor can be computed theoretically under a number of assumptions. This analysis is presented in Appendix B. Theoretical results are close to the simulated ones although they tend to be more optimistic. There are two main effects not accounted for in the theoretical analysis when comparing with simulation results:

In simulations, load is not always equally shared among all the available TSLs. This is due to channel allocation limitations (allocated TSLs have to be consecutive); and, specially, because periodical channel reallocation of the existing TBFs is not performed. This has the effect of worsening the throughput reduction factors in simulations up to 80% TSL utilization.

When interference levels are so high that the TSL capacity cannot be considered constant and independent of TSL assignment dynamics, the theoretical analysis is less exact. In this case, the shape of the trunking efficiency curves changes. This is the case of frequency hopping at high effective frequency load (EFL) levels as mentioned before.

7.3.4 GPRS Performance in a Separate Band with RF Hopping

This section presents spectrum efficiency values for GPRS traffic allocated in the hopping layer. The evolution of the most important network quality indicators is also presented. The LA CIR thresholds have been adjusted for hopping channels. A comparison of the non-hopping and hopping networks will be presented first. Then, a soft-capacity study of GPRS in the hopping layer will be performed.

7.3.4.1 Comparison between FH and Non-FH Deployments

In this section, comparison between frequency hopping and non-hopping networks is presented. In order to be able to produce a fair comparison, the analysis is performed with one TRX configuration and with the same bandwidth. The selected non-hopping configuration is reuse 3/9 and the corresponding FH configuration is 1/1 with nine frequencies. Note also that the comparison is performed with both link adaptation algorithms.

From Figure 7.29, it can be seen that FH does not bring any quality or capacity gain compared with the non-hopping case. It actually degrades the performance, especially in the CS-1–4 case. This behavior is different from that with speech traffic, where significant capacity gains are achieved with FH. The main reasons for this difference are as follows.

Very limited link level gains achieved with frequency hopping when retransmissions and LA are applied. See Section 7.1. In practice, frequency hopping gains are more noticeable with very low BLER, but very low BLER does not make a difference in terms of acknowledged mode throughput.

Highest CS (CS-3 and CS-4) has lower channel coding, so the gain achieved with speech channels becomes an FH loss with CS-3 and CS-4. The lowest CSs (CS-1 and CS-2) have a certain frequency diversity gain (although limited), but interference diversity gain turns negative. Cochannel interference levels in a fractionally reused network can be very high (e.g. with 1/1 reuse, the same frequencies are reused on every site). The limited interleaving depth (four bursts) and limited coding are not enough to cope with the strong cochannel burst hits. For the case of GPRS with CS-1–4, FH results are specially degraded. CS-3 and CS-4 suffer the most with FH, thus significantly limiting the SE.

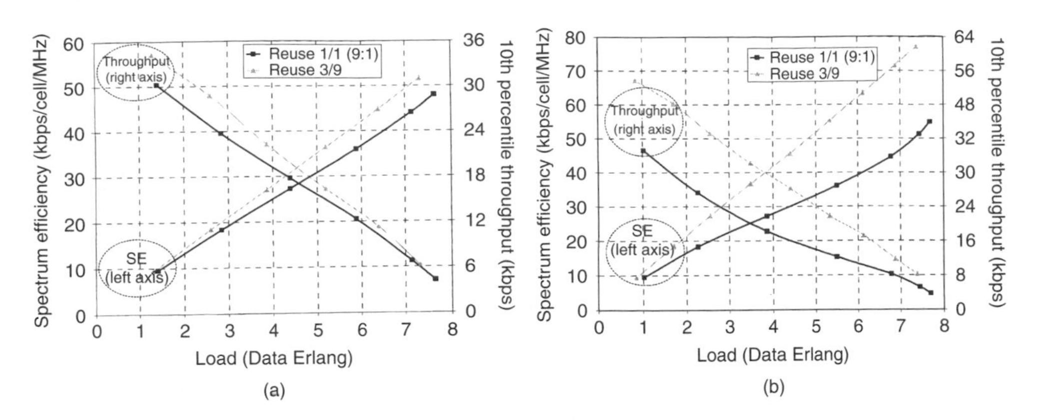

Figure 7.29. Evolution of spectrum efficiency and 10th percentile throughput in both non-hopping and RF-hopping macrocell GPRS networks. (a) GPRS with CS-1–2 and (b) GPRS with CS-1–4. Three TSL-capable terminals. One TRX

With 1/1 hopping reuse, it is possible to introduce up to three TRXs and still avoid cochannel interference from the adjacent cells. By doing this, it is possible to eliminate HW limitations, and make the network purely interference limited. In this case, it is possible to achieve better spectrum efficiency than the reference 3/9 reuse.

In any case, it can be stated that GPRS can be introduced into the hopping network without major problems. Although it is recommended to introduce it first in the BCCH layer, and utilize the hopping layer whenever it is required.

7.3.4.2 Reuse 1/1 Hopping Analysis

In this chapter, a QoS-related soft-capacity analysis is performed. The used FH scheme is reuse 1/1 and the number of hopping frequencies is fixed to 18.

Figure 7.34 shows TSL capacity versus EFL. The analysis is performed with one, two and three TRXs. The same TSL capacity is achieved for the same EFL values, independently of the number of TRXs as soon as there is low TSL sharing among different users. The latter is due to the high probability of RLC protocol stalling that generates a lot of preemptive retransmissions. This effect, of increased TSL capacity when TSL sharing increases is due to preemptive retransmissions and has also been found in non-hopping cases (see Figures 7.25 and 7.26), although with 1/1 hopping reuse, the effect is much more evident. Taking an EFL level of 6%, it corresponds to an average user throughput of $3 \times 12 = 36$ kbps with a three TSL MS. If there is any HW limitation, throughput values should be corrected by the reduction factor. Figure 7.30 represents reduction factor values for one, two and three TRXs. Note also that this analysis is performed with LA utilizing all GPRS coding schemes. Figure 7.31 represents the 10th percentile throughput values and 90th percentile delay in pure soft-capacity situation.

7.3.5 GPRS Spectrum Efficiency with QoS Criterion

Quality-of-service criterion plays an important role in capacity evaluations. Typically, spectrum efficiency increases as a function of input load until the network is fully loaded. At the same time, the quality of calls naturally decreases. It is sensible to assume that

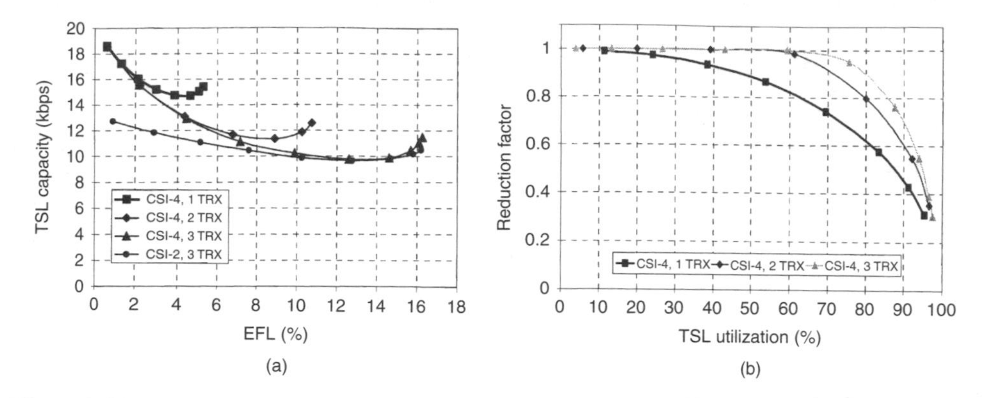

Figure 7.30. TSL capacity and reduction factors for GPRS in the frequency hopping layer. All GPRS coding schemes

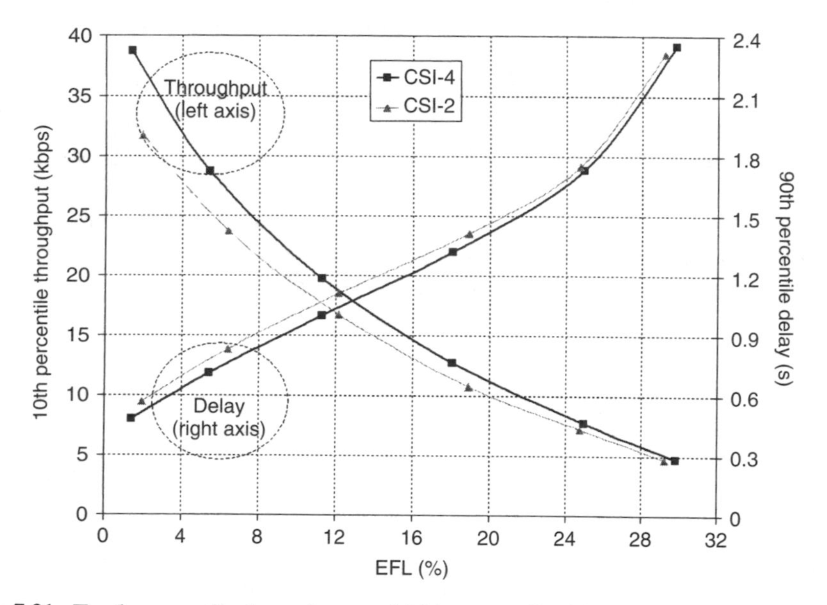

Figure 7.31. Tenth percentile throughput and 90th percentile delay versus EFL for GPRS in a RF-hopping macrocell network with reuse 1/1. Three TSL-capable MS

even for best-effort services, there is certain minimum QoS that must be fulfilled (this is required already by some higher-level protocols like TCP). When simulating spectrum efficiency for a packet data network, offered load is increased until the point where this minimum QoS level is maintained. Since user-perceived quality, not average network level quality, is the indicator to be measured, the QoS criteria must be user-specific. QoS criteria are specified in terms of minimum throughput and maximum delay achieved by 90% of the throughput and delay samples. Figure 7.32 represents GPRS SE figures for different reuses and minimum QoS criteria for three TSL GPRS mobile stations. Four different reuse patterns are considered without hopping, in addition to 1/1 reuse with

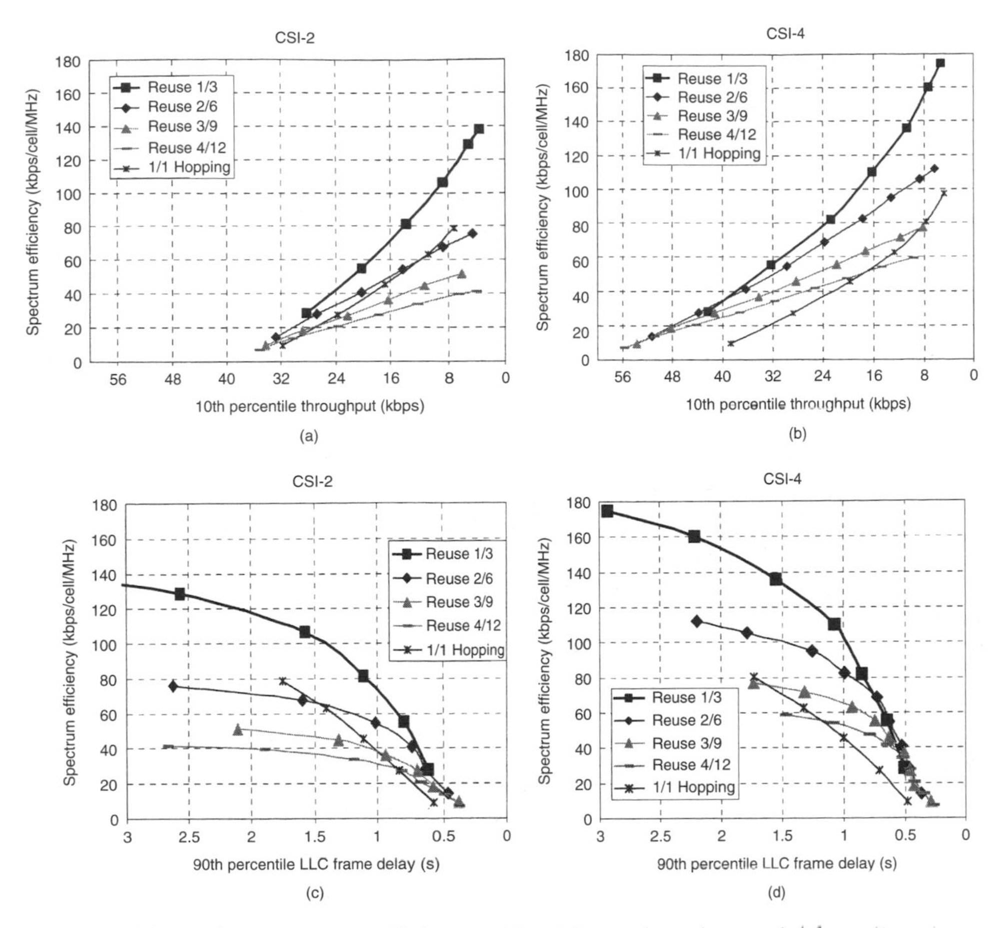

Figure 7.32. Achieved spectrum efficiency with minimum throughput and delay criteria in a macrocell GPRS network with different reuse patterns. (a, c) GPRS with CS-1–2 and (b, d) GPRS with CS-1–4. One TRX

frequency hopping. The SE values are limited by a combination of HW and interference blocking, except in the 1/1 hopping case where there are no HW limitations so that performance is purely interference limited. Hopping performance is degraded also due to RLC protocol stalling, especially in the GPRS CS-1–4 case. Stalling artificially increases the load (up to 30%) and the interference levels due to preemptive retransmissions. With lower user throughputs, stalling is less frequent so that SE improves appreciably.

As can be seen in Figure 7.32, the best spectrum efficiency can be achieved with low user data rates and very tight frequency reuse (reuse 1/3). These kinds of assumptions are very suitable for best-effort data. However, reuse pattern will not be able to fulfill high-quality criteria (high throughput and low delay) in some locations at the cell border. With tighter QoS criteria, i.e. assuming a higher percentage of users above the minimum quality, the spectrum efficiency will be reduced and looser reuse (like 2/6 or 3/9) may give better spectrum efficiency.

7.3.6 *Reuse Partitioning Principle to Increase Spectral Efficiency and QoS Provisioning*

From the previous section, it has been seen that tight reuses provide the highest spectral efficiency, but they suffer from high interference levels at the edge of the cell. Those calls with bad radio link conditions can be moved to a different layer (e.g. BCCH layer) where the reuse is not so aggressive and throughput at the cell border is much better. In this way, two contradictory targets can be met for data services: high spectral efficiency together with high QoS. Very aggressive deployments are also thinkable, like utilizing just one frequency for the high-capacity layer. By making sure that users with good radio link conditions use this layer, spectral efficiency can be boosted. In order to implement these schemes, the reuse partitioning principles presented in Section 6.2 can be used.

7.4 EGPRS System Capacity

7.4.1 *Introduction*

In this section, performance of the EGPRS (Rel'99 specifications) system is evaluated by means of network level simulations. As in the GPRS-related analysis presented earlier, focus is once again on downlink performance. The performance evaluations are done both from the capacity (spectrum efficiency) and quality (throughput, delay) points of view. The analysis concentrating on the evolution of timeslot capacity as a function of EFL is also presented.

7.4.2 *Modeling Issues and Performance Measures*

The modeling of EGPRS service is naturally based on the modeling of GPRS service. In addition to the features already supported in GPRS, EGPRS introduces many new enhancements. The EGPRS-specific modeling issues that are taken into account in the simulations are described below.

7.4.2.1 RRM Algorithms

Link adaptation is similar to that described in Section 7.2.2.3.

7.4.2.2 RLC Layer

Incremental redundancy with receiver memory limitations is implemented. During IR operation, if the incorrectly received RLC block does not fit into MS memory, it will be discarded and it cannot be soft-combined with the old data. MS IR memory considered is such that it is possible to store 61 440 soft values (received raw bits before decoding) per TBF. This is more than enough in order to avoid IR memory problems. Block splitting and block re-segmentation inside the same MCS family as well as header errors are modeled in EGPRS RLC blocks. RLC window size is 384 blocks, which is the maximum value for three-slot EGPRS MS.

7.4.3 *EGPRS Performance with Link Adaptation in a Separate Non-hopping Band*

In this section, performance analysis of the EGPRS system in a separate non-hopping band is presented. Performance of the EGPRS system is analyzed with different reuse patterns and with one TRX configuration. All users are best-effort EGPRS users, i.e. only EGPRS traffic is simulated, and there is no prioritization between different users.

This section is organized in the same way as the corresponding GPRS section. The evolutions of mean net throughput and mean gross LLC frame delay are presented first. Then the evolutions of 10th percentile throughput and 90th percentile delay are studied. After these quality-related sections, the spectrum efficiency values achieved with the EGPRS system will be presented. The analysis also includes the comparison of different reuse patterns with different QoS criteria.

7.4.3.1 Mean Throughput and Mean Delay

The achieved mean throughput values and delays are depicted in Figure 7.33. The achieved values are significantly better than for GPRS. Once again, 1/3 reuse has the worst performance in terms of throughput and delay due to the higher interference levels, while the difference between other reuses is not so large. With low load, it is possible to get quite close to the theoretical maximum bit rates with all reuses. Even with high loads, very good performance values can be achieved compared with GPRS. With the same load levels, average throughput values are 2.5 to 3 times better than with GPRS CS-1–4 and 3 to 4 times better than with GPRS CS-1–2.

7.4.3.2 10th Percentile Throughput and 90th Percentile Delay

Figure 7.33 shows minimum throughput and maximum delay achieved in 90% of the samples (10th percentile throughput and 90th percentile delay, respectively). Note also the relatively larger degradation with 1/3 with respect to other reuses, compared with GPRS results. The absolute throughput and delay levels achieved with certain loads are significantly better than with GPRS. Compared with GPRS, with CS-1–2, the percentile throughput values are 2 to 3.5 times higher. When comparing with GPRS with CS-1–4, gains are of the order of 1.5 to 2.5 times higher. EGPRS gains are lower with tighter reuse schemes since the higher coding schemes provided by the EGPRS bearer are less used.

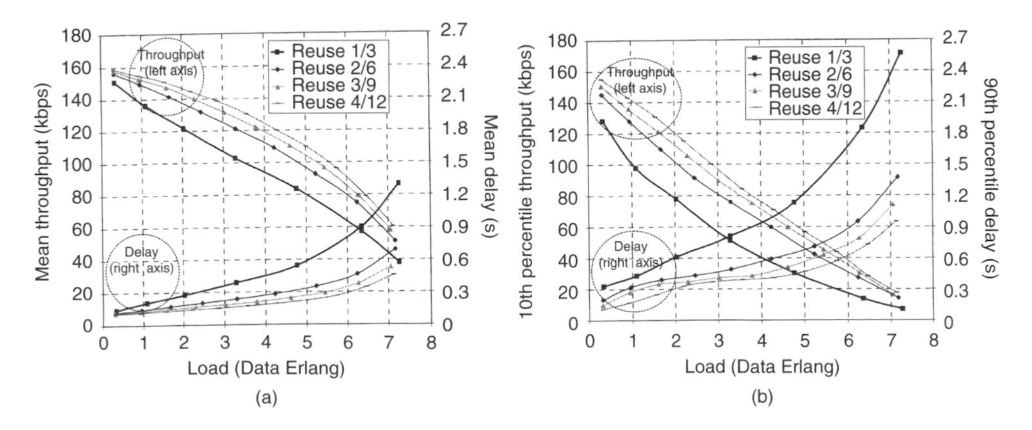

Figure 7.33. Throughput and delay performance of EGPRS in a non-hopping macrocell network with different reuse patterns. Three TSL-capable MS. One TRX

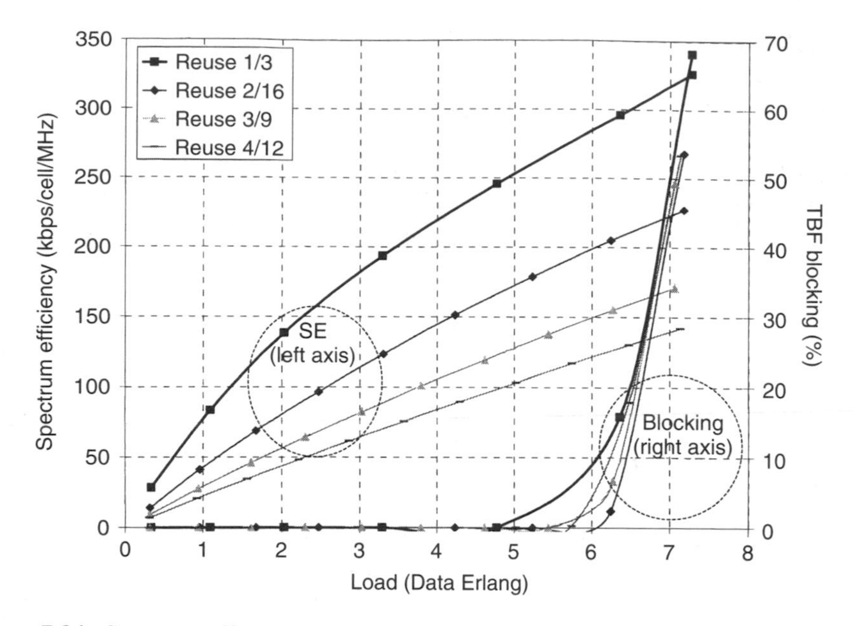

Figure 7.34. Spectrum efficiency and TBF blocking of EGPRS in a non-hopping macrocell network with different reuse patterns

7.4.3.3 Spectrum Efficiency and TBF Blocking

Figure 7.34 shows the spectrum efficiency growth of an EGPRS network as a function of network load. TBF blocking is shown in the same figure in order to see the level of saturation in the network. With EGPRS, the achieved levels of spectrum efficiency are twofold to threefold compared with the same GPRS cases. And as with GPRS, the TBF blocking starts to occur as the load approaches the absolute maximum (8 Erlangs). Although reuse 1/3 has the worst throughput performance, it achieves the highest SE due to the lower number of utilized frequencies.

The simulated SE figures in a fully loaded situation are very much in line with the ones estimated in field trials and presented in Section 7.6.

7.4.3.4 Performance in the BCCH Layer

EGPRS has a similar behavior in the BCCH layer compared with GPRS. Owing to the continuous transmission, lower throughput values are achieved when the network load is low compared with the same reuse but utilizing a non-BCCH TRX. Figure 7.35 shows 10th percentile throughput values and SE with different network loads for different reuses. Seven TSLs are available for both cases. SE is practically the same for both BCCH and non-BCCH cases, the difference is only found in percentile values that defines the minimum quality.

7.4.3.5 TSL Capacity

TSL capacity achieved with EGPRS is presented in Figure 7.36. Owing to the longer RLC window and better retransmission techniques, the probability of stalling is much lower than

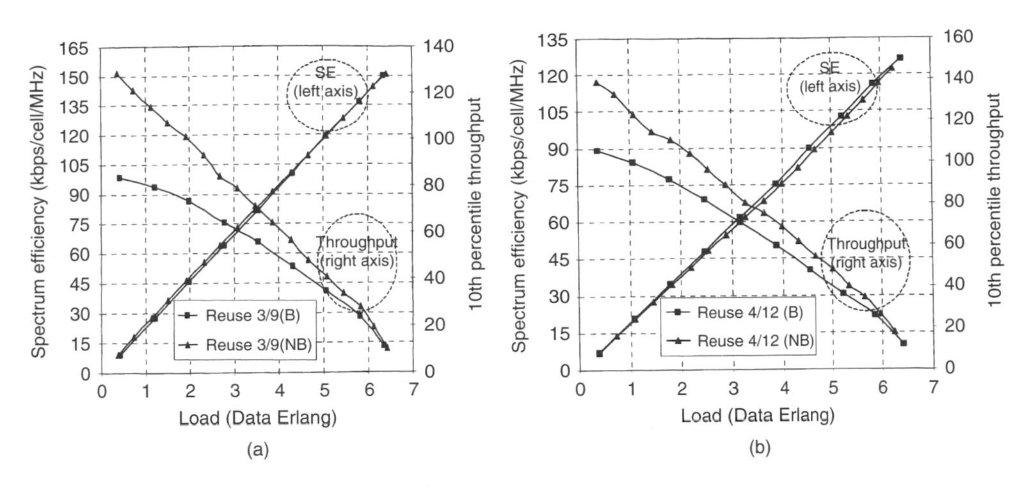

Figure 7.35. Evolution of spectrum efficiency and 10th percentile throughput in a non-hopping EGPRS network with different reuse patterns in both BCCH ('B') and non-BCCH ('NB') layers using EGPRS. (a) reuse 3/9 and (b) reuse 4/12. Three TSL-capable MS. Both graphs are with one TRX and seven TSLs available

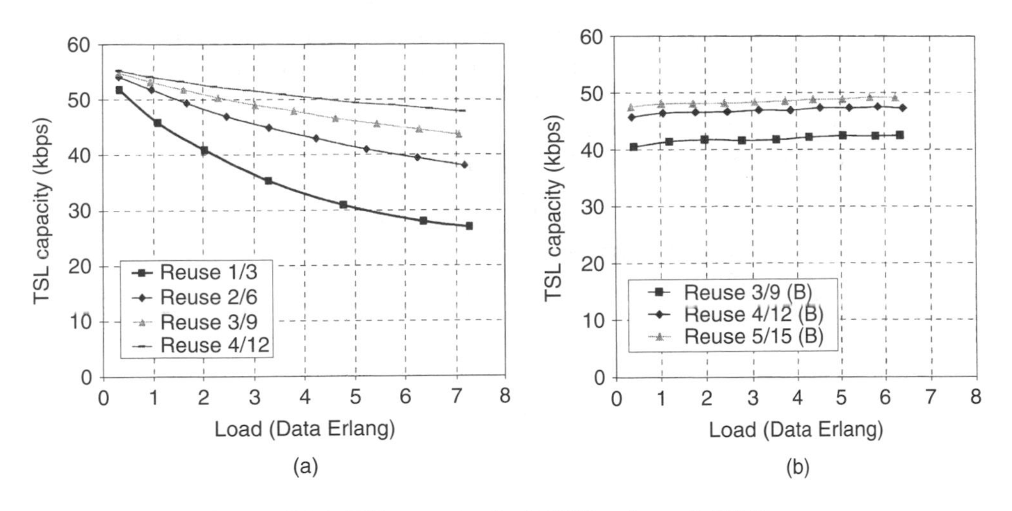

Figure 7.36. TSL capacity for BCCH and non-BCCH layer

with GPRS. Therefore, preemptive transmissions do not influence TSL capacity values very much. TSL capacity values have a larger dynamic range compared with GPRS. For example, reuse 1/3 TSL capacity is almost halved at very high loads. Nevertheless, TSL capacity remains almost constant in the BCCH layer, and therefore BCCH is very suitable for achieving high throughput values. By allocating absolute scheduling priority to a user, the throughput reduction becomes 1 and throughput depends exclusively on the number of allocated TSLs and the TSL capacity. With a four TSL-capable terminal allocated in DL and BCCH reuse of 5/15, high-priority users may experience throughput close to $4 \times 48 = 192$ kbps on average, independently of BCCH traffic load.

7.4.3.6 Trunking Efficiency Analysis

The same trunking efficiency analysis presented in Section 7.3.3 is applicable to EGPRS. Figure 7.37 on the left shows EGPRS reduction factors for one TRX with different reuses. Reuses 2/6, 3/9 and 4/12 have very similar reduction factors. Comparing with GPRS, the reduction factors are also similar to the GPRS reduction factors (found in Figure 7.28).

Reuse 1/3 has lower reduction factors as it has a much larger soft-blocking component. Reduction factors can be seen in Figure 7.37. Reduction factors with two and three TRXs for reuse 1/3 are included in Figure 7.37 (left).

7.4.4 EGPRS Performance in a Separate Band with RF Hopping

This section presents simulated spectrum efficiency values for the EGPRS network with different reuse schemes in the RF-hopping situation. The evolution of the most important network quality indicators is also presented. As in Section 7.3.4, the simulations were done using only non-BCCH hopping TRXs with pure EGPRS traffic.

This section is organized so that a comparison of non-hopping and hopping EGPRS networks is presented first. Finally, a soft-capacity analysis of the EGPRS system is presented.

7.4.4.1 Comparison between FH and Non-FH Deployments

In this section, a comparison between frequency hopping and non-hopping networks is presented. In order to be able to produce a fair comparison, the analysis is performed with one TRX configuration and with the same bandwidth. The selected non-hopping configuration is reuse 3/9 and the corresponding FH configuration is 1/1 with nine frequencies. Figure 7.38 shows that FH does not bring any quality or capacity gain with one TRX.

With nine frequencies and 1/1 hopping reuse, it is possible to increase the number of TRXs up to three (controlling the intrasite cochannel interference; without such control, it can go up to nine). In this way, HW-limitation effects can be minimized, thus networks become pure soft-capacity limited, and better spectrum efficiency than the reference 3/9 case can be achieved.

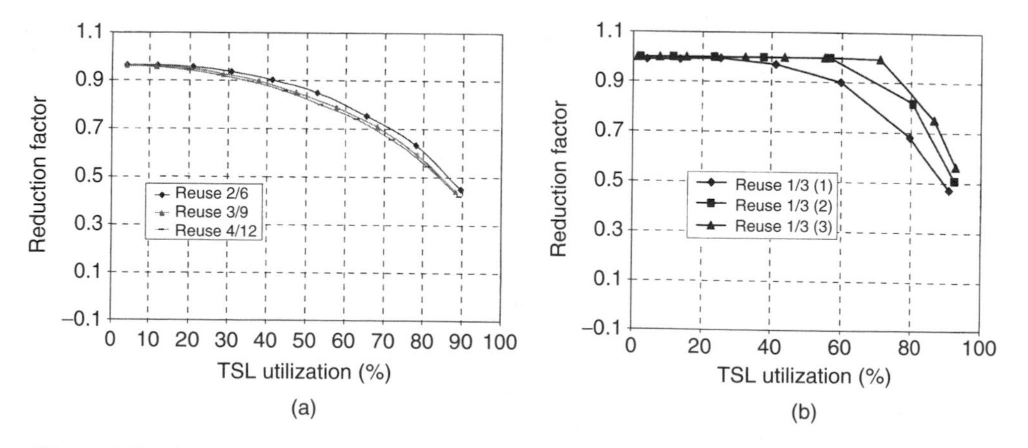

Figure 7.37. EGPRS reduction factor for different reuses. (a) One TRX and (b) one, two and three TRXs with 1/3 reuse. Three TSL-capable MS

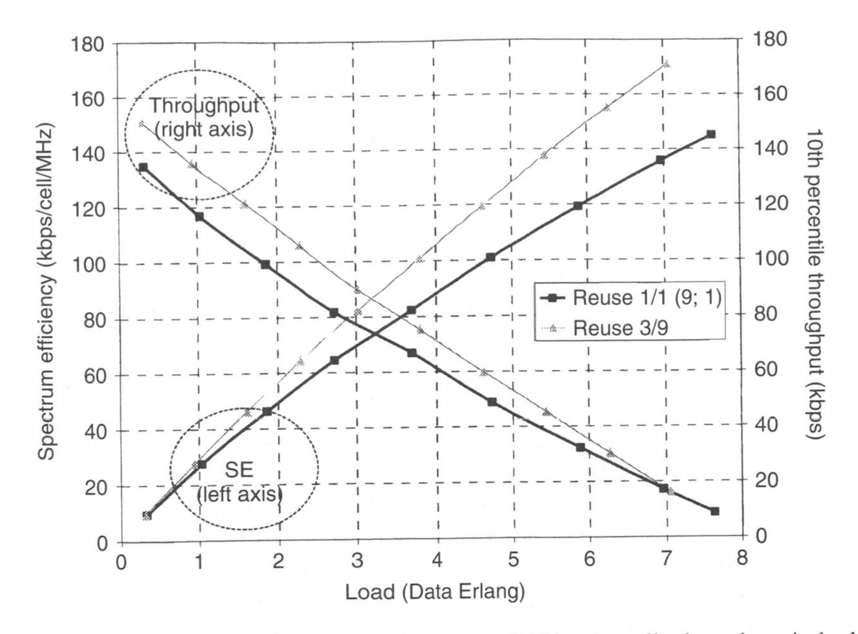

Figure 7.38. Evolution of spectrum efficiency and 10th percentile throughput in both non-hopping and RF-hopping macrocell EGPRS networks. Three TSL-capable MS. One TRX

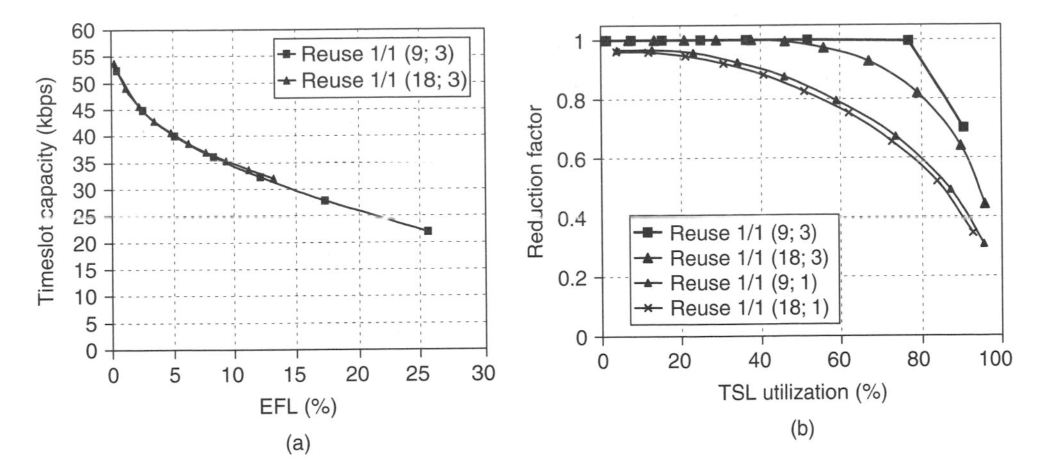

Figure 7.39. Timeslot capacity as a function of EFL for EGPRS in an RF hopping macrocell network with different reuse patterns. Three TSL-capable MS. (X, Y) in legend means X hopping frequencies and Y TRXs

7.4.4.2 Reuse 1/1 Hopping Analysis

TSL capacity is represented in Figure 7.39 as a function of EFL. Results are shown for 9 and 18 frequencies. Note that TSL capacity is the same as soon as the network is loaded with the same EFL. In EGPRS, TSL capacity figures are not affected by preemptive retransmissions in the same way as in GPRS case, as probability of RLC protocol stalling is much lower. Reduction factors for one and three TRXs are also shown in Figure 7.39 (b). Reuse 1/1 with one TRX has a similar reduction factor to that in the non-hopping cases. In case of three TRXs and nine frequencies, the higher soft-blocking component changes the shape of the reduction factor.

7.4.5 *Spectrum Efficiency with QoS Criterion*

Figures 7.40 and 7.41 show spectrum efficiency with minimum QoS criteria for EGPRS traffic with three TSL mobile stations. Different frequency reuses without frequency hopping are considered. Reuse 1/1 results with frequency hopping are also included in the figure. In these simulations, reuse 1/1 is not HW limited (TSLs are not shared by more than one user). Compared with GPRS, the achieved spectrum efficiency for the same quality is twofold with low-quality criteria. EGPRS spectrum efficiency increases with tighter QoS criteria. For example, EGPRS SE is 10 times higher compared with GPRS CS-1–4 with a 10th percentile throughput of 48 kbps. The SE gain of EGPRS is due to two main reasons. First, the much better link performance of EGPRS in acknowledged mode (see Section 7.1.4) makes it possible to achieve same user throughputs with less C/I values. Secondly, the EGPRS has threefold larger peak throughput than GPRS. Reuse 1/1 with frequency hopping is purely interference limited. In this case, network load is increased until the interference level is such that minimum QoS criteria cannot be met. This represents pure soft-capacity performance of EGPRS. Reuse 1/3 without hopping suffers from a combination of HW and interference limitations. But still the achieved

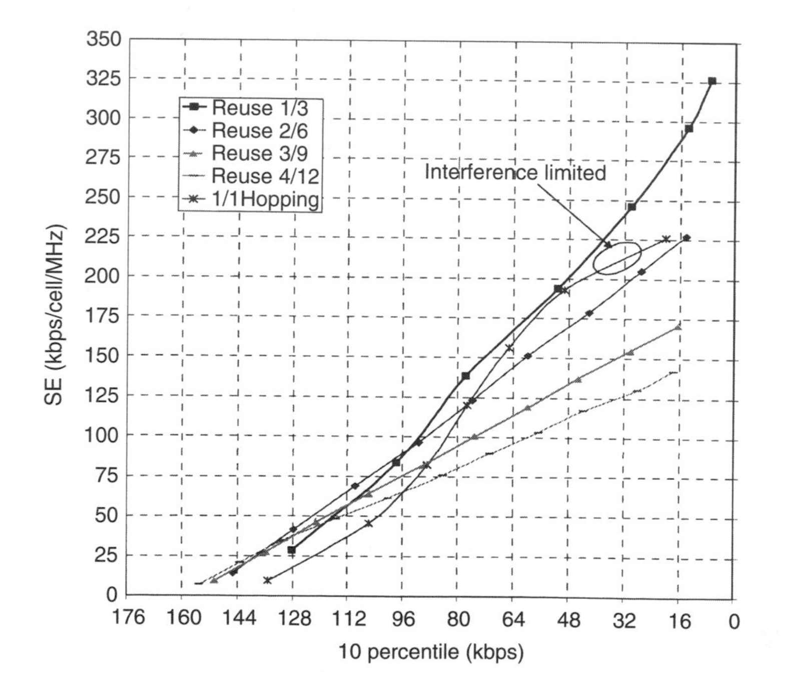

Figure 7.40. Achieved spectrum efficiency values with minimum throughput criteria in a non-hopping macrocell EGPRS network with different reuse patterns. One TRX. Three TSL-capable MS

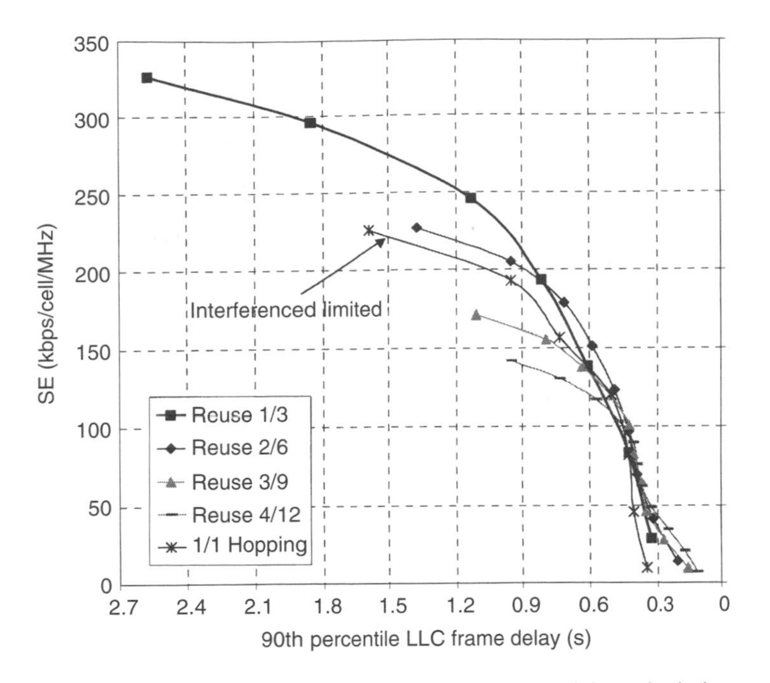

Figure 7.41. Achieved spectrum efficiency values with maximum delay criteria in a non-hopping macrocell EGPRS network with different reuse patterns. One TRX. Three TSL-capable MS

spectral efficiency for 1/3 non-hopping is better than the pure soft-capacity-limited 1/1 hopping case, although the difference is not so dramatic as in the GPRS case. The reason for this is the better performance of EGPRS in FH networks, as described in the Sections 7.3.4 and 7.4.3.

Reuse 1/3 is very suitable for low throughput and delay requirements, but for more restrictive throughput or delay requirements looser reuses are more suitable. Properly designed RRM should follow the envelope of Figure 7.40, maximizing the SE for different QoS requirements. This can be accomplished, for example, with reuse partitioning techniques described in Section 7.3.6.

7.4.5.1 Spectral Efficiency for Different TSL Capabilities

All the previous simulation results correspond to three TSL-capable terminals. Nevertheless, by increasing terminal TSL capability and therefore the number of allocated TSLs per connection, there is more potential data bandwidth allocated to each connection. This is translated into higher throughputs and better spectral efficiency under interference-limited scenarios for a minimum-quality criterion. On the other hand, if the number of allocated TSLs is reduced, there will be a contrary effect.

SE analysis for different allocated TSLs can be performed with existing results. Figure 7.36 can be used to compute TSL capacity values for different network loads (Data Erlangs) and frequency reuses. The number of TRXs available for (E)GPRS can be taken into account considering reduction factors from Figure 7.37. In the case of FH, Figure 7.39 can be used for evaluation of both TSL capacity and reduction factors.

The average throughput values can be estimated as

User throughput = TSL capacity × Number of TSLs × reduction factors.

Changing the number of allocated TSLs affects trucking efficiency, and thus it affects reduction factors. In any case, the average user throughputs will be better with higher number of allocated TSLs. Figure 7.42 shows simulated and ideally estimated average throughput values for one TRX and reuse 3/9. Estimated values (Figure 7.42 (b)) use reduction factors from Appendix B, which assumes that data load is evenly distributed over the TRX. TSL capacity is calculated from Figure 7.39. This figure also shows estimated average throughput values for different number of allocated TSLs.

A higher number of allocated TSLs is directly translated into a higher spectrum efficiency with a minimum-quality criterion. Figure 7.42(a) shows the SE values achieved with minimum-throughput criteria (10th percentile throughput) for one, three and four allocated TSLs. The SE gain is more noticeable when the minimum quality is high, in the interference-limited area. The gain of allocating more TSLs disappears with lower minimum throughput, as the network becomes HW limited. The SE is always better with more allocated TSLs; therefore, it is always better to allocate the maximum TSL capability supported by the MS.

Figure 7.43(b) shows SE versus average and 10th percentile user throughput with 4 TSL-capable mobile stations. 4 TSL is the highest DL configuration possible in the Rel'99 simple mobile station without the need of duplicating the radio parts (see Chapter 1). Rel'5 introduces enhancements that allow the support of 6 TSL in DL with simple terminals (see Chapter 2). The most spectral-efficient reuse is considered (reuse 1/3), therefore the figure represents the maximum EGPRS downlink capacity with simple MS design. The highest SE is achieved with an average data rate of 64 kbps. In order to double the average throughput (up to 128 kbps), the SE is reduced by 30%. In this case, the minimum throughput achieved by 90% of the user is 56 kbps. On the other hand, at the highest spectral efficiency, the minimum throughput is much lower (less then 16 kbps).

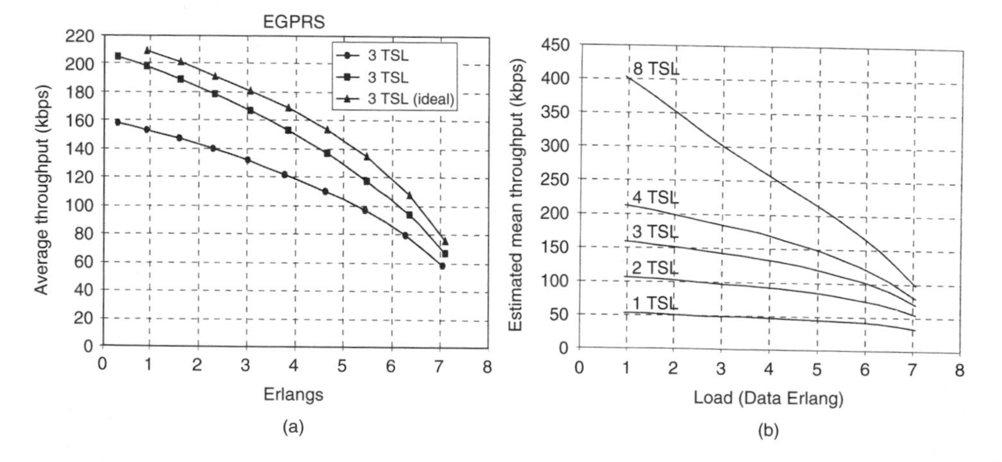

Figure 7.42. (a) Simulated average user throughputs for three and four allocated TSLs. (b) Estimated average user throughput for different number of allocated TSLs. One TRX. Reuse 3/9

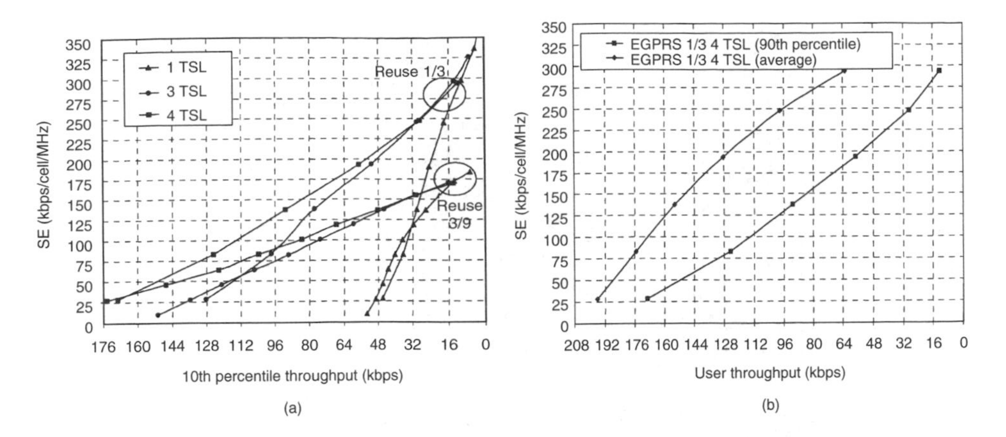

Figure 7.43. (a) EGPRS SE versus 10th percentile user throughput for 1, 3 and 4 TSL-capable MS. 1 TRX. (b) EGPRS SE versus average and 10th percentile user throughput with 4 TSL-capable MS. 1 TRX. Reuse 1/3

7.4.6 Throughput Distribution Analysis

This section will analyze how the throughput experienced by EGPRS MSs varies across the network. The analysis is made on the TBF level, which means that the information generated for each individual TBF is collected and analyzed. Figure 7.44(a), shows a typical throughput snapshot (corresponding to a certain instant), which describes the throughput variation as a function of MS location. Figure 7.44(b) shows the corresponding average throughput distributions in the form of probability density function. The simulation case has been selected in such a way that its load corresponds to the level of 3 Data Erlangs. The achieved mean throughput in this case is about 100 kbps and the 10th percentile throughput about 50 kbps. The quality achieved in certain locations does not only depend on the distance to the serving BTS. The load of the serving BTS might cause queuing and therefore despite excellent radio link quality, the throughput experienced by a certain MS can occasionally be much less than the maximum, even when located very close to the BTS. However, the throughput close to the cell center is naturally still on average better than in the cell border area. No throughput-compensation techniques have been considered for these results. Clever scheduling and channel allocation design may distribute the throughput more equally by allocating more scheduling turn to users at the border of the cells. Note that even with tight reuses like 1/3, a high percentage of users will have high throughputs. At the cell edge, some users may experience low throughput. Those users can be handed over to a different layer with not so tight reuse (e.g. BCCH layer or hopping layer) to improve the throughput if necessary.

7.4.7 Effect of Traffic Burstiness

Earlier results assumed the FTP traffic model where the users access the system to download a file with a fixed length. With some other traffic patterns, EGPRS traffic can be very bursty. For busty traffic, there are two main KPIs to consider: LLC frame delay and TBF throughput. LLC frame delay is applicable to short packets. TBF throughput is still a good quality measure for the download of long files corresponding to long

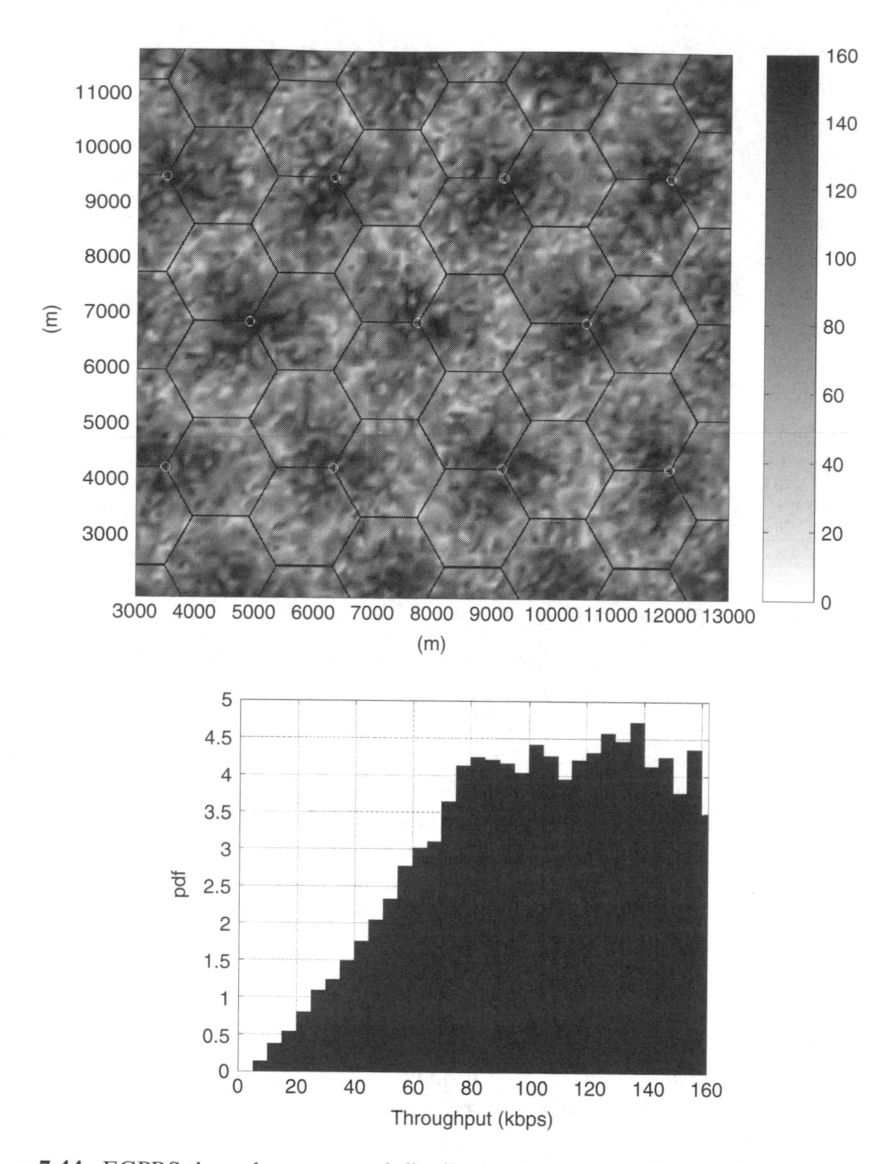

Figure 7.44. EGPRS throughput map and distribution in a macrocell network. Reuse 1/3. One TRX. Three TSL-capable MS

TBF duration. System level-wise, bursty traffic has a tendency to increase protocol and signaling overhead. Some of the effects are as follows.

The number of established TBFs is higher, thus generating more signaling in (P)CCCH channels. Signaling load in Common Control Channel (CCCH) is studied in Chapter 12.

A noticeable delay due to TBF establishment procedure is experienced for short and infrequent packets. A typical TBF establishment delay is on the order of 250 ms. Delayed DL TBF release and Extended UL TBF functionality (see Section 1.5.6) help to reduce this effect.

During the TBF release procedures, the network may generate too many pending ACK block transmissions (preemptive retransmissions) that increase network load, but most of them are unnecessary (they do not contribute that much to SE).

Less protocol efficiency with bursty traffic is traduced into an increase of Data Erlangs for the same cell throughput. This, in practice, will generate higher interference levels and will reduce the spectral efficiency. Different traffic profiles may be more or less bursty and have different dynamic behavior. As a result, for the same HW utilization, the users will experience a reduction factor depending on the traffic profile. For example, consider a cell with 4 TSLs available for GPRS. One user performing a continuous download of a very long file is able to generate 3 Data Erlangs (assuming a 3 TSL-capable terminal) in that cell. As 3 data Erlangs is generated by just one user, the reduction factor is 1, as there is no TSL multiplexing. Assuming now that multiple users with bursty traffic generate 3 Erlangs, there is a certain probability that some of the TSL will be shared, and therefore the reduction factor cannot be 1. Considering the data traffic model from Appendix B, the reduction factor is around 0.55. Even though reduction factor values may change for the same HW utilization, the maximum SE (kbps/cell/Mhz) value achieved by EGPRS is almost independent of traffic burstiness. The only relevant effect in this case is the increased amount of control information (e.g. TBF establishment and release messages) that consumes part of data capacity (up to 8–10% under heavy-load conditions, see Chapter 8).

Figure 7.45 shows network level simulations with simple FTP and HTTP1.1 protocol models. Initial handshake, slow start, TCP ACKs and transmitting windows have been modeled in detail in HTTP1.1 model. Average WWW size is 120 kB. Each new user attempts to download a WWW page. Traffic is bursty by nature due to TCP modeling. HTTP1.1 protocol overhead is more noticeable for high TBF throughputs. End-to-end throughput is degraded compared with simple FTP model due to RTT and dynamic protocol behavior. The maximum SE is close to the one computed in previous sections without HTTP1.1 modeling.

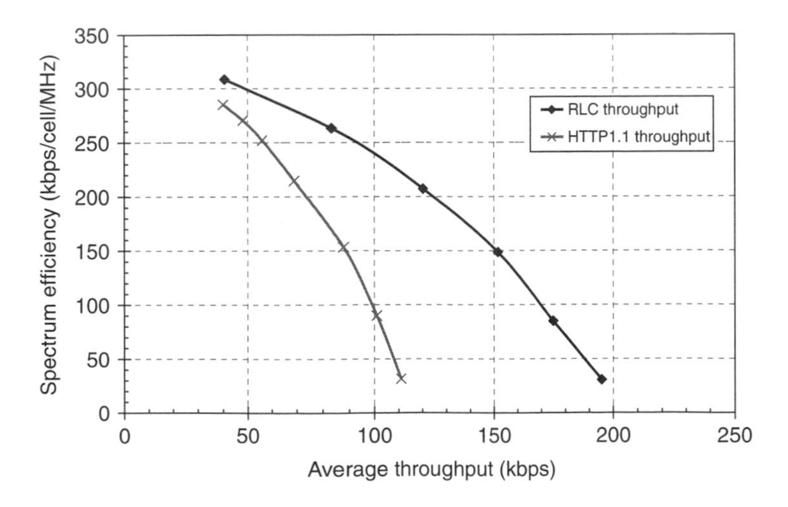

Figure 7.45. Comparison of RLC throughput (simple FTP traffic model) and HTTP1.1 throughput (full model) for 4-TSL capable terminal. Average WWW page size is 120 kB

The performance with bursty traffic can be improved by reducing the radio round trip time to the minimum. By locating RLC/MAC protocol stack as close as possible to the BTS, radio round trip time can be minimized. For short and infrequent packets, the minimum delay depends basically on the overhead delay introduced by the TBF establishment and release procedures, which may be on the order of 500 ms (although this value depends very much on network implementation).

7.4.8 (E)GPRS Deployment

BCCH is the preferred layer for (E)GPRS. In this scenario, the TSL capacity is quite independent of network load and traffic mix (percentage of data and speech users) in the DL direction. This is due to the continuous transmission with constant power over BCCH frequencies in order to enable the MS to measure neighbor cells. As a result, DL interference levels are constant and independent of traffic load. Typical performance figures for TSL capacity, that can be obtained from (E)GPRS traffic-only simulations, are 12, 19 and 43 kbps for BCCH reuse 4/12 with GPRS CS-1–2, GPRS CS-1–4 and EGPRS respectively. An additional characteristic of the BCCH layer is that speech performance in BCCH will not be degraded by (E)GPRS interference in DL.

Whenever the BCCH is fully loaded, new incoming PS traffic can be allocated in a dedicated layer (e.g. reuse 1/3 or 1/1 hopping) so that (E)GPRS spectral efficiency can be maximized. Voice users would be allocated in a separate layer. System level simulations have shown that, although reuse 1/3 provides the highest spectral efficiency, it is more sensitive to network irregularities compared with 1/1 hopping reuse. This is due to the interference randomization effect of frequency hopping. Although 1/3 reuse is still the best from the spectrum efficiency point of view, 1/1 reuse performs quite close to 1/3 in irregular networks. Figure 7.46 shows TSL capacity with regular and irregular networks (traffic hot spots and irregular layout). The average C/I degradation in the simulated irregular network is around 4 dBs with reuse 1/3, while the degradation with 1/1 hopping is only 2 dBs.

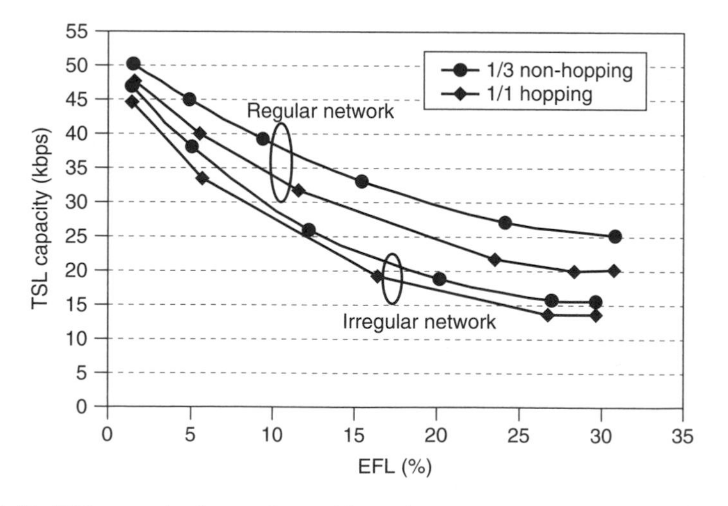

Figure 7.46. TSL capacity in regular and irregular macrocell network with and without frequency hopping

An alternative deployment strategy is mixing PS and CS in the same 1/1 hopping band [9]. In the hopping layer, the interference level depends on how much the network is loaded (EFL operation point of the network). It can also be very dependent on the traffic mix as data services tend to generate more interference than voice. This is because DL PC is not as efficient for (E)GPRS as it is for voice services, and discontinuous transmission (DTX) is not applicable for data services. Total generated (E)GPRS interference can be controlled by limiting maximum transmitted power of (E)GPRS signals. The common hopping-layer approach has the advantage of coping more efficiently with different and dynamic shares of voice and data traffic. With two layers, during certain periods it is possible that (E)GPRS frequencies are very high loaded, while the speech frequencies are low loaded, or the other way around. In this sense, integrating both services in a common frequency-hopping band is a wiser approach as it makes the whole band available for both services, thus providing a better overall performance. Figure 7.47 contains simulation results that compare the common layer (labeled as REF) and separate layer performance. A suitable DL power offset (P0 value, see power control algorithm in Section 7.2) has been applied to (E)GPRS. P0 equal to 6 dB has been selected to have the same voice quality for 50% (E)GPRS penetration. The percentage of bad-quality samples (with FER > 4%) with common layer remains below the separate layer case for (E)GPRS penetration lower than 50%, while average and 10th percentile user throughput is quite similar in both cases. This simulation analysis shows performance benefits of common layer compared with separate layer approach.

7.4.9 *Gradual EDGE Introduction*

Legacy GSM network's BTS equipment may not be able to support EGPRS. An operator with existing legacy BTS equipment, wishing to introduce EDGE has the possibility to do it gradually, which eases the rollout and reduces the required initial investment. In such case, only a fraction of the cells would need to be upgraded to EDGE. This solution is possible if there is a certain degree of cell overlapping, which would be the typical case in any urban deployment. Figure 7.48(a) shows a regular scenario with a

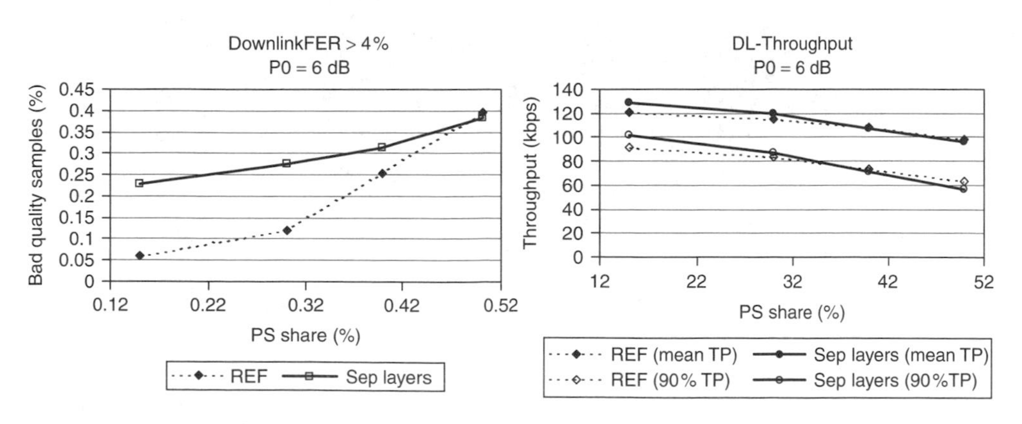

Figure 7.47. Performance of mixed speech and (E)GPRS traffic. 3 TSL-capable EGPRS terminals. AMR speech codecs. (a) Percentage of bad-quality calls (FER > 4%) and (b) average user throughput

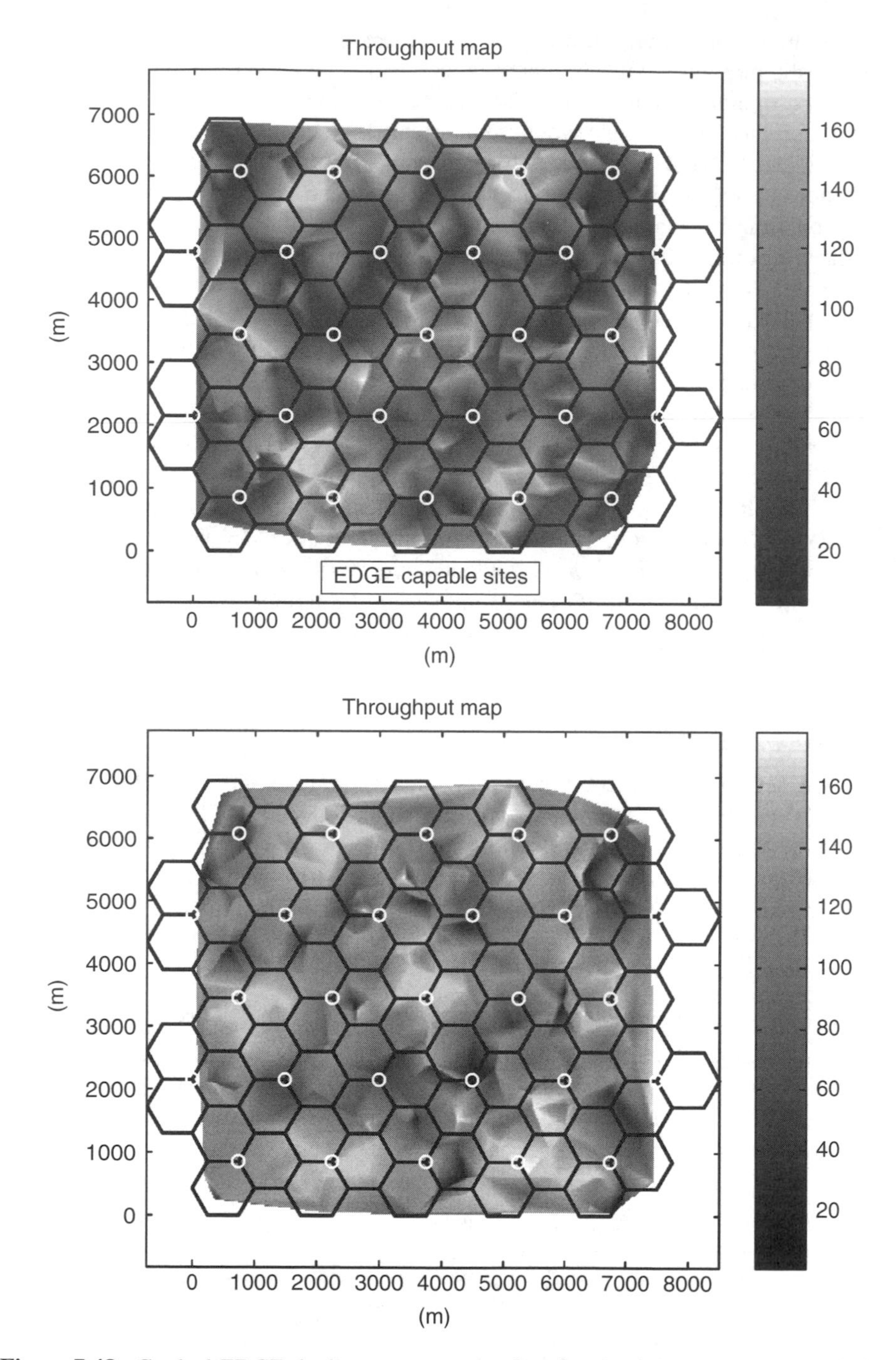

Figure 7.48. Gradual EDGE-deployment scenario. One-fourth of the cells are upgraded to EDGE. Legacy TRX supports only CS-1–2. (a) No cell priorities and (b) increased EDGE-capable cell's dominance area

legacy network supporting GPRS CS-1–2 where 25% of the cells have been upgraded to EDGE. It shows a very irregular throughput map. This situation can be corrected by setting aggressive negative power budget margins, which increases the dominance area of EDGE-capable cells, and therefore bring the EGPRS benefits across the network. The better link performance of EGPRS compared with GPRS CS-1–2 compensates for the lower C/I when extending EDGE cell dominance area, thus EGPRS throughput is still quite high, see Figure 7.48(b).

7.5 Mixed Voice and Data Traffic Capacity

In previous sections, data capacity figures assume (E)GPRS traffic only. Therefore, they represent absolute performance of (E)GPRS systems. In practical deployments, different kind of users (voice and data) may be using the same frequencies and the same HW resources. Different services will interact with each other, affecting the performance of each other. The interaction may be controlled to a certain degree by radio resource management. Different priorities can be applied to different services that have a direct impact on capacity figures. The most important case, since it will be the most widely deployed in initial deployments, is the one in which the (E)GPRS data users use spare resources (timeslots) not utilized by GSM speech users. The average number of free TSLs that can be potentially used by (E)GPRS can be calculated from the average voice Erlangs per cell and the total number of TSLs available on each cell as (assuming full-rate (FR) speech traffic only):

$$\text{Free TSL(FR)} = \text{Total TSL} - \text{Speech Erlangs} - \text{RRM factor}$$

The RRM factor considers practical limitations while managing resources that are shared by PS and CS calls. For example, in case of two separate pools (territories) of PS and CS resources (see channel allocation algorithms in Section 7.2), there is the need to have one or more guard TSLs between the two territories to avoid continuous PS territory upgrades and downgrades. Allocation reconfigurations, which involve resource repacking to make free TSL available to (E)GPRS, introduces (E)GPRS resource allocation delays and also contributes to the RRM factor. Table 7.4 presents the number of free TSLs for different cell configurations and RRM factor equal to 0. The number of occupied speech TSLs in Table 7.4 is calculated with Erlang-B formula with 2% speech blocking probability.

Table 7.4. Available TSLs for best-effort (E)GPRS traffic

Number of TRX	Voice Erlangs at 2% blocking	Free TSLs	Share of TSL available for (E)GPRS (%)
1	2.9	4.1	58
2	9.0	6.0	40
3	14.9	7.1	32
4	21.9	8.1	27
5	29.2	8.8	23
6	36.5	9.5	21

It is assumed that one TSL is reserved for signaling with one and two TRX configurations, while two TSLs are required for the other configurations (signaling capacity dimensioning in mixed cases is covered in Chapter 12). By adding an extra TRX, best-effort cell data capacity can be improved. It is also possible to dedicate some TSLs for (E)GPRS' use exclusively. In this way, a minimum cell data capacity can be guaranteed. The usage of half-rate (HR) speech channel mode reduces the number of TSLs occupied by speech; therefore it increases cell data capacity as follows:

$$\text{Free TSL(HR\&FR)} = \text{Free TSL(FR)} + \text{Speech Erlangs} \cdot \gamma/2 - 0.25$$

where γ is the percentage of HR connections and 0.25 represents the average reduction due to HR connections occupying 0.5 TSL, but making the whole TSL unavailable for (E)GPRS. Tables B.1 and B.2 in Appendix B provides speech Erlangs values for different HR penetration rates. These tables are useful to know the maximum voice capacity (speech Erlangs in previous formula) for a certain speech blocking probability. In order to calculate data capacity, the key figure is the amount of (E)GPRS traffic that can be transmitted in a fully utilized TSL. This concept is called *TSL capacity* and has been described in previous sections. Maximum data capacity per cell can be expressed as:

$$\text{Maximum cell data capacity (kbps/cell)} = \text{Free TSL} \times \text{TSL capacity}$$

TSL capacity depends on average network load (Voice and Data Erlangs). Curves in Figures 7.30 and 7.39 can be used to estimate TSL capacity just knowing the EFL of the hopping layer. Although those curves were derived with 100% data traffic, they can be also applied with mixed voice and data services. In this case, the amount of voice Erlangs should be translated into *Equivalent Data Erlangs*, which represents the amount of data traffic that generates same interference levels than the original voice Erlangs. Erlangs translation factors for mixed voice and data traffic are provided in the next section.

A dimensioning exercise example is useful to better explain the concepts. Consider a hopping layer with 12 hopping frequencies that needs to support 5 voice Erlangs and 200 kbps per cell in busy hour. (E)GPRS is transmitted with maximum BTS power, while voice traffic utilizes DL power control. Assuming an initial typical TSL capacity value of 45 kbps (in order to start the calculation procedure), the number of free TSLs required can be calculated as $200/45 = 4.4$ TSLs. 5 voice Erlangs corresponds to 2.5 equivalent data Erlangs as DL power control is used for voice, while EGPRS is transmitted at full BTS power (see Erlang translation factors later in this section). Figure 7.36 can be used with $\text{EFL} = 100 \times (2.5 + 4.4)/(8 \times 12) = 7.1\%$, giving a TSL capacity of 37 kbps. As this value differs with the initial value (45 kbps), calculations should be iterated. Number of free TSLs in second iteration is $200/37 = 5.4$. EFL in second iteration is $= 100 \times (2.5 + 5.4)/(8 \times 12) = 8.2\%$, which correspond to a TSL capacity of 36 kbps, which is now very close to the previous one (37 kbps), so it can be considered as a valid value. The definitive number of free TSLs required is $200/37 = 5.5$ TSLs.

7.5.1 Best-effort Data Traffic

Maximum cell data capacity is achieved by fully utilizing all the free TSLs (not used by CS) for (E)GPRS. This is only possible with a very high offered load and operating the network close to the (E)GPRS service-saturation point, which leads to very low throughputs. When user throughput is taken into account, the utilization of free TSL (see TSL utilization concept in Section 5.2) does not reach 100%. The amount of free TSLs utilized is a matter of network dimensioning and QoS criteria:

$$\text{Cell data capacity (kbps/cell)} = \text{Free TSL} \times \text{TSL capacity} \times \text{TSL utilization}$$

In order to dimension best-effort data with a QoS criterion it is necessary to define the maximum utilization of free TSLs so that the amount of TSL multiplexing (measured in terms of reduction factor) is such that a minimum-quality criterion is met. Considering average user throughput as QoS criterion, the following equation can be used to check whether a certain amount of free TSLs fulfill the QoS criterion:

$$\text{Average user throughput} = \text{Number Of TSLs} \times \text{TSL capacity} \times \text{Reduction Factor}$$

Reduction factor tables as a function of TSL utilization and number of TRXs can be found in Apendix B.

7.5.2 *Relative Priorities*

Even if the network is close to saturation, it is possible that some users experience high throughput by introducing QoS differentiation mechanisms. QoS differentiation can be done with relative priorities in RLC scheduler (see Section 7.2). Different scheduling priorities change the reduction factor for the same TSL utilization, so that the dimensioning exercise is a matter of estimating reduction factors for different traffic priorities (or traffic classes).

Giving priorities to certain users will degrade low-priority users; therefore the weighted average of reduction factors equals the reduction factor without priorities as follows:

$$\text{RF (no priorities)} = S_1 \cdot \text{RF(TC}_1) + S_2 \cdot \text{RF(TC}_2) + \cdots$$

where S_i is the share ($0 < S_i \leq 1$) and RF(TCi) is the reduction factor for traffic class i.

7.5.3 *Guaranteed Data Traffic*

Guaranteed data traffic (e.g. streaming services) requires that RLC scheduler control the scheduler turns so that a minimum data rate is guaranteed to certain users. The amount of resource required for this kind of traffic can be calculated as:

$$\text{Streaming Data Erlangs} = \text{Nr. of users} \times (\text{Guarantee throughput/TSL capacity})$$

The percentage of satisfied users and the throughput distribution over the network defines the maximum amount of users per cell.

7.5.4 *Erlang Translation Factors*

In the hopping layer, the total amount of traffic determines the interference levels, and therefore the network quality for both speech and data services. For the same amount of traffic (Voice Erlangs + Data Erlangs), different shares of (E)GPRS and speech traffic lead to different interference levels as downlink power control efficiency is different for both kind of services. As it is impossible to have simulation or trial results for all possible shares of traffic, it is very useful to introduce the Erlang translation factors methodology. This methodology establishes equivalent amount of traffic that generates the same amount of interference.

The simplest case to consider is when downlink power control (DL PC) and DTX are not used in voice services, and also DL PC is not used by (E)GPRS. In this case, generated interference levels are equivalent. Therefore, it is possible to translate Data Erlangs into Voice Erlangs. The case of EGPRS with back-off is tackled later in this section. Soft-capacity analysis for data presented in this chapter and for voice, presented in Chapter 6, are fully applicable by considering total EFL as the addition of voice EFL and data EFL. In other words, a Total Erlang is considered as the addition of Voice Erlangs and Data Erlangs. Cell data capacity can be calculated considering Data Erlangs and TSL capacity. TSL capacity versus EFL results from previous sections can be used to calculate TSL capacity. For example, suppose that hopping layer is loaded with 8 Voice Erlangs and 4 Data Erlangs. If the number of hopping frequencies is 24, the network is loaded with 6.25% EFL. On the basis of Figure 7.39, the TSL capacity is 38 kbps giving a total cell data capacity of $4 \times 38 = 152$ kbps.

If DL PC and DTX are used by voice users, a direct conversion between Data Erlangs and Voice Erlangs is not possible. Figure 7.49 shows voice quality versus EFL curves for AMR and enhanced full-rate (EFR) codecs with a quality limit of 2% outage of samples with frame erasure rate (FER) > 4.2%. The EFL curve without power control and DTX would correspond to the speech performance with almost 100% (E)GPRS traffic. The maximum data EFL that can be reached is 4.8%. On the other hand, EFL without power control would correspond to the performance close to 100% voice traffic. The maximum voice EFL that can be reached with DL PC is around 7%, which corresponds with a capacity gain of 45% that is in line with the field trial results presented in Section 6.1.2. An extra 30% gain can be achieved with DL DTX activated (see section entitled *Discontinuous transmission* in Chapter 6), therefore the maximum voice EFL is 9.1%. This is roughly 2 times more Erlangs than in almost 100% (E)GPRS case. From this analysis, the

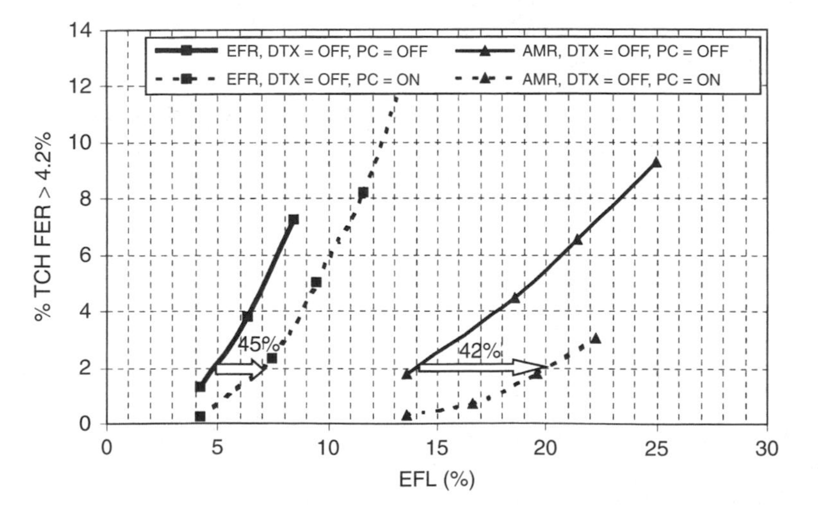

Figure 7.49. Percentage of samples with bad quality (FER > 4.2% and 2% outage) versus EFL for AMR and EFR

following simple rule of thumb can be extracted: *1 Data Erlang is equivalent to 2 Voice Erlangs in terms of voice quality*. The same exercise can be done with AMR voice users, in this case the rule of thumb from simulation results is the same. Note that for EFL calculations, the activation of DTX does not change the Voice Erlang values. Depending on the minimum-quality target and different network topology, different translation factors can be found, so only field trial data will help to fine-tune the translation factor values.

If DL PC is used by (E)GPRS, the analysis requires (E)GPRS DL PC simulations. 8-PSK modulation back-off (2–3 dB in the BTS) can be considered as a special case of DL PC that applies a negative transmission power offset only when 8-PSK modulation is in use. Voice-bad-quality samples will be directly improved if (E)GPRS interference is reduced. Figure 7.50(a) represents voice-bad-quality samples in simulated EGPRS and EFR traffic scenario with 50% data traffic share. It can be seen that it is possible to reach higher EFL values while keeping the same data traffic share when DL PC is applied. Even with aggressive power control settings, the impact in average user throughput is not very relevant in interference-limited conditions. Figure 7.50(b) represents average user throughput with different power control settings.

Figure 7.51 shows data translation factors when DL PC or back-off is applied. One data Erlang with DL PC and back-off is equivalent (in terms of speech quality) to *1/Translation Factor* data Erlangs without DL PC and back-off. Translation factors are almost independent of traffic share unless aggressive power control settings are applied (P0 = 6).

One EGPRS data Erlang with 8-PSK back-off is equivalent to $1/1.5 = 0.66$ data Erlangs (see Figure 7.51(a)) without back-off. Knowing that 1 Data Erlang without back-off is equivalent to two speech Erlangs, it can be concluded that 1 data Erlang with back-off is equivalent to $0.66 \times 2 = 1.33$ voice Erlangs.

Similar equivalences of Data Erlangs with DL power control and voice Erlangs can be done on the basis of Figure 7.51.

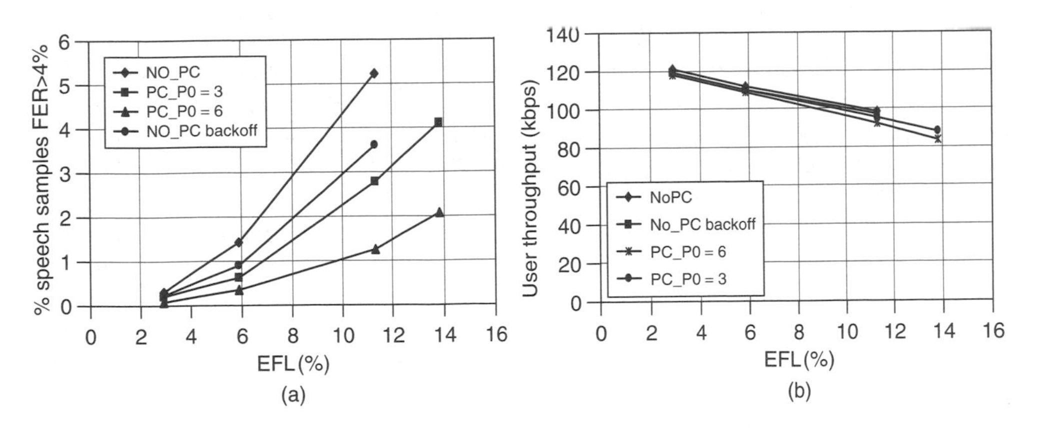

Figure 7.50. Voice-quality evolution with 50% speech and 50% EGPRS data. NO_PC: no EGPRS power control, no 8-PSK back-off. NO_PC back_off: no EGPRS power control, 2 dB 8-PSK back-off. PC_P0 = 3 and PC_P0 = 6: EGPRS power control with P0 = 3 and 6 dB respectively

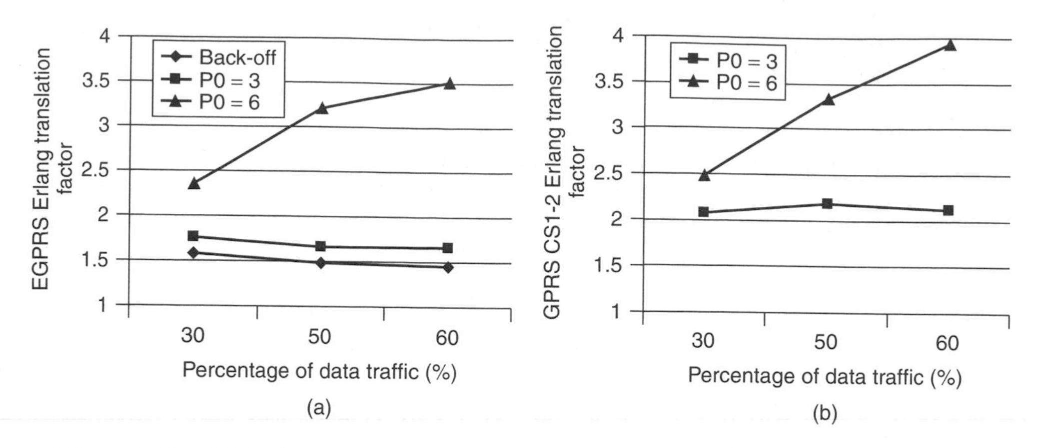

Figure 7.51. GPRS CS1 and 2 and EGPRS translations when DL power control is applied

7.6 (E)GPRS Performance Estimation Based on Real Network Measurements

The performance of (E)GPRS may differ in different networks due to the different topology and propagation environment. Simulation results presented in previous sections show the performance of a regular network with macrocell-propagation characteristics. This section presents a methodology to estimate (E)GPRS performance in a real network based on measurement reports sent back to the network by the MS. This methodology has been used in a field trial carried out in a major European capital (Network 9, in Appendix F). Results provide accurate (E)GPRS performance figures expected in that city.

Any MS with an established voice connection reports periodically (every 480 ms), the received signal level C from the serving cell and the six strongest neighboring cells (N_1, $N_2, \ldots, N_6$). The MS may be detecting and measuring more than six cells but only the six strongest are reported. This will change with the introduction of enhanced measurement report in Rel'99, which largely expands the terminals' reporting capabilities. The reported cells are ordered from the strongest to the weakest. The relative level of signal between serving and reported cell makes it possible to estimate the C/I values, in case the reported cell is a cochannel interferer. Therefore, this information can be utilized to envisage C/I distributions with a given frequency reuse. In order to do this, the cells are replanned with the desired frequency reuse and cochannel interfering cells are defined. Measurement reports are collected from the trial area and stored in a database. The trial area should cover a sufficient number of cells. Each measurement report is processed as follows:

If one (or more) of the six reported cells is a cochannel interferer, the user will experience a throughput that corresponds to $C/I = C/N_i$, where N_i is the sum of cochannel interferer levels.

If there is no cochannel interferer in the six strongest reported cells, it can be said that C/I will be, in the worse case, the one associated with the weakest neighbor (N_6). Throughput versus C/I curves of (E)GPRS tend to saturate for a certain C/I value, $[C/I]_{sat}$, there are two cases to consider:

If $C/N_6 > [C/I]_{sat}$, the MS connection would experience the maximum throughput corresponding to $C/I = [C/I]_{sat}$.

If $C/N_6 < [C/I]_{sat}$, the MS connection would experience, at least, the throughput associated with $C/I = C/N_6$. In this case, it is impossible to know the exact throughput and only worse-case calculations can be done. In order to produce more realistic results, C/I values are compensated with a factor calculated from a static simulator for each reuse. This factor is deduced by calculating the C/I with the information provided by the six strongest cells, and then comparing with the calculated C/I values considering all cells.

If the signal level from the serving cell is too low, the radio conditions can be noise-limited. For each measurement report, the C/N (signal-to-noise ratio) is also calculated as $C/N = C - \text{noise level}$. The noise level is assumed to be −111 dBm. If C/N is larger than C/I, the link is noise-limited and throughput value is calculated with C/N value using throughput versus C/N link level curves.

Figure 7.52 shows TSL capacity distributions. More than 13 million measurement reports have been collected and processed from around 145 cells from one BSC. Microcells have been excluded from the analysis. The results correspond to a fully loaded network as continuous cochannel interference is assumed in the above-described procedure. The procedure also assumes full power transmission; therefore DL PC cannot be taken into account. Frequency hopping cannot be included in the methodology, so non-hopping C/I versus throughput-mapping tables (that can be found in Section 7.1) have been utilized. This provides an accurate analysis in the case of BCCH and worst-case scenario for the non-BCCH TRXs.

TSL capacity can be considered as average user throughput if there is only one user per TSL. This can be achieved with proper HW dimensioning, configuring the available parameters (if they are available to control the maximum number of users per TSL) or by allocating the highest relative priority in the scheduler to some users. TSL capacity and SE figures have been collected in Table 7.5.

Timing advance (TA) information is also included in measurement reports. This information can be considered in the analysis to calculate user throughput figures as a function of the distance from the BTS. Figure 7.53 shows the results. User distance distribution can be found in Figure 7.54.

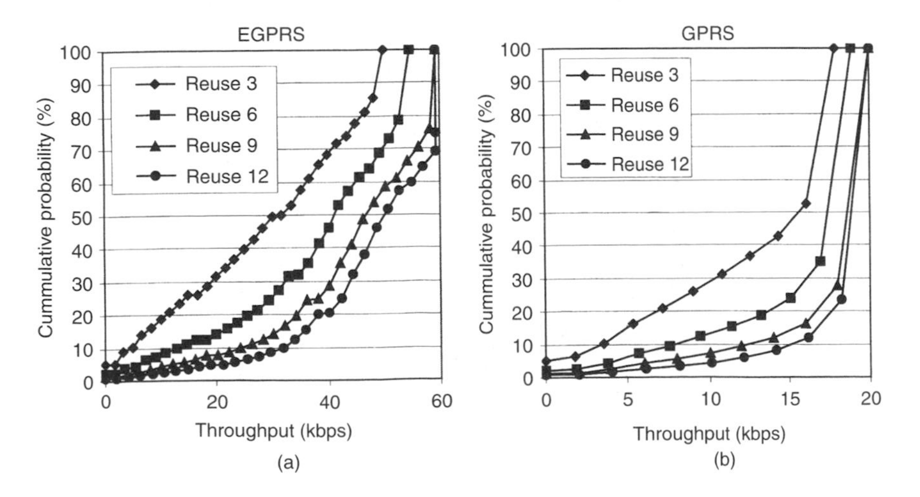

Figure 7.52. EGPRS and GPRS cumulative throughput distributions

Table 7.5. (E)GPRS performance estimated figures from field trial data

	EGPRS			GPRS (CS-1–4)		
Reuse	Minimum throughput/ TSL with 90% location probability	TSL capacity	kbps/MHz/cell	Minimum throughput/ TSL with 90% location probability	TSL capacity	kbps/MHz/cell
3	6	35.4	472	4	14.9	199
6	16	43.1	287	10	17.3	115
9	26	46	204	14	18.2	81
12	32	47.6	159	16	18.8	63

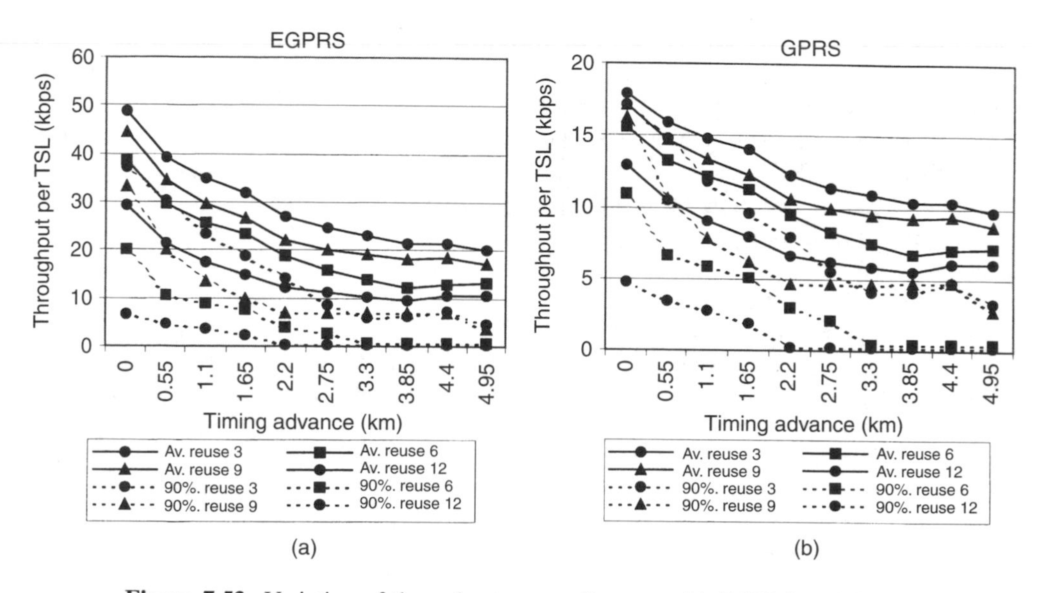

Figure 7.53. Variation of throughput versus distance with EGPRS and GPRS

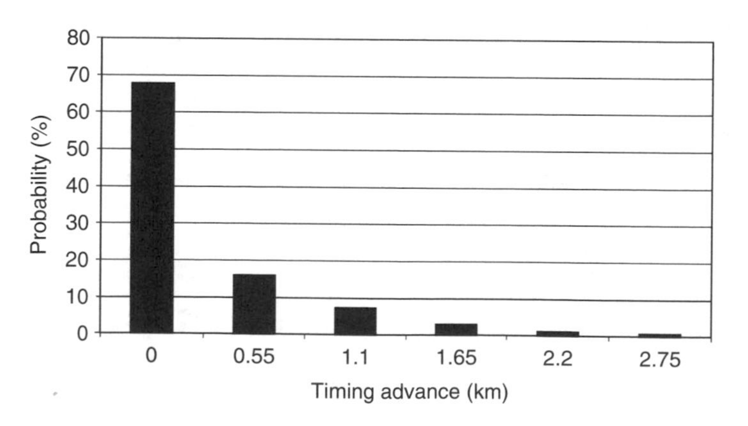

Figure 7.54. User distance distribution

7.7 Application Performance Over (E)GPRS

The basic (E)GPRS service in Downlink direction provided by the BSS is the delivery of LLC data packets to the MS that are coming from the PS core network. Analogous service is offered in UL direction. All performance figures presented in previous sections are measured at LLC- layer level. Nevertheless, QoS experienced by the end users is measured at application level. BSS plays only a part of end-to-end QoS provisioning, according to the QoS architecture described in Chapter 3. The BSS role in the end-to-end QoS picture is called *radio access bearer service* (following Universal Mobile Telecommunications System (UMTS) terminology). The performance analysis presented in previous sections describes the performance of the basic service provided by (E)GPRS bearers. It describes the performance up to LLC layer. The impact of the (E)GPRS bearer over the quality experienced in the application layers is discussed in more detail in Chapter 8.

(E)GPRS BSS provides data link (Layer 2) relay service from the serving GPRS support node (SGSN) to the MS. Internet applications use one of two possible transport protocols: TCP or user datagram protocol (UDP). TCP is a connection-oriented protocol with reliable data transmission, while UDP is not connection oriented and does not provide a reliable data transmission service. WWW, FTP and email applications use TCP protocol, while wireless application protocol (WAP), real-time transport protocol (RTP), Domain Names Server (DNS) and simple network management protocol (SNMP) services use UDP.

TCP is a protocol originally designed and optimized for fixed-PS networks. It may not be optimal for mobile networks without a certain degree of optimization. The main characteristics of TCP to be considered in (E)GPRS networks are described in Chapter 8. A well-known problem of TCP protocol is the amount of overhead to be transmitted over the radio interface. In (E)GPRS networks, the bandwidth is much more scarce and expensive than in fixed networks. Therefore, the protocol overhead should be minimized. The TCP protocol headers, together with IP headers produce an overhead of 40 octets on each TCP/IP packet (IPv4). In order to reduce this overhead, it is better to increase the length of the IP segments to the maximum possible (1500 octets). It is also possible to apply data compression techniques on the subnetwork dependent convergence protocol (SNDCP) layer [10] already in Rel'99. Compression techniques can be applied to the TCP/IP and RTP/IP headers as well as the data part of the IP packets. Van Jacobson's header-compression [11] algorithm can reduce the TCP/IP header length to 4 octets (for most of the packets). Degermark *et al.*'s header compressing [12] can be applied to RTP/IP headers, improving the efficiency for real-time data applications. The data compression algorithm is V.42 bis [13].

Most of the problems with TCP over (E)GPRS and, in general, over wireless networks, come from the existing algorithms to adapt data rates to different speeds. These protocols have been optimized for fixed networks and do not consider the special characteristics of wireless networks. The most important effect of (E)GPRS is the high impact on the overall round trip time (RTT). RTT can be considered as the total time a small packet takes to travel from the service content server through the mobile network (core network and BSS) to the MS and back, see Figure 7.55. The RTT of an (E)GPRS network can change between 0.5 and more than 1 s (depending on the user throughput), while typical RTT of fixed networks can be 1 ms for local area network (LAN), 20 ms for integrated

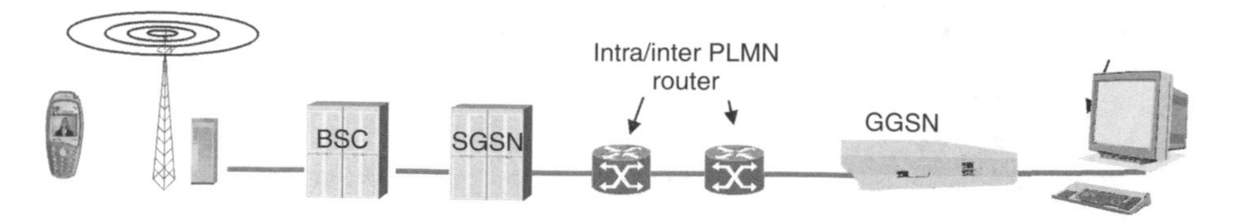

Figure 7.55. Elements affecting the round trip time

services digital network (ISDN), 20 to 200 ms for analogue modem and 125 ms for average Internet.

(E)GPRS cell-reselection delays have also an impact on TCP performance. The typical outage time is a random value that ranges from 0.5 to 2 s. During this time, TCP timers can expire, generating unnecessary retransmissions of TCP segments. GERAN Rel'4 introduces the network-assisted cell change (NACC), which reduces the cell-reselection outage down to 500 ms (max.). Additionally, with the use of the network controlled cell-reselection (NCCR) feature, it is also possible to optimize the cell-reselection performance, so that the impact to short outage in upper layers can be reduced.

WAP is an optimized protocol for wireless network; it can use UDP as a transport, thus avoiding most of the problems associated with TCP. But unfortunately, the majority of the Internet services do not use WAP, but they use instead hypertext transfer protocol (HTTP), which utilizes TCP connections. HTTP 1.1 has some advantages over HTTP 1.0 as it reduces the number of TCP connections (openings and closings) and reuses the same TCP connections for multiple objects.

UDP transport protocol is less problematic than TCP, as it does not require retransmissions, and protocol overhead is much lower. The application layer handles things like possible required retransmissions and reordering of the datagrams. In most data applications, (E)GPRS works in RLC ACK mode, thus guaranteeing reliable transmission over the air interface.

Some streaming and conversational services, like voice over IP (VoIP), uses RTP over UDP. In this case, the (E)GPRS network operates in unacknowledged mode. RRM is then responsible to maintain the negotiated reliability QoS parameter. The most straightforward mechanism to control reliability is the link adaptation (LA) algorithm, that would select the highest (M)CS possible while being under a maximum BLER. It is worthwhile to note that certain streaming services that do not have such tight delay requirements, allow RLC to operate in ACK mode.

A performance enhancement proxy (PEP) [14] is an entity in the network acting on behalf of an end system or a user. The objective of a PEP is to improve the end-user application performance by adapting protocols/information to a bearer. There are two main types of PEP: transport layer and application layer PEP, although a PEP may operate in more than one layer at the same time. Transport layer PEP introduces enhancements to overcome the inherit problem of TCP, especially 'slow start'. There are different mechanisms at this level. Some of them are: TCP ACK handling (TCP ACK manipulation), modified TCP stacks and fine-tuning of TCP parameters.

Application layer PEP solutions include solutions like: caching, information compression and optimization of TCP by the application (fewer TCP connections). See Chapter 8 for further description about PEPs.

7.8 (E)GPRS Performance Measurements

GPRS service has been available for some years in commercial GSM networks. Most of the vendors have decided to implement CS1 and CS2. This chapter shows some (E)GPRS performance measurement in live networks.

7.8.1 TSL Capacity Measurements

TSL capacity figures have been measured in a live network (Network 10, in Appendix F) with data collected from the network management system during the busy hour. Results are presented in Figure 7.56. GPRS traffic is allocated in BCCH layer. CS-1–2 are only implemented in the network and Automatic Link Adaptation is in use for all GPRS MSs. TSL capacity figures are close to the maximum (12 Kbps) in all cells. This is in line with network lcvcl simulation results from Figure 7.26, where the TSL capacity was already very close to 12 Kbps with reuse 12. In Network 10, radio link conditions are good, so almost all the time CS-2 is in use. Control RLC blocks have been excluded from TSL capacity calculations, as they tend to disturb the measurements due to the low GPRS traffic in the network.

Under low data traffic (no TSL multiplexing), a GPRS user receives typically 12 kbps over each allocated TSL in downlink. Other effects like IP overhead, TCP slow start and cell reselections may reduce in practice, the end-to-end throughput, (see Chapter 8 for more detailed analysis).

Simulations have shown that 1/3 reuse is the most spectral-efficient deployment for (E)GPRS. This reuse was tested for GPRS data only in Network 12 (see appendix F). A walking route was defined within the coverage area of a certain cell. The 5 strongest cochannel interfering cells were loaded with around 3 Data Erlangs with 3 TSL-capable GPRS terminals performing continuous data download. During the test, link adaptation never selected CS-1. Note that in the non-hopping case, CS-2 outperforms CS-1 in all radio conditions. The average block error rate in downlink was 5.68%. The TSL capacity can be computed as $12 \times (1\text{-BLER}) = 11.2$ kbps/TSL, which is in line with simulation results in Figure 7.26

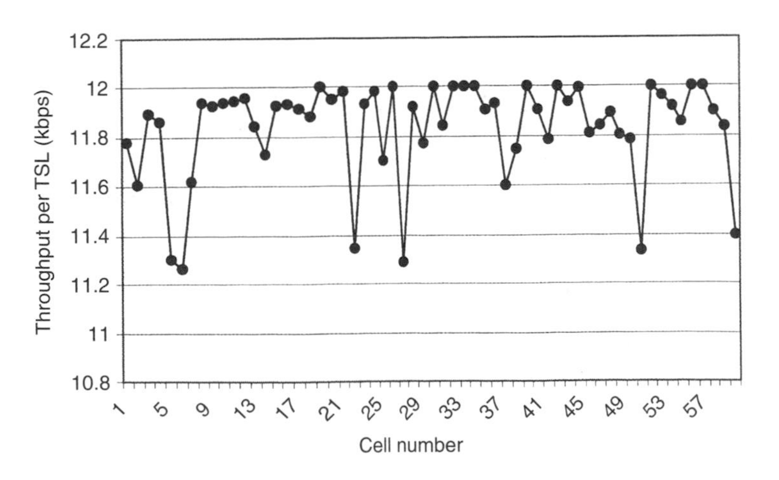

Figure 7.56. TSL capacity measured in a GPRS network

7.8.1.1 Reduction Factors

Throughput reduction factors can be measured in a cell load test with increasing offered data load. The load is increased in this test by increasing the amount of terminals in the cell, while each terminal repeats the same sequence of data accesses. The finite number of terminals utilized limits the maximum amount of TSL multiplexing so that it affects TSL sharing dynamics and, therefore, reduction factor values. For example, two terminals in the same cell generating 1 Data Erlangs in total, will never suffer a throughput degradation larger than half due to TSL multiplexing effects. With a much larger population of user generating same amount of total Data Erlangs, there is a certain probability of multiplexing more than two users and therefore the reduction factor will be lower than in the case of two users.

The infinite population of users assumed in Appendix B cannot be directly applied for comparison with measured reduction factors in this particular test, therefore reduction factor models in Appendix B were modified to consider a limited amount of terminals.

Figure 7.57 shows measured reduction factors versus theoretical modified formulas considering finite number of terminals. Estimated reduction factors tend to be optimistic at high loads (low Reduction Factor values) as theoretical model considers that all available resources are shared in the best possible way, which in practice is difficult to achieve considering realistic implementation of channel allocation algorithms and TSL allocation limitations.

Reduction factors with infinite number of terminals tend to be lower (see Figure 7.53), but represent better the situation in a live network where a large population of terminals generates Data Erlangs.

7.8.1.2 Cell-reselection Delay

Cell reselection involves network procedures that start when a GPRS terminal stops receiving data from the current serving cell and finalize when it resumes data reception on a new cell. The main effects of a cell reselection are the service outage period and potential packet losses. Cell-reselection delay may depend on the GPRS features in use, the most

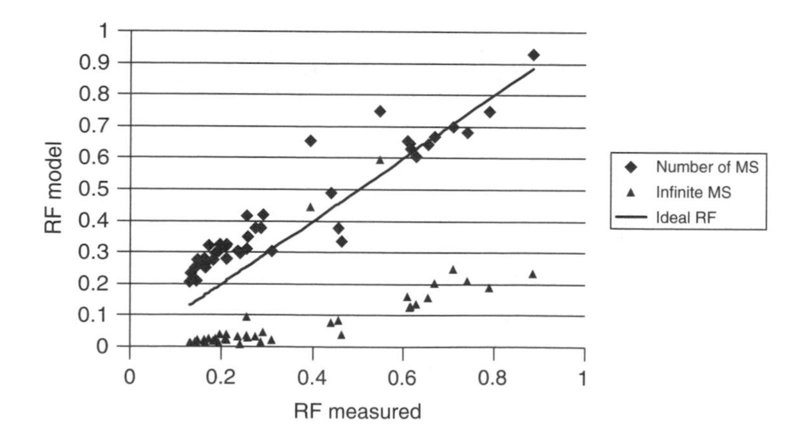

Figure 7.57. Measured versus theoretical reduction factors in cell load test

relevant being:

— PBCCH. The (E)GPRS terminal has to listen to the cell broadcast channel and wait until all needed system information messages have been received before accessing the new cell. As Packet Broadcast Control Channel (PBCCH) is a specific broadcast channel for (E)GPRS, system information can be received faster. With PBCCH, there is also the possibility to utilize the *packet system information (PSI) status* procedure, which enables the MS to access the cell as soon as the MS has received a minimum set of parameters with the rest of system information sent afterwards to the MS over PACCH channel.

— Network-Assisted Cell Change (NACC) is an enhancement introduced in 3GPP Rel'4 that eliminates the need of receiving system information messages over broadcast channels. Following this procedure, the (E)GPRS terminal receives system information messages directly from the serving cell before changing to new target cell.

Most GSM vendors have not implemented PBCCH in their first GPRS system releases. Cell-reselection outage's duration measured in live networks without PBCCH or NACC is quite random. Cell-reselection outages measured in Network 13 in Appendix F ranges from 0.8 and 3.86 s. Cell outage is defined as the time since BCCH is detected from the new cell until the MS receives new data in the new cell. Some cell reselections involve mobility-management procedures, e.g. Routing Area Updates (RAU), before data transmission can be restarted in the new cell. Routing area updates may significantly increase the cell-reselection outage and may take tens of seconds; therefore careful planning of routing areas is very much recommended. Packet losses are possible when (E)GPRS connection is operating in LLC-unacknowledged mode, and may depend on different network implementations. Figure 7.58 shows the average throughput for a 200 kB FTP download, depending on the number of cell reselections during the download. Measurements have been performed with optimized TCP parameters (see Section 7.7 or Chapter 8). TCP slow-start procedure may be triggered during cell reselection due to packet losses. This is reflected in the measured values given in Figure 7.58 where the average measured throughput is shown just considering the cell-reselection outage (no TCP effects case).

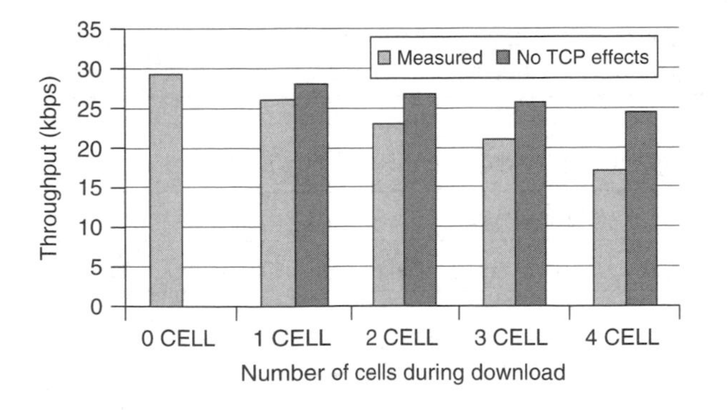

Figure 7.58. Measured average end-to-end throughput in a 200-kB file download with various number of cell reselections during download. Expected throughput, without considering TCP behavior, is also shown. 3(DL) + 1(UL) allocated TSLs

7.8.1.3 End-to-end Data Throughput and Latency Measurements

There are two basic end-to-end performance measurements that characterize the expected data service performance: round trip time and data throughout. Round trip time is defined as the time since a short packet is sent from the MS, received by the application server and finally received back at the MS. It characterizes the end-to-end latency, which is important for time-critical applications (like gaming) and dynamic behavior of Internet protocols. Data throughput is specially important in interactive data services, where the user expects to receive and send data files within a reasonable time.

Round trip time and data throughput have been measured in a field trial in Network 13 (see Appendix F). The current network dimensioning ensures a high probability that the maximum number of TSL supported by the MS can be allocated and low probability of TSL multiplexing. Stationary and driving test were performed.

Two different locations were selected for stationary tests. File downloads with different file sizes were performed with FTP application. The average throughput was dependent on file size due to TCP dynamic behavior. Table 7.6 shows average, maximum and minimum throughput measured during 5 file downloads of the same size. Average, maximum and minimum round trip times were measured from 10 consecutive Ping commands. The test was repeated 5 times for several Ping packet sizes. Results are presented in Table 7.7. Short packet's PING measurements are useful to characterize, for instance, the initial three-way TCP handshake.

A drive test route of around one hour and a half was defined as depicted in Figure 7.59. The test MS was performing continuous 200-kB FTP downloads. The average throughput of each download was recorded. Figure 7.60 shows the statistics collected. 3 TSLs were assigned to the mobile most of the time during the drive test. The lower throughputs compared with stationary tests were mainly due to the effect of the cell-reselection delays

Table 7.6. Application throughput measured in stationary locations

	Stationary 1			Stationary 2		
	Th_mar	Th_average	Th_min	Th_max	Th_average	Th_min
FTP DL 500	31.45	29.9	28.4	31.47	27.16	24.28
FTP DL 200	30.96	27.78	24.77	31.4	27.24	20.29
FTP DL 50	28.98	24.27	18.97	28.98	26.53	20.55
FTP UL 200	10.85	10.64	10.43	9.98	8.25	6.34

Table 7.7. Round trip time measured in milliseconds with Ping command in stationary locations

ping(B)	Stationary 1			Stationary 2		
	Delay_max	Delay_average	Delay_min	Delay_max	Delay_average	Delay_min
Ping(32)	731.3	674	590.6	922	842.88	662.7
Ping(256)	1141.2	1029.86	880.8	1100.4	999.33	921.4
Ping(536)	1528.3	1358.6	1295.7	1599.5	1375.75	1239.1
Ping(1024)	2666.6	2373.2	2048.2	2586	2434.33	2279.5
Ping(1436)	3268.6	2891.56	2644.5	2822.5	2669.2	2537.1
Ping(1460)	3131.3	2917.54	2750.3	2907	2748.85	2647.6

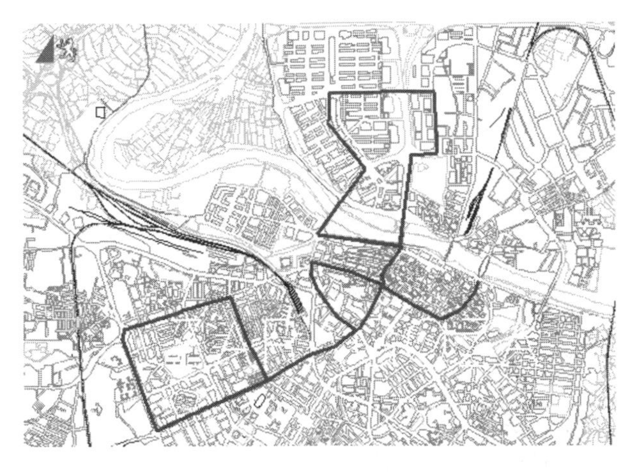

Figure 7.59. Drive test route definition

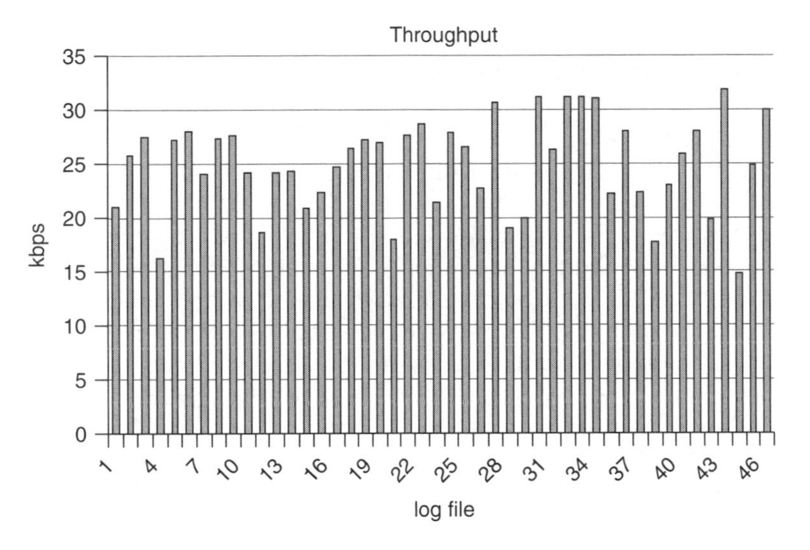

Figure 7.60. Data throughput during 200-kB download in drive test

and Routing Area Updates. In the case of the lowest throughput (14.86 kbps), 2 RAU of 18 and 16 s occurred during the FTP download. The effect of the cell-reselection delays and RAU in the application throughput has already been described in this section.

7.8.1.4 The Effect of Radio Link Conditions

In some user locations, the radio link conditions are not ideal and this will impact the service performance. FTP download, WAP browsing and real-time services have been measured under different radio link conditions in a lab test.

Figure 7.61 shows application level throughput as a function of C/I measured in a field trial. At good radio link conditions, the throughput per TSL is around 10 kbps, slightly worse than CS-2 peak throughput (12 kbps). This represents a 16% overhead introduced

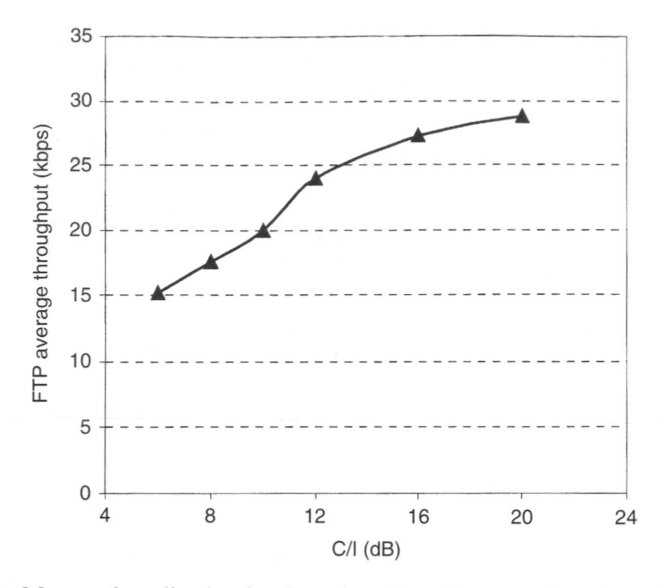

Figure 7.61. Measured application level average throughput as a function of C/I. No FH

by upper layers and GPRS procedures. Note that this figure may depend on how well optimized the TCP/IP is. At C/I of 10 dB, the application throughput per TSL is around 6.7 kbps, while having a look at link performance, the LLC level throughput per TSL is around 10 kbps. Now the overhead is around 33%. No data compression was used at SNDCP later. Similar throughput can be achieved with HTTP for WWW browsing.

Different services like low-resolution movie, slide show and audio have been tested under different C/I conditions. The services have been tested over different radio link conditions in RLC acknowledge mode. Low-resolution movie (moving images plus voice optimized for 30 kbps) has good quality with C/I greater than 20 dB. For worse radio link condition quality degrades very much and becomes unusable with C/I equal to 6. If the movie is optimized for 18 kbps, better quality can be achieved down to 10 dB. Real-time slide show (static images plus voice) at 30 kbps works well for C/I higher than 10 dB, although at 20 kbps, the quality is better and it is maintained down to 8 dB. Audio MP3 monaural at 16 kbps does not show any degradation down to 10 dB; although it can be used with lower quality with lower C/I.

RLC ACK mode proved to be very efficient for real-time services over UDP. There is roughly a 4-dB link level gain for the same application quality with the introduction of LA. Most of the applications have buffers that can cope with the delay introduced by retransmissions.

Table 7.8 shows achieved quality for several real-time services from the quality test performed during a field trial. Mean opinion score (MOS) values are averaged from four different persons.

7.8.2 EGPRS Performance Measurements

Application level throughput has been measured in the lab with a 2 + 1 EGPRS terminal. The maximum MCS supported by the network for this test was MCS-8. Both IR and

Table 7.8. Real-time services' MOS as a function of link quality measured in laboratory trial. GPRS CS-1–2. Three TSL-MS capability

Service	Speed	C/I (dB)					
		20	16	12	10	8	6
Low-resolution movie	30 kbps	4	1.75	2	2	1.75	1
Low-resolution movie	18 kbps	5	4.06	4	3	2.37	1
Slide show	30 kbps	4.75	4.56	3.98	3.56	3.43	2.06
Slide show	20 kbps	4.75	4.56	3.98	3.56	3.43	2.06
Audio MP3/ Monaural	16 kbps	5	5	5	5	3	2

MOS class	Remark	Classification
5	Excellent	Complete relaxation possible, no effort required, no changes from the original
4	Good	Attention necessary, no appreciable effort needed
3	Fair	Moderate effort needed
2	Poor	Considerable effort needed
1	Bad	No meaning understood with any feasible effort

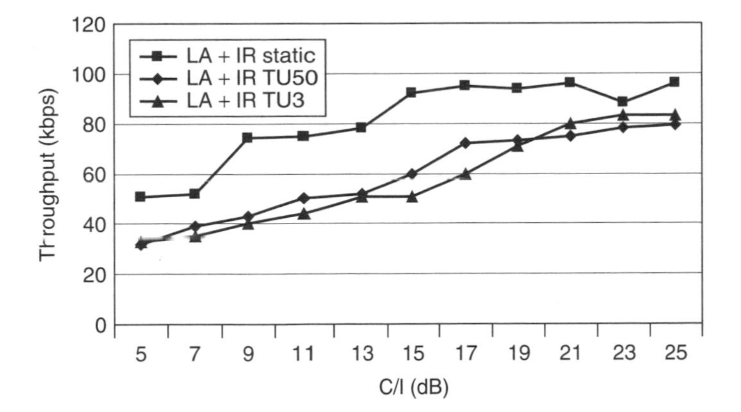

Figure 7.62. Application throughput as a function of C/I. 2 TSL-capable EGPRS terminal. (Source: AT & T Wireless [15])

LA were activated during the test. Different radio channels were tested: static, TU50 and TU3. Frequency hopping was not used. A test terminal was performed FTP download with average file size of 500 kB for different C/I values. Results are shown in Figure 7.62. Owing to the long file download, dynamic protocol effects are not noticeable; therefore, application level throughput is close to data link throughput (see Figure 8.2). Owing to the nature of this test, there is no throughput reduction due to TSL sharing, radio link conditions being the dominant effect. Radio link condition effects were studied in Section 7.1. Maximum throughput per TSL measured ($95/2 = 47.5$ kbps) is relatively close to maximum MCS-8 peak throughput (54.2 kbps).

Link level simulations in Figure 7.2 show that at C/I = 25 dB the ideal throughput per TSL is approximately 50 kbps, that corresponds to 100 kbps with a 2 TSL terminal. Measured throughput at application level is further reduced due to the different effects described in Section 8.1. Neglecting network load effects (not applicable to the EGPRS measurements in figure 7.62) the more noticeable service performance effects are RLC signaling, TCP slow start and headers overhead. In particular RLC signaling is the most dominant one in these measurements. It is due to signaling generated of the PACCH channel associated to the UL TBF allocations necessary to transmit TCP acknowledgments from the MS, see figure 8.3. Assuming an optimized TCP segment size of 1500 octets, the degradation is around 8.3%. This calculation is based on the estimation formula from section 8.1, with:

$$\text{Downloaded bits per UL TBF} = 1500 * 8 * 2$$

$$\text{BitsPerRadioBlock} = 1088 \text{ (corresponding to MCS-8)}$$

$$\text{RLCAcksPerUplinkTBF} = 1$$

Because of the large file (500 kB) used in the test TCP slow start effect should not be noticeable. Optimized TCP segment size minimizes header overhead, which is 3.3% with a segment size of 1500 octets, see Table 8.1. The overall degradation at application level is 11.4%, which corresponds to a total estimated throughput of 88.6 kbps at C/I = 25 dB. The measured throughput at the same C/I (82 kbps) is slightly lower than the estimated one but, certainly, within the expected range of values taking into account possible measurements inaccuracies and the differences possible MS receiver implementation.

Estimated link level throughput at C/I = 5 dB from Figure 7.5 is 15 kbps. It corresponds to 30 kbps with a 2 TSL terminal. Measured service level throughput from Figure 7.62 is around 30 kbps, with is already the same as the simulated values without taking into account service performance effects. This suggests that the sensitivity of the terminal used in the field measurements is better than the one assumed in the link-level simulations.

Application throughput measurements have been also performed in field trials at different distances from the base station. Three positions were selected with high (~25 dB),

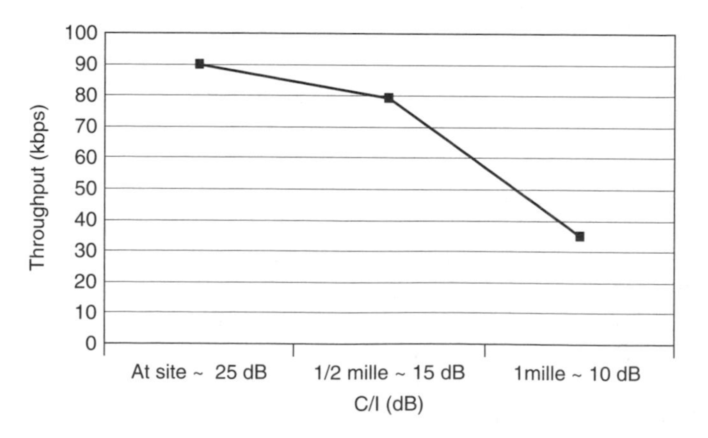

Figure 7.63. Application throughput in field trial. (Source: AT & T Wireless [15])

medium (~15 dB) and low (~10 dB) signal-to-interference ratio. Figure 7.63 shows the application throughput at those locations.

The analysis shown in this section shows that EGPRS field measurements are in line with the assumptions used across this chapter.

References

[1] 3GPP 45.005, Release 1999: Digital Cellular Telecommunications System (Phase 2+); Radio Transmission and Reception.

[2] 3GPP TS GSM 05.05 Version 8.8.0 Release 1999: Digital Cellular Telecommunications System (Phase 2+); Radio Transmission and Reception.

[3] Universal Mobile Telecommunications System (UMTS), *Selection Procedures for the Choice of Radio Transmission Technologies of the UMTS (UMTS 30.03 Version 3.2.0)*, ETSI Technical Report 101 112, 1998–04.

[4] Pons J., Dunlop J., 'Bit Error Rate Based Link Adaptation for GSM', *Proceedings of the 1998, 9th IEEE International Symposium on Personal, Indoor and Mobile Radio Communications, PIMRC*, Part 3, 1998, pp. 1530–1534.

[5] 3GPP 45.03, Release 1999: Digital Cellular Telecommunications System (Phase 2+); Channel Coding.

[6] Rafael S., Julia M., Javier R., Rauli J., 'TCP/IP Performance Over EGPRS Network', *VTC Fall Conference*, Vancouver, 2002.

[7] 3GPP 45.008 Release 1999: Digital Cellular Telecommunications System (Phase 2+); Radio Subsystem Link Control.

[8] Hämäläinen S., Slanina P., Hartman M., Lappeteläinen A., Holma H., Salonaho O., 'A Novel Interface Between Link and System Level Simulations', *Proc. ACTS Mobile Telecommunications Summit*, Aalborg, Denmark, October 1997, pp. 599–604.

[9] Furuskar A., de Bruin P., Johansson C., Simonsson A., 'Mixed Service Management With QoS Control for GERAN—The GSM/EDGE Radio Access Network', *VTC 2001*, Spring IEEE VTS 53rd, Vol. 4, 2001.

[10] 3GPP 3G TS 24.065 Version 3.1.0, Subnetwork Dependent Convergence Protocol (SNDCP).

[11] RFC-1144, Jacobson V., Compressing TCP/IP Headers for Low-Speed Serial Links.

[12] RFC-2507, Degermark M., Nordgren B., Pink S., IP Header Compression.

[13] ITU-T, Recommendation V.42 bis, Data Compression Procedures for Data Circuit-Terminating Equipment (DCE) Using Error Correcting Procedures.

[14] Border J., Kojo M., Griner J., Montenegro G., Shelby Z., *Performance Enhancements Proxies Intended to Mitigate Link-related Degradations*, IETF Informational Document—RFC 3135, June 2001.

[15] Mike B., 'EDGE Update', Presented at *3GSM World Congress*, New Orleans, March 17 2003, Available at www.3Gamericas.com.

8

Packet Data Services and End-user Performance

Gerardo Gomez, Rafael Sanchez, Renaud Cuny, Pekka Kuure and Tapio Paavonen

End-user performance can be defined as the end-user experience of a certain service. While previous chapters have concentrated more on the radio interface performance of voice and data services, this chapter will concentrate on end-user packet data services performance delivered through (E)GPRS networks. Packet data services end-user performance can only be assessed through indicators that are very service dependant and fundamentally different to those presented in Chapter 5, such as response time in web browsing, image quality and number of re-bufferings in video streaming or assurance of delivery within a certain time frame for e-mail.

Section 8.1 provides a general overview of the main factors affecting the end-user performance. Later, Section 8.2 describes a set of packet data services, while Section 8.3 analyses their specific end-user performance. Finally, Section 8.4 introduces possible methods to improve the end-to-end service performance.

8.1 Characterization of End-user Performance

The analysis in this section has been focused on downlink (DL) transmission, as it is expected to be the dominant direction for data applications, but all the findings are also applicable to uplink (UL) transmission.

End-user performance can be characterized through the cumulative performance degradation introduced by each protocol stack layer and network element throughout the system from one terminal to another terminal (or a server). Figure 8.1 below presents the different protocol stack layers and elements in GERAN networks.

In this chapter, a bottom-up approach is followed to characterize the end-user performance. The ideal throughput provided by layer one (physical layer) is considered

GSM, GPRS and EDGE Performance 2nd Ed. Edited by T. Halonen, J. Romero and J. Melero
 ISBN: 0-470-86694-2

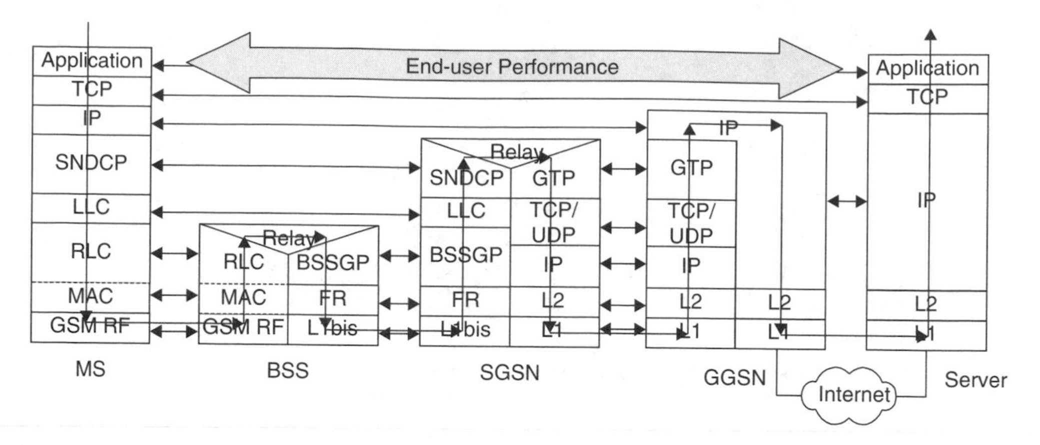

Figure 8.1. Layers affecting end-to-end performance in GERAN

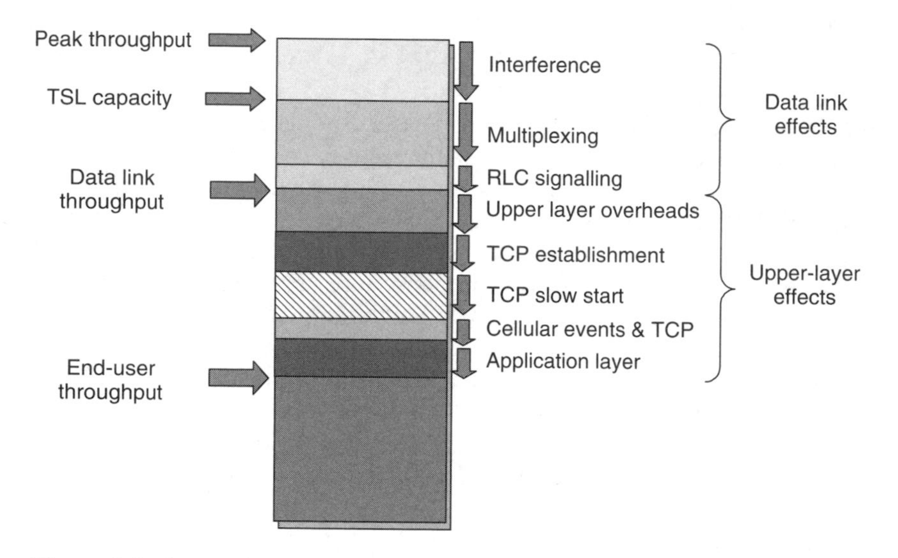

Figure 8.2. Approach used to evaluate end-user performance. TCP transport case

initially, and then, the performance degradation introduced by each protocol stack layer is calculated. Figure 8.2 illustrates the evaluation method for the particular case of services based on Transmission Control Protocol (TCP) transport protocol. Each relevant effect is described below.

8.1.1 Data Link Effects

Data link effects are the set of factors affecting the performance that depend on the network conditions, such as interference, timeslot multiplexing and Radio Link Control (RLC) layer signalling. The resulting performance after these effects is the *Data Link Throughput*, which refers to the payload (throughput) offered by the radio access network to the transport and application layers (as defined in Chapter 5).

8.1.1.1 Peak Air Throughput

In (E)GPRS, the peak air throughput is measured as the amount of data delivered to the Logical Link Control (LLC) layer per unit of time. It is defined after removing the Radio Link Control (RLC) and Medium Access Control (MAC) layers headers. For instance, Coding Scheme 2 (CS2) has a maximum gross throughput of 13.4 kbps, but without the RLC and MAC headers the resulting peak air throughput is 12 kbps (see Table 7.1).

8.1.1.2 Timeslot Capacity

It is the throughput available in a fully utilized Timeslot (TSL) after including the effects of interference. It includes the effect of RLC retransmissions, but not multiplexing and other protocol effects. It depends on network configuration, load and radio link quality. See definition and performance analysis in Sections 5.2, 7.3 and 7.4.

8.1.1.3 Reduction Factor

The available air throughput per user may be further affected by this factor as the existing TSLs are shared between several connections. Reduction factor depends on the amount of TSLs, which is conditioned by the network load and the dimensioning criteria (see Section 5.2 and Appendix B). A good way to avoid excessive timeslot sharing is to carefully design and dimension the network according to the existing speech and (E)GPRS data traffic (see Section 7.5).

8.1.2 Upper Layer Effects

Upper layer effects are the set of factors that depend on the transport and application protocols used by each service. In case of services based on TCP transport protocol like web browsing or File Transfer Protocol (FTP), the main effects are due to header overheads, TCP connection establishment and slow start, specific cellular events over TCP layer, and finally, application layer protocol. Other transport layers such as User Datagram Protocol (UDP), which introduce less control mechanisms and do not require so much interaction between transmitter and receiver will have a more reduced impact on the overall end-user performance.

8.1.2.1 RLC Signalling Overhead

The RLC control blocks used for signalling in the uplink direction also reduce the user throughput. In GPRS, different procedures might require RLC control blocks transmitted in the uplink. The most common case is the transmission of data in the uplink, which requires at least one control block in order to assign the resources during the establishment of one uplinkTemporary Block Flow (TBF). In case the uplink TBF is established when there is already a downlink TBF active, this assignment is sent in the PACCH, which is multiplexed with the downlink packet data traffic channel (PDTCH). In the RLC acknowledged mode, it is also required to send a number of control blocks, transmitted in the PACCH, carrying acknowledgements (ACKs) for the uplink data (as shown in Figure 8.3).

From the description above, it can be derived that this effect is highly dependent on the traffic characteristics, but also on the CS that is in use. With higher CSs each control

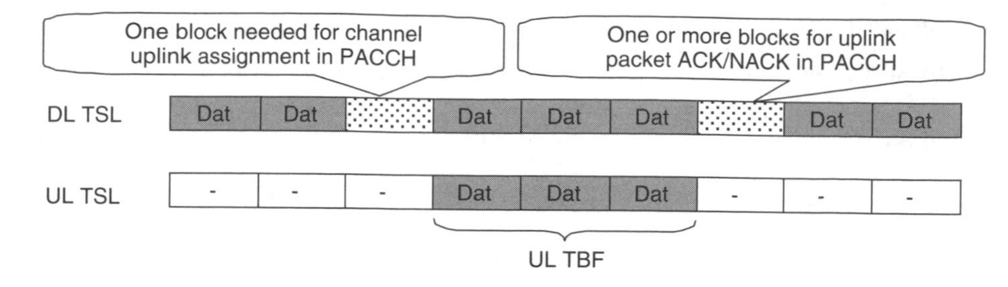

Figure 8.3. Effect of RLC control blocks on transmission

block means more lost bits for data transmission. The following formula tries to describe this degradation in a general case:

$$\left(1 - \frac{\text{DownloadedBitsPerUplinkTBF}}{\begin{array}{c}\text{BitsPerRadioBlock} \cdot (1 + \text{RLCAcksPerUplinkTBF}) \\ + \text{DownloadedBitsPerUplinkTBF}\end{array}}\right)$$

In the case of TCP download, the UL TCP ACKs of 40 bytes are regularly generated per every two downlink TCP data segments, introducing certain throughput degradation.

Other protocols like the UDP, which do not require regular ACKs at transport or application layer level, will suffer much less from this effect.

8.1.2.2 Header Overhead Effects

The data link throughput that the network is able to provide will not be directly perceived at the application level. The maximum achievable throughput at the application level depends on the overhead introduced by each protocol stack layer. In the previous section, the overheads introduced by RLC/MAC layers have been considered. Header overheads introduced by the remaining layers that affect the resulting end-user performance in enhanced GPRS (EGPRS) are shown in Table 8.1 (see also Figure 1.20). The impact of the Internet Protocol (IP) overhead is smaller if the IP packets are large. This is why, in general, it is recommended to use large IP segments (e.g. 1500 bytes) in data transactions in order to enhance spectral efficiency and end-user performance. The IP segment size or MTU (Maximum Transfer Unit) can be configured at IP layer in the terminals and in the servers. Note that there exist different techniques such as header compression to avoid carrying the full IP headers over the air interface (see Section 8.4 for more details). However, these techniques are not always applicable, e.g. when IP security functionality is used end to end, the user IP headers cannot be compressed in the mobile network. It should also be noted that some header compression schemes do not perform well if too many errors or mis-ordering happens between the two compressing nodes. Finally, header compression is a demanding operation in terms of processing power and memory consumption and it should be used only if clear benefits in terms of delay and spectral efficiency can be obtained.

Table 8.1. Overhead in (E)GPRS protocol stack (Rel'97 and Rel'99)

Overhead source	Size	Unit
LLC overhead	6	octets/Packet
SNDCP overhead	4	octets/Packet
IP overhead	20	octets/Packet
TCP overhead	20	octets/Packet

8.1.2.3 Latency (Round Trip Time)

Before continuing with the upper layer effects, described in Figure 8.2, the latency associated to the system must be introduced, as it is a key element when evaluating end-user performance. Although it is not considered as a degradation by itself, it has a direct impact on upper layers behavior. A common way to benchmark the network latency is to evaluate its round trip time (RTT).

The network RTT can be defined as the time it takes to transmit one packet from e.g. a server to a terminal plus the time it takes for the corresponding packet to be sent back from the terminal to the server. Figure 8.4 shows the different elements in a (E)GPRS network that contribute to the RTT.

The RTT is a function of the packet size transmitted in downlink and uplink: RTT (DL,UL). Figure 8.5 shows typical RTTs in GPRS and EGPRS for two cases:

— *RTT (1500 bytes, 40 bytes).* This RTT illustrates common file download scenarios from a server to a mobile terminal, where TCP data segments size (DL) is up to 1500 bytes and the TCP ACKs size (UL) is 40 bytes.

— *RTT (32 bytes, 32 bytes).* This shows the RTT for Ping commands with small packets of 32 bytes. It gives a good idea of the minimum network latency as well.

Different packet sizes RTTs may be used to characterize various performance degradation effects. RTT (32,32) characterizes the minimum network latency and can be used to estimate the performance degradation associated to the transport and application layer establishment protocols. RTT (1500,40) determines the latency associated with the transmission of large packets in DL and small packets in UL and can be used to estimate the impact TCP protocols dynamics (e.g. slow start) have on the end-user performance.

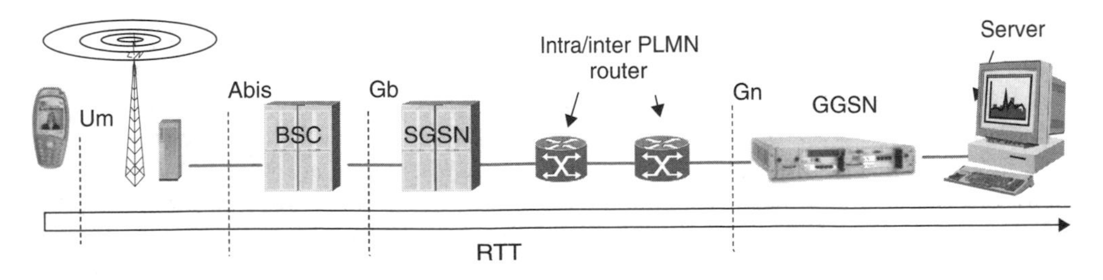

Figure 8.4. End-to-end path in (E)GPRS

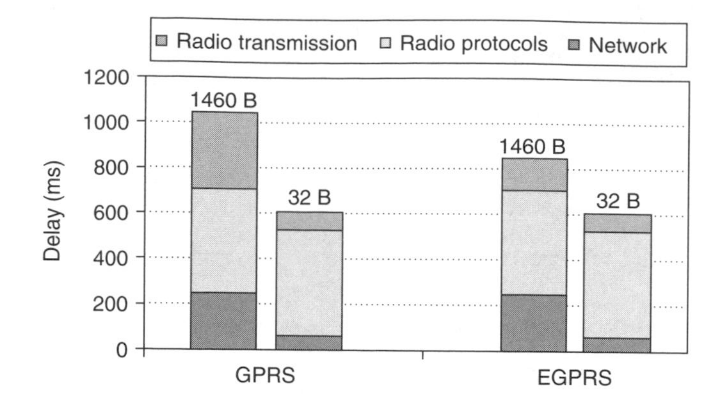

Figure 8.5. Estimated round trip time in (E)GPRS

Figure 8.5, assumes a throughput of 44 kbps for GPRS and 155 kbps for EGPRS in downlink for a 4-timeslots-capable terminal. The Network component includes the delay from Server to the base station subsystem (BSS), while Radio protocols refers to the delays introduced by BSS and MS processing and RLC/MAC layer procedures. Finally, Radio transmission is the time spent through the air interface.

Depending on the traffic characteristics of a particular service, the need of establishing and releasing a TBF may vary, leading to different RTT values. The values shown in Figure 8.5 are estimated for the case of a TCP download in Rel'99, where the Delayed Downlink TBF release feature is supported (no downlink TBF releases). In this case, uplink traffic is bursty and composed by small packets, leading to frequent uplink TBF establishments and releases.

The uplink TBF establishment (part of radio protocol delay) takes a large part of the RTT shown in Figure 8.5. A TBF is established whenever new incoming data needs to be sent over the Um (radio) interface. Such TBF is released when RLC buffers are empty after all available data was sent. In case of bursty traffic, it is common that the TBF is released before the application has sent all its remaining data. For example, the TCP algorithm injects new data in the network only after any segment acknowledging the last window of data has been received. Establishing a TBF takes around 200 to 300 ms, which indeed causes performance penalty if these procedures take place several times during a connection.

In order to decrease the latency, some enhancements were introduced in (E)GPRS specifications: Delayed Downlink TBF release reduces the RTT to about 600 ms (32 bytes packets). This feature is available since Rel'97 (see Section 1.5). Delaying the Uplink TBF release (extended Uplink TBF mode) reduces the RTT to about 420 ms (32 bytes packets). This feature is available in Rel'4 mobile terminals.

Both features are based on the idea of not releasing the resources once there is no more data to transmit. Instead, a timer is started and the TBFs are kept until this timer expires. When traffic is bursty it is quite usual that buffers get empty for small periods, and the time needed for establishing and releasing the TBFs could cause a significant service-performance degradation. The length of this timer shall be set to a suitable value in the range of 1 to 5 s according to specs [1].

Depending on the traffic characteristics of a particular service, the need of establishing and releasing a TBF may vary, leading to different RTT values.

8.1.2.4 Impact of Latency on Service Performance

The RTT has effects on different mechanisms that directly impact end-user performance:

- *Session setup delay.* When a new service is activated, the mobile network may first establish one or several Packet Data Protocol (PDP) contexts in order to reserve resources (note that in case of always on PDP context, this procedure is not required). Additionally, the application signalling procedures needed in most IP-based services causes additional latency.
- *TCP performance.* The establishment of a TCP connection and transmission rate are directly affected by the RTT as explained in more detail in the following section.
- *Service interactivity.* Some services (such as voice or real-time gaming) that require small end-to-end delays may not be well supported over packet-switched technology if the round trip time is large. However, when the traffic pattern is not bursty, the value of the RTT that applies would not be affected by frequent TBF establishment and releases (i.e. third column in Figure 8.6).

Earlier in this section we explained that usually large IP segments are recommended in order to minimize the effects of IP headers overhead. Although this is true, in slow systems where the latency can be high, large IP segments may take considerable time to be transmitted and introduce noticeable delays for the end user. For example, the latency when accessing a remote computer (e.g. using Telnet) should be kept small. In general interactive response time should not exceed the well-known human factor limit of 100 to 200 ms [2]. So a tradeoff should be found between efficiency and latency in the design of a sub-network.

8.1.2.5 TCP Performance

Transport Control Protocol (TCP) [3] is currently used in 80% to 90% of data transactions in fixed networks. In mobile networks, the use of TCP is expected to grow significantly

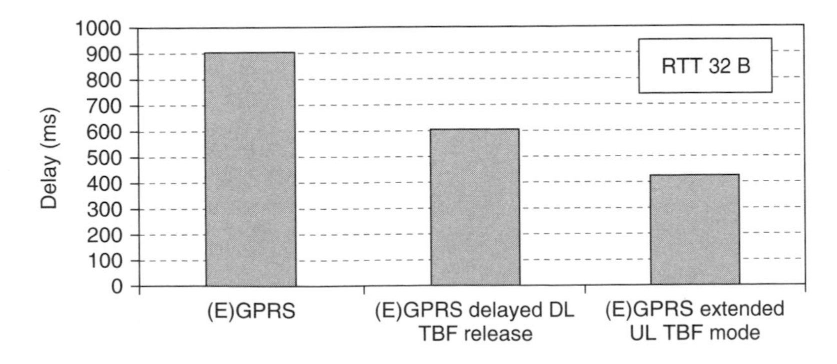

Figure 8.6. Round trip time evolution in (E)GPRS. Typical delays

as packet data applications usage increases. This foreseen growth will be caused partly by the introduction of Wireless Application Protocol (WAP) 2.0, which can use the TCP stack instead of UDP stack (which was mandatory in earlier WAP versions). Also, the deployment of solutions for wireless corporate access mobile office applications will probably increase the amount of TCP traffic in mobile networks.

TCP provides reliable data transfer by means of a retransmission mechanism based on acknowledgments and retransmission timers.

TCP performance impact in cellular networks is largely affected by the RTT. Large round trip delay makes initial data rates slow due to TCP long connection establishment. TCP uses a three-way handshake for connection establishment (see Figure 8.7). Therefore the time required for establishment procedure is $1.5 \times$ RTT, which means a bit less than one second, assuming 600 ms RTT for small packets (40 bytes) as shown in Figure 8.6.

TCP also includes a congestion control mechanism [4] that aims at avoiding Internet congestions occurring frequently. This congestion control mechanism has a big influence on the actual throughput an end user will obtain. The amount of data a TCP can send at once is determined by the minimum value between the receiver's advertised window and the congestion window.

The receiver's advertised window is the most recently advertised receiver window and is based on the receiver buffering status and capabilities.

The congestion window size is maintained by the TCP sender and is increased or decreased according to standard recommendations [4, 5]. Figure 8.8 illustrates the evolution of a TCP congestion window during a connection: At the beginning of the transaction, the TCP sender can transmit only few segments (the maximum number of segments depends on the maximum segment size: MSS). The reasoning is that because the network congestion status is not known, it is considered not safe to start sending at full speed. Each time an ACK is received by the sender, the congestion window is increased by one segment, which leads to an exponential increase until the slow start threshold is reached. After this threshold, the congestion window grows linearly until the advertised window is reached. It is important to mention that TCP algorithm interprets every lost packet as

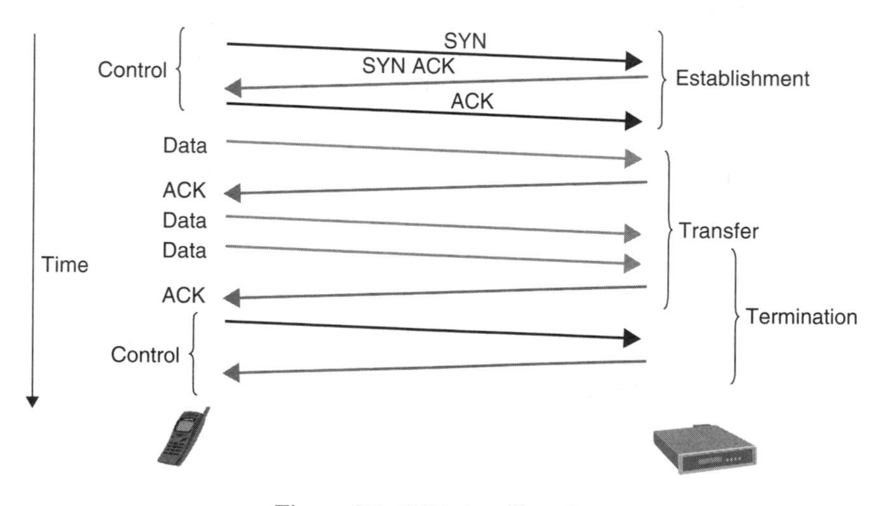

Figure 8.7. TCP signalling chart

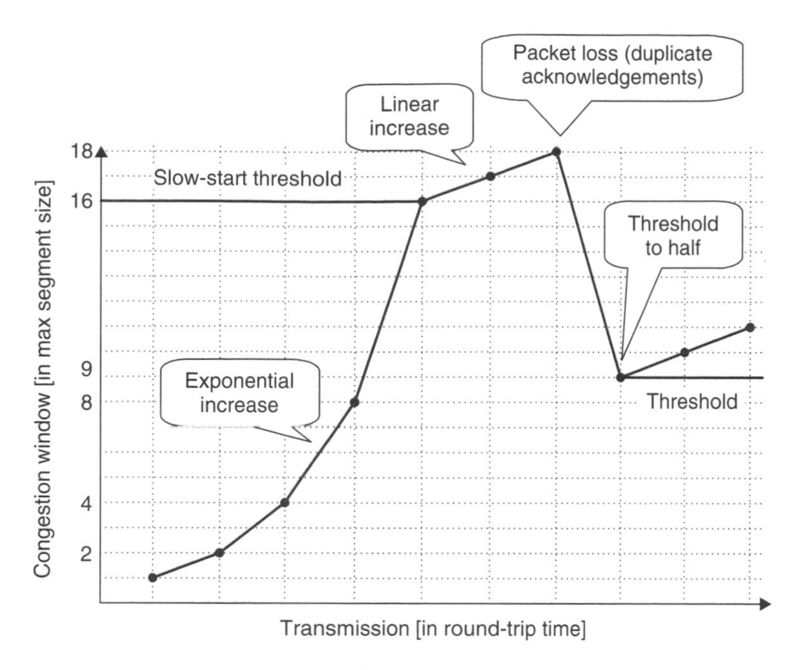

Figure 8.8. TCP congestion window

a sign of network congestion and therefore drastically reduces its sending rate when it detects such an event. The 'fast recovery' mechanism was introduced to recover faster from packet losses and works as follows: if a single packet has been lost in a window of segments, a sequence of duplicate ACKs will be received by the data sender. These duplicate ACKs are generated by the packets that follow the lost packet, where the receiver acknowledges each of these packets with the ACK sequence value of the lost packet. In this case the sender immediately retransmits the lost packets and reduces its congestion window by half. After this, TCP linearly increases its congestion window by one segment per RTT ('congestion avoidance' phase).

Fast changes in RTT may have bad side effects on the TCP performance. The TCP sender also maintains a timeout timer for every packet sent. If no ACK is received after the expiration of this timer, the congestion window drops to one segment and the oldest unacknowledged packet is retransmitted, as illustrated in Figure 8.9.

A TCP timeout timer (Retransmission Timeout (RTO)) is calculated as follows:

$$\text{RTO} = A + 4D$$

where A = average RTT, D = standard deviation of average RTT.

One potential problem in wireless networks is that the round trip delay may vary greatly depending on packet size, or sudden buffer growth. Also, cell reselections introduce a gap in transmission (see Section 7.8). A few seconds of radio transmission gap may mean, for example, 10 s of interruption at the application layer (i.e. before sending new data), if the TCP timeout timer expires. These situations are very undesirable as they lead to unnecessary retransmissions and heavy throughput reduction. Some methods to avoid such situations are presented in Section 8.4.

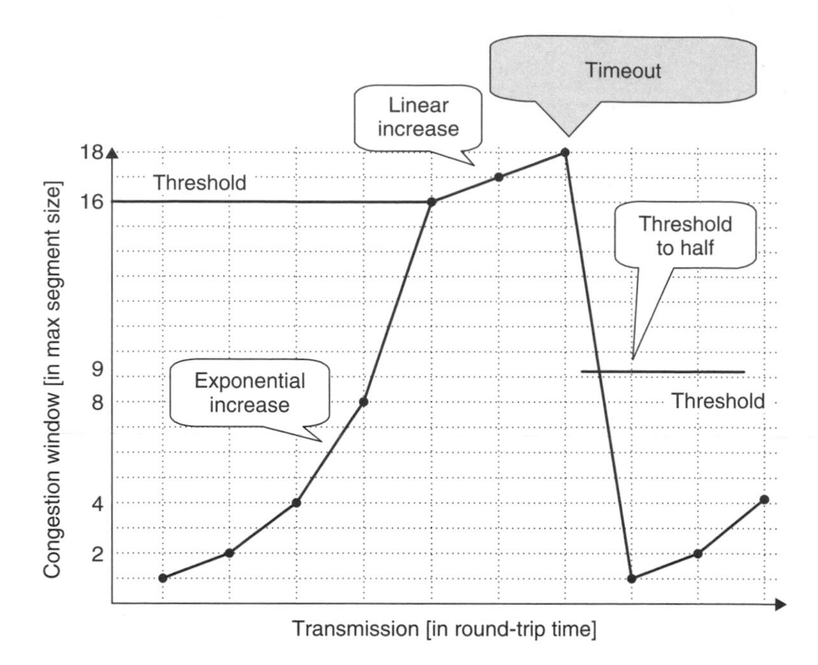

Figure 8.9. TCP timeout effects on the congestion window

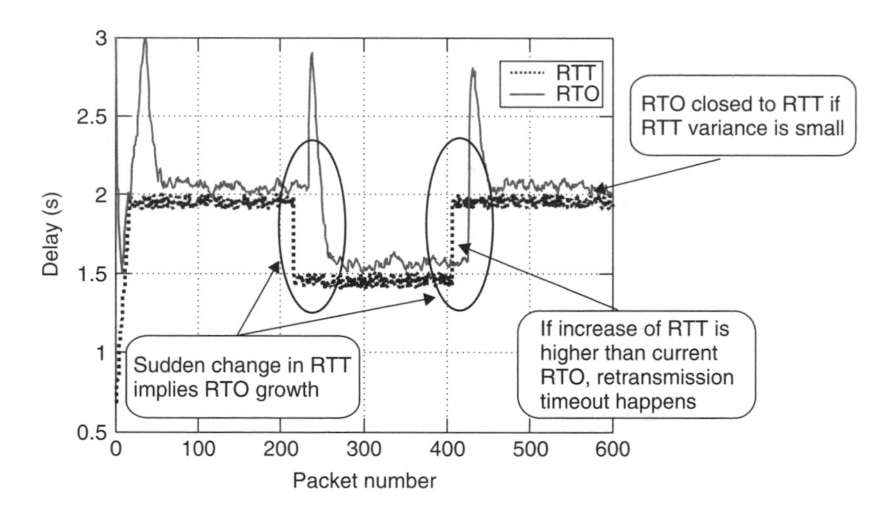

Figure 8.10. RTO estimation versus RTT evolution

An example of RTO evolution following RTT is shown in Figure 8.10. It should be noted that the RTT estimated for the characterization of the network, Figure 8.5, is only valid at the beginning of a TCP transmission. As the congestion window grows, the total RTT measured by the TCP layer will be the addition of the initial RTT and the transmission delay of the congestion window.

A more complex and accurate formula that calculates the TCP throughput depending on packet loss rate is presented below [2]:

$$\text{TCP\_throughput} = \frac{\text{MSS}}{(\text{RTT} \cdot \sqrt{1.33 \cdot p} + \text{RTO} \cdot p \cdot (1 + 32 \cdot p^2) \cdot \min(1; 3 \cdot \sqrt{0.75 \cdot p}))}$$

where:
MSS: is the segment size being used by the connection
RTT: is the end-to-end round trip time of the TCP connection
RTO: is the packet timeout (based on RTT)
p: is the packet loss rate for the path (i.e. 0.01 if there is 1% packet loss).

This formula assumes that the advertised window is large enough to maximize the throughput (see Bandwidth Delay Product (BDP) in the next section). It is applicable to standard TCP Reno implementation, although enhancements and different implementations may add some changes.

As presented in this section there are a few issues related to the efficiency of basic standard TCP in wireless networks. Some standard recommendations for improving the performance of TCP in slow links or in links with errors are provided by the IETF: see [6, 7]. Also some well-known non-standard methods are available and may be applied in special cases. In practice, each implementation of a TCP stack can include its own solutions and the basic behavior may be different from one system to another. Section 8.4 presents in more detail the possible ways to improve IP traffic in wireless environment.

It should be noted that the WAP forum defined a standard protocol to be used in cellular networks: Wireless Application Protocol (WAP). The motivation for developing WAP was to extend Internet technologies to wireless networks, bearers and devices. WAP is currently available in most cellular networks in order to transmit data over packet switched technology (see Section 8.2 for more details on WAP).

8.1.2.6 Bandwidth Delay Product

TCP performance not only depends on the available data link throughput but also on the product of this throughput and the round trip delay. The BDP measures the amount of data that would fill the transmission 'pipe', that is the amount of unacknowledged data that the TCP must have in transmission in order to keep the pipe full. Maximum TCP throughput depends directly on the size of the congestion window and the RTT. When there is no congestion, the maximum congestion window size is determined by the advertised window, and the throughput is limited by the following formula:

$$\text{TCP\_throughput} < \frac{\text{advertised\_window (bits)}}{\text{round\_trip\_time (seconds)}}$$

The bandwidth delay product (BDP) has therefore some implications when selecting the proper maximum window in a TCP receiver (see Figure 8.11). The maximum throughput will be limited during a transmission either by slowest link in the path (i.e. GPRS link) or the quotient between congestion window size and RTT, whichever is smaller.

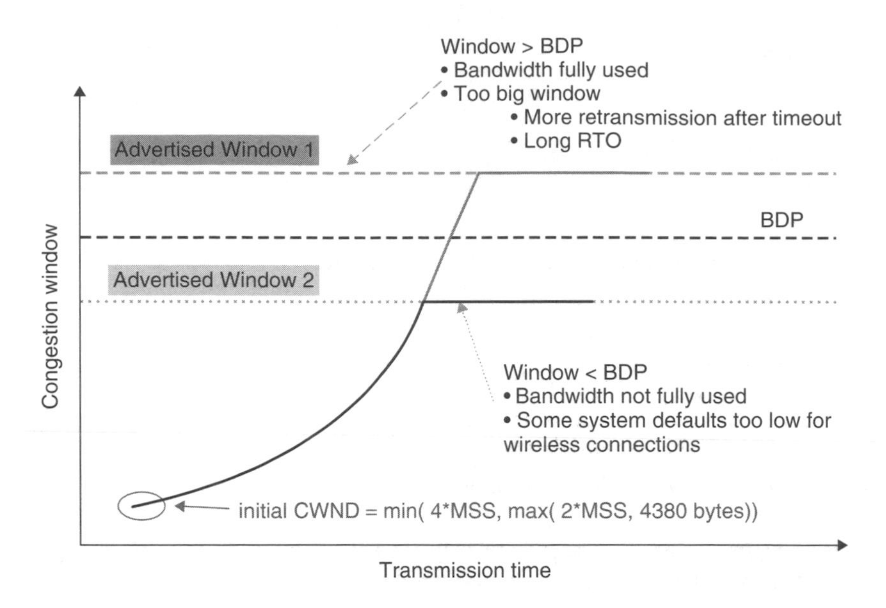

Figure 8.11. Advertised window and bandwidth delay product

8.1.2.7 UDP Performance

User Datagram Protocol (UDP) transport protocol is less problematic than TCP in wireless, as it does not require retransmissions and the protocol overhead is significantly lower. Anyhow, in most data applications (E)GPRS works in the RLC ACK mode, thus guaranteeing reliable transmission over the air interface.

Some streaming services, such as Voice over IP (VoIP), use Real-Time Protocol (RTP) over UDP. In this case, the (E)GPRS network may operate in negative acknowledge (NACK) mode. The Radio Resource Management (RRM) is then responsible to maintain the negotiated Reliability QoS parameter. The most straightforward mechanism to control reliability is the Link Adaptation (LA) algorithm, which would select the highest (M)CS possible while being under a maximum block error rate (BLER). It is worthwhile noting that certain streaming services, which do not have such stringent delay requirements, allows RLC to operate in the ACK mode.

8.1.2.8 Application Layer

The application protocol also plays an important role in the final packet data services end-user performance. A session establishment procedure for the application is usually required and, as stated earlier, the application overhead also reduces the end-user throughput.

For instance, Web services use Hyper Text Transfer Protocol (HTTP) protocol to transfer the HTML files. The client first sends a request to the server with the address of the main page. After the main page has been downloaded and decoded, the client sends a new request for the embedded objects. HTTP version 1.0 establishes a separate TCP connection for each embedded object in the page whereas HTTP 1.1 may use only one TCP connection for all the objects.

As illustrated in Figure 8.14, HTTP version 1.1 can use the *persistent connection* feature where only one TCP connection is established for each connected Web server. Also this version may use pipelining where multiple requests for objects can be sent without the need to wait until the previous request has been served. HTTP version 1.1 is highly recommended for the wireless environment (HTTP 1.1 is currently the dominant version in use).

8.1.3 Performance Characterization Example. HTTP Performance in GPRS

As an example of the characterization of packet data services end-user performance, web browsing (HTTP) over GPRS is analysed. The assumptions taken are as follows:

- GPRS radio with CS-1 and CS-2
- TSL Capacity = 11.1 kbps
- Reduction factor = 99%, that is, no significant timeslot multiplexing
- Round trip time (1460,40) is estimated around 1 s for Rel'99: TCP application with MSS = 1460B, and ACK packet = 40B. Uplink TBF establishment is considered (see Figure 8.5)
- HTTP 1.1 with pipelining is used to minimize the application layer degradation
- Analysis is done considering downlink transmission, and assuming 4 TSLs in downlink and 2 in uplink[1].

The results of this analysis have been benchmarked with simulation results and field measurements. This benchmarking has proven this methodology to be an excellent approach to estimate end-user performance and how it is affected by different conditions of the network.

As illustrated in Figure 8.12 the ideal throughput is about 54 kbps. Then each effect described in this section reduces this value by either adding overhead or adding delay. Each line in the figure shows the obtained throughput after one additional effect is considered. It is interesting to notice how the effect of TCP Layer, in absence of retransmissions, introduces a constant degradation, which highly depends on the amount of data that is downloaded, having more impact for small-sized files.

8.2 Packet Data Services

This section focuses on the description of a subset of the most relevant mobile data applications in current and future (E)GPRS networks, which includes Web Browsing (WWW), Multimedia Messaging (MMS), Streaming, on-line Gaming and Push to Talk over Cellular. Special emphasis is given to the application layer signalling of the different services.

In order to analyse the end-user performance, it is important to understand the implications of each network element along the path between terminal to server or terminal to terminal, as well as the protocols involved in each scenario.

[1] Class 1 terminals cannot provide 4 TSLs in downlink and 2 in uplink at the same time, but it is possible to dynamically use the 4 + 1 and 3 + 2 possible configurations.

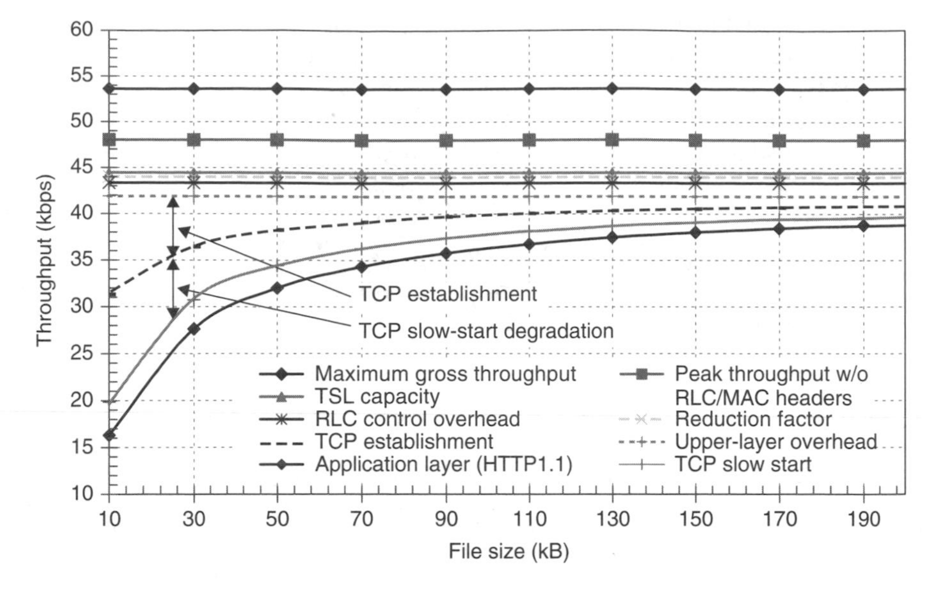

Figure 8.12. Web-browsing performance in GPRS

8.2.1 Web Browsing

Web browsing constitutes the dominant traffic in fixed IP networks, and HTTP is the protocol used by browsers to access web pages and surf through the Internet. This service is also expected to become one of the critical packet data services in (E)GPRS networks.

Typically a web browsing session consists of one or several accesses to different web pages. Between the accesses to each page, some time is usually used to read the downloaded information. A Web page can be described as a main page containing the text and structure of the page as well as a set of Uniform Resource Locator (URL) addresses corresponding to a certain number of embedded objects (e.g. pictures, tables, etc.). Several studies have been made to describe the distribution of web page sizes through the Internet. However, these kinds of studies tend to become out of date quite soon, as the possibility of improving the quality and contents of the pages tend to increase its average size with the years as described in the next section. The average size of a today's typical web page is around 150 kB [8]. Figure 8.13 shows an example of a typical message chart produced during a web page download using HTTP protocol. It must be noted that some browsers may open two or more connections in parallel. The following actions are taken during the page download:

A more detailed description of the procedures shown in Figure 8.13 is presented below:

- PDP context establishment. This is only needed when establishing a new session. It negotiates the parameters of the wireless connection between the terminal and the network. After this PDP context has been established, transmission of upper layer data can start. It is up to the terminal implementation to use an 'always-on' PDP Context suitable for web traffic.

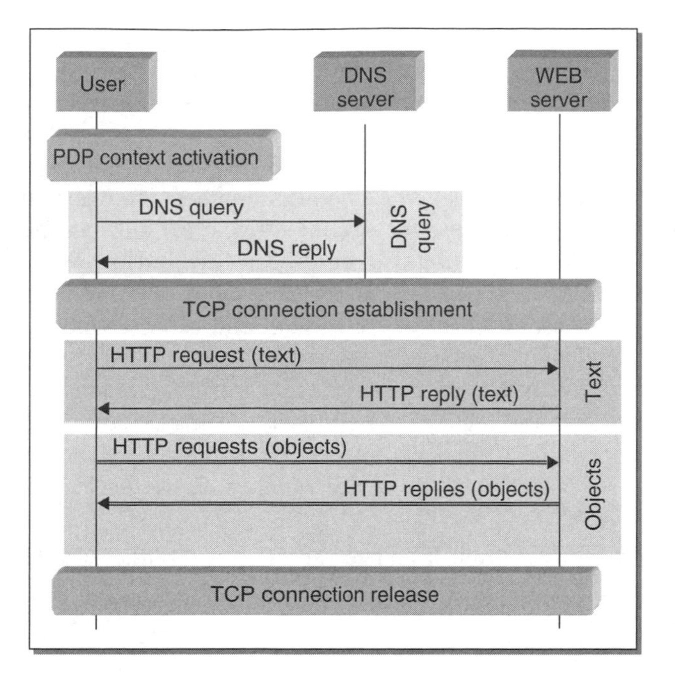

Figure 8.13. Web browsing message sequence chart

- Domain Name Server (DNS) query. Before establishing a connection with a new server, its IP address needs to be identified. A DNS query is sent from the terminal to the DNS server including the domain name of the web server (e.g. www.google.com) and a DNS reply is returned with the IP address (e.g. 216.239.39.101).
- TCP connection establishment. HTTP protocol uses TCP as transport layer. Depending on the HTTP version and browser implementation, one or several TCP connections may be needed. In any case, for the download of the main object, a TCP connection must first be established.
- First an HTTP request is sent to the server requesting the primary object of the web page. Once this element has been received the client will be able to show the skeleton and text of the web page. It also gets the URL addresses of the rest of the components (pictures, tables, etc.) of the page.

After the main object has been received, the behavior varies depending on the version of the HTTP protocol.

- *HTTP 1.0.* A different TCP connection is established and released to download every element of the web page. This typically causes some overload because of the multiple TCP establishments and associated slow start phases. In order to improve the performance of this scheme, commercial browsers establish several TCP connections in parallel (normally four).

- *HTTP 1.1.* The main improvement with respect to the HTTP 1.0 is the usage of the *persistent connection* concept. In this case several objects can be downloaded through the same TCP connection. Once one object is fully received, the request of the following object is sent. Since the objects contained in the web may be stored in different servers, at least one TCP connection is needed for each server to which the browser is connected. The *pipelining* option is another improvement to the protocol that allows sending the request for several objects as soon as they are available, instead of waiting for the previous object to be received. However, *pipelining* is not mandatory and therefore not widely used at the moment by some popular browsers because it can increase the load of the servers and cause congestion unnecessarily. Instead, two parallel TCP persistent connections are normally used. Although this strategy increases the overheads, the multiplexing effect gives the benefit of reducing the probability that radio link is not used because there is no upper layer data to be transmitted. Slow start and application layer effects are reduced in this case. Anyhow, it is up to browser implementers to decide whether *pipelining* is implemented or not.

HTTP 1.1 can only be used if client and server support it. Otherwise HTTP 1.0 is used by default. Figure 8.14 in the previous section illustrates the differences between HTTP 1.0 and 1.1.

8.2.2 WAP Browsing

The WAP is an open specification that allows wireless devices to easily access and interact with information and services. The standard WAP configuration has a WAP terminal communicating with a WAP gateway, which is then connected to the web. However, the WAP architecture can quite easily support other configurations. WAP 1.x is based on the WAP protocol stack (Wireless Session Protocol (WSP), Wireless Transport Protocol (WTP), Wireless Transport Layer Security (WTLS) and Wireless Datagram Protocol (WDP)) over UDP, while WAP 2.0 uses wireless TCP and HTTP protocols.

An example of the WAP (1.x) browsing scenario is depicted in Figure 8.15.

As in the case of web browsing, one PDP Context is required to transmit the GPRS data. After the PDP Context activation, the WSP is responsible for establishing the session

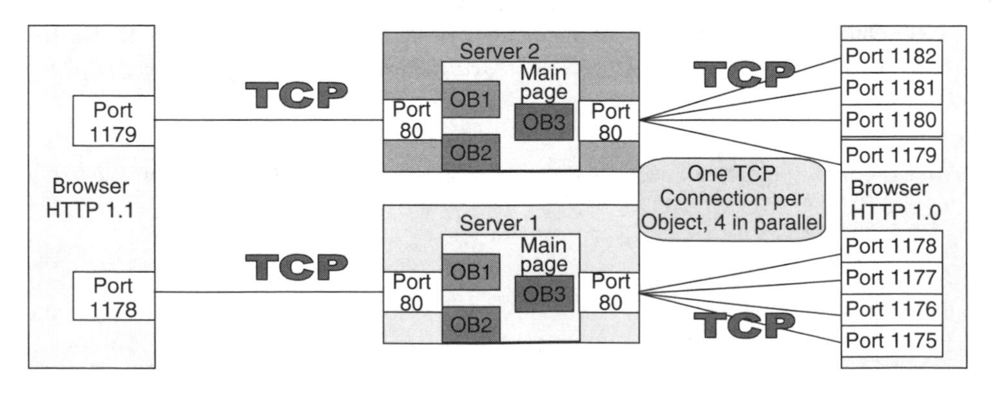

Figure 8.14. HTTP 1.1 versus HTTP 1.0

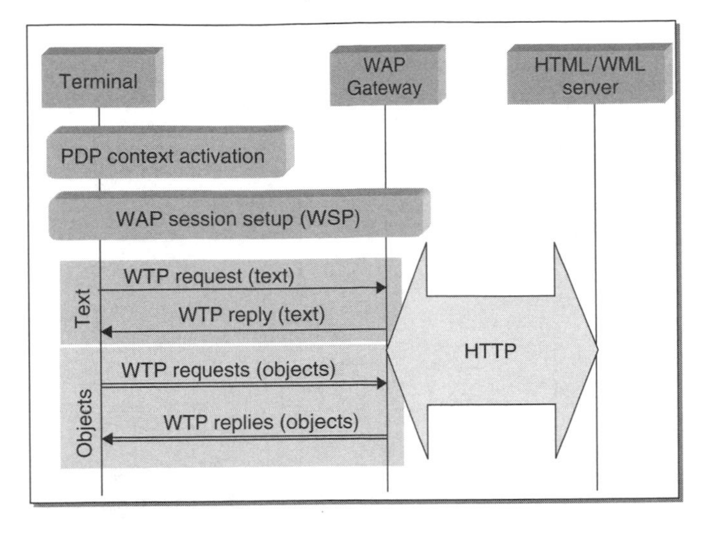

Figure 8.15. WAP browsing message sequence chart

between the mobile device and the WAP gateway, and it provides a generalized session management capability such as the ability to suspend and resume sessions.

The terminal sends its request to the WAP gateway, which then requests the content from the HTML/WAP Mark-up Language (WML) server. Note that although the connection between terminal and WAP gateway is based on WAP protocol stack, the connection between the WAP gateway and HTML/WML server is normally based on the HTTP protocol stack, where a TCP connection must be established before requesting the content from the HTML/WML server. A previous DNS query is performed by the WAP gateway in order to get the IP address of the HTML/WML server.

The server replies to the request with WML files, which the gateway can then optimize for transmission over the low bandwidth link to the terminal. A modified approach, which allows the browsing of standard web pages, is to use a HTML filter to convert the HTML pages to WML.

Once the main element of the WML page has been received, the client will be able to request the rest of the objects, as in the case of web browsing.

8.2.3 Multimedia Messaging Service

Multimedia Messaging Service (MMS) is a natural evolution of the Short Message Service (SMS), extending the text content with capability of transmitting images (pictures, image streams) and sound (voice, music) as a part of the message, in addition to textual content.

Different protocols can be used as MMS transport mechanism, such as WAP, HTTP or Session Initiation Protocol (SIP) (the selection is implementation dependant). In this section, WAP's WSP is assumed as the MMS transport mechanism, which is the most common approach used nowadays. Figure 8.16 shows an example of a signalling chart illustrating the PDP Context activation, message origination, delivery and delivery report in the WAP-based MMS scenario.

In the first step of the sequence chart, the MMS sender originates a Multimedia Message (MM) by sending a send request to the MMS Server.

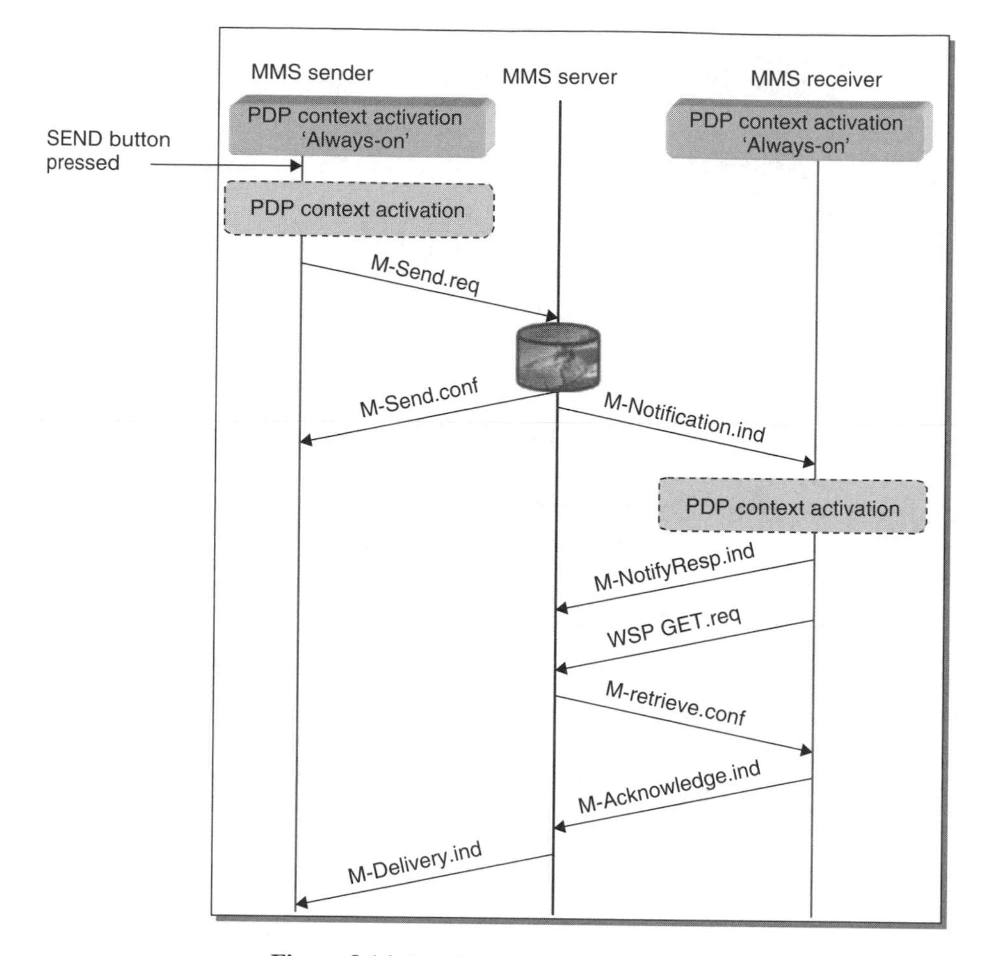

Figure 8.16. MMS message sequence chart

Then, once the MMS Server has stored the MM, it sends a notification indication to the remote user. Normally, from 1 to 3 SMSs are needed to notify the remote user of MMs available for retrieval. Included, as a data component, is the URL of the MM that the MMS receiver is to use for the retrieval. The number of SMSs depends on the length of the notification message, which is text-based. This length can be different depending on the URL and the optional headers. Measurements have shown that typically two SMSs are used.

These SMS notifications can be done over GSM or GPRS. The number of SMSs as well as the use of GSM or GPRS technologies is an important aspect from a service performance point of view, since generally the SMS over GPRS delay is smaller. Once the remote user has received the notification, he is able to get the MM from the server.

One PDP Context per terminal is required to transmit the MM through the GPRS network. A suitable traffic class for this PDP Context could be *background*, which may be also shared by other traffic with similar QoS requirements (see Chapter 3). The use of an 'always-on' PDP Context for conveying the MMS traffic is an implementation issue for the terminal. Therefore, in case no 'always-on' PDP Context is used, an extra delay caused by the activation of a new PDP Context must be taken into account for both transmitter and receiver. The activation of the PDP Context will take place after the send button is pressed at the origin and once the SMS notification is received at the other terminal.

The MMS Server may request information that would permit knowing that the MM was actually received by the remote user. One approach would be for a distinct acknowledge-indication message to be passed from the MMS receiver to the MMS Server. Finally, the MMS Server is responsible for supporting an optional delivery report back to the originator MMS user.

8.2.4 Streaming

Multimedia streaming services are receiving considerable interest in the mobile network business [9, 10]. When using a streaming service, a server opens a connection to the client terminal and begins to send the media at approximately the bit rate at which the decoder needs the information, or playout rate. The playout rate is defined by the compression algorithms, color depth, terminal screen size and resolution and frames per second of the stream. While the media is being received, the client plays the media with a small delay or no delay at all, depending on the buffering scheme used.

The 3GPP multimedia streaming service is being standardized [11] based on control and transport IETF protocols such as Real-Time Streaming Protocol (RTSP), Real-Time Transport Protocol (RTP) [12] and Session Description Protocol (SDP).

RTSP is an application level client-server protocol that uses the TCP as a transport layer [13]. This protocol is used to control the delivery of real-time streaming data and allows to set-up one or several data streams with certain characteristics. It also allows sending control commands during the transmission (e.g. play, pause, etc.), but does not convey the data itself. The media stream is transmitted with RTP protocol, which is supported over UDP.

A presentation description defines the set of different streams controlled by RTSP. The format of the presentation description is not defined in [13], but one example is the session description format SDP, which is specified in [14]. The SDP includes information of the data encoding (e.g. mean bit rate) and port numbers used for the streams. Each stream can be identified with an RTSP URL. This URL points out to the media server that is responsible for handling that particular media stream.

The application layer signalling exchange between the streaming client and server is outlined in Figure 8.17. A detailed description of the Primary and Secondary PDP Context activation needed for the streaming service initiation can be found in Section 3.2.

A primary PDP context is activated for the RTSP signalling between the terminal and the streaming server. After creating a TCP connection to the streaming server, the terminal sends an RTSP DESCRIBE request to the server. This request indicates that the server should send the information about the media that it is going to send. This information includes the encoding of the media and the corresponding UDP port numbers.

The streaming server sends a 200 OK response containing a presentation description in the form of an SDP message. The SDP describes the streaming media the terminal is about to receive. It should be noted that the IETF RTSP specification [13] does not mandate

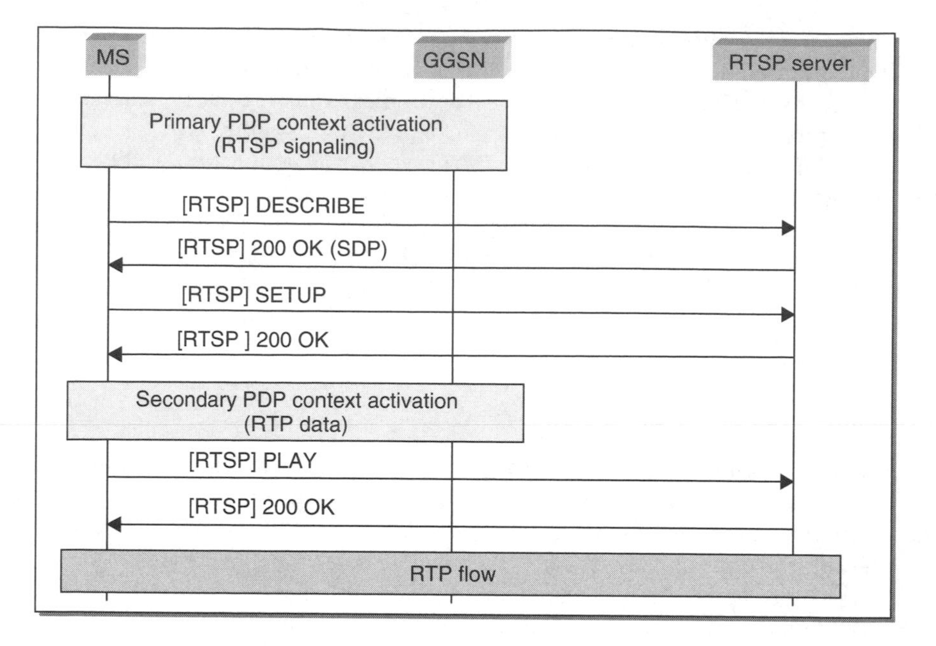

Figure 8.17. Streaming service initiation sequence chart

the use of the DESCRIBE method for this media initialization phase. However, the 3GPP standard [15], which defines the protocols and codecs for the transparent end-to-end packet switched streaming service in 3G networks, mandates the use of the DESCRIBE method for the conveyance of the media description.

After receiving the RTSP confirmation, the terminal sends a SETUP request to the server. This message indicates the transport information of the stream including the UDP port numbers the terminal is going to use for the RTP stream. The port numbers indicated in the DESCRIBE and the corresponding 200 OK messages are recommended values that can be overridden by the values given in the SETUP request and the corresponding 200 OK response. The server acknowledges the SETUP request by sending a 200 OK response back to the terminal.

A secondary PDP context for the streaming media (RTP flow) needs to be activated. During this procedure, the QoS requirements of the RTP flow are negotiated with the network. Guaranteed bit rate, including the effect of RTP/UDP/IP headers, among other QoS parameters described in Chapter 3 are considered. Optionally, its related control protocol, called Real-time Transport Control Protocol (RTCP), can be used to monitor the quality of the RTP session. Once the resources for the media are successfully reserved, the terminal sends the streaming server a PLAY request in order to start the transmission of the stream. After the ACK (200 OK response), the server can start sending the stream as an RTP flow.

The application level in the terminal does not start to deliver the user data from the beginning of the transmission. Instead, the data is stored in an application buffer, which is used to compensate delay variations in the network. This way, the initial delay experienced by the user will be increased by the buffering delay. If the buffer length goes under a certain threshold during the transmission, a re-buffering phase is triggered, in which the application stops delivering data to the end user.

A good way to indicate that a streaming user experiences a good quality is that no re-bufferings are needed during the whole session.

8.2.5 Gaming

On-line gaming represents an attractive business opportunity for operators, manufacturers and developers. In order to be able to make the most of this opportunity, it is crucial for the game developers and network operators to gain enough understanding of the characteristics and quality-of-service requirements of different game types.

There exist several different types of network games with varying requirements for network support. Real-time gaming traffic will be highly sensitive to the network latency, whereas turn-based games will not. Hence, different type of games will have distinct QoS requirements, e.g. Quake (real-time action game) versus Chess (turn-based game).

In general, three broad groups of network games can be considered [16]:

- *Action games.* Usually contain virtual persons moving (most likely shooting at each other) in real-time virtual environment (e.g. Quake II). These types of games normally have real-time requirements, meaning that a certain delay and bit rate need to be guaranteed by the network in order to provide an acceptable perceived end-user quality.
- *Real-time strategy games.* Typically involve 2D or 3D terrain that contains several units moving and performing tasks in real time (e.g. Age of Kings). Although this type of game has an interactive nature, normally higher delays can be accepted compared to action games.
- *Turn-based games.* Two or more participants (computer or end-user players) make their moves sequentially in turns (e.g. Panzer General 3D). Turn-based games have the loosest delay requirements, allowing even several seconds between any interaction between players.

Two different scenarios (game server in the network and peer-to-peer gaming) might imply different service performances due to longer RTTs in the last case, as Figure 8.18 depicts.

8.2.6 Push to Talk over Cellular (PoC)

Push to talk over cellular (PoC) is a novel service that introduces a direct one-to-one and one-to-many voice communication service in the cellular networks. The principle of communication behind the service is simple—just select a person or a group and press the button to talk. This makes the connection instantaneous, as the receiver does not even have to answer the call.

Present cellular voice calls are based on a conversational circuit switched bearer, which implies that resources are reserved for the duration of the call and released when terminating the call. This is full duplex communication, in which the delay requirements are very strict, since for acceptable end-user perception the end-to-end delay must be less than 400 ms.

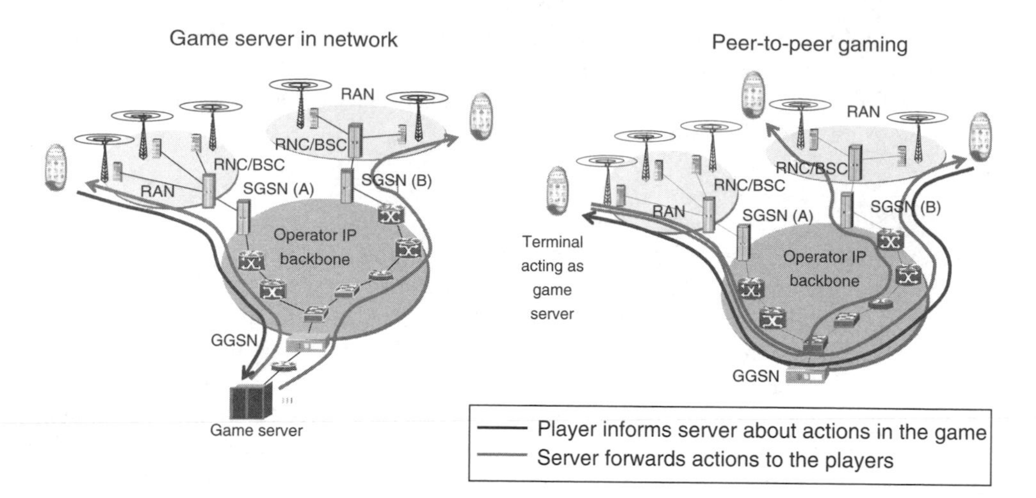

Figure 8.18. Network gaming scenarios

Push-to-talk (PTT) has many differences compared to conventional full duplex telephony. Whilst in traditional full duplex calls, resources are reserved in both directions during the call, in PTT call resources are reserved only when needed and in one direction at a time as seen in Figure 8.19. This has an analogy, the walkie-talkie, where one person is talking while other people are listening. This unidirectional nature of half-duplex communication relaxes the delay requirements of the concept, as the user has to press the button first to access the channel and start talking just when hearing indication on granted channel.

From system viewpoint, as resources have to be reserved and released based on activity, a packet data bearer such as (E)GPRS, will provide a useful platform to build the service. The key enablers for the service are quick access to PDTCHs together with the efficient use of radio and system resources. In fact, in a typical push-to-talk usage profile, an average subscriber uses few speech turns during the session, each speech turn lasting only a few seconds.

Figure 8.20 describes the differences between Circuit-switched (CS) and push-to-talk (PTT) calls. In the CS Call one timeslot is reserved in UL and DL directions until the end of the call. If half rate codec is used one half of the slots in UL and DL are used. In the

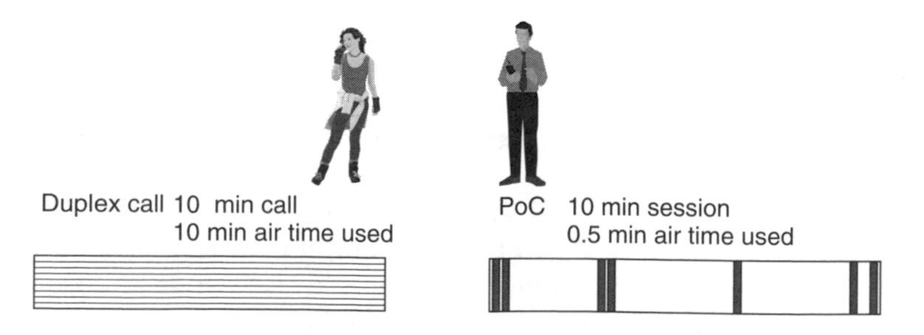

Figure 8.19. Call profiles of circuit-switched call and PoC

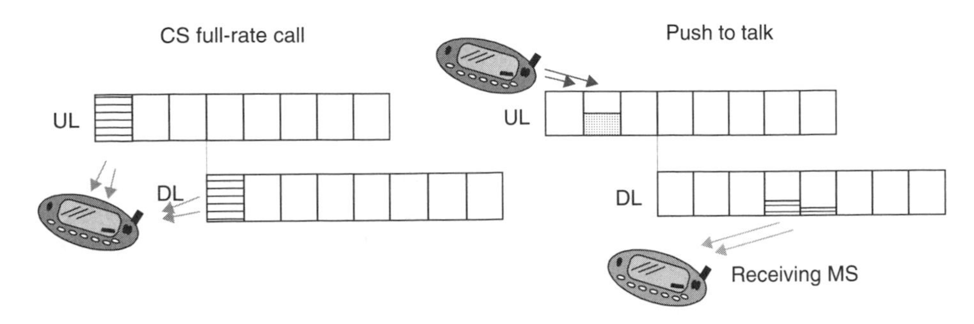

Figure 8.20. Resource allocation of circuit-switched call and PoC

case of push to talk, only one direction is reserved at a time. When LA is in use, the TSL can be shared by several users. If multiple timeslots can be used, additional multiplexing gain can be achieved from the system's point of view.

From the architectural point of view, for push to talk there are two end points, the push-to-talk server and the terminal. Signalling between the terminal and the server is performed with SIP [17] while the speech data is transferred on top of RTP/UDP/IP.

The push-to-talk server terminates the session with the terminals, i.e. that all terminals will have sessions with the server and that the push-to-talk call will consist of two sessions, one session for the originating user and the other for the receiving user. The servers' responsibility is to check the rights related to calls and finally establish a connection to the receiving end. While in the call, the server will route the speech to the receiving end.

Figure 8.21 describes how the PoC communication goes through the GPRS system. When the push-to-talk (PTT) key is pressed the terminal sends IP packets (first signalling, and later voice frames) via GPRS system to the PoC server. The server makes the needed access, checks the packets and forwards them to the recipient. In the case of group communication, the server additionally multiplies and distributes the IP packets for each recipient. In the later 3GPP versions, the multicast feature may be used to reduce the

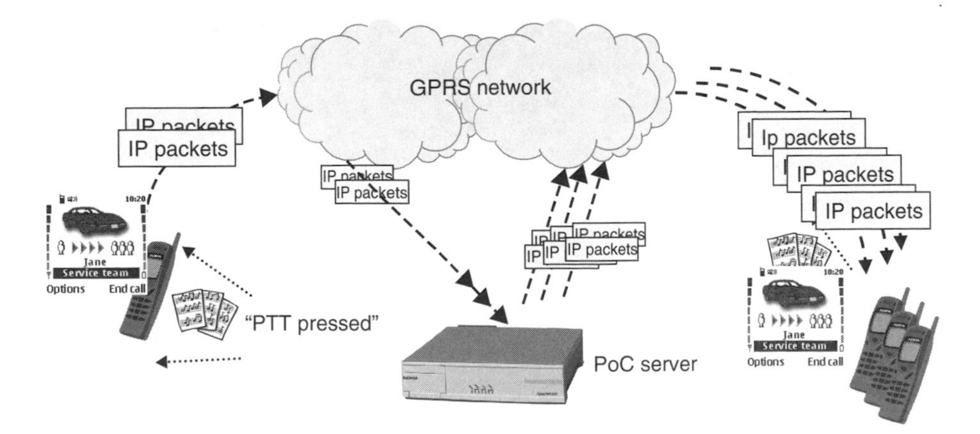

Figure 8.21. The PoC communication between two terminals via GPRS network and PoC server

amount of forwarded IP packets. In the next sections, some cases where a PoC system is used are presented.

8.2.6.1 Personal PoC Communication

Personal communication enables a user to have a voice communication with another user, where the users speak one at a time e.g. by holding a talk button or equivalent when speaking. The voice is heard at the called party end without any user actions.

8.2.6.2 Group PoC Communication

Group communication enables a user to have a voice communication with a multiple number of other users, where the users speak one at a time, for example, by holding a talk button or equivalent when speaking. The voice is heard at the called party end without any user actions.

8.2.6.3 Personal Communication Call Set up

The call set up in personal communication is the phase when the calling party is informing the PoC Server about the called party, the call rights are checked and the communication parameters are negotiated. From the called party point of view, the call set up is the phase, when the PoC Server is informing the called party about the incoming personal communication and the communication parameters are negotiated. The call is established automatically without any called party user actions. Personal communication phases are described in Figure 8.22.

8.2.6.4 Joining a Group

Joining or activating a group is a specific action from the point of the user. The user can do this, for example, by selecting the URL of the group in the user interface of the terminal. During the joining procedure the communication rights are checked and the

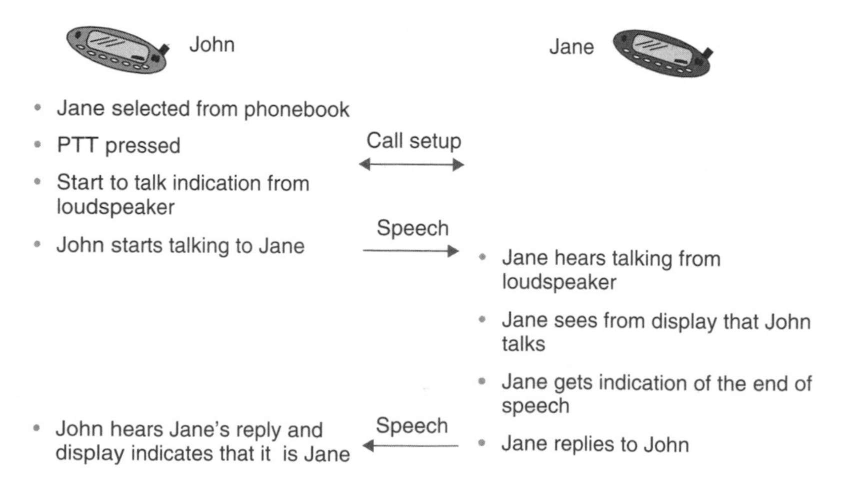

Figure 8.22. Personal communication phases

communication parameters are negotiated between the user and the PoC server. Typically, joining to the group is made in good time before talking.

8.2.6.5 Talk Burst Start and End

After the call set up (in Personal Communication) or joining (in Group Communication) phase, the user will initiate voice transmission (talk burst) by pressing the PTT key (PTT) or using some equivalent control.

Only one user is allowed to speak in a group at a time. The system has the means to arbitrate between concurrent attempts to transmit to the same group. In case of two or more simultaneous transmissions in a group, only one should be forwarded to the recipients. Users shall get an indication if their speech is not being delivered to the group.

The talk burst starts and the talking party can be indicated to the listening user. Also the end of talk burst can be indicated to the listening user.

8.2.6.6 PoC Service Signalling

PoC service uses two parallel and always-on PDP contexts. An interactive class PDP context is used for control signalling purposes and a streaming class PDP context for media purposes. For optimizing the call set-up time both PDP contexts may be always on.

When switched to the PoC service the terminal establishes the primary PDP context with interactive traffic class, makes a login procedure to the PoC server and in case group communication is used, it also declares its group identity to the PoC server. When registered to the PoC service the mobile establishes the secondary PDP context with the streaming traffic class for the media transfer purposes. The PDP context establishment takes time and causes additional delay in call set up. The two parallel PDP contexts are described in Figure 8.23.

8.2.6.7 Call Set-up Times in PoC

Personal communications need a call set up and talk burst start. However, the group communication does not have a similar call set up because the subscriber joins the group call after making the login to the PoC service. Group and personal communication talk burst starts are quite similar, but in the case of group communication there are several receiving parties. In the case of call set up, more information may be transferred between mobile stations and PoC servers. Also the PoC server has to go through more extensive access

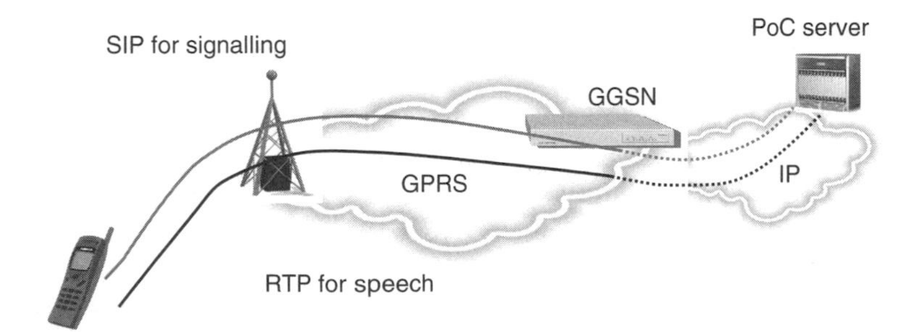

Figure 8.23. Signalling and speech PDP contexts

procedures for the call set up than for the talk burst control. Typically, the called party is in the idle mode most of the time to save the battery, which requires the called party to be paged. However, the next talk burst start messages are normally sent immediately after the previous talk burst, and paging would not be needed.

In the next section delay elements are described. The values of such delays may vary depending on the selected scenarios.

8.2.6.8 One-to-one Call Set-up

The call set-up time in the PoC service includes two important elements:

- The earliest time to start to talk
- The earliest time the voice is through to the called party.

When the Push-to-talk Key (PTT) is pressed, the terminal will establish a data link through the radio system to the PoC Server, i.e. the uplink temporary block flow (UL TBF) between MS and base station controller (BSC) and the packet flow context (PFC) between BSC and serving GPRS support node (SGSN) are established. The earliest time to indicate to the user that talking is allowed to be started without having a voice buffer in the transmitting terminal is the moment when the uplink connection is established to the GPRS system. This moment may be indicated to the user e.g. by an audio tone. The user starts talking after human reaction time. Parallel to human reaction time, voice coding and audio frame packetizing, the call set-up signalling can be transferred over the radio link to the PoC Server as described in Figure 8.24. If talking starts before receiving an ACK from the called party, it sometimes happens that the called party is not available anymore and the system has to indicate this to the user by sending a Stop-to-Talk message.

The earliest time to get voice through to the called party includes the delay introduced by all the procedures described in Figure 8.24, including extra delays caused by retransmissions, in case that the RLC ACK mode is used. The jitter buffer shall be long enough to cover most of the packet delays caused by the IP network, load on the packet scheduler and lower layer packet losses and repetitions.

8.2.6.9 Talk Burst Start

The talk burst start in PoC also includes the same two important elements as described earlier in the call set up:

- The earliest time to start to talk
- The earliest time the voice gets through to the called party

The Talk Burst Start message flow is quite similar to the call set up, although the messages may include less information, because most of it is already transferred during the call set up phase. The only difference is that typically the receiving party is in the ready state and there is no need to be paged before sending data packets. Also the processing time in the PoC server may be essentially shorter.

In the case of the first talk burst in the group communication, the called parties may be in the idle mode and therefore they have to be paged before sending any signalling or voice packets.

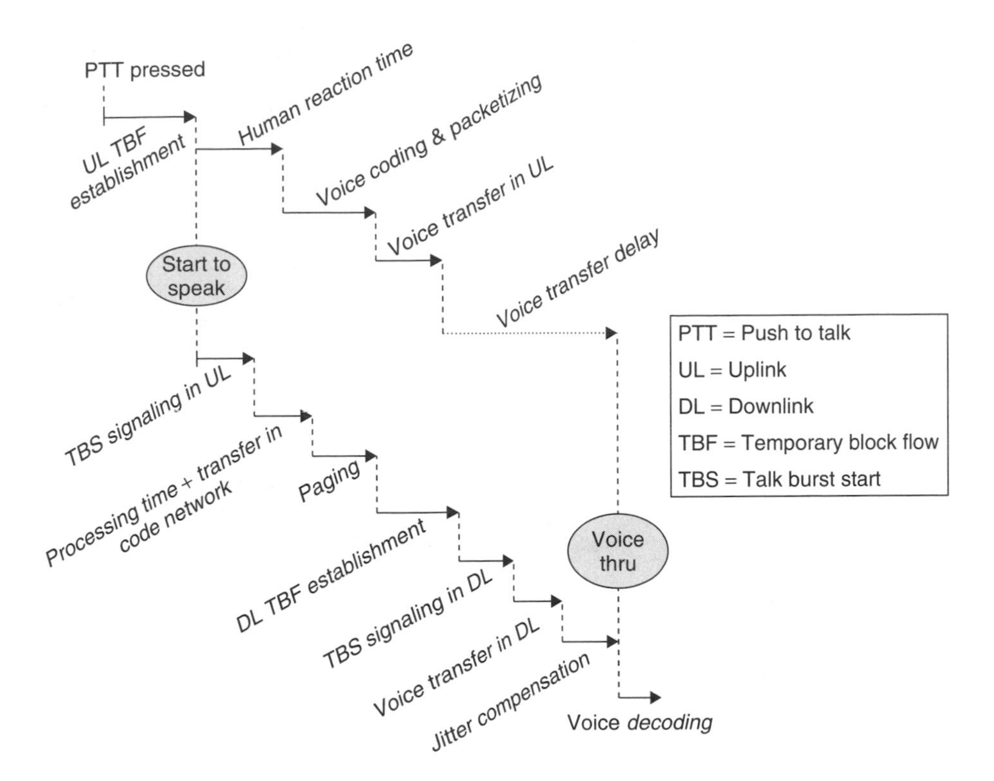

Figure 8.24. Call set up diagrams

8.3 End-user Performance Analysis

In this section, a performance analysis of the packet data services described in the previous section is provided. This analysis illustrates the main contributions to the overall service performance degradation, and provides a performance comparison between GPRS and EGPRS technologies.

The content of the services is a key aspect in the service performance. Because of the evolution of the networks, enabling higher capacities and lower delays, the support of more advanced services with higher content sizes becomes feasible. The content size evolution along the years is something that can be observed in current Internet, where web pages are becoming more sophisticated, with more complex and an increasing number of objects. Figure 8.25 provides an estimation of the content size evolution up to 2008 for MMS, Web/WAP pages and video files [8].

In this section, the generated performance figures assume content sizes approximately corresponding to the year 2002 as well as terminals capable of 4 TSLs in downlink and 2 TSLs in uplink (multi-slot class 10 assumed where $4 + 1$ and $3 + 2$ are dynamically supported according to connection requirements). Considering typical loaded network conditions (up to 150 kbps/MHz/sector), this configuration achieves an effective downlink data link throughput (as described in Section 8.1) of 44 kbps for GPRS (assuming only CS-1–2) and 155 kbps for EGPRS, as Figure 8.26 shows (Chapter 7 provides a more detailed analysis of data link throughputs achieved at different loads for (E)GPRS).

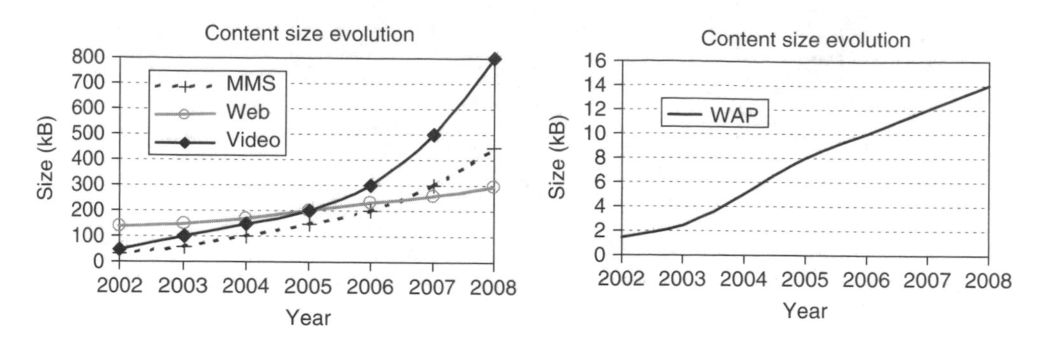

Figure 8.25. Content size evolution

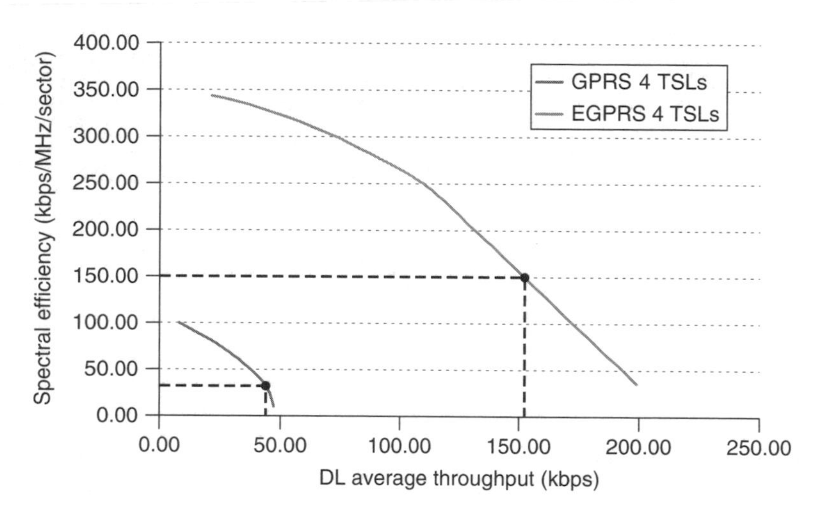

Figure 8.26. Data link throughput assumptions for GPRS and EGPRS

8.3.1 Web Browsing Performance

In a web browsing session, the end-user performance is characterized by the click-to-consume delay, which is critical to the end-user perception due to the high interactive nature of the service.

The following assumptions have been considered in the web performance analysis:

- HTTP1.1 (over TCP/IP) is used, assuming *pipelining* and *persistent connection* features.
- 100 kB web page (10 kB correspond to the text and 90 kB to the objects).
- No compression in network elements considered.

Delay results are shown in Figure 8.27, where *first page* results corresponds to the delay needed when downloading the first page within a web session, including the PDP Context activation delay, DNS query, TCP connection establishment and all the HTTP signalling transactions (as explained in Section 8.2). However, secondary and successive pages can

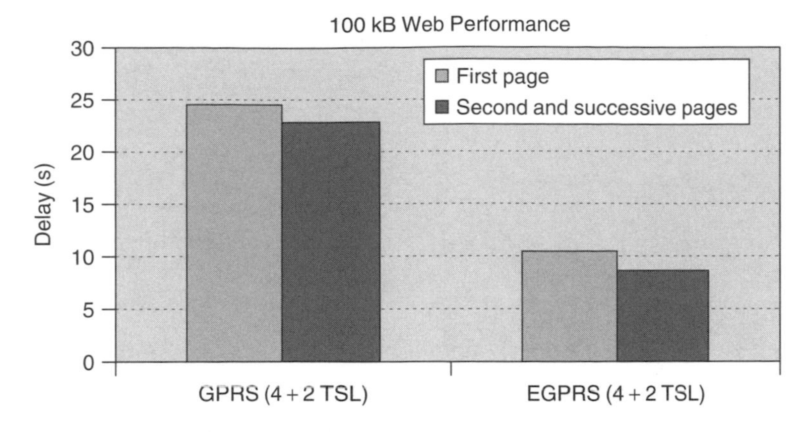

Figure 8.27. Web browsing delay. MS class 10

be downloaded a bit faster since PDP Context is already established for the web traffic so PDP Context activation delay is not considered. Note that in case an 'always-on' PDP Context is used for web browsing, the delays needed when downloading the first page would be similar to the second and successive pages, since PDP Context activation delay would not be considered.

From the delay breakdown shown in Figure 8.28, it can be observed that the main contribution to the overall delay comes from the retrieval of the objects & text, corresponding in this example to 71.9 and 15.7% for GPRS and 53.3 and 21.1% for EGPRS, respectively. Higher bit rates in EGPRS considerably decreases the downloading time of the objects (from 17.6 to 6.5 s approximately). PDP Context activation delay also causes an important contribution to the download delay of the first web page.

Figure 8.29 shows the delay performance and mean throughput at application layer for different web page sizes. Low application throughput for small file sizes is because of the higher contribution of the fixed components of the delay breakdown over the whole download time. For big file sizes, the throughput tends to converge to the maximum achievable throughput after considering the protocol overhead effect on the data link throughput.

In the delay figure, a slight non-linear behavior can be observed for small page sizes due to the TCP slow start effect. This effect is more significant in the EGPRS case since the time needed to reach the transmission window size that fulfils the BDP condition

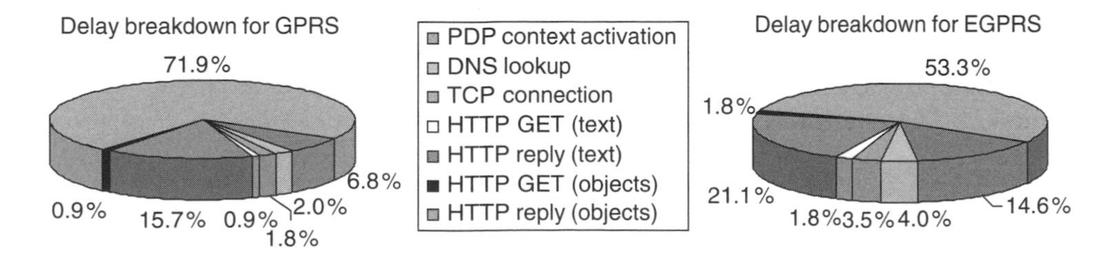

Figure 8.28. Web browsing delay breakdown

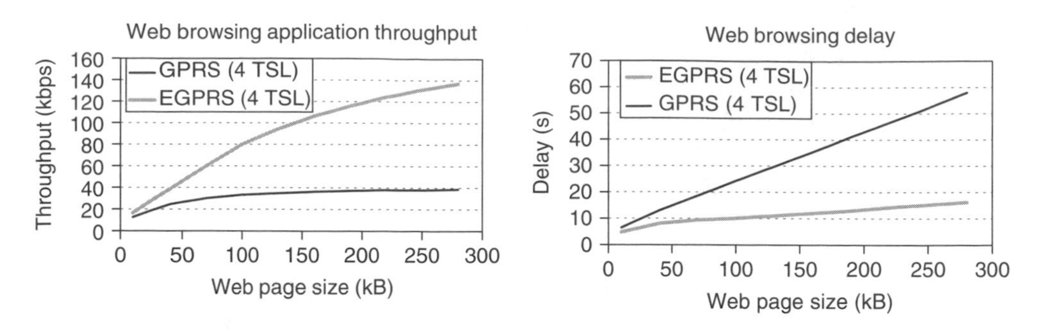

Figure 8.29. Web browsing performance versus page size. MS class 10

is higher. In this case the theoretical BDP for GPRS is in the order of 6 kB while for EGPRS it is 17 kB.

Web browsing end-user performance is defined by the comprehensive measure of the performance a user experiences when interacting with a web service and the perception this user carries after the transaction. In general terms, web-browsing services require a certain level of fluent interaction, e.g. a user could consider an acceptable maximum delay of around 15 s before receiving any response in his terminal, and could decide to discontinue the session if delays were over 30 s. EGPRS will be a significantly better solution for web browsing since it substantially improves the end-user experienced delays.

8.3.2 WAP Browsing Performance

The key indicator to define the end-user experience in a WAP browsing scenario is the same as in the web browsing case, that is, the delay from click to consume.

The following assumptions have been considered in the WAP performance analysis:

- WAP 1.x protocol stack (over UDP) is used.
- 1.4 kB WAP page (400 bytes correspond to the text and 1 kB to an object).

Delay results for GPRS and EGPRS are shown in Figure 8.30, including the PDP Context activation delay, WAP session set-up and all the WTP signalling transactions needed to download the WAP page (as explained in Section 8.2.

Figure 8.31 shows the delay breakdown for GPRS and EGPRS technologies. Since WAP pages in general (1.4 kB in this example) are relatively small, compared to typical web pages, the transmission of the text and objects through the air is not an issue. Actually, the higher contribution in this example is because of the PDP Context activation delay (higher than the 50% for both radio technologies). As explained for other services, it would be possible to use an 'always-on' PDP Context in order to improve the end-user experience. This improvement would be only visible for the first page, since second and consecutive pages can reuse the already activated PDP Context.

The delay performance for different WAP page sizes is depicted in Figure 8.32. The evolution is more linear than the web performance case, since UDP is used in the GPRS network. The TCP connection used between the WAP gateway and the server

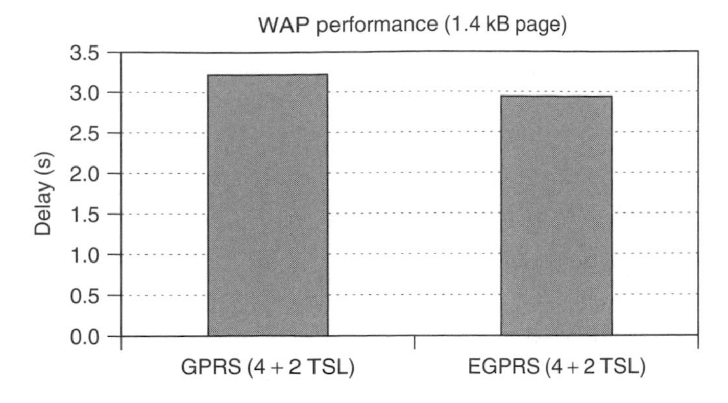

Figure 8.30. Web browsing delay. MS class 10

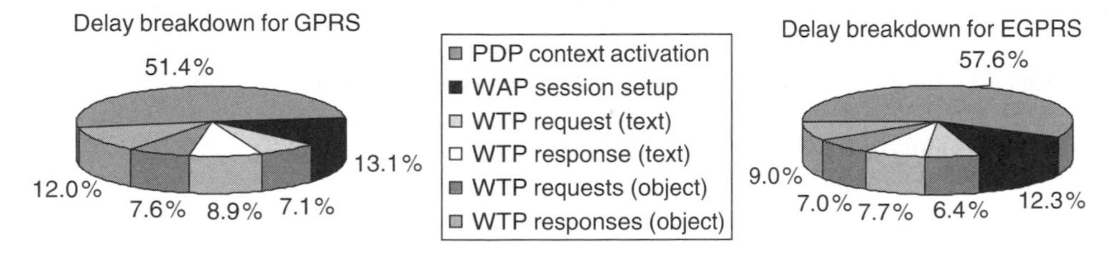

Figure 8.31. WAP browsing delay breakdown

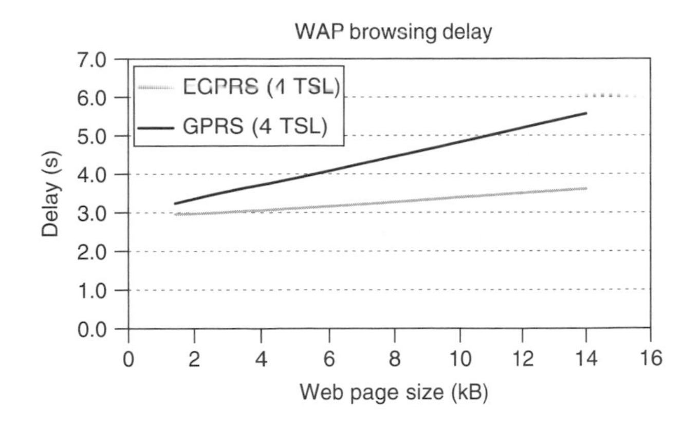

Figure 8.32. WAP browsing performance versus page size. MS class 10

does not affect the performance because of the higher throughputs and lower RTTs in fixed networks.

Although the delay analysis made for 1.4 kB does not show a big improvement with EGPRS (around 9%), for larger WAP page sizes this improvement is quite relevant (e.g. a 14 kB WAP page can be downloaded 35% faster with EGPRS as compared with GPRS).

8.3.3 Multimedia Messaging Service Performance

MMS end-user performance is mainly defined by the terminal-to-terminal delay and delivery assurance. Although generally the MMS delay requirements for the end user may not be very critical, it might depend on the interactivity and urgency of the scenario considered.

The following assumptions have been considered in the MMS performance analysis:

- MMS over WAP1.x/UDP/IP
- MM size of 30 kB
- 2 SMSs (over GSM) are considered.

Two cases have been analysed: the end-to-end MMS delay with and without 'always-on' PDP Context (see Figure 8.33).

An analysis of the main delay contributions for GPRS and EGPRS is displayed in Figure 8.34. A major impact on the performance is caused by the notification indication to the remote user from the MMS server. Around 17 s are required for the transmission of 2 SMSs over GSM although the use of SMS over GPRS would improve the performance around 6 to 8 s. Quite relevant is also the transmission and reception of the MM to/from the server, mainly limited by the UL radio bandwidth.

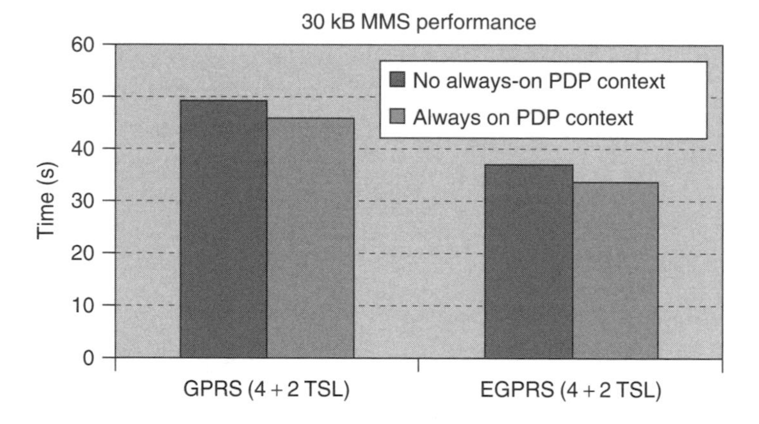

Figure 8.33. MMS delay. MS class 10

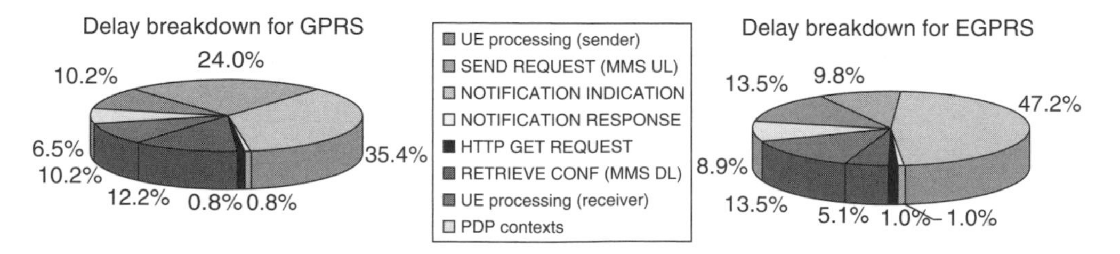

Figure 8.34. MMS delay breakdown

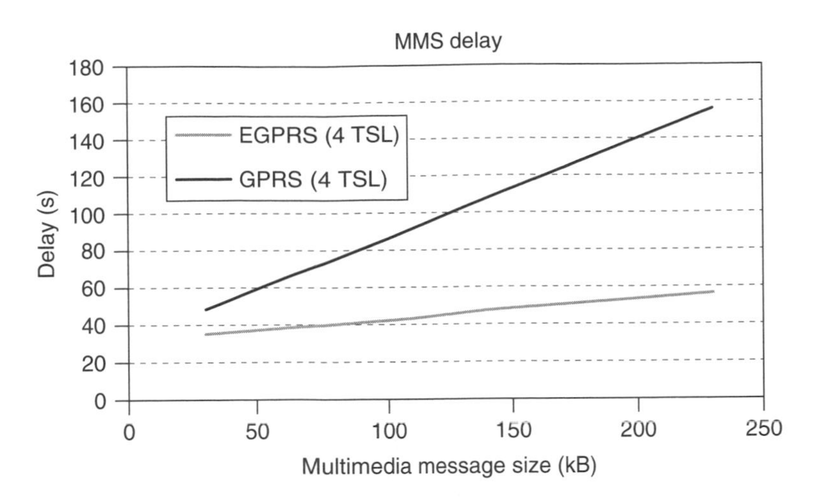

Figure 8.35. MMS performance versus multimedia message size

Delay results for different MM sizes are shown in Figure 8.35. In this case, we can observe a linear performance with the size of the message, since UDP is used instead of TCP and therefore, the slow start mechanism does not affect the performance.

8.3.4 Streaming Performance

As mentioned in Section 8.2, the end-user performance for streaming services is mainly defined by the set-up delay (click-to-consume delay), audio/video quality and number of re-bufferings.

The time from 'click' to consume is mainly affected by the buffering delay in the terminal, which depends on terminal memory, expected throughput and potential network events that cause an interruption of the data flow. For instance, a 40 kbps video stream that should cope with a 5 s outage will require 25 kB terminal buffer memory, computed as:

$$\text{BufferMemory (kB)} = \text{Bit rate (kbps)} \times \text{Time (s)}/8$$

In the performance figures provided in this section, 13 s for GPRS and 8 s for EGPRS are assumed as recommended application buffer values.

Figure 8.36 provides a comparison between a streaming session set-up delay for GPRS and EGPRS. The delay values include the Primary PDP Context activation delay (used for RTSP messages), the RTSP negotiation delay, the Secondary PDP Context activation delay (used for RTP streaming data) and finally, the initial buffering delay in the application. As occurs with previous services, if RTSP signalling is using 'always-on' PDP Context (implementation-dependant), the Primary PDP Context activation delay would not be considered.

In this example, the difference in the performance between GPRS and EGPRS is mainly due to the initial buffering delay. The RTSP negotiation delay is not much lower in EGPRS due to the small packet sizes of RTSP messages (from 100 to 600 bytes) where the radio link throughput is not so relevant.

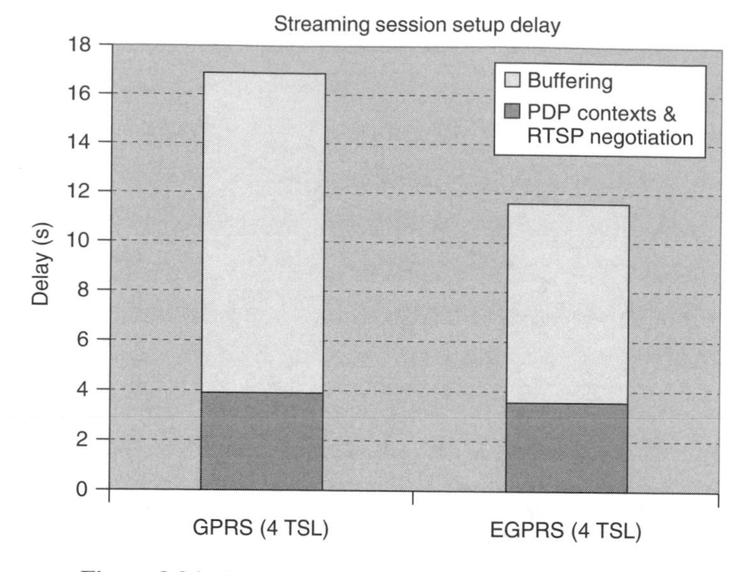

Figure 8.36. Streaming session set-up delay. MS class 10

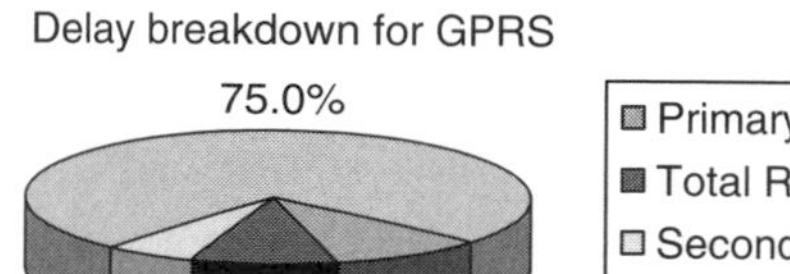

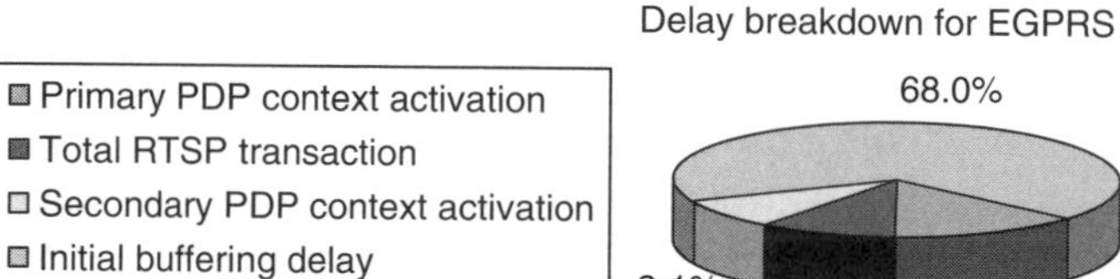

Figure 8.37. Streaming session set-up delay

	Audio + Video: QCIF	Audio + Video: TV resolution
Picture size	176 × 144	720 × 576, 720 × 480
Framerate	8–15 fps	Upto 30 fps
Audio	Mono narrowband	Stereo
Network bitrate	28–64 kbps	Upto 1 Mbps

	Audio: CD Quality
Sample freq	44.1 kHz/channel
Quantisation	16 bits/sample
Audio	Stereo
Uncompressed bitrate	1.4 Mbps
Network bitrate	128 kbps

Figure 8.38. Examples of streaming requirements

More detailed information of the delay contributions to the overall streaming session set-up is given in Figure 8.37.

Throughput requirements of the streaming service depends on many things, such as the terminal screen size and resolution, compression algorithms, streaming content, color depth and frames per second. Some examples of streaming requirements are presented in Figure 8.38, for audio/video QCIF, audio/video TV resolution and audio CD quality.

During the streaming service session, the network should be able to fulfil the streaming requirements as much as possible so that no re-bufferings occur. However, a combination of many factors can lead to a re-buffering, such as cell reselections, bad radio link conditions, or high load (which causes high interference and shared timeslot capacity to the streaming connection).

8.3.5 On-line Gaming Performance

Section 8.2 provided a general description and characterization of networks games. It is evident that network delays can have a major impact on the players perceived performance of the game. The amount of delay that is good or just acceptable varies depending on the type of game [16].

This section provides an overview about the on-line gaming performance for three different game types: action games, real-time strategy games and turn-based strategy games. An estimation of the RTTs for 32 bytes (in the two scenarios shown in the Figure 8.18) is compared with different thresholds that define the maximum delays to experience a good or acceptable quality from the end-user point of view (see Figure 8.39).

As explained in Section 8.1, RTTs may be different depending on the TBF status. The values shown in Figure 8.39 consider both Delayed DL TBF release (Rel'99) and Extended Uplink TBF (Rel'4) features. However, the traffic characteristics for different gaming categories may trigger the release of the downlink TBF, particularly for those games with high mean inter-arrival times between packets such as turn-based games. In that case, an additional delay corresponding to the downlink TBF establishment procedure must be considered.

As a conclusion of the analysis, it can be observed that in Rel'99 the peer-to-peer gaming scenario would only support turn-based games, while in the scenario using a game server in the network, also real-time strategy games and even action games could achieve an acceptable quality.

When considering Rel'4 (Extended Uplink TBF), the peer to peer gaming would support the real-time strategy games delay requirements and in the network game server scenario, action games end-user experience would be significantly improved.

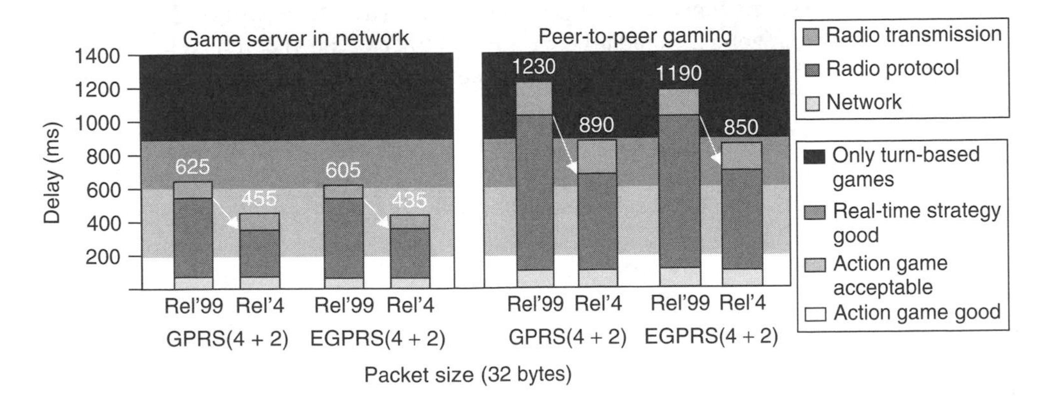

Figure 8.39. Round trip times for different gaming scenarios

8.3.6 Push to Talk over Cellular Performance

In order to monitor the end-user perception when PoC is in use, the most important indicators, in addition to the voice quality, are the delays experienced during set-up phase and ongoing communications. Two main parameters are studied: Start to Speak time and Turn Around Time (TAT).

- The Start to Speak (StS) Time is measured from the PTT press to the indication (e.g. beep) of the speak permission. It is the time the user has to wait before starting to speak.
- The Turn Around Time (TAT) is measured from the PTT release to the moment when the answer arrives. It is the time the user has to await the answer for the question, statement, etc.

It is obvious that users experience better quality when delays are shorter. It is estimated that the start-to-speak delay is acceptable up to 3 s. To sustain the synchronization of conversation the TAT should be less than 4 s.

8.4 Methods to Optimize End-user Performance

Section 8.1 described the performance degradation introduced by all protocol layers and network elements from the server to the terminal. It also explained the performance issues involved when using Internet protocols over mobile wireless networks. This section will present the different techniques that can be used to improve the efficiency of data transactions in the cellular environment. There exist several standardized optimization techniques proposed by the IETF and many of them are already in use in current networks. Examples of such used methods are Header compression, signalling compression, content compression, TCP parameter optimization for wireless, TCP for short transactions (T/TCP). There is also a set of non-standard methods that can be used, but their deployment needs to consider other issues such as security.

8.4.1.1 TCP Optimizations

There are several ways to enhance TCP performance in wireless networks. These can be grouped into three categories: Standard TCP improvements and recommendations, TCP for short transactions, buffer congestion management features and other non-standard improvement features that may be implemented in some parts of the mobile network.

Standard TCP improvements and recommendations. RFC 3481 [18] recommends considering certain TCP parameters and options when using TCP over 2.5 and 3G networks.

- Appropriate receiver window size
- Increased initial window (RFC3390, [5])
- Limited transmit (RFC3042, [19])
- IP MTU larger than default
- Selective acknowledgement (RFC2018, [20])
- Explicit congestion notification (RFC3168, [21] and RFC2884, [22])

- Timestamp option (RFC1323, [23])
- Acknowledge every segment.

An example of the parameter settings effect on end-user performance is given in Figure 8.40 that displays the TCP receiver's advertised window values versus application throughput in EGPRS environment. These results have been obtained by running a real TCP application over a dynamic EGPRS simulator (described in Appendix E).

As displayed above an optimal throughput cannot be achieved if the receiver advertised window is too small (as explained in Section 8.2). On the other hand, too long receiver windows when sending large files may fill the buffers in the bottleneck routers, causing not only longer delays but also risks of congestion and packet discarding. The recommended receiver advertised window value for EGPRS is between 25 kB to 30 kB, whereas for GPRS the value should be around 10 kB depending on the number of TSLs terminal capability. Correct TCP parameters setting for (E)GPRS as well as for other radio technologies (e.g. WCDMA) is very important in order to achieve the desired network and end-user performance.

- *TCP for transactions*. Section 8.1 presented the TCP establishment procedure that uses a three-way handshake to complete connection establishment, and uses acknowledged FIN signals for each side to close the connection (as illustrated in Figure 8.41). This TCP reliability aspect impacts application efficiency, because the minimum time to complete the TCP transaction for the client is $2 \times$ RTT. TCP for Transactions (T/TCP) [24] was designed to improve the performance of small transactions. T/TCP places the query data and the closing FIN in the initial SYN packet. This can be seen as trying to open a session, pass data and close the sender's side of the session within a single packet. If the server accepts this format, it replies with a single packet, which contains a SYN response, an ACK of the query data, the server's data in response and the closing FIN. All that is needed in order to complete the transaction is for the query system to ACK the server's data and FIN. For the client, the time to complete this T/TCP transaction is one RTT. T/TCP requires changes in both the sender and the receiver in order to operate. The protocol allows the initiating side to back off to use TCP if the receiver cannot respond to the initial T/TCP packet. T/TCP is not very much deployed in the Internet because the limited handshake procedure makes

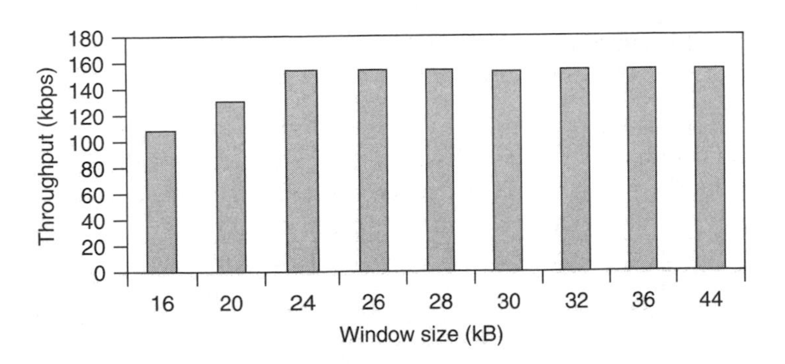

Figure 8.40. TCP receiver window size effect on throughput

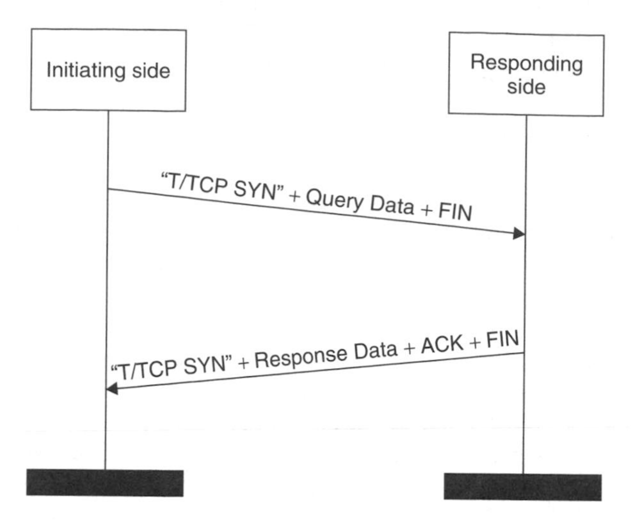

Figure 8.41. T/TCP operation

it more vulnerable from a security perspective. This is illustrative of the nature of the compromises that occur within protocol design, where optimizing one characteristic of a protocol may be at the expense of another benefit of the protocol.

Buffer congestion control mechanism. As explained before, buffer congestion is a risk, especially in the downlink buffers of the bottleneck mobile network elements (e.g. 2G-SGSN in (E)GPRS). Therefore, appropriate congestion control methods should be implemented in such network elements. There exist several methods to control congestion:

- The most well-known standard buffer management method is probably the Random Early Detection (RED). The principle in RED is to increase the packet-dropping probability of incoming packets if the buffer occupancy grows. Ultimately, if the buffer length exceeds a pre-defined threshold, every incoming packet will be dropped. The advantage of starting dropping packets before the buffer is completely full is to force some of the TCP sources to reduce their sending rate before congestion actually happens. Another advantage is to avoid global synchronization of Internet routers. Global synchronization happens when the router buffers load oscillate from almost empty to congested, and is caused mostly by TCP dynamic congestion control mechanism of different connections acting at the same time.
- *Fast-TCP.* The idea here is to start delaying slightly the TCP acknowledgments in the UL buffer of a router if the downlink buffer exceeds a defined threshold. As a result, the time at which the TCP sources will send their next window of data will be also delayed which gives more time to router buffers to recover from congestion. Furthermore, delaying slightly the TCP acknowledgment will slow down the growth of the congestion window in the TCP sender, which is also beneficial in case of congestions. The delay to apply to uplink TCP ACKs should be carefully computed since too large delays would cause TCP timeout in the TCP senders, which is not the objective.

- *Window pacing*. This scheme proposes to reduce the advertised window of some TCP ACKs in uplink if the downlink buffer exceeds some threshold. As for RED, the intention is to force the TCP senders to decrease their sending rate. The advantage as compared to RED is that packets are actually not dropped in the downlink direction, preventing throughput degradation for the end user. The drawback is of course that modifying the TCP header is a heavier task than just dropping incoming IP packets. Also if IP security is used, this feature cannot be applied since the TCP headers cannot be read anymore.

Some network elements may implement more than one congestion control method and thus define several sets of thresholds.

Optimization features to apply in the mobile network. A collection of other schemes can be implemented in the mobile network to improve TCP performance over wireless links. These methods, which are typically non-standard related, should be used with care and their implications towards end-to-end functionalities such as security should be well understood by network operators. RFC 3135 [25] presents a general IETF view on such network features underlining the most important issues to be considered when using such schemes. Some of the most relevant features are:

- *Redundancy elimination*. This functionality monitors the latest TCP acknowledged sequence number in a router, which enables that unnecessary retransmitted TCP segments are discarded before the radio link.
- *Split TCP approach*. One TCP connection is established from the server to a gateway and another one is established from a gateway to the mobile terminal where e.g. TCP parameter optimization can be applied. This approach although powerful in some cases, is in general not recommended by the IETF since it breaks the end-to-end TCP semantics.
- *Deletion of duplicate acknowledgement*. This technique suppresses the duplicate ACKs on their way from the TCP receiver back to the TCP sender, while detecting that the lost packet can be retransmitted to the TCP receiver with link layer mechanism, thus avoiding a fast retransmit at the TCP sender and preventing the congestion window from shrinking.
- *RTT adjustment*. This feature delays some acknowledgments in the wireless router depending on the size of the corresponding data segment sent over the air interface. As explained in Section 8.2, the RTT in wireless networks may vary greatly with the size of the packets. Delaying the ACK allows to have control over the RTT variation. This avoids spurious retransmissions at the TCP sender due to timer expiration.

This list is non-exhaustive and a few other TCP optimization methods could be designed to improve overall end-user performance.

8.4.1.2 Header Compression

The motivation to reduce the size of IP headers (or simply remove them) in wireless networks is two fold:

- Save capacity (radio and transport)
- Reduce the Round Trip Time (RTT).

Header compression is a key feature for real-time services in wireless networks where radio capacity is scarce and RTT can be fairly large. Header compression benefits are maximized for small packets because the related capacity saving rates are high. In (E)GPRS the services that would take advantage of header compression include Push to Talk (PoC) and video traffic in case the video stream needs to be transferred within a short time (e.g. See What I See type of service or video conference).

A downside of header compression is that it requires processing power and memory in the network elements and therefore the decision to apply this scheme should be made depending on the traffic characteristics and delay requirements. Some standard header compression methods are provided by the IETF [26–28].

8.4.1.3 Application Layer Optimizations

Application layer optimization techniques can be very powerful and in some cases almost mandatory in order to allow acceptable end-user experience. Mobile office type of application transactions can be, for instance, greatly optimized for wireless allowing much better performances. These techniques can be sorted in the following categories:

- Content compression by applying standard or non-standard data compression schemes.
- Content and image optimization:
  - *Filtering.* Content that increases download times can be removed from Web pages.
  - *Downgrading.* Image resolution is reduced before it is sent across the wireless link, which decreases download times.
- *Server caching.* When a client has downloaded a particular document, and it is still in the cache, upon reload the server is asked to check the newest document available from the origin server against the version cached by the client. For many documents the differences between versions are small and it is enough to transmit only these small differences to reconstruct the newer version.
- *Business application support.*
  - Less signalling, for example, email application
  - Support for other applications including folder sharing.

8.4.1.4 Performance Enhancing Proxies

A performance enhancing proxy (PEP) is an entity in the network acting on behalf of an end system or a user. The objective of a PEP is to improve the application performance by adapting protocols/information to a bearer. A PEP may apply part or all of the optimization methods presented so far in this section.

A PEP can be located behind the gateway GPRS support node (GGSN) (and before the internet backbone) or inside Intranet corporate premises depending on whether it

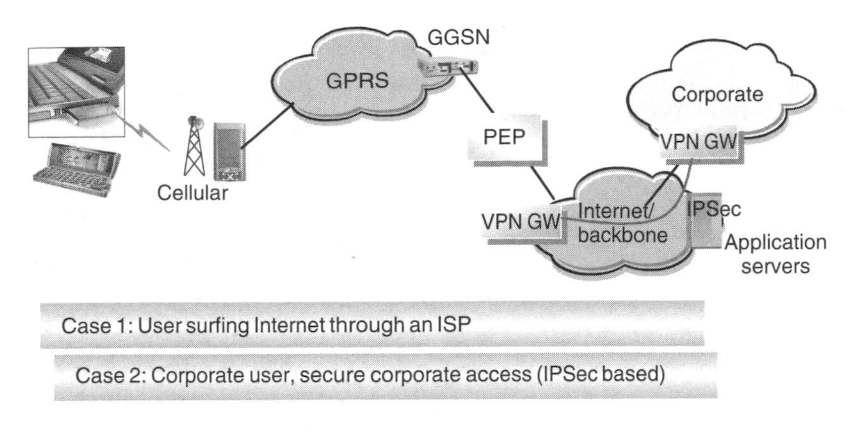

Figure 8.42. PEP near GGSN

is controlled by the mobile network operator or by the enterprise network. Figure 8.42 presents a scenario where the PEP is inside the mobile network operator's domain.

In this case, the PEP is used by mobile phone users for e.g. for surfing the Internet or by corporate users. For the latter, the assumption is that the mobile network security level is considered high enough.

Some enterprises may want to ensure end-to-end security for their transactions. In this case the PEP can be hosted in their network as illustrated in Figure 8.43. IP security is then used from the server in the corporate network to mobile terminal.

PEP clients are added to the mobile terminals (e.g. laptops) allowing increased optimization capabilities. For example, wireless profiled TCP or content compression may require the application of a PEP client. Finally, it should be noted that optimization does not always require a PEP as a separate physical entity, but can be implemented in standard mobile network elements as well. Actually, some of these methods are more efficient near the radio interface (e.g. congestion methods, redundancy elimination) and should preferably be placed there. Furthermore, reducing the number of network elements also reduces the RTT and network management complexity.

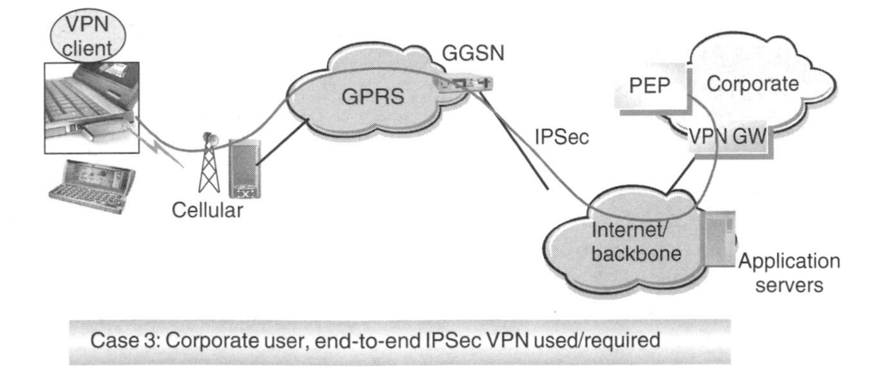

Figure 8.43. PEP in a corporate network

References

[1] 3GPP TS 44.060, Radio Link Control/Medium Access Control (RLC/MAC) Protocol, Release 4, v5.0.0, February 2002.
[2] Karn P., Advice for Internet Subnetwork Designers, Internet Draft, February 2003.
[3] RFC 793, Transmission Control Protocol DARPA Internet Program Protocol Specification, September 1981.
[4] RFC 2581, TCP Congestion Control, April 1999.
[5] RFC 3390, Increasing TCP's Initial Window, October 2002.
[6] RFC 3150, End-to-end Performance Implications of Slow Links, July 2001.
[7] RFC 3155, End-to-end Performance Implications of Links with Errors, July 2001.
[8] Dr. Yrjö Neuvo (Executive Vice president, CTO, NMP), Mobile Phone Technology Development, Nokia Year End Strategy Update, December 2002, http://www.corporate-ir.net/ireye/ir_site.zhtml?ticker =NOK&script =1010&item_id =619407.
[9] Montes H., Gomez G., Paris J. F., Cuny R., 'Deployment of IP Multimedia Streaming Services in Third-Generation Mobile Networks', *IEEE Wireless Commun.*, **9**(5), 2002.
[10] Montes H., Gomez G., Fernandez D., 'An End-to-End QoS Framework for Multimedia Streaming Services in 3G Networks', *Personal, Indoor and Mobile Radio Communications, 2002*, Vol. 4, 2002.
[11] 3GPP TS 23.140; TSG Terminals; Multimedia Messaging Service (MMS); Functional Description; Stage 2 (Release 6), v6.0.0, December 2002.
[12] Schulzrinne H., Casner S., Frederick R., Jacobson V., *RTP: A Transport Protocol for Real-Time Applications*, IETF RFC 1889, 1996.
[13] Schulzrinne H., Rao A., Lanphier R., *Real Time Streaming Protocol (RTSP)*, IETF RFC 2326, April 1998.
[14] Handley M., Jacobson V., *SDP: Session Description Protocol*, IETF RFC 2327, 1998.
[15] 3GPP TS 26.234; TSG Services and Systems Aspects; Transparent End-to-End Packet Switched Streaming Services (PSS); Protocols and Codecs, Release 4, v4.5.0, December 2002.
[16] Lakkakorpi J., Heiner A., Ruutu J., *Measurement and characterization of Internet Gaming Traffic*, Paper Presented at Helsinki University of Technology, Networking Laboratory in February 2002. Available on request.
[17] Handley M., Schulzrinne H., Schooler E., Rosenberg J., *SIP: Session Initiation Protocol*, IETF RFC 2543, March 1999.
[18] RFC 3481, TCP Over Second (2.5G) and Third (3G) Generation Wireless Networks, February 2003.
[19] RFC 3042, Enhancing TCP's Loss Recovery Using Limited Transmit, January 2001.
[20] RFC 2018, TCP Selective Acknowledgment Options, October 1996.
[21] RFC 3168, The Addition of Explicit Congestion Notification (ECN) to IP, September 2001.
[22] RFC 2884, Performance Evaluation of Explicit Congestion Notification (ECN) in IP Networks, July 2000.
[23] RFC 1323, TCP Extensions for High Performance, May 1992.
[24] Braden R., *T/TCP—TCP Extensions for Transactions*, RFC 1644, July 1994.

[25] Border J., Kojo M., Griner J., Montenegro G., Shelby Z., *Performance Enhancing Proxies Intended to Mitigate Link-Related Degradations*, RFC 3135, June 2001.
[26] RFC 3095, RObust Header Compression (ROHC): Framework and Four Profiles: RTP, UDP, ESP, and Uncompressed, July 2001.
[27] RFC 2507, IP Header Compression, February 1999.
[28] RFC 2508, Compressing IP/UDP/RTP Headers for Low-Speed Serial Links, February 1999.

9

Dynamic Frequency and Channel Allocation

Matti Salmenkaita

Dynamic channel allocation (DCA) is a wide definition for functionality that aims to optimise the usage of radio resources by applying intelligence to the channel selection process. DCA schemes can be based on many different variables and different applications provide varying degrees of freedom in the channel selection. The DCA schemes are commonly categorised as traffic adaptive (TA-DCA) or interference adaptive (IA-DCA) [1].

Traffic adaptive DCA algorithms allow dynamic usage of a common channel pool thus avoiding the lack of flexibility that is characteristic of a fixed channel allocation. The primary aim of the TA-DCA schemes is to adapt to the fluctuations in the traffic distribution, thus minimising the call blocking probability due to the lack of available channels. The TA-DCA algorithms avoid interference between cells by considering the channels used in the nearby cells and by utilising some form of pre-planned compatibility matrix or exclusion zone to govern the process. Examples of TA-DCA schemes can be found in [2–4].

Interference adaptive DCA algorithms are similarly able to minimise the blocking probability by allowing dynamic usage of a common channel pool. However, instead of a fixed compatibility matrix the IA-DCA schemes utilise real-time interference measurements to avoid interference. The usage of real-time measurements allows higher flexibility and adaptability in the channel selection process and therefore generally leads to better performance with the expense of increased complexity [5–7].

This chapter presents a dynamic channel allocation scheme called *dynamic frequency and channel allocation* (DFCA[1]). DFCA is a centralised IA-DCA scheme based on the estimation of carrier to interference ratio (CIR) on possible channels. The aim is to optimise the distribution of interference taking into account the unique channel quality requirements of different users and services. DFCA utilises cyclic frequency hopping for

[1] Nokia proprietary functionality.

GSM, GPRS and EDGE Performance 2nd Ed. Edited by T. Halonen, J. Romero and J. Melero
 ISBN: 0-470-86694-2

maximum frequency diversity gain. With adequate base transceiver station (BTS) support DFCA can freely choose the mobile allocation (MA) list, mobile allocation index offset (MAIO), timeslot (TSL) and the training sequence code (TSC) that will be used for each connection regardless of the physical transceiver. The basic requirement for DFCA is that the air interface is synchronised, as synchronisation provides timeslot alignment and time division multiple access (TDMA) frame number control, which are needed for timeslot level CIR estimation with cyclic frequency hopping.

9.1 Air Interface Synchronisation

9.1.1 GSM Synchronisation Basics

The basic time unit in the global system for mobile communications (GSM) air interface is a TDMA frame consisting of eight timeslots. The placement of logical channels in timeslots as well as the current frequency selection in the frequency hopping sequence are determined by the TDMA frame number. In an unsynchronised network, the frame clocks in base stations are initialised independently of each other leading to random frame alignment and numbering. This is illustrated in Figure 9.1. The air interface can be considered synchronised when the TDMA frame alignment and frame numbers of all BTSs are known and controlled. An example of this is presented in Figure 9.2.

9.1.2 Implementation of Synchronisation

Typically, the base stations that are physically located in the same BTS cabinet are using the same frame clock source and therefore they are always synchronised relative to each

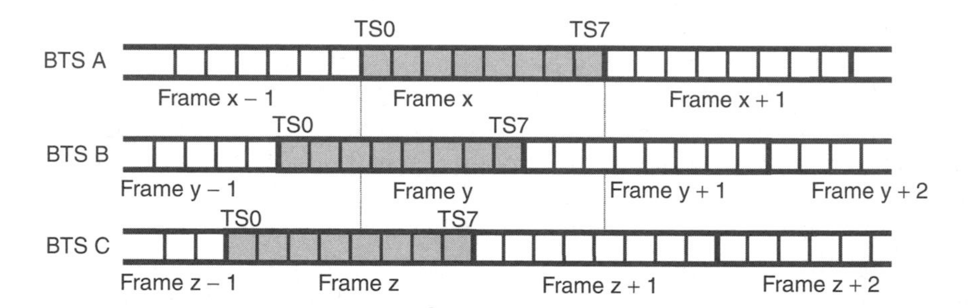

Figure 9.1. Random TDMA frame alignment

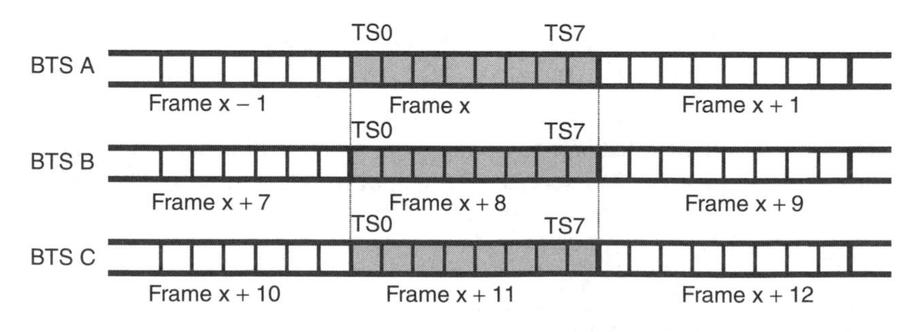

Figure 9.2. Synchronised TDMA frame alignment

other. It is also quite straightforward to chain a common clock signal to all the BTS cabinets in the same site, and this way achieve local synchronisation even if the BTSs are located in different cabinets.

Area level synchronisation is more difficult to achieve, as the common master clock signal must be provided in all BTS sites, which may be located several kilometres apart. A common solution to this problem is to use the stable and globally available clock signal from the global positioning system (GPS) satellites.

A GPS receiver in every BTS site provides a frame clock signal that is then distributed to all BTS cabinets in the site. This frame clock is calculated from the GPS clock signal according to the same rules in all the sites leading to common TDMA frame alignment and numbering in all synchronised base stations.

9.1.3 TDMA Frame Number Considerations

A common TDMA frame number is not desirable at area level as it leads to coincident broadcast control channel (BCCH) and traffic channel (TCH) multiframes in all the base stations. For BCCH, this results in coincident transmission of synchronisation bursts in every 10 TDMA frames. The synchronisation burst carries the base station identification code (BSIC) of a cell, which is needed together with the frequency correction burst for a mobile station (MS) to identify the neighbouring cell and to synchronise with it. Synchronisation with a common TDMA frame number leads to a situation where in 9 out of 10 TDMA frames there is no synchronisation burst transmission from any base station and in 1 out of 10 TDMA frames all the base stations transmit the synchronisation burst simultaneously due to coincident BCCH multiframes.

The mobile stations in-active-mode attempt to search for the neighbours only during idle frames that occur once within each TCH multiframe of 26 TDMA frames. On the other hand, frequency correction burst or synchronisation burst transmission occurs 10 times within one BCCH multiframe of 51 TDMA frames as illustrated in Figure 9.3. Thus, in a synchronised network with a common TDMA frame number only $10/51 = 19.2\%$ of the TCH idle frames can be used to search for the neighbouring cells. Because of this, the mobile stations would need significantly longer time to identify the neighbouring base stations leading to decreased BSIC decoding performance.

The second problem with the common frame number is that the slow associated control channel (SACCH) transmissions that always occur in the same instant within the TCH multiframe would occur simultaneously all over the synchronised area. Since SACCH is always transmitted regardless of the voice activity, it would not benefit from discontinuous transmission (DTX) at all.

These problems can be avoided by applying different frame number offsets throughout the synchronised area. Each base station can have a fixed frame number offset that is added to the global frame number. This is illustrated in Figure 9.2 where the BTSs B and C have relative frame number offsets of 8 and 11, respectively, towards BTS A. It is also possible to apply the frame number offsets on the site level so that all base stations in one site use the same frame number.

9.1.4 Synchronisation Accuracy

Synchronisation has to be accurate enough to ensure that proper timeslot alignment is maintained as network features relying on synchronisation such as DFCA assume nearly

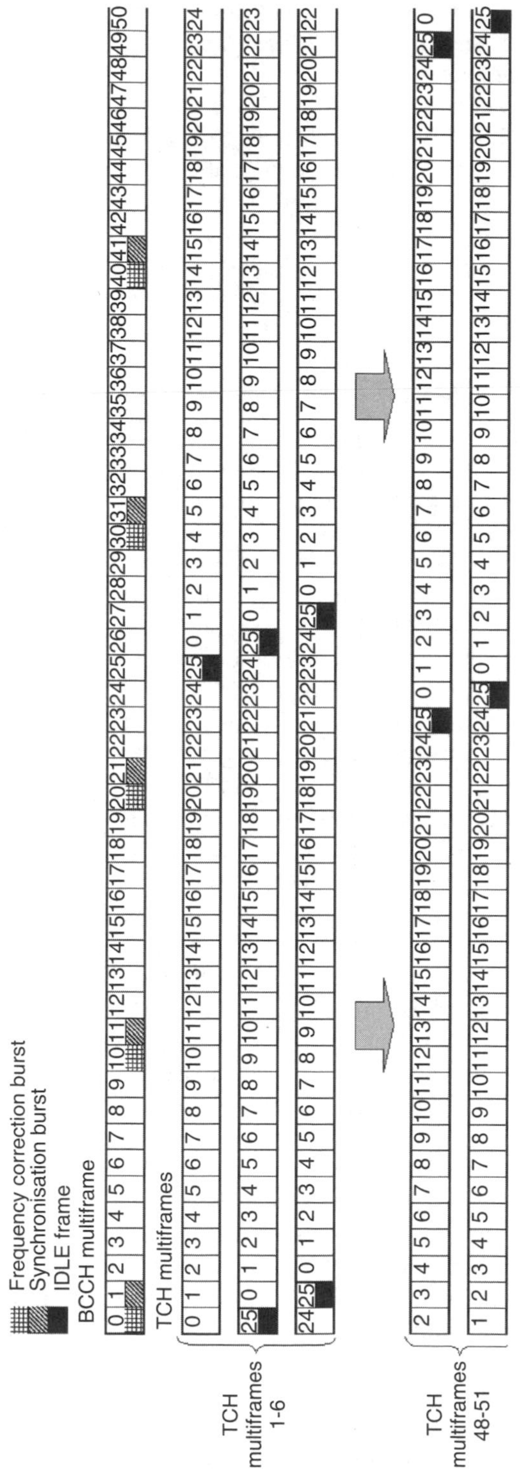

Figure 9.3. Relation of the BCCH and TCH multiframes

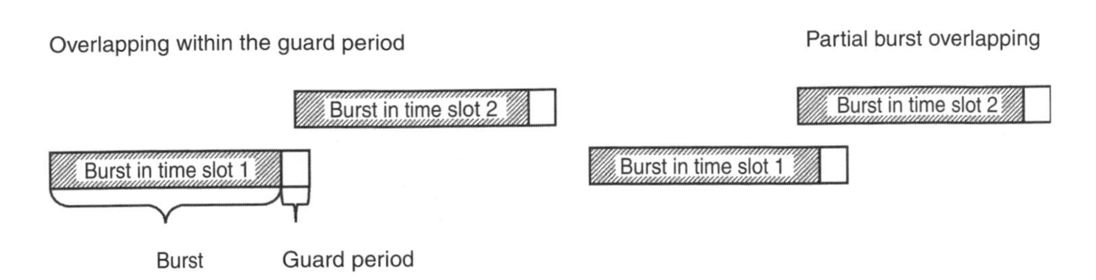

Figure 9.4. Burst overlapping

perfect timeslot alignment so that there is no interference between different timeslots. However, the guard period that is located at the end of the bursts provides some room for overlapping without detrimental effects (Figure 9.4).

All bursts, except the access burst, have a guard period lasting 8.25 bit periods that corresponds to 30.46 μs and about 9.1 km in propagation distance. Thus, assuming an ideal situation with the mobile station located right next to the serving base station, the interfering transmission can originate up to 9.1 km away without burst overlapping between the adjacent timeslots. This distance is referred to here as the *synchronisation radius* (D_0). Generally, the synchronisation radius is the maximum propagation distance difference between the serving and the interfering signal. If the propagation distance difference exceeds D_0, then partial burst overlapping starts to occur. The synchronisation radius is illustrated in Figure 9.5.

Typically, the high serving cell signal level close to the base station ensures that the CIR is sufficiently high to avoid quality degradation. The most sensitive location is the cell border area where the serving cell signal level is the lowest. Two different cases can be identified depending on the relative location of the mobile station. These are illustrated in Figure 9.6.

In both cases, the propagation distance for the serving cell signal C is denoted as d_c and for the interfering signal I as d_i. Additionally, in the uplink cases, the propagation distance of the uplink signal that is being interfered with is denoted as d_{c1}. In the uplink, the timing advance is used to compensate the signal propagation delay; so, ideally from a timing point of view d_c and d_{c1} appear to be zero. Consequently, propagation delay difference as observed in the base station corresponds to $d_i - d_c$, which is also the propagation distance difference in the downlink direction. So, ideally the uplink and the downlink situations are

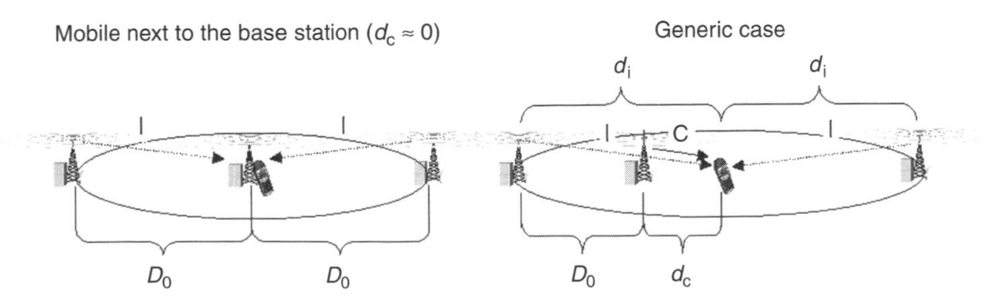

Figure 9.5. Propagation distances

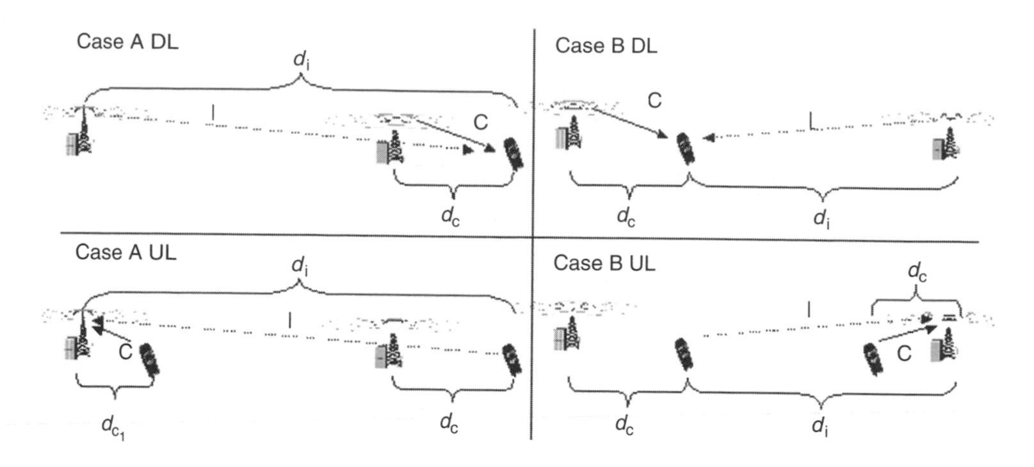

Figure 9.6. The two studied cases considering both downlink and uplink

similar. However, the granularity of the uplink-timing advance introduces an additional error source that does not exist in the downlink direction.

In order to avoid burst overlapping between successive timeslots, the propagation distance difference ($d_i - d_c$) must be below the synchronisation radius D_0. In this analysis, the impact of the interfering base stations outside the synchronisation radius is considered, so in the worst case the d_i can be expressed as

$$D_0 = d_i - d_c \Rightarrow d_i = D_0 + d_c$$

In the ideal case, D_0 is about 9.1 km as presented before. However, multi-path propagation introduces delayed signal components that effectively reduce D_0. The delay and magnitude of these signal components can be estimated by studying the commonly used multi-path channel models presented in [8]. It is assumed that the components with more than 10 dB of attenuation can be excluded. Typical delays after which the multi-path components have been attenuated by more than 10 dB are presented in Table 9.1 for typical urban (TU), rural area (RA) and hilly terrain (HT) models. For each model, also the corresponding D_0 is presented.

At this stage the impact of the multi-path propagation is estimated, but the accuracy of the synchronisation is not yet accounted for. The synchronisation accuracy depends on the stability of the timing source, the way the synchronisation signal is distributed and the calibration of the used equipment. In practice, with GPS-based synchronisation accuracy

Table 9.1. Impact of multi-path channels on D_0

Channel model	No multi-path	Typical urban (TU)	Rural area (RA)	Hilly terrain (HT)
Approx. delay for > 10 dB attenuation	0 μs	5 μs	0.4 μs	16 μs
D_0	9.1 km	7.6 km	9.0 km	4.3 km

of the order of 2 μs is achievable. The granularity of the uplink timing advance setting is 1 bit period corresponding to approximately 3.69 μs. This granularity introduces an additional error source that is only applicable in the uplink direction. Therefore, for the uplink an additional error of 3.69 μs must be added. Figure 9.7 presents D_0 as a function of the synchronisation error for different multi-path channels.

Burst overlapping may occur when an interfering transmission originates from a base station that is located more than $D_0 + d_c$ away from the mobile station. However, if the D_0 is high enough the propagation loss has already attenuated the interfering signal sufficiently so that the interference impact is negligible. The worst-case situation is when the mobile station is located in the cell border where the serving cell signal strength is the lowest. Figure 9.8 presents the cell border CIR as a function of the cell radius d_c assuming 2 and 6 μs synchronisation errors and the worst-case situation where $d_i = D_0 + d_c$ and burst overlapping starts to occur.

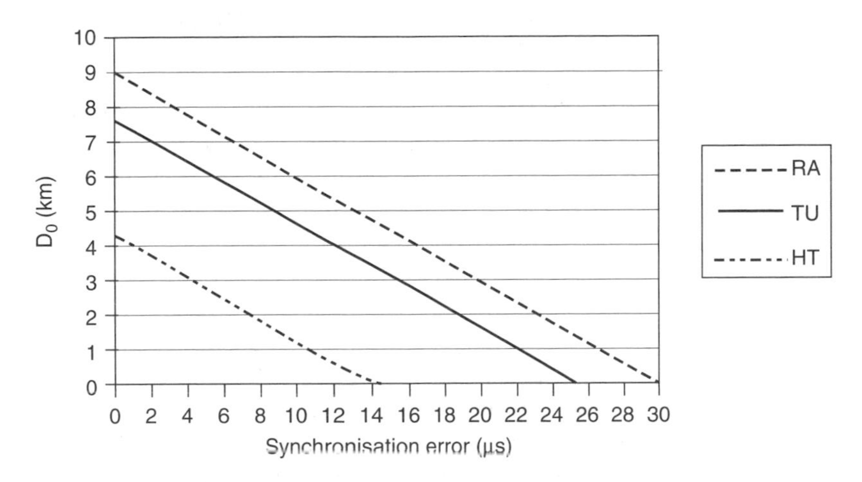

Figure 9.7. D_0 as a function of the synchronisation error

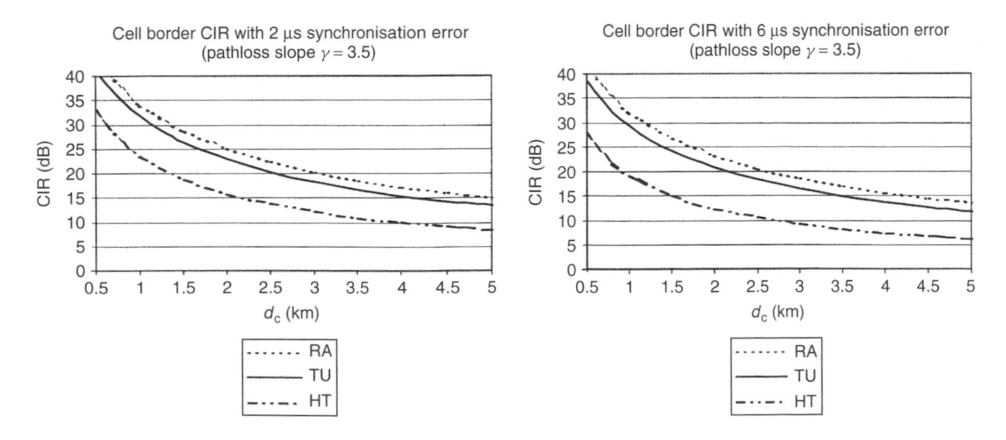

Figure 9.8. Cell border CIR as a function of cell radius of d_c

Figure 9.8 shows that in TU and rural environments the cell radius can be up to several kilometres without significant interference from overlapping bursts. Typically urban area cells are so small that the interference from the cells outside the synchronisation radius is negligible. Since the rural cells are usually much larger, the overlapping bursts may cause some degradation of the part of the burst that is overlapped. However, small overlapping is not likely to cause significant quality degradation as the training sequence is located in the middle of the burst and the channel coding is in most cases able to correct the errors. HT is the most restrictive as the strong delayed signal components shrink the synchronisation radius.

9.2 DFCA Concept

DFCA is based on the timeslot alignment and TDMA frame number control provided by network level synchronisation. The timeslot alignment ensures that the GSM air interface timeslots are coincident throughout the network. This makes it possible to take all the interference considerations to the timeslot level.

As GSM uses the combination of frequency division multiple access (FDMA) and time division multiple access (TDMA), the radio channel is determined by the frequency and the timeslot. When a channel assignment needs to be performed as a result of a newly initiated connection or handover, DFCA will evaluate all the possible channels and then choose the most suitable one in terms of CIR for the assignment. For this reason, an estimate of the CIR is determined for each available radio channel.

9.2.1 CIR Estimation

The estimation of CIR is based on the downlink signal level measurements of the serving cell and the neighbouring cells for which the BCCH frequencies are measured. These measurements are continuously performed and reported to the network by the mobile stations in active mode. This is a standard GSM functionality mainly used to provide path loss information for handover decisions. If one of the reported neighbouring cells would use the same frequency as the serving cell, the potential CIR can then be calculated as follows:

$$\mathrm{CIR_n} = \mathrm{RXLEV_{servcell}} + \mathrm{Pr_{servcell}} - \mathrm{RXLEV_{neigbcelln}}$$

where RXLEV is the GSM parameter corresponding to the measured signal level expressed in dBm and Pr is the current power control power reduction in the serving cell.

With legacy handsets the maximum number of reported neighbouring cell RXLEV values is limited to six. However, the GSM standard Rel'99-compatible handsets are able to report more by using enhanced measurement reporting (EMR). In most cases, reporting only six neighbours is sufficient for handover evaluation purposes but for interference estimation the potential CIR has to be determined for all the surrounding cells that may be significant interference sources. Typically, in an urban environment, there may be more than six such cells. The CIR estimate towards such unreported cells can be obtained by collecting long-term neighbouring cell RXLEV measurement statistics from all the mobile stations in a cell. From these statistics, a CIR with a given outage probability can be determined for all the surrounding cells. This way the network is able to have a CIR estimate for all the potentially interfering cells even if some of them were not included in the latest measurement report. The directly measured CIR is available for the neighbouring cells that are included in the latest measurement report and for the rest of the surrounding cells the statistical CIR estimation can be used. Since the downlink path loss is generally equal to the uplink path loss, the CIR statistics can also be used to determine an estimate of the uplink CIR.

The method described above provides the means to determine a potential co-channel CIR towards each of the reported neighbouring cells, but the CIR is only realised when the same or adjacent frequency is being used in a neighbouring cell in the same timeslot. In case of adjacent channel interference, the resulting CIR is scaled up by a fixed 18-dB offset as this is typical adjacent channel attenuation in GSM. In order to determine the real CIR, the network must examine the timeslots and frequencies used by the active connections in the potentially interfering cells.

Power control typically leads to reduced transmission powers. The current transmission power reduction of the interfering connection is taken into account by scaling up the estimated CIR accordingly. By processing this information, a matrix showing the estimated CIR for all the possible frequency and timeslot combinations can be generated.

9.2.2 *Combination with Frequency Hopping*

The air interface synchronisation makes it possible to control the frame numbers in each of the cells. This allows utilisation of cyclic frequency hopping while still keeping the interference relations fixed in the network. The frequency band dedicated to DFCA operation is divided into one or more MA lists. A frequency in the non-hopping case corresponds to a combination of a MA list and a MAIO in the cyclic frequency hopping case. Two connections using the same MA list and MAIO are co-channel interferers to each other if the cells have the same frame number. This property is illustrated in Figure 9.9. Offset in the frame number means similar offset (modulo MA length) in the MAIO of the interfering connection.

In case of cyclic hopping, a radio channel can be uniquely defined by the used MA list, MAIO and the timeslot. Thus, the radio channels are identified by these variables in the CIR matrix as presented in Figure 9.10. The CIR matrix contains the estimated CIR for each available radio channel. Some timeslots may not always be available as the availability of a timeslot depends on base station hardware resources that might already be in use by other connections.

9.2.3 *Radio Channel Selection*

In the channel assignment, the CIR matrix is searched for a radio channel that provides a CIR that fulfils the CIR requirement of the user. However, the new assignment may cause excessive interference to some ongoing connections in the neighbouring cells. To avoid this, a similar CIR evaluation is performed for all the ongoing connections sharing the same or adjacent radio channel. For each possible radio channel and each affected user, the margin between the predicted CIR and the user CIR target is calculated. The channel allocation algorithm chooses the channel taking into account the estimated CIR for the new connection as well as for all the affected existing connections. The effect of a new allocation on other users is thus taken into account. The expressions presented below

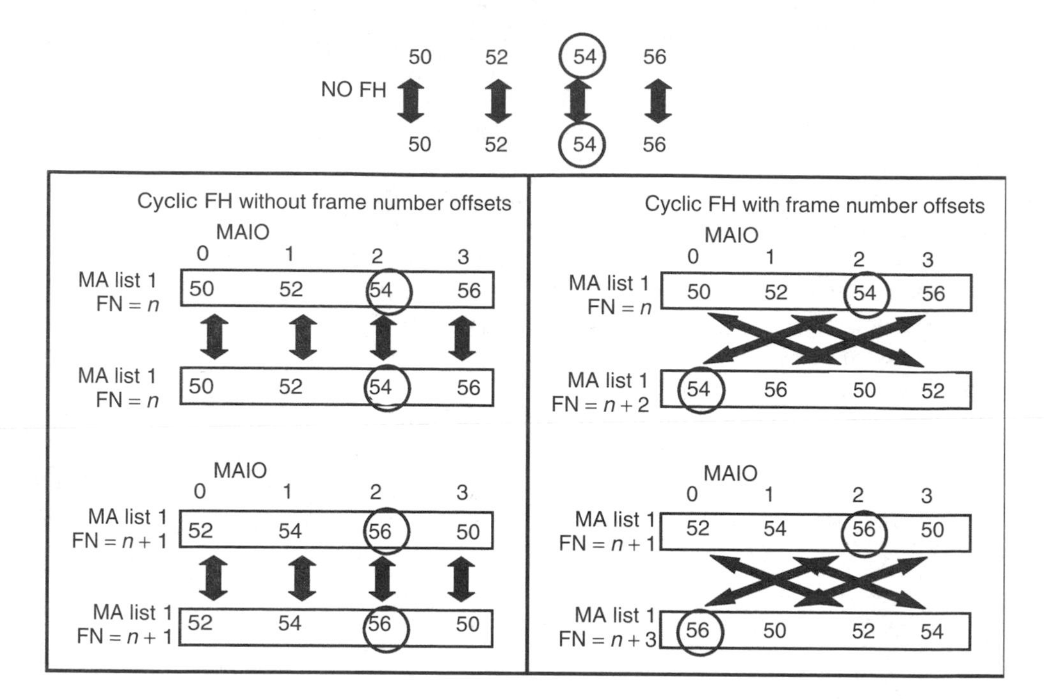

Figure 9.9. Fixed interference relations with cyclic hopping

MA, MAIO	Time Slot 0	1	2	3	4	5	6	7
1, 0	5	24	Not available	13	23	Not available	45	Not available
1, 1	32	8		26	14		9	
1, 2	11	6		18	34		33	
1, 3	27	13		22	6		14	
2, 0	36	21		16	13		37	
2, 1	25	10		14	24		17	
2, 2	9	7		20	31		22	
2, 3	21	9		11	15		24	

Figure 9.10. Example CIR matrix

show how the minimum CIR margin is determined for each candidate radio channel. Finally, the radio channel with the highest positive CIR margin is selected for channel assignment. The evaluation of the radio channels based on the CIR margin is illustrated in Figure 9.11.

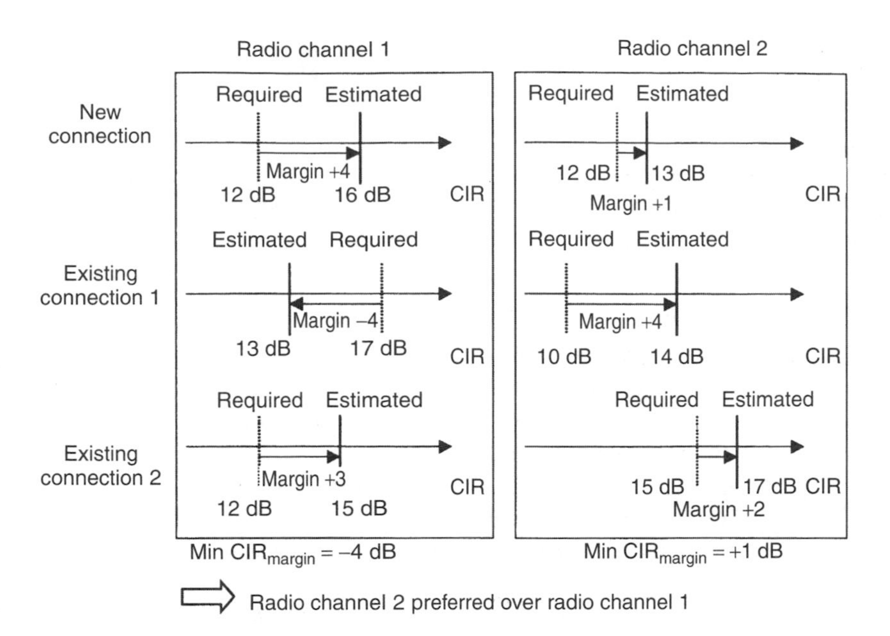

Figure 9.11. Example of DFCA radio channel selection

$$CIR_{margin_{C_{own}}} = \min\left(CIR_{est_{C_{own-DL}}}\, CIR_{target_{C_{own-DL}}},\ CIR_{est_{C_{own-UL}}}\, CIR_{target_{C_{own-UL}}}\right)$$

$$CIR_{margin_{C_{other_n}}} = \min\left(CIR_{est_{C_{other-DL_n}}} - CIR_{target_{C_{other-DL_n}}},\ CIR_{est_{C_{other-UL_n}}} - CIR_{target_{C_{other-UL_n}}}\right)$$

$$CIR_{margin_{min_C}} = \min\left(CIR_{margin_{C_{own}}},\ CIR_{margin_{C_{other_n}}}\right)$$

As a result of the above procedure, the channel with the most CIR margin above the CIR targets will be chosen. Thus, the DFCA algorithm aims to maximise the CIR while maintaining the relative CIR margins between users with different CIR requirements. As the network load increases, the provided CIRs will gradually decrease. The CIR differentiation between user classes is achieved because the CIR margin is calculated on the basis of user class specific CIR targets. Optionally, a soft blocking CIR limit can be defined for each user class. Whenever the channel assignment would lead to an estimated CIR to fall below the soft blocking CIR limit, the assignment is rejected. This procedure ensures that the new or any of the existing connections will not be exposed to intolerable interference.

9.2.4 Information Exchange

As described before, the DFCA algorithm needs to have access to mobile station (MS) measurement reports and real-time channel occupancy and transmission power information in the neighbouring cells. Owing to these requirements, it is natural to locate the DFCA algorithm in the network element that handles the radio resource management. In the standard base station subsystem (BSS) architecture, this network element is the base station controller (BSC). One BSC can handle only a limited number of base stations that typically cover one geographical area. Normally, each of the radio resource management (RRM) functions is only aware of the channel occupancy situation within the area it controls. The same applies also for the availability of the MS measurement reports. To allow transparent DFCA operation across the border areas, the RRM functions have to exchange channel occupancy, MS measurement report and current transmission power information.

9.2.5 DFCA Frequency Hopping Modes

The optimum DFCA operation requires a DFCA frequency hopping mode where a base station has to be able to use any DFCA MA list and MAIO in any of the timeslots. This means that the base station must support timeslot level synthesised frequency hopping and wideband combining. Wideband combining that is required for synthesised frequency hopping may not be feasible in some base stations due to increased RF losses and older base station hardware may not support DFCA hopping mode. However, DFCA can be used in the RRM even if the DFCA hopping mode is not available. The following operating modes are possible:

- DFCA frequency hopping
- synthesised frequency hopping (RF hopping) with DFCA RRM
- baseband frequency hopping (BB hopping) with DFCA RRM.

DFCA hopping mode provides full radio channel selection freedom for the DFCA algorithm and therefore the optimum performance. The synthesised frequency hopping with DFCA RRM consists of a BTS that is configured for synthesised hopping with a fixed DFCA MA list and fixed pre-planned MAIOs for each transceiver (TRX). Because of the fixed MA list and the fixed MAIOs, the DFCA algorithm has limited freedom to choose the most suitable channel for each connection. Each timeslot in every TRX has already a fixed MA list and MAIO that cannot be dynamically changed. Thus, the DFCA algorithm can only select the most suitable timeslot from all the free timeslots. In low load situation, the number of free timeslots is high so DFCA is able to optimise the performance quite effectively. However, in high load situation only a few timeslots may be available severely restricting the possibilities to find a suitable radio channel and therefore degrading the performance. The baseband frequency hopping with DFCA is very similar to the synthesised hopping case. Also, with baseband hopping the MA list and the TRX MAIOs are fixed. The only additional restriction with baseband hopping is that the MA list must contain as many frequencies as there are DFCA TRXs in the base station. If the number of the DFCA TRXs is low, this will limit the frequency diversity gain and therefore cause further performance degradation.

With both synthesised frequency hopping and baseband frequency hopping, a basic frequency planning must be performed. In synthesised hopping, the MAIO reuse must be planned and in the baseband hopping the allocation of DFCA MA lists to different BB hopping BTSs must be decided. The quality of this underlying frequency plan determines maximum achievable quality during the busy hours.

If the share of DFCA hopping capable BTSs is sufficiently high and they are equally distributed among the RF and/or BB hopping BTSs, then the full channel selection freedom of the DFCA hopping capable BTSs provides very effective adaptation to the surrounding RF/BB hopping BTSs. This decreases the importance of the frequency planning of the RF/BB hopping BTSs and at the same time enables increased robustness also during the busy hours maintaining the DFCA gains.

9.3 Application of DFCA for Circuit-switched (CS) Services

9.3.1 Multitude of CS Services

Optimising the network performance has become an increasingly difficult task as the variety of circuit-switched (CS) services has increased. Nowadays there are basic full-rate (FR) and half-rate (HR) services, advanced multi-rate (AMR) codecs for both full-rate and half-rate channels as well as high-speed circuit-switched data (HSCSD) services with the option of utilising multiple timeslots for one connection. There are some differences in channel coding between all these services and as a result the robustness against interference varies significantly. For example, an AMR connection can tolerate almost 6 dB higher interference level than a connection using enhanced full-rate (EFR) codec while still maintaining comparable performance in terms of frame erasure rate (FER) or Mean Opinion Score (MOS). Half-rate channels have generally less room for strong channel coding and the interleaving depth of HR channels is four instead of the eight that applies for full-rate (FR) channels. Thus, FR channels in general tolerate a higher interference level than HR channels. These link performance issues have been covered in more detail in Chapter 6.

The minimum CIR a connection can tolerate sets the lower limit for the frequency reuse distance. However, with a mixture of services in the network, the least robust of the used channel types determines the minimum tolerable reuse. Adaptive power control helps project the gains of the robust connections to the more vulnerable ones by aggressively reducing the transmission powers of the robust connections. However, simulations have shown that power control alone is insufficient for balancing the performances of different services.

For better balancing of performances, the different CIR requirements have to be taken into account already in the channel assignment. One possible solution is the usage of the reuse partitioning concept introduced in Chapter 6, where the frequency band has been divided into two separate pools that are then allocated to TRXs providing two different layers with different frequency reuse factors. The robust AMR connections can be primarily assigned to the underlay with low reuse factor and high spectral efficiency while the less robust connections are assigned to the overlay with conventional reuse factor where sufficiently high CIR can be guaranteed. This approach is reasonably powerful but it suffers from lack of flexibility as increase in the AMR-capable handset penetration requires constant review of the frequency band split between underlay and overlay bands and a change in frequency bands generally requires adjustments in the frequency plan. Since legacy mobiles cannot be guaranteed sufficient quality on the underlay, the overlay must be conservatively dimensioned so that instances where a legacy handset needs to be assigned to underlay due to overlay blocking are minimised.

Table 9.2. Examples of CIR targets

Connection type	Target CIR (dB)	Soft blocking threshold (dB)
FR/EFR	15	10
HR	18	13
AMR FR	9	4
AMR HR	15	10
HSCSD	18	Not used

9.3.2 The DFCA Way

DFCA is ideal for managing users with different CIR requirements. With DFCA different CIR targets can be assigned according to the robustness of the connection. Examples are presented in Table 9.2.

With DFCA, each connection is automatically assigned a channel that aims to satisfy the interference requirements associated with the connection type. Furthermore, the same requirements are evaluated every time a new potentially interfering channel assignment is being done, thus aiming to keep the CIR above the target during the whole connection. An ongoing connection is never allowed to fall below the soft blocking limit as a result of the establishment of an interfering connection. Thus, DFCA automatically optimises the usage of the frequency resource without need for network planning changes as a consequence of growing AMR penetration and changes in the geographical traffic distribution. Growing AMR usage can simply be seen as increasing network capacity as higher effective frequency loads (EFLs) can be achieved while still maintaining sufficient quality for both EFR and AMR connections.

9.4 DFCA Simulations with CS Services

The performance of DFCA has been evaluated in different scenarios by network level simulations. The used simulation tool and the simulation parameters are described in Appendix E. A complete DFCA RRM algorithm was implemented into the simulator. In all scenarios, the DFCA performance is compared with a reference case that utilises random frequency hopping with 1/1 reuse strategy and fractional loading taking advantage of both power control and DTX. Only the traffic on the non-BCCH TRXs is considered. The connection quality is studied by observing the FER in 2-s intervals. FER over 4.2% is considered an indication of unsatisfactory quality during the past 2 s. The network performance is benchmarked by calculating the percentage of these bad-quality instances. In the graphs, this percentage is called an *outage*.

Most of the simulations have been performed in an ideal regular three-sectorised network layout. In order to verify the DFCA performance in a more typical scenario, a real city centre network layout was implemented into the simulator with varying antenna orientations, cell sizes and cell traffics.

9.4.1 Performance in Ideal Network Layout

The ideal network layout consists of 75 cells in regular three-sectorised configuration. The traffic density is even and the cells have equal number of transceivers.

9.4.1.1 EFR

In the first simulation case, the performance of DFCA is studied with all EFR handsets. The limited robustness of EFR channel coding requires low effective frequency loads. In standard interference-limited scenarios, downlink is the limiting link mainly due to application of uplink diversity reception and the regularity of the network layout. In this scenario, DFCA is able to give a significant boost to downlink performance and with 2% outage level an overall network capacity gain of 66% can be observed (Figure 9.12).

9.4.1.2 AMR

In the second case, the DFCA performance is studied assuming that all the handsets in the network are AMR capable (Figure 9.13). Both AMR FR and AMR HR channels are considered. In high load situation, usage of AMR HR is preferred because of the efficient usage of hardware resources. In the reference case, some connections in marginal conditions have to be kept in the FR channel due to higher robustness of the FR channel, whereas in DFCA the best performance in high-load situation is achieved with full AMR

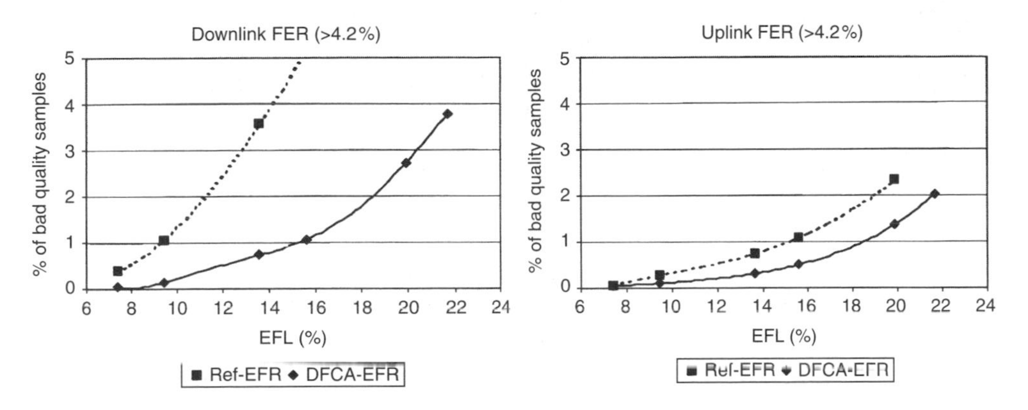

Figure 9.12. DFCA performance with EFR users

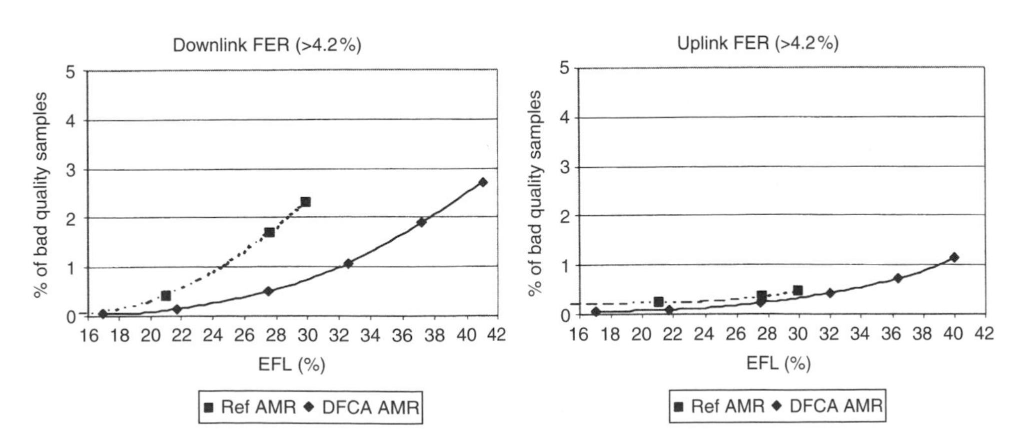

Figure 9.13. DFCA performance with AMR users

HR usage. Fixed codecs with 5.9-kbps bit rate are used for both AMR FR and AMR HR. At 2% outage level, the DFCA gain relative to the reference case performance is 35%.

9.4.1.3 Partial AMR Penetration

Since the AMR support requirement for handsets has only been introduced in a relatively recent release of the GSM specifications (Rel'98), it is common that the share of AMR-capable end users is growing gradually as old handsets are replaced with newer ones. In this case, full AMR gains are not available until 100% AMR-capable handset penetration is achieved. The challenge for an operator is how to extract the full benefit of the AMR-capable handsets while still providing good quality of service for the legacy handsets.

The following simulations have been performed to examine this problem and to evaluate the impact of the CIR provisioning capability of DFCA. The DFCA algorithm can be defined to allow the assignment of radio channels with lower CIR for the AMR-capable handsets while providing sufficiently high CIR for the legacy handsets. Naturally, more aggressive power control settings are applied for the AMR FR connections. The quality experienced by the EFR users is examined separately as this is typically limiting the maximum tolerable network load. As an example, a case with 50% AMR-capable handset penetration is examined (Figure 9.14). In DFCA case, HR channels are used as DFCA provides the best performance with HR channels. In the reference case, the quality experienced by the EFR users has improved from the pure EFR case presented in Figure 9.12 because of the lower power levels used by the interfering AMR users, but the quality gap between the EFR users and the AMR users is significant. It is therefore clear that the performance experienced by the EFR users is the limiting factor that defines the maximum achievable capacity.

DFCA CIR provisioning proves to be very effective in equalising the qualities experienced by these two user groups. With DFCA controlled channel assignment the quality gap disappears as the EFR users experience virtually identical quality with the AMR users. At 2% outage level, a DFCA gain of 77% is achieved when the quality experienced by the EFR users is considered.

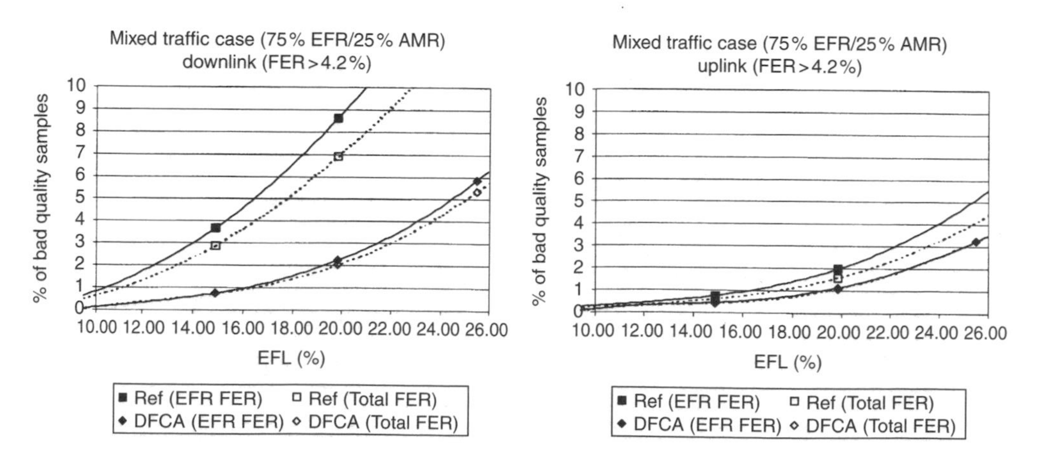

Figure 9.14. DFCA performance with 50% AMR penetration

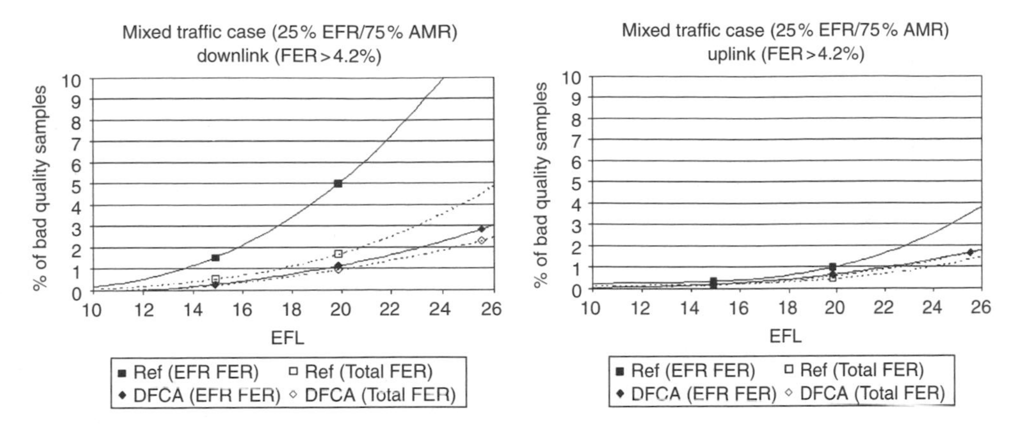

Figure 9.15. Spectral efficiency with variable AMR penetrations in ideal network

Figure 9.15 presents the achieved spectral efficiency separately for EFR users (x-axis) and AMR users (y-axis) with different AMR terminal penetrations. In these simulations, for each AMR terminal penetration rate, the network was loaded to the point where the EFR users' percentage of bad quality samples exceeds 2%. At this point, the carried spectral efficiency in terms of Erl/MHz/cell was determined separately for EFR and AMR users. The overall network spectral efficiency can be determined by summing EFR and AMR spectral efficiency numbers in any of the simulation points. The results indicate 63, 77 and 78% DFCA capacity gains for 25, 50 and 75% AMR terminal penetrations, respectively.

9.4.1.4 Narrowband Performance

Narrowband deployment is presented in Chapter 10, where general narrowband specific issues are presented in detail. In this section, the impact of narrowband deployment on DFCA performance is presented.

In order to evaluate the behaviour of DFCA in narrowband cases, two narrowband scenarios are studied. The BCCH TRXs are modelled in the simulations, but only the quality of the traffic on the hopping TRXs is considered. In these simulations, the frequency band available for DFCA is very limited. The small number of frequencies causes three detrimental effects on the DFCA performance.

Firstly, the small number of frequencies available for DFCA limits the number of available channels. This degrades the DFCA performance, as the probability of finding a suitable channel from a small pool of channels is smaller. This can be viewed as a consequence of degraded trunking efficiency in the channel selection.

Adjacent channel interference coming from the BCCH frequency band can also have a significant negative impact of DFCA performance, if the number of frequencies in the DFCA MA list is small. With DFCA that is using cyclic frequency hopping the adjacent channel interference coming from the BCCH band impacts regularly every nth burst where n is the number of frequencies in the DFCA MA list. Since BCCH is always transmitted with full power in the downlink direction, the impact of this interference can be significant for some users especially if the DFCA MA list is short. Furthermore, DFCA cannot take this interference contribution into account in the channel selection, so in some cases this additional uncounted interference may be critical for the connection quality.

The small number of frequencies can also limit the possibilities of dividing the DFCA frequency band into multiple MA lists. With only one DFCA MA list the frequencies in the list are consecutive. Adjacent frequencies have typically significant fast fading correlation that degrades the frequency diversity gain especially when cyclic frequency hopping is used. The fast fading correlation between the frequencies is modelled in the simulation tool as presented in Appendix E.

3.6-MHz Band

The first evaluated narrowband case is the extreme case of 3.6-MHz total frequency band. Out of this band 12 carriers have been allocated for BCCH and the remaining 6 carrier frequencies have been allocated to DFCA. The DFCA band is directly adjacent to the BCCH band.

The results in Figure 9.16 show about 26% capacity gain, but the gain reduction is obvious when compared to the wideband DFCA case where 66% gain is achieved.

4.8-MHz Band

In this case, the total available frequency band is 4.8 MHz corresponding to 24 carrier frequencies. Again 12 of these are allocated to the BCCH. The remaining 12 form the DFCA frequency band that is again adjacent to the BCCH band.

The doubling of the DFCA frequency band provides a significant improvement in DFCA performance. Although the overall performance is still slightly below the performance of the ideal wideband case, the DFCA gain is already 55% at the 2% outage level (Figure 9.17).

It can be concluded that although the DFCA performance degrades in the extremely narrowband case, the gains have returned to normal level in the 4.8-MHz case.

9.4.1.5 Impact of MS Measurement Errors

The CIR estimation that is done within the DFCA algorithm is based on the serving cell and neighbouring cell signal level measurements that are performed and reported by

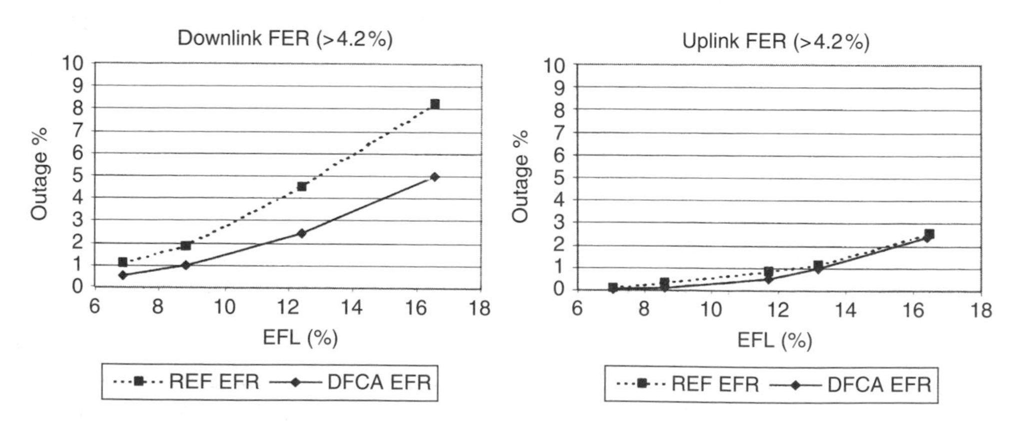

Figure 9.16. DFCA performance in 3.6-MHz narrowband case

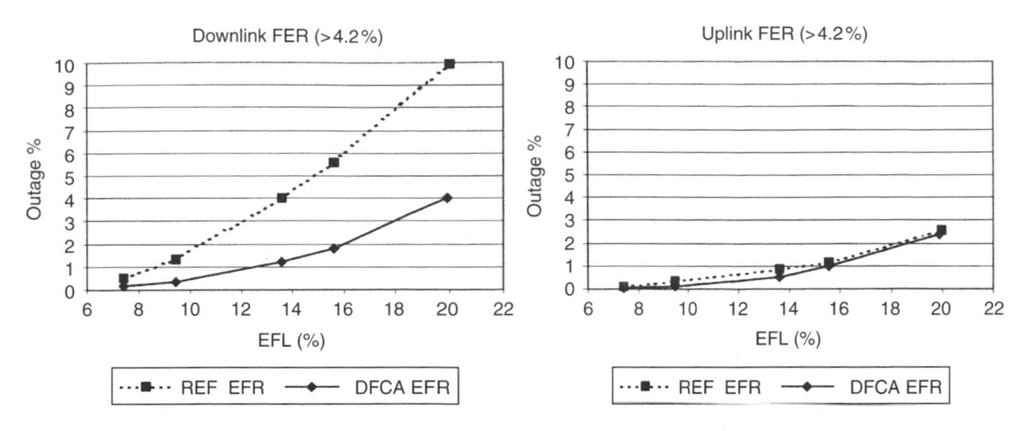

Figure 9.17. DFCA performance in 5-MHz narrowband case

the mobiles. The inaccuracies in these measurements can cause errors in the DFCA CIR estimations and therefore degrade the performance. Each GSM handset must satisfy the minimum signal level measurement accuracy requirements that are stated in the GSM specifications [9].

The absolute signal level accuracy requirement is ±6 dB in both normal and extreme temperature conditions. If the signals are in the same frequency band and at least 14 dB over the MS reference sensitivity level then the relative signal level accuracy requirement is ±2 dB. In these simulations, the occurrence of these errors is assumed to be normally distributed with 99% probability of the error being within the above-mentioned limits. The cumulative density functions of the errors are presented in Figure 9.18.

The simulations were run assuming EFR users in a regular network layout. Figure 9.19 shows that DFCA is very robust against the impact of these measurement errors. In downlink performance, no difference can be noticed and in uplink the impact is hardly noticeable.

DFCA never relies on the absolute reported signal level values. Instead, it always calculates the potential CIRs from each measurement report and uses them for the background CIR statistics and also as a direct input in the CIR estimation. This direct CIR calculation makes DFCA immune to the absolute MS measurement errors. Only relative

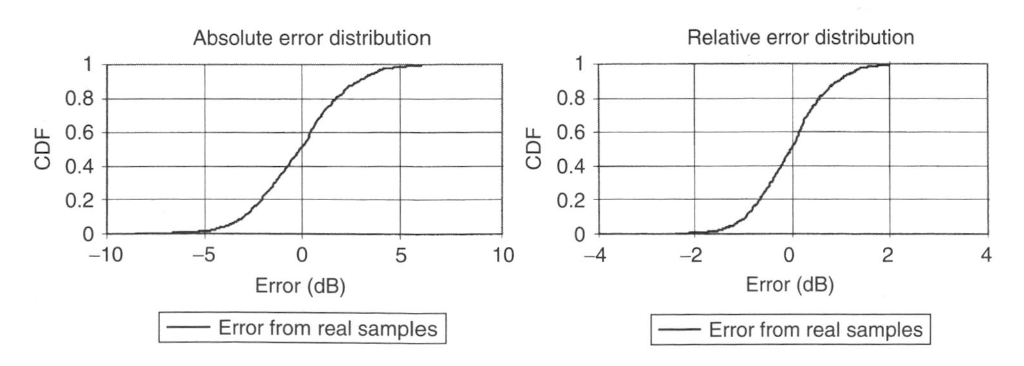

Figure 9.18. Modelling of MS measurement errors

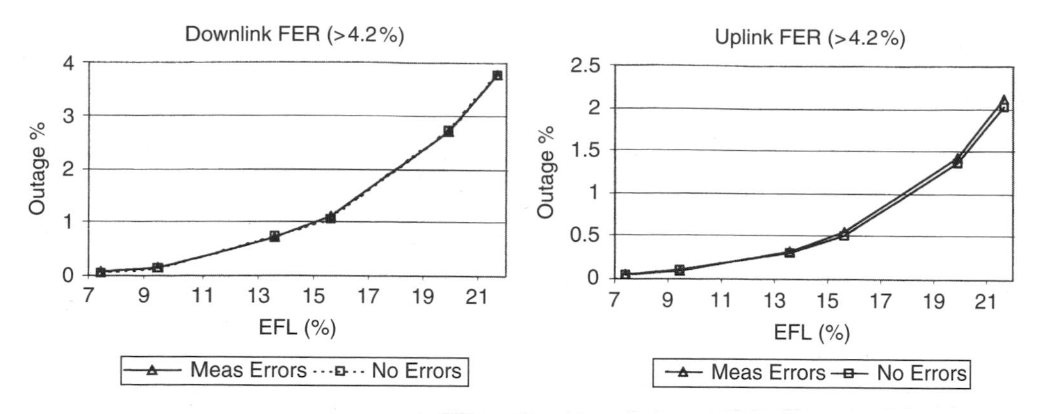

Figure 9.19. DFCA EFR performance with measurement errors

error has effect. But this effect is limited as it is averaged out in the CIR statistics that are used to determine the statistical CIR values. The relative error impacts only the part of the CIR estimation that uses the potential CIR values that have been determined directly from the latest measurement reports.

9.4.1.6 Impact of the Number of Reported Neighbours

Since the DFCA CIR estimation is based on the reported serving cell and neighbouring cell signal levels, the number of the neighbouring cells that are included in the measurement reports has some impact on the DFCA performance (Figure 9.20). In all the previous simulations, it was assumed that each measurement report contains the signal levels of six strongest neighbours that correspond to the capabilities of legacy handsets. However, this may not always be the case. Sometimes high BCCH interference level may not allow successful BSIC decoding in which case the neighbour cannot be reported. Also, the newer handsets are able to utilise EMR that allows reporting of many more neighbours in one report and by using interleaving of the reported cells, even more can be reported over successive reports.

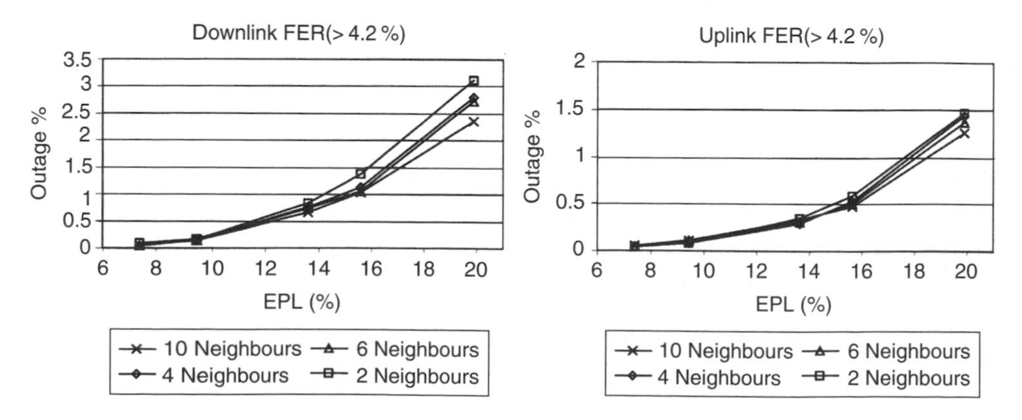

Figure 9.20. Impact of the number of reported neighbours

The results show that the number of reported neighbouring cells has virtually no impact in lower network loads. In higher loads the impact is clearly distinguishable, but not dramatic. The results show that about 5% extra gain can be achieved if the measurement reports contain 10 neighbours instead of 6.

In low-load situations, the DFCA algorithm keeps the interfering connections in far away cells as this maximises the CIR. With higher loads the interfering connections must be allowed in closer neighbouring cells. For DFCA, the importance of the latest downlink measurement reports is significant only when there is a possibility that an interfering connection may exist in the reported neighbours. In low-load situation, this is not the case and therefore the impact of the number of reported neighbours is very small. In low-load situation, the interference estimation relies heavily on the background interference matrix (BIM) that is relatively insensitive to the number of neighbours included in each downlink measurement report.

9.4.2 *Performance in Typical Network Layout*

In the previously presented simulation results, DFCA performance has been evaluated in ideal network layout with even traffic distribution, equal cell sizes and standard antenna orientations. This kind of approach is justifiable in order to be able to get easily comparable and reproducible results in a highly controlled standard environment. However, real networks have always a varying degree of irregularities. The performance of DFCA has also been benchmarked in a typical scenario since it was expected that the highly dynamic nature of DFCA should enable good adaptation to the irregularities.

The typical scenario is created by importing the macro-layer network layout of a real network in a city centre and its surroundings. The modelled network is the same as Network 1 in Appendix F. An uneven traffic distribution is created in order to achieve the cell loads that were observed in the real network. As a result, the cell traffic is highly variable. The number of TRXs per cell ranges from two to six according to the real network configuration. A map of the modelled city centre and the cell dominance areas of the whole area are presented in Figure 9.21.

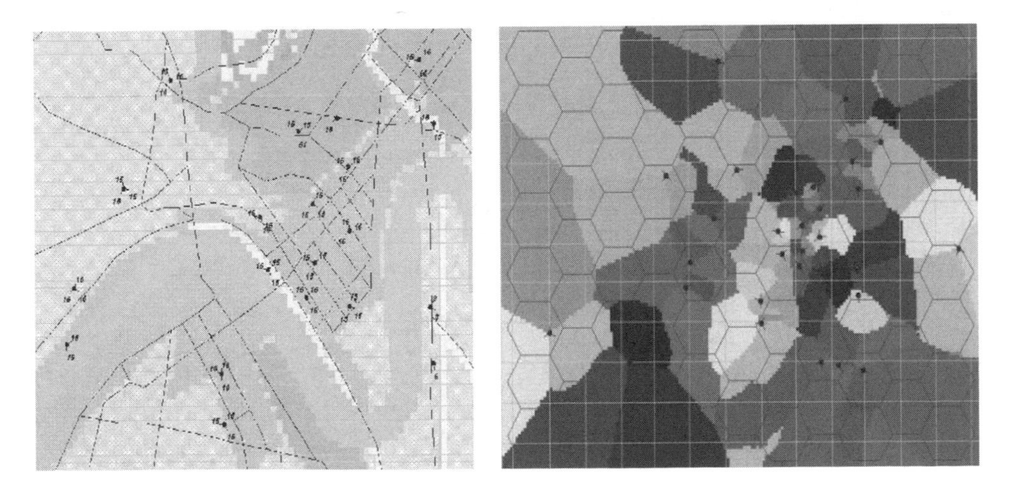

Figure 9.21. Irregular network layout with cell dominance areas

9.4.2.1 EFR

At first, the performance with EFR users is examined. A comparison with the same case in the ideal network layout scenario in Figure 9.22 reveals that the irregularities cause a clear relative degradation in uplink performance. However, with a small margin downlink is still the limiting link. The DFCA gain has been clearly improved. At 2% outage level, the DFCA capacity gain is about 72%.

9.4.2.2 AMR

The AMR simulations in the typical irregular network scenario have been performed with the same assumptions used in the corresponding ideal network scenario. In the case of AMR users, the irregular network is clearly uplink-limited. When the uplink performance is studied at 2% outage level a capacity gain of 62% can be observed (Figure 9.23).

9.4.2.3 Partial AMR Penetration

The performance with partial AMR-capable terminal penetrations in the typical network scenario has been evaluated the same way as was done in the corresponding ideal network

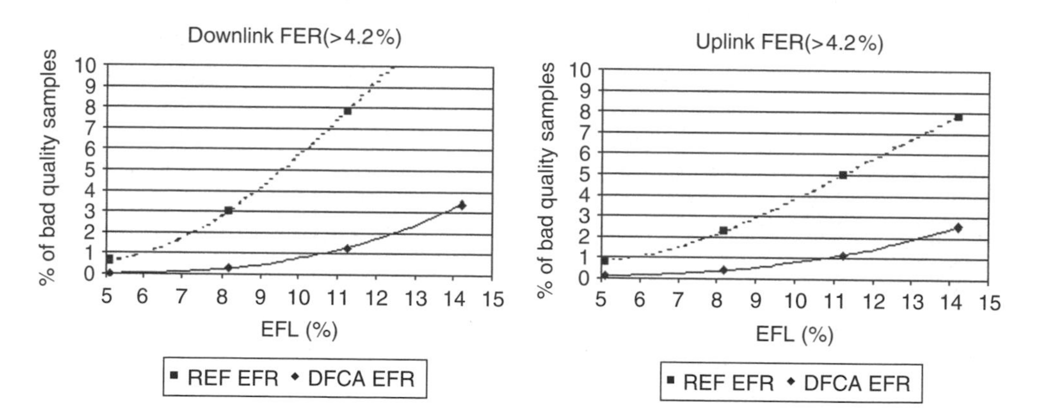

Figure 9.22. DFCA performance with EFR users

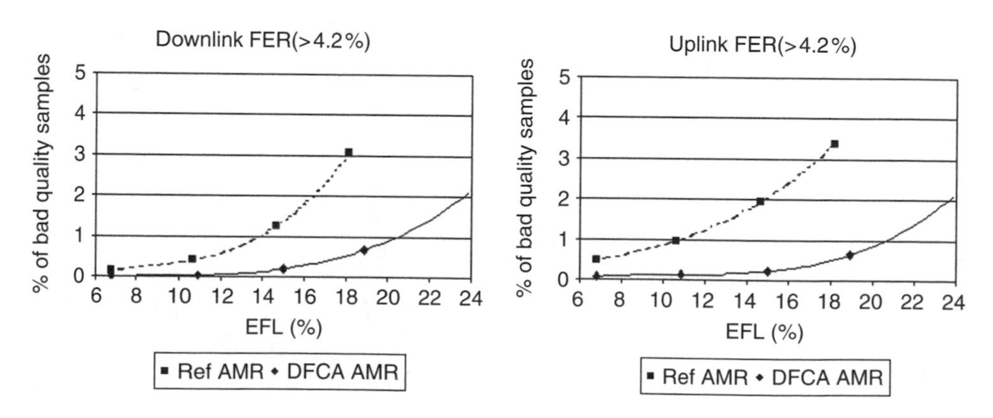

Figure 9.23. DFCA performance with AMR users

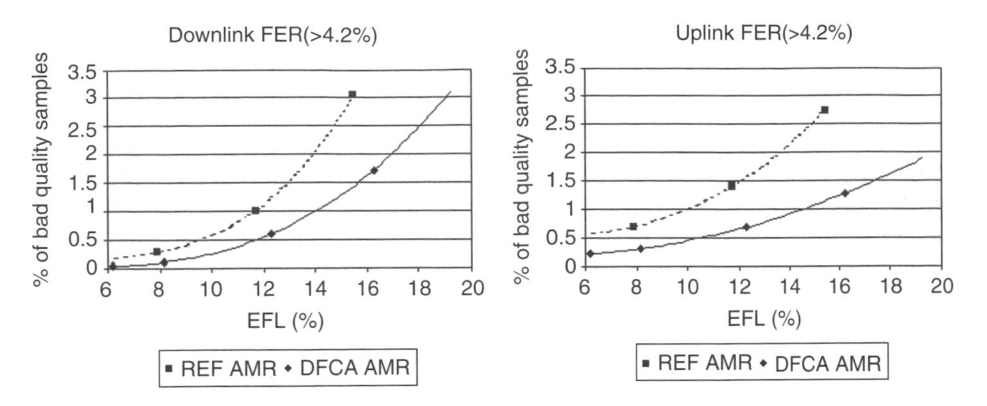

Figure 9.24. Spectral efficiency with variable AMR penetrations in typical network

scenario. The results shown in Figure 9.24 indicate DFCA gains of about 80% for 25, 50 and 75% AMR penetrations.

9.4.2.4 AMR Narrowband Performance (3.6 MHz)

The impact of narrowband deployment in the typical network scenario has been studied with the same assumptions that have been used in all AMR simulations in this chapter. The used frequency band is 3.6 MHz out of which 2.4 MHz is used by the BCCH layer. This leaves only 1.2 MHz available for DFCA operation.

As with the previous wideband results, now also the uplink is the limiting link in the reference case. However, at 2% outage level DFCA case is downlink limited. DFCA gains of 20 and 45% can be observed for downlink and uplink, respectively (Figure 9.25).

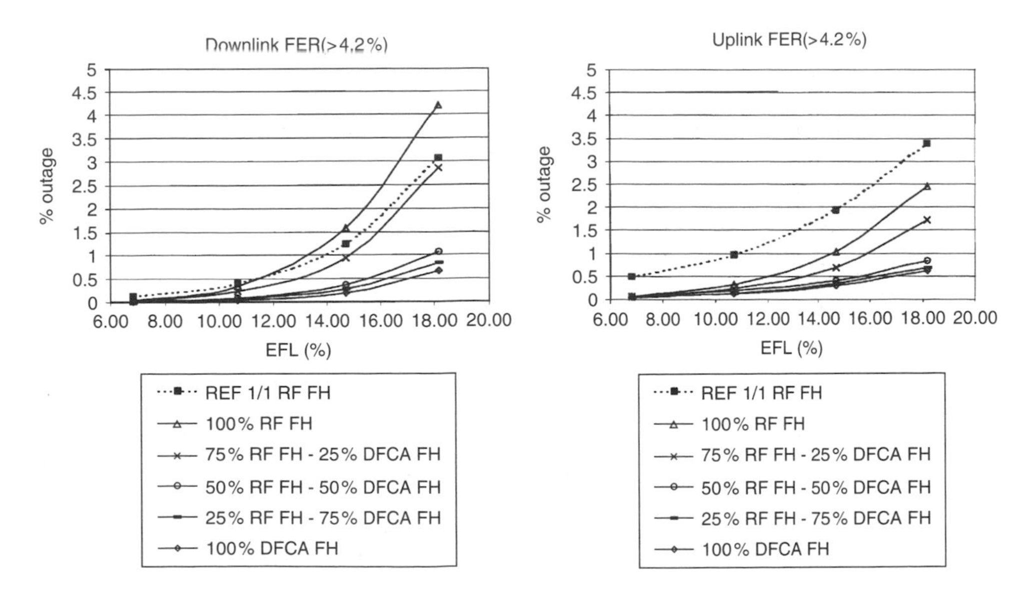

Figure 9.25. DFCA performance with 3.6-MHz band and AMR users

9.4.2.5 Impact of Frequency Hopping Modes

As explained in Section 9.2.5, a complete DFCA implementation requires that the base station is capable of selecting the radio transmission and reception frequency independently for every timeslot in each DFCA TRX. This allows the DFCA algorithm to freely choose the radio channel for each connection. In DFCA, the radio channel is defined by the selection of the timeslot, MA list and MAIO.

However, DFCA can be used together with conventional synthesised and baseband frequency hopping modes. With these hopping modes, each TRX is hopping over a fixed MA list using a fixed MAIO in all the timeslots. Therefore, the radio channel selection is limited to the fixed radio channels that have been pre-assigned to the available timeslots. As a consequence, the radio channel selection freedom available for the DFCA algorithm is directly dependent on the number of free timeslots in the cell. In low-load situation, the freedom is the greatest but as the load and timeslot usage increases, the radio channel selection freedom decreases.

In this simulation, the performance of the synthesised frequency hopping mode with DFCA RRM is compared with the performance of the DFCA frequency hopping mode that provides full radio channel selection freedom. Also, some intermediate cases are studied where 25, 50 and 75% of the base stations are DFCA frequency hopping capable while the rest use synthesised hopping with DFCA RRM. All these cases are compared with the reference case that uses fractionally loaded 1/1 reuse strategy with random synthesised frequency hopping with conventional RRM (Figure 9.26).

As expected, a reduction in the share of the DFCA frequency hopping capable base stations leads to a reduction in the DFCA performance. However, if at least 50% of the base stations are capable of DFCA frequency hopping, then only a small performance degradation can be observed indicating that the surrounding DFCA frequency hopping capable base stations are able to adapt efficiently to the limitations of the synthesised hopping base stations. Therefore, it is evident that almost optimum performance can be

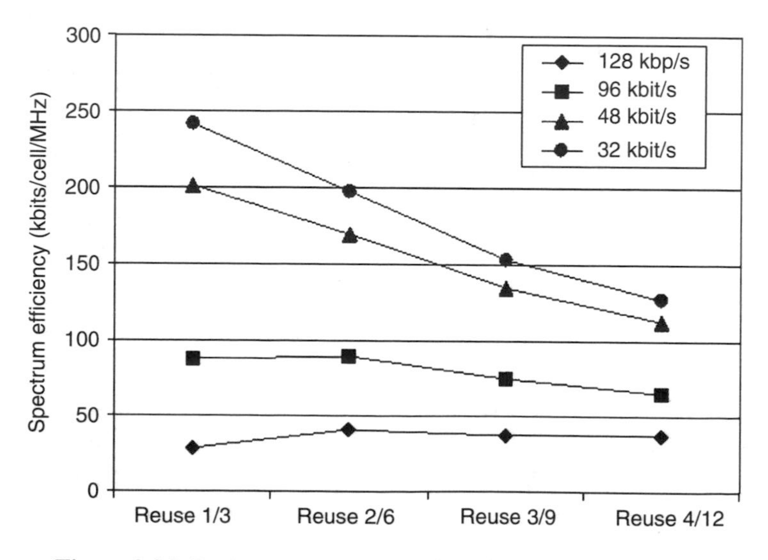

Figure 9.26. Performance with partial DFCA hopping capability

achieved if at least 50% of the base stations have DFCA hopping capability and they are equally distributed among the synthesised hopping base stations.

In all but the peak load situations also, the more heavily restricted DFCA cases provide significant quality gains. This is clearly visible especially in uplink, where the 100% synthesised hopping with DFCA RRM as well as the case with 25% of DFCA hopping capable base stations provide over 100% reduction in the FER outage. This is almost equal to the quality gain provided by the cases with a higher share of DFCA hopping capable base stations and proves that the channel selection freedom in low loads is sufficient for good DFCA performance.

9.4.3 *Summary of the Simulation Results*

A summary of the simulation results is presented in Table 9.3.

9.5 DFCA for Packet-switched (PS) Services

Packet-switched (PS) data services can be used by a multitude of different applications that can have very different quality-of-service (QoS) requirements. The most important of these is the throughput. Some of the services such as applications using streaming data can be very demanding requiring a high and guaranteed throughput, whereas some other services such as the short message service or background email download may not have specific throughput requirements. The complete QoS framework has been presented in Chapter 3.

The user throughput depends on the interference level on the used radio channels as well as on the amount of radio resources assigned to the user. The radio resources are defined by the number of timeslots and the amount of scheduling turns within these timeslots. The radio resources can be efficiently managed by conventional means presented in Chapter 6, but without DFCA the interference level on the radio channels cannot be accurately controlled. For example, in a random frequency hopping network, the frequency load defines an average level of interference, but large random deviations remain depending on the timeslot and the location of the user. Furthermore, the highly variable burst-by-burst CIR caused by random frequency hopping is somewhat detrimental to the enhanced GPRS (EGPRS) performance as shown in Chapter 7. All this makes the estimation of

Table 9.3. Summary of the simulation results

Simulated scenario	Gain in ideal network (limiting link in parentheses) (%)	Gain in typical network (limiting link in parentheses) (%)
EFR	66 (DL)	72 (DL)
AMR	35 (DL)	62 (UL)
75% EFR/25% AMR	63 (DL)	80 (DL)
50% EFR/50% AMR	77 (DL)	80 (DL)
25% EFR/75% AMR	78 (DL)	80 (DL)
3.6-MHz EFR	26 (DL)	N/A
4.8-MHz EFR	55 (DL)	N/A
3.6-MHz AMR	N/A	20 (DL)

the interference levels prior to the channel assignment inaccurate. DFCA provides means to estimate and control the level of interference in timeslot level and as DFCA utilises cyclic frequency hopping without relying on interference diversity, the burst-by-burst CIR variations are smaller allowing better EGPRS performance. Thus, with DFCA all the variables that define the achievable throughput can be kept in close control.

The available time slots in a cell are divided into dedicated time slots and shared time slots. The shared time slots are used for PS services and shared signalling channels only. They are able to serve multiple PS connections in one time slot, whereas the dedicated time slots can only serve one or two connections per time slot for full-rate and half rate services respectively. The dedicated time slots can carry all circuit-switched (CS) services as well as packet-switched (PS) services in the dedicated medium access control (MAC) mode. The time slots in each territory are kept consecutive within each TRX and the territory borders can move dynamically on the basis of the current need. An example of the territories is presented in Figure 9.27.

The DFCA algorithm chooses a common radio channel parameter (MA list, MAIO and TSC) for the whole shared territory. Multislot connections require that the same radio channel be used in all the assigned timeslots. The bursty character of the PS traffic makes TBF-based radio channel selection very challenging to realise. Using a common radio channel for each shared territory significantly simplifies the channel assignment task and it ensures maximum scheduling freedom when multiple users are sharing the same time slot. The selection of the radio channel parameters for the shared territories can be made automatically on the basis of the BIM during low traffic hours and the selected shared territory radio channel parameters are kept unchanged throughout the day.

As the radio channel selections for the connections on the dedicated territory are done in real time by the DFCA algorithm based on the CIR estimates, the maximum allowed interference level affecting each of the shared territory time slots can be controlled as well. This can be achieved by taking the used shared territory radio channels into account when the dedicated territory DFCA channel assignments are done. The required minimum CIR for each of the shared territory time slots can be defined automatically based on the throughput requirement. Thus, DFCA can enable high throughputs by providing high CIR to the shared territory time slots where it is needed. Since the high CIR is only provided where it is needed, the spectral efficiency of the network is still maximised.

On the other hand, the speech traffic on the dedicated territory is very vulnerable to the interference coming from the PS traffic. When performing a channel selection for a speech connection, DFCA can take the interference from the PS connections into account and thus protect the speech calls from excessive interference. The level of allowed interference is determined by the used channel type (EFR, HR, AMR HR, AMR FR...).

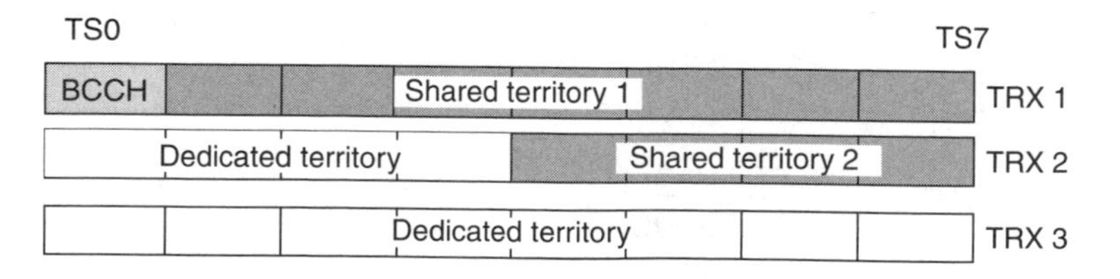

Figure 9.27. Shared and dedicated territories

DFCA is therefore able to manage a mixture of PS and CS connections with variable CIR requirements in a common frequency band without sacrificing the spectral efficiency.

Owing to the different characteristics of background best-effort traffic and high bit rate guaranteed throughput traffic, different approaches are employed. The background traffic is very suitable for multiplexing many users into the same radio channels and utilising packet-scheduling function to distribute the available capacity to the users. This approach maximises the time slot utilisation and is very effective in equalising capacity that is offered to different background data connections.

A high bit rate guaranteed throughput connection can be assigned either to a shared channel within the shared territory or to a dedicated channel in the dedicated territory. In most cases, a shared channel is the preferred option as it allows better utilisation of the radio resource as the excess capacity can be provided to background connections whenever the guaranteed throughput connections do not fully utilise the capacity. However, if the required throughput cannot be provided in the shared territory, e.g. because of high load or CIR restrictions, a guaranteed throughput connection can be assigned to a dedicated channel where the DFCA algorithm can pick the used radio channel individually for the connection. These connections are treated in a similar way as the CS connections, so one or more time slots are assigned exclusively for the connection with a minimum CIR requirement that is calculated on the basis of the required throughput and the number of time slots.

Connections with a guaranteed and high throughput requirement need high and constant CIR. DFCA can provide this by making sure that in a time slot that is used by such connection the same radio channel is reused so far away that the required CIR is achieved. Since the CIR estimation is done by utilising the DL measurement report sent by the mobile, the required minimum CIR can be accurately provided also in the cell border areas making it possible to provide high bit rates everywhere in the network. Once the connection is proceeding, the DFCA algorithm will make sure that new connections in the surrounding cells do not cause the CIR to drop below the target. This is the key to being able to guarantee a certain throughput.

9.6 Simulations of DFCA in Mixed CS and PS Services Environment

The performance of DFCA controlling both CS and PS traffic has been studied by network level simulations. Again, the performance has been benchmarked in the ideal network scenario where the network layout is regular and cell traffic is equal as well as in the typical network scenario with irregular network layout and variable cell traffic loads. These are exactly the same scenarios that were used in the DFCA simulations with only CS traffic presented in Section 9.4.

Multiple simulations have been performed with different CS and PS traffic loads. In each simulation, the maximum network load that still allows both the CS and PS traffic quality requirements to be met was found. The CS traffic quality requirement was the same 2% outage of FER > 4.2%, which has been commonly used throughout this book. The packet-switched (PS) traffic quality requirement was the average logical link control (LLC) level connection throughput of 70 kbps. The resulting maximum network performance for best-effort file transfer protocol (FTP) traffic and enhanced full-rate EFR speech traffic is presented in terms of spectral efficiency in Figure 9.28. The spectral

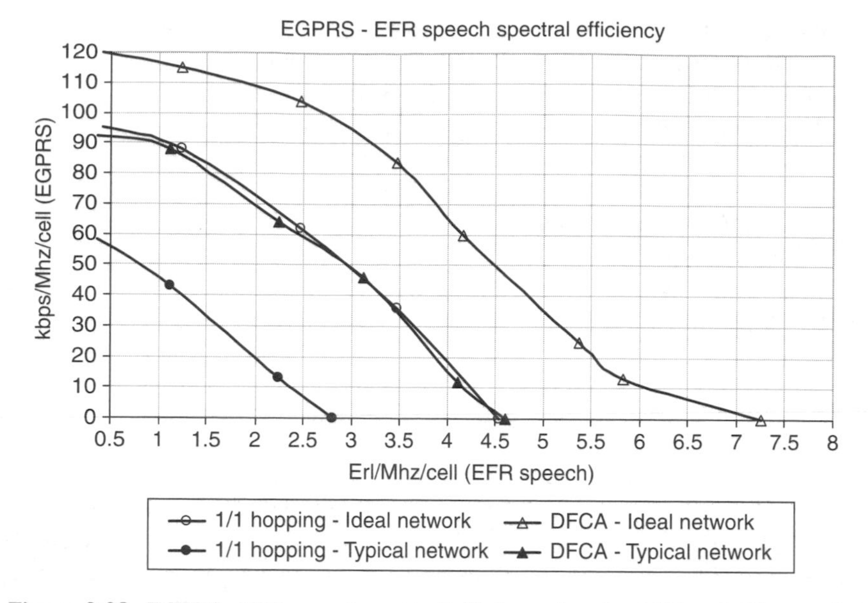

Figure 9.28. EGPRS–EFR speech spectral efficiency in mixed CS and PS scenarios

efficiency is determined separately for EFR speech traffic (x-axis) and EGPRS data traffic (y-axis) in each simulation point. In order to limit the interference from EGPRS traffic to the speech traffic, the EGPRS downlink power control with a fixed power reduction of 6 dB has been used in all simulated cases. The BCCH layer is not included in these simulations.

The results confirm the DFCA gains in case of pure CS speech traffic that have been presented in the previous sections. When there is no EGPRS traffic at all DFCA can achieve over 60% capacity gain in an ideal network scenario and in typical irregular network scenario the gain is even slightly higher. The results also show that considerable gains are maintained when more and more EGPRS traffic is introduced. It can also be seen that when the speech traffic approaches 0, the EGPRS spectral efficiency is still clearly higher in the DFCA case. This can be explained by the cyclic frequency hopping mode used in the DFCA cases that is more suitable to the EGPRS traffic as random burst collisions with very low CIR are avoided. More information can be found in [10, 11]. Note that the overlapping of the DFCA performance curve corresponding to the typical irregular network and the 1/1 hopping performance curve corresponding to the ideal network is just a coincidence.

9.7 Summary

As demonstrated in the previous sections, DFCA brings several important benefits. For an operator with high traffic and limited spectrum, the capacity gains provided by DFCA are very important. Another significant benefit is the excellent performance with HR channels that enables very high HR usage without compromising performance. Higher HR usage is reflected in higher BTS hardware utilisation and thus reduced BTS and transmission expenditure. It also helps in minimising the combining losses and the number of BTS antennas. DFCA can also efficiently manage a mix of CS and PS traffic in a common frequency band enhancing both CS and PS traffic performance.

On top of these network performance gains, DFCA simplifies network planning and operation. The DFCA TRXs do not have pre-assigned frequencies, frequency hopping parameters or training sequences as DFCA assigns the radio channel parameters automatically. This reduces the network planning tasks and simplifies the capacity upgrades as new TRXs can be introduced with minimum planning requirements.

Air interface synchronisation is required in order to be able to use frequency hopping with DFCA and to make accurate time slot level interference estimations possible. However, the benefits of synchronisation are not limited to DFCA. The interference cancellation techniques that can be used in both base station and mobile terminal receivers typically perform significantly better in synchronised environment, where the interference situation remains relatively static for the whole duration of the burst. Therefore, air interface synchronisation can provide significant network performance boost when terminals employing interference cancellation techniques are introduced in the network. Moreover, the DFCA channel selection algorithm can be enhanced to optimise the interference conditions for the terminals utilising interference cancellation. The current interference cancellation techniques can normally cancel the interference coming from the most dominant interference source. For the connections utilising interference cancellation, the DFCA channel selection algorithm can favour channels that have a suitable combination of both carrier to interference ratio CIR and the dominant to rest interference ratio (DIR) rather than basing the channel selection purely on the CIR. This way the gains of the interference cancellation techniques can be more effectively exploited. But even without these SAIC specific enhancements, DFCA is very beneficial in a network utilising SAIC as simulations indicate that the gains from SAIC and the gains from the basic DFCA are cumulative. More information of these interference cancellation techniques can be found in Chapter 6.

References

[1] Beck R., Panzer H., 'Strategies for Handover and Dynamic Channel Allocation in Micro-Cellular Mobile Radio Systems', *IEEE Vehicular Technology. Conference*, 1989, pp. 178–185.

[2] Cimini L. J., Foschini G. J., Chih-Lin I, Miljanic Z., 'Call Blocking Performance of Distributed Algorithms for Dynamic Channel Allocation in Microcells', *IEEE Trans. Commun.*, **42**(8), 1994, 2600–2607.

[3] Jordan S., Khan A., 'A Performance Bound on Dynamic Channel Allocation in Cellular Systems: Equal Load', *IEEE Trans. Veh. Technol.*, **43**(2), 1994, 333–344.

[4] Baiocchi A., Priscoli F. D., Grilli F., Sestini F., 'The Geometric Dynamic Channel Allocation as a Practical Strategy in Mobile Networks with Bursty User Mobility', *IEEE Trans. Veh. Technol.*, **44**(1), 1995, 14–23.

[5] Cheng M., Chuang J., 'Performance Evolution of Distributed Measurements-Based Dynamic Channel Assignment in Local Wireless Communications', *IEEE J. Selected Areas Commun.*, **14**(4), 1996, 698–710.

[6] Borgonovo F., Capone A., Molinaro A., 'The Impact of Signal Strength Measures on the Efficiency of Dynamic Channel Allocation Techniques', *IEEE Int. Conf. Commun.*, **3**, 1998, 1400–1404.
[7] Verdone R., Zanella A., Zuliani L., 'Dynamic Channel Allocation Schemes in Mobile Radio Systems with Frequency Hopping', *12th IEEE International Symposium on Personal, Indoor and Mobile Radio Communications*, Vol. 2, September/October 2001, pp. 157–162.
[8] GSM 05.05, Radio Transmission and Reception, 3GPP, Release 1999.
[9] GSM 05.08, Radio Subsystem Link Control, 3GPP, Release 1999.
[10] Salmenkaita M., Gimenez J., Tapia P., Fernandez-Navarro M., 'Analysis of Dynamic Frequency and Channel Assignment in Irregular Network Environment', *IEEE Vehicular Technology Conference*, Fall, 2002, pp. 2215–2219.
[11] Gimenez J., Tapia P., Salmenkaita M., Fernandez-Navarro M., 'Optimizing the GSM/EDGE Air Interface for Multiple Services with Dynamic Frequency and Channel Assignment', *5th International Symposium on Wireless Personal Multimedia Communications*, 2002, pp. 863–867.

10

Narrowband Deployment

Angel-Luis Rivada, Timo Halonen, Jari Hulkkonen and Juan Melero

Bandwidth is a scarce resource. Some networks have to be initially deployed with a very limited amount of frequency spectrum. A potential limitation in these narrowband deployment scenarios is the number of frequencies required for the broadcast control channel (BCCH) layer. Traditionally, global system for mobile communications (GSM) networks have been deployed with loose BCCH reuses owing to the permanent full-power transmission of BCCH transceivers (TRXs) in downlink (DL). In narrowband scenarios, this could lead to limiting the number of available frequencies for the hopping layer deployment, effectively limiting its performance and the number of additional deployed TRXs per cell. Therefore, in narrowband cases, the tightening of the BCCH reuses is critical to maximize the overall system performance and capacity.

In this chapter, the factors that differentiate narrowband networks from wideband scenarios are identified. Also, deployment guidelines and solutions are presented in order to achieve the maximum spectral efficiency with the scarce spectrum resources available. Radio Frequency (RF) hopping design is assumed, resulting in two effective layers with separated frequency band, the BCCH and the hopping layer. BCCH and RF hopping deployment schemes will be analysed, and functionality to tighten the BCCH reuse will be described, including BCCH underlay and transmit diversity techniques. Additionally, a specific adaptive multi-rate (AMR) analysis is presented as a tool to tighten BCCH reuse. Finally, the common BCCH functionality is presented as a mechanism to enhance the spectral efficiency in narrowband deployment scenarios.

10.1 What is a Narrowband Network?

Networks having an allocated spectrum smaller than 5 MHz per link (10-MHz full-duplex transmission with 5 MHz per link) are considered narrowband networks in the analysis performed in this chapter. Extreme examples of these scenarios can be found across the United States, where, in some cases, GSM has been initially deployed with 3.6 MHz or even less bandwidth. In other countries in Europe and Asia, narrowband scenarios are less

GSM, GPRS and EDGE Performance 2nd Ed. Edited by T. Halonen, J. Romero and J. Melero
 ISBN: 0-470-86694-2

usual due to the different spectrum licensing policies of the local regulatory bodies. However, in some specific situations, other issues limit the number of available frequencies. There are some typical scenarios in which the deployment bandwidth is limited:

- Frequency spectrum re-farming, e.g. technology migration from existing time division multiple access (TDMA) IS-136 or code division multiple access (CDMA) IS-95 to GSM/enhanced data rates for global evolution (EDGE).
- Narrow licensed frequency spectrum due to specific regulations of the country, i.e. 5-MHz allocations in US Personal Communication System (PCS) 1900 band.
- Microcell deployment (band splitting).

10.1.1 Frequency Spectrum Re-farming. Technology Migration

The term spectrum re-farming is used to describe reorganisation of certain frequency spectrum. The reasons behind re-farming are often related to a radio technology migration (e.g. from TDMA IS-136 or IS-95 to GSM/EDGE). However, re-farming is a time-consuming process and it does not happen overnight, mostly because of the fact that the terminal migration is a lengthy process. In most of the cases, wireless operators have to maintain both the old and new radio technology systems using a common frequency spectrum. Therefore, operators have to utilise minimum amount of spectrum for the initial deployment of the new radio technology in order to cope with existing customers in the old system during the migration period. In these circumstances, the deployment of GSM/EDGE requires specific attention.

In the case where a cellular operator decides to perform a technology migration from IS-95 (cdmaOne) or IS-136 (US-TDMA), with a limited available bandwidth, to GSM/EDGE, this will take place using a phased approach. In the first phase, at least a minimum amount of frequencies have to be re-farmed in order to deploy the GSM BCCH TRX layer. The second phase already provides enough capacity for majority of the traffic to be migrated to GSM/EDGE and in the last phase the whole frequency spectrum is re-farmed. Phases one and two are usually going to be GSM narrowband scenarios. Figure 10.1 illustrates an extreme example of IS-95 migration to GSM/EDGE. In this case, IS-95 is deployed in 5 MHz per link with one or two carriers being deployed per sector. In the first phase, only the BCCH layer is deployed, providing sufficient capacity for initial introduction of GSM/EDGE. During this first phase, the existing traffic is shared between the BCCH TRX and the IS-95 carriers. When the traffic carried by the GSM/EDGE system is such that the second IS-95 carrier becomes redundant, the second phase starts with the migration of the spectrum that is initially assigned to the second IS-95 carrier. Finally, when the terminal migration is completed, the remaining spectrum is migrated to GSM/EDGE.

In these technology migration scenarios, the potential cross system interference has to be taken into account. Appendix D includes an analysis of the cross system interference issues to be considered when different radio technologies are deployed within the same band.

10.1.2 Narrow Licensed Frequency Spectrum

Most European GSM operators have large spectrum allocations with dual band availability (900 and 1800 MHz). Also, most of the Asian operators have deployed the 1800

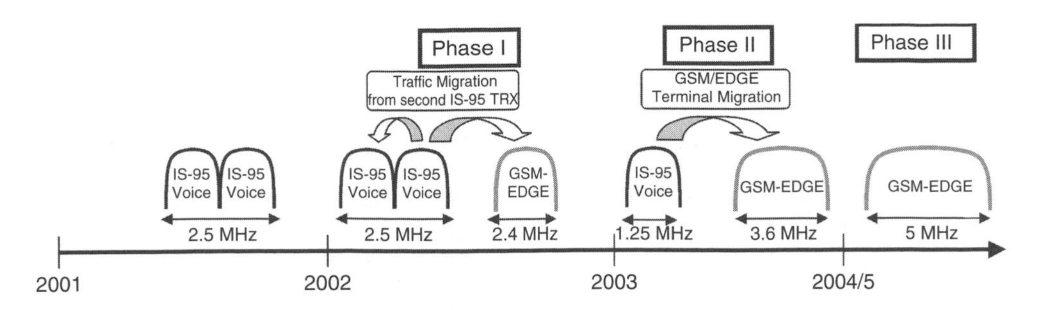

Figure 10.1. Frequency band re-farming example from IS-95 to GSM/EDGE

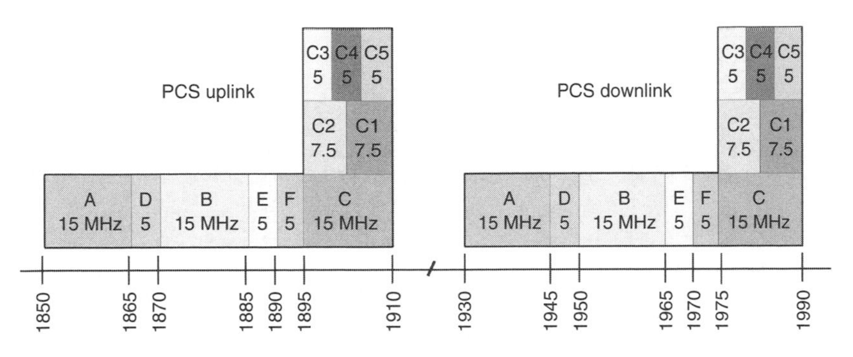

Figure 10.2. US PCS 1900 licensed bandwidth allocations

band to get more frequency spectrum. However, there are still operators facing GSM spectrum limitations. In America, especially in the United States, nation-wide operators may only be licensed 5 MHz in some of their markets (in the United States, the spectrum is not licensed nation-wide). Figure 10.2 shows the fragmented US PCS 1900 licensed bandwidth allocation.

Figure 10.3 illustrates the distribution of population according to the bandwidth availability of each market covered by the top five US operators. It shows that almost 20% of markets have to deploy cellular networks under narrowband conditions (10 MHz = 5 MHz in uplink and 5 MHz in downlink).

10.1.3 Microcell Deployment

Another situation in which many operators have to cope with narrow spectrum availability is in traffic-dense urban areas, where it is necessary to deploy dense microcellular networks. In this case, frequency spectrum is often split into two separate bands, one for macrocells and another one for microcells, both requiring separate BCCH allocations. A similar situation occurs in case of indoor systems deployment. Therefore, even with relatively large amounts of spectrum allocated, the operator has to sometimes deal with the defined narrowband scenario.

In order to overcome the potential performance problems associated with narrowband deployment, several solutions are described in the following sections.

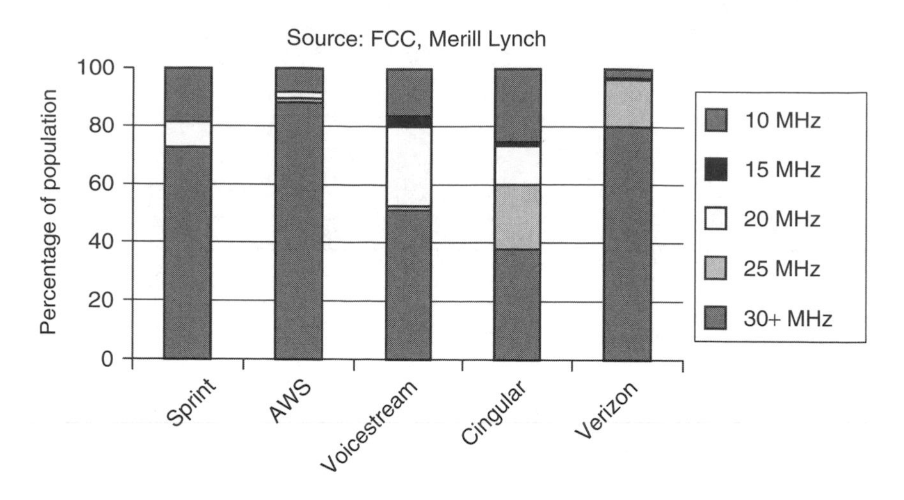

Figure 10.3. Spectrum distribution of top five US wireless operators (this allocation assumes that an agreement is reached regarding Nextwave 1900 band spectrum, with such spectrum reassigned according to 2001 spectrum re-auction results)

10.2 Performance of Narrowband Networks

An efficient BCCH deployment is critical in the narrowband scenarios. Owing to the nature of the GSM BCCH channel, GSM operators have traditionally reserved 15 frequencies (3 MHz) or more for the BCCH TRX deployment. This means that in the case of 3.6-MHz deployment (18 frequencies), only three frequencies would remain for additional transceivers. Not only does this limit the number of additional TRXs deployed but also seriously impacts the performance of the hopping layer since the frequency diversity gains would be substantially limited.

One option to enhance hopping layer performance is to tighten the BCCH reuse to release more frequency channels to become available for the hopping layer. However, the performance of BCCH layer may get degraded and since the traffic served by the BCCH TRX is likely to be a relatively high percentage of the total amount of traffic carried by the cell, this may have a significant impact on the overall performance. Therefore, for narrowband scenarios, in order to obtain the maximum spectral efficiency with the spectrum allocated, both BCCH and hopping layer performance have to be very carefully optimised. In order to benchmark the performance of the network, the overall bad quality samples (BQS) need to be considered. If a certain outage T of BQS or unsatisfied users is allowed in a network, in order to see the maximum capacity achievable in the network, the effect of BCCH performance can be expressed as follows:

$$\text{BQS}_{\text{BCCH}} \cdot x + \text{BQS}_{\text{hopping}} \cdot (1 - x) = T$$

where

- BQS_{BCCH} is the outage of BQS for BCCH TRXs
- $\text{BQS}_{\text{hopping}}$ is the outage of BQS for hopping TRXs

- x is the absorption of BCCH TRXs, i.e. the percentage of traffic carried by BCCH TRX (accordingly, $1 - x$ is the percentage of traffic carried by hopping TRXs).

This simple equation illustrates that the overall performance depends on how much traffic is served by each layer and its quality. For example, in narrowband cases, if tight reuses for BCCH are deployed without any other advanced feature, more frequencies would be used for hopping and the performance of hopping would be better (lower $BQS_{hopping}$), but the traffic served by BCCH TRX would be degraded (higher BQS_{BCCH}). Hence, the amount of traffic served by BCCH has to be considered when defining the best possible deployment strategy in narrowband scenarios.

10.3 Allocation of BCCH and Hopping Bands

In narrowband scenarios, the BCCH deployment will determine the frequency hopping–layer efficiency and capacity. This efficiency depends on how much uncorrelated the fading profiles of the bursts that make up a speech frame are. By using frequency hopping, these bursts use different frequencies and are affected by different fading conditions (frequency diversity). The higher the number of frequencies available to hop over and the higher the frequency separation among them (to reduce the partial correlation of the fading profile), the higher the gain of frequency diversity [1]. The use of hopping also randomises the interference signals received, having an averaging effect (interference diversity). Again, this effect is more efficient when there are more frequencies to hop over. These concepts are described in detail in Chapter 5.

In narrowband cases, the number of frequencies in the hopping layer limits the overall hopping layer performance. One approach to make them more efficient from the frequency diversity point of view would be to increase the frequency spacing as much as possible in order to increase the effective hopping bandwidth. This could be done by interleaving the BCCH and hopping frequencies. However, the BCCH TRXs transmit at full power in the DL and this would create adjacent interference in the hopping layer that might degrade the performance of the hopping layer even more. Another option would be to use separate bands to isolate the effect of adjacent channel interference from BCCH, but this would suffer limited frequency diversity gains. These configurations and some others (with and without guardbands) are analysed in the next sections for different bandwidths.

10.3.1 BCCH Reuse for Narrowband Scenarios

As previously mentioned, traditionally, practical GSM BCCH deployments have used a reuse of 15 or higher since the available spectrum usually did not impose limitations in the BCCH reuse to be deployed. However, in narrowband scenarios the need to tighten the BCCH reuse in order to release more frequencies for the hopping layer forces the careful analysis of the most efficient BCCH-deployment strategy. With efficient network planning (site locations, antenna locations and height, downtilts, beamwidth, etc.), a reuse of 12 frequencies for BCCH is achievable. In this chapter, narrowband networks with BCCH reuse of 12 have been considered. Some methods to even lower the BCCH reuse without compromising the performance will be covered along the chapter.

Using 12 frequencies for BCCH leaves only 6 frequencies for hopping transceivers in the case of 3.6-MHz bandwidth per link and 13 hopping frequencies in the 5-MHz case (one less in case 100-kHz guardband is required at both sides of the 5-MHz licensed band).

10.3.2 Narrowband BCCH and Hopping Deployment Strategies

Initially, two basic strategies can be used to allocate the available band between BCCH and hopping layers: either select a group of contiguous frequencies for BCCH and another one for hopping frequencies (hereinafter called 'non-staggered'); or interleave hopping frequencies with BCCH ones ('staggered' strategy). Both criteria are shown in Figure 10.4.

10.3.2.1 Impact on Frequency Diversity Gain

Figure 10.5 displays the link performance impact of the different allocation strategies for TU3 fading profile. Curves for non-hopping and uncorrelated hopping frequency configurations are also plotted. The 5-MHz case has a frequency diversity gain over 1 dB higher (at 1% frame error rate (FER)) than that of the 3.6-MHz case.

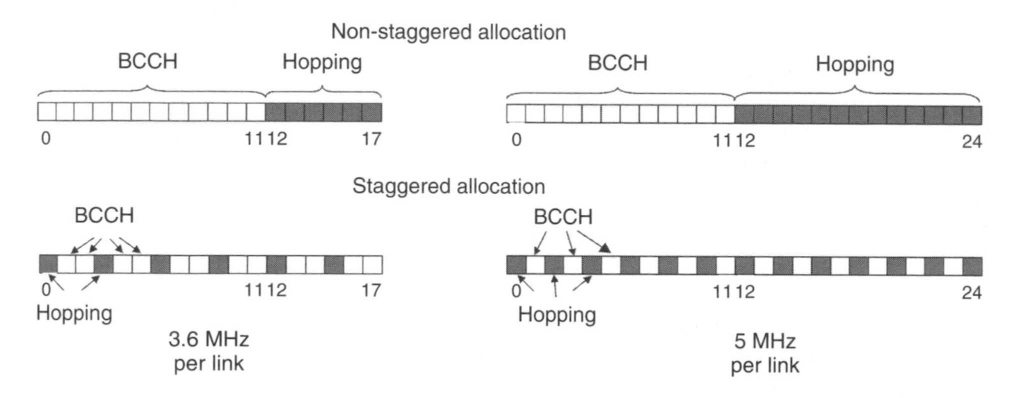

Figure 10.4. Basic BCCH band allocation strategies

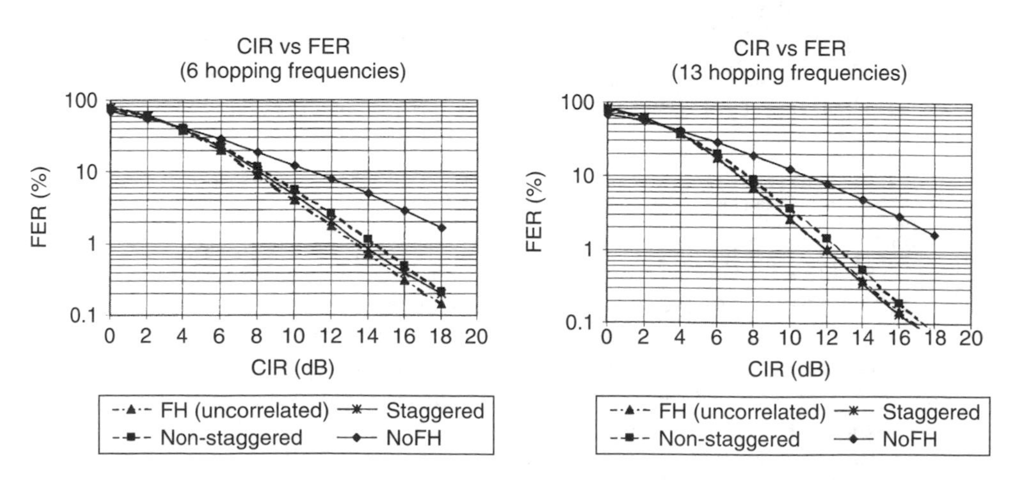

Figure 10.5. Impact on frequency diversity gain based on simulations

Additionally, the effect of the correlation of the hopping frequency fading profiles is also displayed in Figure 10.5. When using the non-staggered strategy, the hopping frequencies are closer to each other and their fading profiles are more correlated than in the staggered case, where there exists more separation among the hopping frequencies. The effect is that both strategies experience different diversity gains. For 3.6 MHz, the frequency diversity gain at 1% frame error rate (FER) is around 0.6 dB higher when the staggered configuration is used instead of the non-staggered one. For 5 MHz, the difference rises to 0.8 dB at 1% FER. In this last case, the staggered configuration reaches the maximum frequency diversity associated with 13 frequencies, i.e. the uncorrelated frequency diversity case.

10.3.2.2 System-level Performance: Influence of Adjacent Channel Interference

Figure 10.6 presents the overall system performance associated with the different configurations presented. In the x-axis, effective frequency load (EFL) is used as a measure of traffic load (as presented in Chapter 5). It can be seen that the non-staggered strategy outperforms the staggered one, despite the higher diversity gain of the latter one. The staggered case suffers from the permanent adjacent channel interference coming from BCCH frequencies in DL (always transmitting at full power). Simulations show that the power control gain is limited by continuous adjacent channel interference. This figure also shows that the performance with 13 hopping frequencies is better than the one with 6 hopping frequencies due to higher diversity gain.

10.3.3 Need of Guardband

According to the analysis from the previous section, the best performance would be achieved with an allocation strategy that provides high diversity gain without suffering degradation due to adjacent channel interference from BCCH. One way to reduce this

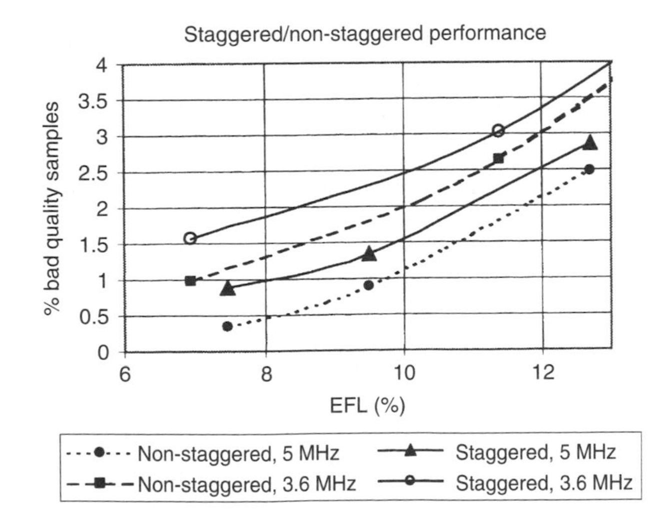

Figure 10.6. Network performance for different allocation strategies and bandwidths. Typical urban 3 km/h (TU3) and 1500 m inter-site distance, DL performance, hopping layer simulations

interference is the use of guardbands between BCCH and hopping frequencies. In the non-staggered case, using only a guard channel between the BCCH band and hopping band might reduce the adjacent interference coming from the BCCH frequency adjacent to the hopping pool of frequencies and therefore this could potentially improve the performance. However, this implies the loss of one hopping frequency with the consequent degradation in diversity gain and capacity. This and other strategies, which are displayed and described in Figure 10.7 and Table 10.1 respectively, are studied in this section.

Figure 10.8 displays the results for the studied configurations. In this case, the staggered case (CONF 5) has an equivalent performance to that with one list with contiguous frequencies (CONF 1, non-staggered). This is due to the larger inter-site distances used,

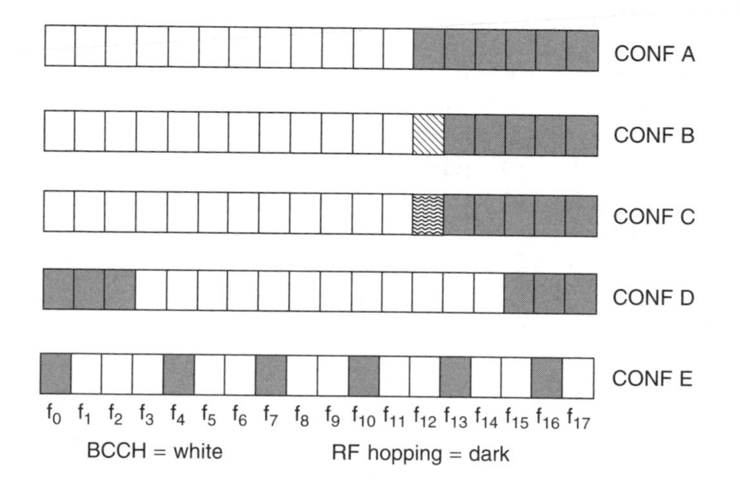

Figure 10.7. BCCH allocation strategies

Table 10.1. Description of different configurations tested

CONF A	Non-staggered case: MA-list = $\{f_{12}f_{13}f_{14}\ldots f_{17}\}$, BCCH = $\{f_0\ldots f_{11}\}$
CONF B	CONF A using guardband when BCCH carrier of the cell is adjacent to MA-list i.e. removing f_{12} from MA-list of those cells using f_{11} as BCCH carrier.
CONF C	1 list of contiguous frequencies using a guardband channel permanently, i.e. MA-list: $\{f_{13}f_{14}\ldots f_{17}\}$; f_{12} not used.
CONF D	MA-list built from 2 sub-bands at the beginning and end of whole-band MA-List: $\{f_0f_1f_2\} + \{f_{15}f_{16}f_{17}\}$. BCCH band is located in between those sub-bands
CONF E	Staggered case

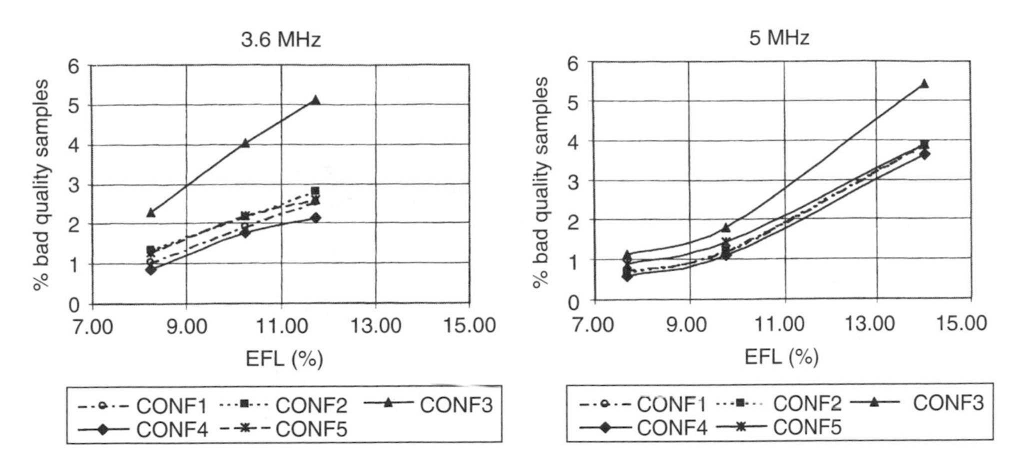

Figure 10.8. Network performance for different BCCH allocation strategies. TU3 and 3000 m inter-site distance, DL performance simulations

which makes the network to be more coverage limited and, the effectiveness of power control is reduced as higher power has to be transmitted to cover the cell area. From the displayed results, it is clear that guardbands between BCCH and hopping layer degrade the overall performance. Finally, the best performing case is CONF 4, even though its gain is only significant in the 3.6-MHz case. This configuration uses one Mobile Allocation (MA) list with frequencies taken from two sub-bands: one at the beginning of the total available band, and another one from the end of such band, which leads to an increase in the effective hopping band (high frequency diversity) with a low impact of the BCCH adjacent interference in the hopping layer.

The main conclusion from this study is that the non-staggered configuration outperforms the staggered one, not requiring a protecting guardband between the BCCH and hopping bands. CONF 4 yields the best performance, especially in very narrow deployment scenarios, where frequency diversity degradation is more significant.

10.4 BCCH Underlay

As mentioned before, in narrowband scenarios, the reduced number of frequencies to be deployed in the hopping layer limits the system performance and capacity. Therefore, the tightening of the BCCH reuse is critical. This section presents a functionality, capable of tightening BCCH reuses without degrading the overall performance, hence releasing frequencies to be deployed in the hopping layer and increasing the system performance and capacity. This solution is called *BCCH underlay*, and belongs to the reuse partitioning category presented in Chapter 6.

Traditionally, the limiting factor for the minimum BCCH reuse has been the performance of the traffic channels carried in the BCCH carrier (usually from timeslot (TSL) 1 to 7). As BCCH reuses tighten, the interference distribution will worsen and the performance of these traffic channels will degrade below the acceptable level. On the other hand, common signalling channels carried by TSL 0 of the BCCH carrier show a more robust link performance due to the coding and retransmission schemes they use (see

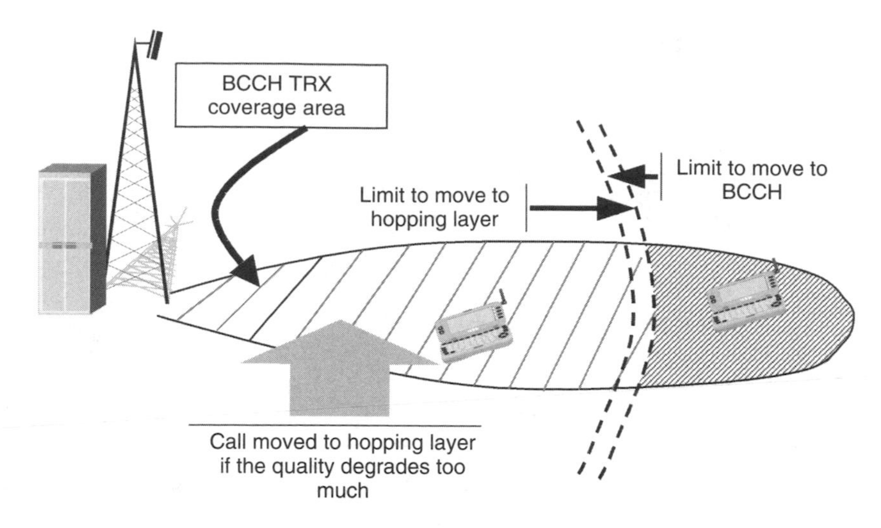

Figure 10.9. BCCH underlay principle

Chapter 12). *BCCH underlay* functionality will deploy a channel allocation strategy that ensures that the BCCH traffic channels' performance is adequate. Under these circumstances, the new limiting factor for the minimum BCCH reuse will be the common control channels' performance and this will allow the deployment of tighter BCCH reuses.

10.4.1 Description of BCCH Underlay Concept

BCCH underlay is a non-standardised functionality based on the reuse partitioning concept. Its basic target is to improve both coverage and capacity of radio networks. It ensures that the BCCH carrier is used only by connections whose performance can be guaranteed, while the other hopping TRXs not only cover the whole cell area but also increase the effective coverage due to their frequency hopping diversity gain.

Figure 10.9 illustrates the concept. The effective area where BCCH TRX serves traffic can be modified in order to ensure adequate performance of the calls allocated in the BCCH carrier while maintaining a high absorption. Therefore, the cell consists of two different logical layers: the BCCH TRX layer with a limited effective coverage area, and a hopping layer that serves the whole cell, providing extended coverage at the edge of the cells.

Finally, the signalling needed to set up and maintain the calls is properly established at the border of the cells, thanks to the robustness of signalling channels due to the coding and retransmission schemes used by these channels.

10.4.2 BCCH Underlay Simulation and Trial Results

10.4.2.1 Performance in 3.6-MHz Networks with 100% Enhanced Full Rate (EFR) Users

In order to understand the benefits of *BCCH underlay*, an analysis is performed for the scenario of 3.6 MHz, when only 18 frequencies are available. Two cases have been covered: a standard deployment with 12 BCCH and 6 hopping frequencies, and a *BCCH*

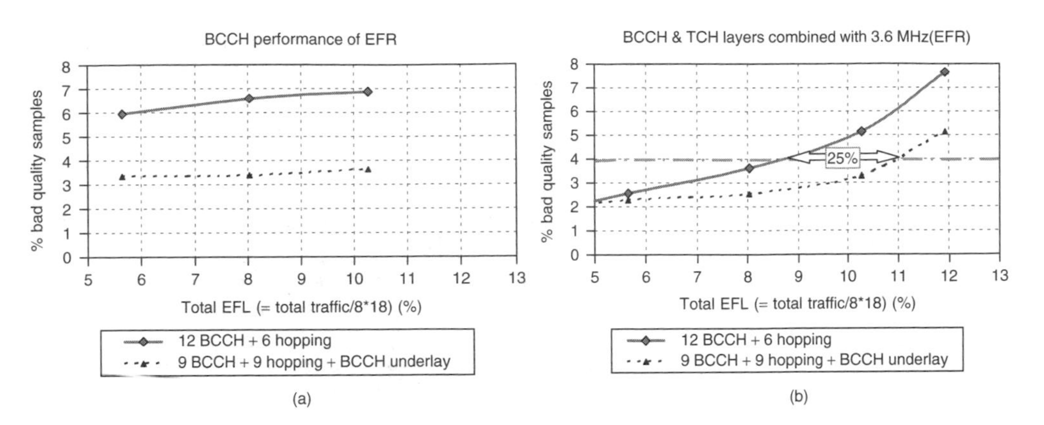

Figure 10.10. Performance of *BCCH underlay* for 3.6 MHz. TU3 and 1500 m inter-site distance, DL performance simulations

underlay deployment with 9 BCCH and 9 hopping frequencies. Figure 10.10 presents the DL performance results. The EFL used in the x-axis corresponds to the '*total EFL*', taking into account the traffic served and frequencies used by both BCCH and hopping layers.

Figure 10.10(a) displays the performance of the BCCH layer when *BCCH underlay* is used. The BCCH band allocation strategy used is the non-staggered one. Since in this strategy, BCCH frequencies are separated from hopping frequencies, the performance of BCCH in DL is almost traffic load independent, especially for the standard case. The *BCCH underlay* case shows an improved BCCH performance even if the number of BCCH frequencies deployed is lower. Figure 10.10(b) shows the overall performance benefits of *BCCH underlay*, including both BCCH and hopping layers. The use of *BCCH underlay* provides a capacity gain of 25%.

There are some important effects when BCCH underlay is used:

- The extended hopping list provides higher–frequency diversity gains and enables further flexibility in order to deploy efficient MAIO management strategies.
- The quality of BCCH layer is improved by the channel allocation scheme in use.
- The performance of hopping layer when using BCCH underlay improves when Total EFL is considered (Figure 10.11(a)). This effect is more noticed the narrower the band of the network is.
- However, when using BCCH underlay, the performance of the hopping layer suffers the effect of not taking full advantage of power control gain since the BCCH layer carries the calls closer to the cell site. In order to display such behaviour, the performance of hopping layer is displayed versus real hopping EFL in x-axis in (Figure 10.11(b)).

Apart from the controlled performance of BCCH layer, the gain of *BCCH underlay* is due to the higher number of frequencies available to hop over. The relative enhancement of *BCCH underlay* is therefore higher for very narrowband deployment cases, such as the ones used in migration scenarios. Table 10.2 below summarises the expected capacity gains for different reference case scenarios.

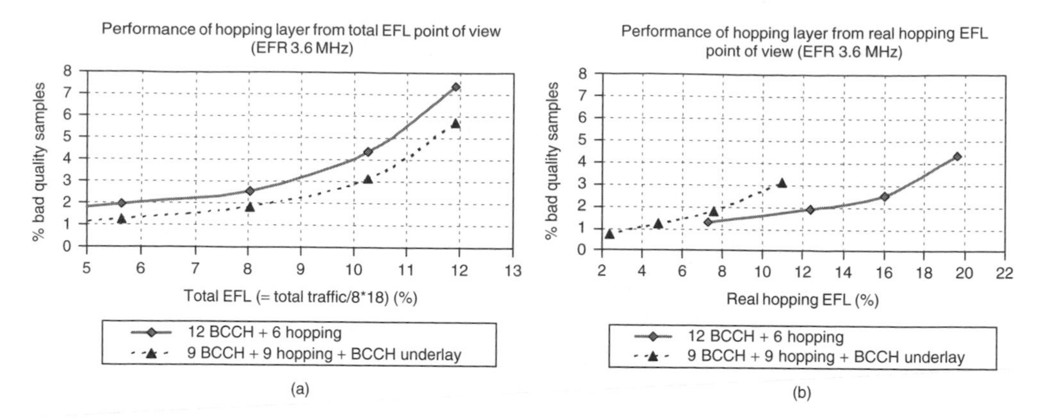

Figure 10.11. Performance of hopping layer for 3.6-MHz case with BCCH underlay (it has been added for this specific case, the performance of other cases to clarify the effect of BCCH underlay in hopping layer)

Table 10.2. Relation between the increase on hopping frequencies with BCCH underlay and the effective capacity gain

Reference case	BCCH underlay	% Hopping frequencies (%)	Capacity gain (%)
3.6 MHz: 15 + 3	12 + 6	100	>25
3.6 MHz: 12 + 6	9 + 9	50	25
5 MHz: 12 + 12	9 + 15	25	<10
5 MHz: 15 + 9	12 + 12	33	25–10

10.4.2.2 Performance of Narrowband Networks with 100% AMR Users (2.4 MHz and 3.6 MHz). BCCH Reuse 9

As already presented in Chapter 6, AMR Full Rate (FR) codecs bring higher robustness and better speech quality under degraded radio conditions than Enhanced Full Rate (EFR). This allows lower CIR values maintaining acceptable speech quality. For narrowband scenarios and 100% AMR penetration, that higher robustness could be exploited deploying tighter BCCH reuses and higher hopping frequency loads. It should be noticed that when using 100% penetration AMR terminals, the performance of the reference case with 9 frequencies is similar to the one measured when 12 frequencies and EFR codecs are used (see Chapter 6). However, realistic deployments have to count either on high percentages of legacy or roaming GSM EFR terminals, which might yield to a more conservative deployment. Anyhow, for this analysis a 100% AMR penetration is considered in order to understand its potential benefit.

The scenarios under analysis in this case are:

- 2.4-MHz network with 2 TRX/BTS.
- 3.6-MHz network with 4 TRX/BTS.

Figure 10.12(a) shows the performance of BCCH layer and Figure 10.12(b) shows the total performance. From them, it can be seen that BCCH performance for EFR and 12

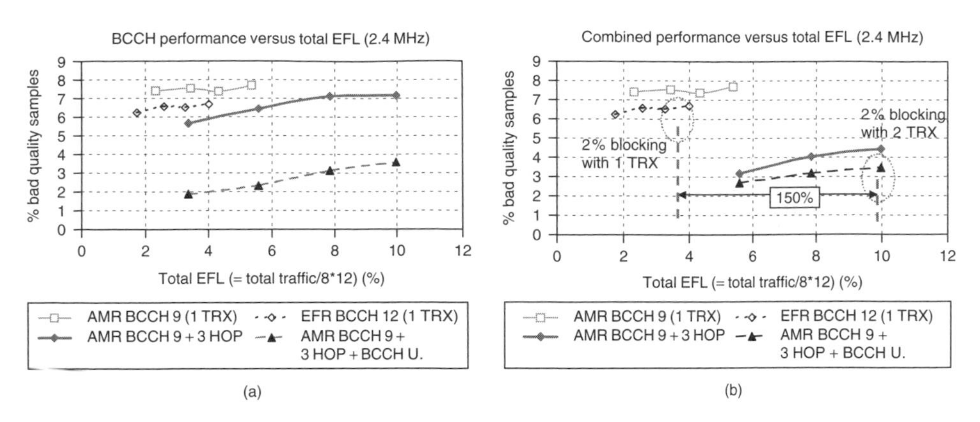

Figure 10.12. Performance using AMR + BCCH 9 in a 2.4-MHz network

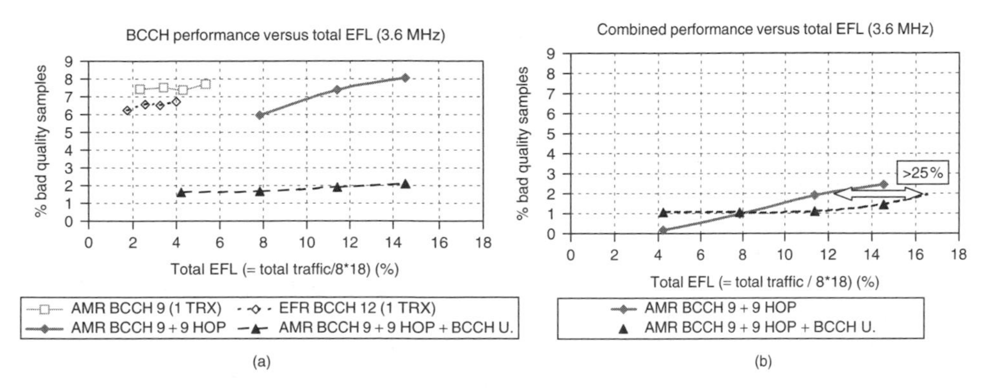

Figure 10.13. Performance using AMR+BCCH 9 in a 3.6-MHz network

BCCH reuse is quite similar to AMR-FR with reuse 9 when using only 1 TRX in total (BCCH TRX). However, the performance of BCCH when also having hopping TRXs in the simulations is a bit different. This effect is due to the different traffic distribution when having more than 1 TRX.

The first effect that we can see of using BCCH reuse of 9 with AMR-FR is that already a second TRX can be added for extremely narrowband cases of 2.4 MHz, increasing quite much the capacity when compared with the case of full EFR penetration (150% capacity gain taking traffic at 2% hard blocking).

Secondly, the impact of BCCH underlay even when using same BCCH reuse in reference case and BCCH underlay case of 9 can be analysed. From Figures 10.13 (a, b), it can be seen that with underlay BCCH, it is possible to control the quality of the BCCH and keep it under a certain limit. This way, the combination of AMR codec and underlay BCCH algorithm provide the best results for high total EFL values (higher than 25% capacity gain at 2% outage of BQS for 3.6 MHz).

10.4.2.3 Field Trial Results of BCCH Underlay

This feature has been tested in trials in Network 6 (see Appendix F). The improvement in performance in terms of drop call rate (DCR) is displayed in Figure 10.14. The network

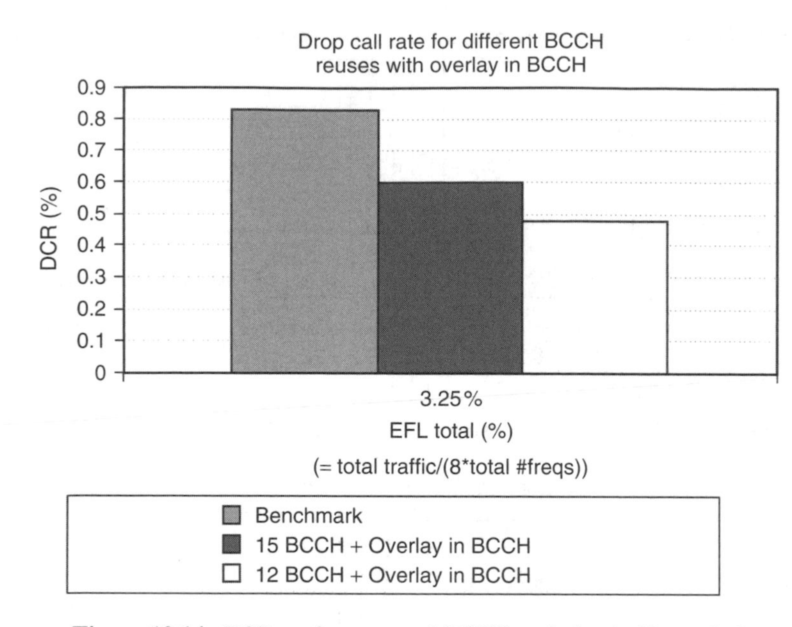

Figure 10.14. DCR performance of BCCH underlay in Network 6

had 22 frequencies available with around two TRXs per cell. The benchmark case did not have hopping in use. By using *BCCH underlay*, BCCH and hopping layers were defined, using both frequencies. The gains over the benchmark case are partly due to the use of hopping. However, it can be seen that the overall performance improves with the use of tighter BCCH reuses owing to the use of the *BCCH underlay* functionality.

According to the performance results presented in this section, *BCCH underlay* is an adequate solution to improve the performance of narrowband deployment scenarios, especially for extreme cases with very few frequencies available to hop over (3.6 MHz), which are likely in migration scenarios. For bandwidths higher than 5 MHz, other solutions, such as the reuse partitioning techniques presented in Chapter 6 provide better performance.

10.5 Transmit Diversity Gains

Chapter 11 provides a detailed description of various transmit diversity techniques, which can be used to improve DL performance. Suitable methods for GSM/EDGE are, e.g. delay diversity, phase hopping, and antenna hopping. The link performance gains of these transmit diversity techniques are most significant when the channel diversity is otherwise poor, like in the non-hopping BCCH layer. Simulation results in Chapter 11 show that by employing a combination of delay diversity and phase hopping (DD/PH), or antenna hopping, with two transmit antennas, the BCCH reuse factor could be tightened, maintaining the reference quality. Furthermore, transmit diversity also improves the hopping layer performance when the frequency hopping frequency diversity gain is limited in narrowband scenarios. A method to model transmit diversity in a system-level simulator, as well as examples of system-level capacity results, are presented in [2].

Figure 10.15 presents the impact of DD/PH on spectral efficiency for different bandwidths. The results are based on simulations that were performed with BCCH reuse factor of 12 in the reference one–transmit antenna simulation and reuse factor of 9 after the introduction of DD/PH. Similar results are achievable with antenna hopping in the simulated interference-limited network (see Figure 11.18 in Chapter 11) (figure refers to *BCCH reuse improvement with transmit diversity methods*). Released frequencies from BCCH band are available to the hopping layer (frequency hopping with 1/1 reuse), thus enabling a higher traffic in the network. As the results in Figure 10.15 show, transmit diversity significantly improved spectral efficiency when the available frequency spectrum was limited to 3.6 MHz. By using DD/PH only on the BCCH band, a network with 3.6-MHz total available band achieved the same spectral efficiency as it had 4.8 MHz with one transmit antenna. Furthermore, the usage of DD/PH in the hopping band further improved the efficiency. The usage of transmit diversity improves performance in all the simulated bandwidths, but it is especially important in the narrowband networks since it improves their spectral efficiency to close to that of the wider bandwidth scenarios.

Usage of the above-mentioned transmit diversity methods improves only DL quality. However, as the most significant gains were achieved on the BCCH layer and the hopping layer is generally DL limited, it is assumed that the results translate directly into the system capacity gains. In addition, uplink performance can be further enhanced by various multi-antenna techniques as described in Chapter 11.

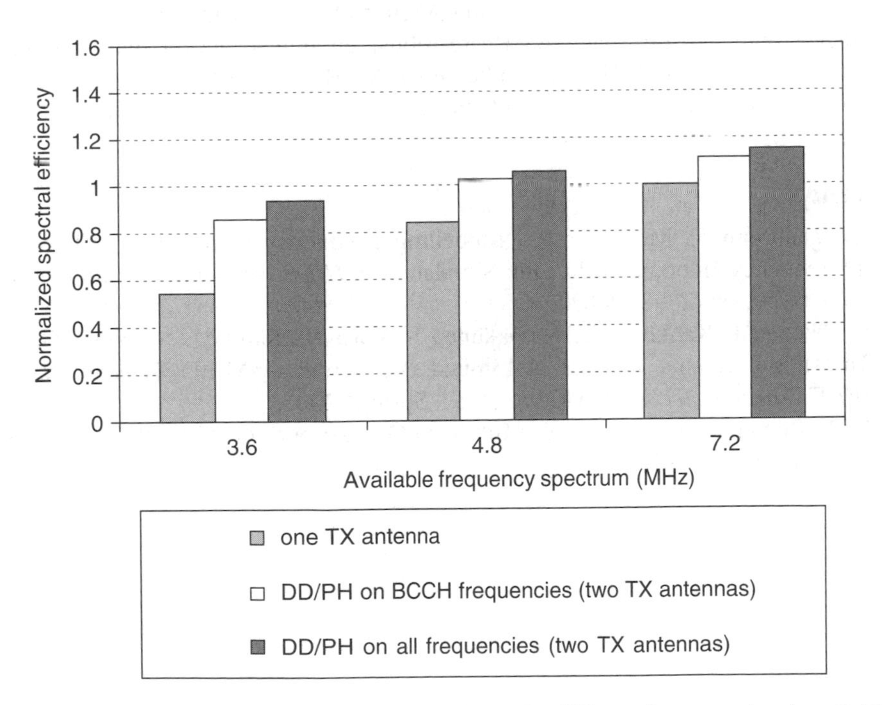

Figure 10.15. Impact of DD/PH on spectral efficiency for different frequency band availability. Downlink simulations for GSM-EFR speech

10.6 Common BCCH

Many operators have access to two or more frequency bands (generally 900/1800 in Europe and Asia Pacific (APAC), and 800/1900 in America). For those operators, and especially if one of the licensed frequency bands is narrow, common BCCH is a competitive solution.

In standard deployment of GSM/EDGE in multiple frequency bands, each band would require a separate BCCH layer, thus limiting the overall performance of the system since the BCCH layer is less spectral efficient than the hopping layer. Common BCCH functionality is introduced in the European Telecommunications Standards Institute (ETSI) GSM specifications [3]. The solution consists of the deployment of a single BCCH layer for both frequency bands. The main advantages of the common BCCH functionality are improved trunking gain, higher spectral efficiency, better quality due to decreased handover rate and better neighbour reporting performance. Common BCCH can be configured in only one of the bands of operation when resources across all bands are co-located and synchronised.

10.7 Other Strategies to Tighten BCCH Reuse

In realistic deployments, scenarios with mixed EFR, AMR and (E)GPRS traffic have to be considered. As has been explained through the chapter, in narrowband scenarios, there is a need to ensure good performance of both BCCH and hopping layers. One deployment strategy to tighten BCCH reuses without the need to use the BCCH underlay functionality would be to ensure that only (E)GPRS and AMR traffic is assigned to the BCCH TRX. Both packet-switched traffic and AMR are robust enough to accept tighter reuses. An efficient way to get all BCCH traffic to be either AMR or (E)GPRS is to give priority in the channel allocation for these types of traffic to be placed in the BCCH.

References

[1] Klingenbrunn T., Mogensen P., 'Modelling Frequency Correlation of Fast Fading in Frequency Hopping GSM Link Simulations', *IEEE Vehicular Technology Conference*, 1999, pp. 2398–2402.

[2] Hulkkonen J., Kähkönen T., Hämäläinen J., Korpi T., Säily M., 'Capacity Gain from Transmit Diversity Methods in Limited Bandwidth GSM/EDGE Networks', *57th IEEE Vehicular Technology Conference*, Spring, 2003.

[3] GSM Recommendations 03.26, Multiband Operation of GSM 900/1800 by a Single Operator, ETSI, 1998.

11

Link Performance Enhancements

Riku Pirhonen, Matti Salmenkaita, Timo Kähkönen and Mikko Säily

Radio link performance sets one of the most important baselines for the overall network performance. The basic link performance is determined by the GERAN specifications that define the radio channel structure and transmission techniques such as modulation, channel coding and interleaving. The specifications also define the minimum performance that base and mobile station transceivers must achieve. It is however possible to enhance performance with a number of additional radio link techniques that are feasible at the network side and supported by standard terminals. This chapter reviews some of these techniques by analysing their feasibility and performance, as well as presenting the basics of radio link performance.

First, Section 11.1 introduces key link performance measures, minimum performance requirements and link budget calculations. Section 11.2 gives an overview of the enhanced radio link techniques and presents link-level simulation results. Finally, Sections 11.3 and 11.4 discuss achievable capacity and coverage gains based on network simulations and field measurements.

11.1 Basics of Radio Link Performance

The GERAN radio link performance depends on the prevailing radio environment as well as on the performance of radio transmission and reception. This section focuses on the effects of transceiver equipment. The underlying GSM radio system and general propagation effects have been described in Chapter 5.

For a transmitter, the most important performance measures include transmit power, output spectrum, frequency tolerance and modulation accuracy. While the maximum available output power is pivotal for coverage-limited links, the dynamic range and accuracy of power control are more important in interference-limited situations. Output spectrum

GSM, GPRS and EDGE Performance 2nd *Ed.* Edited by T. Halonen, J. Romero and J. Melero
 ISBN: 0-470-86694-2

is also more critical in interference-limited operations as it affects the amount of interference between adjacent frequency channels within a network as well as between different networks. On the other hand, frequency tolerance and modulation accuracy have a direct impact on the signal detection at the receiver, and are thus important in any environment. Especially for 8-PSK, high modulation accuracy is crucial, and because of the variable envelope of the 8-PSK signal, modulation accuracy implicitly measures linearity of the power amplification. To achieve sufficient accuracy with practical power amplifiers, 2 to 4 dB back-off from the maximum output power is usually required for 8-PSK.

In coverage-limited situations, the most important receiver characteristic is sensitivity, which is the minimum received signal level at which the maximum error rates do not exceed the specified values for a given radio propagation condition. The baseline error rate measure is the raw bit error rate, which is the bit error rate before error correction. In addition, the frame erasure rate (FER) (or block error rate (BLER)) and residual bit error rate must be below the value specified for the logical channel in question. The FER is the ratio of frames that the receiver rejects and residual bit error rate is the ratio of bit errors within non-rejected frames. In case of speech channels, there is a trade-off between these two measures for the best speech quality. In case of data transmission, high reliability is required in error detection and BLER is the ratio of blocks including bit errors after error correction.

In interference-limited environment, it is important for a receiver to perform properly when the antenna captures interfering signals in addition to the wanted signal. Co-channel interference is most harmful, as the front-end filtering does not attenuate it. In the presence of signals using adjacent frequencies, the performance degradation is mainly caused by the overlapping output spectrum. For larger frequency offsets, the characteristics of radio frequency equipment (such as transmitter wideband noise, receiver and transmitter intermodulation) have a more significant role. The receiver performance in the presence of co-channel and adjacent channel interference separated by 200, 400 and 600 kHz is measured as the interference ratio (for co-channel C/I_c, for adjacent channel C/I_a) at which the specified error rates are not exceeded.

11.1.1 Minimum Performance Requirements

The operators, regulators and equipment manufacturers jointly define the minimum performance requirements for base and mobile stations as part of the standardisation process. The requirements for GERAN are given in the 3rd Generation Partnership Project (3GPP) GERAN specification series [1]. For example, the maximum error rate requirements for sensitivity and interference ratio measurements are specified for different services, radio channel conditions, frequency bands, mobile speeds, frequency hopping operation mode and different equipment classes. The maximum allowed error rates as such do not represent the minimum quality for a certain service, but the test cases are defined so that the piece of equipment that passes the conformance tests has at least the reference performance on which network operators can rely in the network deployment and end users when purchasing terminal equipment. In addition, for example, the output spectrum requirements are necessary in order to prevent interference to other users and networks, or to other services.

The various test parameters stated above for sensitivity and interference ratio measurements make a large set of test case combinations. As an example, Table 11.1 shows

Table 11.1. Minimum sensitivity requirements for TU50 (Ideal frequency hopping) channel on the GSM 850 and 900 band[a]

Service	Error rate			Reference sensitivity level	
	FER	Class 1b RBER[b]	Class II	Base station (dBm)	Mobile (dBm)
TCH/AFS12.2	2.4%	1.5%	—	−104	−102
TCH/AFS10.2	0.85%	0.15%	—	−104	−102
TCH/AFS7.95	0.045%	0.032%	—	−104	−102
TCH/AFS7.40	0.069%	0.016%	—	−104	−102
TCH/AFS6.7	0.017%	0.022%	—	−104	−102
TCH/AFS5.9	<0.01%	0.001%	—	−104	−102
TCH/AFS5.15	<0.01%	0.001%	—	−104	−102
TCH/AFS4.75	<0.01%	0.001%	—	−104	−102
TCH/AHS7.95	20%	2.3%	5%	−104	−102
TCH/AHS7.40	16%	1.4%	5.3%	−104	−102
TCH/AHS6.7	9.2%	1.1%	5.8%	−104	−102
TCH/AHS5.9	5.7%	0.51%	6%	−104	−102
TCH/AHS5.15	2.5%	0.51%	6.3%	−104	−102
TCH/AHS4.75	1.2%	0.17%	6.4%	−104	−102
PDTCH/MCS-1	BLER < 10%			−103	
PDTCH/MCS-2	BLER < 10%			−101	
PDTCH/MCS-3	BLER < 10%			−96.5	
PDTCH/MCS-4	BLER < 10%			−91	
PDTCH/MCS-5	BLER < 10%			−97	−94
PDTCH/MCS-6	BLER < 10%			−94.5	−91.5
PDTCH/MCS-7	BLER < 10%			−88.5	−84
PDTCH/MCS-8	BTS BLER < 10, MS < 30%			−84	−83
PDTCH/MCS-9	BTS BLER < 10, MS < 30%			−80	−78.5

[a] Handset value for 'small handset' category.
[b] RBER = residual bit error rate.

the maximum error rates in the sensitivity measurement for typical urban (TU) radio propagation conditions. These specific test cases for adaptive multi-rate (AMR) traffic channels (TCH) and packet data traffic channels (PDTCH) assume ideal frequency hopping, mobile speed of 50 km/h and 850- or 900-MHz band. The exact technical characteristics and methods of conformance tests are given in [2] for mobile stations and in [3] for base stations.

In order to pass the conformance test, the measured sensitivity must be less than a specified limit, called the *reference sensitivity level*. The reference sensitivity level depends on the type or class of the mobile or base station. In addition, different levels have been specified for packet-switched channels as can be seen in the example test case of Table 11.1. Similarly, the actual interference ratio must be less than the specified interference ratio, which varies according to the equipment type and test parameters discussed above.

The performance figures stated in the technical data of radio equipment products are usually given as 'typical', 'guaranteed' or 'better than'. The 'guaranteed' or 'better than' figures take into account possible variations in performance introduced in the manufacturing and installation processes. Typically, the actual sensitivity of high-end equipment is several decibels better than the reference sensitivity. Moreover, in the sensitivity and interference ratio tests for conformance, the receivers equipped with diversity are tested by disabling the diversity input. As will be shown in Section 11.2, diversity further improves sensitivity by several decibels.

In addition to basic performance requirements, enhanced general packet radio system (EGPRS)-capable mobile stations have to meet a specific performance requirement for incremental redundancy (IR). IR is a type II hybrid automatic repeat request (H-ARQ) retransmission scheme that sets, for example, additional memory requirements for the receiver as described in Chapter 7. The test is carried out in MCS-9 mode in a static channel at a signal level of −97 dBm. In these conditions, the long-term throughput per timeslot (TSL), when measured between the logical link control (LLC) and radio link control (RLC)/medium access control (MAC) layers has to exceed 20 kbps.

11.1.2 Radio Link Power Budget

The radio link between a base station and a mobile station is best described by a radio link power budget, which numerically defines the relation between the radio link characteristics and the maximum service distance from the base station, i.e. the *cell range*.

In order to define the power budget, the minimum acceptable signal level at the receiver input needs to be specified. In interference-free spectrum, the main cause of transmission errors is the system noise that is present in the signal detection. By neglecting the amplification in the receiver that also applies to the wanted signal, the noise level can be calculated as the product of Boltzmann's constant k, receiver temperature T, receiver noise bandwidth B and receiver noise factor NF. Then, by assuming certain propagation conditions, the signal-to-noise ratio (SNR) that is required for a particular service to achieve a certain minimum quality of service can be determined. Finally, the required signal level in the receiver input is simply the noise power plus SNR, in decibel scale. Table 11.2 shows a numeric example of these calculations, which apply when the interference level can be assumed to be well below the system noise level. Separate calculations are usually needed for uplink (UL) and downlink (DL) directions. When capacity requirements demand a low frequency reuse, interference typically overrides noise, which must be taken into account by adding an adequate interference degradation margin to the signal

Table 11.2. Calculation of required signal strength

A	Thermal noise	kT	−174.0 dBmW/Hz	
B	Noise bandwidth	B	52.6 dB Hz	180 kHz
C	Noise factor	NF	5 dB	Estimate
D	Noise[a]	N	−116.4 dBm	D = A + B + C
E	Signal-to-noise ratio	SNR	6 dB	Specific for the service
F	Isotropic signal power		−110.4 dBm	F = D + E

[a] At the receiver filter output.

level. In this case, apart from the constantly transmitted broadcast control channel (BCCH) frequencies, the actual cell range depends on the prevailing traffic load.

When the minimum acceptable signal level has been specified, the remaining task is to derive the cell range starting from the maximum output power of the transmitter, as illustrated in Figure 11.1. This must be done, of course, separately for uplink and downlink directions. In the downlink direction, the losses caused by base station equipment such as duplexer, combiner, filter, connector and cable, are first subtracted (in decibels) from the output power measured. Then the isotropic antenna gain of the transmitter is added to yield the effective isotropic radiated power (EIRP), which is equivalent to the power transmitted towards the receiver if the signal was radiated equally in all directions. The difference between EIRP and the power level in the receiver input then includes the propagation path loss, the isotropic antenna gain of the receiver, and cable and connector losses in the receiver in case of a base station. In addition, in case of a handheld mobile station, the absorption, detuning and mismatch of the antenna by the human body causes a so-called *body loss*, which must be included in the power budget. These calculations are exemplified in Table 11.3 for a voice service using AMR 12.2 and 5.15 kbps codecs, and for a packet data service assuming minimum bit rate requirements of 16, 32, and 48 kbps. The data rates are per time slot, so that, for example, a multislot mobile station capable of receiving or transmitting data on three time slots would be able to provide three times higher data rates.

Table 11.3 illustrates the maximum allowed propagation path loss in dB for each service. For any particular environment, the actual cell range can then be calculated with a propagation model tuned for the environment. Nevertheless, however accurate the

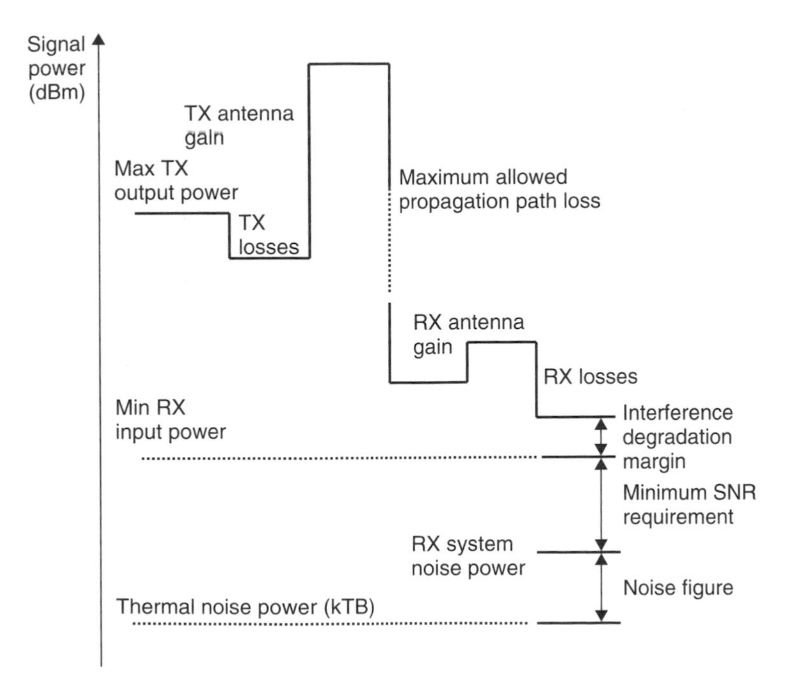

Figure 11.1. Link budget elements

Table 11.3. Example radio link power budgets for voice and data services

	AMR 12.2		AMR 5.15		16 kbps/1 TSL		32 kbps/1 TSL		48 kbps/1 TSL	
	Uplink	Downlink	Uplink	Downlink	Uplink	Downlink	Uplink	Downlink	Uplink	Downlink
Transmitter										
TXTransmitted power[a]	30 dBm	45 dBm	30 dBm	45 dBm	26 dBm	43 dBm	26 dBm	43 dBm	26 dBm	43 dBm
Combiner losses	0	2.4 dB	0	2.4 dB	0	2.4 dB	0	2.4 dB	0	2.4 dB
Cable and connector losses	0	2 dB	0	2 dB	0	2 dB	0	2 dB	0	2 dB
Body loss	3 dB	0	3 dB	0	0		0	0	0	0
TX antenna gain	0 dBi	18 dBi	0 dBi	18 dBi	0 dBi	18 dBi	0 dBi	18 dBi	0 dBi	18 dBi
Receiver										
Antenna gain	18 dBi	0 dBi	18 dBi	0 dBi	18 dBi	0 dBi	18 dBi	0 dBi	18 dBi	0 dBi
Body losses[b]	0 dB	3 dB	0 dB	3 dB	0 dB	0 dB	0 dB	0 dB	0 dB	0 dB
Link										
SNR requirement	7 dB[c]	7 dB[c]	2.2 dB	2.2 dB	8.5 dB	8.5 dB	15.6 dB	15.6 dB	27 dB	27 dB
Receiver sensitivity[d]	−109.4	−104.4	−114.4	−109.2	−107.9	−102.9	−100.8	−95.8	−89.4	−84.4
Max. allowed path loss	154.4	160	159.2	164.8	151.9	159.5	144.8	152.4	133.4	141

[a] 8-PSK transmission is assumed to require 2 dB back-off in base station and 4 dB back-off in the mobile.
[b] Body loss is estimated at 3 dB for voice services due to the position of the mobile. The assumption is that, for data services the terminal will be operated in a position that does not introduce body loss.
[c] Estimated performance for 1% FER in TU3 multipath channel with frequency hopping.
[d] Base station receiver sensitivity is system noise level −116.4 dB plus required signal-to-noise ratio in terms of energy per symbol-to-noise ratio (E_s/N_0). For the mobile, assumption is 5 dB weaker implementation.

path loss prediction is, it usually can match only with the average of path losses for a certain distance between transmitter and receiver. The remaining uncertainty can be often modelled as a slow fading component following a lognormal distribution. Then the cell range can be derived from the maximum path loss for a certain coverage criterion called *location probability*. For example, it can be required that the probability of getting at least the minimum quality of service is more than 95% over the cell area. For this an additional slow fading margin needs to be included in the link budget. The margin depends on the path loss factor and standard deviation of the fading. For example, if the standard deviation of the slow fading component is 7 dB and the path loss slope is 3.5, then the fading margin would be 7.3 dB for a location probability of 95%.

For voice service, it is practical to balance the link budget so that the cell ranges for uplink and downlink are equal. Otherwise, the better direction would have good performance unnecessarily. In the examples of Table 11.3, downlink allows 5.6 dB higher path loss than uplink. In typical coverage-limited deployments, uplink performance is further improved by 4 to 5 dB with diversity reception, which would almost balance these link budget examples. The packet data link budgets in Table 11.3 have an additional 2 dB difference between uplink and downlink for the same bit rate because the mobile station is assumed to have 2 dB greater power amplifier back-up for 8-PSK modulation to achieve the modulation accuracy requirements. On the other hand, data services typically set a higher bit rate requirement for downlink, so that link budget with different SNR requirements would need to be compared.

For additional information about radio link budgets and cell planning in GERAN, see the recommendations in [4].

11.2 Overview of Radio Link Enhancements

GERAN specifications define a number of advanced techniques for efficient radio transmission. As discussed in the preceding section, these standardized techniques as well as specified minimum performance requirements for physical equipment set the baseline for radio link performance. Performance can be enhanced beyond this baseline by using state-of-the-art equipment and introducing additional transmission techniques that need not be standardized.

11.2.1 Uplink Diversity Reception

In order to compensate the link budget imbalance between uplink and downlink, it has been a common practice to have diversity reception at the base station. By using multiple antennas, more signal energy can be collected, and this improves SNR when the signals from the antennas are coherently combined. In addition, it is possible to achieve diversity gain in multipath environment, which is based on the idea of having independently faded radio signals from MS to two or more receiving antennas at the base station. With two antennas, this can be achieved by having sufficient horizontal spatial separation between the antennas or antennas with orthogonal polarisations, and with four antennas by having two horizontally spaced pairs of cross-polarized antennas.

The signals from the diversity branches can be combined before symbol detection algorithm (pre-combining), within detection, or after detection (post-combining). When the signals are perturbed only by system noise, optimal combining has a simple implementation

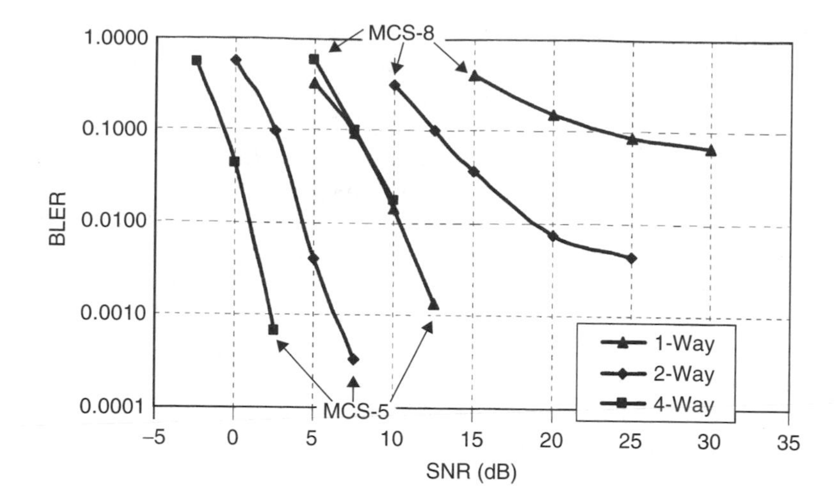

Figure 11.2. Maximum ratio combining performance for two-way and four-way diversity reception

as an SNR-weighted pre-combination, called *maximal ratio combining* (*MRC*). Figure 11.2 illustrates MRC performance by showing simulated block error ratios as a function of SNR for two coding schemes on the packet data channel. For these results, frequency hopping was used, the channel profile was typical urban at 50 km/h, and signal fading in the antennas was assumed uncorrelated.

Exact diversity gains will depend on the environment and the service. In Figure 11.2, the weaker channel coding scheme MCS-8 gains more from diversity than the more robust MCS-5. As MCS-5 has pretty strong channel coding, frequency hopping already provides diversity gain that is seen in terms of BLER performance. Therefore there is about 2-dB gain on top of the theoretical 3-dB power gain increase that two receiver antennas provide. MCS-8 in turn has weak channel coding, so frequency hopping does not provide that much gain. Therefore, the diversity gain from multiple antennas contributes more to the performance than in the MCS-5 case. For voice service, it is usually safe to assume 4- to 5-dB overall gain from the diversity reception in the link budget.

11.2.2 Uplink Interference Rejection

When the main source of errors is interference, typically coming from co-channel and adjacent channel transmitters, an additional performance improvement can be achieved by interference rejection methods. One of the simplest methods, from the algorithm point of view, is a pre-combining diversity algorithm called *interference rejection combining* (*IRC*). It decorrelates the interfering signals from the antenna branches so that their total power in the combined signal will be minimized. Figure 11.3 shows interference performance with IRC for two packet data MCSs in terms of block error rate. The results are for TU3 channel with frequency hopping and assume a single co-channel interfering signal that is synchronized with the wanted signal on the time slot level. Because practical IRC algorithms rely on the estimates of the interfering signals taken from the training sequence part of the received burst, their performance degrades when the signals

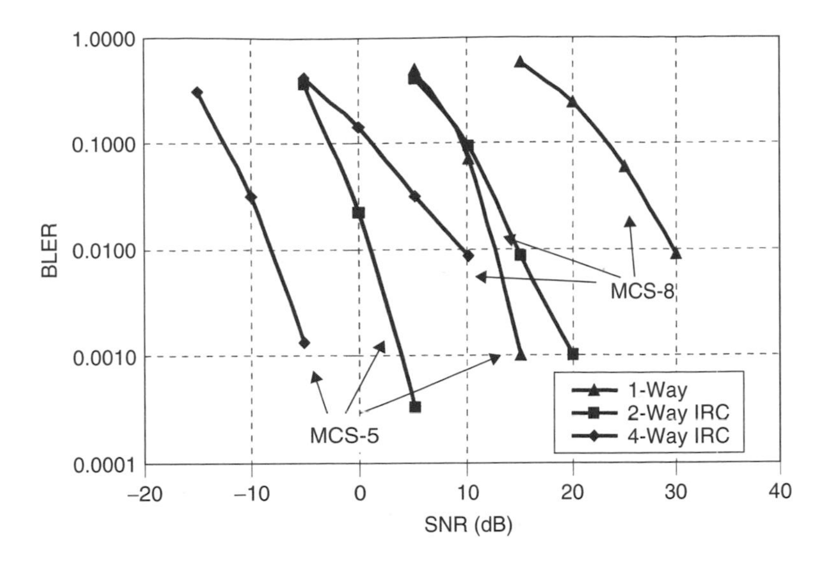

Figure 11.3. Interference rejection combining performance for two- and four-way diversity reception

change during the burst. Therefore, synchronisation is an important prerequisite in order to achieve maximum gains.

Generally speaking, IRC is able to efficiently reject $M/2$ interfering signals, where M is the number of antenna branches. In a realistic case where two antennas capture more than one interfering signal, or one signal close to the system noise level, link performance depends on the dominant to rest interference ratio (DIR). Figure 11.4 includes TCH/FS frame erasure rate results in case of three different DIR values for IRC in case of synchronous interference. Clearly when DIR approaches zero in linear scale (i.e. interference becomes noise-like), IRC performance approaches that of MRC.

As the results above show, interference rejection combining is a very powerful method to reduce the harmful effects of co- and adjacent channel interference. Nevertheless, in order to significantly benefit in the system spectrum efficiency, it is necessary to also improve interference performance in the downlink, for example, with single antenna interference cancellation (SAIC) techniques considered in Chapter 6.

11.2.3 Mast Head Amplifier

When the mobile station belongs to the lowest power class of 0.25 W ($\approx$24 dBm) or downlink performance is improved, uplink diversity may not be enough to balance link budgets. An additional way to improve uplink performance is to place a low noise amplifier between the antenna and feeder cable as shown in Figure 11.5. Mast head amplifiers (MHA) are used to reduce the composite noise figure of the base station receiver system. Assuming that the amplifier gain is considerably higher than the receiver noise factor plus cable loss, MHA compensates the feeder cable loss in the power budget in uplink direction. Moreover, as the noise figure of MHA is better than that of the base station receiver, MHA further improves the sensitivity of the base station. In doing so, the uplink link budget is improved and service coverage increases.

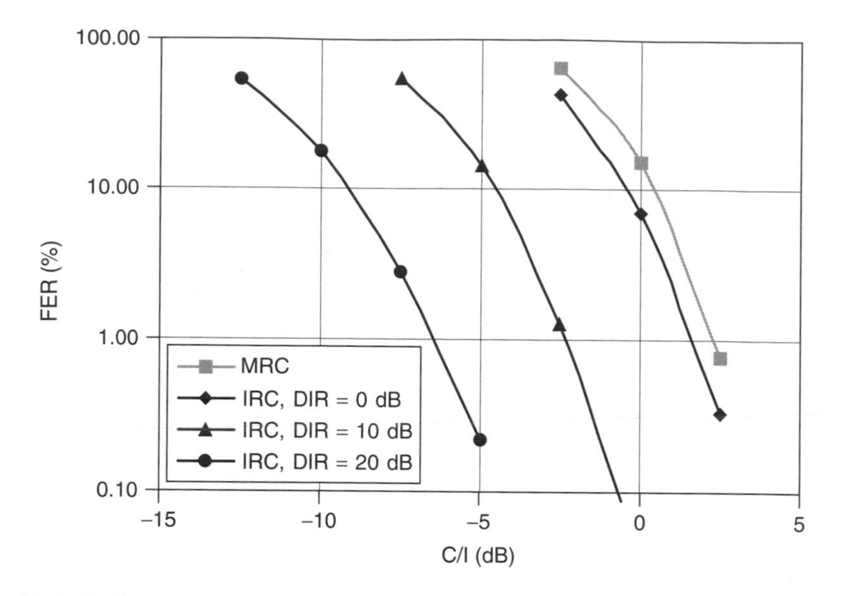

Figure 11.4. Uplink interference rejection performance in different interference environments. Interference is assumed to be synchronous on the timeslot level

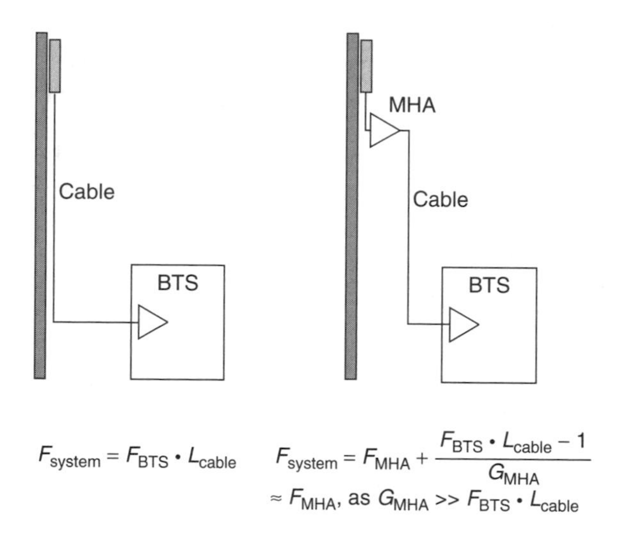

$$F_{system} = F_{BTS} \cdot L_{cable} \qquad F_{system} = F_{MHA} + \frac{F_{BTS} \cdot L_{cable} - 1}{G_{MHA}}$$
$$\approx F_{MHA}, \text{ as } G_{MHA} >> F_{BTS} \cdot L_{cable}$$

Figure 11.5. Mast head amplifier compensates cable loss and improves sensitivity. F_{system} = composite system noise figure, F_{BTS} = BTS noise figure, F_{MHA} = MHA noise figure, G_{MHA} = MHA gain, L_{feeder} = feeder loss

11.2.4 Downlink Transmit Diversity

The diversity antennas at the base station can also be used for transmission to improve downlink performance. Suitable transmit diversity techniques for GERAN are, for example, delay diversity (DD), phase hopping (PH) and antenna hopping (AH). As these techniques require no changes in mobile stations, full gain will be reached with the existing terminal base in the network. As with uplink diversity, in order to achieve significant downlink diversity gains, the antenna signals must be as decorrelated as possible either by cross-polarisation or spatial separation. Another class of open loop transmit diversity techniques includes space–time block codes, but they require changes to the standard, both at the base and mobile station side. Feasibility of closed loop techniques, which use channel information in adjusting the phase and amplitude of transmitted signals, is restricted by the lack of fast feedback from the receiver in GERAN.

11.2.4.1 Delay Diversity

Delay diversity refers to a transmission scheme in which delayed copies of the modulated signal are transmitted over multiple antennas (Figure 11.6). By using sufficient antenna spacing or cross-polarized antennas at the base station, independent fading can be realized on the received signals. Ideally, the receiver constitutes an equalizer that coherently combines the signals. Thus, delay diversity would be able to achieve the same diversity gain as MRC reception. Delay is needed in order for the equalizer to be able to separate the signals. In practice, delay is induced in digital domain after which the signals are amplified separately in analog domain ideally giving 3 dB extra transmission power (an equal improvement in SNR). Delay diversity is especially suitable for GERAN as the specified performance requirements necessitate receivers to have equalizers that cope with delay spreads up to five symbols. Optimal delay depends on the propagation conditions, channel coding and exact equalizer implementation, but usually 1 symbol delay gives close to optimal performance.

11.2.4.2 Delay Diversity with Phase Hopping

In practice, the modulation process and radio channel make it impossible for a receiver to totally separate the delayed signal components in delay diversity. Therefore further diversity gain can be achieved if the transmitter also rotates the phase of the second antenna. The idea behind phase hopping is to prevent continuous, destructive summing of the received signals. The different effects of delay diversity and phase rotation to the signal power are illustrated in Figure 11.7. The diversity gain of phase hopping scheme comes through channel coding and interleaving similarly to that of terminal movement and frequency hopping. Phase hopping is therefore most beneficial when the terminal

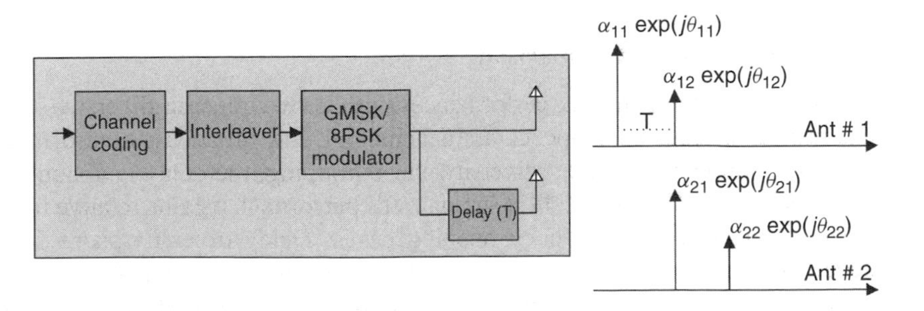

Figure 11.6. Transmitter with delay diversity

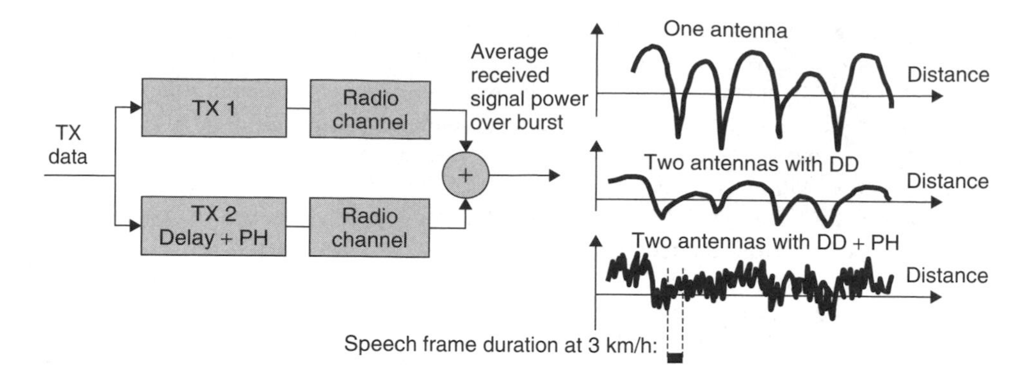

Figure 11.7. Effects of delay diversity and phase rotation to the signal power

moves slowly and frequency hopping is not used. Hopping can be periodic or random. In periodic hopping, it is preferable to change the phase in steps of 360/N degrees so that 360-degree phase rotation happens during one coded data block interleaved over N bursts. In the case of random hopping, the phase for each burst is selected from a Mary Phase Shift Keying (MPSK) constellation with uniform sampling. Since random hopping does not guarantee maximal interleaving depth for a given radio block, it cannot outperform periodic hopping.

Figure 11.7 shows that whereas delay diversity (DD) equalizes the fading profile of the received signal, phase hopping (DD + PH) increases the fading rate, which gives diversity gain through channel coding and interleaving similarly to terminal movement and frequency hopping.

11.2.4.3 Antenna Hopping

Antenna hopping means a scheme in which the transmitter alternately uses the two antennas to transmit bursts (Figure 11.8). The idea is to decorrelate the fading profile of each alternate burst, which is translated into an improved BLER performance through interleaving and channel coding. Antenna hopping can improve link performance on carriers where frequency hopping is not in use or is ineffective because of high correlation between the frequencies. It provides ideally the same diversity gain as pure periodic phase hopping, but lacks the power increase that is achieved from simultaneous transmission through multiple antennas using two power amplifiers.

11.2.4.4 Link Performance Comparison

Figures 11.9 and 11.10 compare link performances with above transmit diversity schemes using two antennas in case of a speech traffic channel and broadcast control channel without frequency hopping. For delay diversity/phase hopping, the delay is 1 symbol and power increase is not included. Table 11.4 summarizes performance gains relative to single antenna transmission also for frequency hopping cases. Delay diversity/phase hopping gains include no power increase. The two AMR modes in this comparison are half-rate 7.4 kbps codec (AHS/7.40) and full-rate 12.2 kbps codec (AFS/12.2). Two different cases with frequency hopping are assumed and can be compared with non-hopping case

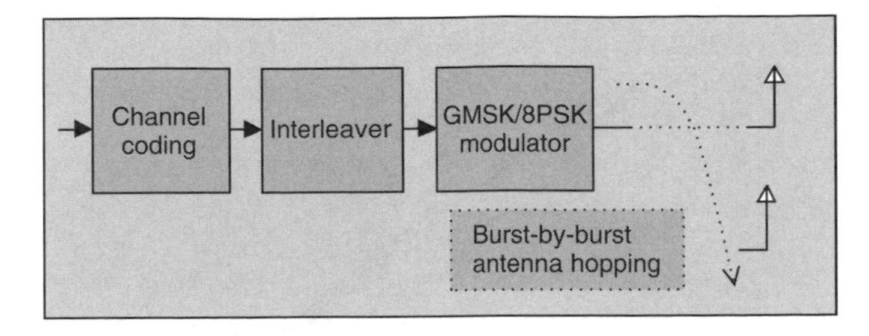

Figure 11.8. Antenna hopping principle

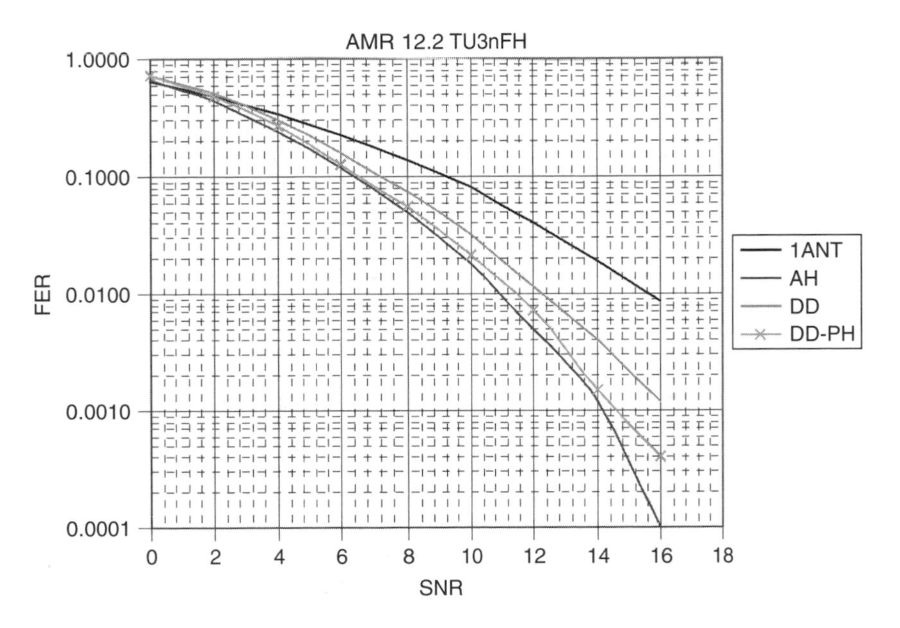

Figure 11.9. Downlink enhancement performance comparison for voice service (normalized power)

Table 11.4. Downlink diversity gain for two AMR modes at 1% FER

Gain (dB)		nFH	6FH	iFH
AHS/7.4 kbps	One antenna	—	2.6	4.3
	AH	1.5	3.4	4.6
	DD + PH	3.0	5.4	5.9
AFS/12.2	One antenna	—	6.6	8.7
	AH	4.7	7.4	8.5
	DD + PH	4.3	8.2	9.2

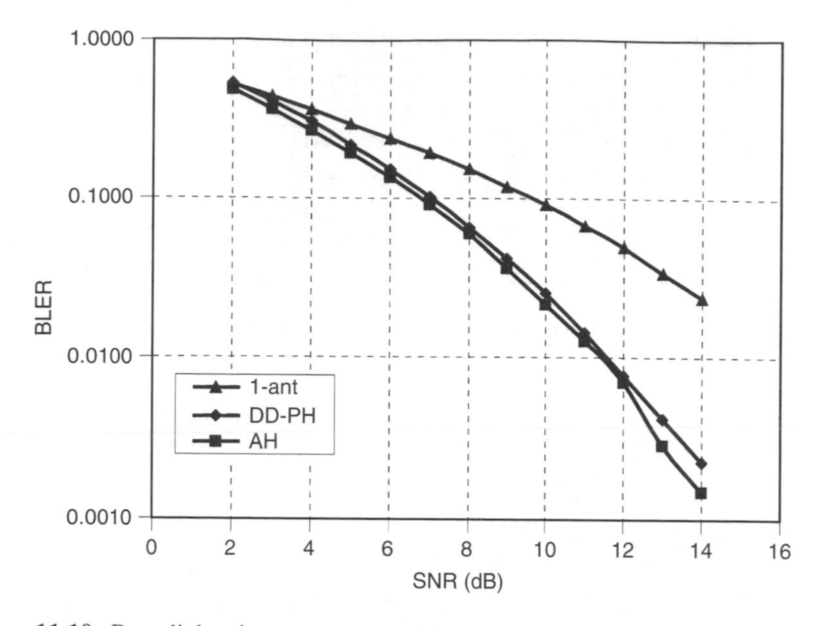

Figure 11.10. Downlink enhancement potential comparison for broadcast control channel (normalized power)

(nFH). The reduced frequency hopping (6FH) case refers to the narrowband deployment scenario with only six consecutive frequencies to hop over. The ideal frequency hopping (iFH) case assumes full decorrelation of fading between bursts belonging to the same frame, which is achieved only when the hopping bandwidth is sufficiently wide. Owing to its strong channel coding and long interleaving period, full-rate speech service gets most of the achievable diversity gain from frequency hopping. Therefore, gains from transmit diversity techniques are smaller with than without hopping. As half-rate mode has less channel coding and shorter interleaving depth, it gets considerably higher gain from delay diversity in the hopping cases. All these results assume independent fading between received signals coming from different antennas.

11.2.5 Macrodiversity

Macrodiversity refers to a spatial diversity where transmission or reception is performed simultaneously at two or more base station sites. The large physical separation of reception sites means that macrodiversity can combat slow fading, as the propagation paths can be largely uncorrelated for shadowing effects. In principle, macrodiversity can also be used to combat fast fading, but local spatial or polarisation diversity implemented within one base transceiver station (BTS) site is usually sufficient for this purpose.

In GSM, the first requirement for the implementation of macrodiversity is air interface synchronisation. Synchronisation is needed in order to achieve timeslot alignment that is required in order for the reception to be possible in multiple sites. Air interface synchronisation is presented in more detail in Chapter 9.

11.2.5.1 Uplink Macrodiversity

Macrodiversity resembles soft handover in code division multiple access (CDMA) systems in a sense that when macrodiversity is applied, the user is connected to multiple base stations simultaneously. It is used for users in marginal field strength or carrier-to-interference ratio (CIR) condition where the slow fading effects can cause significant degradation of link quality. When such a condition is detected, the diversity reception is activated in one or more surrounding BTSs for which the field strength and CIR are found to be acceptable. The main received signal and the diversity signals are combined in the base station controller (BSC), see Figure 11.11. The combining can be realized as a simple selection combining based on signal quality criteria or a more advanced scheme such as multiply detected macrodiversity (MDM) [6] or log-likelihood ratio (LLR) [7] can be used. However, macrodiversity is not a replacement for a handover. A power budget handover to another BTS should happen normally when the network has observed that the path loss from another BTS is permanently less than the path loss from the serving BTS. More complete introduction to macrodiversity can be found in [8].

If macrodiversity is used, more receivers have to be installed to provide the extra hardware capacity. The required increase in the hardware capacity is directly proportional to the percentage of users for whom macrodiversity is being applied at any one time, and the average number of base stations involved in the reception of one user. Naturally, a transmission link to the BSC must also be provided for each extra receiver.

In order to effectively follow the shadow fading conditions, the reaction time for the application of macrodiversity must be fast enough. Typically, the radio link conditions must be evaluated at least every slow associated control channel (SACCH) period of 0.48 s but even more frequent evaluations are needed for optimum performance. This reaction time requirement leaves no room for changes in the radio channel parameters such as frequency, timeslot or timing advance in conjunction with the application of macrodiversity. Thus, macrodiversity is transparent to the mobile station and changes in the radio channel configuration are not permitted when macrodiversity is applied. The fact that the timeslot cannot be changed leads to increased probability of macrodiversity

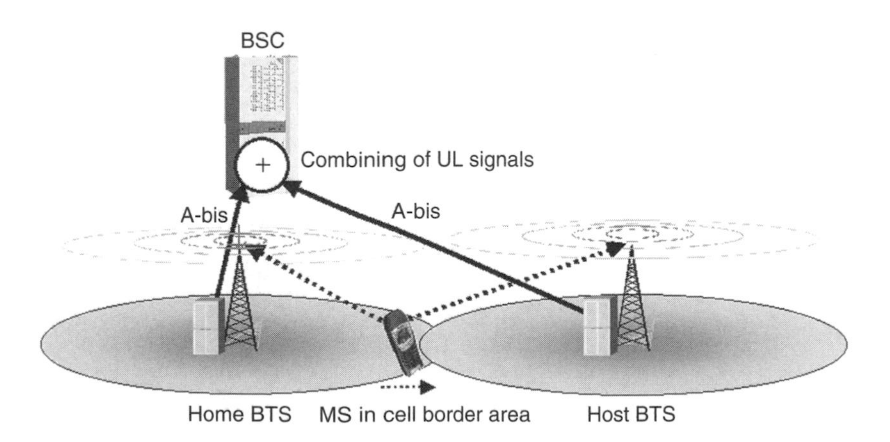

Figure 11.11. Uplink macrodiversity

blocking even if the BTS hardware capacity has been increased proportionally to the share of users applying macrodiversity.

11.2.5.2 Downlink Macrodiversity

In principle, macrodiversity could also be applied to the downlink direction. Whenever the user is subjected to marginal field strength or CIR condition, the downlink diversity transmission could be initiated in another BTS simultaneously with the uplink diversity reception. The user would then have two or more downlink signals available each having different propagation paths, which therefore increase the probability that the signal strength available to the user is sufficiently high despite slow fading. Figure 11.12 schematically represents the downlink macrodiversity operation. Downlink macrodiversity can be based on either a simultaneous downlink transmission from multiple base stations or it is possible to use a switching principle where the downlink transmission is always performed from one BTS only. The BTS can be selected on the basis of the path loss measurements reported by the mobile station.

Downlink macrodiversity has some problems and limitations that are not present in uplink macrodiversity. In the uplink direction, all BTS receivers see a normal transmission from a single mobile station. For a mobile station receiver, the diversity transmission causes additional multipaths to cope with. In order for them to be beneficial in reception, the main multipath components must fit the equalisation window of the mobile receiver. The performance requirements associated with radio channel models defined in GSM specifications implicitly require the window to be about 5 km in propagation distance, and it should therefore be ensured that the propagation path distance differences should be clearly below 5 km. In the case of switching downlink macrodiversity, the propagation distance difference introduces a sudden change in downlink signal timing whenever a switch occurs, which also limits the maximum allowed propagation path distance difference. Another problem with the downlink macrodiversity arises if the mobile station moves fast in different directions relative to the host and home BTS. The movement would thus introduce different Doppler shifts in the received signal components that could not be compensated by the receiver. Also, as downlink macrodiversity involves fast

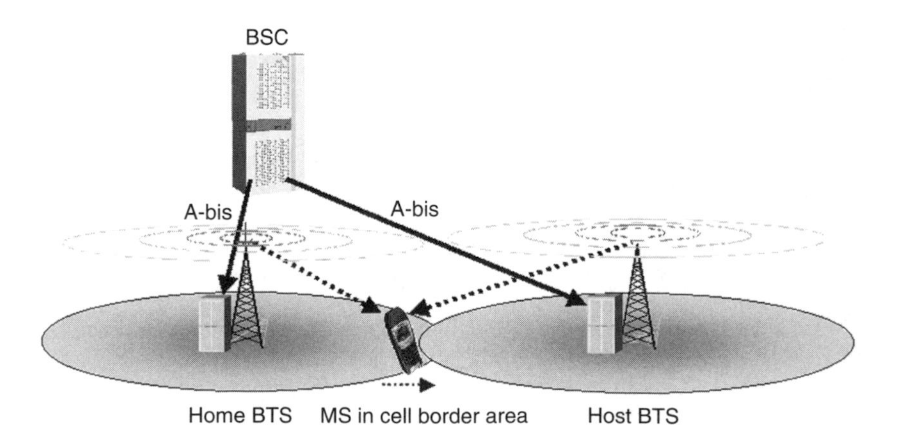

Figure 11.12. Downlink macrodiversity

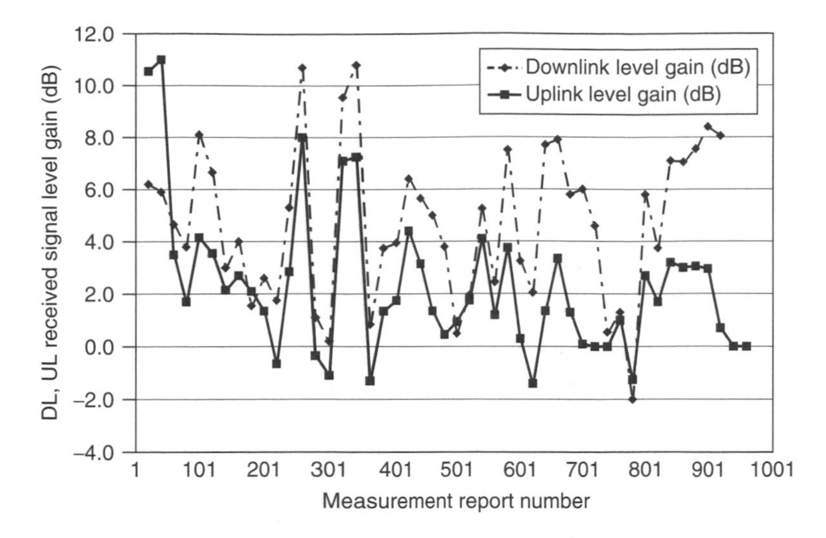

Figure 11.13. Downlink- and uplink-received signal level gain using transmitter and receiver diversity at the base station in rural propagation area

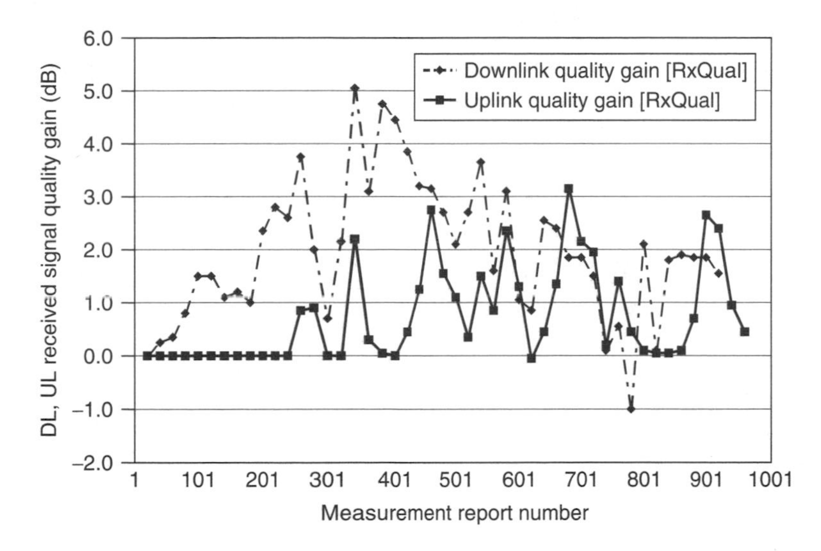

Figure 11.14. Downlink- and uplink-received signal quality gain using transmitter and receiver diversity at the base station in rural propagation area

2.0 and 0.8 quality classes respectively. Using global positioning system (GPS) data of the measurement system to the point where the calls were dropped yields 1.2 km or 42% increase in coverage radius. When average RxQuality exceeds 5, the coverage improvement was 0.66 km or 33% in coverage radius. Corresponding improvements in coverage efficiency, as defined above, are 102 and 77%, and reductions in the number of BTS sites 50 and 43%.

11.4 Capacity Improvements

In case a radio link technique provides an almost constant interference performance improvement that is independent of the nature of interference, then the attainable capacity enhancement can be roughly estimated by using the method of Appendix C. This is not the case with the techniques considered in this chapter. Therefore, results from network simulations that have included detailed models of the evaluated techniques are presented in this section. The considered simulated network (see Appendix E) is purely interference-limited with practically perfect coverage both in uplink and downlink.

11.4.1 Uplink Diversity Reception

In order to evaluate the capacity enhancement from uplink diversity reception with MRC and IRC receiver algorithms, enhanced full-rate (EFR) voice service in a synchronized GSM network has been simulated with different traffic loads. Since the BCCH reuse is clearly downlink limited, only the quality of the hopping layer is of interest. For a fair comparison, the downlink quality is also taken into account so that a connection will be classified as having bad quality if either direction has average FER over the quality threshold of 1%. This relatively tight threshold is used in order to see the performance gains at moderate traffic loads.

Figure 11.15 presents simulation results for the hopping layer, which included 12 frequencies with 1/1 reuse. For the same overall quality, the diversity reception with MRC allows almost 40% more load than the single antenna reception. On the other hand, IRC outperforms MRC with only about 10% in capacity. The main reason for this is evident in Figure 11.16, which separately shows downlink and uplink performance. Owing to quality handovers, downlink performance is not completely independent of uplink performance, but variations are quite small. Downlink performance shown in Figure 11.16 is for the IRC case.

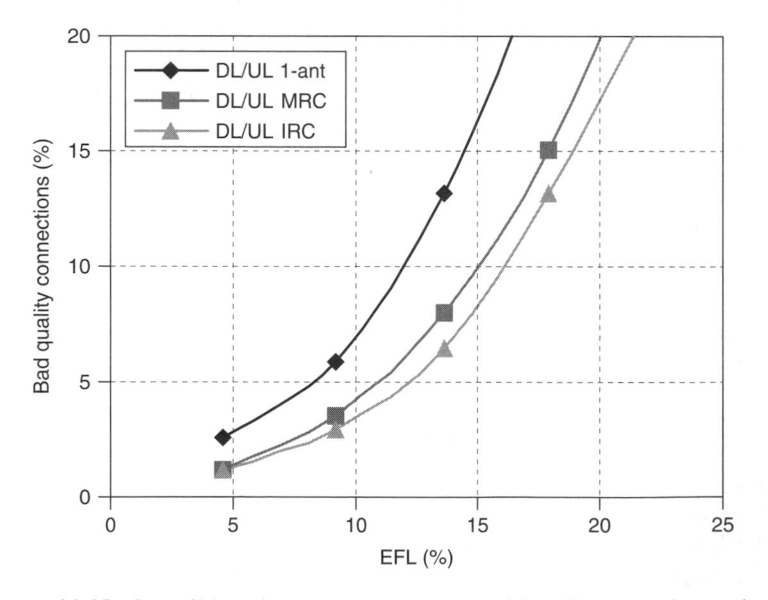

Figure 11.15. Overall hopping layer performance with uplink diversity methods

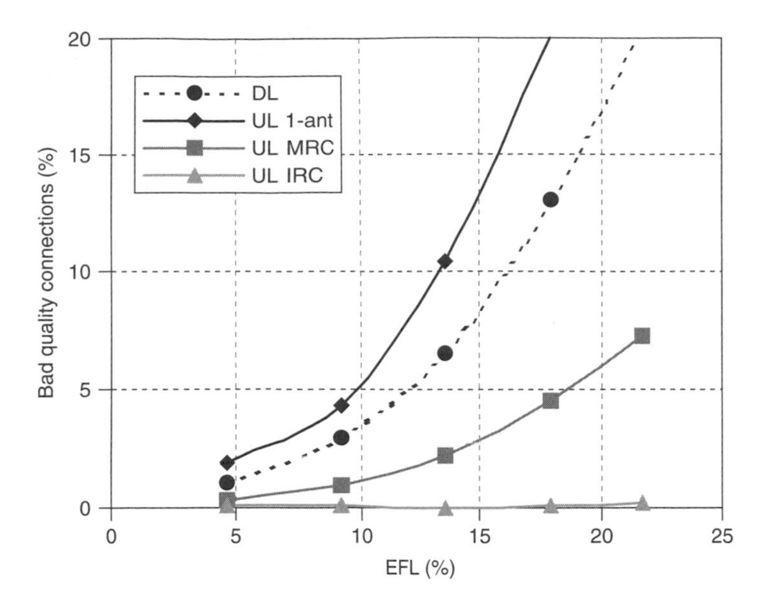

Figure 11.16. Downlink and uplink hopping layer performances with uplink diversity methods

When uplink performance improves, the network becomes more and more downlink limited, and the effect of uplink improvement decreases. Indeed, the overall DL/UL IRC performance is practically identical to downlink performance, which means that the network is purely downlink limited. Most of the existing GSM networks are using UL diversity, therefore the-downlink quality in interference-limited areas tend to be worse in most of the cells.

Single antenna interference cancellation (SAIC, see [5] and Chapter 6) is a promising capacity enhancement technique that applies only to DL, therefore it is important to evaluate with simulations whether uplink can be a limiting factor for SAIC gains. A SAIC receiver model was introduced in the simulator. The overall performance results are shown in Figure 11.17. MRC allows over 100% more load than single antenna reception, and IRC outperforms MRC by 25%. Downlink and uplink performances are shown separately in Figure 11.18. Both DL and UL become relevant limiting factors with MRC when SAIC is considered. Uplink with IRC still significantly outperforms downlink, and the network is still practically purely downlink-limited.

11.4.2 Downlink Transmit Diversity

Link performance results in Section 11.2 showed that transmit diversity provides most significant gains when frequency hopping is not used. This is always the case with BCCH control and TCH assuming that baseband hopping is not used. In addition, BCCH carriers are continuously transmitted in downlink so that their minimum reuse is clearly downlink-limited. Therefore transmit diversity methods, such as delay diversity/phase hopping and antenna hopping, can be used to improve BCCH link performance and tighten BCCH reuses. In order to evaluate this improvement, BCCH layer network simulations were performed with different reuses with and without transmit diversity. Figure 11.19 shows the percentage of connections with average FER over 1% in these simulations (typical

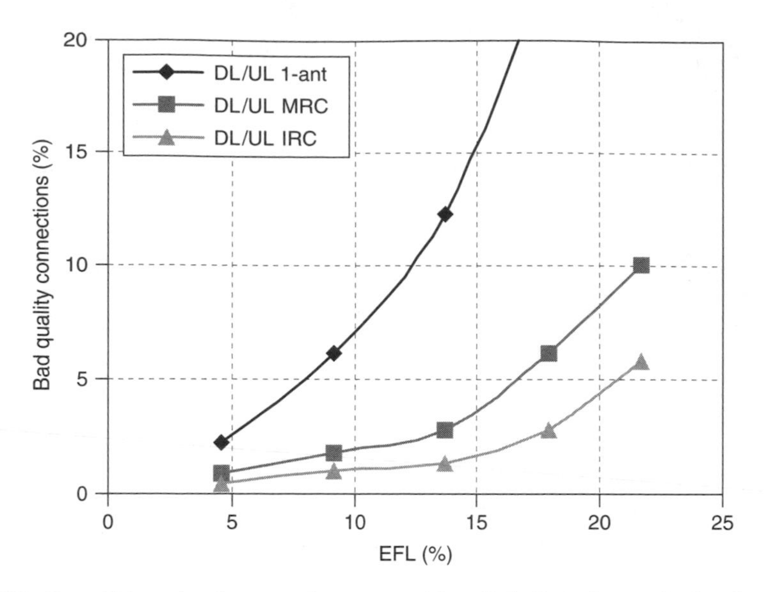

Figure 11.17. Overall hopping layer performance with uplink diversity methods when terminals use a joint detection algorithm for interference cancellation

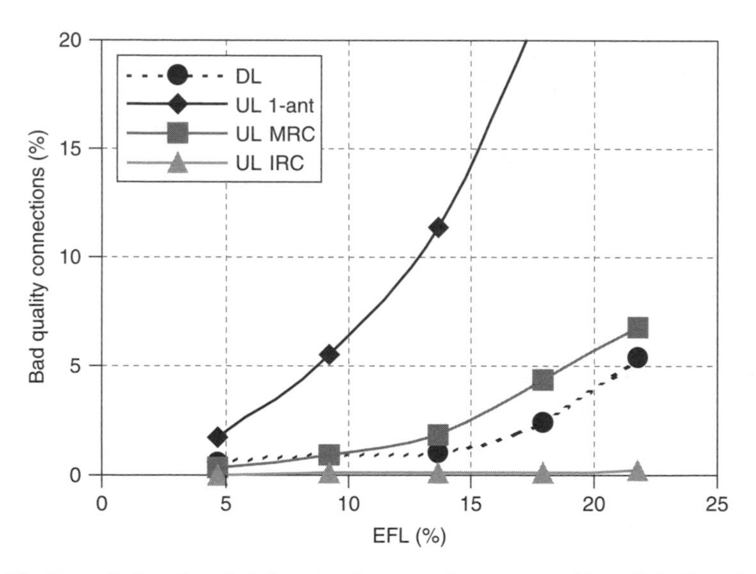

Figure 11.18. Downlink and uplink hopping layer performances with uplink diversity methods when terminals use a joint detection algorithm for interference cancellation

urban 3 km/h radio channel). When transmit diversity is in use, the BCCH reuse factor could be reduced from 15 to 12, or from 12 to 9 having similar quality. This means that three BCCH frequencies could be moved to the hopping layer in order to increase the overall capacity. In narrowband deployments, due to increased frequency diversity, the additional frequencies also improve spectrum efficiency in the hopping layer. This effect,

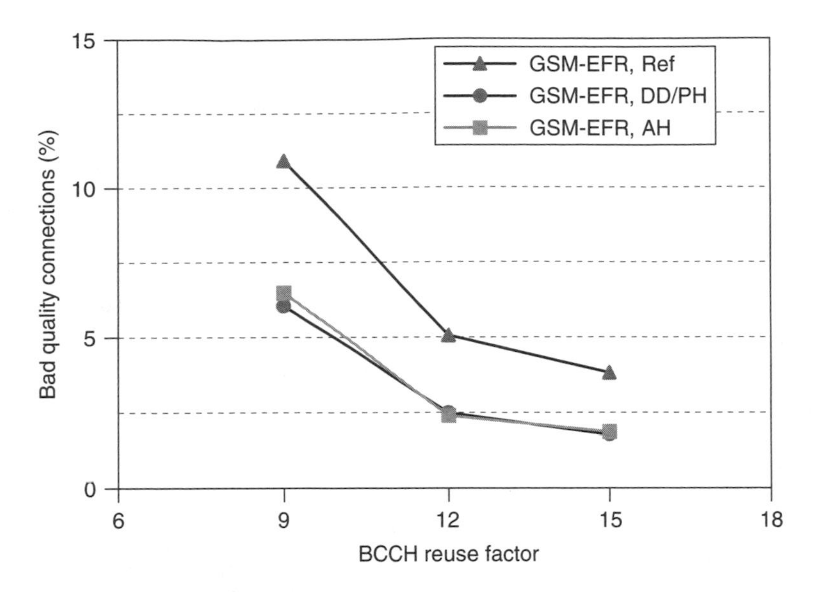

Figure 11.19. BCCH reuse improvement with transmit diversity methods

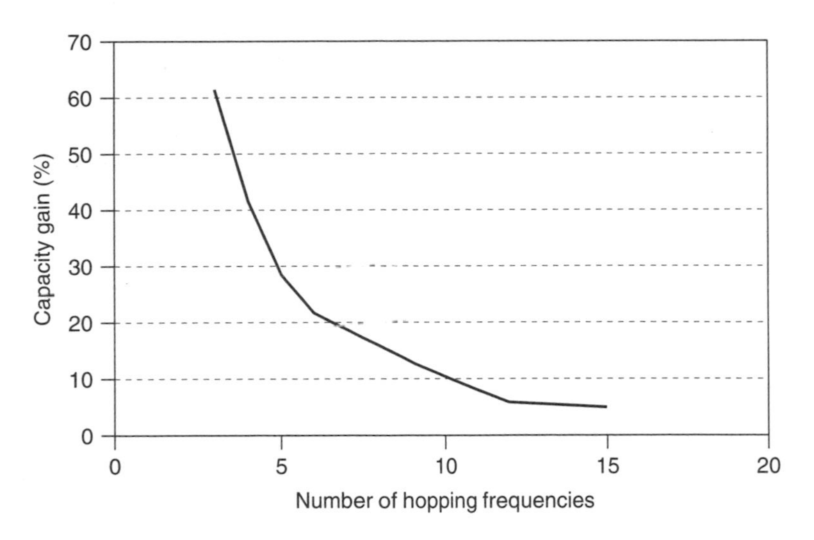

Figure 11.20. Hopping layer capacity improvement with DD/PH

which significantly depends on the frequency correlation properties of the radio channel, is illustrated in Figure 5.40 for frequency hopping with 1/1 hopping reuse.

Transmit diversity methods can also be applied to hopping layer in narrowband scenarios to compensate for the limited frequency diversity gain. Capacity gains with delay diversity/phase hopping are shown in Figure 11.20. These results assume frequency hopping according to 1/1 hopping reuse. Transmit diversity gain is high when frequency hopping is inefficient. The overall capacity gains from transmit diversity in networks of different bandwidths have been considered in Chapter 10 where simulation results show that capacity gains from 10 to 70% can be achieved. Highest gains are achieved in the networks with less than 5-MHz spectrum. In these narrowband networks, BCCH reuse improvement has the most significant effect and also the performance of the hopping layer can be improved the most.

11.4.3 Macrodiversity

The performance of a switching macrodiversity scheme has been evaluated with network-level simulations. In these simulations, the uplink-received signal level was the trigger for a macrodiversity switch. The link was transferred to another base station whenever the uplink path loss was found to be lower in the other BTS. The uplink path loss evaluation was performed every 120 ms. If a switch was needed; it was performed without any additional delay. In this simulation, both uplink and downlink were switched on the basis of uplink path loss measurements. Power budget handovers were performed normally whenever the 6-dB path loss hysteresis was satisfied in a periodical check (Table 11.5).

The simulation results in terms of FER performance are shown in Figures 11.21 and 11.22. The graphs show the share of FER samples that indicate higher than 8% FER during the past second. This share of 'bad' FER samples is referred to as an outage. This benchmark has been selected in order to capture even short periods of time when the user would have been expected to experience bad speech quality.

As can be expected, the results show that macrodiversity provides more gain with fast moving mobiles. This is due to the fact that fast moving mobiles move across cell border areas more frequently and they are therefore able to benefit more from macrodiversity. With fast moving mobiles, the gain of downlink macrodiversity is clearly larger than that of uplink macrodiversity. With slow moving mobiles, the uplink gain disappears in low load situations and in the downlink, losses can be observed.

Table 11.5. Simulation parameters

Cell layout	3-sectorized hexagonal
Site-to-site distance	1500 m
Frequency	900 MHz
Propagation model	COST-Hata
Standard deviation of slow fading	6 dB
Path correlation factor	0.15
Voice activity factor	50%
Power control UL and DL	RXLEV and RXQUAL
Power budget HO margin	6 dB
Power budget HO averaging	10 SACCH periods
Frequency plan	1/1 hopping
Speech codec	EFR
Macrodiversity evaluation and switching period	120 ms
Number of candidate BTSs for macrodiversity switching	6
FER sampling period	1 s

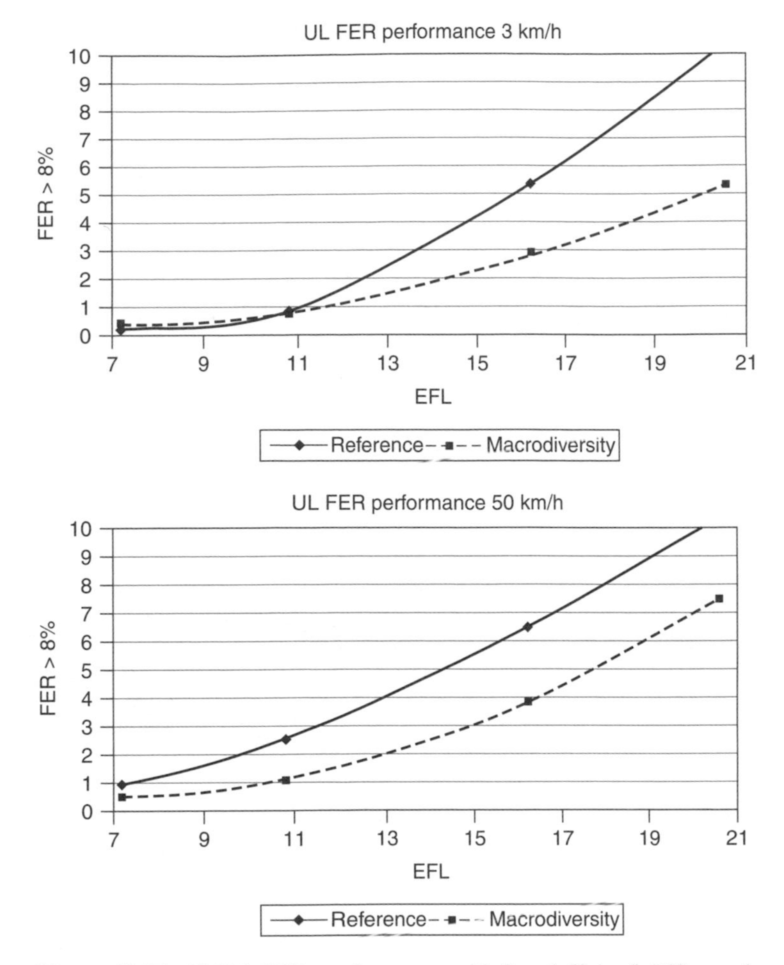

Figure 11.21. Uplink FER performance with 3 and 50 km/h MS speeds

The drawback of downlink macrodiversity is that since the downlink transmission is switched away from the home BTS without changing the radio channel, it brakes the frequency plan and causes sudden changes in the interference situation. For example, in the 1/1 reuse case with random frequency hopping, a switch of downlink transmission to a BTS in another site causes intracell co-channel collisions because the switched connection uses a different hopping sequence number than the rest of the traffic in the cell. With slow moving mobiles and low loads, this results in overall loss.

The impact of AMR speech codecs with more robust channel coding is expected to reduce macrodiversity gain. With AMR, the effective frequency reuse distance can be extremely low leading to very high frequency load. With such high loads, the intracell co-channel collisions are more frequent when the uplink reception or downlink transmission point is changed to another BTS using different hopping sequence number. This can cause a significant increase in the interference level, limiting possibilities to apply macrodiversity reception.

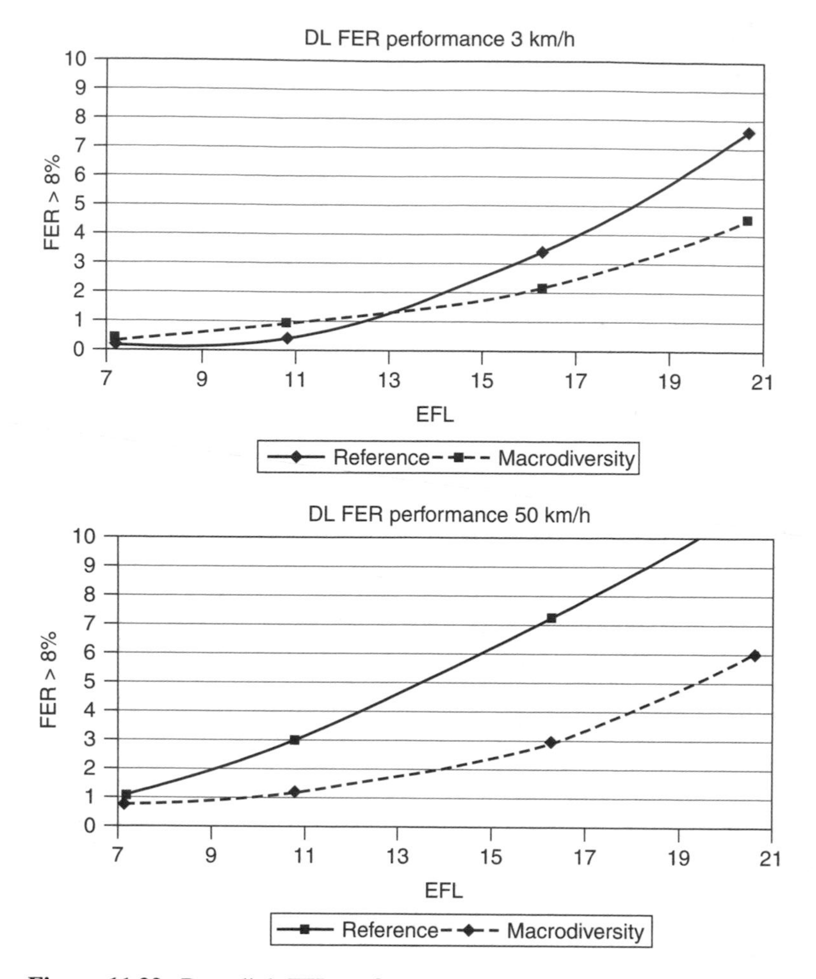

Figure 11.22. Downlink FER performance with 3 and 50 km/h MS speeds

Table 11.6. Macrodiversity gain summary (limiting link in parenthesis)

0.5% outage		1% outage		2% outage	
3 km/h	50 km/h	3 km/h	50 km/h	3 km/h	50 km/h
−30% (DL)	N/A	0%	29% (DL)	14% (DL)	32% (UL)

The simulated macrodiversity gain is summarized in Table 11.6.

Macrodiversity system level gains are very dependent on mobile speed, becoming marginal if the users are slow moving. The assumption that all users are fast moving cannot be maintained in most of capacity-limited scenarios. On the contrary, most of the users tend to be slow moving and the macrodiversity gain turns to be quite limited.

As downlink macrodiversity involves transmitting the signal from two or more base stations either simultaneously or in a switched manner, there are some limitations related to timing and Doppler shift that can complicate the implementation. Further complications can be expected related to the shared GPRS channels and the transmission of the uplink state flag. The implementations of location services that rely on timing measurements performed by the mobile stations can also be problematic if the point downlink of transmission is allowed to be changed during the location measurement. Furthermore, downlink macrodiversity cannot be used with DFCA, which provides superior interference management and performance gains as presented in Chapter 9.

References

[1] 3GPP TS 45.005, Radio Transmission and Reception Specifications.
[2] 3GPP TS 51.010, Mobile Station (MS) Conformance Specification.
[3] 3GPP TS 11.21, Base Station Subsystem (BSS) Equipment Specification; Radio Aspects.
[4] 3GPP TR 43.030, Radio Network Planning Aspects.
[5] Ranta P. A., Honkasalo Z.-C., Tapaninen J., 'TDMA Cellular Network Application of an Interference Cancellation Technique', *1995 IEEE 45th Vehicular Technology Conference*, Vol. 1, 1995, pp. 296–300.
[6] Haas Z. J., Li C.-P., 'The Multiply-Detected Macrodiversity Scheme for Wireless Cellular Systems', *IEEE Trans. Veh. Technol.*, **47**, 1998, 506–530.
[7] Valenti M. C., Woerner B. D., 'Iterative Multiuser Detection, Macrodiversity Combining and Decoding for the TDMA Cellular Uplink', *IEEE J. Selected Areas Commun.*, **19**(8), 2001, 1570–1583.
[8] Weiss U., 'Designing Macroscopic Diversity Cellular Systems', *IEEE 49th Vehicular Technology Conference*, Vol. 3, 1999, pp. 2054–2058.
[9] Harri Holma, Antti Toskala, *WCDMA for UMTS*, Wiley, Chichester, 2000.

12

Control Channels Performance and Dimensioning

Timo Kähkönen and Jorge Navarro

The performance analysis of GSM/GPRS control channels presented in this chapter supplements the performance analysis of preceding chapters where the focus has been on the traffic channels.

The first part covers potential synchronization and signalling problems caused by performance limitations of control channels in demanding radio conditions. The potential problems include, for example, call set-up and handover signalling failures, which directly degrade end-user quality. This analysis is necessary when considering system coverage and capacity enhancements that use radio link techniques, such as adaptive multi-rate (AMR) speech coding, for improving the performance of traffic channels while leaving control channels unchanged.

In the second part, the signalling capacity of different control channel configurations is evaluated. The main objective is to find the required configuration for cells with certain voice and data traffic loads. This analysis will provide guidelines for proper control channel dimensioning.

12.1 Introduction to Control Channels

12.1.1 Physical and Logical Channels

GSM allocates the physical resources of the air interface, that is, frequency channels and timeslots by using two levels of channels: physical and logical. Physical channels are described both in frequency and time domain in such a way that one channel occupies one fixed timeslot in every time division multiple access (TDMA) frame but may change the frequency channel according to its frequency hopping sequence (see Chapter 6) [1]. A physical channel can be allocated to several logical channels according to a multiframe structure consisting of 26, 51 or 52 TDMA frames. The logical channels divide into

GSM, GPRS and EDGE Performance 2nd Ed. Edited by T. Halonen, J. Romero and J. Melero
 ISBN: 0-470-86694-2

traffic and control channels that are used either in circuit- or packet-switched mode. Normal traffic channels (TCH) carry circuit-switched speech or user data, and packet data traffic channels (PDTCH) carry packet-switched user data. The control channels are used for synchronization, radio link control and upper-layer signalling procedures. In this chapter, the control channels that have been introduced for packet-switched services are referred to as *packet control channels*, and those that were included in the original GSM specifications are called *GSM control channels*. The PDTCH and packet control channels are dynamically multiplexed on physical channels that are called packet data channels (PDCH), while the TCH and GSM control channels are mapped into fixed timeslots.

According to the use of radio resources, the GSM control channels can be divided into the following three groups: broadcast channels, common control channels (CCCH) and dedicated control channels. The broadcast channels are the frequency correction channel (FCCH), synchronization channel (SCH) and broadcast control channel (BCCH) on which base stations continuously broadcast frequency reference, time and system information. The common control channels are the random access channel (RACH), paging channel (PCH) and access grant channel (AGCH). They are used to establish a radio resource (RR) connection, after which the network can privately change signalling and radio link control messages with the mobile station on the dedicated control channels. The dedicated control channels are the stand-alone control channel (SDCCH), fast-associated control channel (FACCH) and slow-associated control channel (SACCH). The BCCH, RACH, PCH and AGCH have counterparts in packet control channels whose abbreviated names are PBCCH, PRACH, PPCH and PAGCH. Other packet control channels are the packet notification channel (PNCH), packet-associated control channel (PACCH) and packet timing advance control channel uplink/downlink (PTCCH/U, PTCCH/D).

Transmission on the logical channels is based on four types of bursts. Frequency correction bursts are transmitted on the FCCH and synchronization bursts on the SCH. On the RACH, PRACH and PTCCH/U, only access bursts are used. On other channels, normal bursts are used with two exceptions: on the FACCH and PACCH, access bursts are used to transmit the first message in the uplink when the timing advance required for a transmission of a normal burst is unknown. Modulation of all the control channels is Gaussian minimum shift keying (GMSK).

12.1.2 Control Channel Configurations

The GSM control channels are mapped on to physical channels according to a 51-multiframe structure. The physical channel that carries the broadcast channels is called a *BCCH channel*. It always occupies timeslot 0 at a fixed carrier of the cell, and the broadcast channels use 14 frames in each of its 51-multiframe (a frame includes one timeslot in a TDMA frame). The remaining frames are allocated to the common and dedicated control channels, for which the operator can use either a combined or a non-combined configuration, depending on the number of carriers in the cell. With the combined configuration, the BCCH channel includes four SDCCH subchannels with their SACCHs and three blocks of four timeslots for the CCCH (Figure 12.1). In the downlink, the PCH and AGCH share these CCCH blocks on a block-by-block basis. The operator can, however, reserve a fixed number of blocks for the AGCH that may not be used as the PCH. With the non-combined configuration, all the blocks left from broadcast channels are reserved for the CCCH, and one or more additional physical channels must be allocated for the

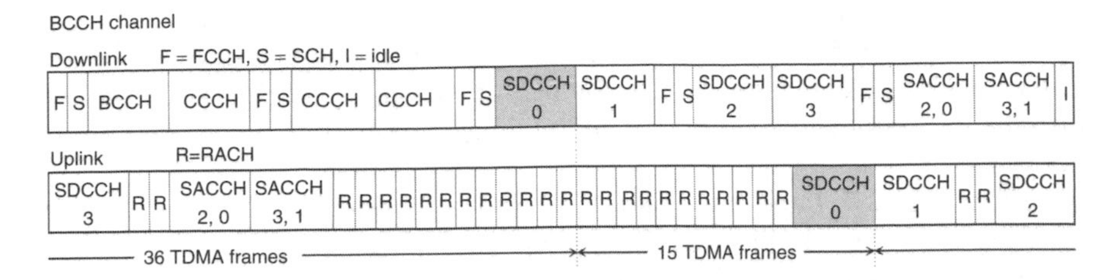

Figure 12.1. The combined configuration of GSM control channels

BCCH channel
Downlink F = FCCH, S = SCCH, I = idle
F S BCCH CCCH F S CCCH CCCH F S CCCH CCCH F S CCCH CCCH F S CCCH CCCH I
Uplink R = RACH
R R R R R R R R R R R R R R R R R R R R R R R R R R R R R R R R R R R R R R R R R R R R R R R R R R R
Any physical channel
Downlink I = idle
SDCCH 0 SDCCH 1 SDCCH 2 SDCCH 3 SDCCH 4 SDCCH 5 SDCCH 6 SDCCH 7 SACCH 4, 0 SACCH 5, 1 SACCH 6, 2 SACCH 7, 3 I I I
Uplink I = idle
SACCH 5, 1 SACCH 6, 2 SACCH 7, 3 I I I SDCCH 0 SDCCH 1 SDCCH 2 SDCCH 3 SDCCH 4 SDCCH 5 SDCCH 6 SDCCH 7 SACCH 4, 0
36 TDMA frames
15 TDMA frame

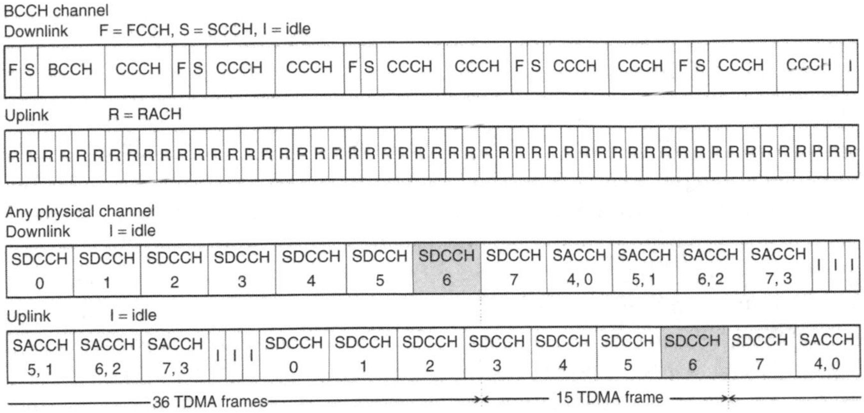

Figure 12.2. The non-combined configuration of GSM control channels

SDCCH usage (Figure 12.2). Additional physical channels can, of course, be allocated for the SDCCH also when the combined configuration is used. In the uplink, the RACH has 27 timeslots in the combined configuration, and all the 51 timeslots in the non-combined configuration.

An SDCCH subchannel transfers a complete data link layer (L2) frame in its block of four timeslots in every 51-multiframe, whereas its SACCH occupies only every other 51-multiframe. As indicated in Figures 12.1 and 12.2, an SDCCH subchannel always has an offset between the downlink and uplink frames, making it possible to receive the response for a transmitted L2 frame before the next transmission possibility and thereby minimizing transmission delays. The network has a longer response time of 36 TDMA frame periods compared with 15 of mobile stations because L2 frames carry network layer (L3) messages, which have transmission delays between base transceiver station (BTS), base station controller (BSC) and mobile services switching centre (MSC).

(E)GPRS can use the existing BCCH and CCCH for broadcast and common signalling or accommodate packet common control channels (PCCCH) with the PBCCH on PDCHs. A PDCH is independently shared in the downlink and uplink directions according to a 52-multiframe structure, which is divided into 12 radio blocks of 4 consecutive timeslots. The first one to three blocks of one PDCH that contains the PCCCH are allocated to the PBCCH in the downlink. Other downlink blocks are allocated on a block-by-block basis for the PAGCH, PNCH, PPCH, PACCH or PDTCH, the actual usage being indicated by the message type. The network can, however, define a fixed number of blocks in addition to PBCCH blocks, which may not be used as the PPCH. On an uplink PDCH containing a PCCCH, all 12 blocks in a multiframe can be used as the PRACH, PACCH or PDTCH. The operator can optionally reserve a fixed number of blocks that can only be used as the PRACH.

The FACCH, SACCH and PACCH share the same physical channel with their traffic channels, except when one-block PACCH allocation is used in a physical channel containing the PCCCH during a PDTCH allocation. An L2 frame transmitted on the FACCH steals the bits of one 20-ms speech frame on a full-rate TCH and of two 20-ms speech frames on a half-rate TCH. The transmitter indicates this to the receiver by setting the stealing flag bits to one.

12.1.3 Usage of Control Channels

The purpose of the FCCH and SCH is to enable mobile stations to synchronize to the network both in the idle and the dedicated mode. All the bits that modulate a frequency correction burst on the FCCH are zero, and thus, by the characteristics of Gaussian minimum shift keying (GMSK) modulation, transmitted bursts are periods of sine waves. By detecting these bursts, a mobile station can find the BCCH channel of the cell and remove the oscillator frequency offset with respect to the base station. During or after adjustment, it attempts to decode SCH information contained in a synchronization burst. SCH information includes the current reduced TDMA frame number and the identity code of the base station. Once a mobile station has successfully decoded SCH information, it starts to decode system information on the BCCH. System information includes the channel allocation in the cell, the BCCH frequencies of neighbour cells and a number of control parameters for the subsequent communication within the cell. Having this information, the mobile station can decode the CCCH, transmit on the RACH and also synchronize to the neighbour cells for the cell reselection. In cells in which PCCCH exists, packet data–specific system information is transmitted on the PBCCH, otherwise it is also transmitted on the BCCH.

Two basic types of signalling procedures using common and dedicated control channels are illustrated in Figure 12.3. In mobile originating procedures, the mobile station starts an RR connection establishment by sending channel request on the RACH. This 8-bit message can be repeated on 3–50 access bursts to increase decoding success probability in bad radio conditions [2]. As several mobile stations may simultaneously attempt

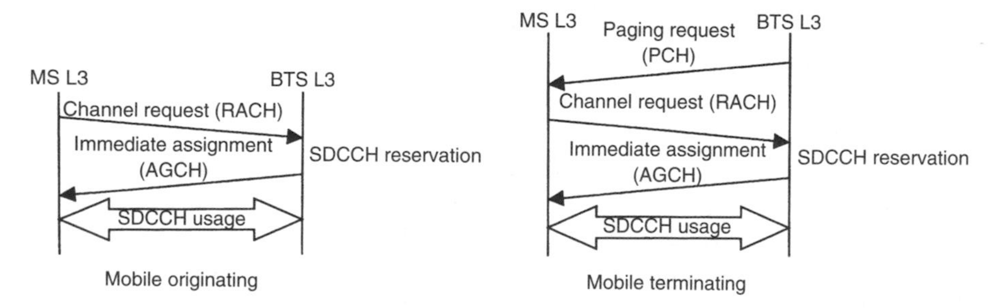

Figure 12.3. Basic signalling procedures for circuit-switched services

to transmit their channel requests, a Slotted Aloha protocol is applied on the RACH. Once the network has received a channel request, it reserves an SDCCH subchannel for subsequent signalling and sends immediate assignment on the AGCH. If all the SDCCH subchannels are reserved, network queues the request or rejects it by sending immediate assignment reject. In mobile terminating procedures, the network calls a mobile station by broadcasting paging request on the PCHs of the cells within the location area of the mobile station. By the receipt of paging request, the called mobile station tries a random access as in mobile originating procedures. While using the SDCCH, a mobile station, being in dedicated mode, regularly receives and transmits radio link control messages on the SACCH. The messages contain power control and timing advance commands in the downlink, and measurement reports and the current Tx-power level in the uplink.

The SDCCH usage depends on the signalling need, which can be a location update, a call establishment, or a short message service (SMS) message transfer. On the SDCCH, L2 frames are transmitted according to a send-and-wait automatic repeat request (ARQ) protocol as illustrated in Figure 12.4 [3]. Response frames carry acknowledgements, which can also piggyback on information frames when L3 messages are transmitted in consecutive blocks in the uplink and downlink. The first pair of frames establishes and the last pair releases an L2-acknowledged mode link.

Once a traffic channel has been allocated, the FACCH can be used for signalling. The principal purpose of the FACCH is to carry handover signalling, but it can also be employed in call establishments. As on the SDCCH, L2 frames on FACCH are transmitted in acknowledged mode. While a traffic channel is in use, its SACCH carries radio link control messages every 0.48 s in both directions.

The (P)CCCH subchannels are used in (E)GPRS to establish an RR connection between a mobile station and the network for signalling, or a unidirectional temporary block flow (TBF) for data transfer. The establishment procedures are presented in Figure 12.5 [4]. While a mobile station has a TBF, it is in a packet transfer mode; otherwise it is in a packet idle mode (see Chapter 1). The establishment procedures follow the principles described above for circuit-switched services except that in a mobile-terminated access, the paging need depends on the mobility management (MM) state. A mobile station enters an MM READY state when it starts transmitting or receiving data, that is, when it changes to packet transfer mode. It stays in this state until it stops sending or receiving data, that is, until it changes to packet idle mode, and a certain time elapses, after which it enters an MM STANDBY state. In the MM READY state, a mobile station performs cell updates, and no paging is needed. In the MM STANDBY state, the network knows only the

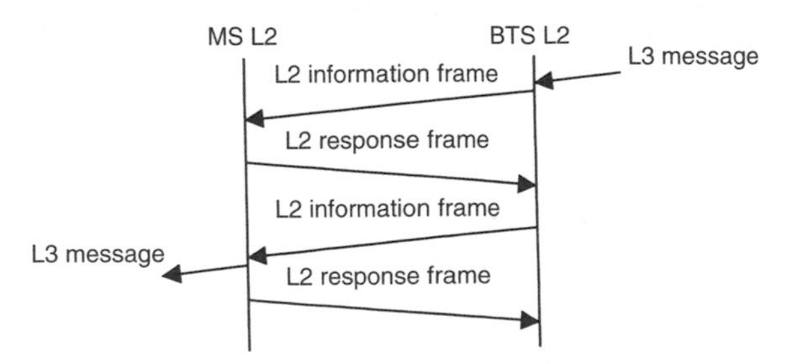

Figure 12.4. Acknowledged mode signalling on the SDCCH and FACCH

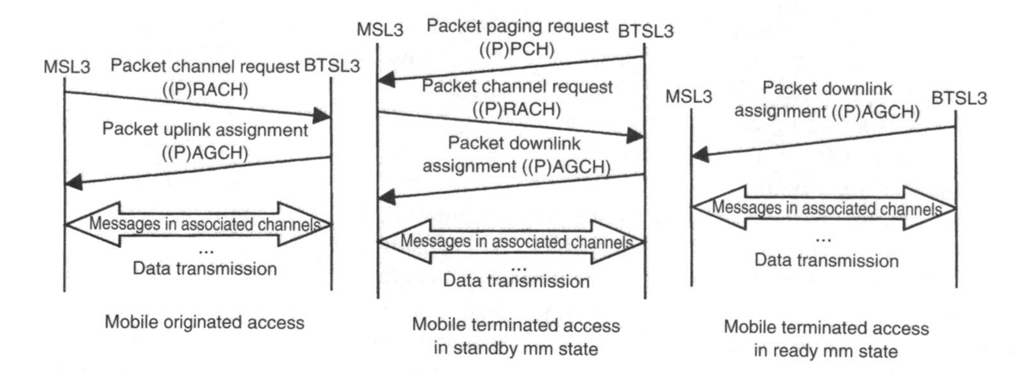

Figure 12.5. Basic signalling procedures for packet-switched services

routing area of the mobile station and it has to send packet paging request in all the cells within the routing area in order to contact the mobile station. Once a TBF is allocated, the PDTCH is used for data transfer and the PACCH for polling and acknowledgement of messages.

12.1.4 Channel Coding and Interleaving

For the detection and correction of transmission errors, channel coding and interleaving are performed for the signalling messages at the physical layer as outlined in Figure 12.6. A message includes 8 bits on the RACH, either 8 or 11 bits on the PRACH and 18 bits on the SCH. Other channels transfer 184 bits in a message, which is an L2 frame (link access protocol for the Dm channel (LAPDm) frame in GSM and radio link control (RLC) block in (E)GPRS) [5].

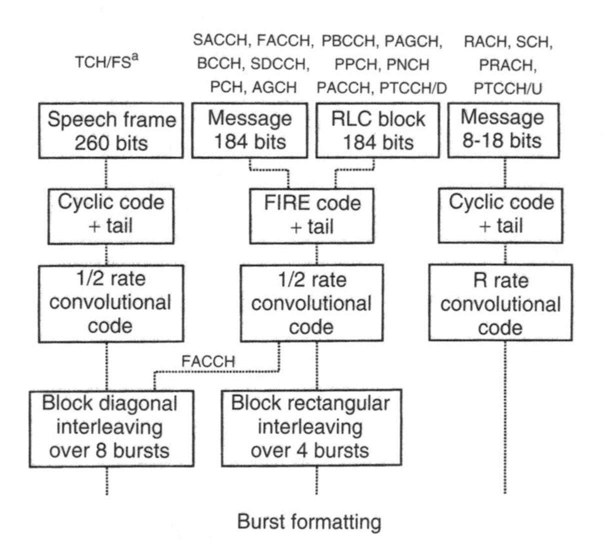

Figure 12.6. Channel coding and interleaving on the control channels. [a]GERAN Rel'5 terminology is used. This corresponds to a full-rate TCH

On all the control channels, a half-rate convolutional code is used for error correction. A small exception in regard to the code rate is the PRACH, on which 6 coded bits are not transmitted in the case of an 11-bit message. The RACH and PRACH use a cyclic code of 6 parity bits and the SCH uses a cyclic code of 10 parity bits for error detection. Other channels use a FIRE code with 40 parity bits for enabling error correction besides error detection.

The BCCH, SDCCH, PCH, AGCH and the packet control channels, except PRACH and PTCCH/U, carry a message in four normal bursts in consecutive TDMA frames. As these channels have the same channel coding and interleaving, their physical layer reliabilities are equal. When the SACCH is associated with an SDCCH subchannel, it is identical to the above channels, but when it is associated with a TCH/FS, it has a longer interleaving over four bursts in consecutive 26 multiframes. As the RACH, SCH and PRACH carry messages in single bursts, they do not have interleaving. The GSM full-rate speech channel (TCH/FS), which is used as a reference channel in Section 12.2, uses a cyclic code of 3 parity bits for the 50 most significant bits produced by the speech encoder during a period of 20 ms. For error correction, the bits are protected with a half-rate convolutional code, and the coded bits are interleaved over eight normal bursts. In the case of full-rate speech, the FACCH has the same interleaving as the TCH. While in the case of half-rate speech, the interleaving is over four bursts on the TCH and over six bursts on the FACCH.

12.2 Physical Layer Reliability

In this section, the reliabilities of control channels on the physical layer are compared on the basis of radio link simulations. The principal measure of physical layer reliability is the frame erasure rate (FER), which gives the ratio of detected, uncorrectable, erroneous frames on a logical channel in certain radio conditions. The upper-layer protocols must take into account the possibility of undetected errors, but it is small enough to be discarded in this analysis. The focus is on coverage-limited conditions, where a radio link has a low SNR and experiences fading. In interference-limited conditions, where a radio link has a low C/I, the unwanted signal also varies in time and therefore error rates drop more slowly when the average signal quality improves. As this holds for all the logical channels, their respective performances are similar regardless of whether coverage- or interference-limited conditions are considered. Only GSM control channels will be named in the analysis, but as one can see from Figure 12.6, results also apply to packet control channels.

12.2.1 Simulation Model

The physical layer simulator that was used performs channel coding, interleaving, burst formatting and modulation according to specifications. Modulated bursts are transmitted over a typical urban (TU) channel model [6], which brings about multipath fading and time spread. In the simulations, a carrier frequency of 900 MHz and mobile speeds of 3 and 50 km/h (TU3 and TU50 channels) were used. The results apply, for example, to the 1800-MHz band when the speeds are halved. When interleaving is applied and frequency hopping is not used, reliability improves at the same speed when the carrier frequency increases. When frequency hopping is used, it is assumed to be ideal so that the radio channel is uncorrelated between bursts. In the case of coverage-limited links, the bursts are corrupted by independent Gaussian noise samples according to a given SNR per bit (E_b/N_0) value. In the case of interference-limited links, a continuous co-channel signal is transmitted over a similar, uncorrelated radio channel and added to the wanted signal before reception. The simulated receiver follows a realistic implementation with floating-point arithmetics, adaptive channel estimation and 16-state log-MAP equalizer without antenna diversity. The FER was calculated over 20 000 frames on the SCH and RACH and over 5000 frames on the other channels.

12.2.2 Comparison of Channels

The frame erasure rates in the TU3 channel are shown in Figure 12.7. In this channel, the mobile station moves half a wavelength in about 40 TDMA frames, and the signal power is almost unchanged during the shortest interleaving period of four TDMA frames. Therefore, the signal power varies between frames on the channels like BCCH using this period, and the frame erasure rate decreases slowly as the signal quality (E_b/N_0) improves. The same also applies to the SCH and RACH, on which frames are transmitted in single bursts. The longest interleaving period makes the frame erasure rate drop most rapidly on the SACCH when E_b/N_0 improves. The TCH/FS and FACCH are intermediate cases that slightly benefit from interleaving. Because the TCH/FS has a shorter frame length and poorer error detection than the FACCH, it has lower frame erasure rates at low values of E_b/N_0. On the other hand, the FACCH performs better at high values of E_b/N_0 due to its FIRE coding, which enables the receiver to correct errors that get through the convolutional coding.

Figure 12.8 shows the results for the TU50 channel. Now half a wavelength takes only 2.4 TDMA frames and the number of bursts over which frames are interleaved has more

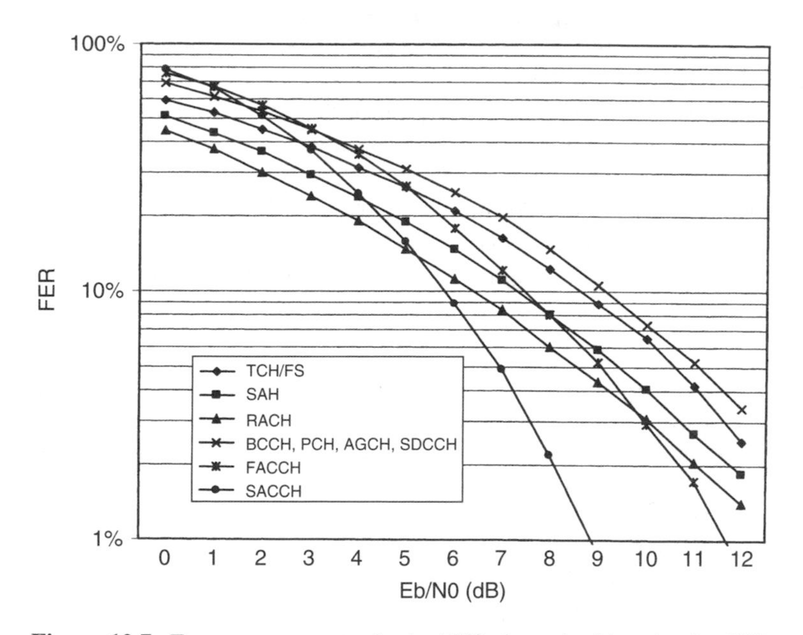

Figure 12.7. Frame erasure rates in the TU3 channel with noise (no FH)

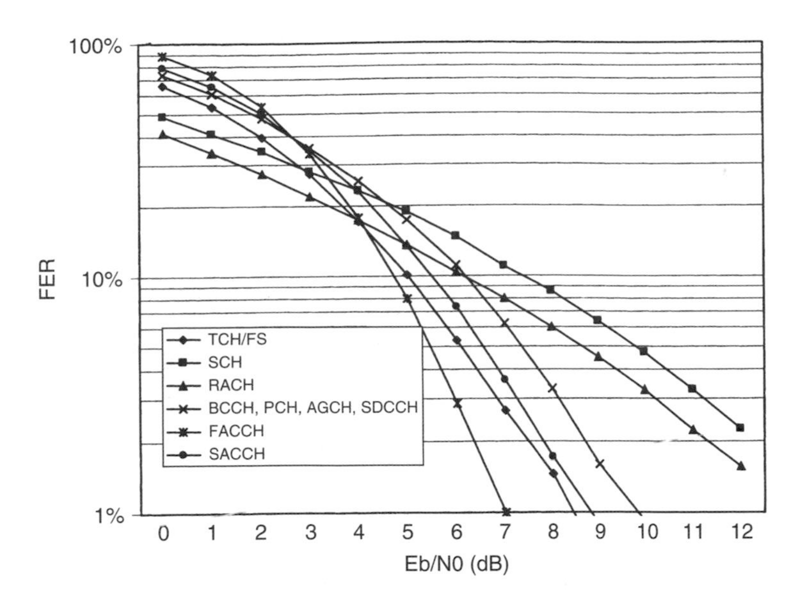

Figure 12.8. Frame erasure rates in the TU50 radio channel with noise (no FH)

significance than the interleaving period. Therefore, TCH/FS and FACCH have lowest error ratios above an E_b/N_0 of 4 dB.

By comparing Figure 12.7 with Figure 12.8, one can see that on all the channels, except the SACCH, SCH and RACH, physical layer reliability improves when the mobile speed increases. This is due to increased time diversity. The TU channel provides some frequency diversity within signal bandwidth due to its time spread, but more effective frequency diversity is achieved with frequency hopping, except for the channels that are mapped on the BCCH channel. This can be seen from the frame erasure rates in the TU3 channel with ideal frequency hopping shown in Figure 12.9. To indicate the difference between E_b/N_0 and C/I results, Figure 12.10 shows frame erasure rates in the TU3 channel with co-channel interference and ideal frequency hopping.

Because the FCCH is not decoded, it is not as simple a reliability measure as other channels. The algorithms that use the FCCH for finding the BCCH channel and correcting the frequency offset are implementation dependent, but usually they consist of burst detection, offset estimation and frequency correction. The convergence time of frequency correction depends on the initial frequency offset and the quality of received bursts. During initial synchronization, a mobile station receives over 20 frequency correction bursts per second, which is more than enough in practical algorithms even if the E_b/N_0 was 0 dB. The frequency offset that remains after convergence depends on the algorithm and SNR. Its effect on the detection of bursts increases with the burst length so that the loss in SNR is most considerable on the channels using the normal burst. The receiver can, however, employ a frequency correction algorithm also in the burst detection to reduce the effect. When a mobile station has a connection to another base station, it can receive the FCCH only every 0.6 s, on average. In this case, as the mobile station is already synchronized to the network, it must only detect the FCCH to be able to find SCH and no frequency correction is needed.

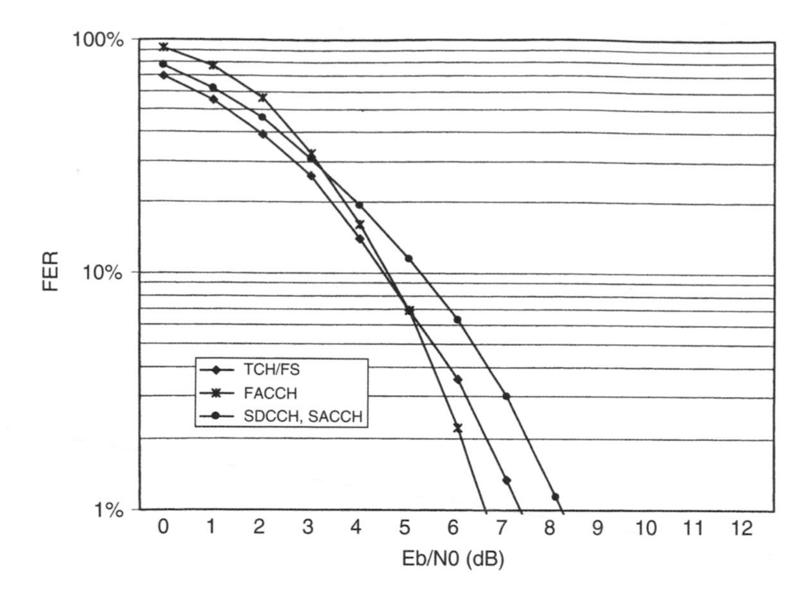

Figure 12.9. Frame erasure rates in the TU3 channel with noise (ideal FH)

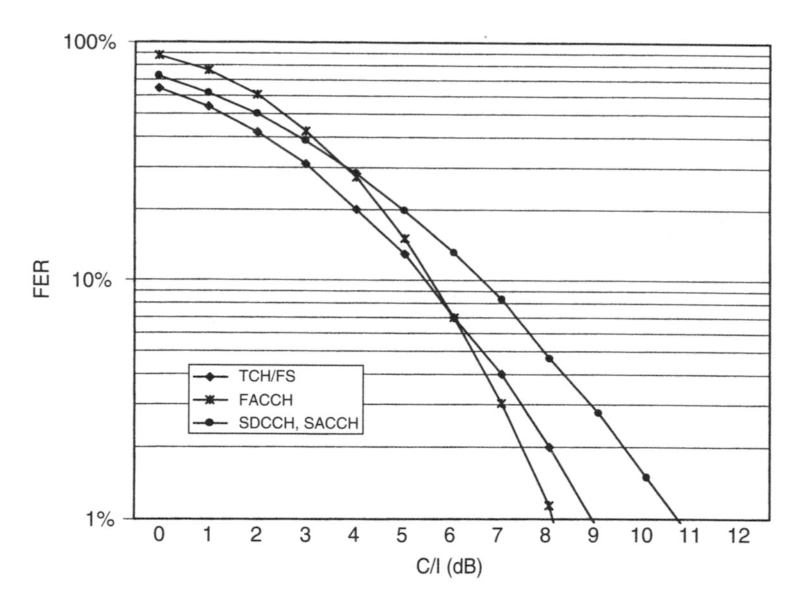

Figure 12.10. Frame erasure rates in the TU3 channel with co-channel interference (ideal FH)

12.3 Signalling Reliability and Delays

In this section, transmission reliability and delays of control channels are analysed taking the data link layer into account besides the physical layer. For simplicity it is assumed that the radio channel is statistically stationary during signalling procedures that take only a few seconds at maximum (i.e. the time average and correlation of C/N are constants). The message flows of the L3 procedures that are analysed have been obtained from signalling traces of a real network. The focus is on the signalling of circuit-switched services, but the results that consider broadcast and common signalling also apply to (E)GPRS.

12.3.1 Probabilistic Models

On the broadcast channels, L3 messages are transmitted repeatedly without acknowledgements at the data link layer. Assuming that the radio channel is uncorrelated between transmissions, the success probability after N-transmissions is

$$1 - p^N \tag{1}$$

where p is the frame error probability. By discarding undetected errors, the frame erasure rate obtained from link simulations can be used as an estimate of p for a given average C/N. Thus, the success probability is a function of time and C/N. Transmissions can be assumed uncorrelated when the mobile stations move half a wavelength or more between them. Otherwise, the frame error probability depends on the success of previous transmissions, in which case the success probability can be estimated from the frame error sequences logged in physical layer simulations.

A successful RR connection establishment on the common control channels requires that paging request, channel request and immediate assignment are all decoded successfully. The network repeats paging request after a certain time without receiving paging response. Assuming that the average C/N is constant in both directions and transmissions are uncorrelated, the probability of a successful establishment is

$$1 - (1 - (1 - p)(1 - (1 - P_k(1 - p))^{M+1})^L \tag{2}$$

where p is the frame error probability on the PCH and AGCH, P_k is the decoding success probability of the channel request that is spread into k bursts (3–50), M is the maximum number of retransmissions for channel request (1, 2, 4 or 7) and L is the maximum number of paging attempts (not specified).

As described in Section 12.1, the data link layer increases transmission reliability with a send-and-wait ARQ protocol on the SDCCH and FACCH. That is, an information frame is retransmitted until both the frame itself and its acknowledgement are received correctly. Suppose that a signalling procedure consists of K information frame transmissions when all the frames are correctly received, and that a separate response frame carries the acknowledgement of each information frame. Then, the success probability of the procedure after N information frame transmissions ($N \geq K$) is

$$1 - \sum_{i=0}^{K-1} \binom{N}{i} [(1 - p_\mathrm{u})(1 - p_\mathrm{d})]^i [1 - (1 - p_\mathrm{u})(1 - p_\mathrm{d})]^{N-i} \tag{3}$$

where the frame error probabilities in the uplink and downlink are constants p_u and p_d. This requires that the average C/N is constant and all transmissions are uncorrelated. The expected number of information frame transmissions is then

$$\frac{K}{(1 - p_\mathrm{u})(1 - p_\mathrm{d})}. \tag{4}$$

Formulas (3) and (4) are inapplicable when acknowledgements piggyback on the information frames, which is possible if L3 messages are alternately transmitted in the uplink and downlink and they fit to single L2 frames. In this case, probabilities and average times can be estimated with simulations. Nevertheless, (3) gives the lower limit of the success probability and (4) the upper limit of the expected number of transmissions.

The restrictions on the number of retransmissions that protocol specifications determine for noticing link failures can make signalling procedures fail. In an L2 acknowledged mode link establishment, the maximum number of retransmissions is five [3]. Since the establishment consists of one information frame and its response frame, the success probability is

$$1 - (1 - (1 - p_\mathrm{u})(1 - p_\mathrm{d}))^6. \tag{5}$$

If the establishment fails, for example, when starting a call, it blocks the call. After link establishment, the number of retransmissions is limited to 23 on the SDCCH, and to 34 on the FACCH [3]. When these limits are reached, a link failure is indicated to L3, which releases the link. At the other end, the link is released after a timer expiry. For one information frame transmission, the probability of link failure is

$$(1 - (1 - p_\mathrm{u})(1 - p_\mathrm{d}))^N \tag{6}$$

where $N = 24$ for the SDCCH and $N = 35$ for the FACCH. When a signalling procedure comprises K information frame transmissions, and separate response frames carry acknowledgements, the probability of failure becomes

$$1 - (1 - (1 - (1 - p_\mathrm{u})(1 - p_\mathrm{d}))^N)^K. \tag{7}$$

In other cases, the link failure probability can be evaluated by simulation. For example, in the call establishment on the SDCCH, link failure blocks the call, and in the handover signalling on the FACCH, it may drop the call.

Formulas (2) to (7) take into account that a radio link can be in imbalance, in which case frame error probabilities are different in the downlink and uplink. When considering the limiting direction, a balanced link sets the lower limit of C/N or C/I since improvement in one direction allows degradation in the other.

12.3.2 SCH Information Broadcast

A base station broadcasts synchronization information on the SCH with an interval of 10 TDMA frames, or 46 ms. When the mobile speed is over 10 km/h, roughly, transmissions are uncorrelated and the decoding success probability is given by formula (1). In the TU3 channel, correlation between transmissions increases decoding delay as Figure 12.11 shows. The trade-off between the delay and E_b/N_0 requirement is significant: repetitions increase the redundancy and time diversity, which lowers the E_b/N_0 requirement, but the lower the E_b/N_0, the longer the transmission delay.

As the transmission period is short in the initial synchronization, E_b/N_0 can be lower than 4 dB without significant decoding delay. In the pre-synchronization for measurements and handovers, a mobile station can attempt to decode SCH information every 0.6 s, on average. Several decoding failures may thus cause a considerable delay, though with this period transmissions are uncorrelated in the TU3 channel. For example, E_b/N_0 needs to

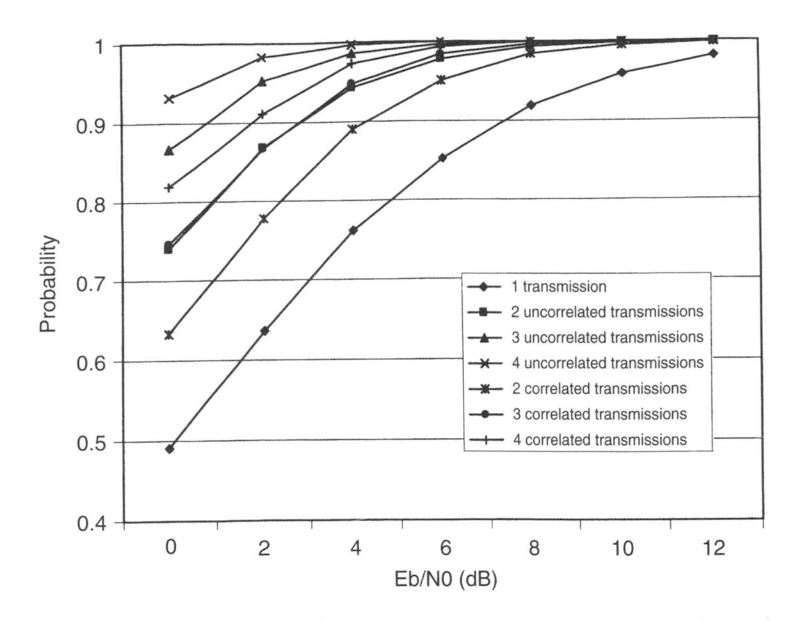

Figure 12.11. The success probability of decoding SCH information (TU3 channel, 900-MHz band)

be 4 dB so that the success probability is close to unity and the decoding delay does not exceed 2.4 s (four transmissions). A delay of a couple of seconds may cause problems, for example, when a high-speed mobile station needs to synchronize to a new cell before a handover.

12.3.3 System Information Broadcast

The system information that is broadcast on the BCCH is grouped into four message types. As the message types are alternately transmitted, the transmission period of a message type is four 51-multiframe periods (0.94 s), which is enough to make transmissions uncorrelated in the TU3 channel. Figure 12.12 shows the decoding success probability of a message type after 1, 2, 3, 4 and 5 transmissions. Since mobile stations decode the BCCH only in the idle mode, decoding delay is not a serious problem for circuit-switched services. In (E)GPRS, a lower delay in the BCCH decoding is more important since a mobile station performs cell re-selections during sessions. For example, with five attempts, the success probability is close to 1 when E_b/N_0 is above 4 dB. This also means that the mobile station may be unable to decode the (P)PCH for 5 s when E_b/N_0 is 4 dB or lower in the new cell.

12.3.4 RR Connection Establishment

The success probability of a mobile-terminated RR connection establishment, which is performed on the common control channels, can be evaluated with formula (2). Before that, it must be examined how the number of bursts used to spread the channel request transmission impacts on the decoding success probability. Figure 12.13 shows the decoding success probabilities when using 3 uncorrelated bursts, and when using 3 and 10

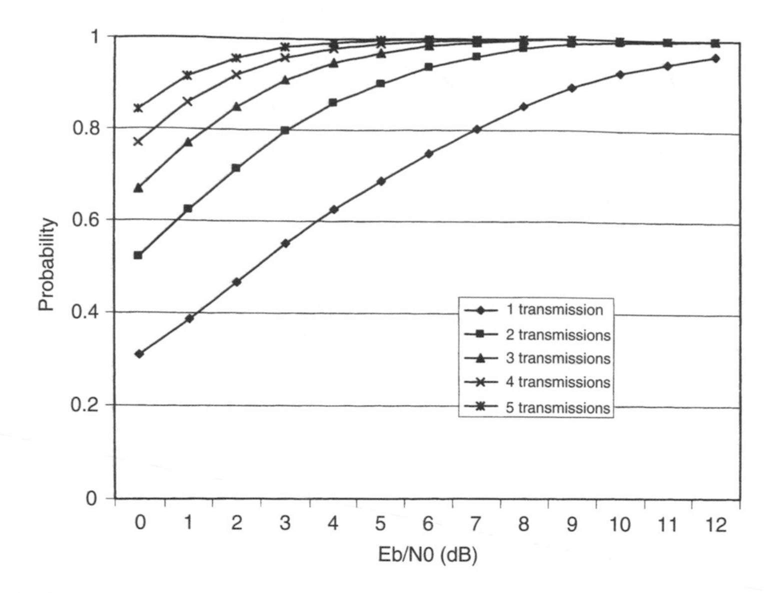

Figure 12.12. The success probability of decoding system information on the BCCH (TU3 channel, 900-MHz band)

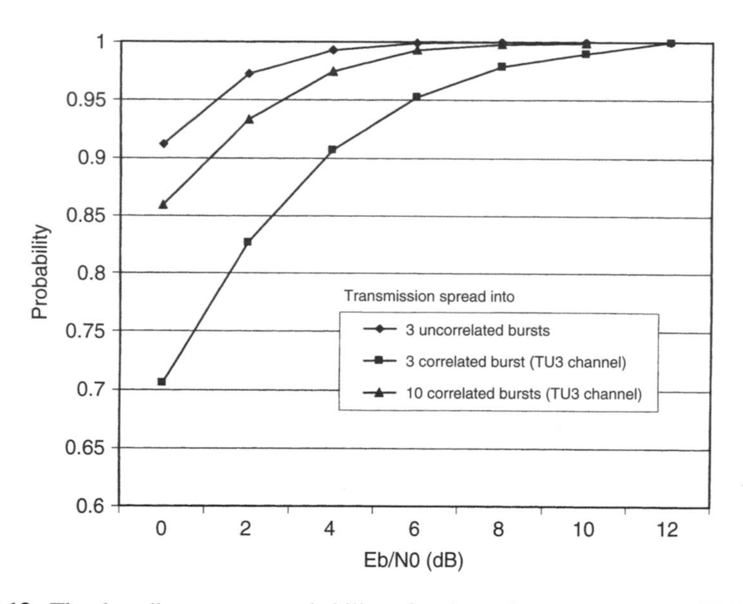

Figure 12.13. The decoding success probability of a channel request message (TU3 channel, 900-MHz band)

consecutive bursts in the TU3 channel. It can be seen that in the case of 3 bursts in the TU3 channel, E_b/N_0 must be over 10 dB in order that the success probability of the first transmission is close to 1. By using 10 bursts, the requirement drops to below 6 dB. The results show that the success probability substantially depends on the correlation between transmissions. This suggests that, in order to maximize both the reliability and capacity of the RACH, using bursts for retransmissions is preferable to spreading single transmission. On the other hand, spreading messages into a maximum of 50 bursts minimizes the delay if the capacity is unlimited.

To evaluate the success probability of the whole RR connection establishment, it is assumed that channel request is spread into three bursts, the maximum number of retransmissions for channel request is four, and the collision probability on the RACH is negligible. Retransmissions can be assumed to be uncorrelated in the TU3 channel since the minimum retransmission period for channel request is 41 TDMA frames, or about 190 ms, and the minimum sensible period for paging attempts is about 1 s. It is furthermore reasonable to assume that the period between paging request and immediate assignment is of the order of 51-multiframe. Thus, all transmissions can be assumed uncorrelated and the success probabilities for different number of paging attempts are obtained from formula (2).

The success probabilities are shown as a function of E_b/N_0 for the TU3 channel in Figure 12.14. With a maximum of three paging attempts, the success probability is close to unity when E_b/N_0 is above 11 dB. With 5 paging attempts, the minimum E_b/N_0 requirement drops to 5 dB, and with 10 paging attempts it drops to 3 dB. The success probabilities are, of course, smaller when the RACH has a non-zero collision probability, which would decrease the value of P_k in formula (2). When channel request is spread only into three bursts, as assumed above, the collision probability can be, however, assumed negligible on the basis of Section 12.4.

12.3.5 L2 Link Establishment

The retransmission period of an L2 frame is 235 ms on the SDCCH and 120 ms on the FACCH. With the shorter period of 120 ms, the correlation between transmissions was

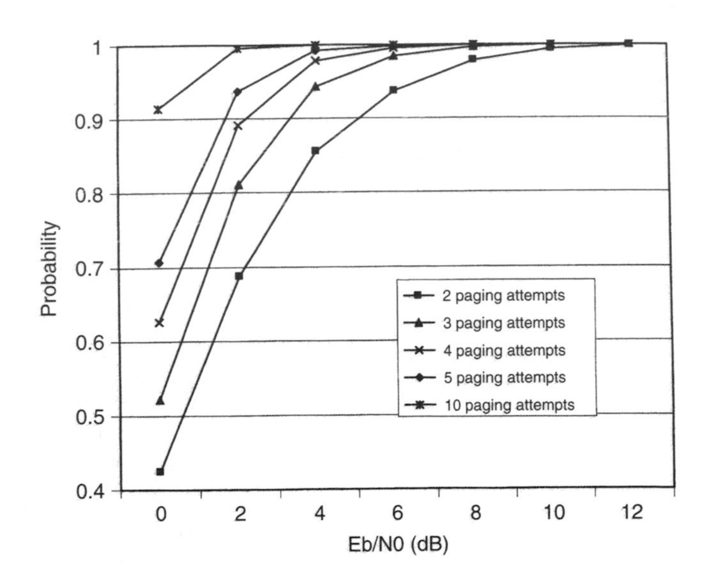

Figure 12.14. The success probability of an RR connection establishment (TU3 channel, 900-MHz band)

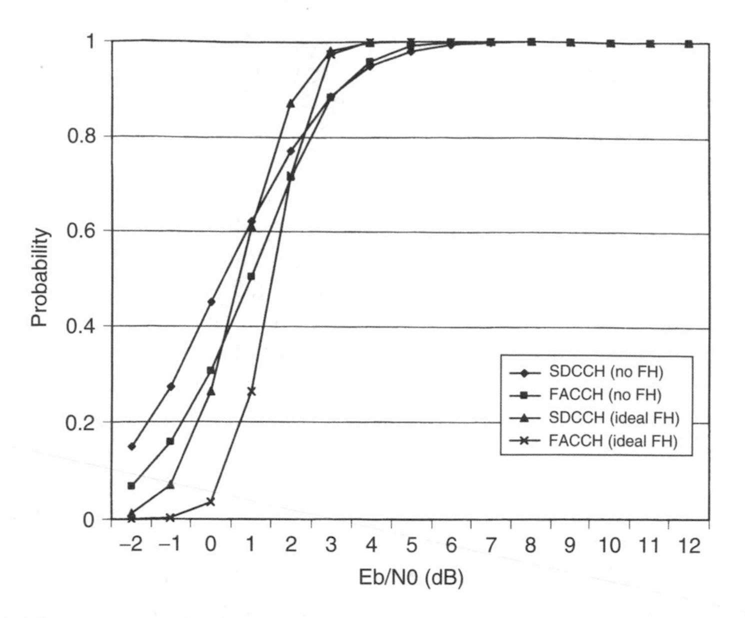

Figure 12.15. The success probability of an L2 link establishment (TU3 channel, 900-MHz band)

found to have negligible impact on the signalling success probability in the TU3 channel. Therefore, transmissions can be assumed uncorrelated. Figure 12.15 shows the success probability of an L2 link establishment in the TU3 channel according to formula (4) with and without frequency hopping. Without frequency hopping, establishment failures are possible when E_b/N_0 is below 8 dB, and with frequency hopping, when E_b/N_0 is below 5 dB.

12.3.6 L2 Link Failure

Figure 12.16 shows the probability of an L2 link failure in the TU3 channel with and without frequency hopping for a call establishment and for a handover command transmission. During a call establishment on the SDCCH, an L2 link failure may occur when E_b/N_0 is below 3 dB, and during a handover command transmission on the FACCH when E_b/N_0 is below 2.5 dB. Frequency hopping decreases these limits by 0–1 dB, but on the other hand, it increases the probability of link failures when E_b/N_0 is below 1 dB.

12.3.7 Call Establishment and Location Update

In a call establishment, the network can choose between three different channel allocation strategies: (1) off-air call set-up, (2) early assignment or (3) very early assignment. These strategies differ in the time when a traffic channel is assigned for the signalling on the FACCH. In strategy (1), the SDCCH is used until the called party answers the call. Because in this strategy, the SDCCH is used for the longest time, it is considered in the following.

Since delays at L3 depend on the signalling needs between MS, BTS, BSC and MSC, the total time of an error-free procedure varies. The simulated case included a total of

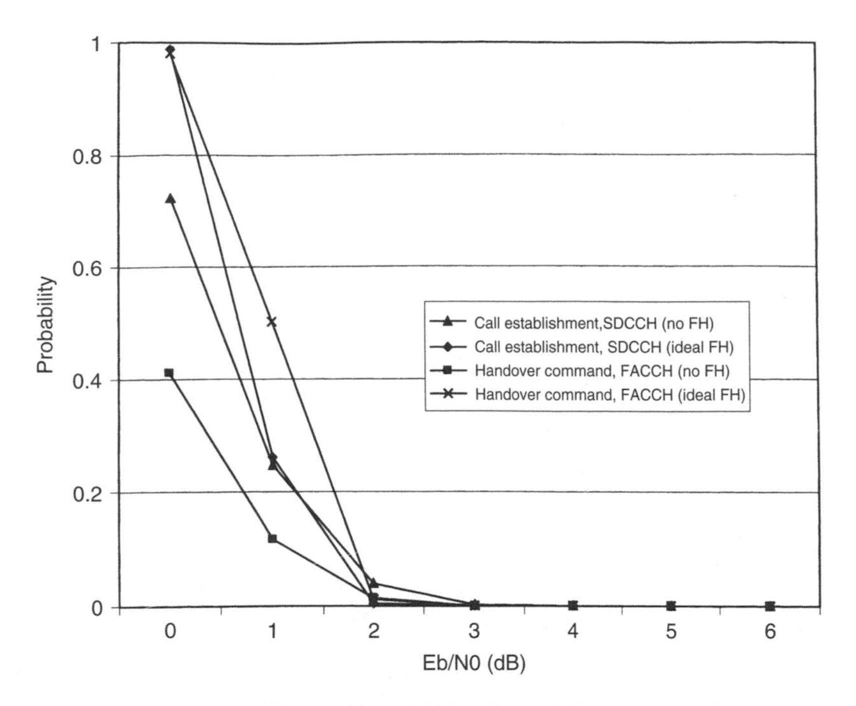

Figure 12.16. The probability of an L2 link failure (TU3 channel, 900-MHz band)

13 information frame transmissions on the SDCCH. If these were transmitted only in one direction with the minimum time between transmissions, the total time would be 13×0.235 s $= 3$ s. The changes in the direction and unused frames increase the time to about 3.5 s. In the simulations, frame erasures were randomized according to the frame error ratios at each E_b/N_0 point and the link was assumed to be in balance. The average time of 10 000 simulated procedures with and without frequency hopping in the TU3 channel is shown in Figure 12.17. Link failures are taken into account, so that only successful procedures are included in the average. It can be seen, for example, that without frequency hopping, the procedure takes 2.7 times longer at 3 dB, on average, than it would take without frame errors.

A location update is also signalled on the SDCCH after initial assignment on the RACH and AGCH. The simulated procedure included authentication signalling and temporary mobile subscriber identity (TMSI) reallocation. The average time of 10 000 procedure executions is shown in Figure 12.17. Because location updates cause a significant load on the SDCCH, several retransmissions are undesirable. The excess load due to retransmissions is about 20% at 8 dB, 50% at 6 dB and 100% at 4 dB in the TU3 channel without frequency hopping.

12.3.8 Handover and Channel Transfer

When performing a handover to another cell, the network sends a handover command to the mobile station on the FACCH using the old channel. The message assigns the new cell and the channel to which the mobile station shall change its transmission. In the case

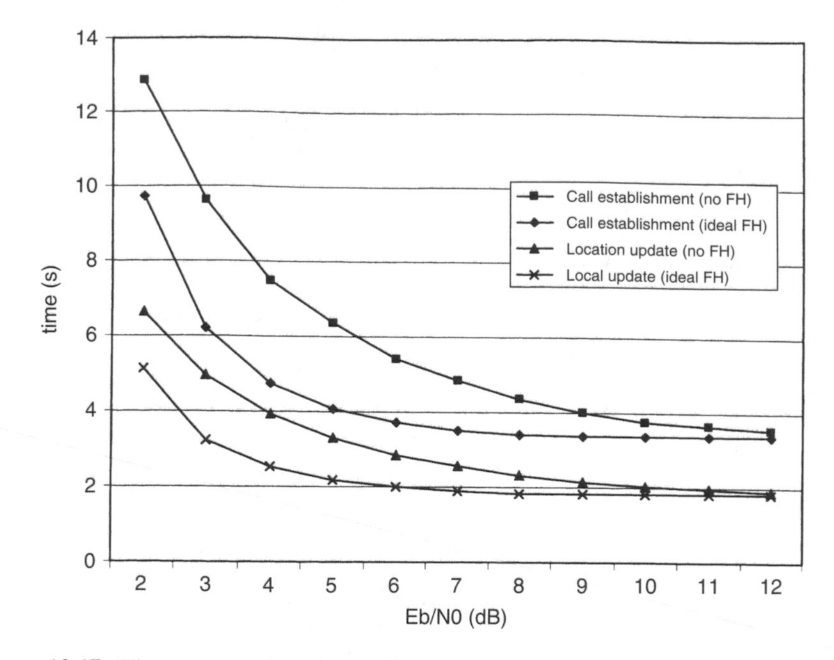

Figure 12.17. The average SDCCH usage time in call establishment and location update procedures (TU3 channel, 900-MHz band)

that is considered, the message fits into two L2 information frames, both of which are acknowledged by a response frame. Thus, formulas (3) and (4) apply with $K = 2$.

Figure 12.18 shows the success probability as the number of retransmitted frames increases from 0 to 15, or the transmission delay increases from 0.24 to 2.04 s. The combination of channel coding and ARQ protocol seems to give sufficient performance, that is, high reliability and low delay, above 6 dB. According to Figure 12.16, a limit appears at 2 dB below which a link failure may occur. However, if C/N drops rapidly, then any additional delay may cause a link failure. Besides transmission of handover command, a successful handover procedure requires successes in measurement report reception (see Section 12.3.4), pre-synchronization (see Section 12.3.2) and in the access and link establishment on the new dedicated channel.

When the network needs to transfer a connection to another physical channel in the same cell, it sends an assignment command to the mobile station on the FACCH using the old channel. This message also fits into two L2 frames and the success probability is the same as that of a handover command shown in Figure 12.18. The mobile station completes the channel transfer by establishing an L2 link and transmitting assignment complete on the FACCH using the new channel.

12.3.9 Measurements and Power Control

The most time critical messages carried by the SACCH include measurement reports in the uplink and power control commands in the downlink direction. Since a message is transmitted every 0.48 s, transmissions may be assumed uncorrelated in the TU3 channel. Decoding success probabilities according to formula (1) for 1–5 transmissions are shown

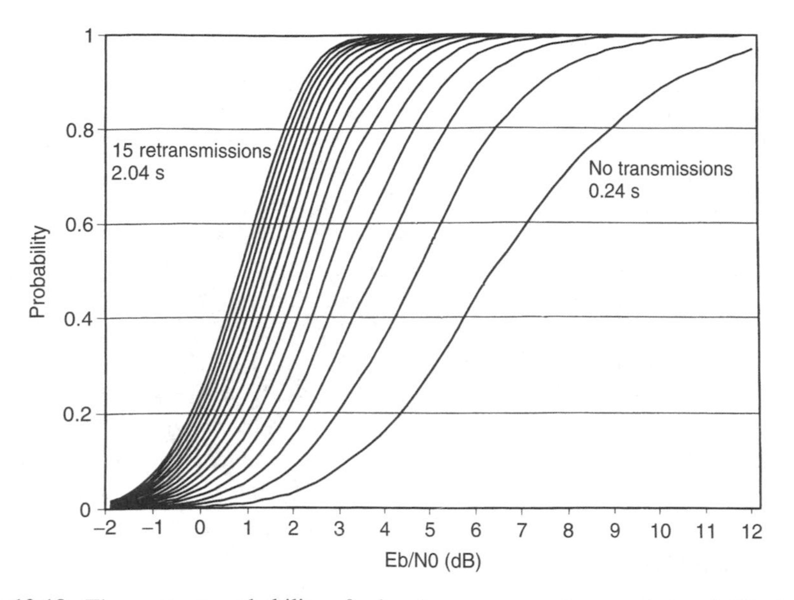

Figure 12.18. The success probability of a handover command transmission (TU3 channel, 900-MHz band, no FH)

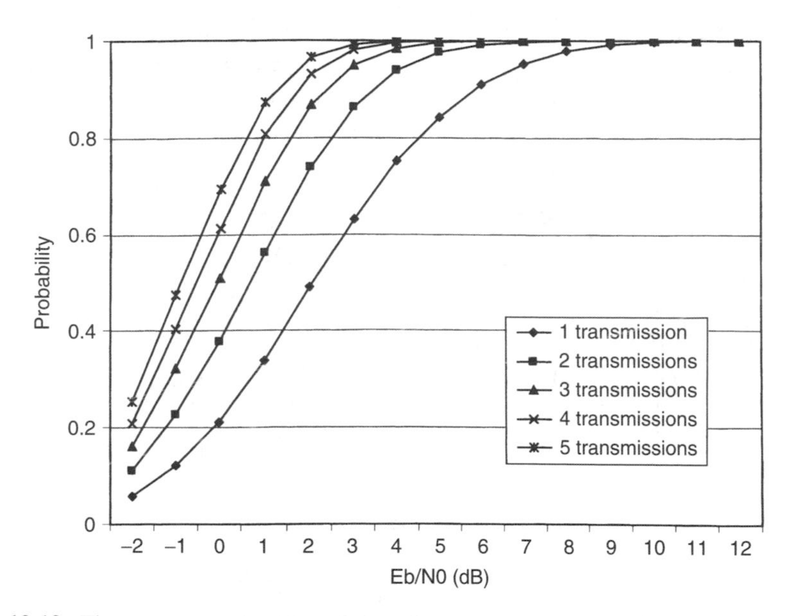

Figure 12.19. The success probability of decoding a message on the SACCH (TU3 channel, 900-MHz band, no FH)

in Figure 12.19. A higher frame erasure rate means longer measurement reporting delays in the uplink and longer power control command reception delays in the downlink. With one decoding attempt, the success probability is close to unity when E_b/N_0 is 11 dB. With 2, 3, 4 and 5 attempts, the corresponding E_b/N_0 values are 7, 5, 4 and 3.5 dB.

12.3.10 Radio Link Failure

To be able to stop transmission and release the physical channel when the signal is irrevocably lost, a mobile station, when using a TCH or SDCCH, updates a radio link counter based on the frame erasures on the SACCH [7]. Every correct frame increases the counter by one and every lost frame decreases it by two, and a radio link failure is declared when the counter reaches zero. The maximum value of the counter is a network parameter, which can take values from 4 to 64 corresponding to a radio link timeout in 2 to 30 s when all frames are lost. The probability of such event can be evaluated as a function of SNR assuming that the counter initially has the maximum value of RLT of and the average SNR is at a certain level for N seconds. Figure 12.20 shows simulation results for TU3 channel with some example values of RLT and N. Clearly, the SNR limit to generate a radio link failure can be decreased by increasing RLT, but this has the drawback that the release time at very low SNR levels increases. On the other hand, the network could optionally induce the release of the channel by stopping transmission based on uplink quality or measurement reports sent by the mobile station.

12.3.11 Conclusions

The preceding analysis considered probabilities of successfully completing signalling procedures as a function of time and C/N. It was assumed that the short-time average of C/N is independent of time. According to the results, it can be concluded that a L2 link establishment on the SDCCH sets the highest E_b/N_0 requirement of 8 dB in the TU3 channel without frequency hopping. The FER with this E_b/N_0 is 13.5%. L2 link establishments are most vulnerable to frame erasures because the number of retransmissions is limited to five and errors may occur in both directions. With ideal frequency hopping, the require-

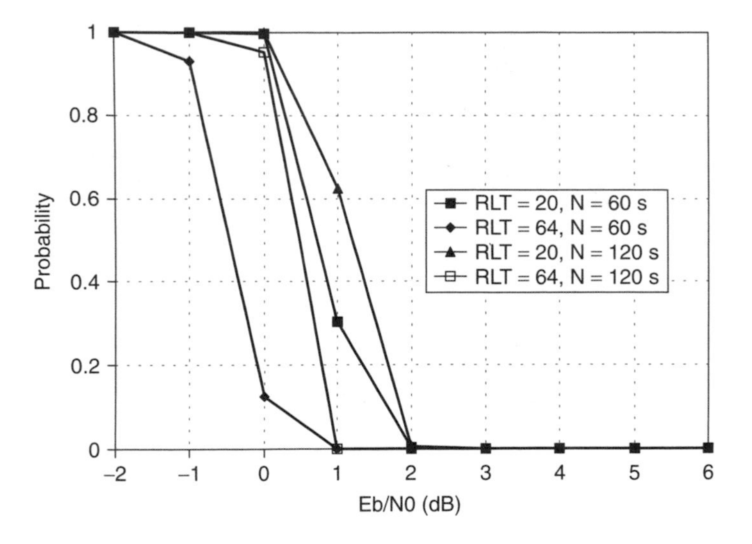

Figure 12.20. Radio link failure probability in a mobile station (TU3 channel, 900-MHz band, no FH)

ment is decreased to 5 dB, which is also met by the other channels if delays of a few seconds are accepted. The GSM full-rate speech channel requires an E_b/N_0 of 7 dB or higher so that the FER is less than 1% with ideal frequency hopping in the TU3 channel. When frequency hopping is non-ideal, the requirement is higher; without hopping, the required E_b/N_0 for a 1% FER is 14.5 dB. Thus, according to the results of this chapter, network coverage and capacity can be potentially improved by enhancing the robustness of speech traffic channels.

12.4 Control Channels versus AMR TCH

As described in Chapter 6, adaptive multi-rate (AMR) speech coding is an efficient technique to decrease minimum C/N and C/I requirements of traffic channels. Its introduction to the network does not, however, improve the performance of control channels. In order to find out if signalling failures or delays limit the system capacity enhancement attainable from AMR, this section compares the performance of AMR full-rate speech traffic channels (TCH/AFS) with the performance of dedicated control channels.

12.4.1 Physical Layer Comparison

Figure 12.21 presents physical layer simulation results for SACCH, SDCCH and FACCH signalling channels and for TCH/AFS of different data bit rates. The results assume TU3 channel with co-channel interference and ideal frequency hopping. According to the results, TCH/AFS12.2 allows 3 dB lower C/I than SDCCH and SACCH for the

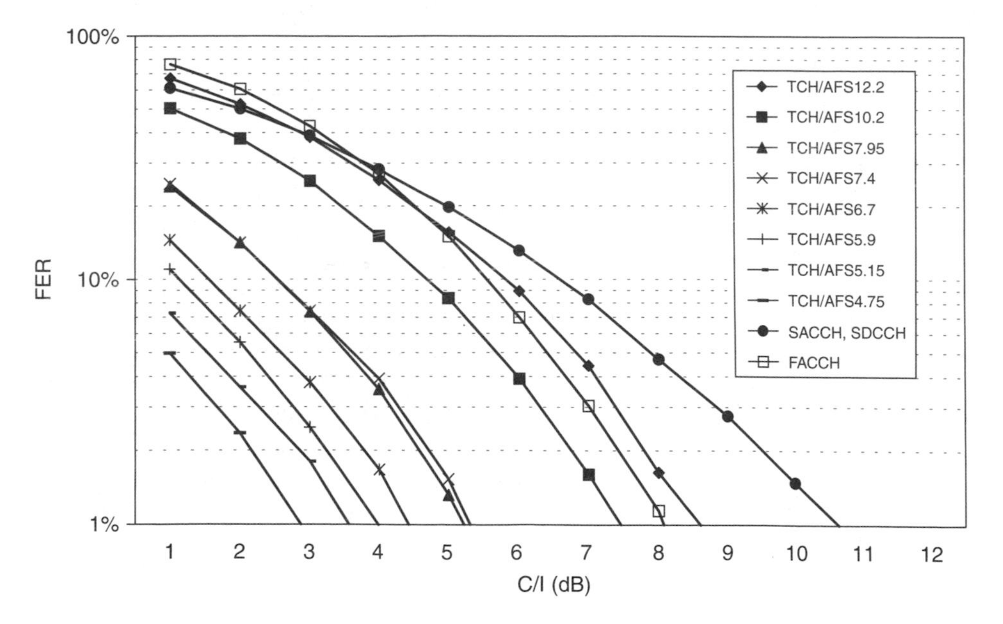

Figure 12.21. Frame erasure rates on the TCH/AFS and dedicated control channels (TU3 channel, ideal FH)

frame erasure rate of 1%. This is mainly due to a half longer interleaving of the full-rate TCH. Lower code rates besides longer interleaving make AFS7.4–4.75 codecs outperform SDCCH and SACCH by 5–8 dB. The FACCH, which has the same interleaving as the full-rate TCH, outperforms AFS12.2 but is less robust than the AFS of lower bit rates.

On the other hand, with the same C/I, the FER is considerably higher on the dedicated control channels than on the low bit rate AFS traffic channels. However, whereas frame erasures decrease speech quality on the TCH, they increase the signalling delay or failure probability on the control channels that use retransmissions to increase reliability. Therefore, in order to analyse the lowest acceptable C/I, the question of how much higher frame erasure rate is allowed on the control channels than on the traffic channels has to be resolved.

12.4.2 System Level Comparison

The analysis of Section 12.3 showed that the SDCCH has the highest C/N requirement of control channels when the link is in balance and statistically stationary during signalling procedures and it is required that the probability of signalling failures is close to zero. The SDCCH FER corresponding to the SNR requirement was 13.5%. According to Figure 12.21, this corresponds to a C/I requirement of 6 dB with ideal frequency hopping in the TU3 channel. As opposed to C/N, C/I is, however, uncorrelated between the downlink and uplink, and in the limiting link establishment signalling, the C/I requirement of the worse direction is smaller if the FER in the other direction is smaller. For example, if the FER is zero in one direction, the C/I requirement of the other direction drops from 6 dB of the balanced link to 4.5 dB. Thus, the C/I requirement of the SDCCH is 4.5–6 dB, depending on the link balance, whereas the C/I requirement is independent of the link balance on the TCH.

In order to evaluate system level performance, the part of SDCCH signalling that is successful when C/I is below 6 dB has to be taken into account as well. Therefore, the signalling performance of an interference-limited network is estimated by weighting link establishment success probabilities of different C/I points with the C/I distribution of the network. The cumulative downlink C/I distribution shown in Figure 12.22 has been obtained from the network simulation corresponding to the highest system capacity with frequency hopping and AMR when the FER on the TCH was used as the quality criterion. The quality criterion was such that frame erasure rates were calculated over 2 s, the samples with a FER greater than 4.2% were classified as bad, and it was required that no more than 2% of the samples are bad (the same quality criteria was used in Section 6.5). Figure 12.22 also shows L2 link establishment success probabilities in the TU3 channel with ideal frequency hopping.

The C/I samples of the distribution are averages over a measurement reporting period of 0.48 s. Using the distribution, the C/I-weighted probability for a link establishment failure was found to be 2.9% when the FER on the uplink and downlink was assumed to be always equal, and 1.7% when one direction was assumed to be always error free. These figures are comparable with the frequency of bad TCH FER samples, which was 2%. These results clearly show that the SDCCH is not robust enough to be allocated on to frequency hopping channels without impacting the service quality when using AMR to maximize system capacity.

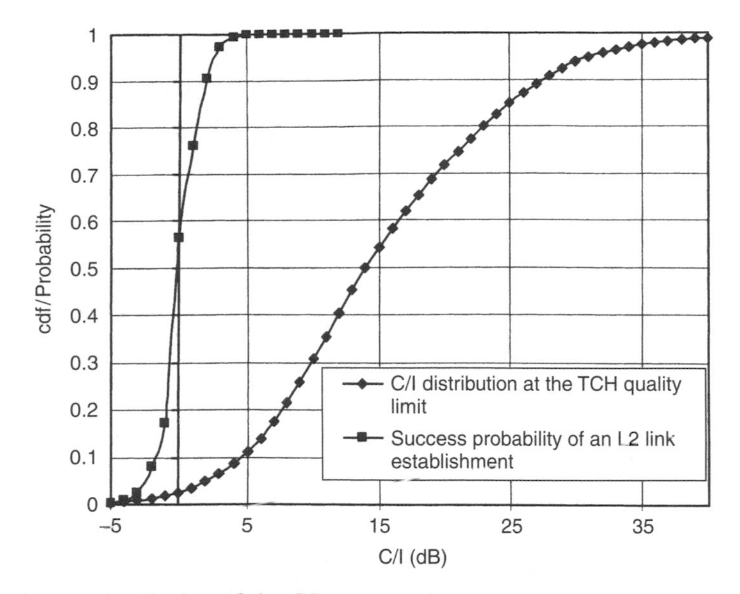

Figure 12.22. The distribution of the C/I average of 0.48 s in an interference-limited network with frequency hopping and the success probability of an L2 link establishment on the SDCCH

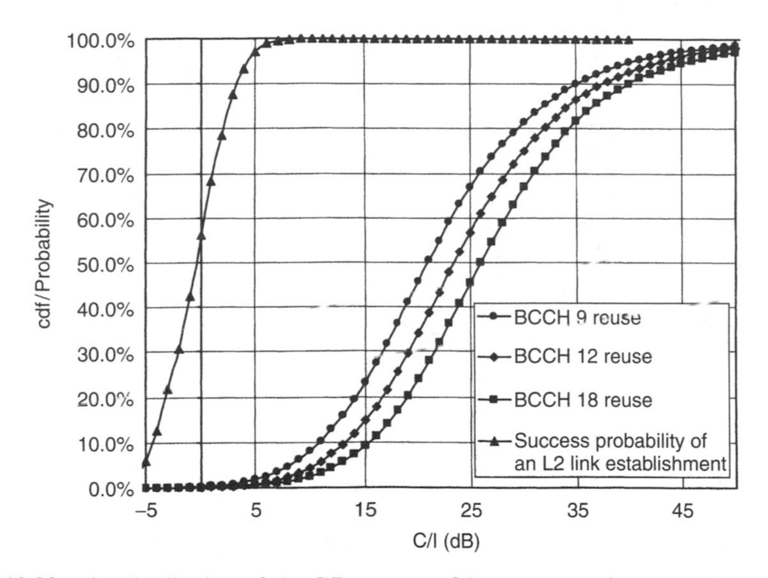

Figure 12.23. The distribution of the C/I average of 0.48 s in interference-limited networks without frequency hopping and the success probability of an L2 establishment on the SDCCH

For a BCCH layer comparison, the downlink C/I distribution of a non-hopping network and frame erasure rates of a non-hopping link were used. Figure 12.23 shows the C/I distributions obtained from network simulations with different reuses and L2 link establishment success probabilities in the TU3 channel without frequency hopping. The AMR TCH qualities of these simulations are presented in Chapter 6.

Table 12.1. The upper and lower limit for the failure probability of an L2 link establishment in networks with reuse 9, 12 and 18 without frequency hopping

Reuse	Upper limit (%)	Lower limit (%)
9	0.5	0.1
12	0.2	0.0
18	0.1	0.0

The C/I-weighted L2 link establishment failure probabilities are shown for different reuses in Table 12.1. The upper limit assumes that all the links are in balance and the lower limit assumes that the uplink is error free. On the basis of these results, it can be concluded that signalling reliability is acceptable with reuse 9 and failures are almost absent with higher reuses.

Although SDCCH is the most limiting control channel based on link failure probabilities in statistically stationary radio conditions, in practice, FACCH and SACCH can be more critical as they are used in handover and power control processes, which are most often performed at a low C/I level. In order that lost frames on the FACCH can cause a link failure when transmitting handover and channel assignment commands in statistically stationary radio conditions, C/I must be below 2 dB (FER = 60%). Because AFS/TCH4.75 FER is moderate (2.4%) at this C/I level, handovers and channel transfers must be performed before speech quality is unsatisfactory. The link failure limit in stationary conditions is, however, insufficient as lost frames affect power control and handover processes, purpose of which is to maintain acceptable signal quality. Figure 12.24 illustrates the delays of these procedures in error-free conditions. The impact of frame erasures on the delays must be considered in order to find out possible limitations of the SACCH and FACCH.

Frame erasures on the SACCH mean lost measurement reports in the uplink and lost power control commands in the downlink direction. Therefore, they increase the delay of power control in reacting on the changes in the radio signal quality. The Rx-quality thresholds of power control found appropriate in network simulations with AMR, correspond approximately to a target window from 8 to 16 dB for C/I. Therefore, it is reasonable to consider delays at a C/I level a little below 8 dB. For example, at 5 dB, the FER on the SACCH is 27% in the TU3 channel with frequency hopping. According to formula (1), with this FER, the probability of losing more than two consecutive frames is 2% and more than four consecutive frames is 0.1%. This means that with a probability of 98%, the delay is below 1.5 s and with a probability of 99.9% below 2.5 s. The same reasoning also applies to the power control commands in the downlink. As a mobile station must be able to change its Tx-power 2 dB in every 60 ms after receiving a command, these delays may not considerably decrease the TCH quality. Lost measurement reports also cause additional delay in the handover process, which is based on the measurements performed by the mobile station as indicated in Figure 12.24.

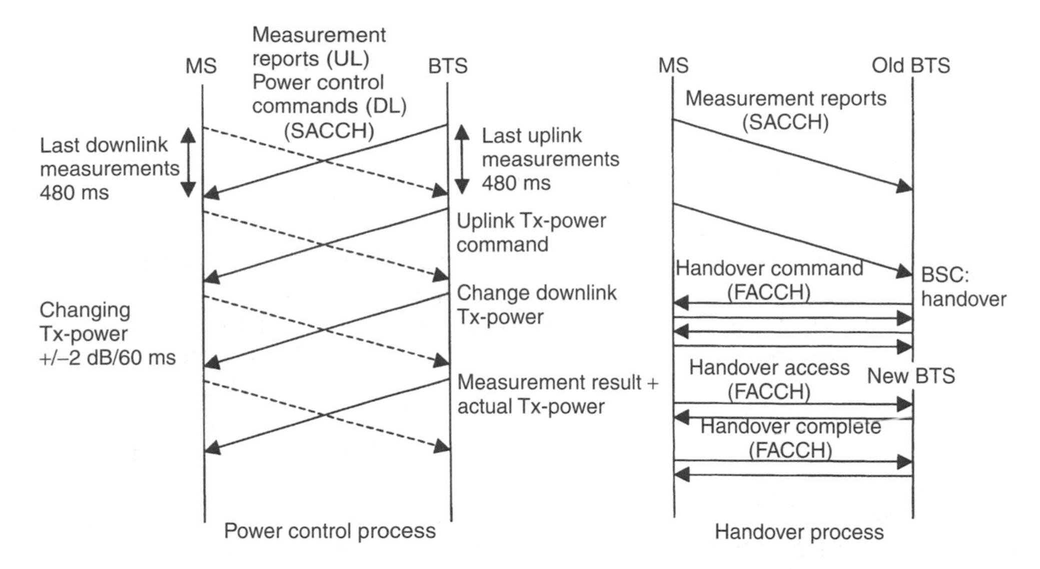

Figure 12.24. Delays in power control and handover processes

The impact of low signal quality on the transmission delay on the FACCH was considered in Section 12.3. It was noted that the signal quality requirement decreases significantly when short additional delays due to retransmissions are accepted. Since the retransmission period is only 120 ms, the delay due to retransmissions is moderate compared with the whole handover process including the downlink measurements and the averaging of measurement reports in the BSC before the handover decision. For example, with an E_b/N_0 of 4 dB, the maximum delay in the transmission of handover command is 2 s in the TU3 channel (see Figure 12.18). This corresponds to a maximum FER of 36% and to a C/I of 3.5 dB with ideal frequency hopping. Accordingly, assuming that a handover decision is made when C/I is 8 dB, the handover command transmission is successful unless C/I drops more than 4.5 dB in 2 s, which plausibly is a rare event.

12.4.3 Conclusions

On the basis of signalling failure probabilities in statistically stationary radio conditions SDCCH is the most limiting control channel. Therefore, its performance must be especially considered when using AMR for network performance improvements. Link level analysis shows that L2 link establishment failures on SDCCH are experienced when the C/I of the limiting direction is below 6 dB in the TU3 channel with ideal frequency hopping. In the hopping layer of an interference-limited network, having acceptable quality for 98% of the full-rate AMR speech samples, the link set-up failure probability was estimated to be 1.7–2.9%. Therefore, if SDCCH is allocated to the hopping channels, link establishment failures may limit the service quality more than traffic channels. This limitation can be avoided by allocating SDCCH to the non-hopping BCCH frequencies in which traffic channels are more limiting than control channels. Also, the performance of SACCH and FACCH is critical, as frame erasures on them increase delays in power control and handover algorithms.

12.5 Signalling Capacity

The way current networks are dimensioned accounts for a certain blocking probability to avoid costly hardware overdimensioning. Thus, the resources are dimensioned to allow a certain blocking probability, which is a design criterion, and the 'hard-blocking' capacity of these resources is defined as the traffic that causes this blocking probability. This section presents a study of signalling channel capacity of typical GSM cell configurations, and includes guidelines about how to optimally configure signalling channels according to the traffic carried by the network. In this section, signalling channels include those control channels in which load varies in time and which are allocated independently from traffic channels, that is, the broadcast and associated control channels are not considered in this section.

The following subsections describe the criteria used to dimension the signalling channels, as well as the signalling procedures and the methodology used to calculate the capacity of these channels. This capacity is shown for GSM voice calls and (E)GPRS data sessions, and finally some conclusions about channel dimensioning are drawn.

12.5.1 Signalling Capacity Criterion

The most common way to dimension the traffic channels for voice calls in GSM is to dimension the network for a 2% blocking probability, which can be calculated using the Erlang-B formulas. However, calls can also be blocked due to lack of signalling resources. Assuming independent blocking probabilities for traffic and signalling channels, the overall blocking experienced by the end users is the sum of both blocking probabilities. Therefore, a criterion is needed in order to obtain the capacity of the channels for signalling purposes. There are some approaches to solve this problem. The first intuitive idea is to use a low blocking probability for the signalling procedures, so all the blocking is caused due to lack of resources for traffic channels. In this way, a blocking probability 10 times less can be used (0.2%). The second criterion is to allow more blocking in order not to overdimension the resources for signalling purposes. This makes it reasonable to check the capacity of signalling channels for a blocking probability of 1 and 2%.

Consequently, the capacity of signalling channels has been obtained for a blocking probability of 0.2, 1 and 2%. As the signalling load depends on the call arrival rate but not on the duration of calls, the capacity of signalling channels is determined as the number of calls per hour that can be supported with a given blocking probability, instead of traffic load.

12.5.2 Signalling Capacity for GSM Voice

This section describes the methodology to estimate the capacity of GSM signalling channels, as well as the traffic assumptions taken into account. Then, the capacity obtained is exposed and finally some conclusions are presented.

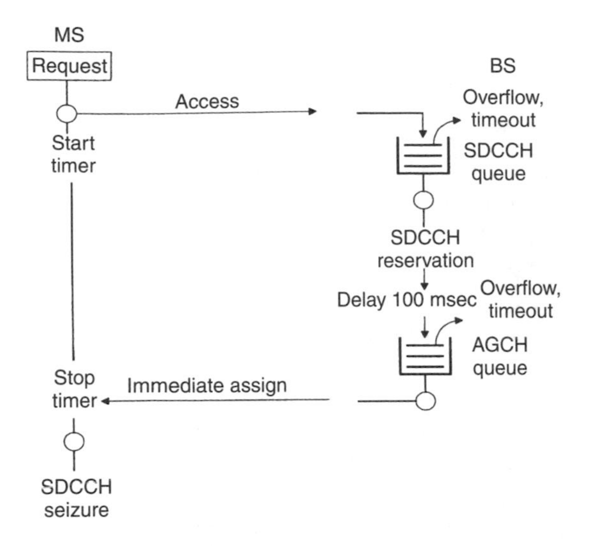

Figure 12.25. Example of signalling procedure modelled with queue systems

The channels used for call establishments in voice services are the CCCH (common control channel) and the SDCCH (stand-alone dedicated control channel) channels. The number of location updates and the SMS traffic also affect the load and therefore the call establishment capacity of these channels.

The configurations of these signalling channels and the procedures they involved were described in Section 12.1. The procedures can be modelled using queue systems because if there are no free SDCCH subchannels, channel request is queued to await subchannel release. Figure 12.25 shows the model for the RR connection establishment procedure.

Since these queue systems are complex including multiple queues, they have no simple analytic solution. Also, the use of timers and reattempts increase complexity. Therefore, simulation analysis is required in order to study the capacity of these channels taking into account all the queues, collisions in the random access, the maximum number of retransmissions and expiring timers.

12.5.2.1 Methodology

The procedure used to find out the capacity of the signalling channel is different for SDCCH and CCCH. The procedure for SDCCH consists of oversizing CCCH, so only the load of the SDCCH influences the blocking probability. For the CCCH channel, the capacity of each type of subchannel (RACH, AGCH and PCH) is calculated for a blocking probability of 0.1, 0.5 and 1% (half than allowed blocking), assuming that the other CCCH subchannels and SDCCH are overdimensioned. This makes it possible to separately compute the capacity of each subchannel. By using half-smaller blocking probabilities than the blocking criteria, it can be ensured that the overall blocking probability does not exceed 0.2, 1 or 2%. This is also applied for the PCCCH described in Section 12.5.3.

12.5.2.2 GSM Traffic Assumptions

The signalling load in GSM consists of call establishments, SMS messages and location updates (periodic and forced).

The arrival of the voice calls is assumed to follow a Poisson distribution and the duration an exponential distribution with a mean of 100 s. The proportions of mobile originating and mobile terminating calls are assumed to be equal.

In every call, the mobile station attempts to access on the RACH by transmitting channel requests until it gets a response from the network or until it reaches the maximum number of retransmissions (1, 2, 4 or 7). The number of slots between transmissions is a random variable with a uniform distribution in a set $\{S, S+1, \ldots, S+T-1\}$, where T is the number of bursts used to spread a transmission and S is a parameter (from 41 to 217), which depends on T and the channel configuration [4]. In mobile-terminated calls, a paging message is previously sent on the PCH to the mobile station. Then, if there is a free SDCCH subchannel, immediate assignment is sent on the AGCH. The SDCCH is used 2.8 s in call establishments, 3.5 s in location updates and 3.5 s in SMS transfers.

The SMS message transmissions and periodic and forced location updates are also supposed to be Poisson processes. In these procedures, traffic channels are not used, but all the information is carried on the SDCCH. On the basis of real networks, the following traffic assumptions are taken: 0.3 SMS messages are sent per hour per call, 0.3 forced and 0.25 periodic location updates are executed per hour.

12.5.2.3 SDCCH Capacity

Simulation results show that the blocking probability due to SDCCH subchannels follows a behaviour similar to the Erlang-B curves. However, there are some differences in the region of interest (from 0 to 2%). Figure 12.26 compares the evolution of the

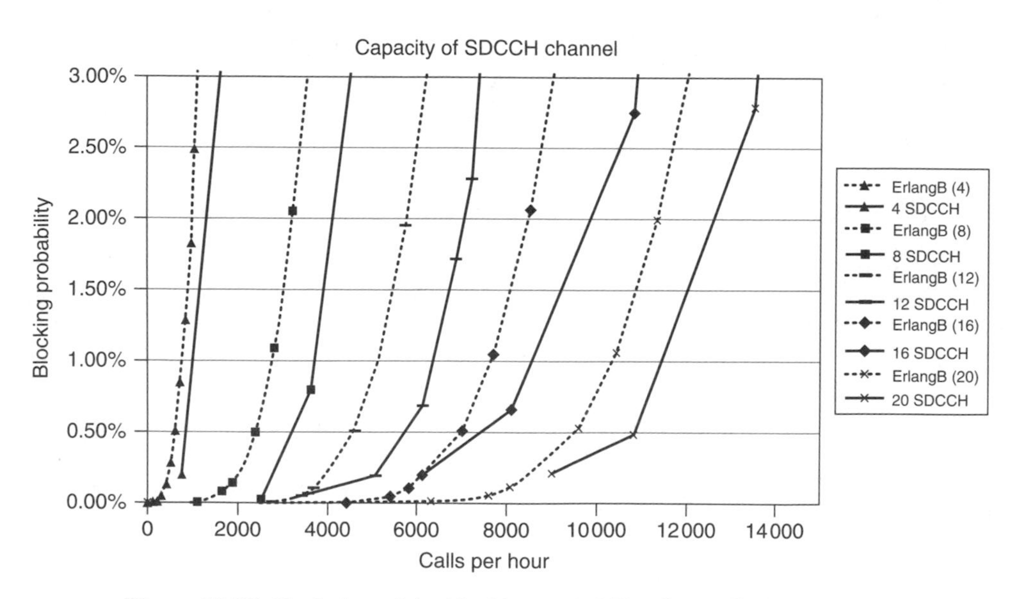

Figure 12.26. Evolution of the blocking probability due to SDCCH channel

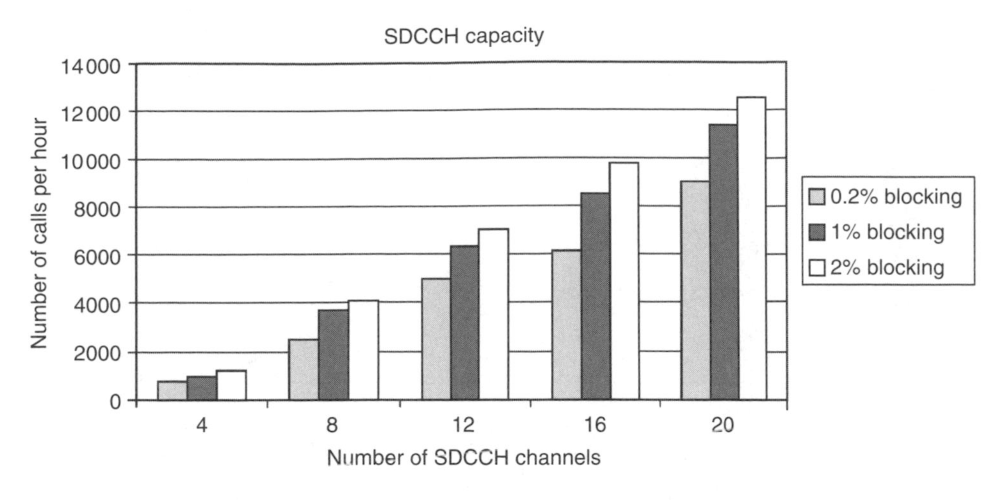

Figure 12.27. SDCCH capacity for different blocking probabilities

blocking probability for some configurations (from 4 to 20 SDCCH subchannels), which was obtained from simulation results, with the Erlang-B curves.

Figure 12.27 exposes the number of calls per hour that the configurations support with a blocking probability of 0.2, 1 and 2%.

SDCCH Capacity for Different SMS Traffic Load

This section studies the impact of different SMS loads on the call establishment capacity of the SDCCH because different operators may have a substantially different SMS load.

The arrivals of calls and SMS messages follow a Poisson distribution, and they use an SDCCH subchannel 2.8 and 3.5 s, respectively. As the SDCCH reservation time is similar for both services, and they follow the same statistical distribution, the capacity of SDCCH depends on its usage time. Assuming that there are x calls per hour, and a SMS messages per call, the SDCCH usage time per hour can be expressed as

$$T_{\mathrm{SDCCHusage}} = x \cdot 2.8 + a \cdot x \cdot 3.5 \text{ [s]}.$$

If the SMS load increases by a factor b, the call rate has to decrease by a factor c to keep the blocking probability. Thus, the SDCCH usage time per hour would be

$$T_{\mathrm{SDCCHusage}} = \frac{1}{c} \cdot x \cdot 2.8 + a \cdot b \cdot x \cdot 3.5 \text{ [s]}.$$

Assuming that the blocking probability remains equal for the same SDCCH usage time, this factor c can be calculated by making both expressions equal. The expression obtained is

$$\frac{1}{c} = 1 + \frac{3.5}{2.8} \cdot a \cdot (1 - b).$$

This expression has been verified by means of simulations. Figure 12.28 shows the relationship between the increase of SMS load and the decrease of call rate needed to maintain the same blocking probability.

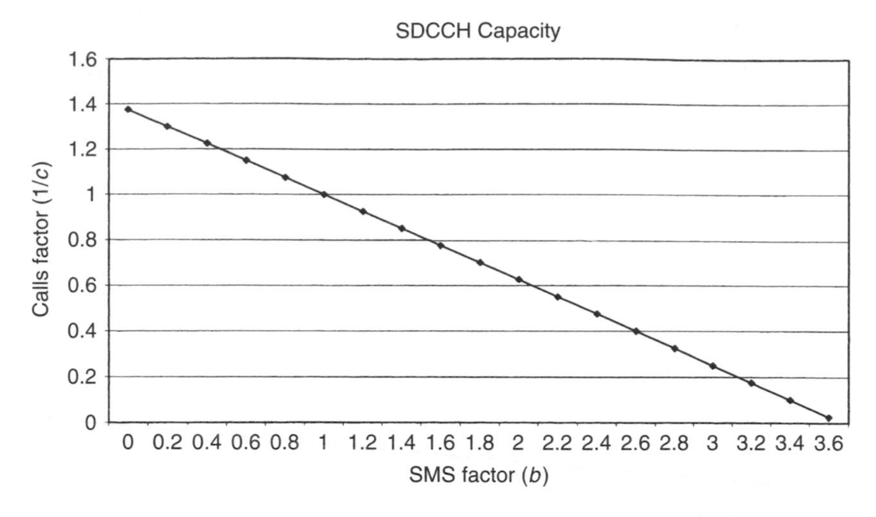

Figure 12.28. Relationship between SMS messages load and calls for a SDCCH capacity

The following example illustrates how this figure and the previous formulas shall be used. Assuming a blocking probability of 0.2% and 12 SDCCH subchannels, the call arrival rate is around 5000 calls (using Figure 12.27) per hour and 0.3 SMS messages are sent per hour per call ($a = 0.3$). If the SMS load was doubled ($b = 2$) then the reduction factor of the call rate would be $1/c = 0.625$. This means that the voice load shall be reduced with 37.5% in order to maintain the same blocking probability (0.2%) when the SMS load increases 100%. The figure makes this calculation easier. The increased factor of the SMS load is represented in the x-axis, and the curve associates this value to the corresponding reduction factor of the call rate. Hence, the value 2 in the x-axis (b) is associated with the value 0.625 in the y-axis ($1/c$).

12.5.2.4 CCCH Capacity

In this section, the capacity of the CCCH subchannels (RACH, AGCH and PCH) is described.

RACH Capacity

The RACH channel is used according to a Slotted Aloha protocol for the random access. However, the theoretical blocking formula cannot be used because it does not take into account GSM particularities: timers (i.e. time between retransmissions if there is no response from the network), retransmissions (the maximum number of retransmissions is a network parameter, typically 4) and the number of slots used for a message. A typical number of slots is 10, but the results exposed in Section 12.3 show that repetition on consecutive slots is not an efficient way of using resources in slow fading radio channels. Therefore, the chosen value for this parameter was 3.

Figure 12.29 displays the evolution of the blocking probability when the load increases, for combined and non-combined configuration. The capacity with different blocking probabilities are shown in Figure 12.34.

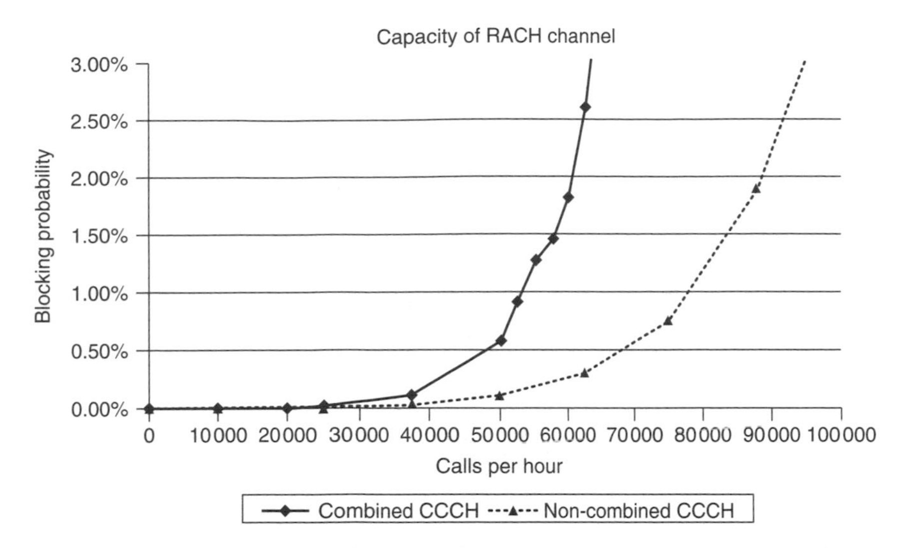

Figure 12.29. Evolution of the blocking probability due to RACH channel

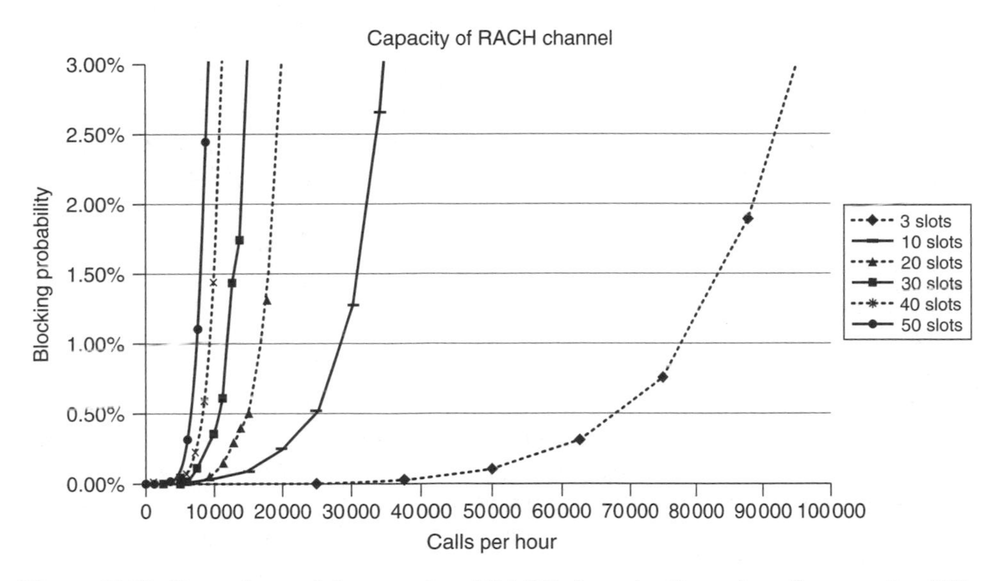

Figure 12.30. Dependence of the capacity of RACH channel with number of consecutive TSLs

Figure 12.30 exposes how the number of consecutive timeslots used for the transmission of channel request (all timeslots contain the same information) affects the capacity of the RACH in a non-combined configuration, showing the evolution of the blocking probability.

AGCH Capacity

With the immediate assignment extended feature, the access grant channel can support two assignments per block, doubling its effective capacity. It is assumed that this feature

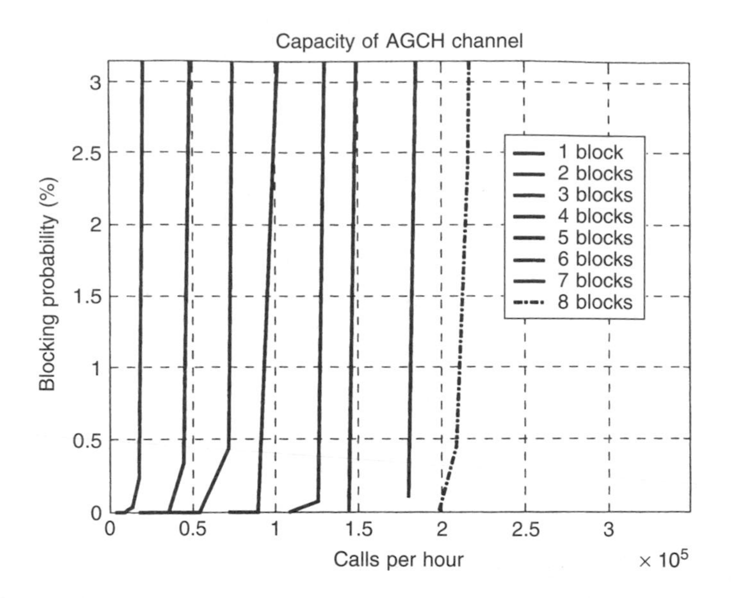

Figure 12.31. Evolution of the blocking probability due to AGCH channel

is used, and the network puts the assignment messages into a queue. In a combined configuration, a 51-multiframe contains three, and in a non-combined configuration, it contains nine blocks shared between the AGCH and the PCH. The operator can reserve 0–2 blocks in the former and 0–8 blocks in the latter case for the AGCH, which are not used for paging, to ensure a mobile station satisfactory access to the system [3]. Thus, the capacity of the AGCH is computed for all the possible number of blocks reserved for the AGCH. Figure 12.31 presents the evolution of the blocking probability due to the AGCH channel. The capacity depending on the number of blocks assigned to the AGCH channel and the allowed blocking probability is exposed in Figure 12.33.

PCH Capacities

There are two paging methods, depending on the type of mobile identification used (IMSI or TMSI), which can provide different number of paging messages per paging block. When using IMSI, only two MS can be paged in a paging message. Instead, up to four MS could be paged in a paging message using the TMSI method. In this section, the average number of paging messages per block is assumed to be three.

If the network does not receive paging response in a predefined time, it repeats the paging request. The maximum number of retransmissions is a network dependent parameter. The value selected in this study is 3 (see the reliability results in Section 12.3.4).

It is assumed that paging requests are produced according to a Poisson distribution, and they are stored in a buffer. The evolution of the blocking probability on the PCH is shown in Figure 12.32, and its capacity is exposed in Figure 12.33.

Summary

The capacity of the CCCH subchannels is summarized in Figures 12.33 and 12.34 by showing the number of calls per hour that the channels support. All the possibilities for

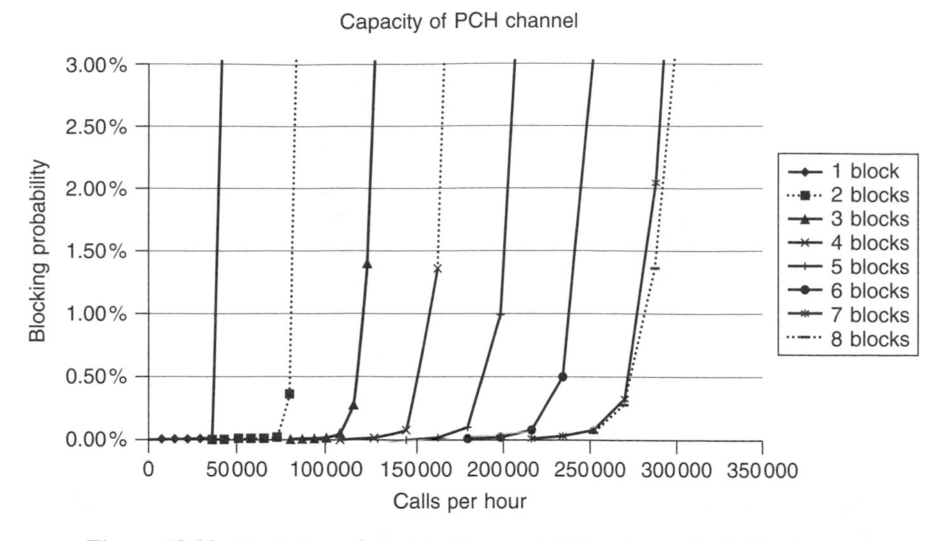

Figure 12.32. Evolution of the blocking probability due to the PCH channel

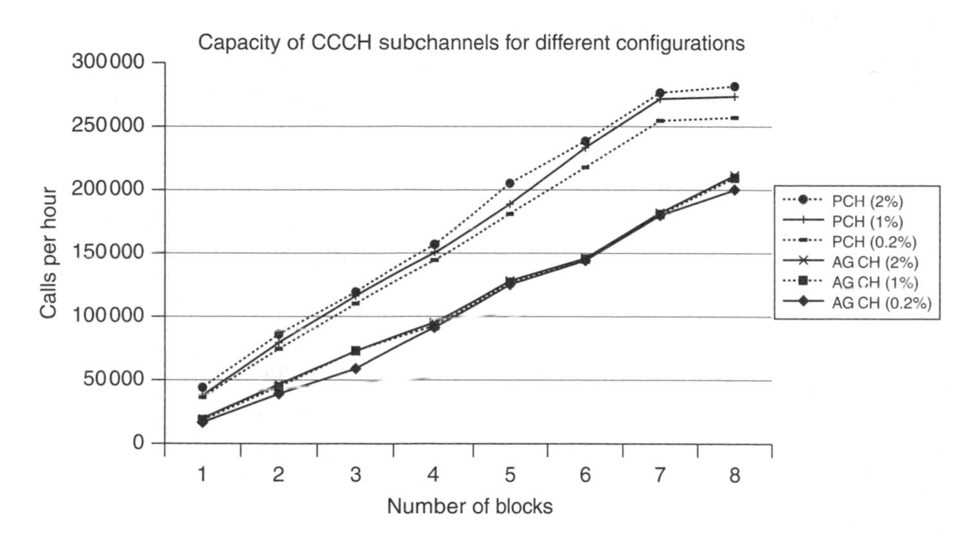

Figure 12.33. Capacity of the CCCH subchannels for different blocking probabilities

the number of radio blocks assigned to each subchannel are represented in the x-axis. The values have been obtained using immediate assignment extended (two assignments per AGCH block) and three paged mobiles per PCH block.

To maximize the capacity of the CCCH, the recommended number of blocks reserved for the AGCH is two in the combined configuration (one block available for the PCH), five in the non-combined configuration (four blocks available for the PCH). These distributions have approximately the same capacity for the PCH and the AGCH channels. In other cases, one of these channels would limit the capacity while the other would be overdimensioned. In a combined configuration, the RACH channel is overdimensioned with respect to

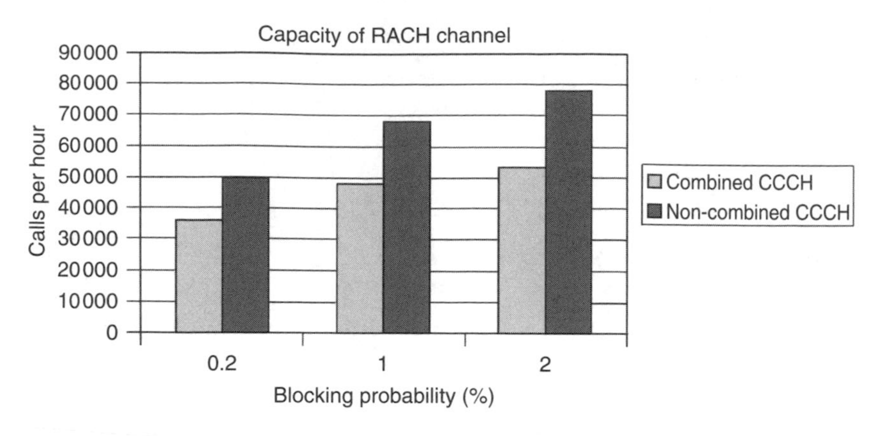

Figure 12.34. Capacity of the RACH channel for different configurations

AGCH and PCH channels, however, in a non-combined configuration, the RACH channel has less capacity than the AGCH and the PCH channels.

12.5.2.5 Number of TRXs Supported

To determine the number of transceivers (TRXs) signalling channels support (with a blocking probability of 0.2%), call arrival rates are translated into traffic loads in Erlangs:

$$\text{Traffic load (Erlangs)} = \frac{\text{Capacity (calls/h)} \times 100 \text{ (s/call)}}{3600 \text{ (s/h)}}.$$

Then, the Erlang-B formula is used to obtain the number of channels needed for a blocking probability of 0.2%. Taking into account that one TRX carries eight physical channels, the number of supported TRXs can be easily calculated by means of the following formula:

$$\text{No. of supported TRXs} = \frac{\begin{array}{c}\text{No. of supported traffic channels} + \text{No. of timeslots}\\ \text{for signalling channels}\end{array}}{8 \text{ (channels/TRX)}}.$$

The obtained results are shown in Figure 12.35. They clearly show that the number of physical channels allocated to the SDCCH determines the signalling capacity in GSM networks as the CCCH is overdimensioned.

12.5.3 Signalling Capacity for (E)GPRS

As explained in Section 12.1.2, there are two possibilities for initiating a packet data transfer in (E)GPRS: using the existing CCCH channel or the new packet common control channel (PCCCH). In this section, the signalling capacities of these two options are evaluated.

12.5.4 (E)GPRS Traffic Assumptions

In order to obtain the blocking due to (E)GPRS data sessions, the following types of communications are considered: email downloading, web browsing and wireless application protocol (WAP). The traffic of data sessions includes these types with the same

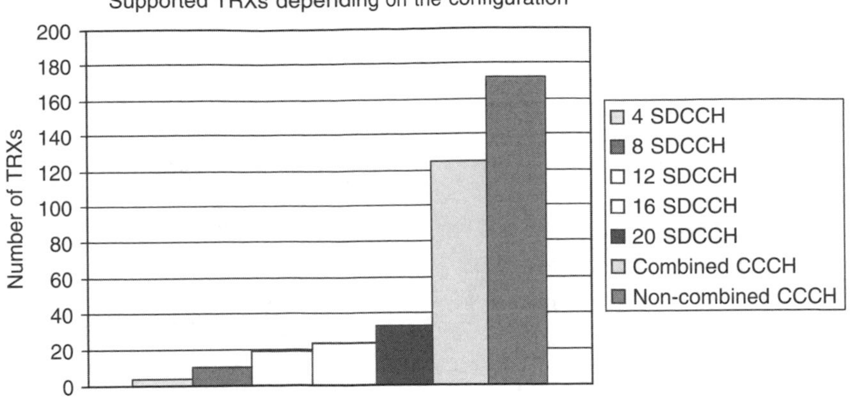

Figure 12.35. Supported TRXs depending on the configuration

probability of occurrence (33.3% for each type of session). The traffic sources send data by using a unidirectional TBF connection between the mobile station and the network. The signalling procedures for TBF establishments, which depend on the data source (the mobile or the network) and on the mobility management state of the mobile, explained in Section 12.1.

Also, the measurements report messages sent for cell reselection are considered because they represent a high percentage of the load on signalling channels. They are transmitted on (P)CCCH when the mobile is in the MM READY state and in packet idle mode, that is, when the mobile is not transmitting data and the READY timer has not expired. The reporting periods are determined by network parameters (NC_REPORTING_PERIOD_T for packet transfer mode and NC_REPORTING_PERIOD_I for packet idle mode).

Other messages like attach, detach, routing area update, and cell update messages are not taken into account because they represent a negligible percentage of the load on (P)CCCH compared with the previous ones. The signalling on the associated channels (PACCH) and the data traffic are not taken into account although they can be transmitted in the (P)CCCH timeslots, because they have lower priority than other (P)CCCH signalling and thus they have no impact on the blocking.

12.5.4.1 GSM and (E)GPRS Sharing CCCH

This section analyses the remaining capacity of the CCCH that can be used for signalling in GPRS without increasing the blocking of GSM calls. This free capacity is obtained for the maximum GSM traffic load of a large cell with 12 TRXs. Using the Erlang-B formula, this load, calculated for a blocking probability of 2% due to traffic channels congestion, is 83.2 Erlangs. Assuming that the duration of a typical call is 100 s, this value can be translated to calls per hour:

$$\frac{83.2 \text{ Erlangs} \times 3600 \text{ s/h}}{100 \text{ s/call}} = 2995.2 \text{ calls/h}.$$

Both, the combined configuration and the non-combined configuration are considered. On the basis of previous simulation analysis, the signalling capacity for (E)GPRS sharing the

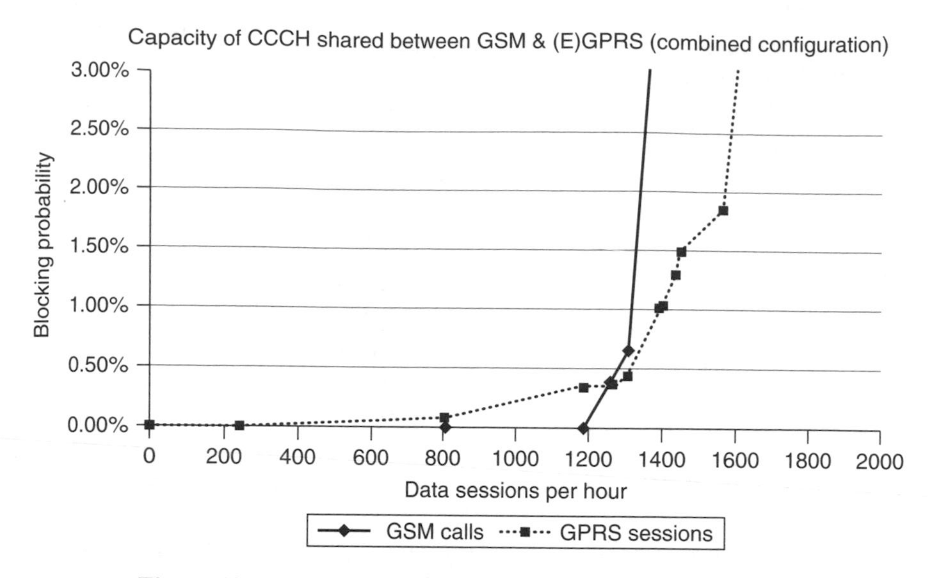

Figure 12.36. Blocking probability for (E)GPRS sharing CCCH

CCCH does not depend on the GSM load, as the signalling load on the CCCH due to voice services is very low.

Combined Configuration

Figure 12.36 presents the evolution of the blocking probability when the number of (E)GPRS sessions increases. The capacity that can be used for the data sessions with this configuration is shown in Figure 12.38.

Non-combined Configuration

The blocking probability for the combined configuration is shown in Figure 12.37 and the values of the capacity with this configuration are shown in Figure 12.38.

Summary

Figure 12.38 exposes the remaining capacity that can be used by (E)GPRS communications with the previous two configurations.

12.5.4.2 Number of TRXs Supported

As it was commented above, the number of TRXs that the CCCH supports for (E)GPRS services is nearly independent of the TRXs used for voice services as their impact on the CCCH load is low.

With the traffic models used, the average amount of data per session is 445 kB. In order to get a first approximation of the number of additional TRXs that can be supported for (E)GPRS sessions, it is assumed that the TRXs are transmitting/receiving all the time and using CS-2 (with a maximum throughput equal to 12 kbps). With this assumption,

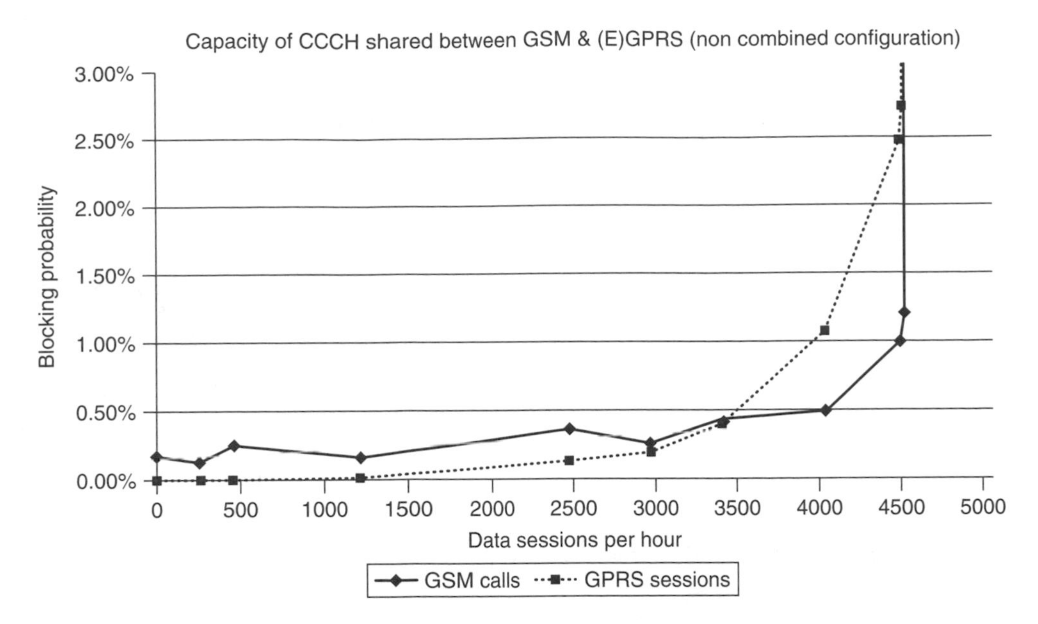

Figure 12.37. Blocking probability for (E)GPRS sharing CCCH

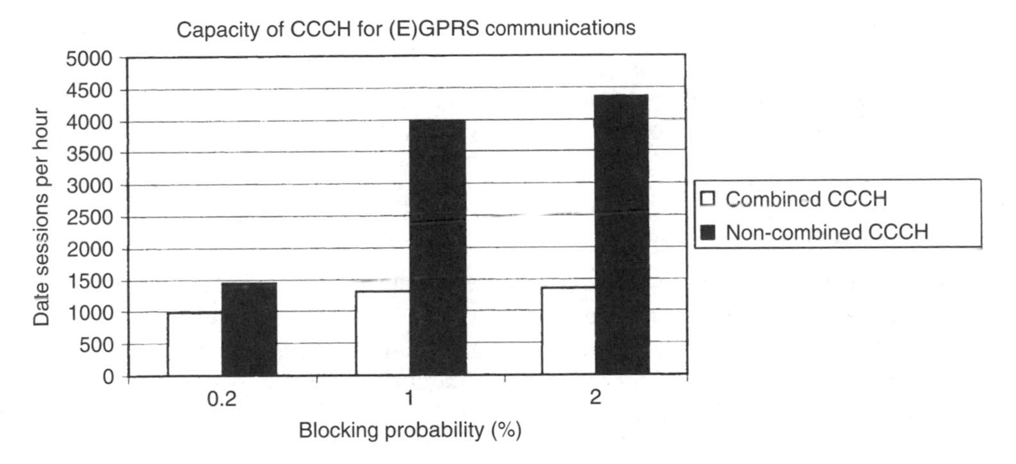

Figure 12.38. Capacity of CCCH for (E)GPRS communications

one TRX can transmit 290 MB per hour. Therefore, one TRX can carry 650 sessions per hour, and so one CCCH supports 1–2 TRXs with the combined configuration and 2–7 TRXs with the non-combined configuration. The exact value depends on the allowed blocking probability.

12.5.4.3 (E)GPRS on PCCCH

As described in Section 12.1, (E)GPRS capable cells can accommodate a PCCCH on one or more packet data channels with a maximum of 12 available blocks in a 52-multiframe.

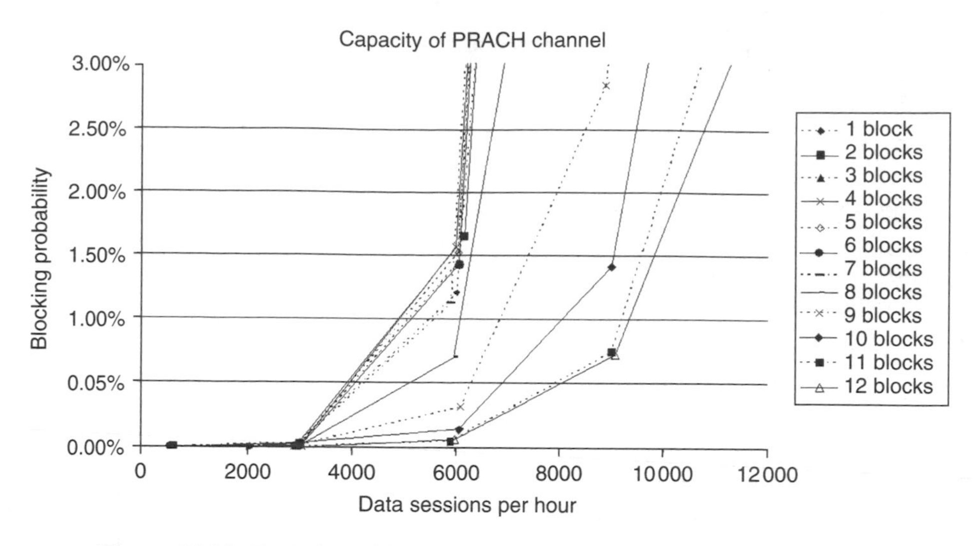

Figure 12.39. Evolution of the blocking probability due to PRACH channel

This section evaluates the capacity of this option.

PRACH Capacity

The capacity of the PRACH was calculated for all possible numbers of blocks (1–12). As for the RACH, the maximum number of retransmissions used is four. Figure 12.39 presents the evolution of the blocking probability for different numbers of blocks reserved for the PRACH.

PAGCH Capacity

As above, all possible values of blocks assigned to the PAGCH channel are taken into account. As opposed to GSM, immediate assignment extended messages cannot be used in (E)GPRS, and thus only one message can be carried per block. Figure 12.40 exposes the blocking probability of the PAGCH as a function of the number of data sessions per hour.

PPCH Capacity

In (E)GPRS, mobile stations are identified by either IMSI, TMSI or P-TMSI and, depending on the identifier from one to four mobile stations, can be addressed in a paging message. As with the PCH, the average number of paged mobiles per paging message is assumed to be three. The evolution of the blocking probability is shown in Figure 12.41.

Summary

The capacity of the PCCCH subchannels for different blocking probabilities is shown in Figure 12.42.

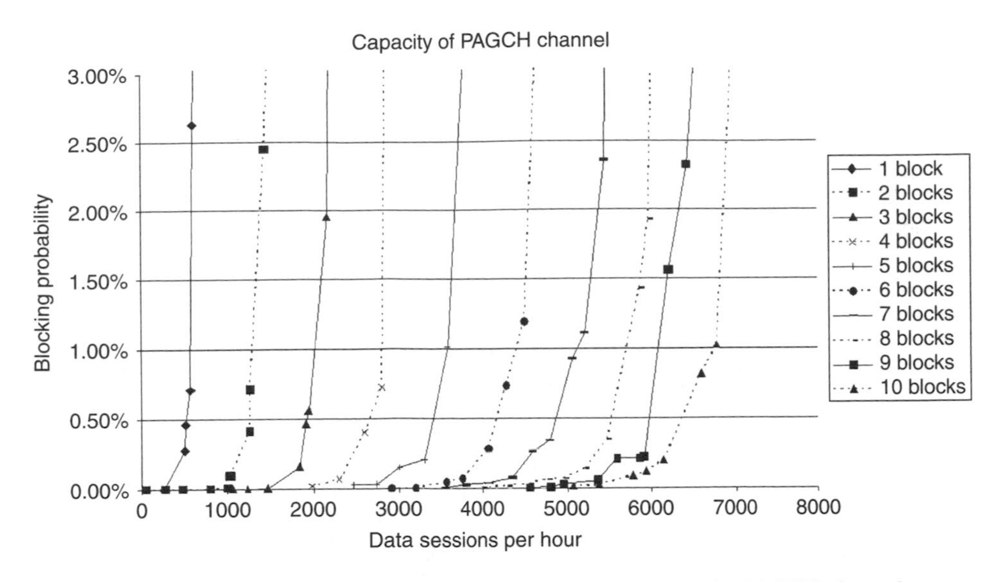

Figure 12.40. Evolution of the blocking probability due to the PAGCH channel

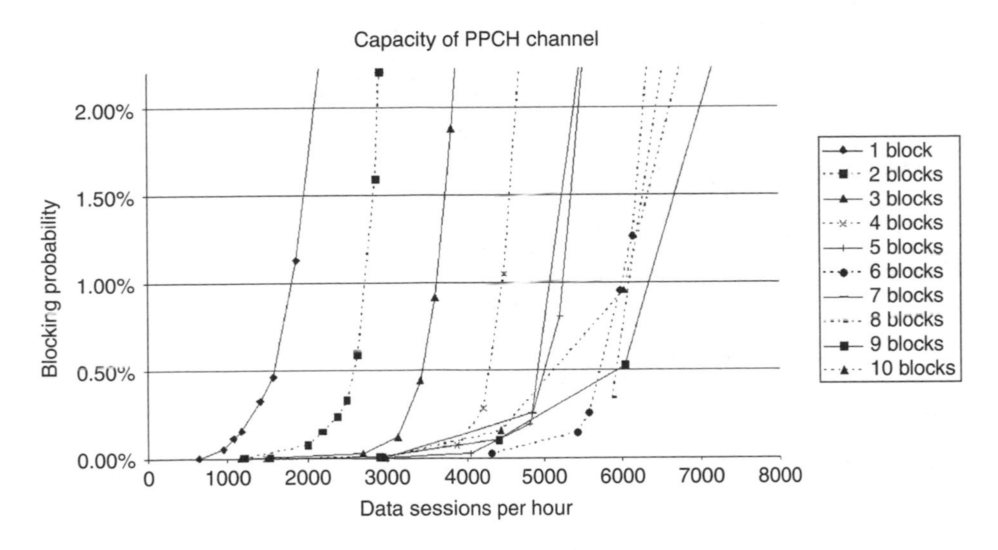

Figure 12.41. Evolution of the blocking probability due to the PPCH channel

Number of TRXs Supported

The recommended distribution of blocks between the PAGCH and PPCH channels is such that it makes the capacity of these channels equal. Thus, according to Figure 12.42, seven blocks should be reserved for the PAGCH, in which case, the remaining four blocks are available for the PPCH channel. Since it was approximated above, that one TRX can carry about 650 sessions per hour, with this block allocation, one PCCCH supports from six to nine additional TRXs dedicated entirely for (E)GPRS.

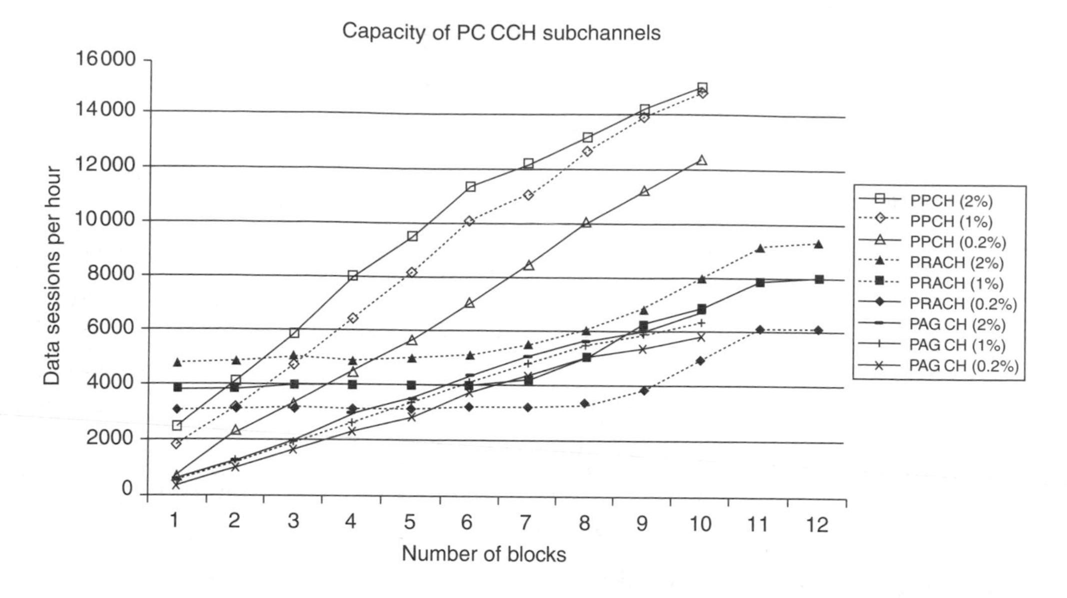

Figure 12.42. Capacity of PCCCH subchannels (data sessions per hour)

12.5.4.4 Comparison between CCCH and PCCCH

Figure 12.43 compares the capacity of the PCCCH and CCCH. It can be seen that the PCCCH has more capacity than the CCCH. However, using a non-combined configuration and a blocking probability of 2%, the capacity of these two channels is similar. Figure 12.43 shows the capacity of both options for (E)GPRS using the recommended configurations.

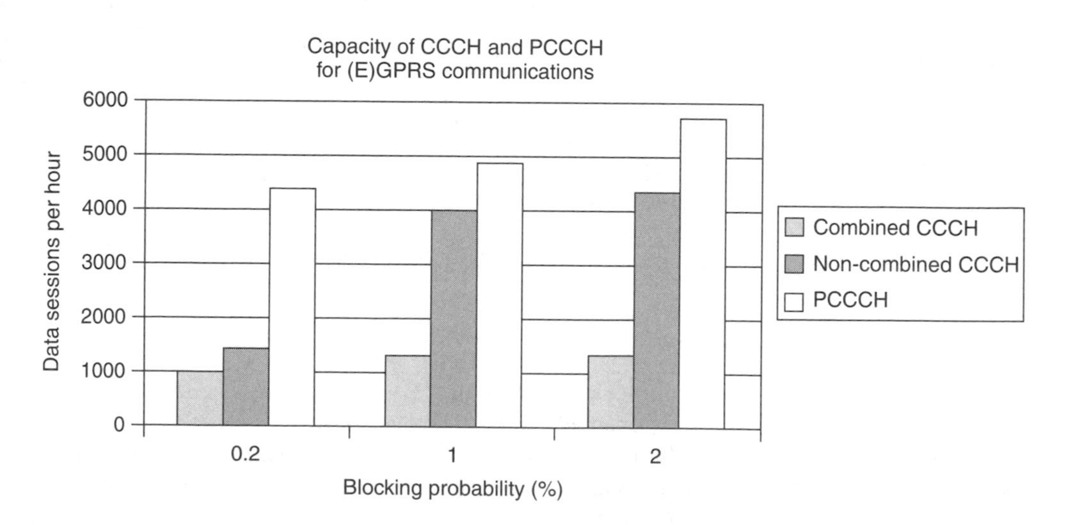

Figure 12.43. Comparison between the capacity of CCCH and PCCCH

12.5.5 Conclusions

For GSM networks, the SDCCH channel is the limiting signalling channel since the CCCH subchannels are overdimensioned for standard TRX configurations. For a blocking probability of 0.2%, a combined configuration, with 2 blocks for AGCH and 1 for PCH, supports 125 TRXs, a non-combined configuration, with 5 blocks for AGCH and 4 blocks for PCH, supports 172 TRXs. With the same blocking probability, 4 SDCCH subchannels support 4 TRXs, 8 SDCCH subchannels 10 TRXs, 12 SDCCH subchannels 19 TRXs, 16 SDCCH subchannels 23 TRXs, and 20 SDCCH subchannels 33 TRXs. A typical high-loaded cell, which must support high GSM load, has 12 TRXs, requires 8 SDCCH subchannels (i.e. one physical channel dedicated to SDCCH). The numbers of blocks for AGCH and PCH and the number of SDCCH subchannels depend, of course, on the traffic assumptions used.

The dimensioning of the location area has a high impact on the number of paging and forced location update messages. With a larger location area, the number of forced location update messages decreases, but the number of paging messages increases because there are more subscribers in the location area. Hence, the appropriate proportion of blocks for AGCH and PCH channels depends on the dimensioning of a location area, which is a compromise between the signalling load due to paging messages and the signalling load due to location updates.

(E)GPRS signalling procedures can use CCCH or PCCCH channels. Assuming cells with 12 TRXs dedicated for GSM speech, the CCCH with combined configuration supports one TRX for (E)GPRS with a blocking probability of .2. The number of supported TRXs increases to two when using the CCCH with non-combined configuration. Since the PCCCH allocated on one PDCH supports more than six TRXs, the required number of PDCHs for the PCCCH can be approximated by dividing the number of TRXs dedicated to (E)GPRS by six.

References

[1] 3GPP TS 05.02 V8.10.0 (2001–08), 3rd Generation Partnership Project; Technical Specification Group GSM/EDGE Radio Access Network; Digital Cellular Telecommunications System (Phase 2+); Multiplexing and Multiple Access on the Radio Path (Release 1999).

[2] 3GPP TS 04.08 V7.10.0 (2000–12), 3rd Generation Partnership Project; Technical Specification Group Core Network; Digital Cellular Telecommunications System (Phase 2+); Mobile Radio Interface Layer 3 Specification (Release 1998).

[3] 3GPP TS 04.06 V8.1.1 (2001–01), 3rd Generation Partnership Project; Technical Specification Group Core Network; Digital Cellular Telecommunications System (Phase 2+); MS–BSS Interface; Data Link (DL) Layer Specification (Release 1999).

[4] GSM 04.60 V8.1.0 (1999–11), European Standard (Telecommunications Series), Digital Cellular Telecommunications System (Phase 2+); General Packet Radio Service (GPRS); Mobile Station (MS)–Base Station System (BSS) Interface; Radio Link Control/Medium Access Control (RLC/MAC) Protocol (GSM 04.60 version 8.1.0 Release 1999).

[5] 3GPP TS 05.03 V8.6.1 (2001–08), 3rd Generation Partnership Project; Technical Specification Group GSM/EDGE Radio Access Network; Digital Cellular Telecommunications System (Phase 2+); Channel Coding (Release 1999).

[6] GSM 05.05 V.8.3.0 (1999–12), European Standard (Telecommunications Series), Digital Cellular Telecommunications System (Phase 2+); Radio Transmission and Reception (Release 1999).
[7] 3GPP TS 05.03 V8.11.0 (2001–08), 3rd Generation Partnership Project; Technical Specification Group GSM/EDGE Radio Access Network; Digital Cellular Telecommunications System (Phase 2+); Radio Subsystem Link Control (Release 1999).

13

Automation and Optimisation

Volker Wille, Sagar Patel, Raquel Barco, Antti Kuurne,
Salvador Pedraza, Matias Toril and Martti Partanen

Owing to the complexity that the third-generation (3G) evolution will bring, planning, optimising and operation of radio networks is a tough challenge that operators have to tackle at a minimum cost. This chapter presents new concepts based on automation and control engineering, designed to deal with all these issues, with functionality capable of fully automating the planning, optimisation and operation of radio networks, which not only decrease the operational expenditure (OPEX) of radio networks but maximise their performance as well.

13.1 Introduction to Radio Network Optimisation

The coming years will bring profound change to the mobile telecommunications industry. This change is due to the changing nature of services offered by mobile operators, the introduction of new air interface technologies as well as the increased level of competition. Most importantly, the business of operators is changing fundamentally as 'mobile phones' will not only be the preferred method of voice communication but also the preferred choice for connecting to the Internet. This change is driven by the desire of an end user to have Internet access at any time regardless of their location.

To enable the convergence of mobile systems with the Internet, new functionality will have to be integrated into existing cellular systems. For operators using global system for mobile communications (GSM) as the second-generation cellular telephony system, this meant adding general packet radio system (GPRS) related hardware (HW) and software to existing networks. Similarly, the core/transmission network will have to be upgraded to facilitate data throughput for the new services. In parallel, operators will be deploying wideband code division multiple access (WCDMA) in order to enable new services that require even higher data rates (see more information in Chapter 14).

However, the addition of new technology increases the complexity of operating the network. When speech telephony was the only service offered, indicators such as dropped

GSM, GPRS and EDGE Performance 2nd Ed. Edited by T. Halonen, J. Romero and J. Melero
 ISBN: 0-470-86694-2

Table 13.1. Growth of operator workforce

1996	1997	1998	1999
4800	5800	7000	7900

Source: D2Vodafone annual reports.

call rate (DCR), speech quality and blocking of call attempts were suitable indicators to give operators an understanding about the performance of the network. With the addition of the packet-switched service, completely new performance indicators such as data transfer speed, delay or availability are required. Thus, the increased complexity of the network makes its analysis and tuning processes far more difficult.

The huge fees that operators had to pay for WCDMA licences have increased their debt burden significantly. In addition, further investment is needed to purchase the WCDMA equipment and finance its installation. Consequently, operators are under increasing financial pressure to produce returns on their investments. At the same time new 3G entrants have been admitted into several markets hence further increasing competition in the mobile industry.

These changes of the network go along with a continuing network roll out even in relatively mature cellular markets. The deployment of additional sites is mainly required to provide increased levels of in-building coverage as mobile users expect to be offered service in all geographical locations. But continuing network growth is also needed to enhance system capacity to cater for the rapid increase in the level of packet-switched traffic.

In the past, operators managed to cope with rapid technological change and growth of the network by increasing their workforce. Table 13.1 gives some indication of the growth of employee numbers experienced by D2Vodafone in recent years. However, due to mounting financial pressures this is not a viable strategy anymore. Therefore, the only feasible option to maintain network quality with the existing workforce whilst also integrating new technology into the network is to increase the level of automation with a view to enhancing the productivity of the employees. This will free resources, which can tackle upcoming challenges brought about by the continuing evolution of the network.

13.1.1 Operational Efficiency

Spectral efficiency is mainly a characteristic of the radio technology and its implementation by the equipment vendor. In addition to spectral efficiency, another crucial factor in providing cost-efficient data transfer is operational efficiency. Operational efficiency is the work-effort that is required to provide a given spectral efficiency at a given grade of service. While it is relatively easy to formulate spectral efficiency by means of a mathematical equation, the formulation of operational efficiency is less straightforward in mathematical terms as diverse aspects such as operator process and the salary of employees have to be considered. However, the validity of the concept of operational efficiency in providing cost-efficient data transfer can be shown with a simple example. If a given feature enables an increase of spectral efficiency by 10%, but the related work-effort has to be increased by 50% to enable this improvement, then it is doubtful if the overall data transfer cost can be reduced by this feature.

13.1.2 Characteristics of Automation

There are two main areas in which automation can be deployed to enhance the operational efficiency of the workforce: the automation of existing manual procedures and the addition of new automatic procedures.

13.1.2.1 Automation of Existing Tasks to Reduce the Workload

When running the network, a number of tasks have to be repeated at regular intervals. Amongst these are frequency and adjacency planning. With the addition of each new cell in the network, it is required to determine which are the most suitable handover (HO) candidates. On all new cells, several cells have to be defined as HO candidates to ensure a satisfactory HO to and from the new cell. On all the existing cells, it is required to remove those HO definitions that will be obsolete through the addition of a new cell. In addition, the new cell also requires a set of appropriate frequencies to enable proper operation, where the frequencies on all existing cells in its vicinity must be taken into account. Providing the means to automate the process of integrating new cells seamlessly into the existing network is just one example of routine tasks that lend themselves to the application of automation. Here, the purpose of automation is to provide the means to increase operational efficiency by reducing the workload associated with existing tasks.

One positive side effect of increasing the level of automation for existing tasks is that the staff is freed from mundane and tedious tasks and can apply their skills in new areas, which bring additional value to the operator. In this sense, automation ensures that the staff is working on challenging tasks, which in turn helps to motivate and retain staff.

13.1.2.2 Automation of New Tasks to Improve the Performance of the Network

Certain tasks cannot be carried out without the aid of automation. Areas that require automation to offer a cost-effective implementation are those involving heavy calculation and/or the evaluation of several input parameters. In contrast to humans, computers can easily carry out analysis of complicated input data. Furthermore, computers are able to repeat this method many times and thus enable the application of a method to all cells in the network. For example, determining optimum parameter values for radio resource algorithms in network elements that are unique for each cell is not viable in large networks without the use of automation. Likewise, in the case of troubleshooting (TS), the most likely reasons for a given network problem are established by combining a large number of factual pieces of information in the form of hardware (HW) alarms, short-term and long-term performance data as well as configuration data such as parameter settings on all surrounding cells. The benefit of automation to these kinds of tasks is an increase in network performance since without the application of automation these tasks could not be carried out on a wide scale, i.e. a large number of cells.

The following sections highlight areas that lend themselves to the application of automation with a view to increasing operational efficiency while also increasing network performance. Thus, it is possible to obtain optimum performance from the existing network at the lowest cost. The technical focus in this chapter is on GERAN but the trend of increasing automation in cellular systems is also applicable to UTRAN networks.

13.1.3 Areas of Automation

Figure 13.1 displays several technical factors contributing to the performance of a GERAN network and their interdependence. Non-technical issues, which also have an influence on network performance and operational efficiency, such as equipment deployment strategy, use of tools and operator processes are not being listed here.

The foundation of a well-performing network is the basic radio platform. Issues such as the location of sites as well as positioning of antennas have a major impact on the distribution of the radio signal and thus on call quality. For example, the closer sites are located near a traffic hot spot, the easier it is to provide good call quality with low signal levels. This will reduce interference to surrounding cells.

In addition, the clearance of faults is also very important in order to ensure that the network operates exactly as designed. If a cell is temporarily non-operational, the performance of all cells in the vicinity is impaired. Ensuring that this cell is speedily brought back into operation is a crucial task to ensure optimum network performance. Methods aimed at improving and automating the fault clearance will be outlined in Section 13.5.

Furthermore, it is the adjacency plan that is the major contributor to network quality. Only with the definition of correct adjacencies will HO to most suitable cells be possible. Without the definition of the optimum adjacencies, HOs occur to cells that are sub-optimal in terms of their radio link performance and subsequently dropping a call might not be avoided. Automating the process of updating adjacency definitions by deleting underused ones and creating missing ones is outlined in Section 13.3. The automation of this task will enhance operational efficiency and, depending on the quality of the existing adjacency management process, also increase network performance.

Similarly, a good frequency plan is of paramount importance for network performance. The better the frequency plan, the less co- and adjacent-channel interference is experienced by subscribers. This means better call quality and subsequently smaller DCR. Thus, frequency planning is a crucial task in building a well-performing network. The application of automation to frequency planning is described in Section 13.2. Here, the emphasis

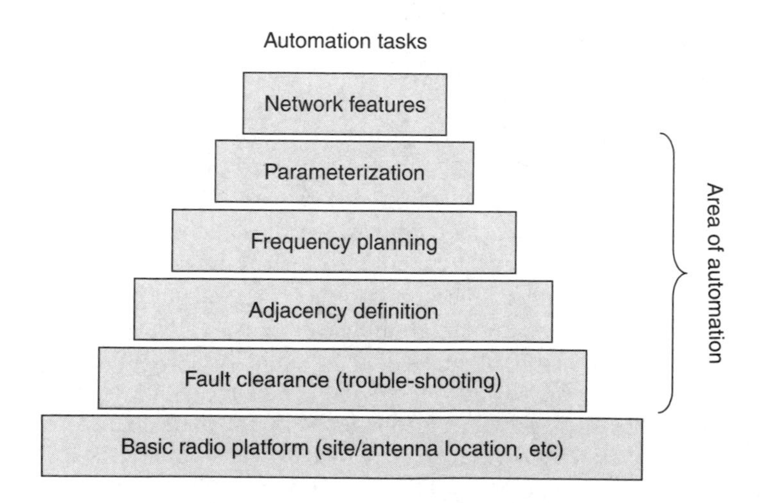

Figure 13.1. Dependency of network planning tasks

is on improving the quality of the frequency plan and reducing the work-effort associated with its creation, i.e. enhancing operational efficiency.

Another important factor that determines network quality is parameterisation of network features. Traditionally, standard parameters have been used network wide, but this is not optimal when trying to enhance network performance. Each cell operates under unique conditions and requires customisation of parameter values to provide optimum performance. Insight into this aspect of automation is provided in Section 13.4.

Automation can be applied to all the tasks shown in Figure 13.1 with a view to enhancing network performance and at the same time reducing the man-effort required to carry out the task. This means that through the application of automation to these tasks, the data transfer cost can be reduced.

On a conceptual level, the structure of the automation process is shown in Figure 13.2. The feedback loop in the figure entails the collection of the information from the network, the analysis of the data and the implementation of the required changes in the network accordingly. An additional task can be integrated into the feedback loop between steps 2 (suggesting changes) and 3 (implementing changes). This step, 'verification of the suggested changes', is optional and is most likely to be used in the initial stages of using an automation feature. Once confidence on the validity of the suggested changes is gained, this task will not be used anymore.

The final decision whether to close the feedback loop and allow the control system to implement changes rests with the operator. Alternatively, the feedback loop might also be open and the control system would only suggest changes, while the verification of these changes and their implementation would still be carried out by operator personnel.

It is worth noting that information to tune the network is provided for free, since most of it is made up of measurement reports that ordinary subscribers are providing free of charge when in call mode. Thus, network planning and optimisation procedures carried out by operators do not require a separate network model where the behaviour of the network is simulated (e.g. network traffic and propagation models used in current frequency planning methods).

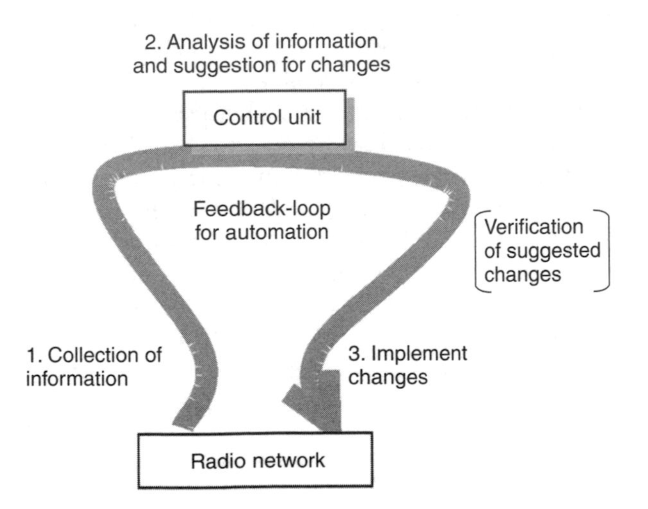

Figure 13.2. Automation feedback loop

13.2 Mobile Measurement-based Frequency Planning

13.2.1 Outline of the Problem

The large number of subscribers in the network and the increasing usage of existing mobile telephony services together with upcoming new services, such as video and audio streaming, will force operators to significantly increase capacity offered by the network. Capacity requirements exist particularly in a city centre environment where the greatest subscriber density is experienced. Sectorisation and cell splitting, two standard methods to increase capacity from macrocells, have generally been employed in GERAN networks. Often operators also utilise advanced network features to reduce the frequency reuse factor and thus further increase network capacity of macro cells. Owing to the limited spectrum and because of HW constraints regarding channel separation required on a cell or a site basis, providing additional capacity from existing macro cells is often impossible without degrading call quality. Owing to the fact that in a city centre environment the distance between macrosites of an operator is typically as low as 300 m and that several operators are competing for new sites, suitable locations to establish additional macro cells are very difficult to obtain. Therefore, network operators also have to use alternative cell types such as micro cells and in-building cells to meet capacity requirements. Distances between micro cells can be of the order of a few tens of metres when deployed on a large scale. Dedicated in-building cells provide tailored solutions in office buildings for corporate customers and in public places such as shopping centres and underground stations. One advantage of the utilisation of small cells is that additional capacity can be provided at 'traffic hot spots' and it is thus possible to target areas with high capacity requirements directly.

The reduced cell sizes also create new challenges. The main challenge is frequency planning. Traditional frequency planning methods rely on the accurate prediction of radio wave propagation. Accurate prediction of propagation is relatively straightforward in non-undulated, rural environments where the signal is transmitted from tall, isolated radio masts towering over all other surrounding buildings. In that scenario, empirical models such as those developed by Okumura–Hata [1] are able to predict signal level with a good degree of accuracy. However, only a minority of cells are deployed under these kinds of conditions. The majority of cells are located in city centres with antennas fixed to ordinary buildings in the midst of taller buildings.

For the prediction of radio wave propagation in microcell environments, other empirical models such as Walfish–Ikegami [2] have been developed. However, even using these bespoke models, signal-level prediction accuracy in dense urban environments is relatively low. The low prediction accuracy is due to several reasons. Firstly, reduced cell sizes require more accurate data to describe topology and morphography of the propagation environment. With macrocell size in the order of around 150 m from the location of the base transceiver station (BTS), it is required to utilise digital map resolution in the order of a few tens of metres. For micro cells with cell sizes of about 50 m, even higher resolution is required, typically, of the order of a few metres (1–5 m). For cost reasons and to speed up computation times, this kind of high-resolution map data is normally not being used.

In addition, propagation models normally determine signal levels about 1.5 m above ground level as this is close to the typical subscriber height. However, a significant share of the subscribers is at elevated locations such as in tall office buildings. The vertical displacement between the location of the signal-level predictions and the actual location of subscribers has an adverse effect on prediction accuracy. Moreover, the fact that these elevated locations are normally indoors further increases prediction inaccuracy. The prediction of signal levels from external cells to indoor environments is particularly difficult, as detailed information on the propagation properties such as penetration loss into the building and propagation loss within the building is generally not available. With the increase of dedicated in-building cells it is also required to frequency plan these accurately in coexistence with outdoor cells. However, commercially available propagation prediction tools do not normally cover these two aspects. Owing to these difficulties in frequency planning, indoor and outdoor cell operators tend to dedicate spectrum for indoor cells when deploying indoor cells on a large scale. In this way, both cell types are separated in the frequency domain. While this approach is easing the planning requirements, it is not optimum from the spectral efficiency point of view.

Deterministic techniques based on uniform theory of diffraction (UTD) have been studied extensively in the academic arena [3, 4] and it is shown that these models can deliver fairly accurate propagation predictions. A prerequisite for obtaining accurate coverage predictions is a high-resolution clutter database. Clutter resolution of the order of 1–2 m is normally required, describing the outline of each building as a polygon. Depending on the use of 2D or 3D prediction methods, the polygon has to describe the buildings in the desired dimension. Obtaining such detailed clutter information can be difficult for some cities and, generally, the cost associated with a high-resolution database is relatively high. In addition, prediction times rise exponentially with clutter resolution and huge memory, and computational requirements exist for the calculation platform. Therefore, UTD-based radio planning tools are currently not widely used by operators.

Owing to these reasons, it is required that a new approach is taken to simplify frequency planning by easing requirements and processes while also improving the quality of the frequency plan. In this novel approach, signal-level measurements provided by ordinary mobile stations (MSs) while in dedicated mode are being used. The reported measurements are sent in regular intervals (usually every 480 ms) and contain the signal level of the serving cell and up to six neighbouring cells. On the basis of this information, which is provided to the operator free of charge, it is possible to create a better frequency plan than is currently possible using standard tools and techniques.

The ability to create a high-quality frequency plan even for complex propagation environments such as city centres and in-building cells enables operators to be more spectrally efficient as less channels have to be used to provide a certain grade of service. These 'saved' frequencies can then be used to increase the capacity of the network. Alternatively, the increased interference prediction accuracy can also be utilised to provide higher network quality.

One additional benefit of using mobile measurement information is the fact that the subscriber distribution within a cell is explicitly included in the information. For example, more signal-level measurements will be reported from locations with a high density of MSs as compared with locations where a low mobile density is experienced. In this way, it is possible that the signal levels from neighbouring cells are weighted depending on the amount of cell traffic that is 'interfered'.

In systems using dynamic frequency and channel allocation (DFCA) (see Chapter 9), the frequency of the BCCH TRXs cannot be dynamically controlled. These TRXs have to be frequency planned using normal, non-dynamic methods. The benefits of the method outlined above can be applied to these TRXs. This means that the reuse of the BCCH can be reduced by the use of this method. Hence, dynamic frequency and channel allocation (DFCA) and mobile measurement–based frequency planning (MMFP) work in harmony to increase spectral efficiency in all layers of the network.

13.2.2 Traditional Frequency Planning

This section provides further detail on current frequency planning by describing the process and methods to carry out frequency planning on the basis of propagation predictions. The new method of MMFP is then outlined in the subsequent section.

13.2.2.1 Frequency Planning Principles

In cellular time division multiple access (TDMA) systems, like GERAN, each TRX in the network requires at least one frequency to operate. As the number of frequencies that an operator has been allocated is far smaller than the number of TRXs in the network, it is required to repeat the use of a frequency several times throughout the network. For example, in a network consisting of 20 000 TRXs using 50 channels, each channel has to be reused about 400 times.

The term reuse factor is normally used to denote the 'distance' at which the use of frequencies is repeated. For example, a reuse factor of 16 implies that in case frequency f_1 is used on a cell A, only on the cell that is 17 cells 'away' from cell A, will f_1 be used again. It is the aim of frequency planning to ensure that all cells that are 'closer' to A in propagation terms, i.e. cells 2–16, will have frequencies different than f_1.

Operators aim to obtain maximum performance from the existing network with the given number of frequencies. This can only be achieved if the available frequencies are reused in an optimum manner. Frequency assignment is the process of providing each TRX in the network with the required number of frequencies while ensuring that interference between cells is minimised, i.e. the network quality is maximised. Frequency assignment is a very crucial task as it has a direct impact on network performance. In order to be able to assign frequencies correctly, it is required to accurately determine interaction between cells within the network in interference terms.

The interaction between cells is normally determined by network planning tools that use propagation prediction methods such as those outlined in the previous section. In planning tools, the geographical area is described by regular squares, called *bins*. A commonly used bin size is 50 m × 50 m. The size of these bins corresponds to the resolution of the map component of the planning tool.

Figure 13.3 displays two pictures that show the same geographical area. The picture in Figure 13.3(a) uses a bin size of 2 m, whereas the picture in Figure 13.3(b) uses a bin size of 50 m. It can be seen in Figure 13.3 that with bins of 2 m it is possible to visually recognise many features of the actual city, such as roads, buildings and open areas. With the larger bin size, on the other hand, such identification is not easily achieved. A bin size of 50 m also means that structural features that are smaller than 50 m disappear or are increased in size to a minimum of 50 m. For that reason, the street width is set to 50 m or the street is not shown at all. Figure 13.3 thus visualises the importance of an accurate description of the propagation environment by a small bin size, as the bin size determines how the propagation prediction tool 'sees' the propagation environment.

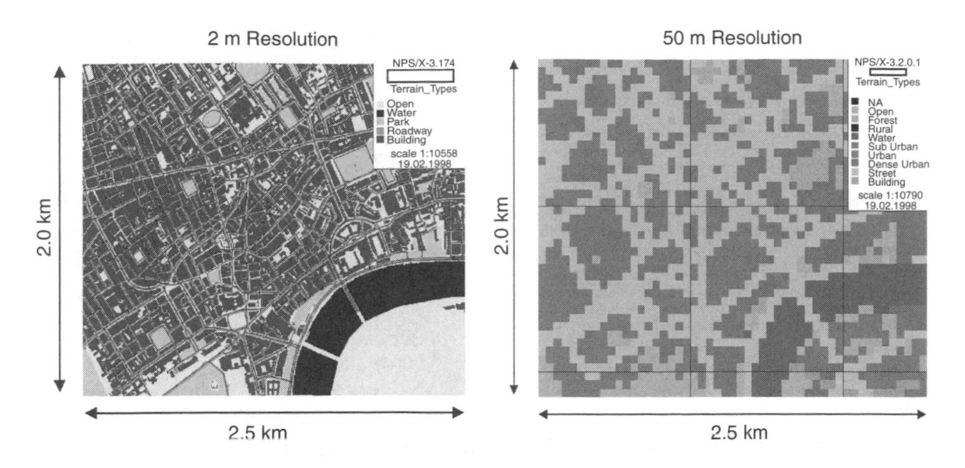

Figure 13.3. Effect of map bin size

Each bin also has a certain attribute that describes its propagation characteristics. In Figure 13.3, this attribute is denoted by the difference in colour. Typical morphographic attributes include water, open, road, forest, rural, suburban, urban and dense urban. The more attributes are being used to describe the propagation environment, the more information can be provided about its morphographic make-up. This means that bin size and number of morphographic attributes of the bin have a significant impact on the propagation accuracy of the planning tool. In general, the smaller the bin size and the larger the number of morphographic attributes, the better the propagation accuracy. However, this increased level of accuracy can only be provided with increased cost for the map component of the planning tool and increased levels of computational power to execute propagation predictions.

On the basis of propagation predictions, the signal strength from each cell is determined within each of the relevant bins. Within each of these bins, signals from several cells are received. Thus, it is possible to determine the interaction between cells within a bin and consequently within a larger area made up of a large number of bins. This interaction between cells is expressed as the *carrier to interference (C/I) ratio* and indicates to what extent cells interfere if they were going to use the same or an adjacent frequency. Analysis is carried out to determine the number of bins in which the C/I between a pair of cells is below a certain threshold such as 9 dB. A threshold of 9 dB is normally chosen since below this threshold, poor speech quality would be obtained, whereas above this value good quality can be provided. By carrying out this analysis for all surrounding cells, it is possible to determine a ranking of the interference obtained from these cells.

The exact formulation of the level of interference between cells is known as *the interference matrix (IM) or inter-cell dependency matrix* [5]. Element x_{ij} in the IM is an indicator of the potential interference produced by cell j on mobiles attached to cell i, assuming that both use the same frequency.

13.2.2.2 Conceptual View of Frequency Planning Process

Figure 13.4 displays the frequency planning process on a conceptual level. It can be seen that the creation of a frequency plan is a sequential process where the final frequency plan

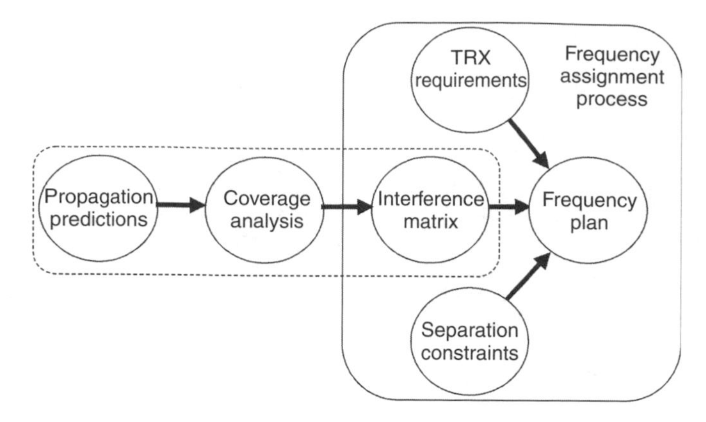

Figure 13.4. The traditional frequency planning process

is based on the results of the previous steps. Propagation predictions are the starting point to create a frequency plan. On the basis of these, the coverage is analysed to establish the level of interference between cells. This is formulated in the form of the IM. On the basis of the IM together with TRX requirement and separation constraints on a cell-by-cell basis, the actual frequency plan is created.

Commercially available network planning tools normally include all this functionality. But it is also becoming more widespread that the tasks associated with creating the IM (shown in dashed box) are located within one tool, whereas the actual frequency assignment is carried out in a dedicated frequency assignment tool (denoted by a solid line).

The difficulty with this sequential process is that the quality of the last stage of the process is still dependent on the quality of the first stage of the process. This means that all deficiencies of propagation predictions are still reflected in the final frequency plan. Hence, even if an ideal frequency assignment process is being assumed, this process is only as good as the input into the process, i.e. the propagation predictions.

13.2.2.3 Support of Different Frequency Hopping Strategies

Over the past few years, a high ratio of operators have started using frequency hopping. Frequency hopping provides two advantages over the non-frequency hopping case. These are frequency diversity and interference diversity (see Chapter 5). Frequency diversity provides higher link resilience as the effect of deep fades (temporal reduction in signal level) is normally only experienced on one of the frequencies and hence the others provide the required link stability. Owing to the effect of interference diversity, it is possible that high levels of interference on one frequency can be offset by low levels of interference experienced on other frequencies. As non-hopping networks lack these benefits, operators do generally apply frequency hopping wherever possible.

There are two types of frequency hopping, baseband (BB) and synthesised/radio frequency (RF). In BB hopping, frequencies are planned as in the non-hopping case, but during the course of a call, the frequency is changed burst by burst, cycling through the frequencies available on the cell, i.e. a call hops from TRX (frequency) to TRX (frequency). In terms of planning BB hopping, there are no differences to the case of non-hopping as each TRX in the network requires one frequency. This means that the above-mentioned process of assigning frequencies is still applicable.

In the case of RF hopping, only the BCCH TRX is non-hopping and thus has to be planned as normal. The other TRXs in the cell are not provided with one frequency, but a set of frequencies through which an active call will be hopping. A possible reuse pattern in RF hoping is 1/1, see Section 6.1.1. This means that each non-BCCH frequency is available to each non-BCCH TRX in the network. Frequency collision (interference) between neighbouring cells is prevented by allowing only a relatively low level of loading of the cell. Owing to the fact that all non-BCCH frequencies are used in each TRX, frequency planning is not required. However, for RF hopping schemes different than 1/1, accurate knowledge of the interaction between cells in the networks is still useful as it is possible to customise the frequency sets that calls are using in RF hopping. Through customisation of these sets performance benefits can be obtained.

One advantage of BB over RF hopping is the fact that remote-tune (cavity) combiners can be used, which have a lower insertion loss than the wideband combiners that are required for RF hopping. Owing to the lower loss of the combiner, higher output powers are achieved with BB and thus larger cell sizes and higher in-building penetration is provided.

RF hopping is the simplest approach to frequency planning as it minimises the planning effort for the traffic channel (TCH) TRX, requiring frequency planning only for the BCCH TRX. RF hopping with regular reuses like 2/2 and 3/3 area also easy to plan and do not require too much planning effort in the non-BCCH layer, see Chapter 6. However, from the spectral efficiency point of view, this is not the optimum approach. For scenarios where building an accurate IM is not possible, this hopping strategy is certainly beneficial. However, the more accurate the ability to determine interference between cells and the more accurate the IM, the more useful it is to create a tailored frequency plan that is making use of the knowledge of interference levels between cells. The knowledge that two cells interfere strongly (or not at all) can be used by the frequency assignment tool to improve the quality of the frequency plan.

13.2.3 The New Frequency Planning Concept

The traditional frequency planning process was described in the previous section and illustrated in Figure 13.4. The new concept described here is called *mobile measurement–based frequency planning* (MMFP). The idea was proposed about five years ago; but in 2002 not many operators really use MMFP. This is due to the fact that without support of the equipment, vendor access to the required mobile measurement data demands huge effort if these data are to be collected on a large number of cells. The main part of the process, shown in Figure 13.4, still applies to MMFP, as it is only the data source that is different. Instead of using radio wave propagation and small-scale *ad hoc* measurements, each call made by the subscribers in the area to be planned contributes to the IM. The idea is useful from an optimisation point of view, as there is no longer a need to predict the radio path or the subscriber distribution, both of which are extremely complex to estimate. The subscriber distribution being implicitly present in Mobile Measurement Reports (MMR) allows really optimising the frequency plan based on the actual traffic hot spots. The IM reflects exactly *where* and *when* subscribers make calls.

Storing the MMR replaces the two phases in Figure 13.4 ‘Propagation predictions’ and ‘Coverage analysis’. There are many different methods to construct the IM from RXLEV values in MMR. These are discussed in Section 13.2.5. After the IM has been built, the MMFP process continues just like the normal frequency planning process. The advantage is that the IM is more accurate than what is obtainable by conventional methods.

Increased accuracy of the measurement-based IM over the prediction-based IM arises from two distinct sources. Firstly, with a conventional method, carrier strength and potential interferer strength are estimated in each map bin as explained in Section 13.2. Each bin contributes one sample, and collecting all samples from each cell service area determines the interference relations for that particular cell. The inevitable error made here is the fact that not each map bin is equally often used for a call by subscribers. This error is avoided when measurement reports are used to construct an IM. This is because each sample now corresponds to a fixed length portion of an actual call. Therefore, those locations from where calls are made more often get to be measured more often and thus contribute more to each cell's interference relations. Secondly, in addition to the fact that the number of samples affecting each cell's interference relations is distributed according to subscriber call locations, each sample is also more accurately estimated in measurement-based method. Each sample is based on measured values of a carrier and a potential interfering cell. Even though there are errors in these measurements, they are of smaller magnitude than errors in propagation predictions.

13.2.4 Signal-level Reporting in GERAN

Radio link measurements done by mobiles were defined for use in HO and RF power control processes. For this purpose, mobiles report not only the serving cell level, but also that of the neighbouring cells, which are defined as HO candidates. It is important to analyse the characteristics of these measurements in order to use them for MMFP.

The signal level from the neighbouring cells can be measured because each BTS continuously transmits a *beacon frequency* (BCCH carrier) at a constant power level containing a 6-bit base station identification code (BSIC). In the specifications [6] it is stated that mobiles should measure the power level of as many beacon carriers as possible, and decode their BSIC in order to determine the cell emitting the signal. After basic processing, mobiles send on the slow associated control channel (SACCH) messages (MMR) to the BTS containing the measured values.

The following process applies: within a period of 120 ms there are 25 intervals in which the mobiles can measure the power of the BCCH carriers indicated in the BCCH allocation (BA) list, normally performing one measurement per interval [6, 7]. This means that the number of samples per SACCH measured on each BCCH carrier will depend on the number of carriers defined in the BA list. For example, if the number of carriers in the BA list is 16 and the MS can measure only one carrier during each interval, the number of samples per carrier in a SACCH multiframe (480 ms) will be 6 or 7(= 100 intervals/16 measurements).

The average of these samples is taken in dBm over one SACCH multiframe and mapped to an RXLEV value between 0 and 63, as follows:

$$\begin{aligned}
&\text{RXLEV} = 0 \Leftrightarrow \text{level} \leq 110\ \text{dBm} \\
&\text{RXLEV} = n \Leftrightarrow -110 + n - 1\ \text{dBm} < \text{level} < -109 + n - 1\ \text{dBm}\ n = 1 \ldots 62 \quad (1) \\
&\text{RXLEV} = 63 \Leftrightarrow \text{level} > -48\ \text{dBm}.
\end{aligned}$$

Then, the MS transmits a measurement report to the BTS, which contains the RXLEV for the BCCH carriers of up to the six cells with the highest RXLEV among those with known and allowed BSIC.

In conclusion, ideal mobile measurements are subject to restrictions:

- Quantisation and truncation (Equation 1);
- The fact that only the six strongest cells are reported;
- The uncertainty in the BSIC decoding procedure.

It is important to analyse the effect of these MMR limitations on MMFP [8]. Figure 13.5 shows the average received power (ARP) (see Section 13.2.5) received in one cell from the surrounding cells (normalised by the total number of MMRs). The figure compares the interference for ideal measurements (Ref) with those of real MMRs including their limitations: firstly, quantisation and truncation (RefQ), secondly only six strongest reported (N6) and finally all limitations considered (N6QBSIC).

It can be observed that quantisation does not have a significant effect when the measured power is inside the margins. Only for cells producing low-level interference (less than -110 dBm), the value is overestimated, because of lower truncation. Upper truncation can be a problem in dense urban areas, where values greater than -47 dBm may be very common.

Another problem observed in the figure is due to the fact that mobiles only provide information about the six strongest interfering cells and therefore other potential strong interferers are not reported. This effect is also clearer in the interferers with low level.

Finally, the BSIC decoding process also has an influence in the interference matrices because the MS has to identify the main interfering cells amongst the ones sharing the same frequency. Decoding can fail if the signal level is too low or there is no dominant interfering cell on that frequency. Two opposite effects can be pointed out. First, if BSIC is decoded, a power level higher than the real one is reported. This is due to the fact that the reported power is not only from the cell whose BSIC is decoded, but the sum of the signals from other cells sharing the same BCCH. This can be observed in the figure on cells producing a high-level interference. On the other hand, if BSIC is not decoded, zero power level is assumed, which is less than the real one. This effect is more important on cells creating low-level interference because the interference level received from cells sharing the same BCCH is usually very similar.

It can also be noticed that even if only the six strongest interfering cells are reported, after a short time several tens of cells are seen. This will allow building accurate IMs (see Section 13.2.5).

In more recent versions of the GSM specifications [6], other types of MMRs, apart from the ones described above, are specified. These are extended measurement reports (ExMR) and enhanced measurement reports (EnMR).

Before sending an ExMR, the MS should receive an extended measurement order (EMO) message requesting the MS to measure the signal level of those frequencies specified in the message. Then, the MS should send one ExMR containing the RXLEV for the carriers specified, considering that BSIC decoding is not required for those frequencies.

It has been found that ExMRs are not very useful for MMR-based frequency planning because measurements are associated with frequencies but not with cells, due to the fact

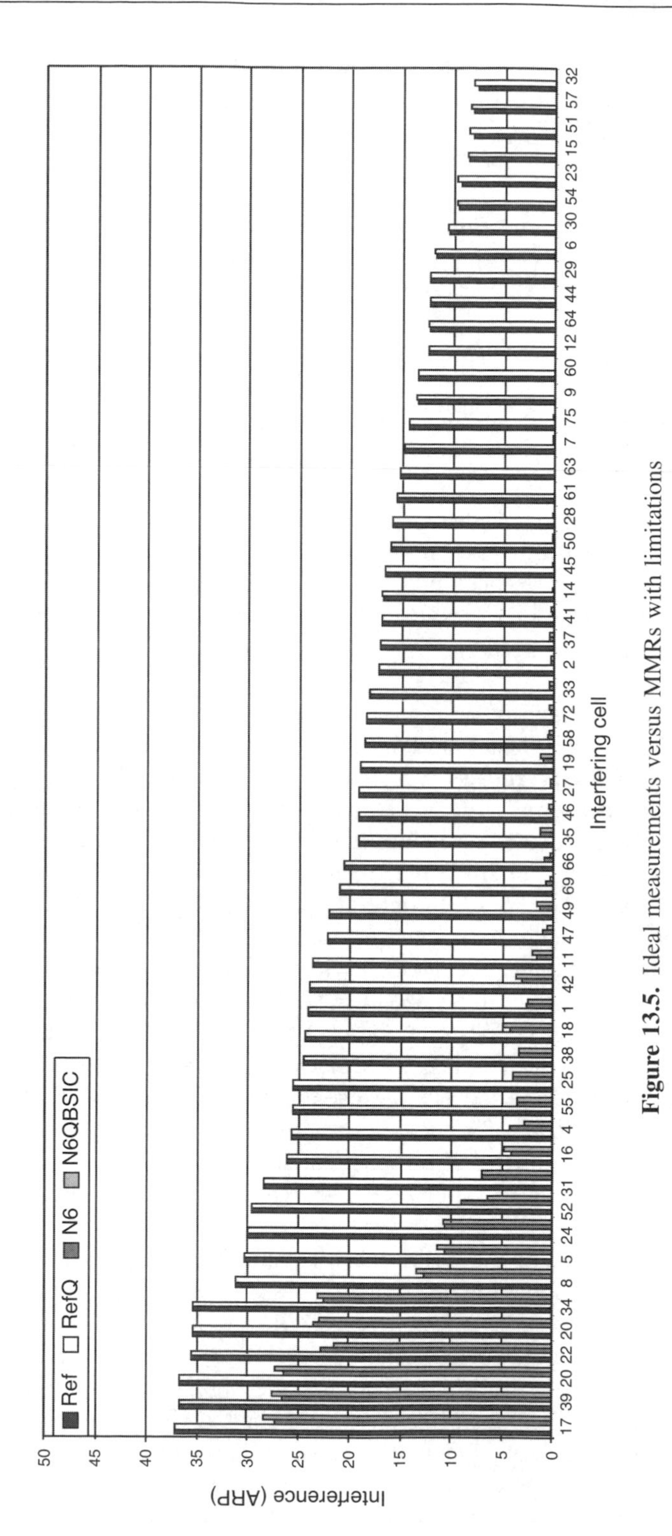

Figure 13.5. Ideal measurements versus MMRs with limitations

that BSIC is not decoded. In order to use these measurements for MMR-based frequency planning, each frequency should be associated with a cell. The first drawback is that only information for a restricted set of cells is available (one for each BCCH frequency). The most logical procedure would be to associate each frequency with the cell with the highest interfering level among the ones sharing that frequency.

Figure 13.6 shows a comparison of the FER_{nw} (see Section 13.2.5) obtained from ideal, real and extended measurement reports for a macro cell scenario. It can be observed that extended measurement reports overestimate the ideal value of interference because they include interference from all cells sharing the same frequency.

The network can also request the MS to send EnMRs. They provide a better estimation of the received signal level from neighbouring cells than conventional MMRs because the maximum number of reported cells in one MMR is not limited to six and also the mapping function from measured power to RXLEV is changed to

$$\begin{aligned}
&\text{RXLEV} = 0 \Leftrightarrow \text{level} \leq -110 \text{ dBm} + \text{SCALE} \\
&\text{RXLEV} = n \Leftrightarrow -110 + n - 1 \text{ dBm} + \text{SCALE} \\
&\qquad < \text{level} < -109 + n - 1 \text{ dBm} + \text{SCALE} \; n = 1 \ldots 62 \qquad (2) \\
&\text{RXLEV} = 63 \Leftrightarrow \text{level} > -48 \text{ dBm} + \text{SCALE}
\end{aligned}$$

where SCALE is an offset indicated in a measurement information message.

As stated before, in micro cells it is common for the signal level to be higher than −47 dBm and therefore the level would be underestimated when using the traditional mapping function. But if the SCALE parameter is incorporated, the maximum reported level can be increased by up to 10 dB.

Another issue to be considered is that, in some scenarios, for EnMRs or multiband mobiles, the number of neighbours reported from several serving cells in different frequency bands could be different to six. Figure 13.7 shows the error between the ideal case (all neighbours reported) and only the 3, 5, 6 or 12 highest neighbours reported for FER_{nw} (see Section 13.2.5). It can be observed that, the higher the number of neighbours reported, the closer the results are to the ideal values. But even when the number of neighbours reported is low the results are still quite accurate. For cells creating small levels of interference (right-hand side of graph), the reduced number of reported cells does not have a significant effect.

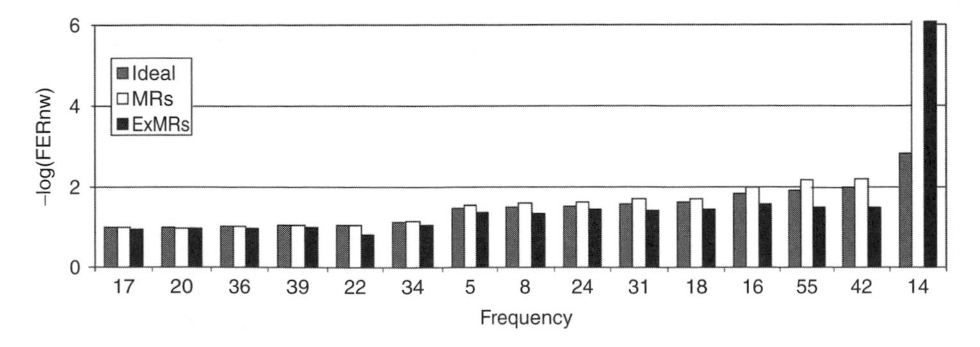

Figure 13.6. Comparison of FER_{nw} for ideal, real and extended MMRs

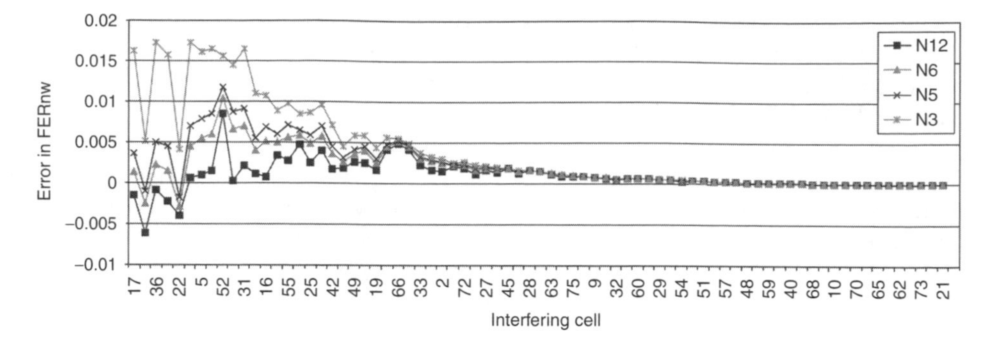

Figure 13.7. Error in $\mathrm{FER_{nw}}$ for different maximum number of reported neighbours

13.2.5 Review of Interference Matrix Types

An inter-cell dependency matrix (IM) is a matrix that describes the interaction between cells. This interaction can be described in several ways, although it normally means radio interference. That is, the element x_{ij} in the matrix is an indicator of the potential interference produced by cell j on mobiles attached to cell i, supposing that both use the same frequency channel. There are several classifications of IMs depending on the aspects considered: radio wave-propagation-prediction–based or measurement-based, uplink (UL) or downlink (DL) IMs, expected values or percentile IMs, etc.

Generally, GSM networks are DL interference limited. Furthermore, in the future, services such as Internet access will be very commonly used, where mobiles will send short requests in the UL, while receiving large amounts of data in the DL. These are the main reasons to concentrate on DL MMR-based IMs. Several types of IMs can be built on the basis of the processing made on the raw MMRs.

Several types of IMs can be built on the basis of the processing made on the raw MMRs. Besides, each of these types can be divided into two groups: traffic-weighted or non-traffic-weighted measurements.

$\mathrm{ARP_w}$ (average received power) represents potential interference from neighbouring cells within the analysed serving area. The non-weighted ARP, $\mathrm{ARP_{nw}}$, is $\mathrm{ARP_w}$ normalised by the total number of MMRs received in the cell.

$$\mathrm{ARP}_{\mathrm{w}_{ij}} = \sum_{k=1}^{N_{ij}} p_{ij}(k) \tag{3}$$

$$\mathrm{ARP}_{\mathrm{nw}_{ij}} = \frac{1}{N}\mathrm{ARP}_{\mathrm{w}_{ij}} \tag{4}$$

where

N_{ij} = number of measurements reported of interfering cell j to serving cell i

$p_{ij}(k)$ = $k$interfering power level (dBm) from cell j reported to serving cell i

N = total number of measurement reports received by serving cell

Another type of IM is the FER (frame erasure rate) matrix [9]. FER is used because it is a very good indicator of speech quality. It represents the percentage of frames being dropped because of its inability to correct bit errors in the vital part (category 1A) of the speech frame, see Section 5.1.2. In order to obtain FER, carrier to interference ratio (CIR) values are calculated for each MMR and then mapped to FEP (frame erasure probability). FEP represents the probability of frame erasure associated with the current MMR. FER is obtained by averaging FEP corresponding to multiple MMRs.

$$\mathrm{FEP}_{ij}(k) = f_{\mathrm{FEP}}(\mathrm{CIR}_{ij}(k)) \tag{5}$$

$$\mathrm{FER}_{\mathrm{w}_{ij}} = \sum_{k=1}^{N_{ij}} \mathrm{FEP}_{ij}(k) \tag{6}$$

$$\mathrm{FER}_{\mathrm{nw}_{ij}} = \frac{1}{N} \sum_{k=1}^{N_{ij}} \mathrm{FEP}_{ij}(k) \tag{7}$$

where f_{FEP} is a mapping function from CIR to FEP [10].

The last analysed IM is the SFER (step FER) matrix [5]. SFER is an estimation of the probability of CIR being lower than a threshold (9 dB). It can be seen as a rough approximation to an FEP–CIR mapping function.

$$f_{\mathrm{SFER}}(x) = \begin{cases} 0 & x \geq 9 \text{ dB} \\ 1 & x < 9 \text{ dB} \end{cases} \tag{8}$$

$$\mathrm{SFER}_{\mathrm{w}_{ij}} = \sum_{k=1}^{N_{ij}} f_{\mathrm{SFER}}(\mathrm{CIR}_{ij}(k)) \tag{9}$$

$$\mathrm{SFER}_{\mathrm{n}_{ij}} = \frac{1}{N} \sum_{k=1}^{N_{ij}} f_{\mathrm{SFER}}(\mathrm{CIR}_{ij}(k)). \tag{10}$$

It is important to point out that FER and SFER provide a better approach to the quality perceived by users than ARP.

13.2.6 MMFP Trial Results

MMFP was tested in a city with a population of 1.5 million (Network 1). Retune area was located in the centre of the city and had 82 cells, all on 900-MHz band. The spectrum used in the trial area and surroundings was about 8-MHz wide. The network was BB hopping with an average of 2.5 TRXs per cell. During weekdays, the average traffic served by trial cells was 3500 Erlang hours per day.

Prior to trial, the average DCR on the trial area was 1.29%. When only drops due to radio reason are counted, RRDCR (radio reason dropped call rate) was 1.16%. The corresponding number of drops per Erlang hour were 0.51 and 0.46. Downlink (DL) and uplink (UL) bit error rate (BER) were 0.41% and 0.32%. The values are based on 8 days of data, from Monday to Monday.

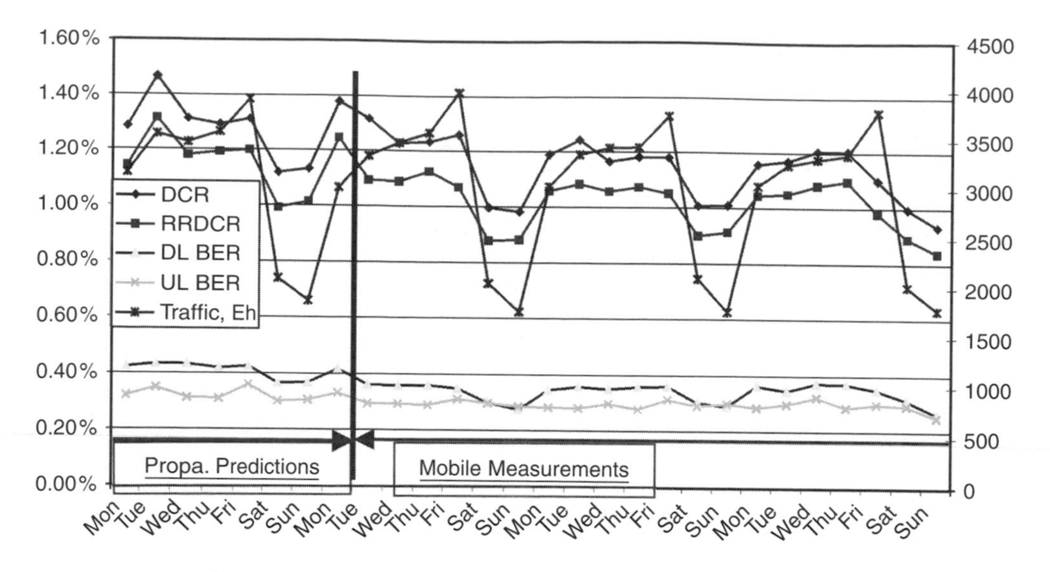

Figure 13.8. Quality in the network before and after MMFP

Figure 13.8 shows quality indicators before and after the retune. The solid vertical line marks the retune. Carried traffic is displayed on secondary *y*-axis as Erlang hours, while the quality indicators have percentage values.

After retuning the trial cells based on mobile measurements, the quality indicators were checked again to compare the quality of different frequency plans. DCR was 1.14% and RRDCR was 1.01%, corresponding to improvement of 12% and 13%. Drops and radio drops per Erlang hour were 0.46 and 0.41, indicating a gain of 10% on both. BERs had decreased to 0.34% for DL and 0.29% for UL, therefore improving by 18% and 10%. Greater improvement in DL was expected, since the mobile measurements only provide DL information. It can be seen however, that using only DL measurements also improves quality in the UL direction. Total number of HOs per carried traffic also decreased after the retune by 2%. This was due to the smaller number of triggered quality HOs.

After the retune, the reason for remaining dropped calls was investigated. It was found out that virtually all of these were due to low signal level. Therefore, frequency planning could not improve DCR in this network any further. The magnitude of the gain obtainable from this method is strongly dependent on the quality of the existing frequency plan. The poorer the performance of the propagation-based plan, the larger the gains that can be obtained from the MMR-based approach.

The trial has proven that despite the measurement limitations inherent to GSM, which were listed in Section 13.2.4, it is possible to utilise these measurements for frequency planning and significantly improve network performance.

13.3 Automated Measurement-based Adjacency Planning

Subscribers can move freely in a cellular network whilst still maintaining traffic connections. This is made possible by the use of HOs. The basic concept involves the subscriber moving from the coverage area of one cell to the coverage of another hence establishing a new connection with the target cell and releasing the connection with the old cell. Such cell change operations are referred to as *handovers* (HOs) and occur while the MSs are in *dedicated mode*, i.e. whilst a call is ongoing.

To perform such HOs, the target or adjacent cells need to be evaluated using adjacent cell measurements. These measurements provide the serving base station controller (BSC) with information about the received power levels from the target cells that are to be considered for HO. The basic reasons for triggering an HO are poor call quality or received signal strength falling below required thresholds and also when the traffic capacity of the serving cell is approaching its limits. Adjacent cell measurements are therefore carried out for all the required target cells and an HO is initiated to the best target cell. Without defining correctly defined cells, call quality and performance are clearly compromised.

To obtain the adjacent cell measurements, the MSs must be informed of the frequencies that are used on all surrounding cells, which need to be measured. These frequencies have to be defined in the BCCH allocation list (BAL) of the cell to be analysed. Two types of BAL can be defined; IDLE, which is used by MSs in idle mode for cell re-selection and ACTIVE, which is used by MSs during dedicated mode for performing HOs. The ACTIVE BAL therefore usually consists of the list of frequencies of the adjacent cell BCCH carrier frequencies normally defined in a neighbour definition list. A maximum of 32 such frequencies or adjacent cells can be defined in a neighbour list in the BSC as per GSM specifications 05.08 in Section 7.1.

The goal of adjacency management is therefore to have the right BCCH carrier frequencies in the neighbour definition list so as to always provide the best target cells to perform HOs. Since any group of adjacent cells, neighbour cells, or target cells for a cell A is a set of cells to which HOs from cell A are permitted, it becomes important that this group of cells is correctly defined. The responsibility of defining these HO target cells is currently a function of network planning tools that operate by simulating the cellular network operations using various definable parameters and constraints. Operators also perform drive tests and through trial and error generate adjacency lists for key areas of the network. Some advanced operators also utilise Abis data to capture adjacency performance metrics and then optimise their neighbour definition lists.

The techniques mentioned are, however, laborious and the complexity of managing adjacencies increases as the network evolves and it's behaviour alters. An automated and accurate adjacency management method would be able to react to evolution changes quickly and hence greatly benefit operators, reducing operational efforts and increasing the efficiency of the network.

The following sections provide a description of how capable a measurement-based adjacency management technique is in performing the task of always providing the best possible candidates as HO targets.

13.3.1 Maintaining Adjacencies

Network performance and quality depend on various operational factors but one of the first stages in a network optimisation loop involves frequency and adjacency cell planning. To maintain and manage the adjacency relations involves the definition of the correct adjacencies for HOs and eliminating the rogue or underperforming adjacencies. Both these steps aim to reduce drop calls and congestion, improve call quality, support the traffic density patterns and hence provide better network performance. However, this is generally an extremely tedious task. In a network of 30 000 cells, where each cell may have up to 20 adjacencies, it requires handling 600 000 adjacency relations. Assuming that on average it takes 5 min to manage one adjacency, the total time required for 600 000 adjacencies is around 5 man years. This provides a synoptic view of the resource and time required within the daily operational tasks of running a cellular network. The present technique most frequently employed by operators is to accurately try and model their network using off-line planning tools utilising propagation algorithms. Although this can prove to be somewhat successful, modelling of the actual network is a painstaking, laborious and resource-consuming task. The constantly changing network means plans generated may quickly become outdated and efficiency is lost due to the operational speed of such tools.

13.3.2 Solution Description

In light of the problems and failures of the existing methods and solutions, an automated mobile measurement–based adjacency management technique has been developed. A major advantage of using measurements comes from the fact that they intrinsically take into account all propagation effects, subscriber density and movement conditions. Planning tools tend to be inaccurate especially in complex radio propagation environments, such as city centres, where the radio wave propagation is difficult to predict. A measurement-based technique can therefore be seen to offer advantages in terms of more accurate adjacency definitions than possible with a theoretical-based approach.

Tests using measurement data show the differences between a prediction-based approach and a measurement-based one. The data shown in Figure 13.9 comes from real network tests conducted in a busy city environment where mobile measurement data from several days were analysed. The results are best analysed by displaying the 32 strongest predicted adjacencies ranked against the measured signal strengths. It can be seen that the predicted ranking is quite different from that obtained using measurements and that some of the cells predicted to be good target cells were not even measured by mobiles in a dedicated mode. The results show that, for example, the sixth strongest predicted adjacency was not even measured and is therefore incorrectly defined as an adjacency. This adjacency should therefore be recommended for deletion to reduce the chance of drop calls occurring.

The various performance metrics of adjacencies, such as HO attempts, HO successes, etc. are normally stored in the network management database. Such data can easily be utilised for deletion purposes and is readily available without having to facilitate for any kind of special data collection. However, for the more difficult task of identifying missing adjacencies, it becomes necessary to measure all the surrounding cells and then deduce the best candidates for HOs.

Any method to generate IM (interference matrix) values described in Section 13.2.5, can also be used to decide whether a cell should be added to the neighbour definition list of another cell. If the IM value corresponding to cells A and B is high, it indicates a high degree of coverage overlap between the two cells. Therefore, the likelihood is high that HOs will be made between the cells.

13.3.3 A Description of the Adjacency Management Process

This process has been identified as having two parts; deleting obsolete adjacencies and creating missing ones with both processes exploiting MS measurements. The entire process is repeated on a regular basis to maintain optimal adjacency performance in the network.

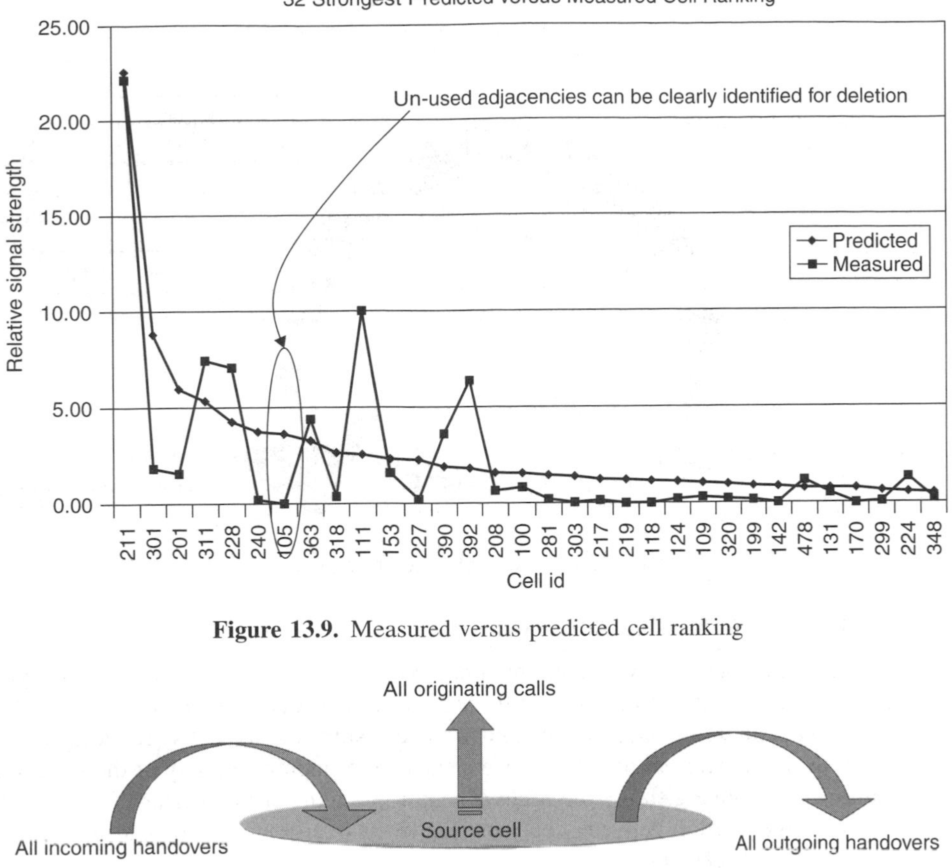

Figure 13.9. Measured versus predicted cell ranking

Figure 13.10. Handover traffic percentage

13.3.3.1 Deletion of Obsolete Adjacencies

It is important when looking at the deletion process to first identify and benchmark the performance criteria that are required. Factors like network growth and strategy are important and contribute towards the benchmarking of the performance requirement. Any existing adjacencies that do not then meet the required performance criteria are deleted from the network. An example performance metric applied to the HO performance data is shown in Figure 13.10, where data from existing adjacencies are compared with a value known as HO traffic percentage. This is evaluated as being the ratio of the calls set-up and handed into cell A, which are then later directed successfully to cell B to the total successful HOs of all defined adjacencies of cell A. This together with the HO success percentage to a target cell metric provides an effective benchmark to allow detection of badly and underperforming adjacencies. The overall deletion process can be outlined as follows and is shown in Figure 13.11.

The majority of the steps highlighted in the deletion procedure are automated. The benchmarking and verification steps, however, need to be carefully assessed and implemented in the procedure.

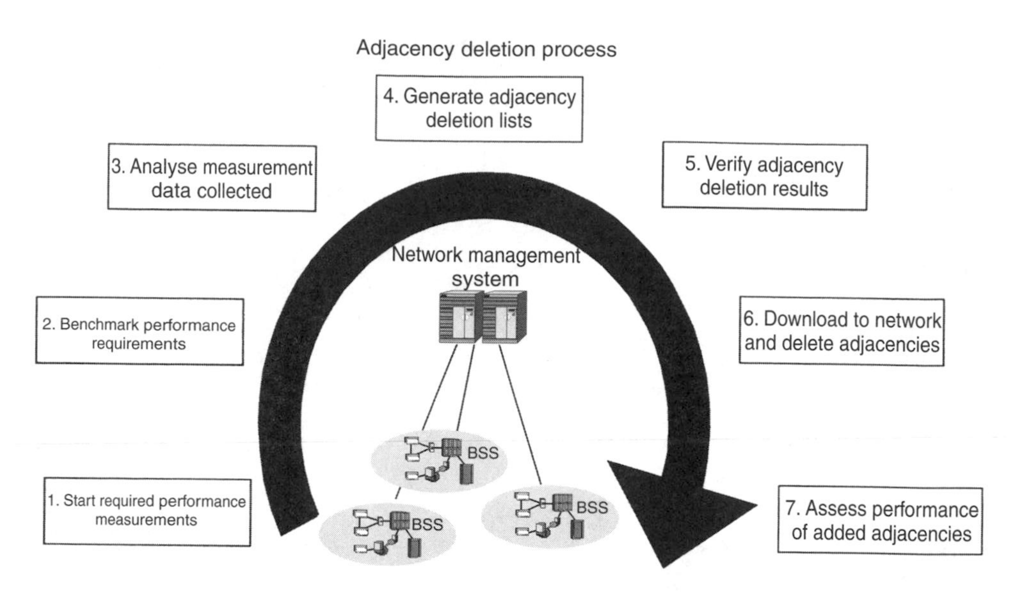

Figure 13.11. Adjacency deletion process

13.3.3.2 Creation of Missing Adjacencies

The other part of the adjacency management procedure concerns the aspect of identifying missing adjacencies. This process is a little more complex than the deletion of obsolete adjacencies and requires measurement data from all surrounding cells. Ranking of the measured cells as shown in Figure 13.12 is a clear way of identifying any of the measured target cells. Such a ranking shows the relative signal strengths of the measured cells or the extent of coverage overlap between cells and the most overlapping cells can be selected as the new target adjacencies.

As with the deletion process, a benchmark must be applied that allows the selection of the right HO candidates from such a ranking. An example benchmark for target cell selection could be determined from the product of the average signal strength value and the average number of samples of a measured cell. The creation process can be highlighted as is shown in Figure 13.13.

Most of the creation process steps can be automated and this also includes restoring the BAL situation back to normal. Always having 32 BCCH frequencies to be measured by mobiles can contribute to some degree of network performance degradation, which is not desirable. Once the newly created adjacencies are created, they must be assessed by their performance in the network and must be seen to take HOs. The whole adjacency management process has to be repeated at regular intervals, such as on a monthly basis, to keep up with network evolution and changes.

13.3.4 Network Test Results

To test the functionality of this new approach, a trial was designed and implemented during the early part of 2001 using a network management system (NMS) as a central feature control system. An entire area managed by a BSC in a busy city environment was the subject of the trial where the trial objectives were to

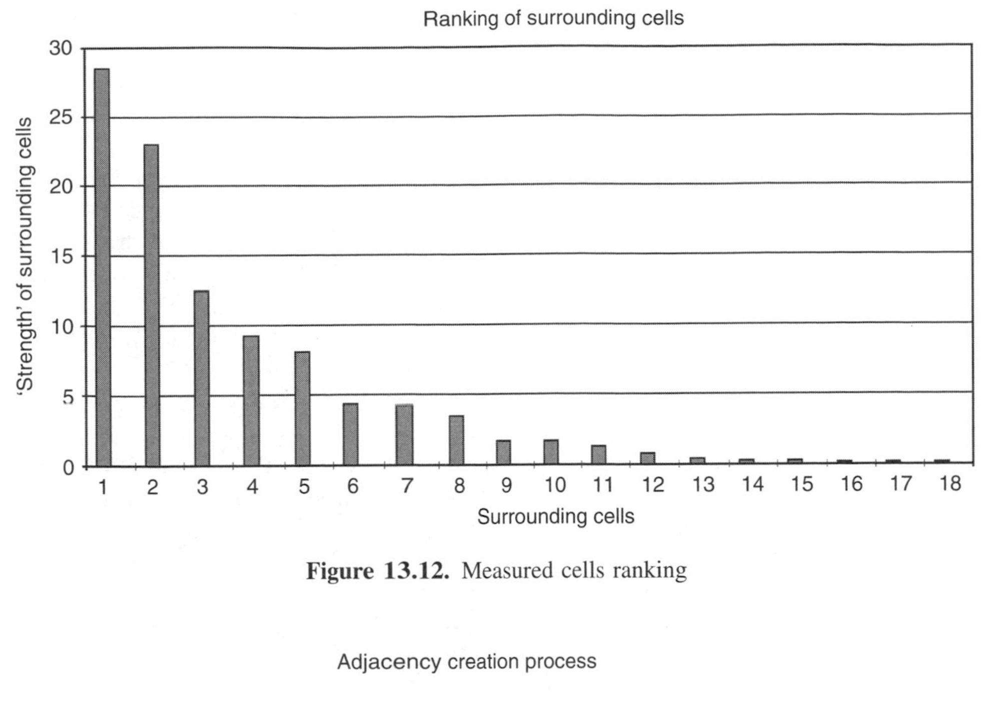

Figure 13.12. Measured cells ranking

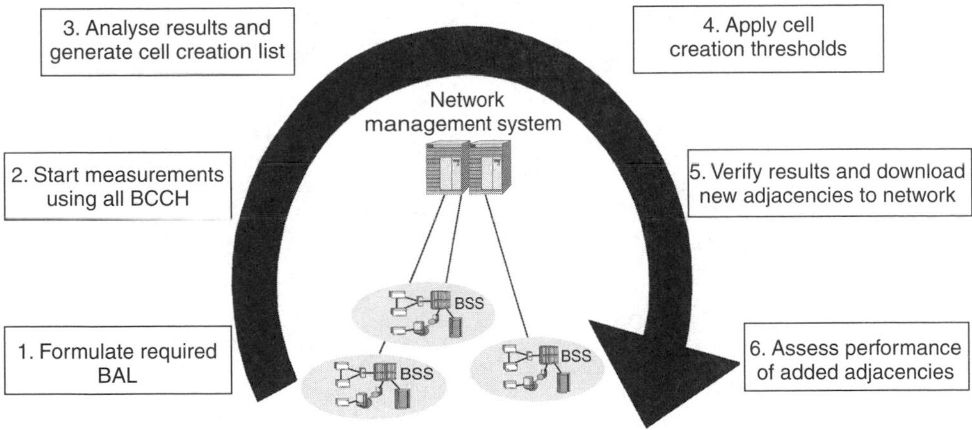

Figure 13.13. Adjacency creation process

- determine and delete unused adjacencies for a BSC,
- determine and create missing/required adjacencies for the same BSC and
- verify the network performance improvements related to the new adjacency plans.

The trial was also focused into looking at the 900-layer part of the dual band network and an important secondary objective of the trial became the mapping of 900-layer adjacencies to 1800-layer cells in the network to determine 1800-layer adjacencies. The possibility of using a mapping technique successfully was based on the fact that all the 1800-layer cells were co-sited to the trial subject 900-layer cells and were all in a high cell density area.

The deletion process was performed first and allowed the selection of obsolete adjacencies. A performance target was determined and any adjacencies failing to meet the criteria were deleted.

The creation process required measurements from all neighbouring cells and these were obtained using all 32 of the operators' 900-MHz BCCH frequencies. The measurements yielded the BCCH, BSIC, number of samples and average measured signal strength and example data shown in Figure 13.14. A threshold was generated based on performances of the existing adjacencies to select suitable candidate adjacencies. Results obtained this way were verified before being downloaded to the network and performance data studied to determine the success of the newly added adjacencies.

Figure 13.14 also shows the results and performances of the newly created adjacencies in case of a measured microcell.

The data shown contains the target cells identified by their BCCH and BSIC, their measured coverage overlap in terms of the sum of DL signal strength and the number of times the measured target cell was seen. Once these measured cells were benchmarked and created as new adjacencies, their performance was monitored. The data presented shows the newly created adjacencies are clearly handling a large percentage of the overall HOs in terms of successful attempts made in both directions. The newly created adjacencies are highlighted so their performance can be easily compared to the existing adjacencies and are seen to perform better than the best existing adjacencies.

The HO utilisation of the added adjacencies clearly show that using a measurement-based technique is valuable, providing a viable solution, which can also be easily automated.

CASE 1:

Mobile Measurements

SOURCE-BTS	SUM_DL_SIG_LV	SAMPLES	T-BCCH	T-BSIC	T-CI
073	2835662	67630	113	53	134
073	2034602	60600	530	53	434
073	674451	24639	103	51	234
073	373939	15562	106	50	334

Performance Statistics

BTS ID	Att_HO =>	<= HO_Att	Target BTS
073	1550	378	434
073	1165	2256	134
073	657	317	889
073	384	74	737
073	324	647	293
073	231	378	393
073	92	139	192
073	54	110	221
073	50	8	612
073	48	73	334
073	25	95	234
073	17	0	534
073	15	42	312
073	2	0	634
073	1	0	534

Figure 13.14. ADCE trial results—case1 (Source: microcell)

13.4 Automated Parameter Optimisation

13.4.1 Outline of Problem

Current procedures for network design estimate the radio network performance on the basis of empirical propagation models, whose accuracy is largely dependent on the precision with which the actual environment has been modelled in the tool. Adjustments of planning parameters, such as base locations, antenna tilt and orientation, BTS output power or frequency plans, are normally carried out during the design phase following a trial-and-error approach. During the subsequent deployment phase, some limited further field-based tuning of the previously mentioned parameters is performed on the basis of the feedback data obtained from drive tests or key performance statistics. Finally, during the operational phase, operators fix parameters in radio resource algorithms to common sets of default values shared between cells, even if not optimum performance in terms of quality/capacity is achieved, as the large set of parameters makes the detailed parameter planning process on a cell-by-cell basis a time-consuming task. Unfortunately, this homogeneity hypothesis may be far from reality, as propagation, interference and traffic distribution vary both in space and time throughout the network. As a consequence, degradation of system performance is observed.

In cases where existing specific features are extremely sensitive to network peculiarities (e.g. traffic management in complex urban scenarios), tailoring of parameters is sometimes performed by means of field trials carried out on isolated areas, which define the default values to be applied across the network. To account for non-homogeneity in the space domain, few advanced operators extend parameter optimisation to different scenarios (e.g. rural, urban, tunnel, indoors) or layer/band (e.g. Macro900/1800, Micro900/1800, Pico1800, Motorway900). Nonetheless, cell-by-cell parameterisation is rarely considered because of time and cost constraints, so parameter retuning is normally carried out only on those cells where performance problems exist, so the potential of parameters that can be adjusted on a cell-by-cell (or even adjacency-by-adjacency) basis is not fully exploited.

Even if optimum parameter values were reached to deal with spatial peculiarities, network behaviour is still changing with time. Temporal variation results from a combination of both long-term trends (i.e. user growth) and periodical changes (i.e. season, day and hour periods). Whilst slow trend variation can be easily coped with by means of network re-planning, no automatic instantaneously reactive process is currently in use, still remaining as a challenging optimisation task.

From this analysis, it can be concluded that the flexibility emerging from the existing large set of parameters included in various network algorithms cannot be currently fully utilised because of the complexity in current networks and a lack of tools to fully support this activity. Obviously, a great potential gain can be obtained from *automatic parameter tuning and optimising processes* that take advantage of cell and adjacency parameters. Such strategy would help operators during the deployment and operational stages, providing benefit from cost savings and improved performance despite changes of the environment both in space and time.

Automatic optimisation has clear and direct benefits for the operators, since it aims at maximising radio performance for a given network infrastructure through a minimum manual effort. Keeping the network in an optimum point continuously, operators can minimise traditional overdimensioning factors, thus reducing the infrastructure costs to have a certain grade and QoS. At the same time, network radio performance is improved, increasing revenue and user satisfaction.

The increased complexity involved in parameter consistency checks would be the main disadvantage derived from this cell-basis optimisation approach. Hence, only a subset of radio parameters must be subject to optimisation, while the rest are set to appropriate default values.

13.4.2 Control Engineering for Automatic Parameter Optimisation in Mobile Networks

The present section focuses on the application of traditional control techniques to increase network performance in cellular networks. Once general control techniques are put into the context of mobile telecommunication systems, an architecture for the control system is proposed.

13.4.2.1 Control Engineering Principles

Automatic control has played a vital role in the advance of engineering and science [11], providing the means for attaining optimal performance of dynamic systems, improving productivity and relieving the drudgery of many routine manual operations.

Among the different approaches to control system design, open-loop control systems refer to those systems in which the output has no effect on the control action. In other words, in an open-loop control system the output is neither measured nor fed back for comparison with the input. In this case, the control system is a static mapping function from target space into control action space. This type of control can only be used when the behaviour of the system under control is very well known and no major disturbances affect it. Unfortunately, from a system-engineering point of view, the radio access network (RAN) in a telecommunication network is a complex system (i.e. large number of cells, non-linear behaviour, space and time non-homogeneities). Consequently, construction of precise analytical models of current mobile networks is a difficult task, thus preventing this approach from being applied extensively.

In contrast to the previous technique, the principle of feedback presented in Figure 13.15 proves essential in the design of control systems where no model of the system under control is obtained, as the one under consideration. Feedback control, or close-loop control, refers to an operation that, in the presence of disturbances, aims to reduce the difference between the output of a system and some reference input and that does so on the basis of this difference, the error. Here, only unpredictable disturbances are so specified, since predictable or known disturbances can always be compensated for within the system.

As depicted in Figure 13.16, the RAN performance optimisation problem can be formulated as a control problem in which (a) the inputs of the system are the radio parameters and (b) the outputs are statistical counters. In addition, (c) the customers using the resources constitute the unpredictable disturbance. (d) The inputs to the optimisation system are those operator definable internal indicators that permit the determination of the network performance: costs of different events in the systems (e.g. such as a block call or a drop call), gains derived for system utilisation (e.g. such as revenue of speech call and packet data calls) and restriction in key performance indicators (KPIs). (e) The outputs

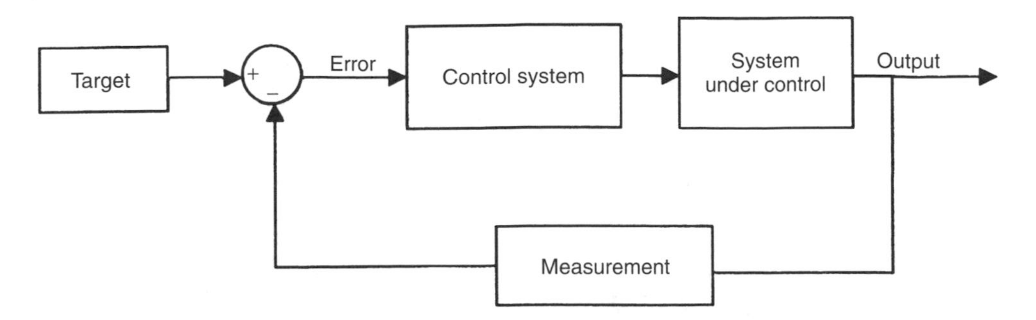

Figure 13.15. Feedback control system diagram

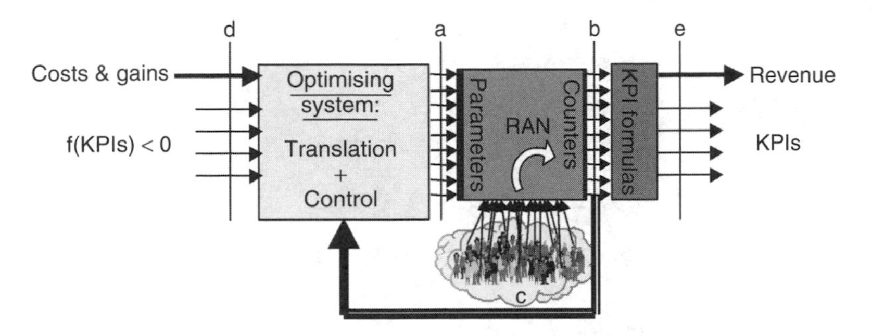

Figure 13.16. Control of telecommunication radio access system

of the system, which measure the overall performance, are the revenue and the actual KPI values, which are obtained from raw counters. The optimisation system defines the optimum set of parameter values performing a translation from operator inputs and with feedback control from counters.

13.4.2.2 Control System Architecture

The great size and high number of different dynamics involved in the system make it impossible to develop a unique control system that tackles optimisation of all parameters. Instead, a modular analysis of the system is required. In Figure 13.17, a control hierarchical architecture and multiple control system is presented, which will be later mapped into network elements.

The definition process of the controller architecture is as follows:

- Firstly, the RAN is split into smaller subsystems that can be analysed and controlled separately. Those subsystems should be as small as possible and at the same time as separate as possible to simplify the control system design. Some examples of those subsystems could be, as shown in Figure 13.17, traffic management, power control, GPRS dimensioning, etc.
- Then, a control subsystem is designed for each radio subsystem (first layer in control architecture). The control approach selected in this phase depends on the subsystem characteristics, linearity, dynamic involved, etc.

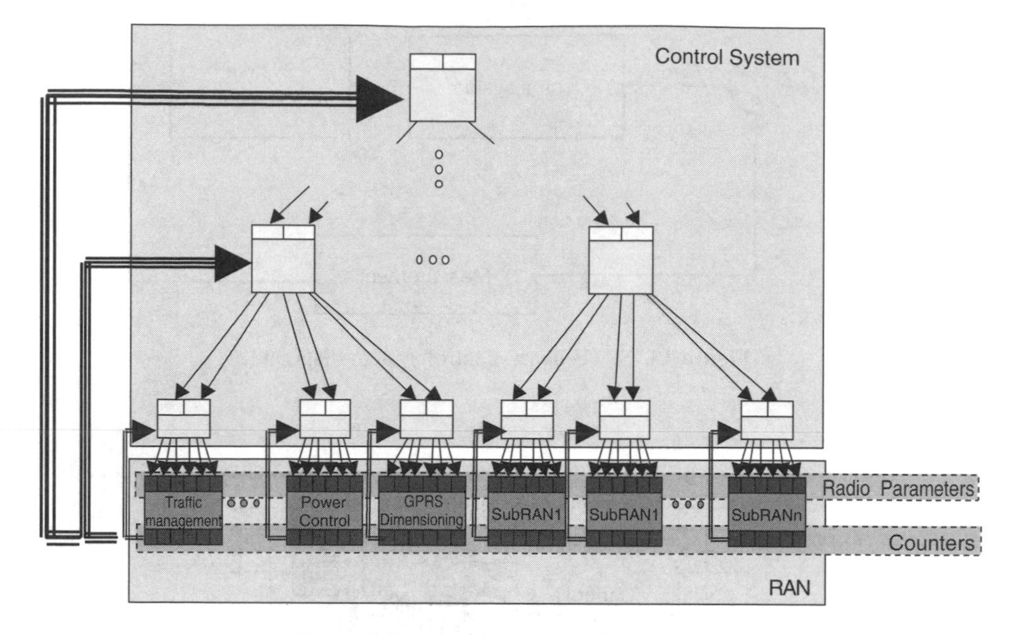

Figure 13.17. Control system architecture

Finally, the algorithm is integrated in the overall control architecture. Splitting the system into subsystems, dynamic independence has been assumed. Actually, soft interaction between subsystems exists, which is accounted for in higher layers of the control architecture. Synchronisation and communication between algorithms is also done in those levels. The main characteristics of this hierarchical control structure are as follows:

- *Scalability*. The control system can be built in different phases with minimum impact to other parts of the system. Whenever a new control system is developed, it can be integrated relatively easily into the existing system.
- *Maintainability*. As the control system is split into smaller subsystems, any change in one system has no effect on other systems and thus the maintenance effort is minimised.
- *Reusability*. Control subsystems developed for a RAN subsystem can easily be adapted to similar problems.
- *Capacity to distribute*. The control system can be distributed through different elements along the network (BTS, BSC, NMS, etc.) in order to minimise the information flow hence the control system can be sparse.
- *Multisystem*. Such a control system architecture deployed for RAN of telecommunication networks can be applied for any radio access technology (GSM, EDGE, WCDMA, WLAN, etc.) and even for other network subsystems (core, IP network, etc.).

Final implementation of this hierarchical structure requires a mapping process of control subsystems to network elements. In this process, control algorithm requirements, both in time response and amount of statistical information, must be considered. Agile algorithms (i.e. seconds timescale) will sit on elements close to the radio resource management algorithm (i.e. BSC in GERAN or radio network controller (RNC) in WCDMA networks). This ability to track instantaneous changes makes this approach extremely suitable to deal with changes in the network environment in the course of time (e.g. temporary traffic blocking or interference peaks). Likewise, those algorithms that require area or network level statistics will be placed on elements with wider and longer visibility (i.e. **operations support system** OSS). This information completeness, together with its inability to deal with fast variations, makes this approach suitable to deal with steady cell peculiarities.

13.4.3 Applications of Radio Network Parameter Optimisation

The present section will give the reader an understanding about how previously discussed control techniques may be applied to increase the performance of current cellular networks. Once the main areas for radio network parameter optimisation are identified, several examples of control elements for different GERAN radio subsystems will be presented. It is worth noting that the application set presented here is not meant to be comprehensive, but rather illustrative of the underlying capabilities of these automation processes.

13.4.3.1 Areas of Radio Network Parameter Optimisation

The starting point of any controller design is the definition of the parameters to be controlled and the state variables that lead the control actions.

As a starting point, it could be stated that nearly any parameter that defines the behaviour of radio resource algorithms in the network is a candidate for adjustment to account for temporal or spatial non-uniformity in the network. Nonetheless, access control, handover and power control procedures are envisaged as the main areas where parameter optimisation may lead to significant performance gains, based on their prevalent role in the quality experienced by the mobile user.

From the measurements collected, radio resource algorithms evaluate potential target cells on the basis of the comparison against some restrictive condition (e.g. signal-level threshold). Valid cells are subsequently prioritised on the basis of some specific figure of merit (i.e. absolute signal-level received, power budget between serving and adjacent cell). Consequently, parameters involved in thresholding and priority biasing actions are obviously suitable for optimisation in a period, cell and/or adjacency level.

Equally important is a good understanding of the main reasons for space and time unevenness in mobile networks in order to identify the main drivers of the optimisation process. The first cause of cell differences is the uneven radio propagation environment. Deployment of several basic cellular structures (e.g. macro cell, micro cell, overlay, underlay) to accommodate high traffic density while still keeping high QoS translates into differences in signal propagation and interference. Hence, adaptation to local interference and propagation conditions is desirable.

Likewise, traffic is another main driver for network adaptation. Owing to user mobility, small areas with instantaneous high user density tend to be rather usual but not spatially fixed (e.g. hot spot, events). In order to cope with these large peaks of capacity demands, networks may be overdimensioned in the design phase, leading to an inefficient usage of resources. To solve this problem, several strategies that take advantage of the overlapping area among neighbouring cells have been proposed in the literature. In this direction, both off-line and real-time adjustment of HO parameters prove to be instrumental in relieving persistent and temporary congestion problems.

In the following, several optimisation procedures intended for automatic off-line (i.e. non-real-time) adaptation of parameters within radio resource algorithms are presented.

13.4.3.2 Optimisation of Signal-level Thresholds in GSM/GPRS

Access to a cell that is unable to provide an adequate connection quality should be prevented both in cell (re)selection and HO procedures. Minimum signal-level thresholds take charge of that responsibility, and tuning of these parameters is therefore a key step in the optimisation of those processes.

Information about the concrete signal level that will ensure acceptable call quality may be derived from cell statistics related to connection measurement reports (MRs) [6]. Collection of these measurement results (named as *Level-quality Statistics* on the following) on a cell basis provides a means to obtain the relationship between received level and perceived connection signal quality in any particular cell. Thus, interference and propagation peculiarities are highlighted so that existing signal-level thresholds present in cell (re)selection and HO processes can be easily upgraded into signal-quality thresholds.

An example of such statistics from a cell in a real network is depicted in Figure 13.18. This 3D illustration shows a 2D histogram of (*RXLEV*, *RXQUAL*) samples, where x-axis, y-axis and z-axis represent received signal level, perceived signal quality and number of occurrences respectively. It is worth noting that discretisation of signal-level values is normally applied in order to keep the amount of information in network database to a reasonable limit.

The construction process of the mapping function between signal level and quality is summarised in Figure 13.19, while the intermediate variables in the reckoning process are

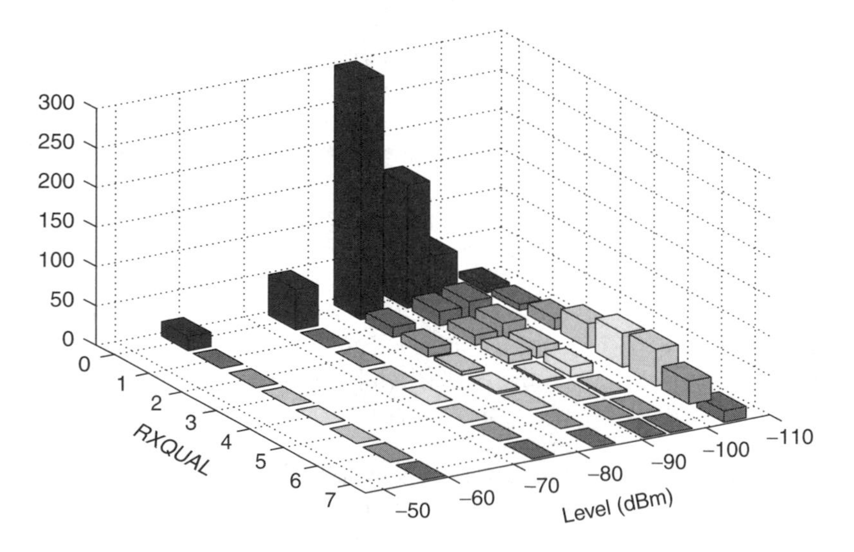

Figure 13.18. Level-quality statistics of a cell in a real network

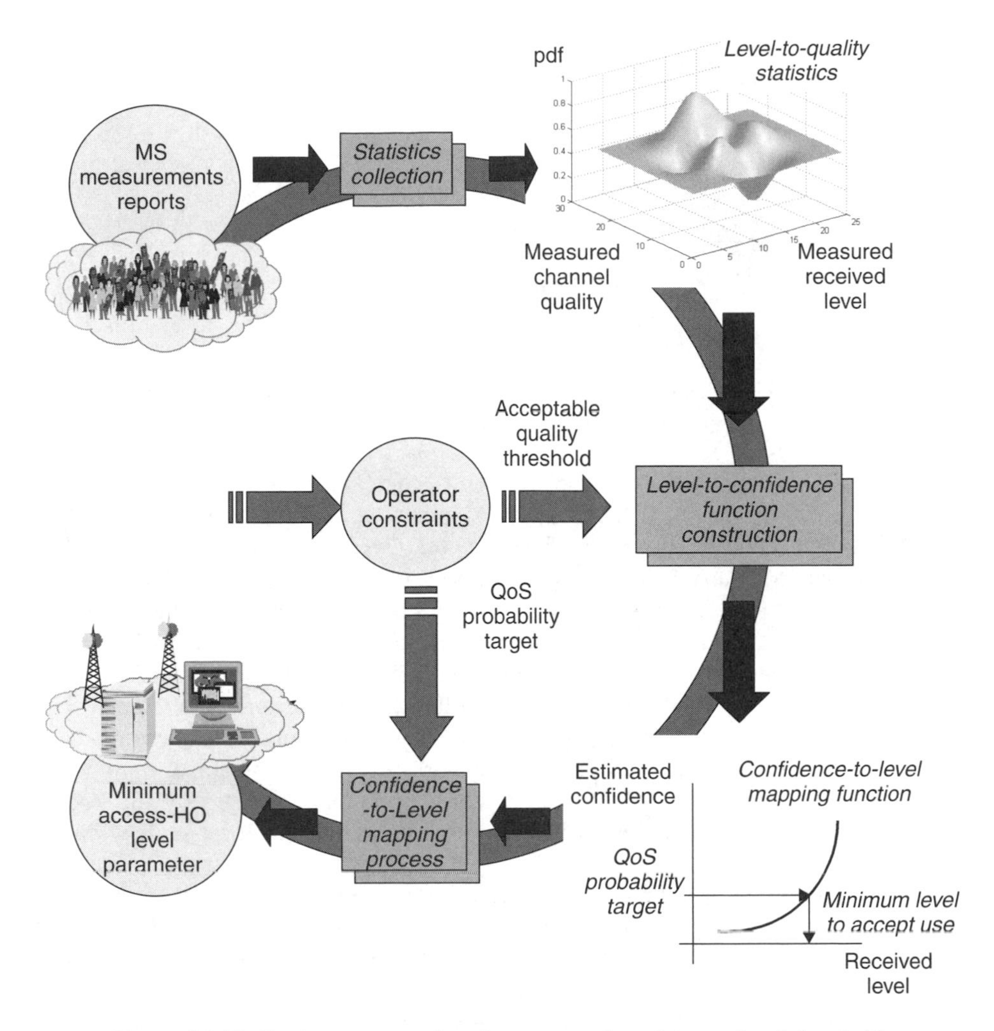

Figure 13.19. Basic structure of tuning process for minimum level thresholds

expressed in (12). First, the 2D joint probability density function (p.d.f.) of *RxLev* and *RxQual* random variables (i.e. p.d.f. $_{RxLev,RxQual}$) is estimated through normalisation by the total number of samples. Subsequently, the p.d.f. of *RxQual* conditioned to *RxLev* (i.e. $p.d.f._{RxLev/RxQual}$) may easily be obtained through normalisation by the number of MRs in every level band. By means of this function, the probability of perceiving a certain connection quality, given that a certain level is received, is determined. Finally, the c.d.f. (cumulative probability density function) of *RxQual* conditioned to *RxLev*, which is the core of the optimisation process, is extracted by the cumulative sum in the quality axis.

$$
\begin{aligned}
p.d.f._{RxLev,RxQual} &= \mathrm{P}(RxLev = RXLEV, RxQual = RXQUAL) \\
p.d.f._{RxLev/RxQual} &= \mathrm{P}(RxQual = RXQUAL/RxLev = RXLEV) \\
c.d.f._{RxLev/RxQual} &= \mathrm{P}(RxQual < RXQUAL/RxLev = RXLEV)
\end{aligned}
\tag{11}
$$

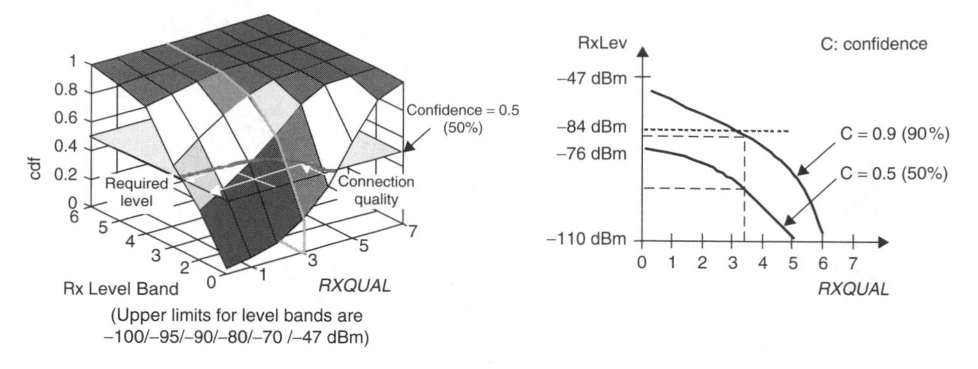

Figure 13.20. Cumulative Quality–Level distribution and Confidence influence on Quality-to-Level mapping process

Figure 13.20 depicts in detail the construction of the *Quality-to-Level* relationship for a cell. Figure 13.20 (a) depicts the $c.d.f._{RxQual/RxLev}$ surface that represents the probability to perceive a connection quality better than *RxQual* given a certain level *RxLev*. When a certain connection quality value is fixed as a requirement, a vertical plane parallel to the *xz* plane is defined (e.g. $RxQual = 3$). The dotted line formed by the intersection between the previous plane and the surface under study represents the probability (or confidence) of reaching the predefined quality target for every level band. On the other hand, whenever a certain confidence is desired in the decision, a horizontal plane parallel to the *xy* plane is defined (e.g. $c.d.f._{RxQual/RxLev} = 0.5$). In this case, the curve represents the quality that is assured for every level band with a certain confidence. Figure 13.20 (b) represents how constraints related to the confidence may be used by the operator to tailor the deterministic relationship between signal quality and level built within the algorithm.

Once this *Quality-to-level* function has been determined, the mapping of connection signal-quality constraints set by the operator into minimum signal-level requirements is rather straightforward. By means of this explicit relationship, signal-level constraints in cell (re)selection or HO algorithms may be easily upgraded into signal-quality constraints on a cell-by-cell basis. Simulation results presented in [12] show that the proposed optimisation procedure is instrumental in scenarios where severe local quality or coverage problems are found, whenever enough cell overlapping is ensured.

13.4.3.3 Optimisation of Handover Margins in GSM/GPRS

In cellular networks, the HO process takes charge of ensuring that any MS is always connected to the most suitable cell. In GSM/GPRS networks, power budget (PBGT)handover (HO) assures that, under normal conditions, any MS is served by the cell that provides minimum path loss. Such a process is initiated whenever the averaged signal level from a neighbour station exceeds the one received from the serving base station (BS) by a certain margin, provided a certain minimum signal level is assured. Both conditions can be formulated as in (13)

$$PBGT_{i \to j} = RXLEV\_NCELL_j - RXLEV\_DL \geq HoMargin_{i \to j} \quad (12)$$

$$RXLEV\_NCELL_j \geq RxLevMinCell_{i \to j}$$

where $PBGT_{i \to j}$ is the power budget of neighbour cell j with respect to serving cell i, $RXLEV\_NCELL_j$ and $RXLEV\_DL$ stand for the received signal levels from neighbour cell j and serving cell i, $HoMargin_{i \to j}$ and $RxLevMinCell_{i \to j}$ are the parameters that define the relative HO margin and the absolute HO threshold considered for that adjacency.

Optimisation is first focused on the requirement for a cell to be included in the candidate list, which is built on the basis of minimum signal-level constraints. Thus, signal-level thresholds may be individually tuned to adapt to local interference and propagation conditions, as explained in the previous section.

Subsequently, margins provide a powerful mechanism to easily control the operational area of a cell. Modification of margins causes displacement of the crossing point where HO between adjacent cells occurs, as depicted in Figure 13.21. Thus, both cell size and shape adjustment is attained, provided a certain degree of cell overlapping is present in the network.

The modification of cell operational areas through this cell resizing effect may be finally utilised to level (or balance) some concrete problem between adjacent cells (e.g. congestion, high drop call rate or interference). Equation (13) presents a possible computation of the margins as a weighted sum of differences of problem indicators between adjacent cells,

$$\begin{aligned} HoMargin_{i \to j} &= \sum_k W_k{}^*[f_k(Problem\ indicator_{k_j}) - f_k(Problem\ indicator_{k_i})] \\ &= W_1{}^*[f_1(Cong_j) - f_1(Cong_i)] + W_2{}^*[f_2(Qual_j) - f_2(Qual_i)] \\ &\quad + W_3{}^*[f_3(Interf_j) - f_3(Interf_i)] \end{aligned} \tag{13}$$

where $HoMargin_{i \to j}$ stands for the power budget margin from serving cell to adjacent cell, W_x are the relative weights for problem prioritisation, *Problem indicator* is any metric related to congestion ($Cong_x$), quality ($Qual_x$) or interference ($Interf_x$), and $f_x(.)$ is the mapping function defined by the operator to derive problem severity from performance indicators.

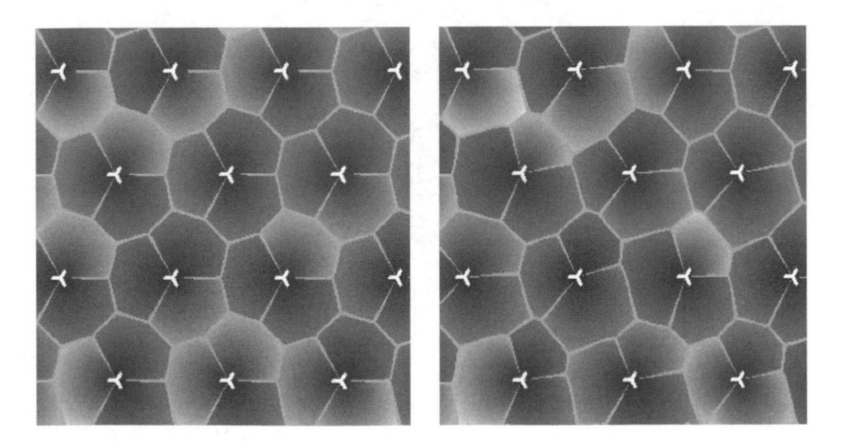

Figure 13.21. Cell breathing as a consequence of margin parameter tuning

As a particular case of the previous rule, several references in the literature have proposed the modification of HO margins to relief permanent localised congestion problems caused by the uneven appearance of traffic in a cellular network both in time and space [13–15]. Such a problem is counteracted in the long term with planning strategies such as TRX expansion or cell splitting, but parameter optimisation remains the only solution in the short term for those cells that cannot be upgraded quickly. Traffic balancing through permanent adaptation of HO margins on an adjacency-by-adjacency basis can greatly minimise congestion without the need for any HW upgrades, thus providing a cost-effective method to increase network capacity.

A possible implementation of the adaptation process is depicted in Figure 13.22. In the figure, an incremental controller is launched on an adjacency-by-adjacency basis in order to suggest changes in current power budget margin values. The direction and magnitude of parameter changes are computed from traffic indicators in both cells of the adjacency.

In the first block, congestion performance indicators are mapped into a meaningful metric related to problems (i.e. namely 'cost' in the optimisation terminology). The congestion-to-cost mapping process takes the form of a non-linear function on the basis of problem threshold detection, controlled by the operator strategy. Subsequently, comparison of the cost metric provides information about the deviation from the equilibrium situation in which problems in adjacent cells are equally severe. Consequently, parameter correction is proportional to the deviation from the equilibrium (or balance) situation. The aggressiveness parameter takes charge of speed and stability issues by controlling the feedback loop gain. Finally, the constraint control block enables the operator to set limits to the adaptation process to avoid excessive increase of co-channel interference caused by cell enlargement. Likewise, selective activation of the optimisation on an adjacency basis is governed.

It is worth noting that the proposal for changes in both directions of the adjacency (i.e. source > target and target > source) are calculated simultaneously, resulting in the same magnitude and opposite direction. Hysteresis region is thus maintained in order to avoid unnecessary HOs due to the ping-pong effect.

A practical example of the previous architecture applied for off-line adaptation of HO margins to deal with congestion problems caused by operator tariff policy is presented

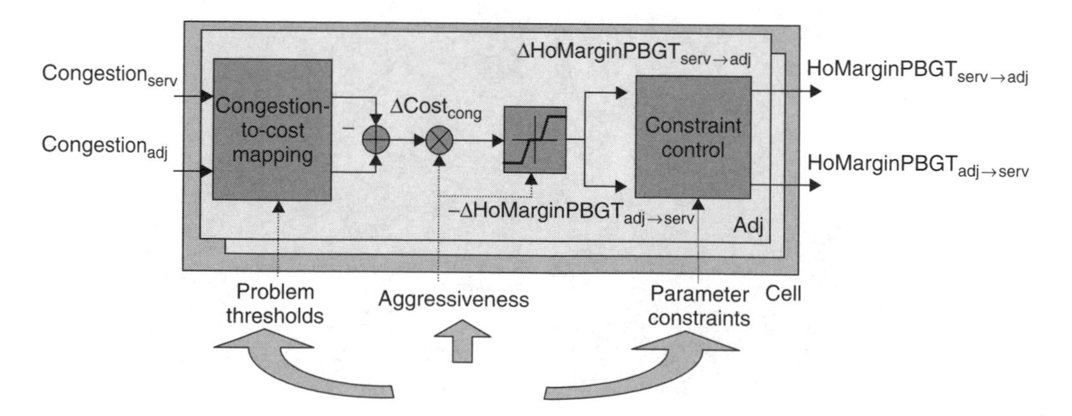

Figure 13.22. Basic implementation of tuning algorithm for traffic balance

in [15]. Equation (14) summarises the concrete optimisation rule employed,

$$HoMargin_{serv \rightarrow adj} \equiv K^{*}[CR_{adj} - CR_{serv}] = -HoMargin_{adj \rightarrow serv} \quad (14)$$

where $HoMargin_{serv \rightarrow adj}$ is the PBGT HO margin, CR_x is the congestion rate (i.e. percentage of time where congestion occurs during the busy hour) and K is the gain of the feedback loop. In addition, several considerations during the development stage were included in order to reduce the processing load and the download time required for parameter re-configuration in a real network. First, only significant adjacencies in terms of HO traffic share were considered in the computation process. Likewise, proposal of subtle parameter changes was neglected. Moreover, for those cases where congestion problems disappear (e.g. due to TRX expansion or cell splitting in the vicinity), a slow return to the homogeneous default settings is achieved by means of a procedure that steps back gradually to the initial situation.

Field trial results showed that more traffic could be carried in the network while the blocking of call set-up attempts could also be reduced when this optimisation procedure is applied. Moreover, congestion relief in the network led to a higher HO success rate, thus improving the performance of macroscopic diversity (i.e. multiple base station) from cell area overlap, and consequently reducing the dropped call probability in the trial area.

13.5 Automated Troubleshooting of Cellular Network Based on Bayesian Networks

13.5.1 Introduction

Troubleshooting (TS) in cellular systems—the identification and correction of problems in the network—is an integral part of the ongoing network operation and optimisation process. As a cellular system consists of several thousand elements, even with highly reliable HW not all of these elements are always performing in an optimum manner. Furthermore, in a growing network, sometimes it is difficult to control that all elements interact as they are expected to or that the configuration is the right one. To ensure that the network is providing maximum performance, continuous TS should be applied.

The TS of a cellular network has three different aspects: first, identification of cells that are facing problems in terms of performance; second, establishing the reasons why a particular cell is not performing well; and finally, implementing a solution to the problems. Of these three tasks, establishing the cause of the fault is by far the most difficult and time-consuming one since a large number of causes could result in detrimental performance. Being able to identify the cause of poor performance in a cell can be one of the most difficult issues, especially if there are several problems masking the real cause.

Currently, TS is largely a manual process, in which the person looking into the reasons for the fault has to carry out a series of checks in order to establish the causes of the problem. This process is characterised by eliminating likely problem causes in order to single out the actual one. In the process of TS, several applications and databases have to be queried to analyse performance data, cell configurations and HW alarms.

The speed of detecting problems is dependent on the level of expertise of the troubleshooter, the level of information available and the quality of the tools displaying the relevant pieces of information. This means that, in addition to a good understanding of the possible causes of the problem, a very good understanding of the tools available to access the sources of information is also required. Owing to the complexity of the system, it is almost impossible for newcomers to carry out this task in a proficient manner. Therefore, some of the most experienced members of staff are involved in rectifying problems in the network rather than looking after its further development. Hence, several operators consider TS as one of the most time- and resource-consuming tasks associated with the operation of the network.

The growing size of cellular networks, together with the increasing complexity of the network elements, creates an apparent need for an automated troubleshooting tool (TST). With the help of a TST, the time required to locate the reason for a fault that is causing a problem will be greatly reduced. This means that network performance will be enhanced as the downtime and the time with reduced QoS will be limited significantly. In addition, by automating the TS process, fewer personnel and thus fewer operational costs are necessary to maintain a network of a given size. By creating an application that is based on the knowledge of the expert troubleshooter, the process of fault rectification will be de-skilled as the majority of problems can be rectified with the help of the automated TS system. The knowledge of this highly experienced staff, which is released from the TS work, can then be utilised for other aspects of the network optimisation, thus further increasing network performance. Thereby, the automated TS system will ensure that operators will increase returns on their investment.

The aim of such an automated tool is then to de-skill and speed up the TS process by automatically gathering the required information from the data sources and making reasoning with the information related to the faulty cell. The TST can guide the less-experienced engineer in the direction of the problem or directly indicate the reason. It is important to point out that the efficiency of the TST largely depends on the skill level of the TS expert who 'taught' the system because the reasoning is based on his knowledge. The higher the level of expertise of the 'teacher', the higher the proficiency of the TST.

One additional benefit is that the knowledge from different experts can be stored in the tool, therefore not being dependent on the staff working for the company at any time. Furthermore, with the TST, the work is more systematic than when it was a manual procedure where the knowledge was only in the head of the experts.

The gains of the automated TST for an operator are significant as fewer TS personnel with a lower skill level can resolve more network problems in less time. For an average size network, the gain that can be obtained through increased network performance, which is brought about by automated TS, is in the order of several million euros per year.

In summary, automatic TS of problems in cellular networks has an important role in network operations to provide maximum QoS with minimum downtime. That means that TS will contribute to an increase in the network performance, whereas cost will be highly reduced, therefore increasing operational efficiency. On the other hand, once the number of active faults in the network is greatly reduced by TS, the performance benefits of adaptive parameterisation of the network will be significantly increased.

13.5.2 Troubleshooting Process

Automatic TS consists of the following three steps:

- Fault detection: automatic detection of bad performing cells based on performance indicators, alarms, etc.

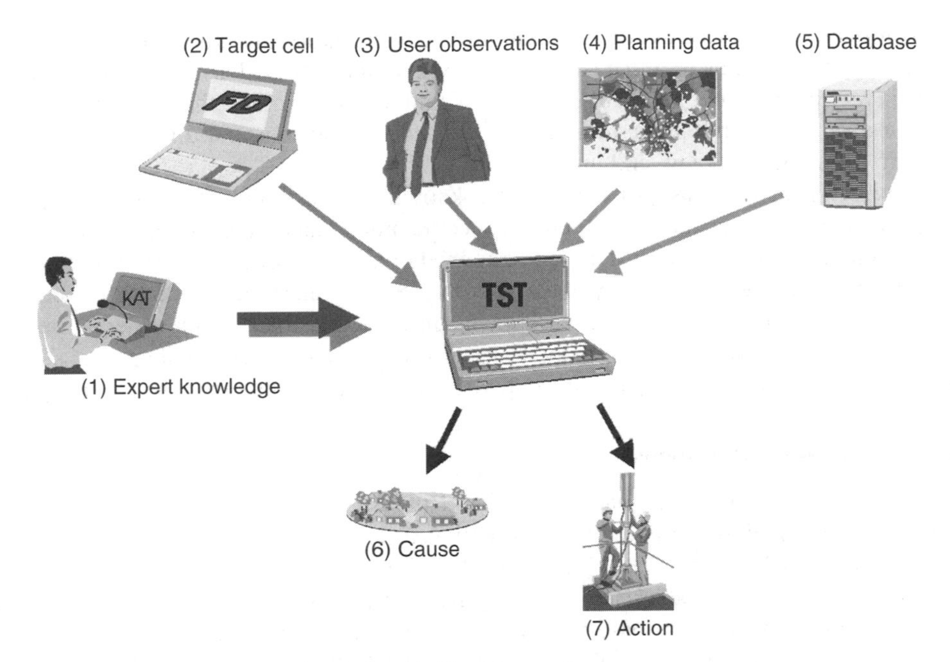

Figure 13.23. Troubleshooting process

- Cause identification: automatic reasoning mechanisms to identify the cause of the problems and the best sequence of actions to solve them.
- Problem solving: execution of the actions to solve the problems.

Hereafter, when speaking about TS, it will be understood that we refer to cause identification, keeping in mind that TS is a wider process. Figure 13.23 explains the cause identification procedure. First, (1) the knowledge of the troubleshooters has to be transferred to the system using a knowledge acquisition tool (KAT). (2) The fault detection subsystem indicates which are the cells with problems and the type of problem. The TST chooses the proper model and makes the reasoning on the basis of (3) user observations (e.g. antenna down tilt), (4) planning data (e.g. frequency reuse) (5) and indicators from databases (e.g. measured interference level, HW alarms, etc.). (6) The output of the TST is a list of possible causes with the probability associated with each cause. (7) The TST also recommends the most cost-efficient action to solve the problem.

13.5.3 Decision Support Systems

When there is a problem in the network, e.g. high amount of dropped calls, TS experts analyse some KPIs that show the status of the network. On the basis of their previous experience, they can find the cause of the problem and decide how to proceed. One significant issue is that the number of variables to be taken into account can be very high and so the process can take a long time and a lot of effort. Besides, there are only very few specialists in the company who actually have the knowledge to determine the right cause. Therefore, the TS performance can be very much dependent on these experts, who may not always be available for TS or might also change jobs.

Traditional expert systems for fault diagnosis try to extract the knowledge of one or several experts to build a tool that can be used by non-specialists. This allows newcomers to effectively analyse and solve the majority of faults. Furthermore, the knowledge from different experts is stored in the tool in an organised manner, not in a diffuse set of rules that only a few people know and understand.

An expert system consists of two parts: a knowledge base and an inference engine. The knowledge base contains the domain-specific knowledge, which should be transferred from the experts to the system (e.g. knowledge about mobile networks TS, medical knowledge of diseases and their symptoms, etc.). The inference engine performs reasoning to obtain conclusions based on the knowledge base and the domain observations. It is important to point out that these expert systems are actually decision support systems, which do not try to completely replace experts, but to support less experienced people.

13.5.3.1 Rule-based Systems

The first expert systems were based on rules. Rules have the form: 'IF A THEN B', where A is an assertion and B can be an assertion or an action. Some rules for TS could be: 'IF there are HW alarms THEN there is a HW fault' or 'IF number of HO failures is too high THEN check whether calls are being handled by expected adjacencies'. The set of rules forms the knowledge base and the inference engine can select the rules relevant to the particular case under consideration and suggest further assertions or deductions.

In several cases, operators have flowcharts describing their TS procedure, which can be directly converted into rules. The main advantage of rule-based systems is that they provide a logical way of reasoning for human beings, and therefore it is easy to build models based on the flowcharts. But rule-based systems also have several drawbacks: they are not suitable for large and complex domains, they are only valid for deterministic cases, they model the TS procedure rather than the domain, etc.

Furthermore, normally experts have some uncertainty in their reasoning, that is, conclusions are not usually completely certain, e.g. 'IF there are HW alarms THEN *very probably* there is an HW fault'. There are many schemes for treating uncertainty in rule-based systems (certainty factors, fuzzy logic, etc.), but normally combining uncertainties from different rules is very complex, so that the resulting systems may lead to wrong conclusions.

13.5.3.2 Bayesian Networks

One alternative to rule-based expert systems are Bayesian networks (BNs) [16, 17], which use classical probability calculus and decision theory to cope with uncertainty. BNs got their name from the eighteenth-century Reverend Thomas Bayes, the eighteenth-century mathematician who first used probability inductively and established a mathematical basis for probability inference. He formulated the Bayes' theorem, which in the BN is used to infer in the opposite direction of the causal one. For example, if there is an interference problem, very probably there will be a high number of interference HOs. Thus, there is a causal relation from interference problem to number of interference HOs. But it is also possible to do reasoning in the reverse direction. For instance, if there is a high number of interference HOs, then the probable cause is interference in the network. Using Bayes' theorem, it is possible to calculate the second probability from the first one.

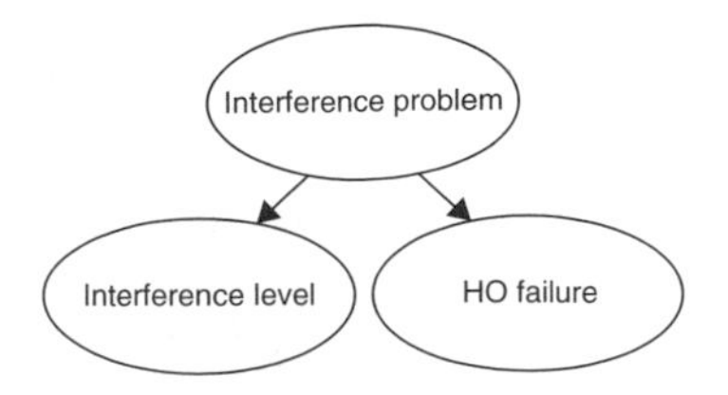

Figure 13.24. Example of Bayesian network

A BN consists of a set of variables or nodes (which have different exclusive states) and a set of arrows between these variables. Arrows normally reflect cause–effect relations within the domain. The strength of an effect is modelled as a probability. For example, Figure 13.24 shows a very simple Bayes model. If there is an interference problem in the cellular network, this will cause HO failures and the interference level will probably be high. In this case, there are three variables with the causal relations shown in the figure. The states of the variables are interference problem: yes/no; interference level: high/medium/low; and HO failure: yes/no. The meaning of these states should be defined; for example, high interference level can be specified so that for 95% of the calls the interference level is higher than −85 dBm.

The conditional probabilities associated with the links have to be specified too, that is, the probability of the interference level being high, medium or low, given that there is an interference problem and likewise the probability of the HO failure being in state 'yes' or 'no' given that there is an interference problem. Furthermore, it is necessary to define the prior probability of having an interference problem.

A BN has several advantages compared with rule-based expert systems such as the following:

- Reasoning can be made even if not all the required information is available, which is the most common situation in mobile networks;
- It can cope not only with deterministic cases, as it was the case with rule-based systems, but also with uncertainty using probability theory;
- It models the domain, whereas rule-based systems modelled the expert's way of reasoning;
- It can learn with each new case;
- It can learn from data available from the domain.

A clear drawback of a BN is that it is relatively difficult to build models, especially if several probabilities have to be specified.

13.5.4 Bayesian Network Models

When creating a BN there should be a compromise between accuracy and simplicity. On the one hand, the Bayesian model should reflect the domain as accurately as possible in order to assure that the right cause of the problem is found in each case. On the other hand, if the model is too complex the knowledge acquisition will be extremely difficult and the algorithms to find the causes will be far more complicated. Indeed, one of the major problems when trying to build models for mobile networks has been obtaining knowledge from the experts. Therefore, the aim is to define a model as simple as possible, but sufficiently accurate.

In order to create a model, the important variables have to be identified and their exclusive states have to be defined. Then, it is necessary to specify the dependency between the variables. Finally, probabilities have to be assigned.

For simplicity, variables will be divided into three types: causes, evidences and actions. *Causes* are the possible faults that are causing problems in the network. *Evidence*s, which are variables related to the causes, can be divided into two classes: symptoms and conditions. *Symptoms* are indications of the causes. *Conditions* are variables that could have caused the problem or could have an impact on some cause. *Actions* are the activities that may solve the problem once the cause is known.

Figure 13.25 shows an example to illustrate these definitions. For example, frequency reuse may have an influence on having interference, i.e. if the frequency reuse is very tight it is more probable to have an interference problem than if co-channel cells are further apart. On the other hand, if there is an interference problem, very probably the interference level will be high and there will be HO failures.

Once the variables, their states and the complete structure are defined, then the probabilities have to be specified. When a node has several parents, this becomes a cumbersome task because probabilities for each combination of the parent nodes have to be set. For example, even for an expert it is very difficult to define which is the probability of having a HO failure given that there is an interference problem and, at the same time, a coverage problem but not a HW problem.

In order to simplify, two plain models are proposed: naïve model and noisy-OR model.

The *naïve model* shown in Figure 13.24 has been used in many diagnostic systems [17]. When this model has been used in medicine, C represented a set of alternative diseases and A_i were potential symptoms of the diseases. This model can also be used for TS, C

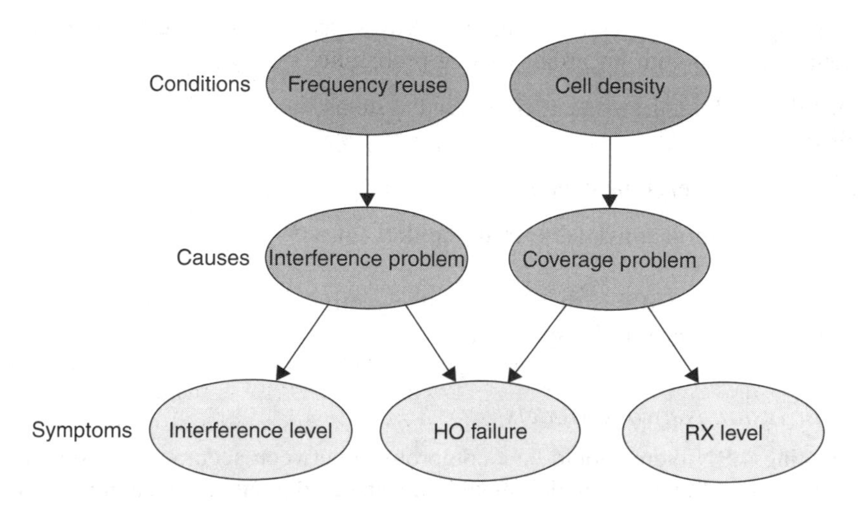

Figure 13.25. Example of Bayesian model for troubleshooting

would be the possible causes of the problem and A_i the possible symptoms, conditions and actions.

The following assumptions are implicitly inherent in this model:

- Single fault assumption: the states of C are mutually exclusive, that is, only one cause can be present at a time.
- The evidences A_i are independent given the cause C. It implies that once the cause C is known, information about some evidence A_i is of no relevance to predict the value of the other A_i.

To overcome the single fault restriction, which is the main limitation of the naïve model, the *Noisy-OR model* or other *causal-independence models* can be used instead [18]. As was explained before, one of the major problems in building a Bayesian model with multiple faults is the difficulty of providing conditional probabilities for a symptom given each possible combination of the causes. For example, in Figure 13.25 the probability of node S should be specified for each possible value of the causes C_i. That means specifying 32 probabilities, whereas only 6 probabilities have to be defined if Noisy-OR is assumed.

Noisy-OR supposes that C_i causes S unless an inhibitor prevents it. For example, if there is either a coverage problem or an interference problem, then there will be a high number of quality HOs unless something prevents it. That means, that almost certainly there will be a high number of quality HOs, although there is a slight possibility that the symptom is not present even if any of the causes is occurring. Considering that the inhibitors are independent, the probabilities are very easy to calculate.

13.5.5 Knowledge Acquisition

Converting the expert knowledge into a BN can be a complex process. If the domain expert not only knows about TS mobile networks but also understands BNs, he would be able to build accurate models. But this is not usually the case; normally, a knowledge engineer should work with the expert and try to reflect his knowledge in the form of BNs. Frequently, this is a troublesome task because of the different languages used by the BN and the TS expert, and it can take a long time to build even a simple model.

The TST will never be able to work better than the expert who transferred his knowledge to the system. This means that the knowledge acquisition is a crucial task because the TS will only give accurate results if the experts have a deep understanding of the domain and if this is correctly converted into BN models.

Eliciting knowledge from experts should include the following:

- *Variables*. what are the important variables to consider? What are the states of these variables?
- *Structure*. what are the causal relations between the variables? Are there any dependencies or independencies between variables?
- *Parameters*. what are the probabilities needed in the defined structure?
- *Utilities*. what is the usefulness of the different possible actions?

These stages can take place iteratively, being a difficult and time-consuming process. In this context it would be very useful to have a tool acting as the knowledge engineer, i.e. it should guide the experts through a sequence of questions and it should automatically translate the answers into Bayesian models. This tool will be called the *knowledge acquisition tool* (KAT).

Nevertheless, it can be extremely complex for KAT to build any BN model. In order to make it simpler, some assumptions can be made, for example, regarding the complexity of the BN structure. In that case, only questions concerning some basic relations, variables, parameters and utilities are required. Besides, if it is assumed that the Bayesian model is any of the previously explained, i.e. naïve or noisy-OR model, only a limited subset of probabilities is needed.

For the simple case shown in Figure 13.25, KAT would ask about the causes (interference, coverage problem), symptoms related to each cause (interference level, HO failure, RXLevel) and conditions related to each cause (frequency reuse, cell density). It would also inquire about the different states of each variable and some probabilities, e.g. probability of each symptom given the related causes. On the basis of the answers to the questions it would automatically build a model like the one shown in Figure 13.25, a naïve model or a noisy-OR model.

When dealing with TS experts, it has been realised that even specifying probabilities for a given structure is quite difficult and that different experts can provide completely different values. This can lead to inaccurate results. Normally, operators have system databases containing the history of most variables in the network, which can be used to train the BN (*learning from data*). This means that it is possible to train the BN on the basis of previous problem cases and the probabilities can then be obtained from those previous experiences instead of being elicited from experts. Furthermore, probabilities can be updated every time there is a new case. This presents the additional value that the system can adapt to different environments, e.g. probabilities can be different for a rural and urban area.

In conclusion, learning from data is an important feature when using BNs. But not only learning parameters, learning the structure of the BN is also possible (although more complex). Therefore, it is not necessary to make simplifications such as done in the naïve model, but the structure is the one that best matches the existing data, and also it adapts to changes in the TS domain.

13.5.6 Troubleshooting Sequence

It is important to point out that the target of TS is not only to determine what is wrong, i.e. what the causes are, but also to solve the problems. At any stage of the process, there are several possible evidences that can be analysed and actions that can be performed. Because these evidences and actions take time and have a cost, it is required to find a sequence of steps (evidences or actions) that minimise the cost (time/money) of solving the problems.

A very simple TS procedure collects all evidences and then makes reasoning based on them, ordering the possible causes according to their probabilities and carrying out the action associated with the most probable cause. This is not an optimum procedure because all evidences are collected, when probably not all of them are required to find out the cause. Furthermore, sometimes it could be more efficient to perform an action different to the one associated with the most probable cause. For example, suppose that the most probable cause was a HW problem, and the next probable one was a configuration problem. In that case, it is perhaps more cost-efficient to make a simple parameter change first to see if the problem was in configuration. Only if it fails to solve the problem, someone has to be sent to the site to change the problematic HW unit, which is always a very expensive and time-consuming action. Without an optimised TS sequence, the HW would have been changed before trying a simple change of a parameter.

An optimum procedure would order the steps (evidences and actions) in a sequence minimising the cost (money/time). In general, some actions will be carried out before collecting all evidences and very probably the problem will be solved by one of these actions, saving time and money. For example, if after collecting some evidence, the probabilities of faulty HW and bad configuration are very high, the parameter change will be proposed and if the problem is solved, it is not required to collect the other evidences or carry out the other actions anymore.

The procedure consisting of identifying causes of problems and solving them through a sequence of TS steps is called *decision-theoretic troubleshooting*. Some algorithms to choose the best sequence of steps are proposed in [16, 19, 20].

13.5.7 Advanced Features

Some issues to be taken into account when modelling with a BN are briefly reviewed in this section.

First, *sensitivity analysis* examines how sensitive the conclusions are to minor changes in probabilities and evidences. As was previously explained, specifying accurate probabilities is a difficult task for the domain experts. Thus, it is important to assure that minor deviations of these probabilities will not lead to big errors in the conclusions (probabilities of the causes). On the other hand, it is important to notice which are the more important evidences, e.g. knowing that the interference level is high may lead to the conclusion that there is an interference problem, independently of the value of the other evidences.

In some cases, it can be required to change some probabilities in a model that is already built. *Tuning* explains how the parameters in the model can be changed to meet the new requirements. For example, let us suppose that there is a high number of quality HOs and this knowledge is incorporated into the model, obtaining the posterior probability of having a coverage problem. But, the expert may think that the obtained probability is not right and prefer to manually set a different value. In that case, all the conditional probabilities previously defined in the model should be tuned in order to obtain the probability required by the expert. Tuning may be especially useful when there is a previously existing model, and then one expert, who may be different to the one that initially built the model, wants to modify certain probability.

Another important topic is verifying the chosen model, i.e. once the model is selected and the probabilities are specified, it is important to check the validity of the model. This can be done by processing several cases and examining that the conclusions obtained are the right ones.

13.5.8 Interaction with the Network Management System

The network management system (NMS) is an essential part of cellular networks providing operational tools for the operator to run the network efficiently. The TS expert system can work independently from the NMS, but most of the benefits are achieved when it is an integrated part of it. This integrated solution will provide direct access to information required in fault analysis as well as access to the operators' fault management system. Also, some operators can have special automatic fault-detection tools in their NMS that can highlight cells that are performing badly. With an integrated solution, it is possible to start up TS analysis automatically on the basis of the output of the fault-finding tool.

An integrated solution is also beneficial in case of TS of multi-vendor networks as well as with networks where new network system layer is added, e.g. extension from 2G network to 3G. The TS expert system based on BN methods is system independent itself, so it is flexible to support multi-vendor networks and 3G networks with minor modifications.

The access to the fault management system in the NMS guarantees that work done with the TST is very well synchronised with the whole fault management process of the operator. Thus, all relevant TS cases can be automatically directed to the TS expert system, and if it finds the solution, the case is cleared, reported and filed. If the problem is not found by the expert system, it can be redirected to the specialists for further analysis and the final conclusions can be incorporated into the knowledge of the expert system.

References

[1] Okumura T., Ohmori E., Fukuda K., 'Field Strength and its Variability in VHF and UHF Land Mobile Services', *Rev. Elect. Commun. Lab.*, **16**(9–10), 1968, 825–873.

[2] Walfish J., Bertoni H. L., 'A Theoretical Model of UHF Propagation in Urban Environments', *IEEE Trans. Antenna Propagation*, **AP-36**, 1988, 1788–1796.

[3] Athanasiadou G. E., Nix A. R., McGeehan J. P., 'A Ray Tracing Algorithm for Microcellular and Indoor Propagation Modelling', *ICAP'95*, IEE Conference Publication No. 407, 1995, pp. 231–235.

[4] Brown P. G., Constantinou C. C., 'Prediction of Radiowave Propagation in Urban Microcell Environments using Ray Tracing Methods', *ICAP'95*, IEE Conference Publication No. 407, 1995, pp. 373–376.

[5] Timmer Y., Bergenlid M., 'Estimating the Inter Cell Dependency Matrix in a GSM Network', *VTC '99*, pp. 3024–3028.

[6] GSM Specification 05.08 v8.4.0, Section 8.4.1.

[7] Mouly M., Pautet M. B., *The GSM System for Mobile Communications*, Cell & Sys., Telecom Publishing, Palaiseau (France) 1992.

[8] Barco R., Cañete F. J., Díez L., Ferrer R., Wille V., 'Analysis of Mobile Measurement-Based Interference Matrices in GSM Networks', *IEEE Vehicular Technology Society Fall Conference, VTC 2001*, 2001.

[9] Kuurne A., *Mobile Measurement Based Frequency Planning in GSM Networks*, Master's Thesis, Helsinki University of Technology, Laboratory of Systems Analysis, Helsinki, 2001.

[10] Nielsen T. T., Wigard J., *Performance Enhancements in a Frequency Hoping GSM Network*, Kluwers Academic Publishers, Dordrecht, 2000.

[11] Astrom K., Albertos P., Blanke M., Isidori A., Schaufelberger W., Sanz R., *Control of Complex Systems*, Springer-Verlag, London, 2000.

[12] Toril M., Pedraza S., Ferrer R., Wille V., 'Optimization of Signal Level Thresholds in Mobile Networks', *Proc. IEEE 55th VTC*, Vol. 4, Birmingham, May 2002, pp. 1655–1659.
[13] Chandra C., Jeanes T., Leung W. H., 'Determination of Optimal Handover Boundaries in a Cellular Network Based on Traffic Distribution Analysis of Mobile Measurement Reports', *Proc. IEEE VTC 1997*, Vol. 1, May 1997, pp. 305–309.
[14] Steuer J., Jobmann K., 'The Use of Mobile Positioning Supported Traffic Density Measurements to Assist Load Balancing Methods Based on Adaptive Cell Sizing', *Proc. IEEE VTC 2002*, 2002, pp. 339–343.
[15] Toril M., Pedraza S., Ferrer R., Wille V., 'Optimization of Handover Margins in GSM/GPRS', *57th IEEE Proc. Vehicular Technology Conference*, Jeju, 2003 (to be published).
[16] Jensen F. V., *Bayesian Networks and Decision Graphs*, UCL Press, New York, 2001.
[17] Cowell R., Dawid A. P., Lauritzen S. L., Spiegelhalter D., *Probabilistic Networks and Expert Systems*, Springer-Verlag, Berlin, 1999.
[18] Kim J. H., Pearl J., 'A Computational Model for Causal and Diagnostic Reasoning in Inference Engine', *Proc. 8th Int. Joint Conf. AI*, 1983, pp. 190–193.
[19] Heckerman D., Breese J. S., Rommelse K., 'Decision-Theoretic Troubleshooting', *ACM Commun.*, **38**(3), 49–56.
[20] Skaanning C., Jensen F., Kjaerulff U., 'Printer Troubleshooting Using Bayesian Networks', *Proc. 13th Int. Conf. Ind. Eng. AI and Expert Systems*, AIE, 2000.

Part 3
3G Evolution Paths

The International Telecommunications Union (ITU) launched the International Mobile Telecommunications-2000 (IMT-2000) program, which, together with the main industry and standardisation bodies worldwide, targets to provide the framework for the definition of the third-generation (3G) mobile systems.

Several radio access technologies have been accepted by ITU as part of the IMT-2000 framework. From the existing IMT-2000-accepted standards, two main differentiated 3G evolution paths are available for current cellular operators. Such paths are the *UMTS 3G multi-radio* and the *cdma2000* evolution paths.

The UMTS Forum, an association with over 250 member organisations, and endorsed by the GSM Association and the 3G Americas, is focused in the successful commercial realisation of universal mobile telecommunication services (UMTS) globally. On the other hand, the CDMA Development Group (CDG) is the equivalent organisation supporting the successful evolution and commercialisation of cdma2000.

The third and last part of this book will introduce the main aspects related to different 3G radio access technologies and evolution paths. Chapter 14 introduces the main code division multiple access (CDMA)-based 3G technologies, analysing their associated performance, both in terms of coverage and capacity. It presents as well a technical comparison between different 3G radio technologies (global system for mobile communications (GSM)/enhanced data rates for global evolution (EDGE), wideband CDMA (WCDMA) and cdma2000). Finally, it introduces the UMTS 3G multi-radio integration concept with its possibilities and performance benefits analysed. Finally, Chapter 15 presents a global market analysis to assist defining the technology evolution strategy within the existing 3G evolution paths in the global frame that characterises the cellular business.

14

IMT-2000 3G Radio Access Technologies

Juan Melero, Antti Toskala, Petteri Hakalin and Antti Tolli

This chapter provides an introduction to the different third-generation (3G) radio technologies, providing a technical benchmark of their voice and data capabilities and presenting the concept of UMTS multi-radio networks.

14.1 IMT-2000 3G Technologies and Evolution Paths

Third-generation radio access technologies aim to provide the mass market with high quality, efficient and easy-to-use wireless mobile multimedia services. IMT-2000 defines a set of technical requirements for the realisation of such targets, which can be summarised as follows:

- high data rates: 144 kbps in all environments and 2 Mbps in low-mobility and indoor environments
- symmetrical and asymmetrical data transmission
- circuit-switched and packet-switched-based services
- speech quality comparable to wire-line quality
- improved spectral efficiency
- several simultaneous services to end users for multimedia services
- seamless incorporation of second-generation cellular systems
- global roaming
- open architecture for the rapid introduction of new services and technology.

GSM, GPRS and EDGE Performance 2nd Ed. Edited by T. Halonen, J. Romero and J. Melero
 ISBN: 0-470-86694-2

Several technologies have been approved by the International Telecommunication Union (ITU) and included in the IMT-2000 3G radio access technologies family. Out of these technologies, wideband code division multiple access (WCDMA), enhanced data rates for global evolution (EDGE) and cdma2000 have gained substantial backing from vendors and operators. Therefore, this chapter will focus on these radio technologies:

- *EDGE.* Enhanced data rates for GSM evolution. This technology provides an enhanced air interface, which includes the adoption of octagonal phase shift keying (8-PSK) modulation, the use of adaptive modulation and coding schemes (AMCS) and incremental redundancy. This technology, fully integrated within the existing GSM radio access network, is an integral part of GSM/EDGE radio access network (GERAN), which, together with the UMTS terrestrial radio access network (UTRAN), constitutes the UMTS 3G multi-radio network. The first two parts of this book describe in depth the characteristics of this radio technology, its evolution and its associated performance.
- *WCDMA (FDD/TDD).* Wideband code division multiple access. It has two modes of operation, frequency division duplex (FDD) and time division duplex (TDD). TDD will not be part of this chapter's analysis due to its expected late introduction and the lack of current market interest. WCDMA FDD is a code division multiple access (CDMA) based technology, specifically designed for the efficient support of wideband multimedia services and implementing an advanced set of CDMA-related functionality. WCDMA is the radio access technology (RAT) of UTRAN. The next section will introduce the main technical principles of this technology.
- *cdma2000.* Code division multiple access 2000. This radio technology is the natural evolution of IS-95 (cdmaOne). It includes additional functionality that increases its spectral efficiency and data rate capability, as described in Section 14.3.

These radio access technologies are grouped under two main categories or evolution paths, which are standardised by different bodies:

- *UMTS multi-radio evolution path.* 3GPP (3rd Generation Partnership Project) is a global project aiming to develop open standards for the UMTS 3rd Generation Mobile System based on evolved GSM core networks. This multi-radio mobile system comprises two different 3G radio access networks, GERAN and UTRAN, which are based on different radio access technologies, GSM/EDGE and WCDMA, respectively. Initially, UMTS defined, within the scope of UTRAN standardisation, a new radio access network architecture, with protocols, interfaces and quality of service architectures specifically designed for the efficient provision of 3G multimedia services. As presented in Chapters 1 and 2, GERAN has adopted all these, hence becoming an integral part of the UMTS 3G frame. Furthermore, 3GPP standards ensure that an efficient integration between UTRAN and GERAN can be accomplished so that they can be merged under a single UMTS multi-radio network. This concept is illustrated in the Figure 14.1.
- *cdma2000 evolution path.* 3GPP2 (3rd Generation Partnership Project 2) forms a collaborative effort of the organisational partners to put together the standardisation of

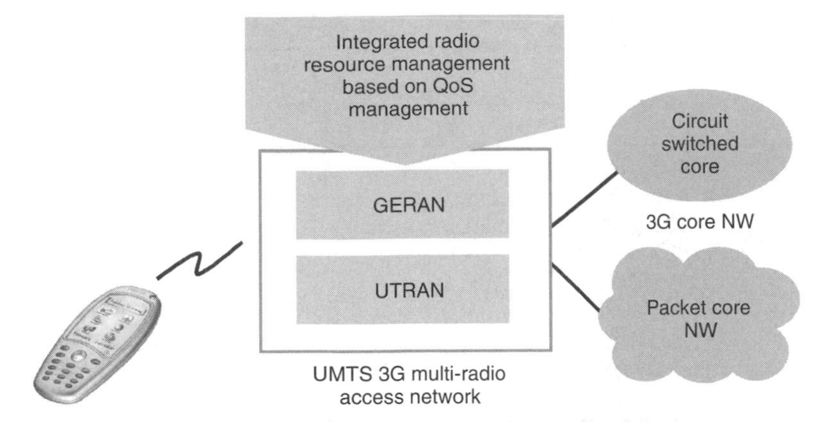

Figure 14.1. UMTS multi-radio network

the cdma2000-based 3G RAT, which is the natural evolution of the IS-95 (cdmaOne) 2G RAT.

14.2 3G Technology Support of Licensed Frequency Bands

An important factor in the 3G-technology evolution path strategy is the support, in the spectrum licensed to operators, of the different radio technologies. Depending on the spectrum licensed for cellular use in a certain region, local operators may have a limited technology choice. Unfortunately, frequency licensing has not been historically uniform across different regions worldwide. Figure 14.2 shows the spectrum allocation in different regions of the ITU IMT-2000 recommended new bands. The recommended ITU spectrum for new bands for 3G deployment has been allocated in some regions for other purposes.

Therefore, operators in different regions will have to try to accommodate the preferred RAT in the specific licensed band. Table 14.1 shows the supported technologies in the main bands allocated worldwide for cellular networks use. Both the UMTS 3G multi-radio (GSM/EDGE and WCDMA) and the cdma2000 technologies are supported in most of the bands. The standardisation of WCDMA support in the 1800 and 1900 bands is being finalised. With this standardisation, and provided infrastructure equipment and terminals

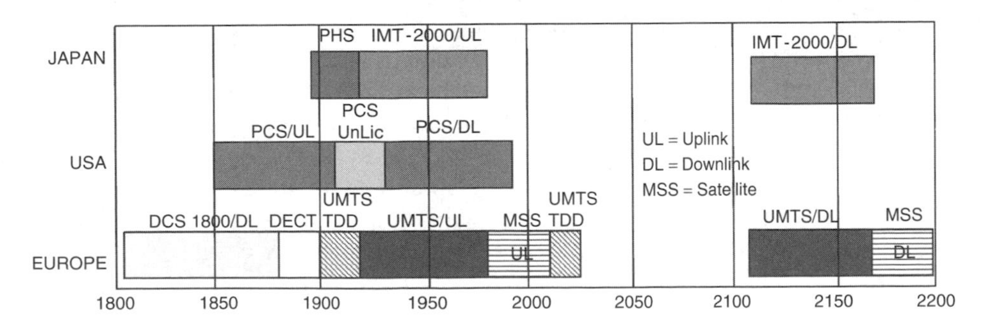

Figure 14.2. Spectrum allocation in different regions of the IMT-2000 recommended new bands for 3G

Table 14.1. 3G RAT support in different bands[a]

Bands (MHz)	MHz	Main deployment regions	GSM/ EDGE	WCDMA	cdma2000
450	450–467	Europe			x
480	478–496	Europe	x		
800	824–894	America	x		x
900	880–960	Europe/APAC	x		x
1700	1750–1870	Korea			x
1800	1710–1880	Europe/APAC	x	x	x
1900	1850–1990	America	x	x	x
2100	1885–2025and 2110–2200	Europe/Japan/ Korea/APAC		x	x

[a] Some of the bands supported in the standards do not have commercial products available.

become available in these bands, wireless operators in America will be able to deploy WCDMA in existing or about to be licensed 1800/1900 cellular bands since new 3G bands have not been licensed yet in this region.

14.3 3G Radio Access Technologies—Introduction

As mentioned before, there are several different radio access technologies defined within ITU, based on either CDMA or TDMA technology. This section will give a technical overview about the CDMA-based technologies, multi-carrier CDMA (cdma2000) and WCDMA. Out of these two radio technologies, WCDMA is the one used in UTRAN, which together with GERAN makes up the UMTS 3G multi-radio network. A more detailed description of these CDMA-based technologies, with specific focus on WCDMA FDD, can be found in [1].

14.3.1 WCDMA Basics

14.3.1.1 Introduction

WCDMA is based on DS-CDMA (direct sequence—code division multiple access) technology in which user-information bits are spread over a wide bandwidth (much larger than the information signal bandwidth) by multiplying the user data with the spreading code. The chip (symbol rate) rate of the spreading sequence is 3.84 Mcps, which, in the WCDMA system deployment is used together with the 5-MHz carrier spacing. The processing gain term refers to the relationship between the signal bandwidth and the information bandwidth as illustrated in Figure 14.3. Thus, the name wideband is derived to differentiate it from the 2G CDMA (IS-95), which has a chip rate of 1.2288 Mcps. In a CDMA system, all users are active at the same time on the same frequency and are separated from each other with the use of user specific spreading codes.

The wide carrier bandwidth of WCDMA allows supporting high user-data rates and also has certain performance benefits, such as increased multipath diversity. The actual carrier spacing to be used by the operator may vary on a 200-kHz grid between approximately 4.4 and 5 MHz, depending on spectrum arrangement and the interference situation.

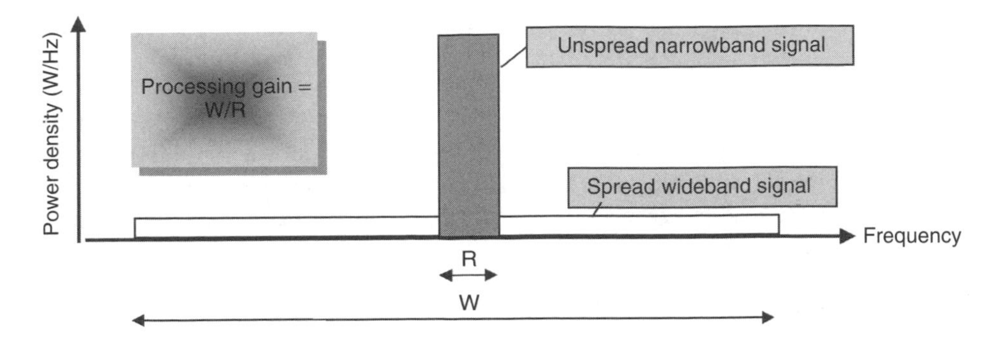

Figure 14.3. Concept of processing gain in CDMA

WCDMA supports highly variable user-data rates, with the concept of obtaining bandwidth on demand (BoD) very well supported. Each user is allocated frames of 10 ms duration, during which the user-data rate is kept constant. However, the data rate among the users can change from frame to frame. This fast radio capacity allocation (or the limits for variation in the uplink) is controlled and coordinated by the radio resource management (RRM) functions in the network to achieve optimum throughput for packet data services and to ensure sufficient quality of service (QoS) for circuit-switched users.

WCDMA supports two basic modes of operation: FDD and TDD, as illustrated in Figure 14.4. In the FDD mode, separate 5-MHz carrier frequencies with duplex spacing are used for the uplink and downlink, respectively, whereas in TDD only one 5-MHz carrier is time shared between the uplink and the downlink. The TDD mode is based heavily on FDD mode concepts and was added in order to leverage the basic WCDMA system also for the unpaired spectrum allocations of the ITU for the IMT-2000 systems. The TDD mode is not expected to be used in the first phase of WCDMA network deployment. The TDD mode is described in more detail in [1] and is not considered further in this book.

WCDMA operates with asynchronous base stations, without the need for chip level timing source. From the signal detection point of view WCDMA uses coherent detection based on the pilot symbols and/or common pilot. WCDMA allows many performance-enhancement methods to be used, such as transmit diversity or advanced CDMA receiver

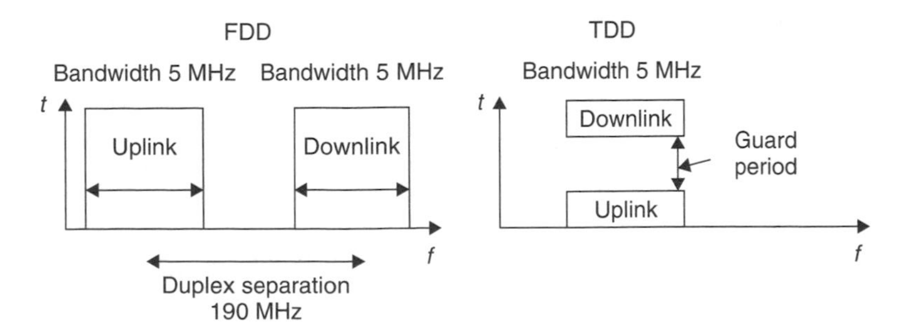

Figure 14.4. FDD and TDD operation principles

concepts. Adaptive antennas can also be efficiently deployed even though its full support with the performance requirements for the terminals and measurements to support RRM is not included in Rel'99 and will be part of the later WCDMA 3GPP standard releases.

As WCDMA is going to be mostly deployed in the current GSM footprint, the support for handovers (HO) between GSM and WCDMA is part of the first standard version. This means that all multi-mode WCDMA/GSM terminals will support measurements from the one system while camped on the other one. This allows networks using both WCDMA and GSM to balance the load between the networks and base the HO on actual measurements from the terminals for different radio conditions in addition to other criteria available. All WCDMA terminals, with the exception of those targeting the Japanese market, are expected to be multi-mode, supporting both WCDMA and GSM.

Table 14.2 summarises the main WCDMA parameters.

14.3.1.2 WCDMA Radio Access Network Architecture

The WCDMA technology has shaped the WCDMA radio access network architecture due to soft handover requirements in any cellular CDMA system with reuse 1. Without support for soft handover, one or more cells would be suffering from the near–far problem when terminals operate near the cell border. In Figure 14.5, the basic WCDMA/UTRAN architecture is shown indicating the connections to the core network for both circuit-switched traffic (Iu-cs) and packet-switched traffic (Iu-ps). The existence of the open, standardised Iur interface between radio network controllers (RNCs) is essential for proper network operation including soft handover support in a multi-vendor environment.

One focus area in the architecture standardisation work in the coming releases is the multi-radio operation. From a service point of view there is going to be a single entity from a core network to provide service to the GSM/WCDMA dual-mode terminal either via GSM or WCDMA radio access. Therefore, it is obvious that optimising the GSM and WCDMA radio access networks (GERAN and UTRAN) jointly is the key for

Table 14.2. Main WCDMA parameters

Multiple access method	DS-CDMA
Duplexing method	Frequency division duplex/time division duplex
Base station synchronisation	Asynchronous operation
Chip rate	3.84 Mcps
Frame length	10 ms
Service multiplexing	Multiple services with different quality of service requirements multiplexed on one connection
Multi-rate concept	Variable spreading factor and multicode
Detection	Coherent using pilot symbols or common pilot
Multi-user detection, smart antennas	Supported by the standard, optional in the implementation

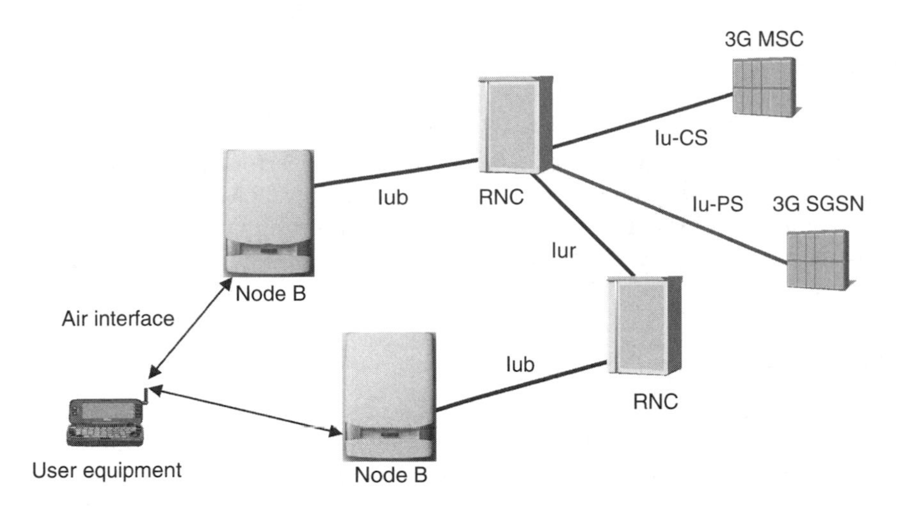

Figure 14.5. The WCDMA radio access network architecture

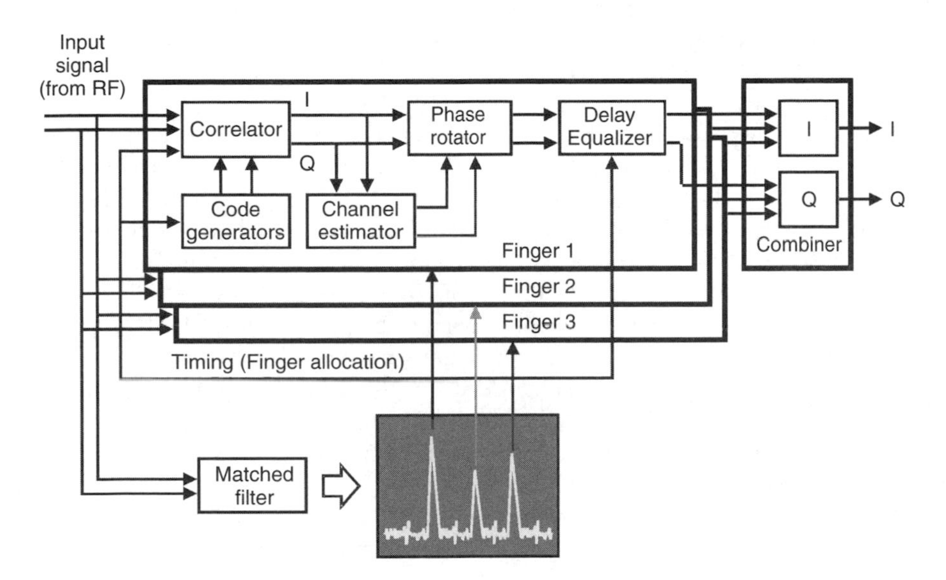

Figure 14.6. Example WCDMA receiver structure

cost efficient multi-radio network operation. The multi-radio architecture concept will be further explored in the Section 14.5 of this chapter.

14.3.1.3 WCDMA Receiver Structure

In WCDMA standardisation, the receiver type to be used is not specified. As in the GSM standards, only minimum performance requirements in particular test cases are specified. In the case of CDMA, the basic receiver solution, known for over 50 years, is the RAKE receiver, which is illustrated in Figure 14.6. The receiver consists of multiple fingers, each one of them being able to receive a multipath component or a different downlink transmission in case soft handover is used. Typically, a RAKE receiver has a searcher, implemented with matched filter, which tracks the delay components from one or more base stations and then allocates the RAKE fingers to different propagation delays. The signal is then combined coherently together for maximum performance. In WCDMA, the common pilot and pilot symbols allow coherent combining. RAKE receivers have been already used in 2G in IS-95 terminals. The first version of IS-95 uplink was non-coherent and thus the base-station receiver principle was slightly different. The later versions of the IS-95 standard also used coherent uplink.

14.3.1.4 Soft Handover + Fast TPC

In WCDMA, since all users share the same time and frequency space, the separation is done in the code domain. The processing gain indicates how much the despreading process can suppress interference present in the whole of the carrier. As the processing gain is limited and reduces further when the data rate increases, the power control is the essential element for a CDMA system operation, especially in the uplink direction. As the pathloss between a terminal and base station varies a lot, it was necessary to use a large dynamic range for the uplink power-control operation. In WCDMA, the typical power class is 21 dBm with minimum requirement for lowest power level of −50 dBm. This results in a 71-dB power control range in the uplink.

The practical power-control operation consists of a simple up/down command sent by the base station and terminal at the rate of 1500 Hz. This causes the transmitter in the normal case to increase or decrease the TX power by 1 dB. This principle is illustrated in Figure 14.7, which shows the uplink power-control operation. Fast power control is used in the WCDMA downlink as well as performance improvements, though in the downlink the near–far problem is not that critical like it is in the uplink.

The operation in soft handover is another important feature for all CDMA systems, including WCDMA (Figure 14.8). All base stations have their own power-control procedure, which ensures that none of them suffers from the near–far problem. In the uplink direction, all base stations will decode the same data and the RNC will then select (on the basis of the error indication from the base station) which frames to forward to the core network. Soft handover support is the key function of the Iur interface, as otherwise the near–far problem would emerge in the border of RNC areas. As WCDMA is designed to operate in an asynchronous fashion, the timing information from the measurements done

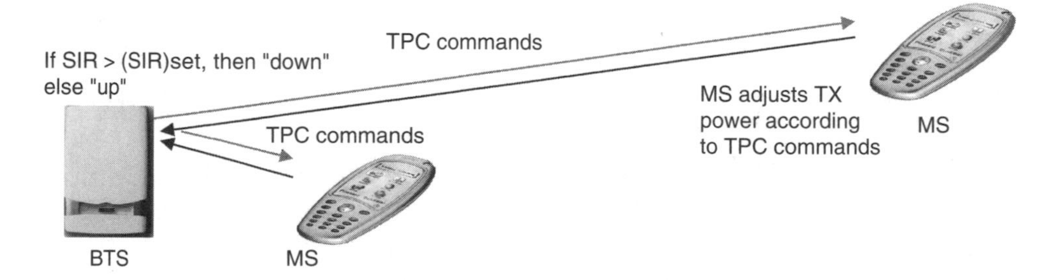

Figure 14.7. Fast closed-loop power-control principle

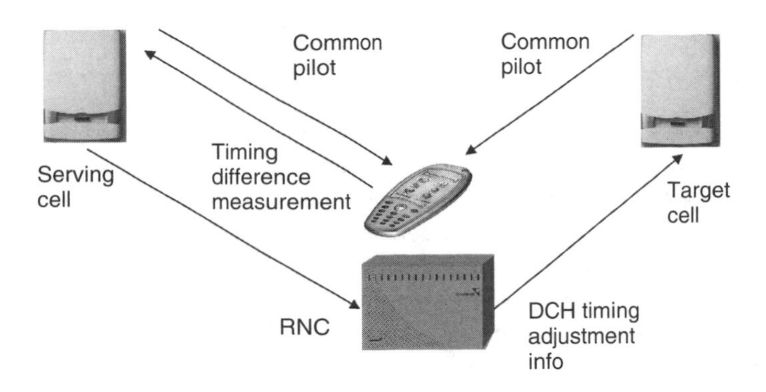

Figure 14.8. Asynchronous WCDMA soft handover operation

by the terminal is provided to the target cells. This allows the target cell to adjust the timing of a new radio link and enables the terminal to combine the links in the RAKE receiver without excessive time difference between the links. In addition to solving the near–far problem, the use of soft handover also provides diversity benefits, especially in the uplink where there is no additional transmission but the same signal is decoded by multiple base stations. In the downlink direction, each new transmission from additional cells also adds interference and does not always have a positive impact on the capacity.

14.3.1.5 WCDMA Packet Data Operation

For packet data transmission in the WCDMA downlink, several alternatives exist, especially in the downlink. There are basically three different channels in Rel'99/Rel'4 WCDMA specifications, completed December 1999 and March 2001, respectively, which can be used for downlink packet data:

- The dedicated channel (DCH) can be used basically for any type of service, and has a fixed spreading factor (SF) in the downlink. Thus, it reserves capacity according to the peak rate for the connection. For example, with an adaptive multi-rate codec (AMR) speech service and packet data, the DCH capacity reserved is equal to the sum of the highest rate used for the AMR speech and the highest rate allowed to be sent simultaneously with full-rate AMR. This can be even used up to 2 Mbps but obviously reserving the code tree for a very high peak rate with low actual duty cycle is not very efficient use of code resources. The DCH is power controlled and may be operated in macro-diversity as well. In the uplink, the DCH SF is variable for the data part while the control channel part has a fixed SF.
- The downlink shared channel (DSCH) has been developed to always operate together with a DCH. This allows defining the channel properties to be best suited for packet data while leaving the data with a tight delay budget to be carried by the DCH. The DSCH in contrast with DCH (or forward access channel (FACH)) has dynamically varying SF informed to the terminal on a frame-by-frame basis with physical layer signalling carried on the DCH. This allows dynamic multiplexing of several users to share the DSCH code resource thus optimising the orthogonal code resource usage in

the downlink. The DSCH is power controlled relative to the DCH power level, and the DSCH is not operated in soft handover. However, the associated DCH can be operated in soft handover.

- The forward access channel (FACH) can be used for downlink packet data as well. The FACH is operated on its own, and it is sent with a fixed SF and typically with a rather high power level to reach all users in the cell. There is no fast power control or soft handover for FACH. The uplink counterpart for FACH is random access channel (RACH), which is intended for short duration 10/20 ms packet data transmission in the uplink. There exists also another uplink option, named common packet channel (CPCH), but that is not foreseen to be part of the first phase WCDMA network deployment.

From the terminal capability point of view, all terminals obviously support DCH and FACH/RACH, but the maximum data rates supported vary. The DSCH support is a terminal optional capability, and the reference classes have optional DSCH of up to a 384-kbps class in Rel'99. In Rel'4, DSCH is part of a 384-kbps reference class. CPCH support is not part of any reference classes.

14.3.1.6 Transmit Diversity

Rel'99 of WCDMA contains a powerful performance enhancement feature, transmit diversity, which is applicable for all downlink transmission. There are two methods included, the open-loop transmit diversity coding, as shown in Figure 14.9, in which the information is space–time coded to be sent from two antennas, and the FB mode transmit diversity, in which the signal is sent from two antennas on the basis of the FB information from the terminal. The FB mode uses phase, and in some cases also amplitude, offsets between the antennas. The use of these two antennas is based on the FB from the terminal, transmitted in the FB bits in the uplink DCH.

14.3.1.7 Service Multiplexing

The simplified WCDMA service multiplexing principle is illustrated in Figure 14.10. In the actual multiplexing chain, there are additional steps such as rate matching or interleaving functions that are necessary as well for proper system operation. A single physical connection (channel) can support more than one service, even if the services have

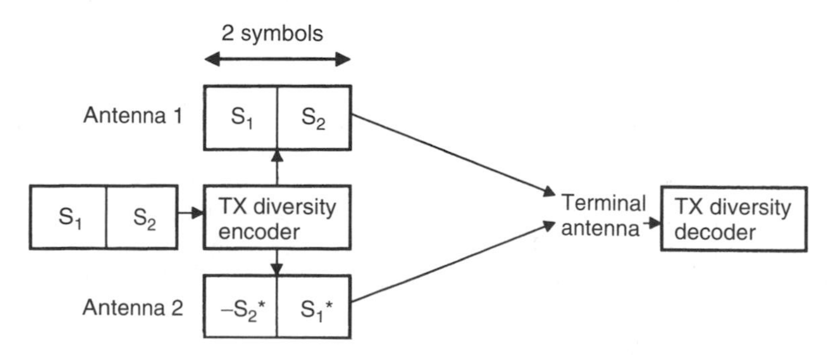

Figure 14.9. Open-loop transmit diversity encoding

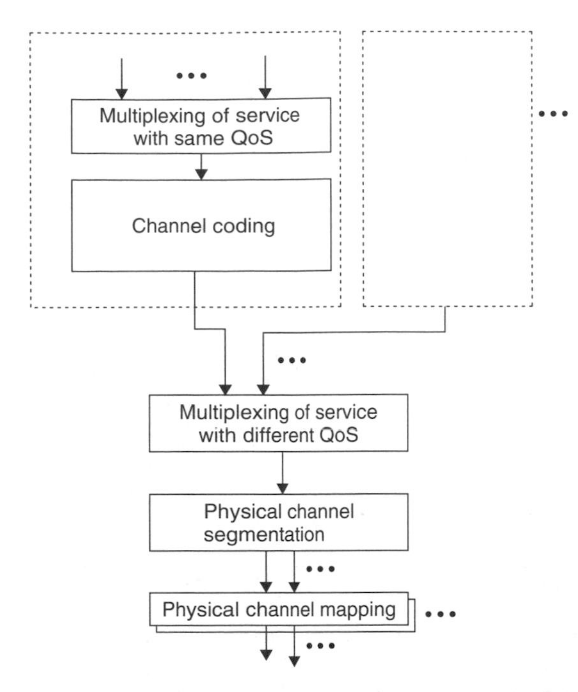

Figure 14.10. WCDMA service multiplexing principle

different quality requirements. This avoids, for example, multicode transmission in case there are two services running simultaneously as they can be dynamically multiplexed on a single physical resource.

14.3.1.8 WCDMA Downlink Packet Data Evolution with HSDPA

In WCDMA Rel'5 specifications, the biggest improvement is the high-speed downlink packet access (HSDPA) concept. In brief, the feature introduces base station (or Node B in 3GPP terminology) based scheduling for the downlink packet-data operation and uses similar methods like GSM evolution such as higher order modulations (e.g. 16QAM in WCDMA) and hybrid automatic repeat request (ARQ) (incremental redundancy and chase combining) to improve packet data performance. In 3GPP, performance improvements up to 100% have been reported compared with Rel'99 architecture packet data performance. The HSDPA principle with Node B based scheduling is illustrated in Figure 14.11. The HSDPA capable Node B will take care of the scheduling functionality and retransmission control, which in a Rel'99/Rel'4 based network is handled solely by the RNC. Terminal or user equipment (UE) in 3GPP terminology, will send positive or negative acknowledgements to Node B, which initiates retransmission at a suitable moment taking into account the channel condition and data in the buffer for different terminals.

14.3.2 Multi-carrier CDMA (cdma2000) Fundamentals

14.3.2.1 Introduction

Multi-carrier CDMA (ITU-R term) or cdma2000 (3GPP2 standards term) is designed to be deployed together with the existing IS-95 (cdmaOne) systems in the same frequency

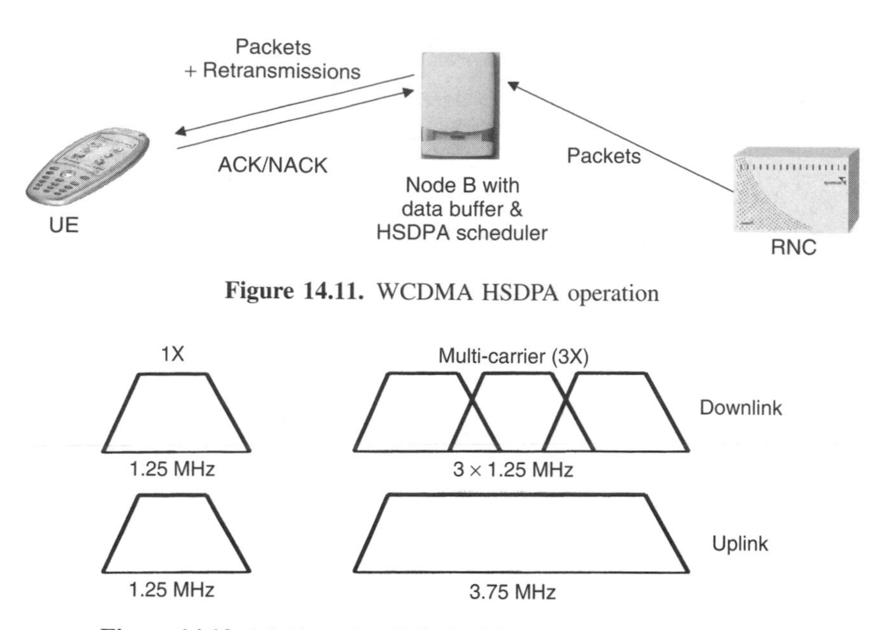

Figure 14.11. WCDMA HSDPA operation

Figure 14.12. Multi-carrier CDMA (3X) and 1X cdma2000 modes

band. With the multi-carrier mode of cdma2000, the downlink transmission consists of multiple (up to three in the IMT-2000 version) parallel narrowband CDMA carriers. The bandwidth in each narrowband carrier is equal to the bandwidth of a single IS-95 carrier.

The multi-carrier mode of cdma2000 is often denoted as 3X to differentiate it from its other mode, the evolved narrowband version of IS-95, 1X. The 3X uplink chip rate is three times the 1X chip rate while in the downlink three parallel 1.2288 Mcps carriers are used. This is illustrated in Figure 14.12.

When compared with WCDMA, the uplink principles are quite similar while the downlink is totally different because of its multi-carrier nature. It seems however that 1X is the focus for practical deployments and thus work on cdma2000 standardisation has also been lately focusing on 1X.

When comparing the WCDMA with 1X there are a number of key differences starting from the synchronous network operation of IS-95 based networks. 1X introduces certain improvements over IS-95 such as downlink fast power control and optional terminal support for downlink transmit diversity. Some of these key differences are summarised in Table 14.3. From the service provisioning point of view, the WCDMA has a more flexible service multiplexing capability, although cdma2000 should be able to support an equivalent set of services.

14.3.2.2 Packet Data Operation with cdma2000

The basic concept of cdma2000 packet data operation with data rates above 14.4 kbps relies on the concept of using a fundamental and supplemental channel. The counterparts in WCDMA are DCH and DSCH, respectively. In cdma2000, the fundamental channel has a fixed SF (as in WCDMA downlink) but it does not contain physical layer information on a frame-by-frame basis as in WCDMA. Instead, the existence of the supplemental

Table 14.3. Key differences between WCDMA and cdma2000 1X

	WCDMA	1X
Carrier spacing	5 MHz	1.25 MHz
Chip rate	3.84 Mcps	1.2288 Mcps
Power-control frequency	1500 Hz, both uplink and downlink	Uplink: 800 Hz, downlink: −800 Hz
Base-station synchronisation	Not needed	Yes, typically obtained via GPS
Inter-frequency handovers	Yes, measurements with slotted mode	Possible, but measurement method not specified
Efficient radio resource management algorithms	Yes, provides required quality of service	Not needed for speech-only networks
Packet data	Load-based packet scheduling	Packet data transmitted as short circuit switched calls
Downlink transmit diversity	Mandatory support of closed and open loop methods	Optional support of open loop method

channel carrying the packet data is informed by higher layer signalling on the supplemental channel. Thus, the operation is similar from the conceptual level but the WCDMA resource-sharing process is more dynamic on a 10 ms frame basis.

14.3.2.3 cdma2000 Network Architecture and Operation

From the radio access network architecture point of view, there is no specification adopted for the radio access network architecture. Depending on the vendor, the level of functionality in the base station will vary and in some implementations the base station can be connected even directly to the mobile switching centre (MSC) while in some other cases there exists a RNC type of entity. There are no open interfaces between different network elements as in WCDMA.

The network design is also based on the use of synchronised base stations. Terminals have been designed accordingly to search for the different base stations with identical timing but different code phase. The timing element needs to be accounted for in the network planning process.

14.3.2.4 cdma2000 Downlink Packet Data Evolution

The evolution of cdma2000 packet data operation is using similar methods as those used in EDGE or WCDMA evolution, such as higher order (adaptive) modulation methods or the combining of retransmissions with hybrid ARQ (HARQ) and faster scheduling based on the FB from the terminal. In the first phase, 1X-EV-DO (DO = data only), already deployed in Korea during 2002, requires packet data to be operated on a separate carrier but the next version, 1X-EV-DV (DV = data and voice), is expected to allow packet data to be combined with speech on the same carrier like in EDGE or WCDMA HSDPA. The target peak data rate of 1X-EV-DV is around 3 Mbps, whereas in WCDMA HSDPA the peak rate targeted is around 10 Mbps.

14.4 3G Radio Access Technology (RAT) Performance Benchmark

This section will benchmark the performance of the relevant 3G radio access technologies, GSM/EDGE and WCDMA as part of the UMTS multi-radio and cdma2000. Different areas will be analysed, such as link-budget performance, technology-data capabilities and spectral efficiency both of voice and data services. The scenario selected for this analysis represents a 10 MHz (10 MHz in downlink + 10 MHz in uplink) deployment in the 1900 band. For GSM/EDGE technology, all the results presented are those resulting from the thorough performance analysis presented in this book. For cdma2000, available information from different sources is used. Finally, for WCDMA, the same methodology and results presented in [1] have been used, adapting the relevant parameters to match the same conditions used in the cdma2000 studies. Future technology evolution has been previously mentioned, although in order to carry out a fair analysis, in this section each technology will be analysed according to the functionality that can be realistically deployed during 2003–2004:

- For GSM/EDGE, AMR, SAIC and dynamic frequency and channel allocation (DFCA) are considered for speech, and EGPRS Rel'99 for data
- WCDMA uses Rel'99 functionality
- cdma2000 uses 1X release 0. Selectable Mode Vocoder (SMV) is included in the speech performance analysis.

14.4.1 Voice Performance

14.4.1.1 Coverage. Link Budget Analysis

The coverage analysis is performed using the link budget calculation. Link budget is defined as the maximum pathloss (dB) between the base transceiver station (BTS) and mobile station (MS). Since the CDMA based technologies are coverage limited in uplink, this is the link analysed. Table 14.4 summarises the uplink link budget for speech services for the three radio technologies deployed using the 1900 band.

Table 14.4. 1900 band speech uplink link budget comparison

	cdma2000	GSM EFR	WCDMA
MS TX power (dBm)	23	30	21
BS sensitivity (dBm)	−124	−110	−124
BS ant diversity gain (dB)	3	4	3
Interference margin (dB)	3	0	3
Fast fading margin (incl. SHO gain) (dB)	4	0	2
BS antenna gain (dBi)	18	18	18
Cable loss incl. MHA gain	0	0	0
Body loss for speech terminal (dB)	3	3	3
MS antenna gain (dB)	0	0	0
Maximum pathloss (dB)	158	159	158

This analysis shows there are no significant differences between the different technologies, with all of them having an approximate maximum pathloss of 158 dB for speech Services.[1]

14.4.1.2 Spectral Efficiency Analysis

This section will analyse the spectral efficiency of the different technologies. As mentioned before, only functionality realistically available in 2003–2004 has been considered in this analysis and a 10-MHz band (10-MHz DL + 10 MHz UL) in the 1900 band has been selected as a deployment scenario. The results are presented in terms of users (voice paths) per sector since this is the common methodology used for the CDMA based technologies. In order to transform these numbers to Erlangs of carried traffic, Erlang B tables should be used.

In Chapters 5, 6 and 9 the baseline performance of GSM and the gains associated with the deployment of AMR, SAIC and DFCA were presented. In Chapter 5, the achievable GSM baseline performance of the hopping layer was defined to be 12% effective frequency load (EFL). In Chapter 6, the hopping layer gain with the introduction of AMR was quantified to be 150%, the Source Adaptation additional gain was estimated to be 25% and finally, the SAIC gain was expected to be at least 40%. Finally, an additional 40% hopping layer gain from DFCA was described in Chapter 9. With this set of functionality, the total GSM capacity reaches 29 users/MHz/sector.

The WCDMA capacity has been calculated using the methodology described in [1]. The following equation is utilised to compute the capacity:

$$\eta_{\mathrm{DL}} = \sum_{k=1}^{K} \frac{(E_{\mathrm{b}}/N_0)_k}{W/R_k} \times [(1-\alpha_k) + i_k] \times v_k$$

where

AMR speech codcc	$R_k = 7.95$ and 12.2 kbps	4.75–12.2 kbps supported in AMR
Effective voice activity	$v_k = 58\%$ (12.2 kbps)	50% activity + DPCCH overhead SF = 128 and slot format 8
E_{b}/N_0 12.2 kbps	$E_{\mathrm{b}}/N_0 = 7.0$ dB (12.2 kbps)	Typical macro cell based on 3GPP Performance requirements [2]
Orthogonality	$\alpha = 0.5$	Typical macro cell
Other cell interference	$i = 0.65$	Typical three-sector macro cell
Load factor	$\eta_{\mathrm{DL}} = 80\%$	80% of pole capacity
Common channels	15%	15% of maximum BTS power
Number of users	k	

[1] GSM pathloss calculation has considered Enhanced Full-Rate (EFR) codec and hence it has not taken into account the benefits associated from the AMR extended link budget (Chapter 11) or Transmit Diversity techniques (Chapter 11).

In the case of 7.95-kbps speech codec and assuming an Eb/No requirement of 7.5 dB, the capacity reaches 88 users per WCDMA carrier. Transmit diversity, which is a standard functionality supported by all terminals, provides an additional 0.5 dB gain, and finally the introduction of Source Adaptation generates an estimated gain of 20%, which brings the total capacity to 23.7 users/MHz/sector.

The cdma2000 1X capacity has been calculated from the available market sources. According to the CDMA Development Group (CDG) [3] the maximum capacity per cdma2000 1X carrier is 33 users. This figure is calculated over the one-path radio channel propagation environment. The equivalent number for the Vehicular B channel profile, which has been selected for all the presented studies since it reflects much better a realistic macrocellular propagation environment, is 24 users per carrier. According to [4] the introduction of SMV will provide an estimated 27% higher capacity without compromising the speech quality. Since seven carriers can be deployed in 10 MHz, the capacity reaches 21.33 users/MHz/sector.

WCDMA, even though based on the same CDMA principles, outperforms cdma2000 1X due to its wideband nature. WCDMA has a higher multipath diversity gain thus reducing the required fast fading margin, a higher effective power due to the reduced BTS power variance associated with the higher number of users per carrier and an important trunking gain in the common channels since there are less carriers for the same bandwidth, which leads to less power required to broadcast the common control channels.

Figure 14.13 displays the speech spectral efficiency results in terms of total number of users per sector. WCDMA and cdma2000 1X have an equivalent performance, while GSM/EDGE clearly outperforms the cdma-based technologies in terms of voice spectral efficiency mostly because of the introduction of advanced downlink interference cancellation techniques (SAIC).

Another aspect to consider is the hardware configuration required to support the maximum spectral efficiency associated with each technology. This can impact the cost of the Capital Expenditure (CAPEX) required to deploy different capacity configurations.

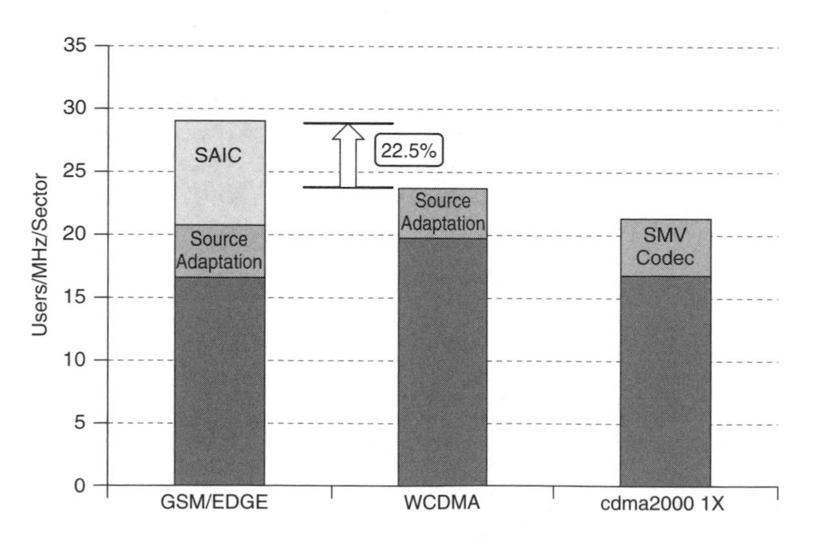

Figure 14.13. Voice spectral efficiency. 10-MHz case in 1900 band

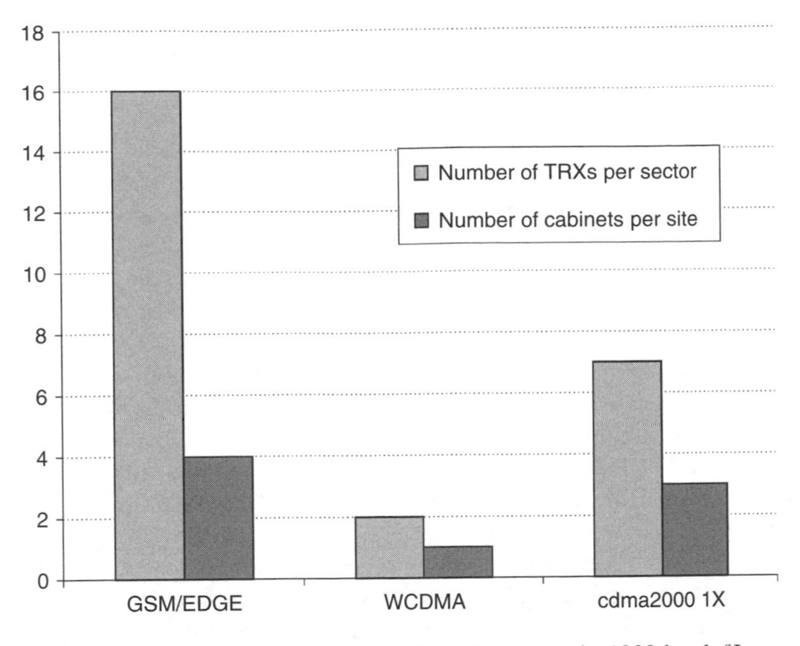

Figure 14.14. Cabinet and TRX configurations[a]. 10-MHz case in 1900 band. [a]It assumes 12 TRXs GSM/EDGE cabinets, 9 TRXs cdma2000 cabinets and 6 TRXs WCDMA cabinets

Figure 14.14 displays the transceiver (TRX) per sector and cabinet per site configurations needed to support a very high capacity configuration (20 users/MHz/sector). The most important element is the cabinet configuration per site. WCDMA is clearly the most efficient technology for very high capacity configurations requiring a single cabinet per site to support its maximum traffic capacity. On the other hand, and because of the lower cost of GSM/EDGE cabinet and TRX units and its granular scalability, this technology will be the most cost efficient for relatively low capacity configurations.

14.4.2 Data Performance

14.4.2.1 Data Capabilities

This section gives an overview of the different areas regarding data capabilities, such as theoretical, realistic and cell edge throughputs, spectral efficiency and QoS capabilities. All the data presented will refer to downlink capabilities since this is the limiting link in the case of data services.

Each technology has a theoretical peak throughput that can be achieved under certain conditions. However, these peak throughputs will probably not be experienced in the field due to the demanding conditions required, the maximum level of interference that can be generated in the systems, or simply due to terminal limitations. In CDMA based technologies, i.e. WCDMA and cdma2000, the maximum throughput is going to be limited by the maximum interference that can be maintained across the network, which can be potentially exceeded by a single user. In EDGE, the peak throughputs are going to be limited by the terminal timeslot configurations supported, which, during 2003, are probably going to be limited to Type 1 configurations. This is due to the cost associated with higher configurations, which would require duplication of the radio architecture of the terminals since they would need to transmit and receive simultaneously. The supported multislot configurations are presented in Chapter 1. The maximum multislot configurations (DL + UL) associated in Rel'99 to Type 1 terminals are 3 + 2, 4 + 1 and 6 + 0. These configurations can be dynamically supported according to the connection requirements. For interactive services such as web browsing, 4 + 1 is the most adequate configuration, providing in downlink a maximum of 236.8-kbps (4 TSL*59.2 kbps/TSL) throughput.

In terms of average throughputs expected in the cell, these are going to be fully dependant on the network load. Table 14.5 displays typical average throughputs for medium loaded networks. The difference between WCDMA and cdma2000 1X throughput values is mostly due to the higher chip rate of WCDMA. These results are in line with the simulation results presented in Figure 14.17 and several cellular operators publicly stated expectations and measured performance.[2]

The data services link budget is conditioned by selected data rates and, in the case of GSM/EDGE, by the MS timeslot configuration capability. A simple approach to this complex problem is to determine what is the available throughput in the cell edge, defining cell edge as the maximum pathloss from the speech link budget calculation. Assuming a typical maximum pathloss at cell edge of 155 dB, and considering the lack of body loss in the data link budget calculation, the selected pathloss has been 152 dB. According to the results presented in Chapter 7, EGPRS 4 + 1 terminals support 128 kbps at the cell edge. WCDMA supports 384 kbps assuming loaded surrounding cells, while cdma2000 1X would support 128 kbps under the same conditions. The CDMA technologies, WCDMA and cdma2000 1X, are capacity limited, that is, in order to support these throughputs at the cell edge a single connection would use all the sector capacity.

Table 14.5 illustrates the maximum throughputs defined in the standards, the typical average throughputs and the achievable throughputs at Cell edge.

14.4.2.2 Best-effort Data Spectral Efficiency

This section presents the capabilities of different technologies in terms of spectral efficiency for data only deployments. The analysis is carried out for packet-switched best-effort data with a simple round-robin scheduling algorithm, which will deliver different data rates to different connections according to their signal-to-interference ratio. The analysis is based on existing simulation results and does not consider application layer throughput impacts.

Table 14.5. Maximum, average and cell edge throughputs

	EGPRS	WCDMA	cdma2000 1X
Absolute maximum (standard)	473.6 kbps	2 Mbps	307 kbps
Average throughput	128 kbps	240 kbps	80 kbps
Cell edge throughput	128 kbps	384 kbps	128 kbps

[2] cdma2000 1X measured/expected average throughput: SKT (Korea) and Bell Mobility (Canada) 60–80 kbps, Sprint (US) 60–70 kbps, Verizon (US) 40–60 kbps. WCDMA measured average throughput: NTT DoCoMo (Japan) 270 kbps.

As presented in Chapter 7, in a loaded network, the TSL capacity of 1/3 reuse converges to 26.25 kbps, which is equivalent to a spectral efficiency of 350 kbps/MHz/sector. The evolution of spectral efficiency with load is illustrated in Figure 14.17. For the BCCH TRX, in the case of reuse 12, the TSL capacity reaches 47 kbps, which is equivalent to 137 kbps/MHz/sector. In a 10 MHz deployment, the total throughput per sector would be 2989 kbps.

According to the methodology presented [1], and using the latest 3GPP performance requirements, the spectral efficiency of WCDMA reaches 161 kbps/MHz for guaranteed throughput power scheduling. The following conditions have been considered for evaluation of the simulations:

E_b/N_0	$E_b/N_0 = 4.5$ dB	Typical macrocell
Orthogonality	$\alpha = 0.5$	Typical macrocell
Other cell interference	$i = 0.65$	Typical three-sector macrocell
Load factor	$\eta_{DL} = 80\%$	80% of the pole capacity
Common channels	15%	20% of max. BTS power

The use of the round robin best-effort data rates scheduling[3] introduces a gain of 40% in the spectral efficiency, increasing its value to 226 kbps per MHz. In a 10-MHz deployment, the total throughput per sector would reach 2260 kbps.

According to [5], for equivalent conditions, cdma2000 1X reaches a capacity of 284.75 kbps per carrier for omnidirectional antenna sites. In sectorised antenna site scenarios, equivalent simulations provide a higher capacity that reaches 230.4 kbps per carrier. As with WCDMA, an equivalent 40% increase is applied for best-effort data scheduling that brings the capacity up to 322.5 kbps per carrier. In the selected 10-MHz scenario, seven carriers can be deployed with a maximum throughput of 2258 kbps per sector.

Figure 14.15 summarises the spectral efficiency results for all technologies. It assumes data only configurations. WCDMA and cdma2000 1X have an equivalent performance. EGPRS is clearly the best performing technology for best-effort data, with a 35% higher capacity than WCDMA or cdma2000 1X. The main reason for this is that EGPRS includes AMCS as well as type II HARQ retransmissions combining. These advanced data functionalities will be only deployed for WCDMA and cdma2000 1X when their data technology evolution (HSDPA and 1X-EV-DV, respectively), are available in commercial products (realistically not earlier than 2005), providing an estimated capacity increase of 75 to 100% [6]. An advantage of HSDPA and 1X-EV-DV over EGPRS is the existence of a link performance fast FB mechanism, which enables fast scheduling algorithms and *multi-user diversity* gains.

1X-EV-DO was first deployed during the year 2002 in Korea. Even though it provides high peak data rates and competitive spectral efficiency, its main disadvantage is that it cannot be mixed with voice in the same carrier so the system trunking efficiency will be

[3] This can be accomplished by either assigning same power to all data users or by using the Dynamic Shared Channel (DSCH).

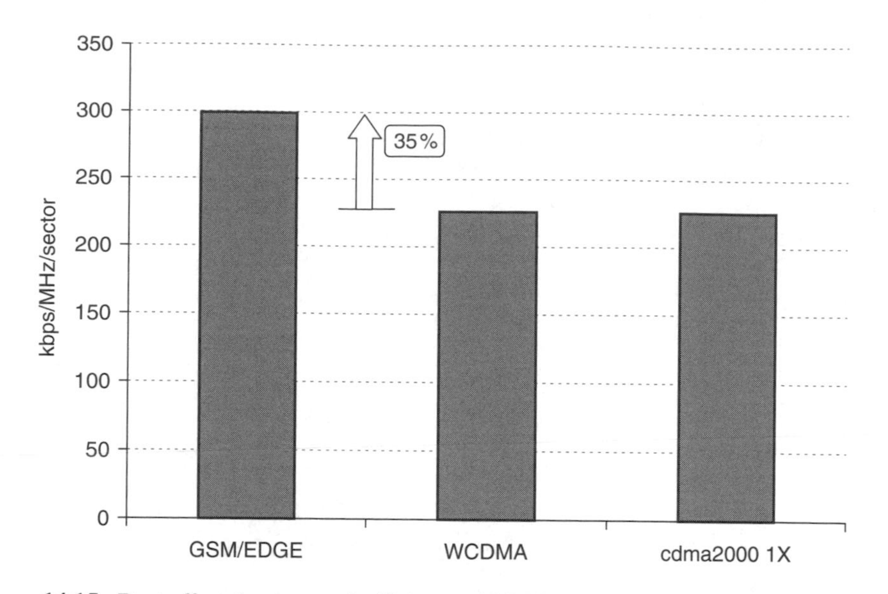

Figure 14.15. Best-effort data spectral efficiency. 10-MHz case (10-MHz DL + 10-MHz UL)

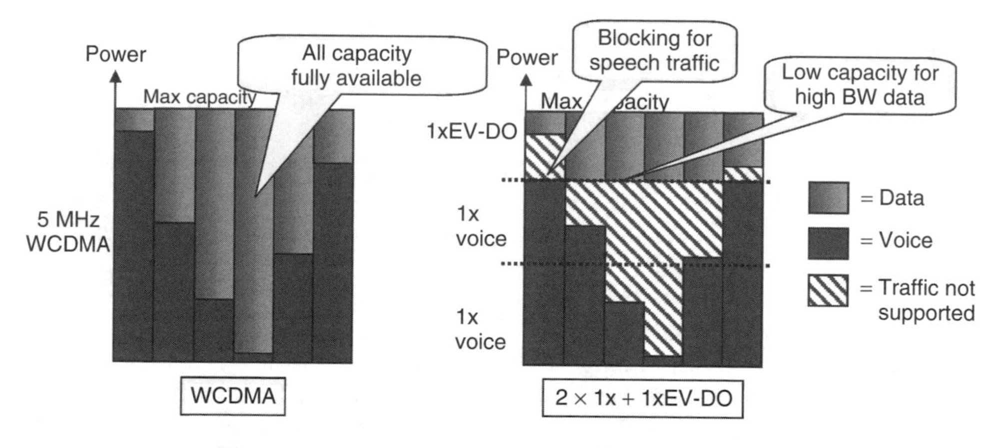

Figure 14.16. 1X-EV-DO trunking efficiency limitation

significantly affected, as displayed in Figure 14.16. Additionally, its introduction requires one full 1.25-MHz carrier to be deployed to support 1X-EV-DO even though initially there will likely be a marginal number of 1X-EV-DO capable terminals. The spectral efficiency for 1X-EV-DO is presented in [7] under ideal conditions. According to [8], with the introduction of realistic interference conditions, the spectral efficiency of 1X-EV-DO reaches 337 kbps per carrier. In the selected scenario this yields a total throughput per sector of 2359 kbps, which makes it slightly more spectral efficient than cdma2000 1X. The main benefit from 1X-EV-DO over cdma2000 1X is the capability to deliver higher throughputs, with peak throughputs potentially as high as 2.4 Mbps.

14.4.2.3 QoS Capabilities and Impact in Performance

As presented in Chapter 3, Rel'99 defines the UMTS traffic classes and basic QoS architecture, which allows for differentiated QoS requirements for various service and user profiles. The GSM/EDGE and WCDMA radio access networks, GERAN and UTRAN, may utilise the available RRM mechanisms (link adaptation, power control, channel allocation and scheduling) in order to ensure that such differentiated requirements are fulfilled. On the other hand, cdma2000 1X release 0 and 1X-EV-DO do not support QoS differentiation for packet-switched data and are effectively best-effort only data technologies. The cdma2000 1X support for an equivalent QoS architecture will come with release A, unlikely to be deployed before 2002.

The spectral efficiency of the radio network is going to be conditioned by the required QoS to be provided to the end users in terms of user throughputs. Figure 14.15 displayed the best-effort Spectral Efficiency for the different technologies under analysis. However, the best-effort figures will decrease as the user-throughput requirements get higher, as displayed in Figure 14.17. This figure display the EGPRS simulation results presented in Chapter 7 and Matlab-based static simulations for CDMA based technologies. The spectral efficiency degradation with high throughputs is, however, significantly different for each technology. Such degradation starts at a certain throughput, which is determined by terminal capabilities. GSM/EDGE is the most efficient technology for average data rates below 128 kbps, while WCDMA becomes more efficient for higher data rates. The combination of GSM/EDGE for relatively low data rates and WCDMA for high data rates provides the most efficient data networks performance.

14.4.3 Conclusions

The speech-link budget analysis, which generally characterises the dimensioned network coverage, showed an equivalent performance for the 3G technologies under analysis.

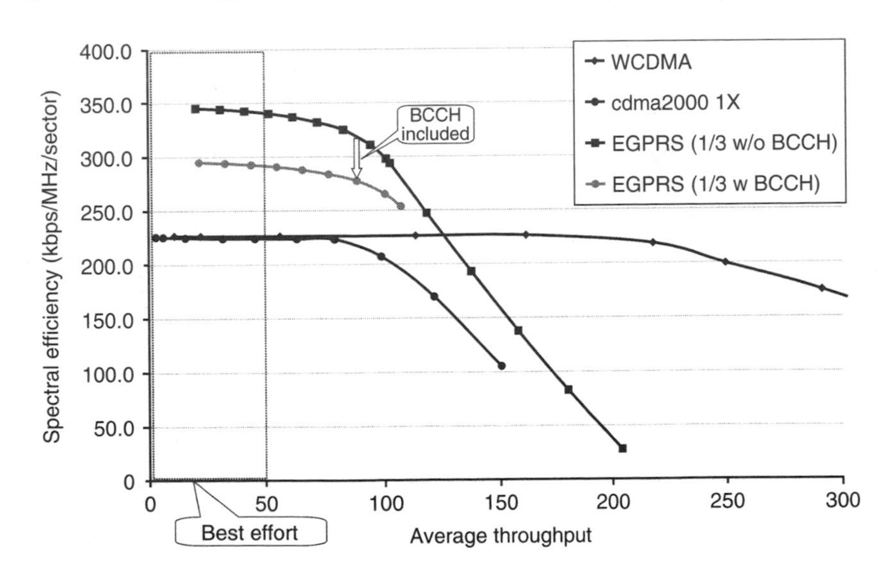

Figure 14.17. Spectral efficiency (load) versus. average throughput. 10 + 10 MHz data only deployment

According to the presented results, in terms of voice spectral efficiency, GSM/EDGE has a 20% higher performance than WCDMA and cdma2000 1X. Similarly, in terms of best-effort data, GSM/EDGE has a 35% higher performance than WCDMA and cdma2000 1X. The spectral efficiency characterisation for different average throughputs show that WCDMA is the most robust technology for throughputs over 128 kbps, due to its wide-band nature. Finally, both GSM/EDGE and WCDMA Rel'99 offer full support of QoS differentiation, whereas cdma2000 1X release 0 and 1X-EV-DO are best-effort only packet-switched data-capable technologies.

3GAmericas endorsed white papers [9, 10] provide a similar analysis as the one delivered in this section reaching equivalent conclusions.

The combination of GSM/EDGE and WCDMA under a seamlessly integrated UMTS multi-radio network should yield the best possible radio performance both for speech and data services. Provided multiple bands are available, GSM/EDGE is the most efficient technology in the already deployed cellular network bands (800, 900, 1800 and 1900) and WCDMA in the new 2100 band (IMT-2000) or other coming bands over 2 GHz (as it is likely the case in the United States). GSM/EDGE can support efficiently speech and narrowband services and WCDMA can offer further capabilities, supporting wide band data services (high quality music or video and high speed internet access) and conversational services (video calls). For operators licensed with multiple bands, a joint operation of GSM/EDGE and WCDMA could be the optimum scenario with a UMTS multi-radio deployment. The next section analyses how these technologies can be efficiently integrated.

14.5 UMTS Multi-radio Integration

14.5.1 Introduction

With the GSM evolution presented in this book, GSM/EDGE effectively becomes a new UMTS RAT since it can get directly connected to the UMTS core network (CN) supporting the same QoS architecture. The UMTS multi-radio networks will make use of two different radio technologies within a single 3G network, GSM/EDGE in existing bands and WCDMA mostly in the IMT-2000 recommended new bands. The advantage of this solution is that these technologies can be seamlessly integrated in order to deliver services through the most efficient radio access network based on QoS management. The service request and QoS negotiation take place between the terminal and the 3G CN, so the radio access network becomes just the means to deliver the requested services to the end user.

In Rel'5, a full integration is achieved with the support of the Iu interfaces that allow GERAN to be connected to the UMTS CN. Following this principle, other radio technologies could be 'plugged' into the UMTS multi-radio network (UTRAN TDD, WLAN ...) as long as they can be connected to the UMTS CNas well.

Figure 14.18 illustrates the current technology evolution towards a seamless broadband network where multiple radio technologies coexist together as part of the same network and are used according to the requirements of the service to be delivered.

14.5.2 UMTS Multi-radio Evolution

In a typical scenario, the integration of GSM/EDGE and WCDMA will take place smoothly. Figure 14.19 shows a typical multi-radio network evolution scenario. The

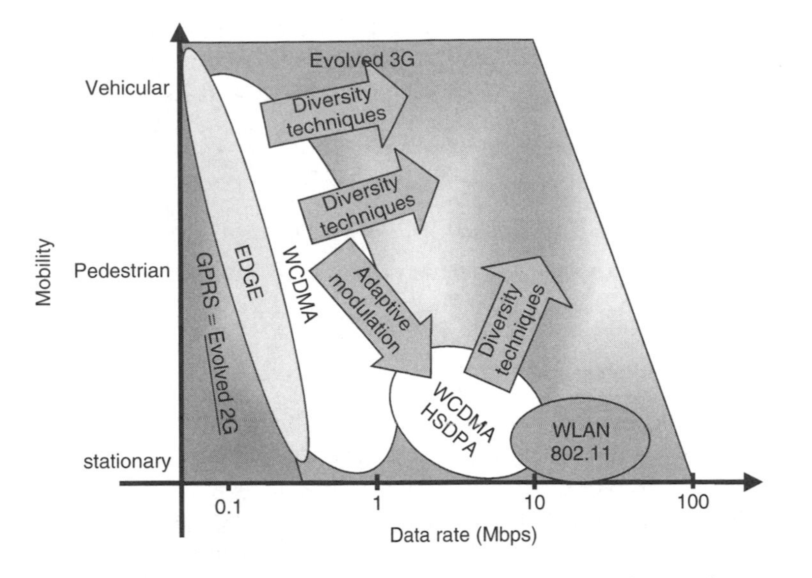

Figure 14.18. Evolution towards a seamless broadband network

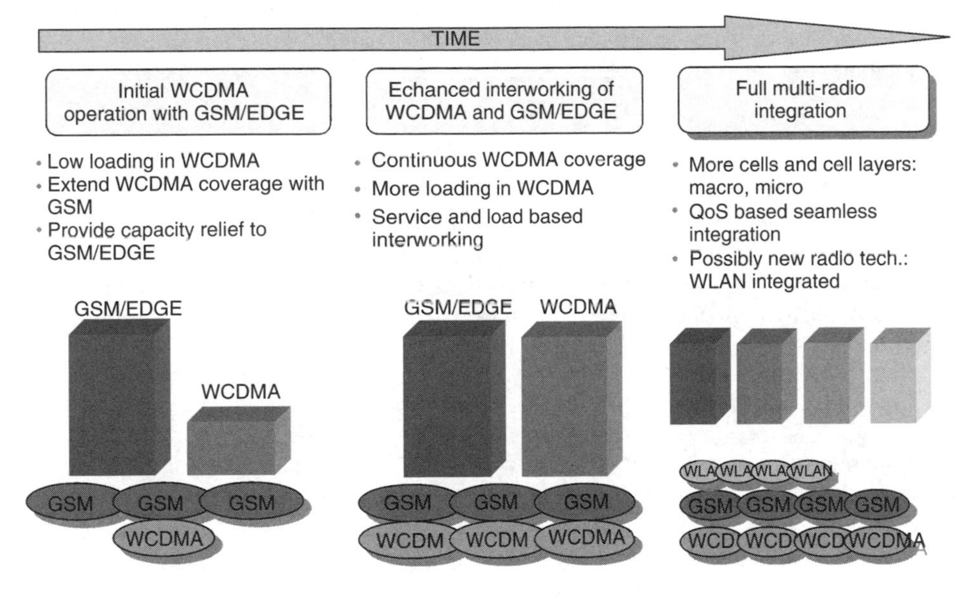

Figure 14.19. Multi-radio evolution

first stages of the multi-system network are characterised by relatively small multi-mode mobile penetration and low initial coverage of the WCDMA network. The level of integration required in this initial scenario is limited to the availability of coverage-based emergency HO.

As the WCDMA coverage grows and the dual-mode mobile penetration increases, the integration of the networks should increase as well providing load and interference balancing between both networks. Load-based HO with traffic differentiation based on service profile, are required at this phase. Mature multi-radio networks with high multi-mode mobile penetration should provide seamless service and RRM integration with the technology selection based on service QoS requirements, radio technology capabilities and resource availability.

14.5.3 Mechanisms for UMTS Multi-radio Integration

This section provides an overview of the most important mechanisms for traffic management between GSM/EDGE and WCDMA.

14.5.3.1 Initial Cell Selection

When the multi-mode terminal is switched on, or following recovery from lack of coverage, the terminal scans the bands for supported radio access technologies (GERAN and UTRAN) and finds the available public land mobile networks (PLMNs). The terminal reads the system information from the strongest cell, in order to find out which PLMN the cell belongs to. It selects the PLMN based on the priority ordered list of PLMN/access technology combinations (GERAN and UTRAN) provided by the subscriber identity module (SIM)/universal subscriber identity module (USIM). In the manual PLMN selection mode, the user may select the desired PLMN. Finally, the terminal initiates registration on this PLMN using the access technology chosen or using the highest priority available access technology for that PLMN.

14.5.3.2 Idle Mode

In general, the traffic management can be divided into idle and active mode traffic management. In idle mode, the terminal is performing the cell selections autonomously based on system information parameters broadcasted on the control channels. Thus, the traffic management flexibility in idle mode is limited as compared with that in active mode in which more information is available. Idle mode parameters can, however, be used to keep idle terminals camped on the preferred technology.

14.5.3.3 Active Mode

Both basic inter-system handover (for circuit switched services) and network controlled re-selection (for packet switched services) between GSM/EDGE and WCDMA systems have been standardised in order to provide the tools for an efficient active mode interworking. Table 14.6 summarises the different possible handovers (HO) and cell reselections (CRS) between different radio access technologies.

14.5.4 Trunking Efficiency Benefits from Multi-radio Integration

One immediate effect of multi-radio integration is the trunking gain resulting from all radio resources being managed as a single pool of resources. Table 14.7 presents the load-sharing gains that could be achieved by the ideal multi-radio resource management of two separate systems, WCDMA and GSM EDGE. In practice, networks will carry a mixture of high and low data rate services, so the overall gain would be a weighted

Table 14.6. Supported integration mechanisms between A/Gb and Iu GERAN modes and UTRAN

Source/target	GERAN A/Iu-cs modes	GERAN Gb/Iu-ps modes	UTRAN CS	UTRAN PS
GERAN A/Iu-cs modes	HO	__	Inter-RAT HO	__
GERAN Gb/Iu-ps modes	__	CRS	__	Inter-RAT CRS
UTRAN CS	Inter-RAT HO	__	HO	__
UTRAN Ps	__	Inter-Rat CRS	__	CRS

Table 14.7. Trunking gains with 2% blocking for 5-MHz WCDMA bandwidth and equivalent GSM/EDGE capacity

	WCDMA/EDGE capacity (Erl)	WCDMA and EDGE separately (Erl)	Common channel pool (Erl)	Trunking gain (%)
Speech	49.7	99.4	107.4	8%
64 kbps	5.1	10.2	13.1	28%
144 kbps	1.6	3.2	5.1	60%

mixture of these gains. Common multi-radio resource management provides more gain at higher bit rates.

The efficient load balance between the available radio technologies will provide an optimum interference distribution as well, which enhances the overall network performance.

14.5.5 QoS-based Multi-radio Integration

The final level of integration comes when the UMTS multi-radio network is able to direct the incoming call or handover to the most appropriate cell and RAT according to its QoS requirements. At the same time, the system should try to maintain the uniform distribution of traffic and interference between different cells, layers and radio systems. The system should evaluate the required QoS and both MS and base station capabilities and then select the most adequate cell/sector and technology.

There can be many information inputs and measurements available from the network to be utilised in the cell/technology selection algorithms. In Table 14.8, a few possible inputs from network elements for the cell selection algorithms are presented.

Table 14.8. Possible inputs for candidate the cell-prioritisation algorithm

Terminal inputs	QoS requirements (RAB parameters), terminal capabilities, measurement report, terminal location
Cell inputs	Cell capability, available maximum capacity, throughput, current UL/DL load for RT, delay for NRT services, interference levels, statistics collected from the idle/active terminals
O&M inputs	Operator preferences, cell capabilities, parameters and settings for algorithms, limits/thresholds, statistics

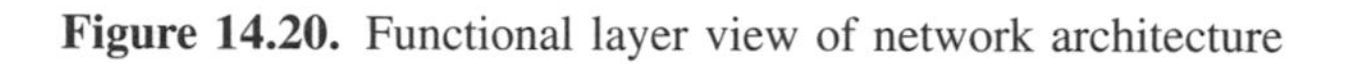

Figure 14.20. Functional layer view of network architecture

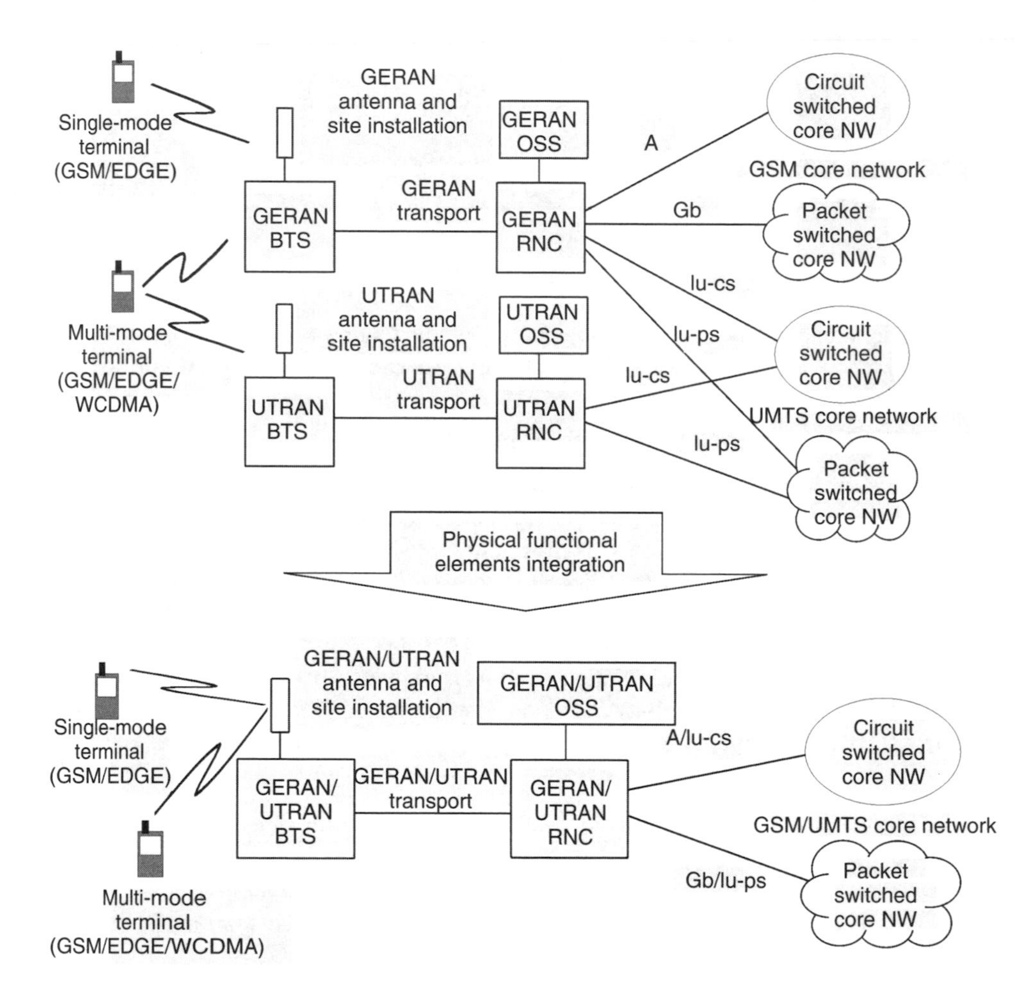

Figure 14.21. UMTS multi-radio network architecture integration

For real-time services (conversational and streaming) the most important indicator, apart from the received signal strength, is the amount of load currently occupied in the cell of interest. The amount of resources dedicated to non-real-time services without any guaranteed bandwidth does not have much effect on new requests of real-time services. Non-real-time NRT traffic is usually scheduled via time-shared channels, and because of lower priority the amount of resources dedicated to interactive and background traffic can be flexibly scaled down if requested by real-time services. Therefore, the occupation of physical resources by NRT traffic is not an optimum criterion for candidate cell prioritisation as a new NRT service request is considered. Instead, a measure of achieved throughput or average delay per NRT connection could be used.

14.5.6 Architecture Integration

Each radio access network has its own functional and physical elements. As the number of radio technologies connected to the UMTS CN increases, the associated complexity and deployment cost of the UMTS multi-radio network increases as well. Figure 14.20 illustrates the main functional entities of each technology.

In order to design an efficient multi-radio solution, these functional and physical entities must be efficiently integrated. As previously mentioned, both radio access technologies, GSM/EDGE and WCDMA, constitute two physical and logically independent radio access networks connected to the same UMTS CN. The required level of architectural integration is achieved as both logical networks merge together into a single physical UMTS radio access network, that is, all the physical elements of the network support both technologies providing the cellular operator a seamless operation of multiple technologies. This concept is illustrated in Figure 14.21.

References

[1] Holma T., *WCDMA for UMTS*, Wiley, Chichester, 2000.
[2] 3GPP Technical Specification 25.101, UE Radio Transmission and Reception (FDD).
[3] CDG, 3rd Generation Mobile Wireless, www.cdg.org., December 2001.
[4] Qualcomm, SMV Capacity Increases, 3GPP2-C11-20001016-010R1, October 2000.
[5] Lin Ma., Zhigang R., 'Capacity Simulations and Analysis for cdma2000 Packet Data Services', *IEEE VTC 2000*, 2000.
[6] 3GPP TR 25.848 V4.0.0. Technical Report 3rd Generation Partnership Project; Technical Specification Group Radio Access Network; Physical Layer Aspects of UTRA High Speed Downlink Packet Access.
[7] Viterbi A. J., Bender P., Black P., Grob M., Padovani R., Sindhushayana N., 'CDMA/HDR: A Bandwidth-Efficient High-Speed Wireless Data Service for Nomadic Users', *IEEE Commun. Magazine*, **July**, 2000.
[8] SKT, R&D 1X-EV-DO Technology Features, www.sktelecom.com.
[9] Voice Capacity Enhancements for GSM Evolution to UMTS, Rysavy Research, http://www.3gamericas.com/pdfs/whitepaper_voice_capacity71802.pdf.
[10] Data Capabilities for GSM Evolution to UMTS, Rysavy Research, http://www.3gamericas.com/PDFs/rysavy_paper_nov19.pdf.

15

3G Technology Strategy and Evolution Paths

Juan Melero

15.1 3G Multimedia Services

15.1.1 Operators' Business Impact

The wireless communications business has experienced an exceptional growth during the last 10 years. This growth has been particularly remarkable in those markets in which GSM was the dominant standard, Europe and Asia Pacific. GSM economies of scale and its fundamental characteristics such as global roaming and competitive short message services (SMS) helped this boost. Current global volumes were not expected in any existing forecast. Figure 15.1 displays the evolution of the cellular subscribers worldwide from 1991 and shows several forecasts for the coming years, which indicate that the cellular subscriber growth rate is expected to continue growing during this decade.

During the 1990s, with the introduction of second-generation (digital) standards, voice telephony went wireless. The 1990s was also the decade of the Internet take off, with Internet-based multimedia services (MMSs) becoming increasingly popular and the web becoming the '*de facto*' world information database. Second-generation (2G) cellular standards were, however, not designed to support these new Internet-based services and applications. Third-generation (3G) standards are meant to become the vehicle for Internet-based multimedia and other data services to go wireless as well. Figure 15.2 shows the projected evolution of the number of devices accessing the Internet. It is expected that the number of wireless devices with access to the Internet will exceed that of fixed devices during 2002.

The take off of these new 3G MMSs will drive a major transformation in the cellular operators' business case. With these services boosting the compound annual growth rate (CAGR) of the global mobile service revenues, the business opportunities for the 3G operators are enormous. Figure 15.3 shows the estimated evolution from 2002 to 2007 of the total global mobile service revenues and the revenue split in terms of different services.

GSM, GPRS and EDGE Performance 2nd Ed. Edited by T. Halonen, J. Romero and J. Melero
 ISBN: 0-470-86694-2

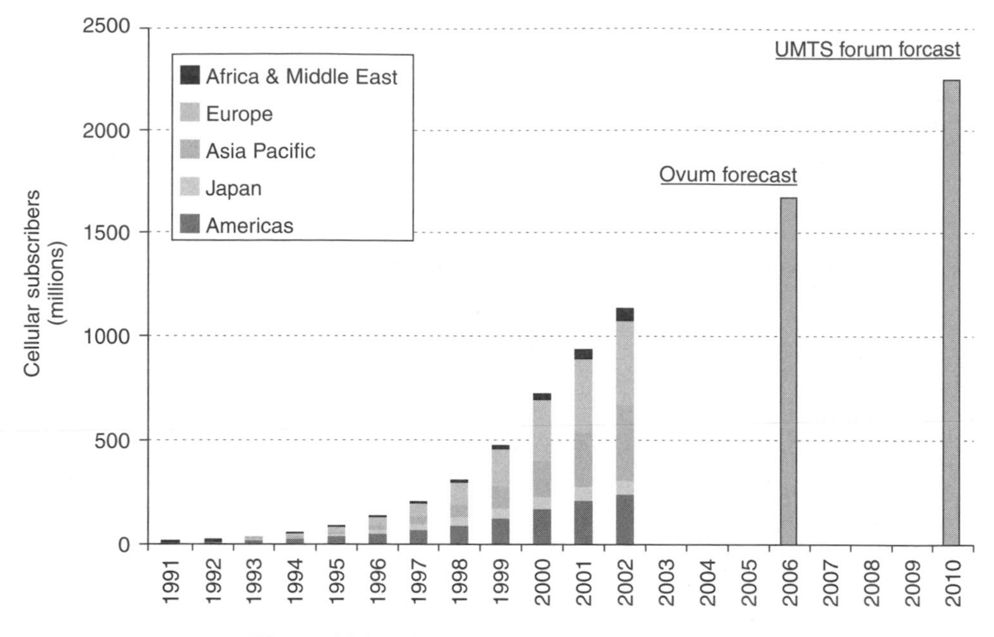

Figure 15.1. Global cellular subscriber evolution

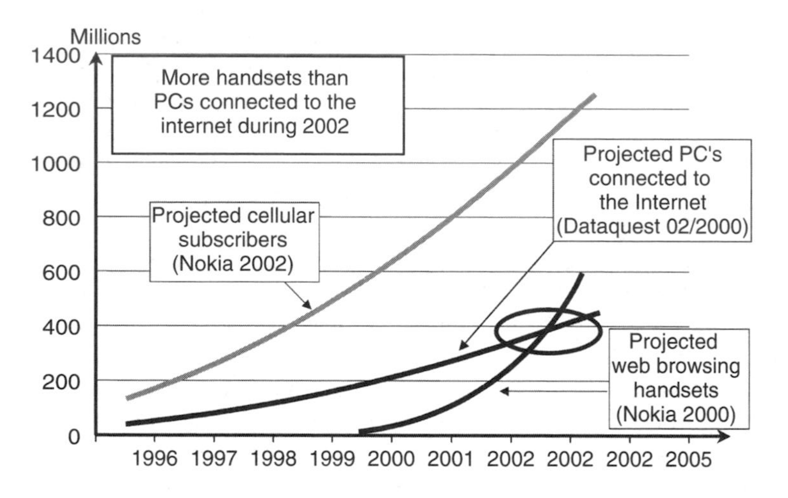

Figure 15.2. Projected penetration of Internet connections

Today, voice services usually account for almost 90% of the total cellular operator revenue, but in the next few years data services are expected to account for more than 30% of the revenues, boosting the total global cellular business revenues to a value almost double its current volume.

During the last few years, the growth experienced, both by the number of subscribers and the voice service usage, was enough for cellular operators to attain a solid revenue growth. However, as these factors tend to stabilise, and the price erosion impacts

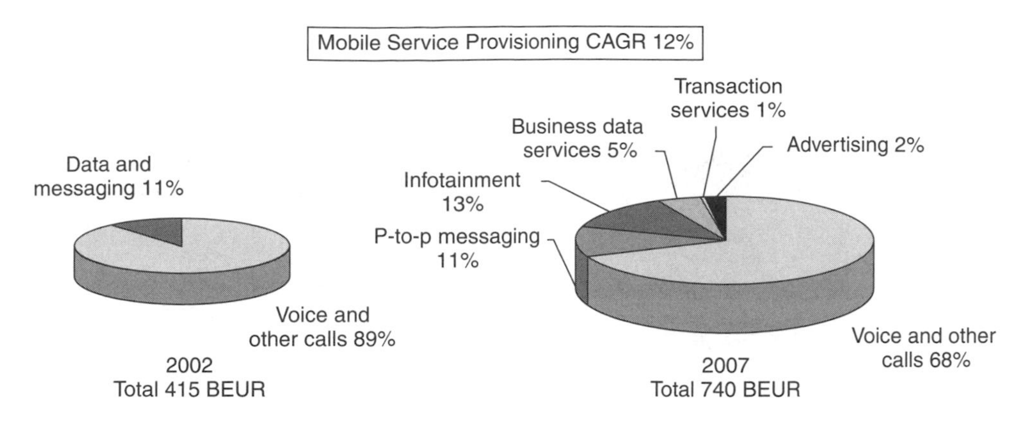

Figure 15.3. Evolution of the global mobile service revenues [1]

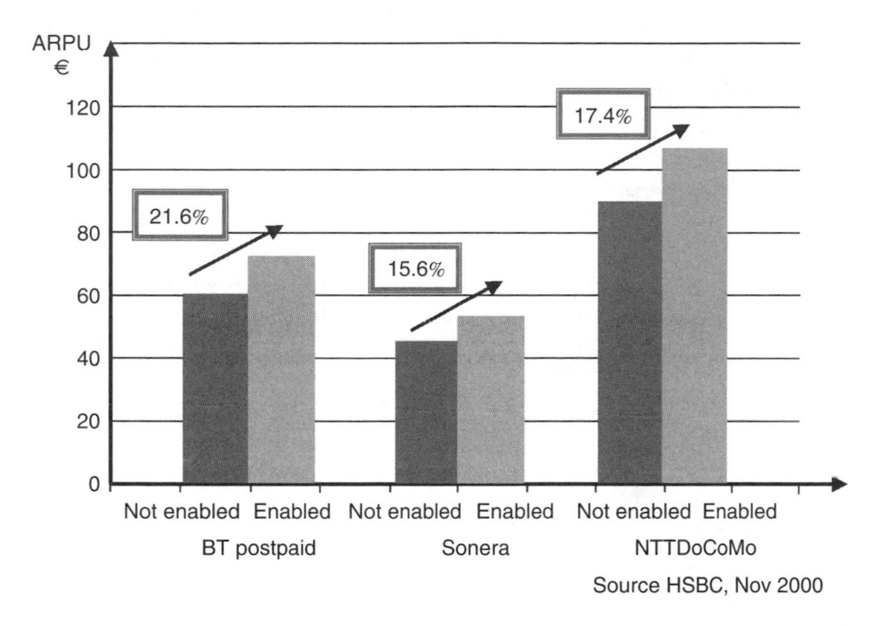

Figure 15.4. Examples of monthly ARPU with the introduction of value-added data services

the margins, the monthly average revenue per user (ARPU) will tend to decrease and the operators' total revenue growth will start to slow down. Figure 15.4 displays some examples of the impact the introduction of value-added data services had on the operators' monthly ARPU. The introduction of the upcoming 3G multimedia services is expected to have a major impact on the revenue per user, thus allowing cellular operators to continue with a sustained revenue growth.

15.1.2 3G Technologies—Requirements

3G radio access networks (RAN) should efficiently deliver the upcoming services providing the quality of service (QoS) expected by the end user. The main characteristic

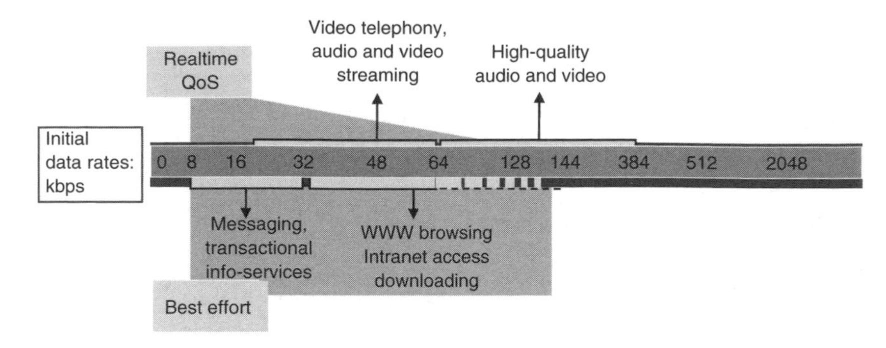

Figure 15.5. 3G service throughput and QoS requirements

to consider when describing data services is the required data rate. Figure 15.5 shows the data rates required by different 3G services. It is important to highlight that most of the foreseen services are well supported with data rates below 128 kbps. As presented in Chapter 7, enhanced general packet radio system (EGPRS) (Rel'99) can already efficiently deliver such data rates throughout the whole cell print by effectively controlling the QoS of the ongoing connections.

QoS management is a key element for the radio access technology to efficiently deliver 3G MMSs, since such services will be fundamentally packet-switched-based. Distinct services and user profiles will demand an important differentiation in terms of the QoS to be provided. Therefore, the RAN should be able to control the QoS provision and ensure that certain guaranteed data rates can be provided throughout the system by dynamically assigning the radio access resources. As presented in Chapter 3, UMTS terrestrial radio access network (UTRAN) and GSM/EDGE radio access network (GERAN) utilise the same UMTS QoS architecture, which was specifically designed for the efficient support of the expected 3G service mix. Even though the cdma2000 first release only supports best-effort data, ongoing standardisation will provide this technology with equivalent QoS architecture.

15.2 Globalisation

3G is about service delivery, and the competitive differentiation operators will seek in order to increase their revenues and market share will be based on when, how and what is offered in terms of services to the targeted end user segment. Provided different technologies are able to deliver a similar service set, the technology utilised in the RAN will be, to a certain extent, transparent to the end user perception. There are, however, a number of aspects of vital strategic importance related to the selection of the right technology, which are linked with the globalisation process of the cellular business.

15.2.1 Technology Globalisation

Today's telecommunications market segment is becoming increasingly global and major players are emerging in this market struggling to take a dominant position. The cellular business is not an exception and, through acquisitions, mergers and strategic alliances,

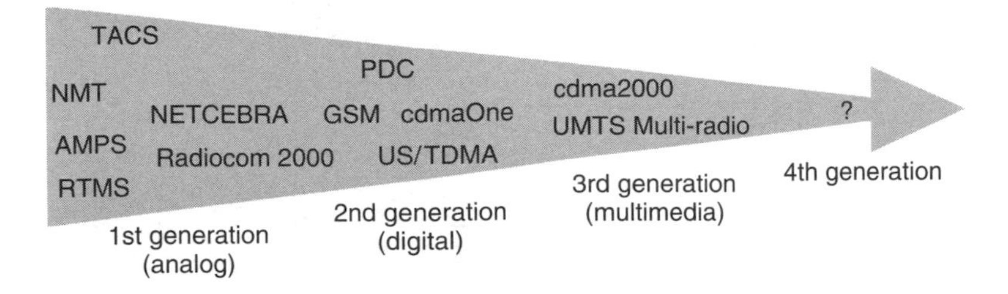

Figure 15.6. Technology globalisation

most operators are linked to one of these global players. This increasingly global business environment has been driving, and will continue doing so, the globalisation of technologies, and in particular, radio access technologies. Therefore, the selection of technology is likely to have an impact on the global strategy of each particular cellular operator. Figure 15.6 displays the technology globalisation evolution and how the number of existing technologies is continuously decreasing with a clear trend to reaching a unique global radio technology in the near future.

15.2.2 Economies of Scale

In this global scenario where operators, technologies, services and marketing strategies are becoming increasingly global, the selection of the right 3G technology is critical to remain competitive. Chapter 14 introduced a technical comparison between the different 3G radio access technologies. Such analysis showed certain advantages of the universal mobile telecommunications system (UMTS) 3G multi-radio access technologies (GSM/EDGE/WCDMA) over cdma2000. But there are other aspects associated with the globalisation trend that will likely be determining factors for the selection of technology. These aspects are related to the benefits associated with 'economies of scale'. The basic principle is related to global business volumes and market shares. The higher the volumes, the lower the technology cost, the faster the technology and its supported services evolve and the easier to deploy true global market strategies.

- *Network infrastructure and mobile handset costs.* The cost of the network infrastructure and mobile terminals is heavily dependent on economies of scale. Higher competitiveness and larger volumes will naturally bring costs down. Hence, the cost impact is likely to be significant if there is a significant difference in the global market share of the technologies. Additionally, the fact that GSM has enjoyed open interfaces and shared intellectual proprietary rights (IPR) has had a significant influence on the infrastructure and terminal costs. This cost impact may become particularly important on the terminal side. Figure 15.7 shows the wholesale prices of GSM and cdmaOne (IS-95) 2G terminals sorted in ascending order, each point representing a terminal price. The impact is of higher relative significance in the low-end section, where unsubsidised prices are roughly double for cdmaOne terminals.
- *Speed of technology development.* Industry analysts estimate that vendors are allocating $200 billion in R&D in order to carry out the specification, design and manufacturing of 3G infrastructure only. The R&D investment in the two differentiated

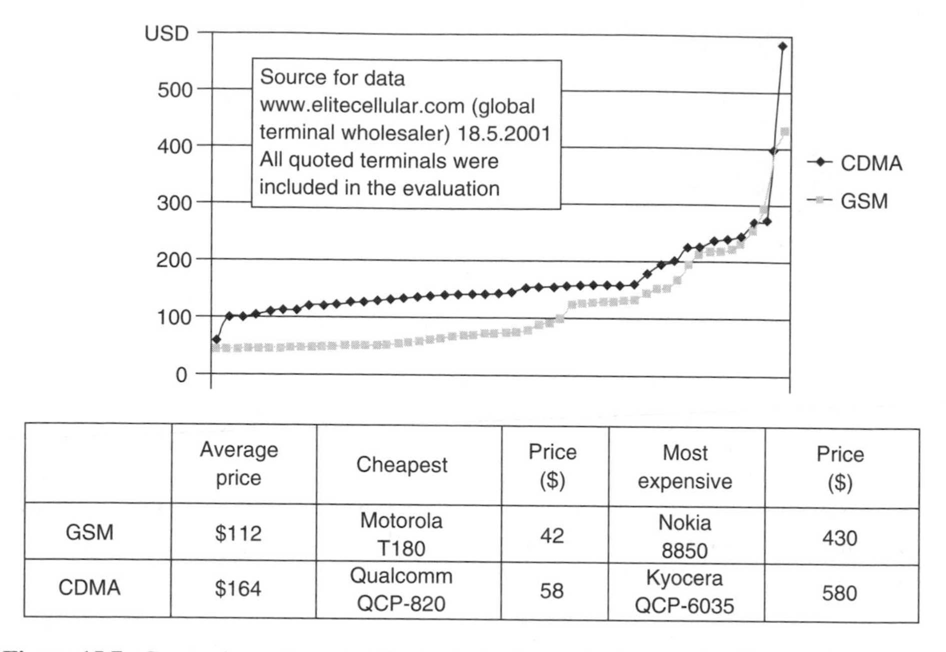

	Average price	Cheapest	Price ($)	Most expensive	Price ($)
GSM	$112	Motorola T180	42	Nokia 8850	430
CDMA	$164	Qualcomm QCP-820	58	Kyocera QCP-6035	580

Figure 15.7. Comparison of unsubsidised wholesale terminal prices for 43 cdmaOne and 54 GSM terminal models [2]

available 3G paths (UMTS 3G multi-radio and cdma2000) will likely be proportional to their respective global market share. Therefore, the choice of the right technology will ensure a faster, more advanced and efficient technology development. Today's products portfolio is a good example of this factor. For instance, in the case of terminals, the number of available GSM models on the market (312) is roughly double that of cdmaOne (159) [3].

- *Roaming and service continuity*. Traditionally, roaming has been one of the competitive advantages that has made GSM technology highly attractive, and recently is becoming increasingly important in parallel with the globalisation process. In the past, GSM subscribers have been able to roam into other GSM networks worldwide. Today, global players are starting to deploy worldwide service continuity strategies, in order to ensure that their customers always stay in their global networks when roaming outside the particular customers' 'home network'. A single global technology will be required in order to realise these targets.

- *Available service portfolio*. Finally, since 3G is, as described before, about delivery of 3G MMSs, the technology with the broader global footprint and pioneering the introduction of true 3G services will have an advantage in terms of service portfolio available. In September 2000, SK Telecom launched its first commercial cdma2000 network, which positioned cdma2000 as the first International Telecommunication Union (ITU) approved 3G technology to be commercially launched worldwide. However, true 3G MMSs, such as video telephony or simultaneous speech and Internet access, with efficient QoS management and data rates up to 384 kbps, were first successfully launched on October 1, 2001 in Tokyo by NTT DoCoMo using UMTS WCDMA technology.

All these items define the competitive benefits coming from economies of scale, and are likely to be determining factors of the selection of 3G technologies. Economies of scale are synonymous with volumes. The current 3G technology paths are based on the evolution of 2G standards and hence the future 3G technologies volumes will be highly influenced by the current 2G technologies volumes and their natural evolution to 3G. The following section will analyse the global trends of the technology evolution.

15.3 3G Technology Evolution Paths. UMTS Multi-radio and cdma2000

Analogue technologies were dominant in the cellular market up to 1997, when their global market share was exceeded by that of 2G digital technologies. As presented before, the last 10 years have seen a phenomenal growth of the cellular penetration worldwide, which has defined the shape of the 2G technologies global market share as displayed in Figure 15.8. GSM technology market share shows a sustained growth, and today it has become the global 2G standard, deployed by more than 460 operators around the world and accounting for more than 70% of the total number of cellular subscribers. On the other hand, despite an important growth during its first years of deployment, cdmaOne market share has now stabilised. During 2001, 81% of all the new subscriptions globally were GSM, whereas only 13% were cdmaOne [4]. The current subscription trend will broaden the already significant relative market share gap between GSM and CDMA.

In terms of revenue generation, GSM SMS has been the most successful 2G data service and has defined a clear frame for the expected highest revenue generator 3G service, the

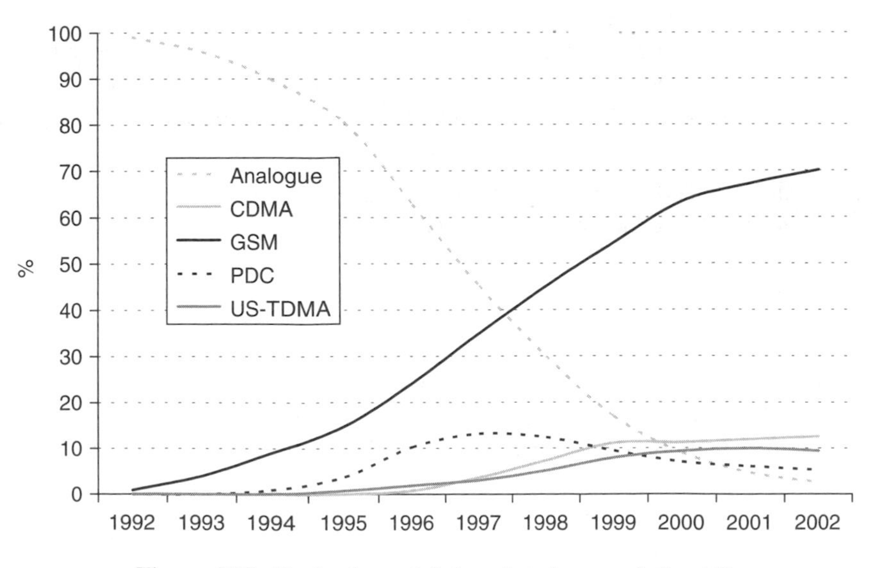

Figure 15.8. Technology global market share evolution [4]

enhanced 3G MMS. According to the GSM Association, 15 billion SMS text messages were sent worldwide during December 2000, which indicates a fivefold increase in the SMS volume in the past year. During October 2001, the GSM Association announced that 1 billion SMS were sent daily worldwide. In the service revenue forecast presented in Figure 15.3, the different messaging services, grouped under the communication category, would account in 2006 for 164 billion a year, which represents a market size equivalent to almost half of the total global market revenues of 2000.

As GSM technology has become by far the most successful 2G standard, UMTS, its 3G multi-radio evolution, made up of the seamless integration of GERAN and UTRAN, is likely not only to match its predecessor but naturally increase its dominance as UMTS multi-radio becomes the evolution path of other 2G technologies.

15.3.1 From 2G to 3G

Today, the convergence of 2G technologies towards the UMTS multi-radio 3G evolution path is clear. The entities, such as operators, global associations and standardisation bodies, which are representing and driving the evolution of three out of the four current most representative 2G technologies, have endorsed the UMTS multi-radio evolution path:

- The GSM Association, representative of the GSM operators worldwide, is a market representative of the 3rd Generation Partnership Project (3GPP) and fully supports the UMTS multi-radio 3G evolution path.
- 3G Americas, a wireless industry association dedicated to the Americas, supports the seamless deployment of GSM, GPRS, EDGE and UMTS throughout the Americas and fully endorses the migration of TDMA operators to the GSM family of technologies. The transition from TDMA to GSM technologies for operators is helped by the GAIT standard functionality (GSM, ANSI-136, Interoperability Team) that enables interoperability between TDMA and GSM technologies. The GAIT terminals are dual mode, effectively supporting TDMA (ANSI-136) and a combination of the GSM family of technologies. With the introduction of such terminals, the US-TDMA operators have a smooth path available to converge into the UMTS multi-radio evolution. Figure 15.9 shows the combined TDMA/GSM/EDGE architecture based on GAIT interworking.
- In Japan, the two most important PDC operators (NTT DoCoMo and J-Phone), which on October 1st, 2001 covered 75.7% of the Japanese subscriber base, have selected UMTS WCDMA technology for their 3G evolution [4].

Table 15.1 summarises the evolution paths associated with the existing 2G technologies.

The alternative 3G evolution path to UMTS is the one endorsed by the CDMA Development Group (CDG), which supports an evolution to 3G based on cdmaOne technology. cdmaOne was initially developed in the United States and soon adopted in other regions. Figure 15.10 shows the market share cdmaOne technology holds in different global regions [4].

Although cdmaOne has been introduced in most of the regions worldwide, only in America, especially in the United States, and to some extent in Asia Pacific, fundamentally in Korea, has it developed an important momentum.

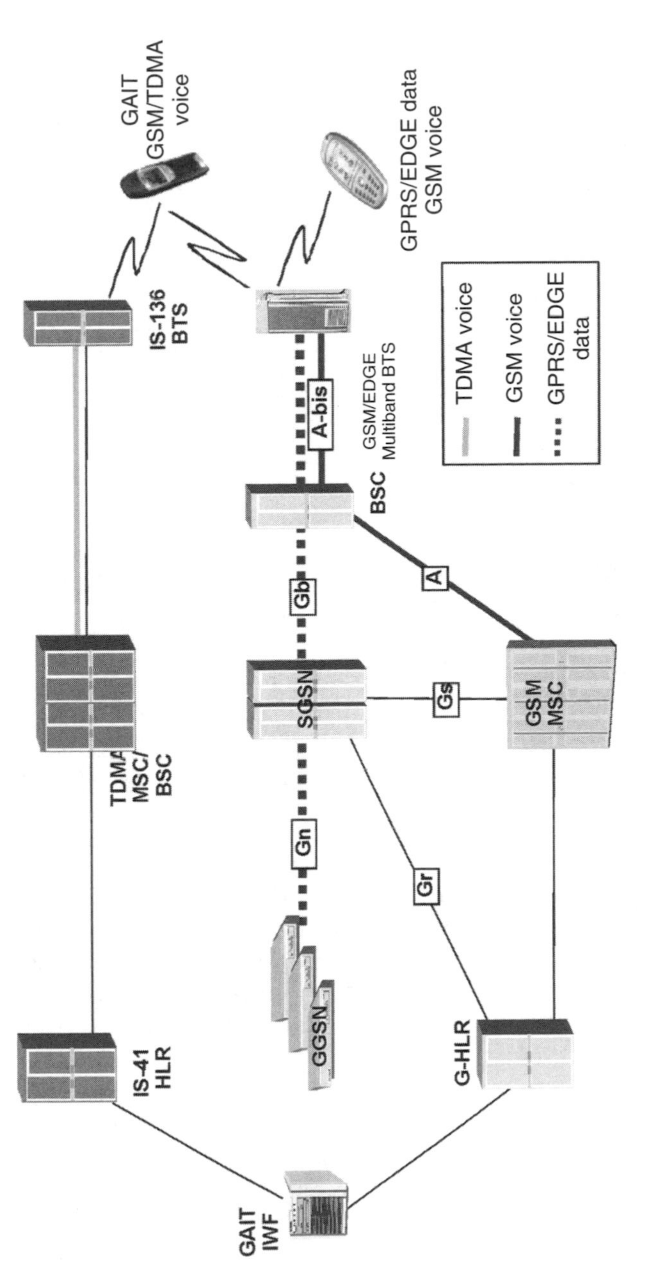

Figure 15.9. TDMA/GSM/EDGE GAIT architecture

Table 15.1. Evolution paths associated with the existing 2G technologies

2G standard	3G standard	Evolution path
GSM	GSM/EDGE/WCDMA	UMTS
PDC (Japan)	WCDMA[1]	UMTS
US-TDMA (IS-136)	TDMA/GSM/EDGE/WCDMA	UMTS
cdmaOne (IS-95)	cdma2000 or GSM/EDGE/WCDMA	cdma2000 or UMTS

[1]The two Japanese operators (NTT DoCoMo and J-Phone) that have selected UMTS WCDMA technology for their 3G evolution represent 92.4% of the existing PDC market [4].

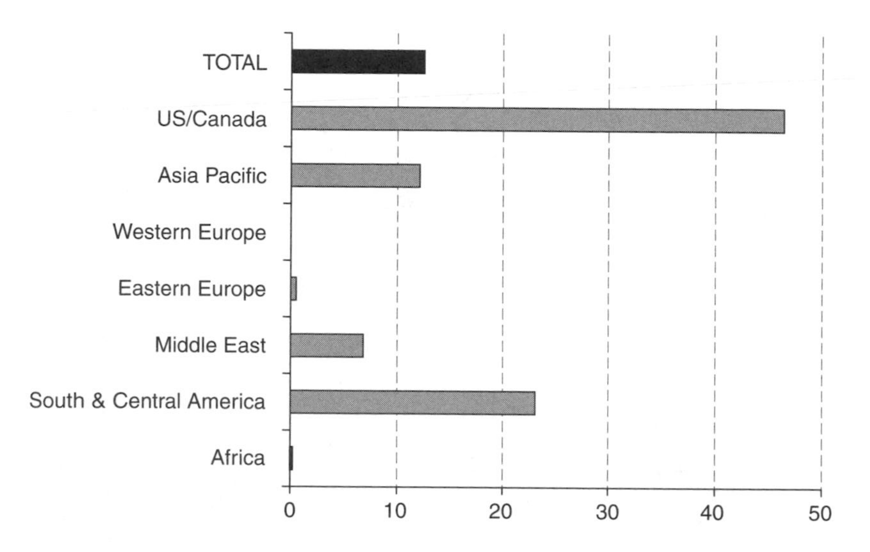

Figure 15.10. cdmaOne worldwide regional market share (cdma2000 terminals included)

South Korea represents 70% of the cdmaOne Asia Pacific customer base. This country accounted for 28.1 million cdmaOne subscribers in June 2001, which was roughly one-third of the cdmaOne subscriber base worldwide. After several acquisitions and mergers during 2001, the three remaining operators defined their 3G evolution path. Two of these operators (SKT and KT) opted for WCDMA hence converging into the UMTS evolution path. These operators hold over 85% of the Korean market share [4]. Therefore Korea, the main cdmaOne stronghold in Asia, is effectively converging towards the UMTS evolution path as well.

In the United States, Verizon Wireless, with 19 million cdmaOne subscribers, accounts for over 40% of the total US cdmaOne subscribers and well over 50% of the potential subscribers if its analogue customer base is counted [4]. Verizon Wireless, which is 44% owned by Vodafone, the largest global wireless operator with 100 million subscribers worldwide [5], has publicly announced the possibility of its UMTS 3G choice by the adoption of WCDMA radio technology. If Verizon Wireless UMTS selection finally takes place, over 40% of the current global cdmaOne customer base (counting SKT, KT and Verizon Wireless cdmaOne subscribers) would evolve towards the UMTS 3G evolution path.

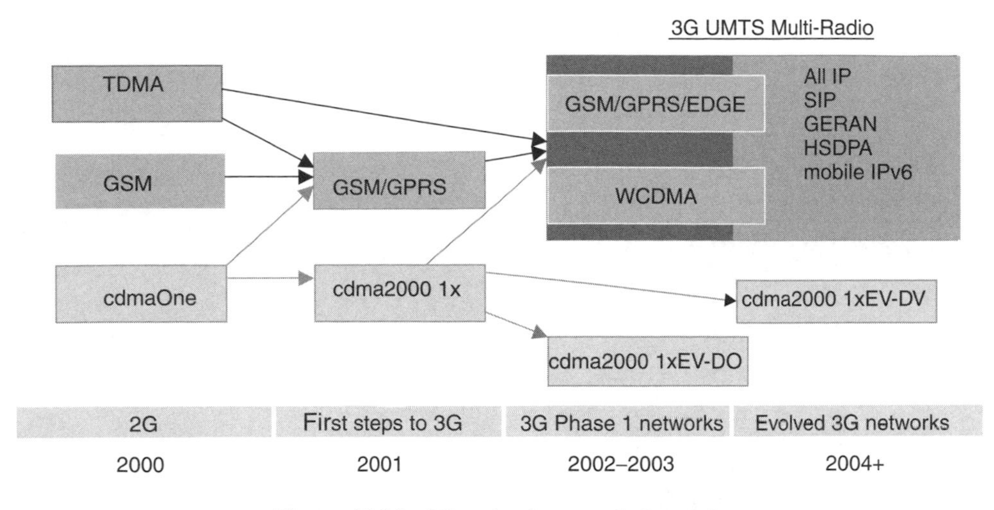

Figure 15.11. 3G technology evolution paths

Additionally, many current cdmaOne operators are analysing the possible migration paths to UMTS. Such migration can be done through the migration to GERAN (GSM/EDGE) with the later integration of UTRAN (WCDMA) or it can be done through the direct introduction of UTRAN. This last option is however dependent on the cellular operator licensed bandwidth, since WCDMA technology is currently not supported in the 800 MHz band, the future availability of dual mode cdmaOne/WCDMA terminals and the integration effectiveness of the technologies. Figure 15.11 summarises the global evolution paths to 3G.

At the end of 2001, the year that UMTS WCDMA was first launched in Tokyo, cdmaOne based technology covered 11.8% of the global market share whereas the technologies to converge to UMTS[1] covered 83.5% of the global market [4]. This constitutes an initial relative advantage of 7 to 1 for the UMTS 3G multi-radio evolution path. With the globalisation dynamics, this gap will surely broaden, possibly ending up in an effective single universal 3G wireless access technology, as it was devised when UMTS was first proposed.

References

[1] Nokia, November 2002.
[2] Elite Cellular (global terminal wholesaler), www.elitecellular.com.
[3] Wireless Tomorrow, www.wirelesstomorrow.com.
[4] EMC Database, January 1 2002.
[5] Vodafone, www.vodafone.com.

[1] Including GSM, PDC and US-TDMA

Appendix A

MAIO Management Limitations

Angel Rivada

When random radio frequency (RF) hopping is used and fractional load is introduced, that is, there are more hopping frequencies than transceivers (TRXs) per sector, there is a certain degree of interference control available in order to avoid undesired co-channels and adjacent channel interference between the sectors of each site. This is achieved by the mobile allocation index offset (MAIO) management functionality, which basically controls the offsets in the Mobile Allocation (MA) list of each TRX within a synchronised site. However, the gains of this functionality get limited as the effective reuse decreases. When the effective reuse is below 6, adjacencies in the hopping band within sites begin to appear. Subsequently, for effective reuses lower than 3, adjacencies within the sector itself and co-channels between the site sectors are unavoidable. Although this is a general problem, narrowband scenarios suffer this limitation for relatively low TRX configurations. This appendix studies the performance degradation as the effective reuse decreases.

A.1 MAIO Management Limitations and Planning

RF hopping networks usually use lists of frequencies (MA lists) that are shared by different sectors to hop over. In order to reduce the impact of interference from sectors using the same MA lists, the usage of the frequencies can be randomised (random hopping). The hopping sequence number (HSN) is a parameter that defines the pseudo-random sequence used. Within one site using a single MA list and HSN, both co-channel and adjacent channel interference can be avoided by controlling the frequencies used in each TRX within the site. To achieve this, the MAIO associated with each TRX must be controlled.

However, the degree of control provided by the MAIO management functionality is conditioned by the site-effective reuse (*number of frequencies in the MA list of the site/average number of TRXs per sector of the site*). If the MA list is conformed by

GSM, GPRS and EDGE Performance 2nd Ed. Edited by T. Halonen, J. Romero and J. Melero
 ISBN: 0-470-86694-2

consecutive frequencies (1/1 hopping reuse), the degrees of control would be:

• Effective reuse ≥ 6	No co-channel or adjacent channel interference within the site
• 6 > Effective reuse ≥ 3	Adjacent channel interference between site sectors
• 3 > Effective reuse ≥ 1	Co-channel and adjacent channel interference between site sectors and adjacent interference within the sectors
• 1 > Effective reuse	Co-channel and adjacent channel interference both between site sectors and within the sectors

Therefore, the MAIO management limitations are not load-dependent but only configuration-dependent (effective reuse-dependent). The different available deployment strategies when the effective reuse goes under the mentioned thresholds are either to maintain the MAIO management in all the sectors within the site, which causes permanent interference between certain TRXs, or start using different HSNs per sector, which increases the overall interference but randomises its occurrence. Figure A.1 shows an example of the performance of different strategies for the case of a single MA list of 12 frequencies using 1/1 hopping reuse with an average of 4 hopping TRXs

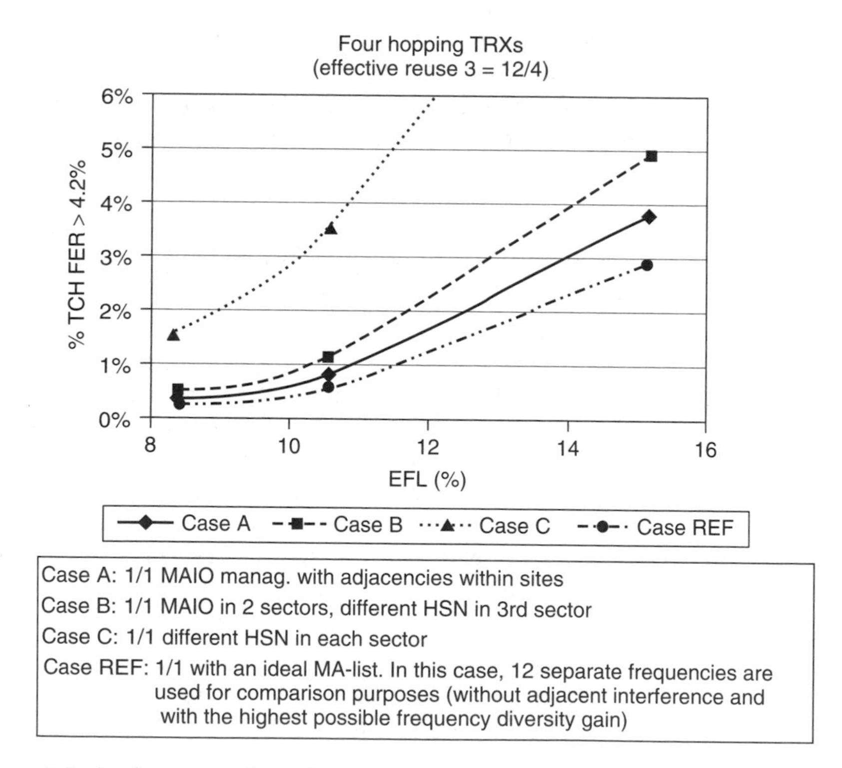

Figure A.1. Performance of four hopping TRXs and different configurations (typical urban 3 km/h (TU3), 3000 m inter-site distance; 65° antenna beamwidth; DL displayed)

per sector, i.e. effective reuse of 3. In this example, adjacencies within the sites are unavoidable. Additionally, in this figure the ideal case with 12 separate frequencies without adjacent interference (reference best performing case) has been added. The 1/1 with MAIO management in all sectors outperforms the other configurations even though existing adjacent channel interference will degrade the performance compared with the interference-free (within the site) case. The same exercise can be applied to higher number of TRXs, where co-channel interference within the site starts to become a problem. In those cases, the co-channel interference should be handled assigning different HSNs between co-channel interfering sectors and applying MAIO management when possible (for example, with effective reuse 2, MAIO management could be applied in two sectors to avoid co-channels and a different HSN in the third one).

A.2 MAIO Management Limitations for Different Effective Reuses and Antenna Beamwidth

Figure A.2 displays the performance of the network when using different effective reuses. The best MAIO management configuration cases are displayed for each effective reuse. Only hopping layer and downlink (DL) results from simulations are displayed. The degradation when going from effective reuse 6 to 4 and 2 can be observed. This is due to the impact of MAIO management degradation. With reuse 4, adjacent channels within sectors of the same site appear; and with reuse 2 co-channels within sites are unavoidable. For a 65° antenna, a degradation of around 5% can be detected when going from reuse 6 to reuse 4, and 10% when going to reuse 2.

Additionally, the effect of using narrower antenna beamwidth is also presented in the same figure. Basically, according to the simulations, in a regular-grid scenario, the use of narrower beamwidth decreases the interference level produced towards sectors of the same site, improving the performance of the interference-limited scenarios. Finally, the MAIO management degradation presented in Figure A.1 will be higher in the case of 90° antennas, as presented in Figure A.2.

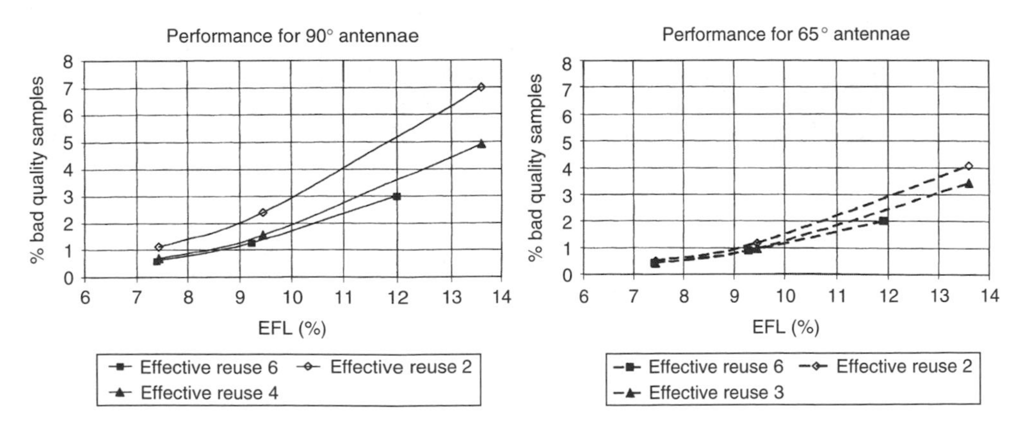

Figure A.2. MAIO management limitations for different antenna beamwidth (DL performance, TU3)

Appendix B

Hardware Dimensioning Studies

Salvador Pedraza

This appendix includes different studies on hardware dimensioning in cases in which the Erlang-B formula cannot be used. The first section covers mixed full and half-rate voice scenarios, while the second deals with packet data-traffic connection sharing the system resources.

B.1 Blocking Probability for Half- and Full-rate Speech Modes

This section is a study about blocking probability (BP) in the GSM system in which two different types of connections can be established: full-rate (FR) connections using one timeslot (TSL) and half-rate (HR) connections using half a TSL. This study is very useful for hardware dimensioning when a percentage of the calls can be converted to HR in order to save resources. Note that this study is purely based on hardware (HW) considerations and does not consider interference issues.

B.1.1 The Erlang-B Formula

The Erlang-B formula has been used for a very long time to dimension the number of transceivers (TRXs) with FR calls. This problem was first studied by Erlang in 1899 [1], who derived the well-known and much used Erlang-B formula for blocking probability (BP):

$$\mathrm{BP}_{\mathrm{ErlangB}}(T, N) = \frac{\dfrac{T^N}{N!}}{\displaystyle\sum_{i=0}^{N} \frac{T^i}{i!}}$$

where
T = traffic generated in the cell
N = number of available channels

Call arrival is assumed to be a Poisson-distributed stochastic process with exponentially distributed call duration. This formula provides the BP in a cell with N channels for a

GSM, GPRS and EDGE Performance 2nd Ed. Edited by T. Halonen, J. Romero and J. Melero
 ISBN: 0-470-86694-2

given traffic (T). It is important to note that T is *potential* traffic and not real carried traffic in the cell, which can be calculated as follows:

$$\mathrm{CT} = T(1 - \mathrm{BP}_{\mathrm{ErlangB}}(T, N)).$$

Although if BP is low it can be assumed that carried traffic is approximately equal to potential traffic

$$\mathrm{CT} \approx T \text{ if BP is low.}$$

The Erlang-B formula is the steady response of the stochastic process of arriving and serving calls assuming the following hypothesis:

- traffic is originated from an infinite number of sources;
- the number of channels in the system is limited (N);
- call requests arrive independently of each other;
- the number of call requests in a time unit is Poisson-distributed with parameter λ;
- the call duration is exponentially distributed with parameter τ;

where
λ = call request arriving rate (e.g. $\lambda = 0.1$ calls/s)
τ = average call duration (e.g. $\tau = 120$ s)

The call termination rate is $\mu = 1/\tau$. Little's formula connects μ and λ:

$$T = \frac{\lambda}{\mu}.$$

B.1.2 Blocking Probability for HR/FR Scenario

Let us assume a combined scenario with a ratio of γ HR connections. There are two factors that may contribute to γ: HR mobile penetration and radio conditions (connections with bad radio conditions cannot be converted to HR). N is the number of available channels (e.g. $N = 15$ for a two TRX cell). The call arrival rate is λ (e.g. $\lambda = 0.1$ calls/s = 1 call/10 s) and the call termination rate is μ (e.g. $\mu = 0.05$ calls/s = 1 call/20 s) or equivalently $\tau = 1/\mu$ is the average call duration.

B.1.2.1 Simple Case: Two Timeslots

For the two timeslot (TSL) case, the problem can be modelled with the state diagram shown in Figure B.1. Here $(PfFhH)$ represents the state in which there are f FR connections and h HR connections. The arrows describe the possible state transitions and the label gives the relative probability for that transition.

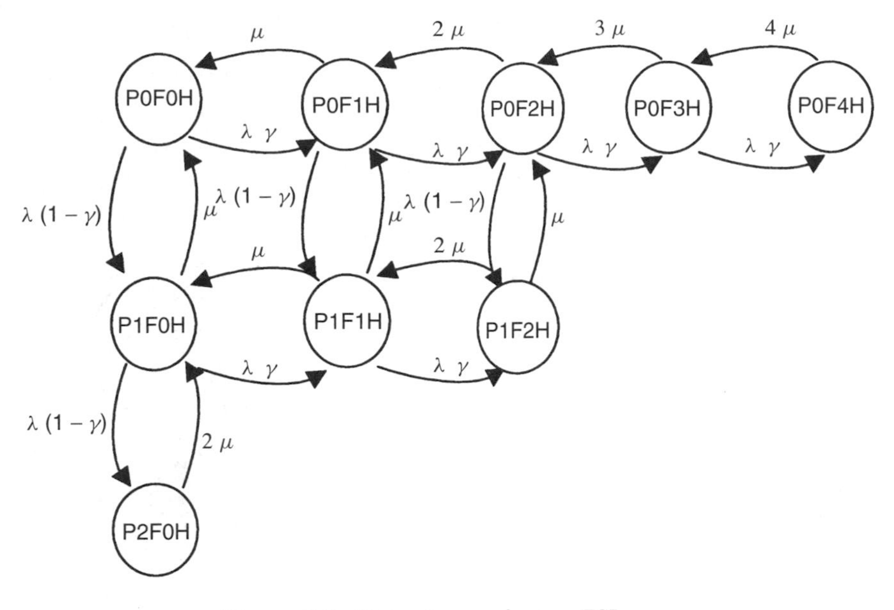

Figure B.1. State diagram for two TSL case

It is possible to derive the steady state equations for Figure B.1, in terms of offered traffic (T):

$$0 = P0F1H + P1F0H - T\,P0F0H$$

$$0 = T\gamma\,P0F0H + 2P0F2H + P1F1H - (T+1)P0F1H$$

$$0 = T\gamma\,P0F1H + 3P0F3H + P1F2H - (T+2)P0F2H$$

$$0 = T\gamma\,P0F2H + 4P0F4H - (T\gamma + 3)P0F3H$$

$$0 = T\gamma\,P0F3H - 4P0F4H$$

$$0 = T(1-\gamma)P0F0H + P1F1H + 2P2F0H - (T+1)P1F0H$$

$$0 = T\gamma\,P1F0H + T(1-\gamma)P0F1H + 2P1F2H - (T\gamma + 2)P1F1H$$

$$0 = T\gamma\,P1F1H + T(1-\gamma)P0F2H - 3P1F2H$$

$$0 = T(1-\gamma)P1F0H - 2P2F0H$$

The solution for BP for two TSLs is

$$BP(T, \gamma, 2) = \frac{T^2(T\gamma^4(T-4) - 8\gamma^3(T-3) + 12\gamma^2(T-3) + 12)}{-4T^3\gamma^2(2\gamma - 3) + T^4\gamma^4 + 12T^2 + 24T + +24}$$

B.1.2.2 General Case: N TSLs

In a general case, there is a similar state diagram with $N^2 + 2N + 1$ states, where a general node has the state transition diagram shown in Figure B.2.

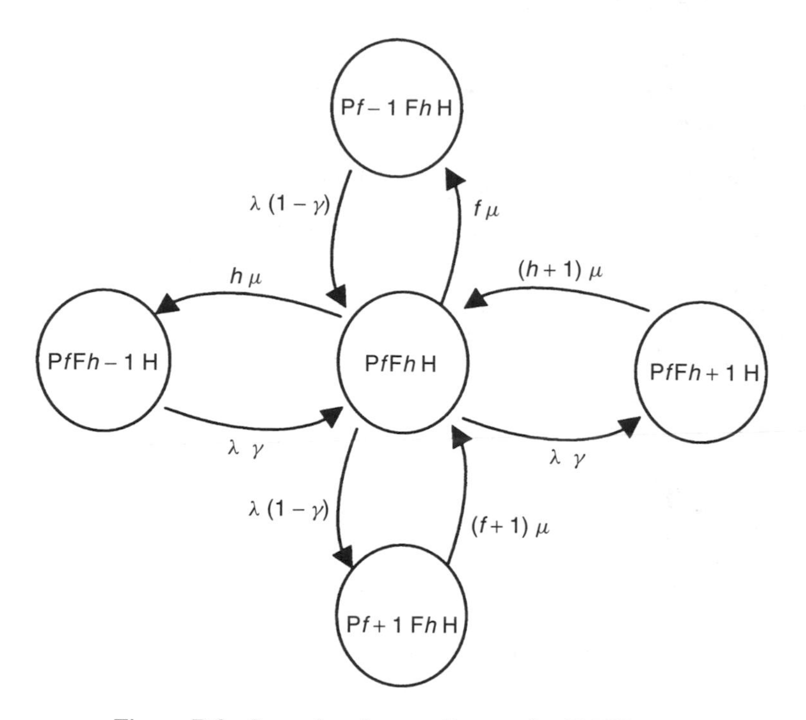

Figure B.2. General node state diagram for N TSL case

Similar to the two TSL case, the following linear equation can be written describing the steady state behaviour of Figure B.2:

$$\begin{aligned}0 = {} & T\gamma PfF(h-1)H + T(1-\gamma)P(f-1)FhH + (h+1)PfF(h+1)H \\ & + (f+1)P(f+1)FhH - fPfFhH \text{ Exist } (P(f-1)FhH) \\ & - T\gamma PfFhH \text{ Exist}(PfF(h+1)H) - T(1-\gamma)PfFhH \text{ Exist}(P(f+1)FhH) \\ & - hPfFhH \text{ Exist}(PfF(h-1)H)\end{aligned}$$

where Exist(PfFhH) is a function equal to 1 if the state exists in the state diagram, and is equal to 0 otherwise. Two terms contribute to the BP equation:

The sum of those states in which the system is totally full:

$$f = \sum_{f=0}^{N-1} PfF(2N-f)H.$$

and when the system has 1/2 TSL free, so that the system is full for FR mobiles, which are $1-\gamma$ of total. So for this case:

$$(1-\gamma)\sum_{f=0}^{N-1} PfF(2(N-f)-1)H.$$

In conclusion, BP can be written as

$$\mathrm{BP} = \sum_{f=0}^{N} PfF(2N - f)H + (1 - \gamma) \sum_{f=0}^{N-1} PfF(2(N - f) - 1)H.$$

Solving the $N^2 + 2N + 1$ dimensional system of equations, BP can be obtained as a function of total traffic generated in the cell (T), capability of HR mobiles (γ) and number of available channels (N):

$$\mathrm{BP} = \mathrm{BP}(T, \gamma, N).$$

A general solution for this equation is not easy to find as in the two TSL case. The next section presents numerical solutions in the form of tables.

B.1.2.3 Numerical Solutions

Similar to the Erlang-B formula, it is more useful for dimensioning purposes to have a table with traffic for a certain grade of service (GoS = blocking probability). Tables B.1 and B.2 have been generated for 1% of GoS and 2% of GoS as a function of the HR capability and number of channels (= TSL).

Notice that the first column ($\gamma = 0\%$) corresponds to the value obtained with the Erlang-B formula for n TSLs and the last column ($\gamma = 100\%$) corresponds to values obtained with the Erlang-B formula for $2n$ TSLs. For example, with eight TSLs, the offered traffic at 2% GoS is 3.62 Erlang if no HR calls can be done ($\gamma = 0\%$), while

Table B.1. Traffic (Erlangs) for 1% GoS with different ratio of HR connections

		Percentage of HR connections										
		0%	10%	20%	30%	40%	50%	60%	70%	80%	90%	100%
	1	0.0101	0.0102	0.0105	0.0111	0.012	0.0133	0.0158	0.0143	0.0279	0.0696	0.1526
	2	0.1526	0.1554	0.1613	0.1722	0.19	0.2144	0.255	0.3097	0.4132	0.5913	0.8694
	3	0.4555	0.4652	0.4875	0.5258	0.5838	0.6615	0.7724	0.9163	1.1335	1.4544	1.909
	4	0.8694	0.8908	0.9398	1.019	1.1312	1.2746	1.4651	1.7013	2.0326	2.4958	3.1277
	5	1.3608	1.3993	1.4826	1.6109	1.7843	1.9964	2.2678	2.5982	3.0452	3.6518	4.4614
	6	1.909	1.9705	2.0931	2.2744	2.5121	2.7939	3.1472	3.5737	4.1381	4.8893	5.8761
	7	2.5009	2.5928	2.7556	2.9877	3.2875	3.6435	4.0839	4.6082	5.2906	6.1866	7.3517
	8	3.1276	3.2512	3.4595	3.7488	4.1152	4.5419	5.0646	5.6873	6.4895	7.5319	8.8753
TSL	9	3.7825	3.9414	4.1969	4.5439	4.977	5.4747	6.0807	6.8034	7.7269	8.9165	10.437
	10	4.4612	4.6604	4.962	5.3627	5.8593	6.4327	7.1275	7.9511	8.9957	10.332	12.031
	11	5.1599	5.4014	5.7503	6.2056	6.7663	7.4159	8.1999	9.1247	10.291	11.774	13.652
	12	5.876	6.1616	6.5584	7.0681	7.6928	8.4196	9.2938	10.32	11.608	13.239	15.296
	13	6.6072	6.9376	7.3835	7.9485	8.6362	9.434	10.394	11.528	12.945	14.728	16.96
	14	7.3517	7.7287	8.2237	8.8434	9.5947	10.469	11.519	12.757	14.299	16.231	18.642
	15	8.108	8.5325	9.077	9.7517	10.567	11.518	12.658	13.998	15.662	17.744	20.339
	16	8.875	9.3477	9.9421	10.672	11.551	12.579	13.81	15.259	17.05	19.281	22.051
	24	15.295	16.16	17.174	18.371	19.787	21.453	23.422	25.687	28.456	31.879	36.109
	32	22.048	23.307	24.767	26.466	28.443	30.714	33.386	36.49	40.267	44.904	50.589
	40	29.007	30.689	32.58	34.747	37.256	40.155	43.567	47.538	52.343	58.198	65.322
	48	36.109	38.227	40.546	43.168	46.196	49.719	53.88	58.77	64.646	71.724	80.221
	56	43.315	45.882	48.621	51.682	55.213	59.355	64.281	70.151	77.148	85.451	95.237

Table B.2. Traffic (Erlangs) for 2% GoS with different ratio of HR connections

		Percentage of HR connections										
		0%	10%	20%	30%	40%	50%	60%	70%	80%	90%	100%
	1	0.0204	0.0207	0.0213	0.0224	0.0243	0.0268	0.032	0.034	0.0559	0.1138	0.2235
	2	0.2236	0.2278	0.2366	0.2527	0.2789	0.315	0.3722	0.4515	0.5828	0.7889	1.0923
	3	0.6024	0.6155	0.6456	0.6965	0.772	0.8729	1.0118	1.1897	1.4419	1.795	2.2759
	4	1.0927	1.1204	1.1828	1.2819	1.4197	1.5938	1.8195	2.0963	2.468	2.9674	3.6271
	5	1.6578	1.7065	1.8084	1.9628	2.1683	2.4169	2.7292	3.1061	3.5987	4.2451	5.084
	6	2.2769	2.3526	2.4985	2.7109	2.986	3.3096	3.7089	4.1874	4.8022	5.5967	6.6147
	7	2.9367	3.0448	3.2375	3.509	3.8538	4.2532	4.7404	5.3216	6.0596	7.003	8.2003
	8	3.6287	3.7733	4.0146	4.3457	4.7601	5.2359	5.8119	6.4968	7.359	8.4517	9.8284
TSL	9	4.3468	4.5311	4.8222	5.2131	5.6969	6.2502	6.9161	7.7053	8.692	9.9344	11.491
	10	5.0864	5.3128	5.6548	6.1058	6.6593	7.2905	8.0469	8.9409	10.053	11.445	13.182
	11	5.8443	6.1143	6.5082	7.0199	7.6432	8.3528	9.2001	10.199	11.436	12.979	14.896
	12	6.6178	6.9206	7.3662	7.9445	8.6452	9.4345	10.372	11.477	12.84	14.534	16.631
	13	7.405	7.7521	8.2508	8.8911	9.6629	10.532	11.561	12.772	14.261	16.105	18.383
	14	8.204	8.5963	9.1486	9.8511	10.694	11.643	12.764	14.082	15.697	17.693	20.15
	15	9.0137	9.4517	10.058	10.823	11.738	12.766	13.981	15.405	17.147	19.294	21.932
	16	9.8328	10.317	10.978	11.806	12.792	13.901	15.208	16.739	18.609	20.907	23.725
	24	16.636	17.497	18.601	19.943	21.515	23.274	25.329	27.724	30.615	34.129	38.392
	32	23.729	24.978	26.522	28.367	30.516	32.932	35.753	39.044	42.985	47.729	53.428
	40	30.998	32.618	34.611	36.982	39.735	42.819	46.408	50.589	55.575	61.548	68.688
	48	38.387	40.358	42.8	45.709	49.082	52.839	57.201	62.278	68.315	75.519	84.1
	56	45.863	48.161	51.051	54.511	58.519	62.955	68.092	74.073	81.164	89.602	99.624

with 100% HR capability 9.82 Erlang can be offered for the same GoS, the same as with 16 TSLs.

B.1.3 Effective Factor

The effective factor (EF) is defined as the ratio between the effective number of channels (the equivalent number of channels required to offer the same traffic at the same GoS with FR mobiles only) and the actual number of channels

$$\mathrm{EF} = \frac{\text{Effective number of channels}}{\text{Actual number of channels}} = \frac{\text{Number of channels with } \gamma = 0}{\text{Number of channels for a certain } \gamma}.$$

This factor shows the resources that can be saved due to HR capability. Logically, the larger the capability (γ), the larger the EF. Tables B.3 and B.4 show the EF for 1 and 2% GoS for different number of TSLs.

From the results with different number of channels, it can be noticed that the EF is not dependent on the number of channels (N) when N is greater than 16. Additionally, the difference between 1 and 2% GoS is relatively small (less than 2%).

Figure B.3 shows the EF for 4, 8, 16 and 56 channels with 2% GoS. It can be noticed that for $N > 16$ the curves are very similar, so that the curve for $N = 56$ can be considered as describing the behaviour for a cell with more than two TRXs ($N > 16$).

Also the respective curves for 1(not shown here) and 2% GoS are very similar.

From Figure B.3, it can be seen that trunking gain increases with the number of channels, although this gain practically disappears for $N > 16$. Additionally, it can be concluded that EF is not a linear function of γ and is always below the diagonal.

Table B.3. Effective number of channels and efficiency factor for 1% GoS

		Percentage of HR connections										
		0%	10%	20%	30%	40%	50%	60%	70%	80%	90%	100%
TSL	1	1.0	1.0013	1.0037	1.0084	1.0168	1.0274	1.0494	1.0362	1.151	1.4747	2.0
	2	1.0	1.0078	1.0234	1.0517	1.0959	1.1516	1.2338	1.3264	1.4578	1.6189	2.0
	3	1.0	1.006	1.0195	1.0417	1.0748	1.1234	1.2153	1.3676	1.5189	1.726	2.0
	4	1.0	1.0118	1.0384	1.0804	1.138	1.2088	1.2995	1.4081	1.5537	1.7479	2.0
	5	1.0	1.0148	1.0462	1.0937	1.1562	1.2304	1.3229	1.4318	1.5743	1.7607	2.0
	6	1.0	1.0179	1.0531	1.1042	1.1697	1.2457	1.3384	1.4476	1.588	1.7693	2.0
	7	1.0	1.0214	1.059	1.1116	1.1783	1.2559	1.3497	1.459	1.5977	1.7753	2.0
	8	1.0	1.024	1.064	1.1187	1.1868	1.2646	1.3581	1.4673	1.605	1.780	2.0
	9	1.0	1.0263	1.0683	1.1244	1.1934	1.2714	1.3647	1.4739	1.6109	1.7837	2.0
	10	1.0	1.0288	1.0719	1.1286	1.1977	1.2763	1.3701	1.4794	1.6156	1.7867	2.0
	11	1.0	1.0309	1.0751	1.1321	1.2014	1.2805	1.3746	1.4839	1.6195	1.7892	2.0
	12	1.0	1.0327	1.0778	1.1351	1.2044	1.284	1.3784	1.4877	1.6228	1.7912	2.0
	13	1.0	1.0343	1.0802	1.1377	1.2069	1.2862	1.3804	1.4903	1.6255	1.7933	2.0
	14	1.0	1.0357	1.0823	1.1399	1.2091	1.2886	1.383	1.4929	1.6279	1.7948	2.0
	15	1.0	1.037	1.0841	1.1419	1.2109	1.2907	1.3852	1.495	1.6295	1.7958	2.0
	16	1.0	1.0381	1.0857	1.1436	1.2126	1.2925	1.3872	1.4973	1.6318	1.7973	2.0
	24	1.0	1.0435	1.094	1.1534	1.2231	1.3044	1.3998	1.5086	1.6405	1.8021	2.0
	32	1.0	1.0457	1.0984	1.1594	1.2299	1.3105	1.4047	1.5133	1.6446	1.8047	2.0
	40	1.0	1.0477	1.1011	1.1619	1.232	1.3126	1.4069	1.5161	1.6475	1.8066	2.0
	48	1.0	1.0492	1.1029	1.1633	1.2328	1.3132	1.4078	1.5184	1.6506	1.809	2.0
	56	1.0	1.0505	1.1042	1.1639	1.2326	1.3128	1.4078	1.5204	1.6541	1.812	2.0

Table B.4. Effective number of channels and efficiency factor for 2% GoS

		Percentage of HR connections										
		0%	10%	20%	30%	40%	50%	60%	70%	80%	90%	100%
TSL	1	1.0	1.0018	1.0052	1.0119	1.0238	1.0389	1.0694	1.0812	1.2082	1.5153	2.0
	2	1.0	1.0084	1.0252	1.0551	1.1014	1.1608	1.2454	1.3462	1.4822	1.7051	2.0
	3	1.0	1.0103	1.0332	1.071	1.125	1.1939	1.2836	1.3941	1.5445	1.7434	2.0
	4	1.0	1.0133	1.0423	1.0876	1.1487	1.2234	1.3176	1.4294	1.5744	1.7618	2.0
	5	1.0	1.0166	1.0505	1.1008	1.1663	1.2438	1.3387	1.4502	1.5919	1.7726	2.0
	6	1.0	1.0199	1.0574	1.111	1.1791	1.2577	1.3526	1.4639	1.6037	1.7799	2.0
	7	1.0	1.023	1.0632	1.1189	1.1885	1.2677	1.3626	1.4737	1.612	1.7851	2.0
	8	1.0	1.0258	1.068	1.1252	1.1956	1.2753	1.3702	1.4811	1.6183	1.789	2.0
	9	1.0	1.0283	1.0721	1.1302	1.2012	1.2812	1.3761	1.4869	1.6232	1.7921	2.0
	10	1.0	1.0304	1.0756	1.1344	1.2057	1.286	1.3809	1.4916	1.6272	1.7945	2.0
	11	1.0	1.0323	1.0785	1.1379	1.2095	1.2899	1.3849	1.4954	1.6305	1.7966	2.0
	12	1.0	1.0326	1.0796	1.1401	1.2126	1.2934	1.3882	1.4987	1.6332	1.7983	2.0
	13	1.0	1.0339	1.0818	1.1426	1.2153	1.2962	1.3911	1.5014	1.6356	1.7998	2.0
	14	1.0	1.0351	1.0836	1.1448	1.2176	1.2986	1.3935	1.5038	1.6376	1.801	2.0
	15	1.0	1.0361	1.0852	1.1467	1.2196	1.3008	1.3957	1.5059	1.6394	1.8022	2.0
	16	1.0	1.037	1.0866	1.1484	1.2214	1.3027	1.3976	1.5078	1.641	1.8032	2.0
	24	1.0	1.0413	1.0937	1.1569	1.2306	1.3124	1.4074	1.5175	1.6493	1.8084	2.0
	32	1.0	1.0435	1.0968	1.1601	1.2336	1.3158	1.4112	1.5219	1.6538	1.8116	2.0
	40	1.0	1.0441	1.0981	1.1621	1.2361	1.3186	1.4142	1.525	1.6566	1.8134	2.0
	48	1.0	1.044	1.0984	1.163	1.2376	1.3204	1.4161	1.5271	1.6586	1.8147	2.0
	56	1.0	1.0434	1.098	1.1633	1.2386	1.3216	1.4175	1.5287	1.660	1.8158	2.0

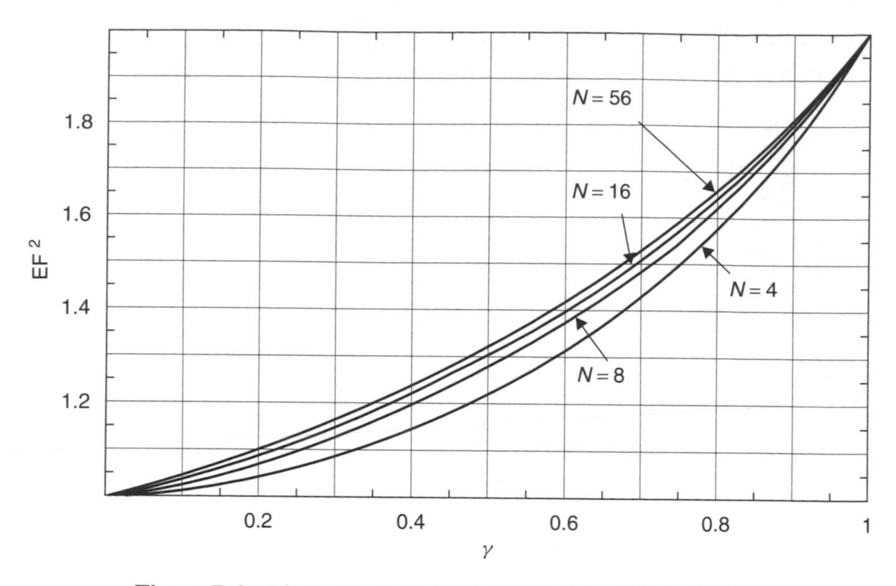

Figure B.3. Effective factor for $N = 4, 8, 16, 56$, 2% GoS

Since the EF does not depend on the number of channels when N is large ($\mathrm{EF}(N, \gamma) \approx \mathrm{EF}(\infty, \gamma)$), it is possible to use the Erlang-B formula to calculate BP or traffic replacing N with a function of capability (γ) as

$$\mathrm{BP}(T, \gamma, N) = \mathrm{BP}_{\text{Erlang B}}(T, N(\gamma)).$$

And

$$N(\gamma) \approx N \times EF(\infty, \gamma)$$

The asymptotic curve $EF(\infty, \gamma)$ can be obtained from the average situation, in which a fraction γ of the calls are half-rate and $(1 - \gamma)$ are full-rate. Half-rate calls take only $\gamma/2$ of the timeslots while full-rate connections take one timeslot each. So,

$$EF(\infty, \gamma) = \frac{1}{\frac{\gamma}{2} + (1 - \gamma)} = \frac{2}{2 - \gamma}$$

Giving:

$$BP(T, \gamma, N) \approx BP_{ErlangB}\left(T, \frac{2N}{2 - \gamma}\right) \text{ Dimensioning factor}$$

A dimensioning factor (DF) can be defined as the rate between the traffic offered for a certain GoS with a capability (γ) and the traffic offered with only FR connections ($\gamma = 0\%$)

$$\mathrm{DF} = \frac{\text{Traffic offered for a certain } \gamma}{\text{Traffic offered for } \gamma = 0}.$$

Tables B.5 and B.6 show the DF for 1 and 2% GoS for different number of TSLs.

DF shows the increase of the offered traffic for a certain number of channels as a function of HR capability. For example, for 16 TSLs and 2% GoS, the offered traffic is

Table B.5. Dimensioning factor for 1% GoS

		Percentage of HR connections										
		0%	10%	20%	30%	40%	50%	60%	70%	80%	90%	100%
	1	1.0	1.0147	1.0419	1.0959	1.1914	1.3126	1.5644	1.4133	2.759	6.893	15.107
	2	1.0	1.0187	1.057	1.1282	1.2454	1.4051	1.6709	2.0293	2.7081	3.8751	5.6977
	3	1.0	1.0213	1.0702	1.1544	1.2817	1.4524	1.6957	2.0116	2.4886	3.193	4.1912
	4	1.0	1.0246	1.0809	1.1721	1.3011	1.4661	1.6851	1.9568	2.3379	2.8707	3.5974
	5	1.0	1.0283	1.0895	1.1838	1.3113	1.4671	1.6666	1.9093	2.2378	2.6836	3.2785
	6	1.0	1.0322	1.0964	1.1914	1.3159	1.4635	1.6486	1.872	2.1677	2.5611	3.0781
	7	1.0	1.0367	1.1018	1.1946	1.3145	1.4569	1.6329	1.8426	2.1154	2.4737	2.9396
	8	1.0	1.0395	1.1061	1.1986	1.3158	1.4522	1.6194	1.8184	2.0749	2.4082	2.8378
TSL	9	1.0	1.042	1.1096	1.2013	1.3158	1.4474	1.6076	1.7986	2.0428	2.3573	2.7594
	10	1.0	1.0447	1.1123	1.2021	1.3134	1.4419	1.5977	1.7823	2.0164	2.316	2.6969
	11	1.0	1.0468	1.1144	1.2026	1.3113	1.4372	1.5891	1.7684	1.9943	2.2819	2.6459
	12	1.0	1.0486	1.1161	1.2029	1.3092	1.4329	1.5817	1.7564	1.9755	2.2531	2.6032
	13	1.0	1.05	1.1175	1.203	1.3071	1.4278	1.5732	1.7448	1.9592	2.229	2.567
	14	1.0	1.0513	1.1186	1.2029	1.3051	1.424	1.5668	1.7353	1.9449	2.2077	2.5358
	15	1.0	1.0524	1.1195	1.2027	1.3032	1.4205	1.5612	1.7265	1.9317	2.1885	2.5085
	16	1.0	1.0533	1.1202	1.2025	1.3015	1.4173	1.5561	1.7193	1.9211	2.1725	2.4846
	24	1.0	1.0566	1.1228	1.2011	1.2937	1.4026	1.5313	1.6795	1.8605	2.0843	2.3608
	32	1.0	1.0571	1.1233	1.2004	1.29	1.3931	1.5142	1.655	1.8263	2.0366	2.2944
	40	1.0	1.058	1.1232	1.1979	1.2844	1.3843	1.5019	1.6388	1.8045	2.0063	2.2519
	48	1.0	1.0587	1.1229	1.1955	1.2794	1.3769	1.4922	1.6276	1.7903	1.9863	2.2217
	56	1.0	1.0593	1.1225	1.1932	1.2747	1.3703	1.484	1.6196	1.7811	1.9728	2.1987

Table B.6. Dimensioning factor for 2% GoS

	Percentage of HR connections										
2%GoS	0%	10%	20%	30%	40%	50%	60%	70%	80%	90%	100%
1	1.0	1.0148	1.0421	1.0965	1.1923	1.3154	1.5664	1.664	2.7409	5.5765	10.95
2	1.0	1.0194	1.0588	1.131	1.248	1.4098	1.6657	2.0206	2.6082	3.5301	4.8878
3	1.0	1.0221	1.0721	1.1566	1.2819	1.4496	1.6802	1.9756	2.3944	2.9808	3.7792
4	1.0	1.0258	1.0829	1.1736	1.2998	1.4592	1.6658	1.9192	2.2596	2.7167	3.3207
5	1.0	1.0298	1.0913	1.1844	1.3085	1.4585	1.6469	1.8744	2.1716	2.5617	3.0679
6	1.0	1.0337	1.0978	1.1912	1.312	1.4542	1.6297	1.8399	2.11	2.4591	2.9064
7	1.0	1.0373	1.1029	1.1954	1.3129	1.4489	1.6149	1.8129	2.0643	2.3857	2.7936
8	1.0	1.0403	1.1068	1.1981	1.3124	1.4436	1.6024	1.7912	2.0289	2.3302	2.7098
9	1.0	1.0429	1.1099	1.1999	1.3112	1.4386	1.5918	1.7735	2.0006	2.2865	2.6448
10	1.0	1.045	1.1123	1.201	1.3099	1.434	1.5828	1.7586	1.9773	2.2512	2.5927
11	1.0	1.0467	1.1141	1.2017	1.3084	1.4299	1.5749	1.746	1.9578	2.2219	2.55
12	1.0	1.0462	1.1136	1.201	1.307	1.4263	1.5681	1.7351	1.9412	2.1972	2.5142
13	1.0	1.0474	1.1147	1.2012	1.3055	1.4229	1.562	1.7256	1.9268	2.176	2.4837
14	1.0	1.0483	1.1156	1.2013	1.3041	1.4198	1.5566	1.7173	1.9143	2.1576	2.4573
15	1.0	1.0491	1.1164	1.2013	1.3028	1.417	1.5517	1.7098	1.9032	2.1415	2.4342
16	1.0	1.0497	1.1169	1.2012	1.3016	1.4144	1.5473	1.7032	1.8933	2.1271	2.4139
24	1.0	1.0521	1.1185	1.1992	1.2937	1.3994	1.523	1.6671	1.8409	2.0522	2.3085
32	1.0	1.0528	1.1179	1.1957	1.2863	1.3881	1.507	1.6457	1.8118	2.0118	2.252
40	1.0	1.0523	1.1166	1.1931	1.2819	1.3814	1.4972	1.632	1.7929	1.9856	2.2159
48	1.0	1.0512	1.1148	1.1906	1.2784	1.3763	1.4899	1.6222	1.7794	1.9671	2.1906
56	1.0	1.0498	1.1128	1.1882	1.2756	1.3723	1.4843	1.6146	1.7692	1.9532	2.1716

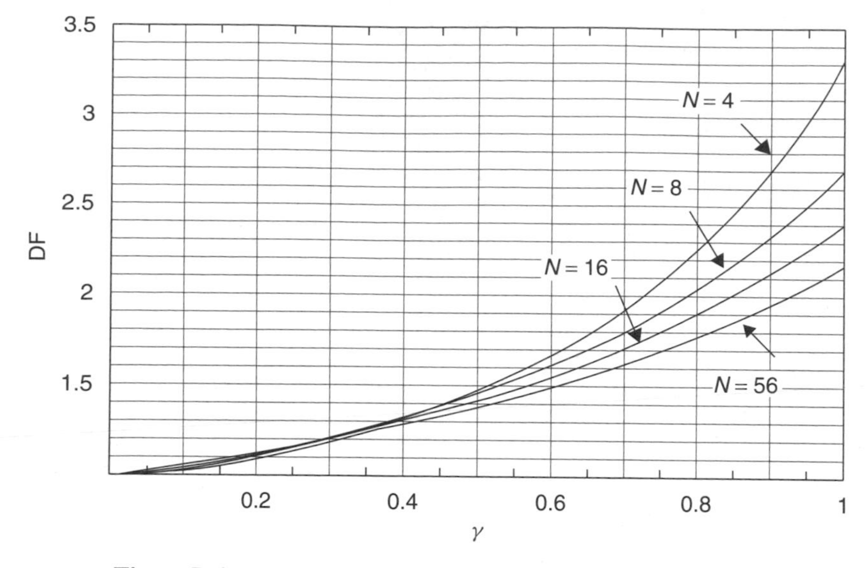

Figure B.4. Dimensioning factor for $N = 4, 8, 16, 56$, 2% GoS

increased by a factor of 1.41 for 50% of HR capability and by 2.41 if all calls are done as HR ($\gamma = 100\%$).

Figure B.4 shows the DF for different number of channels and 2% GoS. It can be seen that offered traffic increases non-linearly and surprisingly, the lines for different number of channels cross for low γ. Additionally, it can be observed how for $\gamma < 50\%$ there is not much difference in DF for different number of channels.

B.2 (E)GPRS HW Dimensioning Analysis

In (E)GPRS, circuit switch (CS) connections use dedicated resources (i.e. TSLs) while packet switch (PS) connections share the resources between the users being served, and consequently, the throughput in PS connections is reduced when the number of active users increases. The throughput experienced by a user can be calculated as (see throughput reduction factor in Section 5.2).

$$\text{User throughput} = \text{Number of allocated TSLs} \times \text{TSL capacity} \times \text{Reduction factor}$$

or

$$\text{User throughput per allocated TSL} = \text{TSL capacity} \times \text{Reduction factor}$$

where TSL capacity is the average throughput per fully utilized TSL. It represents the average throughput from the network point of view. It basically depends on the network carrier/interference (C/I) distribution. Reduction factor (RF) takes into account the user throughput reduction due to TSL sharing with other users. Figure B.5 represents how interference and TSL sharing reduce the EGPRS throughput. In some networks, TSL capacity can be the most limiting factor, while in others TSL sharing can be the dominant effect.

Under the following conditions, it is possible to model the resource utilization problem with a Markov chain [2]:

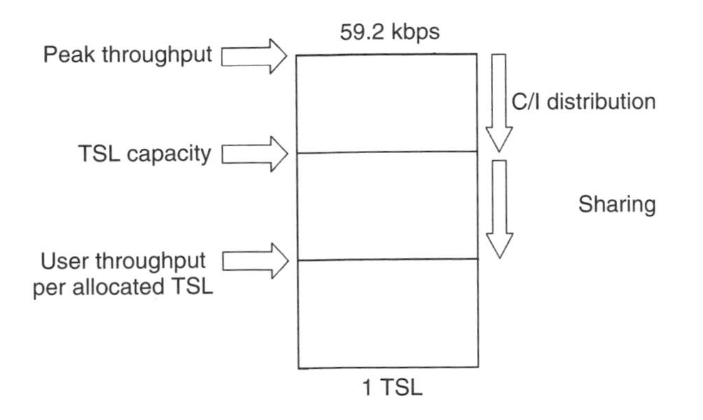

Figure B.5. Effect of interference and HW dimensioning on average user throughput

- Traffic (CS and PS) is originated from an infinite number of sources.
- Lost CS calls are cleared assuming a zero hold time.
- The number of channels in the system is limited (N_s).
- Call requests arrive independently of each other.
- The number of call request in a time unit is Poisson-distributed, with parameter λ_{CS} for CS calls and λ_{PS} for PS connections
- The circuit switch call duration is Exponential-distributed with parameter τ_{CS}. The data call length is Exponential-distributed with parameter ρ.
- The system resources (N_s) are shared in the best possible way, i.e. optimal repacking.
- The data length is much longer than the radio block capacity.
- The connection establishment time is negligible compared to the data transfer time.
- For simplicity, it is assumed that all the PS connections use N_u channels. Radio resource management system assigns maximum number of TSL to the connection for better performance.

Some of these assumptions are difficult to hold for certain kind of packet-switched (PS) traffic. Different results might be expected with different kinds of traffic mix. In any case, the results of this simple modelling are still useful for coarse (E)GPRS HW dimensioning and has been validated in Section 7.8. The assumption that system resources are shared in the best possible way is optimistic, as well, considering realistic implementation of channel allocation algorithm and TSL allocation limitations. Therefore, the analysis represents upper performance limits. The effect of channel allocation limitations is more noticeable with low network load.

Two different scenarios can be considered.

B.2.1 Dedicated PS Resources

In this case, the PS connections use a fixed number of dedicated resources, no CS traffic is considered.

B.2.1.1 Reduction Factor Calculation

The reduction factor is a function of number of TSLs assigned in the connection (N_u), number of TSLs in the system (N_s) and average utilization of resources in the system (U). Note that traffic (Data Erlangs) is $U \times N_s$. More precisely, the reduction factor is a function of N_s/N_u (N), since, for instance, a process in which users take 3 TSLs in each connection in a 24 TSL system can be modelled with a system with 8 TSLs and users taking 1 TSL per connection. Therefore,

$$\mathrm{RF} = f(Ns/Nu, U) = f(N, U).$$

This problem can be modelled as a one-dimensional Markov process and can be solved symbolically.

B.2.1.2 Model Formulation

Let X be a random variable that measures the reduction factor in a certain system state:

$$\begin{cases} X = 0, & \text{if } n = 0 \\ X = 1, & \text{if } 0 < n < N \\ X = N/n, & \text{if } N < n \end{cases}$$

where n is the instantaneous number of connections in the system.

Then, the reduction factor can be calculated as the average of X over the states with at least one connection, as

$$\text{Reduction factor} = \sum_{n=1}^{\infty} X \cdot \frac{P(X = n)}{P(X > 0)} = \sum_{n=1}^{\infty} X \cdot \frac{P(X = n)}{\sum_{i=1}^{\infty} P(X = i)}$$

where $P(X = n)$ is the probability function (i.e. the probability of having n connections in the system).

Under the assumptions considered above, the probability function can be expressed as

$$\begin{cases} P(X = n) = \dfrac{\dfrac{(U \cdot N)^n}{n!}}{\sum_{i=0}^{N} \dfrac{(U \cdot N)^i}{i!} + \sum_{i=N+1}^{\infty} \dfrac{(U \cdot N)^i}{N! \cdot N^{(i-N)}}}, & \text{if } 0 \leq n \leq N \\ P(X = n) = \dfrac{\dfrac{(U \cdot N)^n}{N! \cdot N^{(n-N)}}}{\sum_{i=0}^{N} \dfrac{(U \cdot N)^i}{i!} + \sum_{i=N+1}^{\infty} \dfrac{(U \cdot N)^i}{N! \cdot N^{(i-N)}}}, & \text{if } N < n \end{cases}$$

And finally,

$$\text{Reduction factor} = \frac{\sum_{i=1}^{N} \dfrac{(U \cdot N)^i}{i!} + \sum_{i=N+1}^{\infty} \dfrac{(U \cdot N)^i}{N! \cdot N^{(i-N)}} \left(\dfrac{N}{i}\right)}{\sum_{i=1}^{N} \dfrac{(U \cdot N)^i}{i!} + \sum_{i=N+1}^{\infty} \dfrac{(U \cdot N)^i}{N! \cdot N^{(i-N)}}}.$$

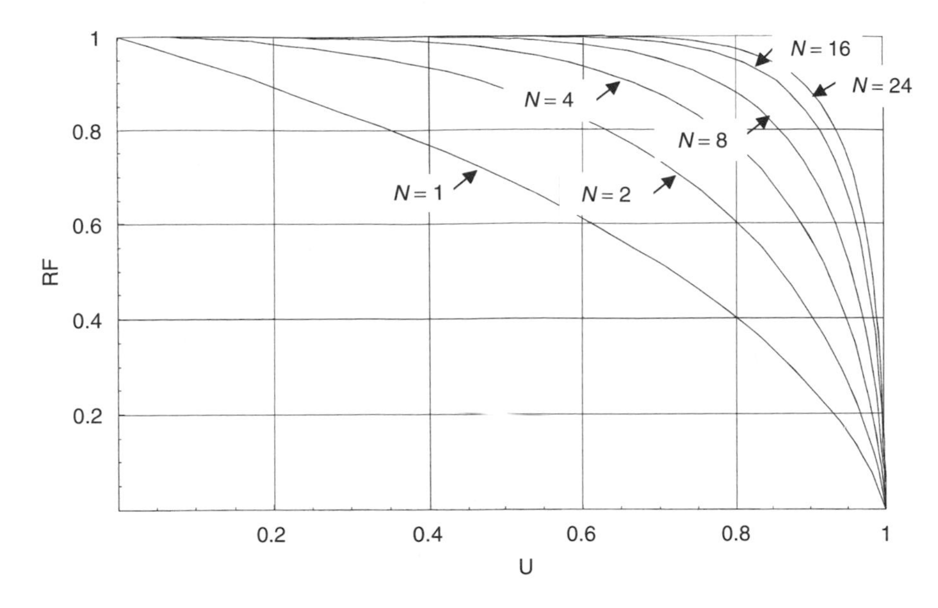

Figure B.6. Reduction factor curves

Figure B.6 graphically represents the above equation.

B.2.1.3 Blocking Probability

In GPRS, there is no blocking as in circuit-switched connections, if a new temporary block flow (TBF) establishment is requested and there are already M users per TSL, the communication is queued in the system to be established later. M is the maximum number of simultaneous TBFs per TSL that (E)GPRS can handle. In this sense, blocking probability means the probability of TBF establishment queuing and this happens when there are more than $M \times N$ active TBFs in the system.

Blocking probability can be calculated as

$$\text{Blocking probability} = \frac{\displaystyle\sum_{i=M\cdot N+1}^{\infty} \frac{(U\cdot N)^i}{N!\cdot N^{(i-N)}}}{\displaystyle\sum_{i=0}^{N} \frac{(U\cdot N)^i}{i!} + \sum_{i=N+1}^{\infty} \frac{(U\cdot N)^i}{N!\cdot N^{(i-N)}}}.$$

Figure B.7 graphically represents the above equation using the typical maximum number of users sharing a TSL, $M = 9$.

B.2.2 Shared PS and CS Resources

In this scenario, it is considered that PS and CS connections share a common pool of resources (TSLs) available. CS calls are considered to have priority with pre-emption over PS, therefore the CS blocking probability is independent of the PS calls and can be easily calculated using Erlang-B formula.

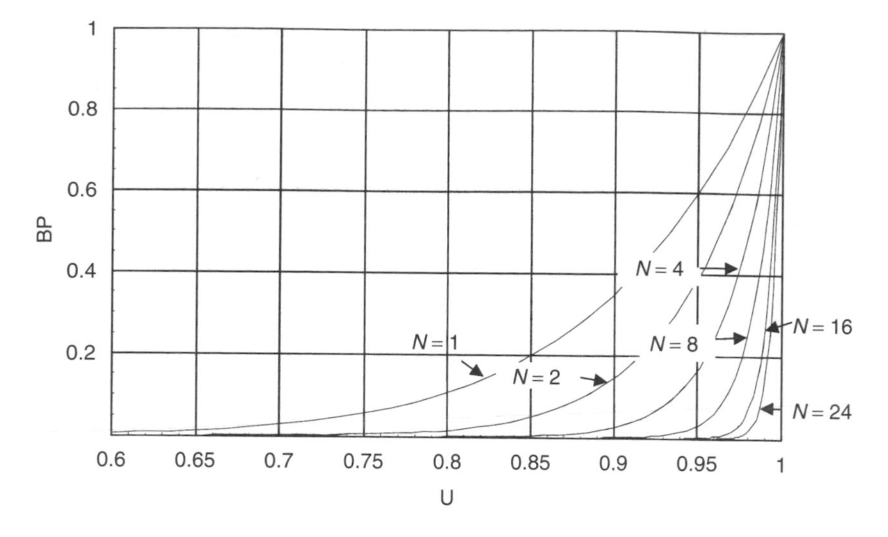

Figure B.7. Blocking probability curves

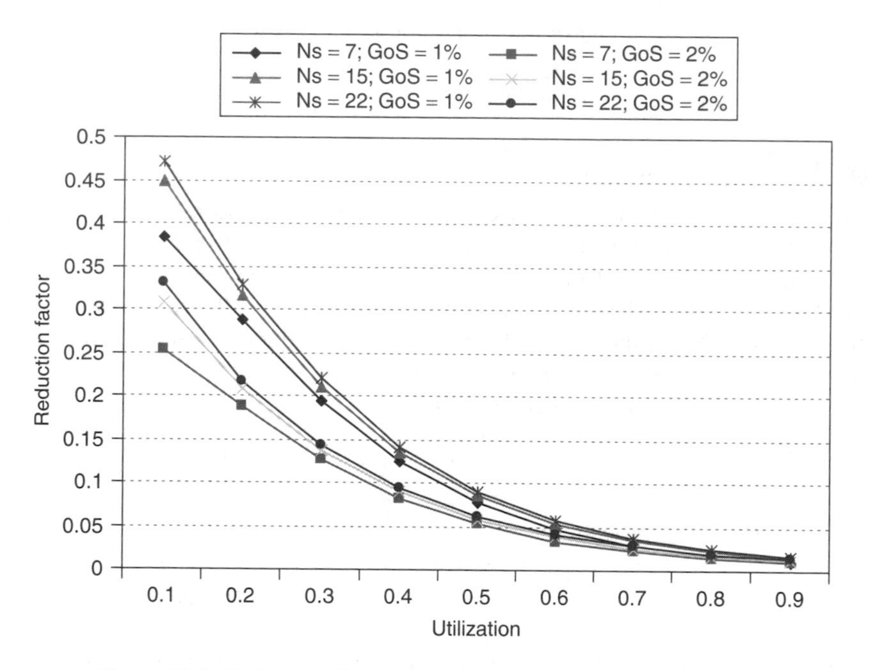

Figure B.8. Reduction factor curves for different N_s and GoS [2]

B.2.2.1 Reduction Factor Calculation

When the same resources are shared between CS and PS connections, the problem becomes more complex and it has to be modelled as a bi-dimensional Markov process. In this case, the solution has been calculated numerically as the steady state solution of a finite approximation of the differential equation system derived from the state Markov process.

Table B.7. Reduction factor for different PS utilization (U), mobile capability (Un) and CS traffic (Tc) for 7 timeslots (1 TRX typical)

	U														
	0.1			0.2			0.3			0.4			0.5		
Tc	Un = 2	Un = 3	Un = 4	Un = 2	Un = 3	Un = 4	Un = 2	Un = 3	Un = 4	Un = 2	Un = 3	Un = 4	Un = 2	Un = 3	Un = 4
0	0.999	0.992	0.968	0.992	0.968	0.920	0.975	0.926	0.858	0.942	0.865	0.780	0.888	0.784	0.688
0.3	0.998	0.989	0.959	0.990	0.961	0.903	0.968	0.913	0.833	0.927	0.843	0.748	0.858	0.747	0.643
0.6	0.996	0.983	0.948	0.983	0.946	0.882	0.952	0.886	0.800	0.892	0.796	0.694	0.784	0.662	0.557
0.9	0.990	0.969	0.927	0.965	0.916	0.844	0.912	0.830	0.737	0.812	0.698	0.596	0.646	0.519	0.425
1.2	0.970	0.935	0.882	0.921	0.851	0.769	0.827	0.722	0.625	0.675	0.550	0.456	0.478	0.363	0.289
1.5	0.923	0.864	0.794	0.834	0.738	0.646	0.693	0.572	0.479	0.512	0.394	0.317	0.331	0.241	0.188
1.8	0.839	0.751	0.665	0.706	0.590	0.498	0.537	0.419	0.339	0.366	0.269	0.211	0.226	0.159	0.122
2.1	0.722	0.611	0.520	0.562	0.444	0.362	0.396	0.295	0.232	0.258	0.183	0.142	0.157	0.108	0.083
2.4	0.593	0.475	0.390	0.431	0.325	0.258	0.289	0.207	0.161	0.184	0.128	0.098	0.112	0.077	0.058
2.7	0.474	0.362	0.289	0.327	0.238	0.185	0.213	0.149	0.115	0.134	0.092	0.070	0.083	0.057	0.043
3	0.375	0.277	0.217	0.250	0.177	0.137	0.161	0.111	0.085	0.102	0.069	0.052	0.064	0.043	0.033

	U											
	0.6			0.7			0.8			0.9		
Tc	Un = 2	Un = 3	Un = 4	Un = 2	Un = 3	Un = 4	Un = 2	Un = 3	Un = 4	Un = 2	Un = 3	Un = 4
0	0.806	0.681	0.580	0.689	0.553	0.458	0.525	0.399	0.320	0.302	0.216	0.168
0.3	0.744	0.614	0.511	0.557	0.430	0.345	0.300	0.215	0.167	0.095	0.065	0.049
0.6	0.605	0.476	0.386	0.368	0.269	0.210	0.163	0.113	0.086	0.052	0.035	0.026
0.9	0.430	0.321	0.253	0.229	0.161	0.124	0.099	0.068	0.051	0.035	0.024	0.018
1.2	0.286	0.205	0.158	0.147	0.101	0.077	0.066	0.045	0.034	0.027	0.018	0.013
1.5	0.189	0.132	0.101	0.099	0.067	0.051	0.048	0.032	0.024	0.021	0.014	0.011
1.8	0.129	0.089	0.067	0.070	0.047	0.036	0.036	0.024	0.018	0.018	0.012	0.009
2.1	0.091	0.062	0.047	0.052	0.035	0.026	0.028	0.019	0.014	0.015	0.010	0.007
2.4	0.067	0.045	0.034	0.039	0.027	0.020	0.023	0.015	0.011	0.013	0.009	0.006
2.7	0.051	0.035	0.026	0.031	0.021	0.016	0.019	0.013	0.009	0.011	0.007	0.006
3	0.040	0.027	0.020	0.025	0.017	0.013	0.016	0.011	0.008	0.010	0.007	0.005

Table B.8. Reduction factor for different PS utilization (U), mobile capability (Un) and CS traffic (Tc) for 14 timeslots (2 TRX typical)

	U														
	0.1			0.2			0.3			0.4			0.5		
Tc	Un = 2	Un = 3	Un = 4	Un = 2	Un = 3	Un = 4	Un = 2	Un = 3	Un = 4	Un = 2	Un = 3	Un = 4	Un = 2	Un = 3	Un = 4
0	1.000	1.000	0.999	1.000	0.998	0.992	0.999	0.991	0.975	0.994	0.973	0.942	0.979	0.938	0.888
0.8	1.000	1.000	0.999	1.000	0.997	0.990	0.997	0.988	0.969	0.990	0.965	0.931	0.969	0.923	0.868
1.6	1.000	0.999	0.997	0.999	0.995	0.985	0.995	0.982	0.959	0.983	0.953	0.912	0.952	0.896	0.834
2.4	1.000	0.999	0.995	0.998	0.991	0.977	0.990	0.971	0.941	0.968	0.927	0.876	0.908	0.834	0.760
3.2	0.999	0.996	0.989	0.993	0.981	0.960	0.974	0.943	0.902	0.923	0.861	0.795	0.800	0.699	0.613
4	0.993	0.984	0.971	0.973	0.948	0.913	0.923	0.867	0.807	0.811	0.718	0.636	0.617	0.501	0.418
4.8	0.969	0.945	0.916	0.911	0.857	0.799	0.801	0.710	0.630	0.626	0.513	0.430	0.418	0.316	0.253
5.6	0.904	0.850	0.795	0.780	0.688	0.608	0.613	0.501	0.420	0.426	0.324	0.260	0.264	0.190	0.148
6.4	0.785	0.695	0.617	0.599	0.488	0.408	0.421	0.320	0.257	0.273	0.197	0.153	0.166	0.116	0.089
7.2	0.633	0.523	0.440	0.428	0.326	0.261	0.279	0.202	0.157	0.175	0.123	0.094	0.108	0.074	0.056
8	0.489	0.379	0.307	0.302	0.219	0.171	0.188	0.132	0.101	0.118	0.081	0.062	0.074	0.050	0.038

	U											
	0.6			0.7			0.8			0.9		
Tc	Un = 2	Un = 3	Un = 4	Un = 2	Un = 3	Un = 4	Un = 2	Un = 3	Un = 4	Un = 2	Un = 3	Un = 4
0	0.944	0.876	0.806	0.875	0.776	0.689	0.742	0.618	0.525	0.493	0.376	0.302
0.8	0.923	0.848	0.773	0.823	0.715	0.625	0.588	0.466	0.383	0.204	0.143	0.110
1.6	0.876	0.787	0.704	0.692	0.572	0.483	0.363	0.267	0.211	0.104	0.071	0.054
2.4	0.762	0.651	0.562	0.490	0.378	0.305	0.212	0.149	0.115	0.066	0.045	0.034
3.2	0.574	0.457	0.377	0.311	0.226	0.177	0.131	0.090	0.069	0.048	0.032	0.024
4	0.382	0.285	0.226	0.194	0.136	0.105	0.086	0.059	0.045	0.036	0.024	0.018
4.8	0.240	0.171	0.132	0.124	0.085	0.065	0.060	0.041	0.031	0.029	0.019	0.015
5.6	0.151	0.105	0.080	0.082	0.056	0.042	0.044	0.030	0.022	0.023	0.016	0.012
6.4	0.098	0.067	0.051	0.057	0.039	0.029	0.033	0.022	0.017	0.019	0.013	0.010
7.2	0.067	0.045	0.034	0.041	0.028	0.021	0.025	0.017	0.013	0.016	0.011	0.008
8	0.047	0.032	0.024	0.030	0.020	0.015	0.020	0.013	0.010	0.013	0.009	0.006

Table B.9. Reduction factor for different PS utilization (U), mobile capability (Un) and CS traffic (Tc) for 22 timeslots (3 TRX typical)

	U														
	0.1			0.2			0.3			0.4			0.5		
Tc	Un = 2	Un = 3	Un = 4	Un = 2	Un = 3	Un = 4	Un = 2	Un = 3	Un = 4	Un = 2	Un = 3	Un = 4	Un = 2	Un = 3	Un = 4
0	1.000	1.000	1.000	1.000	1.000	0.999	1.000	0.999	0.995	0.999	0.994	0.984	0.995	0.980	0.957
1.5	1.000	1.000	1.000	1.000	1.000	0.999	1.000	0.998	0.993	0.998	0.991	0.978	0.991	0.973	0.946
3	1.000	1.000	1.000	1.000	0.999	0.998	0.999	0.996	0.989	0.996	0.986	0.969	0.986	0.961	0.928
4.5	1.000	1.000	0.999	1.000	0.999	0.995	0.998	0.993	0.983	0.992	0.977	0.953	0.970	0.935	0.892
6	1.000	1.000	0.998	0.999	0.996	0.990	0.994	0.984	0.968	0.976	0.950	0.916	0.918	0.859	0.799
7.5	0.999	0.997	0.993	0.993	0.985	0.973	0.974	0.951	0.921	0.917	0.863	0.806	0.776	0.680	0.600
9	0.988	0.980	0.968	0.960	0.933	0.902	0.893	0.835	0.777	0.756	0.659	0.580	0.546	0.435	0.359
10.5	0.942	0.908	0.873	0.847	0.776	0.711	0.701	0.598	0.519	0.511	0.403	0.331	0.327	0.241	0.190
12	0.822	0.744	0.675	0.636	0.529	0.450	0.455	0.352	0.286	0.299	0.219	0.172	0.185	0.130	0.100
13.5	0.645	0.537	0.456	0.421	0.321	0.258	0.272	0.197	0.154	0.172	0.121	0.093	0.108	0.074	0.056
15	0.478	0.370	0.300	0.276	0.199	0.156	0.168	0.118	0.090	0.106	0.073	0.055	0.068	0.046	0.035

	U											
	0.6			0.7			0.8			0.9		
Tc	Un = 2	Un = 3	Un = 4	Un = 2	Un = 3	Un = 4	Un = 2	Un = 3	Un = 4	Un = 2	Un = 3	Un = 4
0	0.981	0.948	0.907	0.942	0.881	0.818	0.850	0.752	0.669	0.627	0.505	0.421
1.5	0.971	0.932	0.886	0.918	0.847	0.778	0.762	0.652	0.565	0.325	0.238	0.187
3	0.953	0.903	0.848	0.851	0.761	0.682	0.550	0.435	0.358	0.164	0.114	0.088
4.5	0.898	0.827	0.758	0.688	0.575	0.490	0.333	0.245	0.193	0.100	0.069	0.052
6	0.757	0.654	0.571	0.465	0.358	0.289	0.198	0.140	0.108	0.069	0.047	0.035
7.5	0.534	0.423	0.348	0.282	0.204	0.160	0.123	0.085	0.065	0.051	0.034	0.026
9	0.326	0.240	0.189	0.168	0.118	0.090	0.081	0.055	0.042	0.039	0.026	0.020
10.5	0.189	0.133	0.102	0.103	0.071	0.054	0.055	0.037	0.028	0.030	0.020	0.015
12	0.111	0.076	0.058	0.065	0.044	0.034	0.039	0.026	0.020	0.023	0.016	0.012
13.5	0.068	0.046	0.035	0.043	0.029	0.022	0.028	0.019	0.014	0.018	0.012	0.009
15	0.044	0.030	0.023	0.030	0.020	0.015	0.020	0.013	0.010	0.014	0.009	0.007

Table B.10. Reduction factor for different PS utilization (U), mobile capability (Un) and CS traffic (Tc) for 30 timeslots (4 TRX typical)

	U														
	0.1			0.2			0.3			0.4			0.5		
Tc	Un = 2	Un = 3	Un = 4	Un = 2	Un = 3	Un = 4	Un = 2	Un = 3	Un = 4	Un = 2	Un = 3	Un = 4	Un = 2	Un = 3	Un = 4
0	1.000	1.000	1.000	1.000	1.000	1.000	1.000	1.000	0.999	1.000	0.999	0.994	0.998	0.993	0.981
2.2	1.000	1.000	1.000	1.000	1.000	1.000	1.000	1.000	0.998	1.000	0.997	0.992	0.997	0.989	0.975
4.4	1.000	1.000	1.000	1.000	1.000	1.000	1.000	0.999	0.997	0.999	0.995	0.988	0.995	0.983	0.964
6.6	1.000	1.000	1.000	1.000	1.000	0.999	1.000	0.998	0.994	0.998	0.991	0.980	0.989	0.971	0.946
8.8	1.000	1.000	1.000	1.000	0.999	0.997	0.999	0.995	0.988	0.992	0.981	0.963	0.967	0.935	0.898
11	1.000	0.999	0.999	0.998	0.996	0.991	0.992	0.983	0.969	0.967	0.939	0.907	0.885	0.820	0.757
13.2	0.996	0.993	0.989	0.985	0.974	0.959	0.952	0.920	0.885	0.863	0.795	0.731	0.683	0.578	0.498
15.4	0.970	0.952	0.932	0.911	0.865	0.819	0.801	0.719	0.648	0.626	0.519	0.440	0.421	0.321	0.259
17.6	0.871	0.810	0.754	0.708	0.610	0.533	0.531	0.424	0.351	0.361	0.270	0.215	0.226	0.161	0.125
19.8	0.687	0.584	0.505	0.455	0.353	0.286	0.299	0.219	0.172	0.192	0.135	0.104	0.121	0.084	0.064
22	0.501	0.393	0.321	0.284	0.206	0.161	0.173	0.121	0.093	0.109	0.075	0.057	0.071	0.048	0.036

	U											
	0.6			0.7			0.8			0.9		
Tc	Un = 2	Un = 3	Un = 4	Un = 2	Un = 3	Un = 4	Un = 2	Un = 3	Un = 4	Un = 2	Un = 3	Un = 4
0	0.992	0.975	0.950	0.970	0.931	0.885	0.904	0.830	0.758	0.713	0.599	0.512
2.2	0.987	0.965	0.936	0.956	0.908	0.857	0.853	0.764	0.686	0.447	0.341	0.275
4.4	0.979	0.950	0.914	0.923	0.861	0.799	0.700	0.589	0.505	0.234	0.167	0.130
6.6	0.955	0.912	0.865	0.822	0.732	0.654	0.467	0.361	0.292	0.140	0.097	0.074
8.8	0.873	0.800	0.732	0.619	0.507	0.426	0.282	0.204	0.160	0.094	0.064	0.049
11	0.684	0.577	0.496	0.392	0.296	0.236	0.171	0.120	0.092	0.068	0.046	0.035
13.2	0.438	0.336	0.271	0.230	0.164	0.127	0.108	0.074	0.056	0.050	0.034	0.026
15.4	0.247	0.177	0.138	0.134	0.093	0.071	0.071	0.048	0.036	0.038	0.026	0.019
17.6	0.135	0.094	0.072	0.080	0.054	0.041	0.047	0.032	0.024	0.029	0.019	0.014
19.8	0.077	0.052	0.040	0.049	0.033	0.025	0.032	0.021	0.016	0.021	0.014	0.011
22	0.047	0.032	0.024	0.032	0.021	0.016	0.022	0.015	0.011	0.015	0.010	0.008

Table B.11. Reduction factor for different PS utilization (U), mobile capability (Un) and CS traffic (Tc) for 37 timeslots (5 TRX typical)

	U														
	0.1			0.2			0.3			0.4			0.5		
Tc	Un = 2	Un = 3	Un = 4	Un = 2	Un = 3	Un = 4	Un = 2	Un = 3	Un = 4	Un = 2	Un = 3	Un = 4	Un = 2	Un = 3	Un = 4
0	1.000	1.000	1.000	1.000	1.000	1.000	1.000	1.000	1.000	1.000	0.999	0.998	0.999	0.996	0.990
2.8	1.000	1.000	1.000	1.000	1.000	1.000	1.000	1.000	0.999	1.000	0.999	0.996	0.999	0.994	0.986
5.6	1.000	1.000	1.000	1.000	1.000	1.000	1.000	1.000	0.999	1.000	0.998	0.994	0.998	0.991	0.980
8.4	1.000	1.000	1.000	1.000	1.000	1.000	1.000	0.999	0.998	0.999	0.996	0.990	0.995	0.984	0.968
11.2	1.000	1.000	1.000	1.000	1.000	0.999	1.000	0.998	0.995	0.997	0.991	0.981	0.985	0.966	0.942
14	1.000	1.000	1.000	1.000	0.999	0.997	0.997	0.993	0.986	0.985	0.971	0.951	0.939	0.897	0.852
16.8	0.999	0.998	0.996	0.994	0.989	0.982	0.978	0.961	0.941	0.924	0.880	0.834	0.787	0.698	0.624
19.6	0.985	0.976	0.965	0.951	0.923	0.893	0.875	0.814	0.758	0.728	0.631	0.554	0.518	0.411	0.339
22.4	0.910	0.866	0.822	0.779	0.694	0.623	0.615	0.508	0.432	0.434	0.334	0.270	0.277	0.201	0.157
25.2	0.733	0.638	0.561	0.504	0.399	0.328	0.340	0.252	0.200	0.221	0.157	0.122	0.141	0.097	0.075
28	0.534	0.424	0.350	0.303	0.221	0.174	0.185	0.130	0.100	0.118	0.081	0.062	0.077	0.052	0.040

	U											
	0.6			0.7			0.8			0.9		
Tc	Un = 2	Un = 3	Un = 4	Un = 2	Un = 3	Un = 4	Un = 2	Un = 3	Un = 4	Un = 2	Un = 3	Un = 4
0	0.996	0.985	0.969	0.981	0.953	0.919	0.931	0.871	0.811	0.765	0.659	0.575
2.8	0.993	0.979	0.959	0.972	0.938	0.898	0.897	0.825	0.757	0.542	0.429	0.354
5.6	0.988	0.969	0.944	0.953	0.908	0.859	0.792	0.694	0.614	0.301	0.220	0.172
8.4	0.976	0.948	0.914	0.892	0.823	0.758	0.580	0.467	0.389	0.180	0.126	0.097
11.2	0.930	0.880	0.828	0.732	0.630	0.549	0.365	0.272	0.217	0.119	0.082	0.062
14	0.790	0.700	0.624	0.498	0.390	0.319	0.221	0.157	0.122	0.084	0.057	0.043
16.8	0.546	0.437	0.362	0.295	0.215	0.169	0.137	0.095	0.072	0.062	0.042	0.032
19.6	0.312	0.229	0.181	0.168	0.118	0.090	0.087	0.059	0.045	0.046	0.031	0.023
22.4	0.166	0.116	0.089	0.097	0.066	0.050	0.056	0.038	0.029	0.034	0.023	0.017
25.2	0.089	0.061	0.046	0.057	0.038	0.029	0.037	0.025	0.019	0.024	0.016	0.012
28	0.051	0.034	0.026	0.035	0.023	0.018	0.024	0.016	0.012	0.017	0.011	0.009

The reduction factor can be calculated as a function of CS Traffic (Voice Erlangs), PS Traffic (Data Erlangs), the call length ratio (W, the ratio between PS and a CS call duration), the amount of TSLs available (N_s) and MS TSL capability (N_u). The average user throughput not only depends on CS and PS traffic but also on the PS traffic burstiness.

Figure B.8 shows RF curves for different amounts of system resources ($N_s = 7, 15, 22$ corresponding to 1, 2 and 3 TRXs) and CS blocking probabilities (1 and 2%). The CS traffic is calculated with Erlang-B formulas for a specific CS blocking probability (GoS). The call length ratio W is 1/120, which corresponds, for instance, to an average CS call of 12 and 1 s of average PS call duration. TSL utilization in x-axis is calculated with different data loads as U = Data Erlangs/(N_s Voice Erlangs).

Reduction factors have been calculated for mixed CS and PS dimensioning purposes in Tables B.7 (1 TRX) to B.11 (5 TRXs). RF have been calculated with W equal to 1/120 and different values of CS Traffic (T_{CS}), PS Utilization (U, as defined before), and mobile capability (N_u).

References

[1] Allen A. O., Probability, Statistics, and Queuing Theory. With Computer Science Application.

[2] Pedraza S., Romero J., Muñoz J., '(E)GPRS Hardware Dimensioning Rules with Minimum Quality Criteria', *IEEE Vehicular Technology Society Fall Conference*, 2002.

Appendix C

Mapping Link Gain to Network Capacity Gain

Pablo Tapia

C.1 Introduction

This appendix presents a simple method to estimate the network capacity associated with link level enhancing functionality. The method converts a link level gain (in dB) to an estimated network capacity increase. This is done by means of a function that has been constructed from extensive network level simulation analysis.

This gain estimation method can be very useful when there is no time or possibilities to model the specific feature in a network simulator, which is usually a highly time-consuming task. However, if resources are available, the best way to achieve reliable capacity numbers is always the feature modelling.

The results presented in this appendix have been obtained from network level simulations. The modelling of the link gain is very simple. It consists of a fixed dB offset that is added systematically after every carrier/interference (C/I) calculation:

$$\text{New C/I (burst } n) = \text{Real C/I (burst } n) + \text{Link offset (dB)}.$$

This offset corresponds to the studied link level gain and generates the same result in the resulting frame error rate (FER) distribution as a link level enhancing functionality with a link offset (dB) gain.

C.2 Theoretical Analysis

In principle, it could be expected that a 3-dB gain (3 dB higher interference tolerated for the same FER) would lead to a capacity increase of up to 100% (since the network can tolerate double the interference). With the same reasoning, a 6-dB gain would translate to a maximum of 300% increase. However, mobile networks are complex systems and

GSM, GPRS and EDGE Performance 2nd Ed. Edited by T. Halonen, J. Romero and J. Melero
 ISBN: 0-470-86694-2

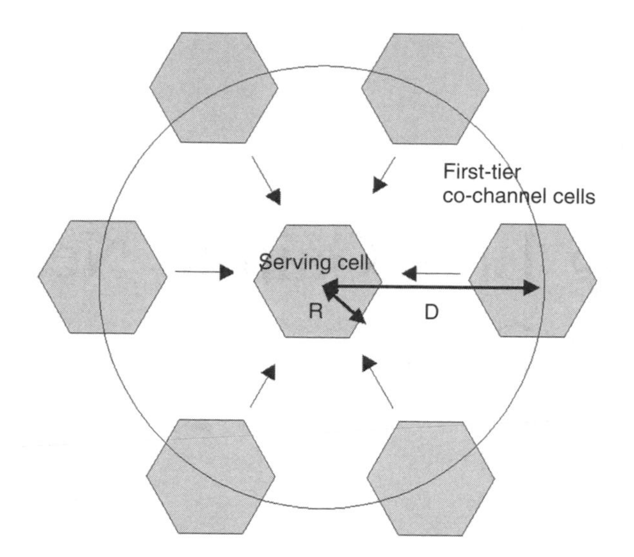

not easy to predict, and these numbers can be considered as *ideal* ones. As it will be shown later, there are several factors that have to be taken into account that will limit the capacity increase associated with certain link level gains. Theoretically, a method could be developed assuming an ideal mobile network to calculate the capacity increase. Considering a generic cellular system with an ideal hexagonal layout, the mean carrier-to-interference ratio (CIR) in the network can be easily estimated on the basis of geometrical considerations (assuming that interference is coming from the first ring of co-channel cells) [1].

Assuming an exponential propagation law, the path loss factor is:

$$Lb = k \cdot r\gamma$$

(k is a constant, r is the distance, and γ the propagation factor)

Considering CIR calculated at the edge of the cell:

$$\mathrm{CIR} = \frac{k_a \cdot R^{-\gamma}}{\sum\limits_{i\,\mathrm{Neighbours}} k_i \cdot r_i^{-\gamma}}$$

Taking into account the relationship between the cell radius (R) and reuse distance (D) in a hexagonal grid:

$$\text{Reuse distance } (D) = \sqrt{3K} \cdot R$$

where K is a rhombic number ($K = 1, 3, 4, 7, 9, \ldots$).

A relationship can be established between the frequency reuse factor (K) and the mean CIR in the network. According to [1], the final expression can be simplified to the following formula, for an omni-directional layout:

$$\mathrm{CIR} = \frac{1}{2} \cdot \frac{1}{(\sqrt{3K} - 1)^{-\gamma} + (\sqrt{3K})^{-\gamma} + (\sqrt{3K} + 1)^{-\gamma}}.$$

Table C.1. Theoretical capacity increase

Link level gain (dB)	Capacity increase (theoretical, $\gamma = 3.5$)
3	43%
6	110%

In the case of sectorised antennas, there is no good simplification for the CIR, and it should be calculated for every specific reuse pattern by geometrical calculations (taking into account also the beamwidth of the antennas: 120°, 65°, etc.). In this theoretical approach, we will just consider omni-directional antennas.

Comparing two different networks, A and B, with the same traffic conditions and different reuse schemes, the capacity difference between them is obtained from the following expression:

$$\text{Capacity increase (\%)} = 100 \times \left(\frac{\text{Reuse A (loose)}}{\text{Reuse B (tight)}}\right) - 1.$$

This capacity increase means that network B is using fewer of the resources than network A, while keeping the same amount of traffic in both systems. This will cause the CIR distribution in network B to be degraded, which will be perceived as a degradation in the quality of the system.

In order to achieve a certain capacity increase by tightening the reuse scheme, it will be necessary to improve the link level performance of the network by a certain factor. This factor will be called *link level gain*, and has to be chosen so that it compensates the CIR degradation caused by the reduction of the reuse:

$$\text{Link level gain} = \frac{\text{CIR}_\text{A}}{\text{CIR}_\text{B}} = \left(\frac{(\sqrt{3K_\text{B}}-1)^{-\gamma} + (\sqrt{3K_\text{B}})^{-\gamma} + (\sqrt{3K_\text{B}}+1)^{-\gamma}}{(\sqrt{3K_\text{A}}-1)^{-\gamma} + (\sqrt{3K_\text{A}})^{-\gamma} + (\sqrt{3K_\text{A}}+1)^{-\gamma}}\right).$$

Table C.1 shows some values calculated as an example.

These results show that introducing some physical limitation (in this case, the propagation factor) reduces the ideal capacity figures. In order to achieve these ideal figures (100% for 3 dB and 300% for 6 dB), a propagation factor of $\gamma = 2$ should be considered, that is, assuming space-free propagation conditions.

It should be noted that these values have been obtained by simplified theoretical calculations. Also, user mobility has not been considered, as well as many other features existing in real networks: frequency hopping, power control, discontinuous transmission (DTX), etc. As a result, this method could only be suitable to calculate the capacity increase on a broadcast control channel (BCCH) transceiver (TRX), under certain circumstances: good network layout, optimal frequency plan, high reuse factors, etc.

C.3 Simulations

In the previous section, a theoretical study presented how the link level gain could be translated into capacity increase of cellular networks. However, real cellular systems are not easy to analyse because of their dynamic behaviour. There are many factors to consider, and the behaviour of each of them separately is not necessarily the same as when they are combined with other effects. With this complexity, the resolution of this problem requires the use of a dynamic network simulator.

The results of the network simulations will be presented here. The simulations have been run over an ideal hexagonal layout, assuming typical urban 3 km/h (TU3) propagation model. The curves present the evolution of FER when effective frequency load (EFL) is increased, calculating the capacity difference between the EFL points with the same quality.

C.3.1 BCCH Layer Performance

Next simulations were configured to emulate BCCH TRX performance. Since BCCH is studied, only one TRX per cell is modelled, assuming full power transmission in all slots, no power control and no frequency hopping. Sectorised antennae with 65° beamwidth were used.

In Figure C.1, three curves are presented: the first one ('no offset') corresponds to a reference network; the second one ('3-dB offset') emulates the performance of a network with 3 dB better link level performance than the reference; and the third one ('6-dB offset') shows the behaviour when the link level gain is 6 dB. The *y*-axis shows the percentage of bad-quality FER samples, that is, those whose FER is above 4%.

The different points in the curve have been obtained by changing the reuse pattern (reducing the number of frequencies). It is the only way to observe differences in the performance on this layer, due to the full-time transmission of the broadcast channels.

When the trend of the curves becomes stable, roughly at 8% outage, it can be seen that the capacity increases for 3- and 6-dB link gain are 40 and 90% respectively. These values are in line with the predicted capacity increase of the theoretical method, which indicated 43 and 110% gain respectively.

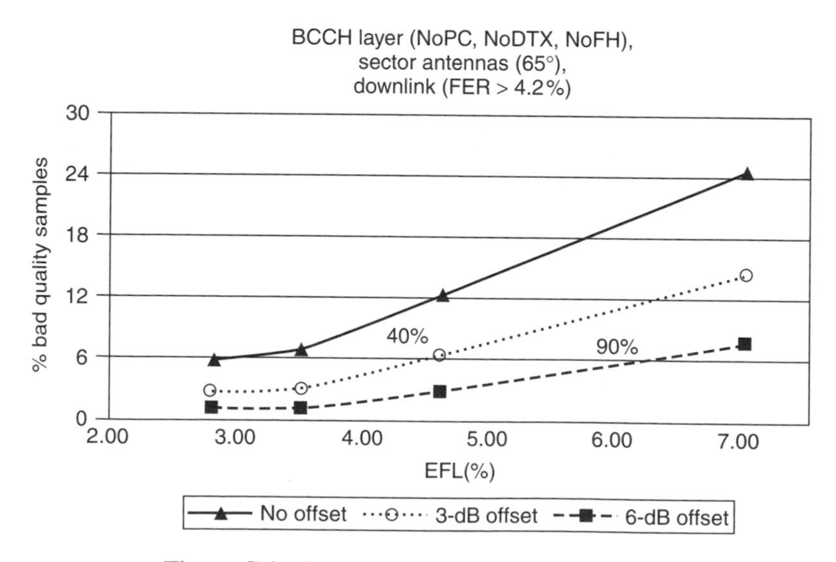

Figure C.1. Capacity increase in the BCCH layer

C.3.2 Hopping Layer

In this book, we have considered the baseline speech global system for mobile communications (GSM) network configuration to include frequency hopping, DTX and power control in the non-BCCH TRXs. The combination of these factors causes the interference to be more equally distributed in the network. This section will include the capacity gain analysis for both hopping and non-hopping configurations in the non-BCCH TRXs.

C.3.2.1 Non-hopping Network

The simulation results in Figure C.2 apply for a non-hopping network, with power control and DTX enabled.

Looking at the results at 5% outage, it can be seen that the capacity increases are 90 and 190% for 3- and 6-dB link level gain respectively. It can be concluded that power control effect is taking better advantage of the potential gains introduced by link level enhancements. This effect will be further analysed in Section C.3.3.

C.3.2.2 Hopping Network

The simulations in Figure C.3 provide the values for a typical GSM network configuration on the traffic channel (TCH) layer. In this case, frequency hopping as well as power control and DTX are switched on. As in the previous section, power control parameters have been tuned to achieve the maximum capacity in every case.

Also, the capacity increase values are now slightly better than in the non-hopping network case: 100% for 3 dB and 215% for 6 dB (at 2% outage).

C.3.3 Effect of Power Control

It has been shown in previous results how the use of power control could be applied to improve the capacity gain due to a link level improvement. While in a network with no

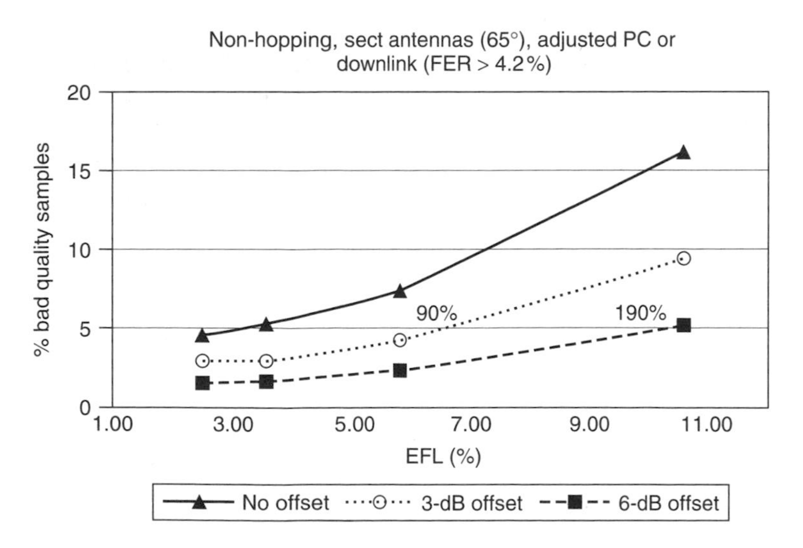

Figure C.2. Capacity increase in the non-BCCH layer (non-hopping, power control, DTX)

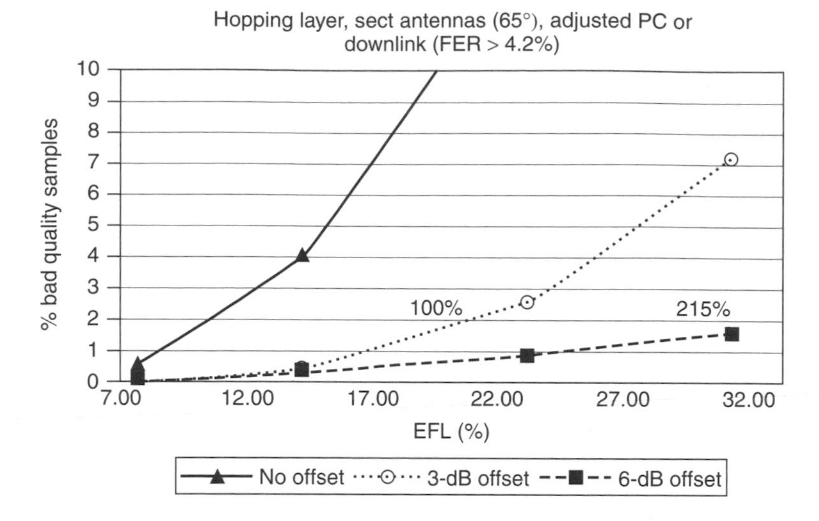

Figure C.3. Capacity increase in the non-BCCH layer (frequency hopping, power control, DTX)

power control the capacity increase with 3 dB was 50%, when power control is activated, this capacity increase rises to 100%.

Power control algorithms are an effective method to limit the waste of spare resources in the network, thus leaving space for new incoming users. As power control reduces the power of some users, there will effectively be more resources that can be utilised to improve the FER of mobiles with poorer quality. This effect can be observed by examining Figure C.4.

This figure represents the *pseudo CIR* that the user with enhanced link level would experience, if the link level gain were translated into an effective reduction of the interference. In reality, gains at link level are provided at the receiver chain; however, these representations can help to understand the effect of power control.

Figure C.4 shows the CIR distribution for the BCCH and hopping layer, both for the reference case and for a 3-dB link level gain. Figure C.4(a) can be taken as an example of a non-power control network, while Figure C.4(b) uses power control.

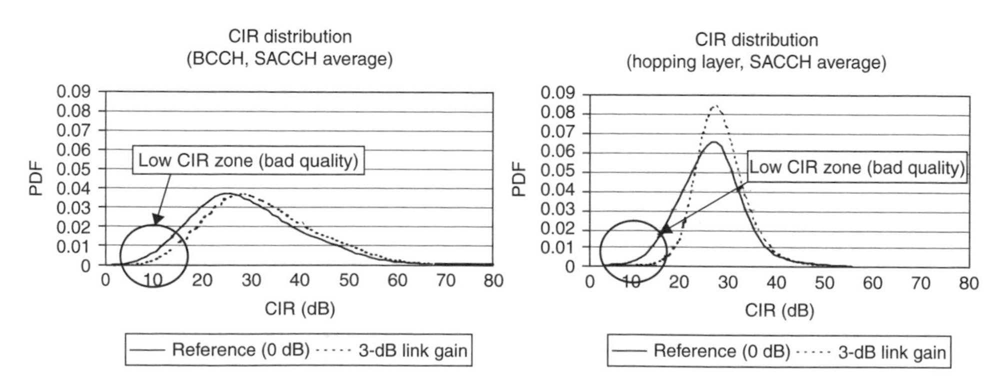

Figure C.4. CIR distribution of BCCH and hopping layer

It can be observed that, when no power control is applied, the shape of the distribution remains almost unchanged (the effect is a right shift of the whole curve). When power control is used, the shape of the distribution changes, and the potential gain from a better link level is used to effectively share the quality in the network; this can be observed by a narrower shape in the 3-dB gain case, which is placed almost completely in the *good CIR zone* of the x-axis (from 15 dB onwards). Thus, both distributions have different impact on the capacity since the difference of the low CIR values is broader when power control is used.

C.4 Final Results and Conclusions

The conclusions of this study can be summarised in Figure C.5, which presents values of link gain versus capacity gain for both typical BCCH and hopping configurations.

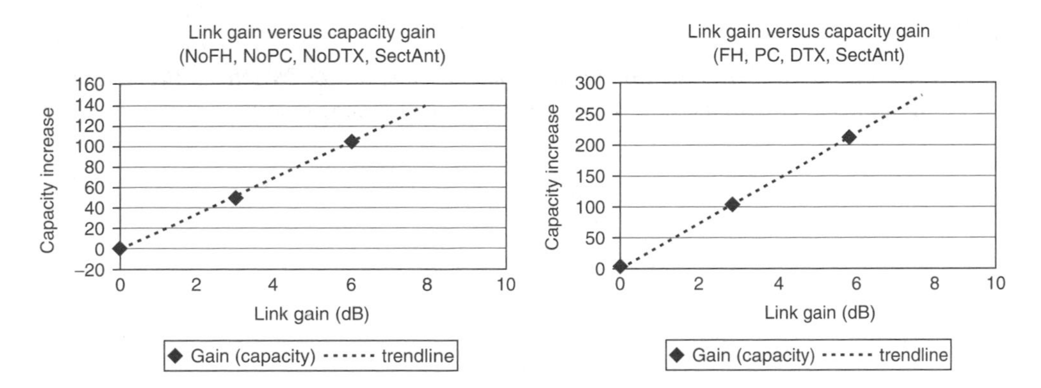

Figure C.5. Link gain versus capacity gain figures (BCCH, hopping layer)

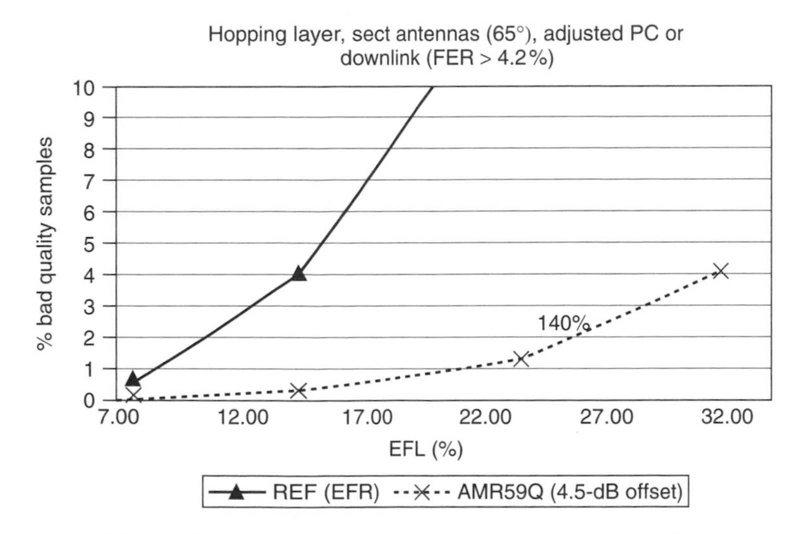

Figure C.6. Simulated performance of AMR FR 5.9 codec

Figure C.5 can be used to calculate the approximate capacity increase of any link level enhancement that achieves a constant gain in the receiver chain, such as theadaptive multi-rate AMR codecs. However, this method should not be directly used to estimate the network gain from those features in which the link gain may change with different propagation channel or instantaneous bit error rate conditions (like, for example, some of the diversity techniques presented in Chapter 11).

For instance, keeping in mind that the link level gain of the AMR 5.90 codec with frequency hopping is 4.5 dB, the capacity increase for this codec would be 150% according to Figure C.5. Figure C.6 shows the results when the codec is modelled in the simulator, where the gain achieved is around 140%.

Finally, note that the values from these figures should only be taken as an *estimation* of the real capacity increase associated with the feature under study. If accurate capacity studies are needed, the feature should be modelled using a dynamic network simulator.

References

[1] Rappaport, T. S., *Wireless Communications, Principles & Practice*, Prentice Hall Communications Engineering and Emerging Technologies Series Copyright 1996, Edition 1999.

Appendix D

Interference between GSM/EDGE and Other Cellular Radio Technologies

Kari Heiska

D.1 Introduction

This appendix describes the main interference sources and the network planning principles associated with the adjacent channel operation of different technologies. The interference sources and basic mechanisms are typically technology independent, but the relative importance of various interference mechanisms is dependent on the implementation of different network elements, locations of the interfered and interfering sites with respect to each other as well as type and size of the cells/sectors. Typically, in order to control the effect of the adjacent system interference, guardbands between the adjacent systems are deployed. The size of these guardbands has to be carefully selected in order to maximise the overall performance.

The interference between global system for mobile communications (GSM)/enhanced data rates for global evolution (EDGE) and other cellular technologies is an important issue to be carefully considered in scenarios when these technologies are deployed in adjacent frequency bands. Typical reasons for these multi-technology deployments would be:

- Technology migration. If cellular operators using IS-136 or IS-95 decide to migrate technology to GSM/EDGE, during the migration period, both technologies are likely to be deployed within the same band and inter-system interference has to be considered for the deployment.
- GSM/EDGE and wideband code division multiple access (WCDMA) multi-radio operation within the same band. GSM/EDGE and WCDMA can be seamlessly integrated

GSM, GPRS and EDGE Performance 2nd Ed. Edited by T. Halonen, J. Romero and J. Melero
 ISBN: 0-470-86694-2

and deployed together using the same 1800 and/or 1900 band. In this deployment scenario, the inter-system interference has to be taken into account.

- Licensed bands using different technologies. In the case in which different technologies are deployed in adjacent bands by different operators, there will be certain inter-system interference to consider. National regulations, such as Federal Communications Commission (FCC) requirements in the United States [1, 2], will ensure that this interference is controlled.

In the first two scenarios, the technologies will typically be co-sited since the same operator deploys them. These cases are under the control of operators' design criteria and will be the focus of this appendix.

The interference sources in the mobile telecommunication systems can be divided into controllable and non-controllable interference. The controllable interference refers to the interference from the sources that can be controlled by the operator itself, that is, the operator is able to move the interfering sites, change the equipment of the site and to reconfigure the network parameters. Non-controllable interference from other operator devices or from background noise is thus more difficult to eliminate.

The easiest way to decrease the inter-system interference is to co-site the interfering systems. By co-siting, it is possible to avoid the near–far effect, which means, for example, that when the mobile of operator 1 is close to operator 2's site and far away from its own site, there could be significant uplink interference from operator 1's mobile to operator 2's base station receiver. Also, when the operator 2's mobile terminal is close to the operator 1's base station, there can be a large downlink interference component from operator 1 to operator 2. Similar interference components occur from operator 2 to operator 1.

Figure D.1 shows the example frequency allocation case in which different systems and different operators have been allocated into the same frequency band. In the general situation, operator 1 and operator 2 are not co-sited and their network structures are not congruent. This means, for example, that when operator 1 utilises micro cells to cover

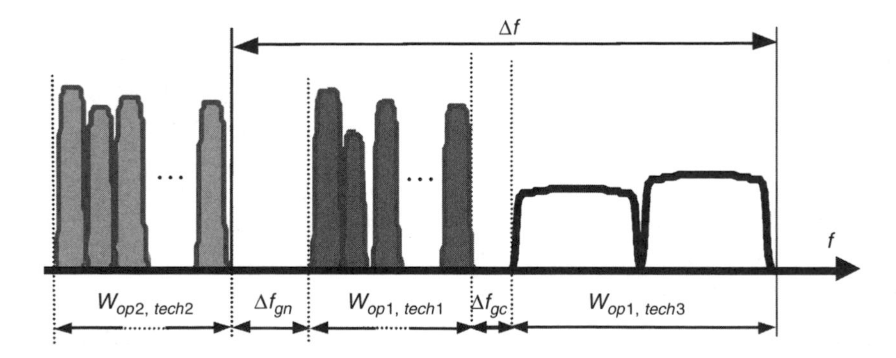

Figure D.1. Frequency allocation example. Operator 1 utilises two different technologies in its frequency band and the adjacent operator 2 uses its own technology. Δf_{gn} refers to the guardband between operators (non-coordinated case) and Δf_{gc} refers to the guardband between technologies (coordinated case)

a certain part of the urban area, operator 2 utilises macro cells. The average cell size for these two operators can thus be rather dissimilar, which increases the inter-system interference.

D.2 Interference Mechanisms

This section describes the most dominant interference mechanisms, which take place in the adjacent channel operation.

D.2.1 Adjacent Channel Power

The adjacent channel interference (ACI) refers to the power coming to the receiver due to finite receive filter attenuation outside the band of interest. Even with the ideal transmitter emission mask there is interference coming from adjacent channels, which depends on the implementation of analogue and digital filtering. Additionally, the ACI is dependent on the power of the interfering system as well as the frequency offset between the interferer and the interfered systems. The effect of the ACI decreases rapidly outside the receive band so that the ACI can be eliminated with an adequate guardband aside the band of interest.

Out-of-band emission refers to all the interference power outside the wanted channel of the interfering system, and can be measured with the spectrum analyser from the output port of the transmitter. When the power density of the transmitter is $p_{\mathrm{e}}(f)$ [W/Hz], its total emitted power, P_{e}, is defined as

$$P_{\mathrm{e}} = \int_{-\infty}^{\infty} p_{\mathrm{e}}(f)\,\mathrm{d}f. \tag{1}$$

The adjacent channel interference ratio (ACIR) is the key parameter describing both the effect of non-ideal receive filtering and the out-of-band emission of the transmitter. L_{ACIR} is then given by

$$L_{\mathrm{ACIR}}(\Delta f) = \frac{\displaystyle\int_{-\infty}^{\infty} p_{\mathrm{e}}(f) \cdot L_{\mathrm{filter}}(f - \Delta f)\,\mathrm{d}f}{\displaystyle\int_{-\infty}^{\infty} p_{\mathrm{e}}(f)\,\mathrm{d}f} \tag{2}$$

$$= \frac{1}{P_{\mathrm{e}}} \int_{-\infty}^{\infty} p_{\mathrm{e}}(f) \cdot L_{\mathrm{filter}}(f - \Delta f)\,\mathrm{d}f \tag{3}$$

where $L_{\mathrm{filter}}(f)$ is the attenuation of the receive filtering at frequency f. Typically, the emission spectrum values have been measured over some finite bandwidth and in some predefined frequency grid. When the carrier separation between two systems is Δf_k, the L_{ACIR} is then given by

$$L_{\mathrm{ACIR}}(\Delta f_k) = \frac{1}{\displaystyle\sum_{k=-M}^{M} p_{\mathrm{e}}(f_k)} \sum_{k=-M}^{M} (p_{\mathrm{e}}(f_k) \cdot L_{\mathrm{filter}}(f_k - \Delta f_k)) \tag{4}$$

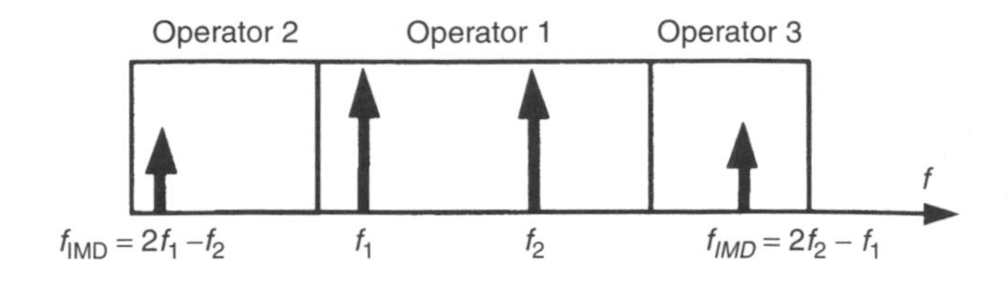

Figure D.2. Third-order intermodulation components falling from operator 1 into the band of operators 2 and 3

where the emission spectrum, $p_e(f_k)$, is negligible if $|f_k| > f_M$. The effect of ACI on the system performance has been studied in [3–5].

D.2.2 Intermodulation Distortion (IMD)

Owing to the non-linearity of the receiver, the receiver's output spectrum may contain frequency components that are not present in the input signal. These new signal components may fall into the band of interest causing intermodulation distortion. The non-linearity takes place in every component of the receiver if the signal level is high enough. The high power, unwanted signal may drive the receiver of the mobile to its non-linear part. When two or more signals are added together in the non-linear element, the resulting outcome from the element includes in addition to the wanted signal frequency also other frequency components. In the radio frequency (RF) design, the third-order intermodulation is particularly problematic, because of its strength and because it falls close to the band of interest. In the case of two interfering signals on frequencies f_1 and f_2, in the proximity of the desired signal, the third-order IMD products are those falling on frequencies $2f_1 - f_2$ and $2f_2 - f_1$. Also, higher-order IMD products exist but they are usually not that strong. The strength of the third-order intermodulation power is given by

$$P_{\mathrm{IMD}}^{\mathrm{in}} = 3 \cdot P_{\mathrm{i}} - 2 \cdot \mathrm{IIP}_3 \tag{5}$$

assuming that the incoming signal powers at frequencies f_1 and f_2 are equal. P_{i} (dBm) is the power at the input of the non-linear component and IIP_3 (dBm) is the third-order input intercept point of the same. The strength of the IMD depends on the input power of the interferer as well as the linearity of the receiver. P_{IMD} is proportional to the third power of P_{i} so that it is large when the receiver is close to the interfering source but decreases rapidly as the path loss increases.

Figure D.2 shows the example of the intermodulation effects between two operators. Operator 1 has two strong signal powers at frequencies f_1 and f_2. These can be originating from one site. If this jamming signal from the adjacent band is high enough, causing saturation at the receiver, the mobiles of operators 2 and 3 might observe IMD power at frequencies $2f_1 - f_2$ and $2f_2 - f_1$ respectively. The performance effects of intermodulation have been described in [6].

D.3 Coverage Effects

This section describes the coverage effect of the ACI and the out-of-band emissions in downlink and uplink.

D.3.1 Downlink

The interference power in downlink is

$$P_r = P_B - L_{min} - L_{ACIR,DL} \tag{6}$$

where P_B is the total transmission power of the interfering base station, L_{min} is the minimum coupling loss from the base station to the mobile station and $L_{ACIR,DL}$ is the ACIR described in Section D.2.1. If the required carrier-to-interference ratio is $(C/I)_{dB}$ and it is assumed that the ACI is the dominant interference mechanism, the needed carrier power is given by

$$C - P_B + L_{min} + L_{ACIR,DL} = (C/I)_{dB}. \tag{7}$$

When the maximum allowed path loss for the interfered system is L_{max} and the maximum transmission power is P_B, the strength of the carrier is given by $P_B - L_{max}$. The needed ACIR in the downlink can then be solved as

$$L_{ACIR,DL} = L_{max} - L_{min} + (C/I)_{dB}. \tag{8}$$

This gives, for example, $L_{ACIR,DL} = 135 - 80 + 20 = 75$ dB, which is a rather large value for the adjacent channel isolation, and might cause dead zones around the interfering site. This was the worst-case calculation, which assumes that the interfering cell is located at the cell edge of the interfered system. This is not a typical case but in an urban area, for example, the areas for highest path loss requirements are located inside the buildings, where there are no adjacent operator sites. Of course, this leads to problems in the case where one operator has an indoor system and another has only outdoor systems. In that case, a large guardband between operators is required. In the case of co-sited base stations, where $L_{max} = L_{min}$, Equation (8) gives $L_{ACIR} = (C/I)_{dB}$.

D.3.2 Uplink

The additional interference power from the adjacent frequency channel decreases the sensitivity of the base station receiver and therefore decreases the coverage area. The uplink interference is large when the interfering mobile station is very close to the base station receiver and is allocated into the closest frequency channel. The maximum required isolation between carriers can be calculated by using this worst-case scenario. This corresponds to the case in which interfered and interfering systems are not co-sited and the interfering system uses high powers in the close proximity of the interfered receiver. The effect of intermodulation has been ignored in this analysis. The received interference power at the receiver is then given by

$$P_r = P_M - L_{min} - L_T - L_{ACIR,UL} \tag{9}$$

where P_M is the mobile station maximum power, L_{min} is the minimum coupling loss and L_T is the average transmit power decrease due to bursty transmission (in time division multiple access (TDMA)). When P_r is set equal to the noise floor, P_N, so that the extra interference causes 3-dB degradation to the link budget, the needed $L_{ACIR,UL}$ can be calculated as

$$L_{ACIR} = P_M - L_{min} - L_T - P_N \tag{10}$$

which gives 30 dBm − 80 dB − 9 dB − (−121 dBm) = 62 dB. Here we have assumed that the system is noise limited where the noise power P_N is given by

$$P_N = 10 \cdot \log_{10}(kTB) + F \tag{11}$$

where k is Boltzmann's constant (1.38×10^{-23} W/Hz/K), T is the temperature of the receiver, B is the system bandwidth and F is the noise figure of the receiver. For the GSM system, (B = 200 kHz), $P_N = -121$ dBm. When the interfered system is limited by its own system the required $L_{ACIR,UL}$ depends on the interference level. If the interference level is, for example, −100 dBm, the needed isolation is 30 dBm − 80 dB − 9 dB − (−100 dBm) = 41 dB.

In the co-sited case, the power control of the interfering system usually decreases the interference problem. If the minimum power of the mobile is 0 dBm, the needed isolation in the thermal noise-limited case is 0 dBm − 80 dB − 9 dB − (−121 dBm) = 32 dB. The GSM power control in downlink is used in the case of traffic channels outside the BCCH transceiver. In the case of control channels, the mobile usually uses full power and therefore the receiver has to tolerate more interference.

If $L_{ACIR,DL}$ is much lower than $L_{ACIR,UL}$, the mobile will be blocked by the base station before it interferes with the base station in uplink. This phenomenon is depicted in Figure D.3.

By inserting the L_{min} from Equation (8) into Equation (9), the uplink interference level can be solved as,

$$P_r = P_M - L_{max} - (C/I)_{dB} - L_T - L_{ACIR,UL} + L_{ACIR,DL}. \tag{12}$$

From this equation it is possible to calculate the uplink interference if $L_{ACIR,UL}$ and $L_{ACIR,DL}$ are known. Figure D.4 shows the uplink interference as a function of difference between $L_{ACIR,UL} - L_{ACIR,DL}$ with different maximum adjacent cell path losses, L_{max}. It can be seen that when $L_{ACIR,UL} = L_{ACIR,DL}$, the uplink interference is above the GSM noise level (−121 dBm) only when the adjacent cell size is rather small. Here, it has also

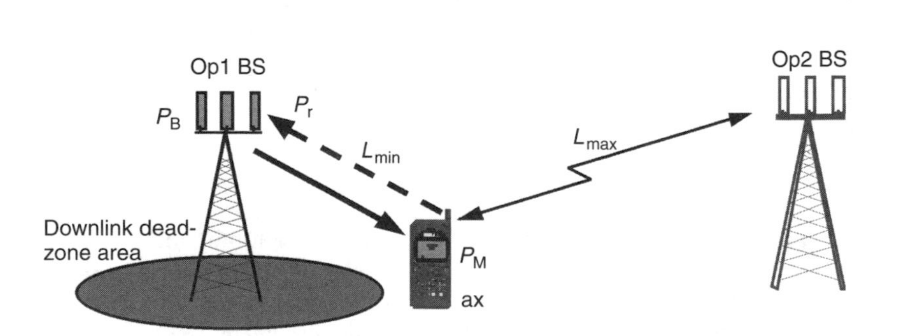

Figure D.3. Interference between operator 1 and operator 2. Operator 1 interferes with operator 2 in downlink and operator 2 interferes with operator 1 in uplink

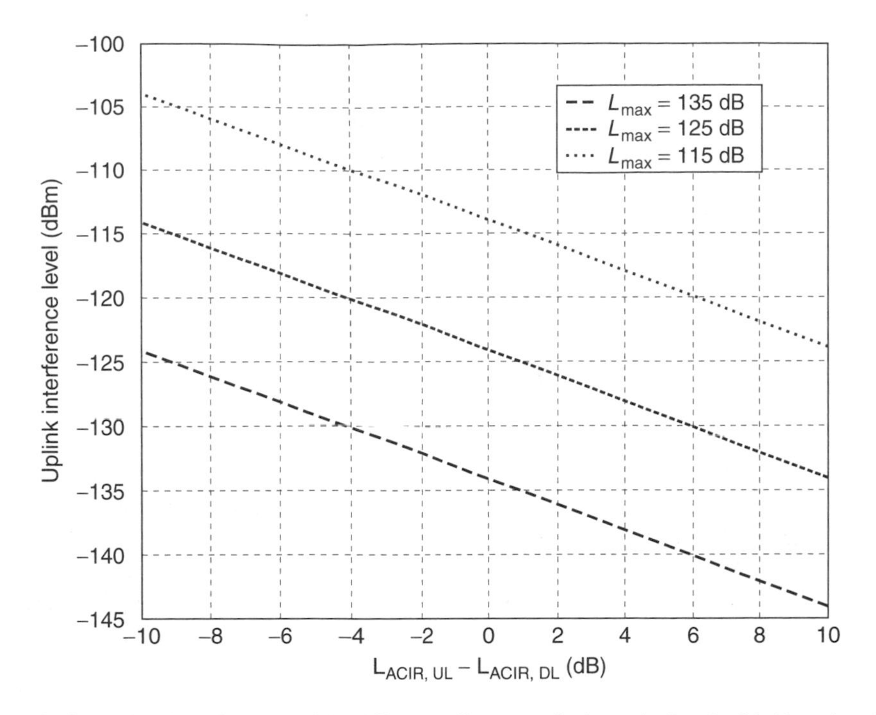

Figure D.4. Uplink interference with different adjacent cell size calculated with Equation (12). $P_M = 30$ dBm, $(C/I)_{dB} = 20$ dB, L_T equals; 9 dB

been assumed that the mobile transmits all the time with the full power. This is, however, not the case when the cell size is smaller and if the uplink power control works properly.

D.4 The Interference from WCDMA to GSM

The capacity loss of the GSM has been simulated in the case where the GSM and the WCDMA systems were deployed at adjacent frequency bands [7]. Other standardisation references handling the GSM and UMTS blocking-related items are [8–11]. The main interference mechanism that affects GSM uplink performance is the out-of-band emission of the WCDMA user equipment (UE) transmitter. The performance degradation is dependent on the transmitting power of the WCDMA UE and thus on the cell size of the interfering WCDMA system. In these simulations, equal cell sizes in both systems were assumed with cell sizes up to 2400 m. The simulation results indicate that the capacity loss in GSM was between 0 to 4% with the carrier separation of 2.7 MHz between WCDMA and the closest GSM carriers. This is in the case in which both systems were deployed with three-sector macrocellular networks and the WCDMA network was fully loaded. This is a worst-case scenario in the sense that the WCDMA base stations were located at the cell edge of the GSM system. The capacity degradation could be even lower when applying the random site locations with frequency hopping and the link adaptation schemes in the GSM system. In the co-sited base stations case, the capacity degradation of the GSM system can be considered to be negligible when deployed with WCDMA system.

D.5 Monte-Carlo Simulation Study (GSM/EDGE and IS-95)

This section describes the example simulation exercise that has been used to estimate the needed guardband between two systems by analysing the effect of ACI in uplink. The selected scenario represents the deployment of IS-95 and GSM/EDGE within the same band. The interference effects of GSM to WCDMA and IS-95 to WCDMA system have not been studied here. These simulation studies can be found in [12, 13]. In this example both systems are co-sited, representing the technology migration scenario. The interference levels in the uplink of IS-95 have been calculated by summing the interference contributions from randomly located GSM/EDGE mobiles. The used carrier frequency, mobile station powers and the path loss of the GSM/EDGE mobile have been generated randomly. The path loss data of realistic network scenarios have been calculated by using appropriate, environmental-specific propagation models. An Okumura–Hata propagation model has been utilised in macro cells and a ray-tracing propagation model [14] in micro cells. The mobile station power has been generated according to polynomial function by mapping the coupling loss value to mobile Tx power value for each mobile in the network. The function has been obtained by fitting the measured mobile station transmission power values from the real GSM/EDGE network. The original experimental data, which have been retrieved from numerous urban and suburban measurements, and the fitted polynomial, are shown in Figure D.5. The used carrier for each mobile has been generated randomly within the total bandwidth of 2.4 MHz (12 GSM/EDGE carriers).

The used emission spectrum of the GSM/EDGE mobile, the IS-95 receive filtering as well as the resulting ACIR function calculated according to Equation (4) are shown in Figure D.6.

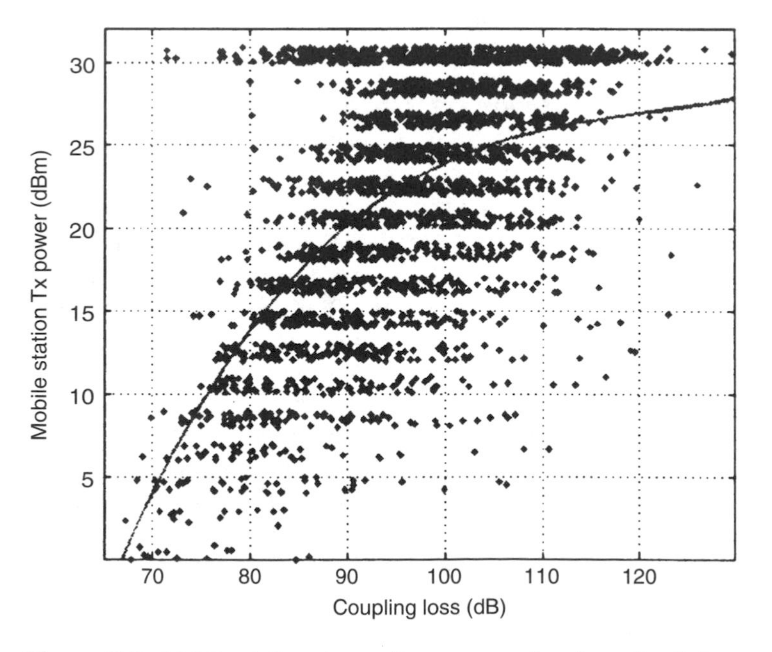

Figure D.5. Mobile stations transmit power as a function of path loss

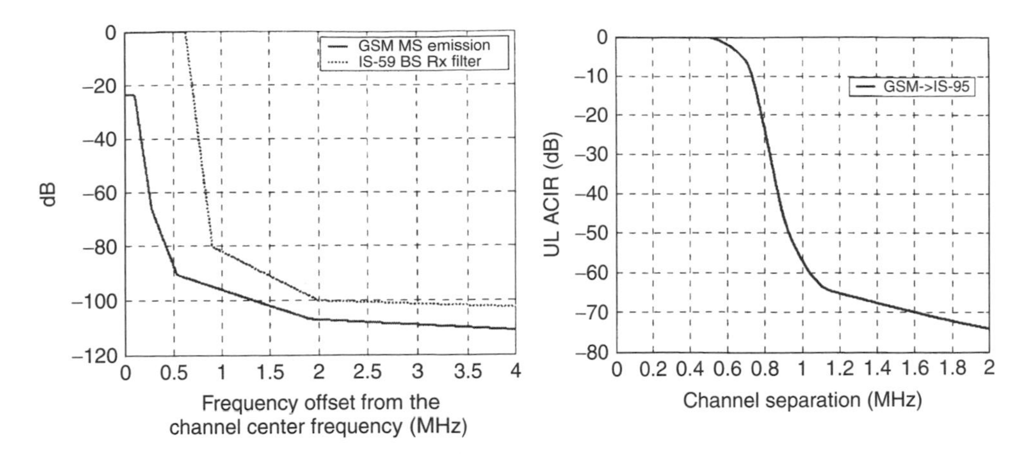

Figure D.6. Filter characteristics of the IS-95 BS and (a) out-of-band emissions of the GSM/EDGE MS and (b) the respective adjacent channel interference ratio (ACIR). These are example functions respecting typical implementation values

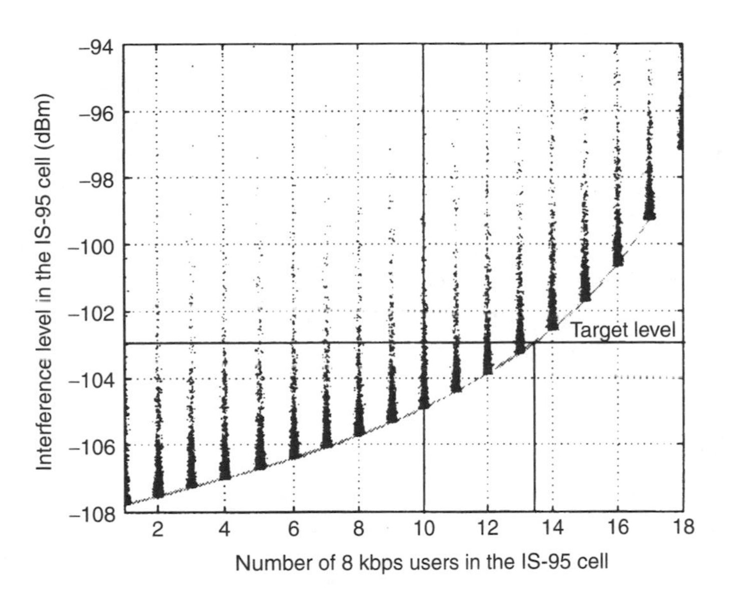

Figure D.7. Interference level as a function of users in a IS-95 cell with (dots) and without (solid line) the adjacent system interference

The target interference level in the IS-95 cell is −103 dBm in these simulations, which means approximately 5 dB noise rise when the noise figure is assumed to be 5 dB. The capacity of the interfered IS-95 network has been defined as the number of users for which the interference is above the target level at maximum 5% of the snapshots. Figure D.7 shows the interference level as a function of number of users for each simulated snapshot. Each simulation case, which means a certain number of IS-95 users, includes 2500 independent snapshots. The solid curve shows the loading-not-interfered network.

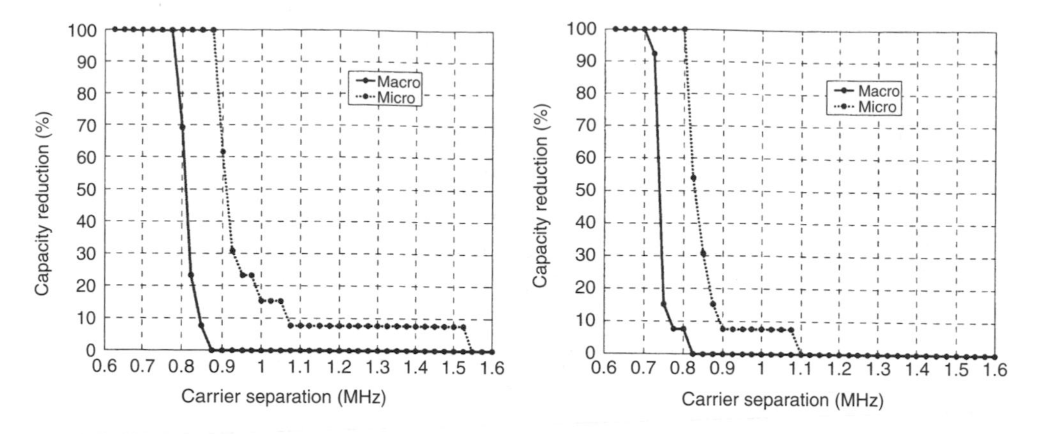

Figure D.8. The capacity reduction as a function of the carrier separation. (a) Seven GSM/EDGE users per IS-95 cell. (b) Two GSM/EDGE users per IS-95 cell

In the figure, the capacity is 10 users with the interference and 13.5 users without the interference. The capacity reduction is thus $100 \cdot (13.5 - 10)/13.5 = 25.93\%$.

Figure D.8 shows the capacity reduction as a function of the channel separation between GSM/EDGE and IS-95 carriers. The GSM/EDGE bandwidth is 0.2 MHz and the IS-95 bandwidth is 1.25 MHz, so that the channel separation without any guardbands is $0.2/2 + 1.25/2 = 0.725$ MHz. It can be seen from the results that the capacity reduction is larger in micro cells than in macro cells. This is because in micro cells, where the BS antenna is close to ground level, the minimum coupling loss is around 60 dB, whereas in macro cells it is around 80 dB. This leads to larger interference levels and also to larger capacity reductions in micro cells. With the example ACIR function, the needed guardband is just around 100 kHz in the macro cell case and 350 kHz in the micro cell case, in order to obtain very low capacity reduction.

D.6 Summary

This appendix presented some radio network planning aspects associated with the coexistence of different systems in the same geographical area. The main reason for ACI is the combined effect of non-idealities in the receive filtering and the out-of-band emissions. The IMD can also be relevant in some cases.

The system has to be able to tolerate a certain amount of additional interference from the adjacent carriers due to limitations in mobile terminal and base station design. The intra-operator interference can be minimised with co-siting, which decreases the near–far problem. One aspect to consider is the GSM/EDGE common control channels planning, since they usually transmit at full power and therefore might cause interference problems. It is preferable to allocate the common channels far away from the adjacent system in order to minimise the interference effects.

The deployment in the same band by a single operator of GSM/EDGE with other radio technologies, such as IS-136, IS-95 and WCDMA, has proven to be feasible without the requirement of significant guardbands in the co-sited scenario. Therefore, operators can deploy multiple technologies in the same site grid without any major performance degradation.

In the non-co-sited case, the near–far problem might exist which means that when the mobile is far away from its own site it can be very close to the adjacent site. In this situation, the interference increases both in uplink and in downlink directions. The downlink interference might cause dead zones around the base station and the uplink interference causes desensitisation in the BS reception. The needed guardband between carriers depends on the particular systems and filter characteristics.

References

[1] National Archives and Records Administration, Code of Federal Regulations: 24.238 CFR, 47, Parts 20–39, http://www.access.gpo.gov/nara/cfr.

[2] http://www.fcc.gov/.

[3] Hamied K., Labedz G., 'AMPS Cell Transmitter Interference to CDMA Mobile Receiver', *Proceedings of Vehicular Technology Conference*, Vol. 3, 1996, pp. 1467–1471.

[4] Schilling D. L., Garodnick J., Grieco D., 'Impact on Capacity to AMPS Jamming CDMA/CDMA Jamming AMPS in Adjacent Cells', *Proceedings of Vehicular Technology Conference*, 1993, pp. 547–549.

[5] Lu Y. Edward L., William C. Y., 'Ambient Noise in Cellular and PCS Bands and its Impact on the CDMA System Capacity and Coverage', Communications, 1995, *ICC '95*, Seattle, 'Gateway to Globalization', 1995, IEEE International Conference, Volume: 2, 1995, pp. 708–712.

[6] Dong Seung K., Heon Jin H., Sang Gee K., 'CDMA Mobile Station Intermodulation Interference Induced by AMPS Competitor Base Station', *Proceedings of IEEE 4th International Symposium on Spread Spectrum Techniques and Applications*, Vol. 1, 1996, pp. 380–384.

[7] GPP TR 25.885: 3rd Generation Partnership Project; Technical Specification Group TSG RAN; UMTS1800/Work Items Technical Report (Release 5 (1900)) 220.

[8] GPP TS 25.104 V4.0.0, UTRA (BS) FDD; Radio Transmission and Reception, 2000.

[9] GPP TR 25.942, V2.1.1, RF System Scenarios, 1999.

[10] 3rd Generation Partnership Project (3GPP), Technical Specification Group (TSG), UTRA FDD, Radio Requirements: TS 25.101 and TS 25.104.

[11] European Telecommunications Standards Institute (ETSI): GSM 05.05, http://www.etsi.org.

[12] Laiho J., Wacker A., Novosad T., *Radio Network Planning and Optimisation for UMTS'*, Wiley, Chichester, 2001, p. 512.

[13] Heiska K., Posti H., Muszynski P., Aikio P., Numminen J., Hämäläinen M., 'Capacity Reduction of WCDMA Downlink in the Presence of Interference from Adjacent Narrowband System', *IEEE Trans. Veh. Technol*, **51**(1), 2002.

[14] Hefiska K., Kangas A., 'Microcell Propagation Model for Network Planning', *Proceedings of Personal Indoor and Mobile Radio Communications, PIMRC '96*, 1996, pp. 148–152.

Appendix E

Simulation Tools

Martti Moisio

E.1 Introduction

This appendix describes first the fundamentals of simulation and network-level simulator (sometimes also called *system* simulators) functionalities. The simulators can be divided into two main categories: *static* simulators and *dynamic* simulators. The properties of these two simulation approaches are discussed and finally, the global system for mobile communications (GSM)/enhanced data rates for global evolution (EDGE) simulator used in this book are described together with the applied cellular models, algorithms and parameters.

E.2 Static Simulations

Static simulators do not model the movement of the mobiles and there is no correlation between consecutive simulation steps. Typical static simulation methodology is the following:

1. Create mobiles in random places in the network. The number of mobile stations (MSs) depends on the wanted load.
2. Select serving base stations (BSs) for each MS according to the received signal level.
3. Adjust transmission powers of links according to, e.g. path loss criteria. After this, the interference situation in the network will be frozen.
4. Execute needed radio resource management (RRM) algorithms, e.g. select coding scheme according to link quality.
5. Collect statistics (carrier/interference (C/I), bit error rate (BER), block error rate (BLER), throughput, etc.).
6. Repeat steps 1 to 5 (typically, a few thousand times) to get enough statistical confidence.

GSM, GPRS and EDGE Performance 2nd Ed. Edited by T. Halonen, J. Romero and J. Melero
 ISBN: 0-470-86694-2

The above procedure is repeated for each wanted operation point (network load). It should be noted that Steps 3 and 4 might also include iterative processes needed, e.g. power control with certain C/I target (typical in code division multiple access (CDMA) simulations [1]). Finally, network-level key performance indicators (KPIs) can be plotted against offered load. BER/BLER/throughput performance indicators are achieved through mapping tables from link simulations, using long-term average C/I versus BER/BLER curves. Normally, fast fading is not taken into account in a static network simulator, its effect is included in the link-level curves.

The main benefit of the static simulation tool is its execution speed. Also, the results are easier to understand and interpret because of simple models and especially lack of dynamics related to radio link control and RRM algorithms. However, these same issues constitute the main drawbacks of the static simulation approach: lack of dynamics, correlation and feedback loops, which form an essential part of cellular system characteristics, especially when dealing with packet data. Typically, a static simulator gives optimistic upper-level results. Especially when studying the performance gain related to radio link algorithms, a dynamic simulator is needed.

E.3 Basic Principles of Dynamic Simulation

The fundamental property of a dynamic simulator is *causality* and *time correlation* between events in the simulated network. When an event (e.g. handover (HO) or PC command) is triggered, it has its cause in the past and consequence in the future. With this kind of modelling of the system dynamics, the effect of feedback loops (input depends on the output) that are present in many algorithms can also be assessed. Maybe the most important building blocks of this kind of dynamics are *simulation time* and *user mobility*, which are not relevant in static simulations. These two features bring the needed determinism in the simulated system—in each simulation step, the time is increased and mobiles move along their paths. However, the movement of mobiles is not in any way necessary for a simulator to be a dynamic one—the important thing is the time correlation of the signal with the evolving simulation time.

Dynamic simulators can be divided into two basic categories: they can be either *event-driven* or *time-driven*. In an event-driven simulator, the functionality is based on events (MS changes position, PC command is executed, etc.) that are processed according to their order of appearance. Events may generate *tasks*, which may be handled only after some time period or after some other event has occurred. This approach resembles the behaviour of typical protocols. It can save simulation time especially if the event's inter-arrival time is relatively long, because the simulator can jump directly to the next pre-calculated event.

The simulator used in this book is a dynamic time-driven simulator, where the simulation time step is fixed (one TDMA frame). This approach is convenient because the time steps can be selected according to the GSM frame structure (time division multiple access (TDMA) frame numbering). Owing to the detailed modelling, high frequency and volume of the events (e.g. receiving the radio burst), the event-based approach would produce more implementation overheads leading probably to longer simulation times. Also, the time-driven approach makes the actual program code more understandable for the software developer. However, event-based methodology is used for certain parts of the program, where it was found to be suitable (PC/HO task handling in base station controller (BSC), general packet radio system (GPRS)/enhanced GPRS (EGPRS) radio link control (RLC) protocol).

An important practical property of the dynamic simulator (actually, any simulator) is that it must be possible to reproduce the results, provided that the input remains the same. Otherwise, the analysis of the system behaviour is almost impossible, not to mention the debugging of the program code. The network simulator should not just be able to produce average performance values for the whole network; often the behaviour of a single mobile is also interesting. It should be possible to select any connection in the network and trace its behaviour. These requirements mean that the simulator must have pseudo-random statistical properties. This means, in practice, the output remains exactly the same with the same input and it can be changed by changing the seed of the random number generator(s). Of course, final simulations must be long enough so that the effect of random number seeds becomes negligible.

A state-of-the-art dynamic simulator should fulfil at least the following requirements:

- Short-enough time resolution to capture the effect of short-term changes in the radio link (multipath fading, frequency hopping, power control, discontinuous transmission, packet scheduling, fast link adaptation using hybrid automatic repeat request (ARQ) schemes, etc.). In GSM, a suitable time step unit is one TDMA frame, i.e. 4.615 ms—this means that every burst in the transmitting link is simulated. It can be assumed that the C/I ratio of the link is fixed for the GSM burst period. Individual bits are not considered in a system-level simulator; bit-level modelling (burst formatting, modulation, channel coding, etc.) is handled in a dedicated link-level simulator. Statistical look-up tables produced by the link simulator are then used to calculate bit and block error rates. It is very important to model the stochastical, non-deterministic processes to produce correct distributions, not just correct long-term averages. For example, when the look-up table produces certain probability for block error, a random experiment must be made against uniform distribution to produce the final result (whether the block was correct or not).
- Multiple cells with enough co-channel and adjacent-channel interfering tiers. To reduce the boundary effects of finite simulation area, some wrap-around methodology [2, 3] is recommended, especially if the number of frequency clusters is small (large reuse factor and/or small number of cells).
- Modelling user-specific quality with the concept of *call*—whether it is a traditional circuit-switched voice call, file downloading or Web browsing session. This is important, since finally the only relevant quality is the user-specific quality. The concept of call includes the important effect of mobility in a cellular environment—a moving MS experiences different signal strength and quality conditions depending on its location and it may be served by several different BSs during its lifetime (hence, the effect of HOs is taken into account). Also, the call may be transferred to a different layer, or it may even have to adapt to different base transceiver station (BTS)/transceiver (TRX) configuration (e.g. some BTSs/TRXs may not support EDGE).
- The software must be efficient enough to allow enough statistically valid calls to be simulated within a reasonable time. As a rule of thumb, the number of simulated calls must be of the order of thousands. Also, to minimize the unavoidable distorting

effects in the beginning and the end of the simulation, the actual simulated real (busy hour) network time should be long enough. If we consider the typical case of call arrivals to the network as a Poisson process with exponential service times, the total simulation length must be long enough compared with the expectation value of the service time. It should be noted that this is not required by the Poisson process itself (which is a memory-less process), but it is required to have enough normally ended calls according to the wanted call length distribution. Traffic warm-up should be used to shorten the time when the level of desired offered load is reached—however, full traffic warm-up is not always possible because the load saturation point is not always known, especially in packet data services. On the basis of the experience with the simulator used in this book, at least 15 min of busy hour time is needed to reach full statistical confidence. It should be noted that a large number of simulated calls do not compensate the requirement for long enough simulation times. Analogously, the simulation time must be further increased with especially low load in order to have high enough number of calls.

E.4 Description of the Simulator and Basic Simulation Models Used in this Book

The simulator[1] used in this book follows the principles and requirements in the previous section. This section describes the simulator functionality and modelling more accurately. For all the people involved in performance analysis, it is important to understand the modelling principles of present-day network-level simulators. Also, this section (as the previous one) might give some hints for those who are interested in developing system-level simulators of their own.

E.4.1 Software and Programming Issues

To ensure easier maintenance, reliability and modularity, C++ [4] object-oriented language was selected. The elements of the GSM network (BSC, BTS, MS, radio block, etc.) form a natural basis for C++ class hierarchy. The fine features of C++ language, like polymorphism and class inheritance are very useful when designing a GSM system simulator. Figure E.1 shows an example of the class structure in the simulator. The base class contains all common GSM-connection (call)-specific functionality and the derived class implements new ones and/or overloads the base class methods whenever needed. Each connection has N pieces of *links* that carries the radio blocks and bursts from transmitter to receiver. For example, a GSM speech connection class has one downlink and one uplink, while a GPRS connection can have several links in one direction.

As is clear for people familiar with object-oriented programming, polymorphism enables run-time dynamic binding (through virtual functions in C++) of the methods. This is a very useful feature when thinking about the modelling of a hierarchical GSM system.

E.4.2 Basic Functionality of the Simulator

Figure E.2 shows a schematic view of the basic functionality of the simulator. For each simulation step, the mobiles are first moved and received powers are calculated for both

[1] The simulator is called SMART, originally developed at Nokia Research Center, Helsinki, Finland.

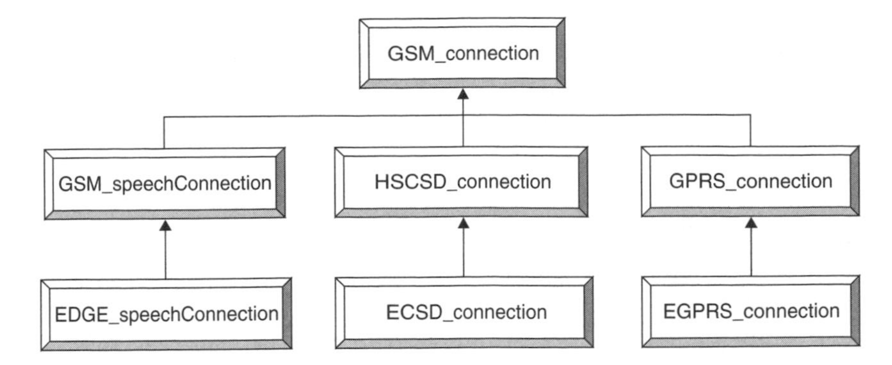

Figure E.1. An example of the hierarchical class structure of the *SMART* GSM simulator

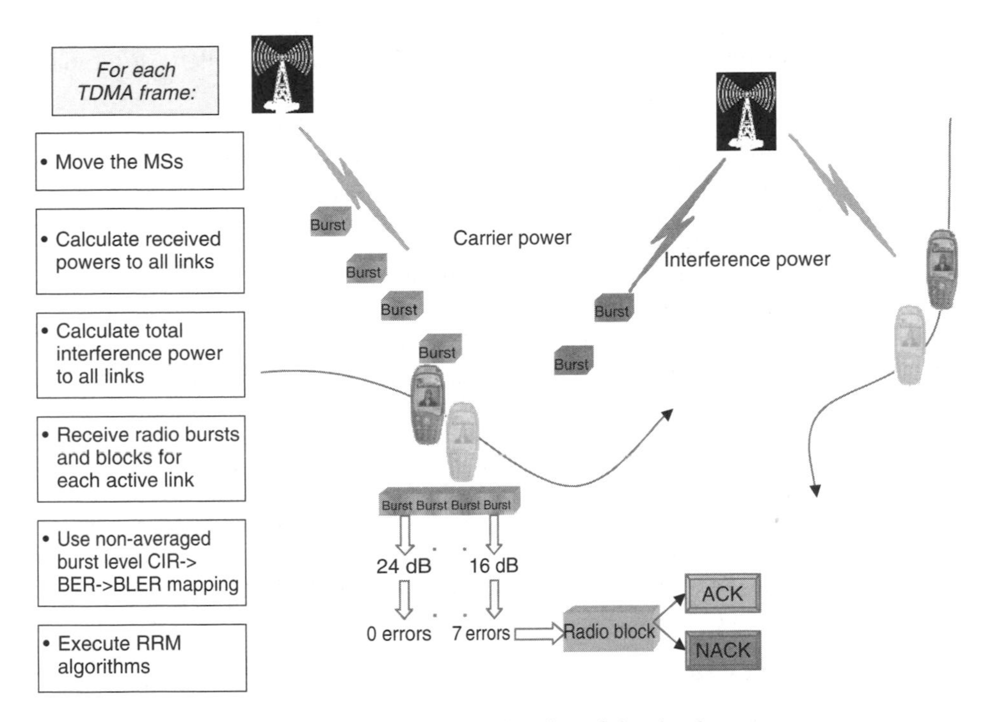

Figure E.2. Basic functionality of the simulator

uplink and downlink. After this, the interference situation in the network is frozen for that time step and C/(I + N) values can be calculated for each active transmission link. The link object binds the receiving and transmitting objects together and contains all the necessary information (antenna gains, transmission powers, MS/BS positions, fading phase, etc.) in order to calculate the power (field strength) in the receiver antenna. All the transmission links in the network are stored in one interference matrix (one for downlink, one for uplink) where each matrix element contains all the links transmitting in that physical channel. This kind of effective and simple data structure is important for the next step in the simulation loop—interference calculation—which is the most important and time-consuming function of a cellular simulator. Received power C (in watts) is calculated for each link according to the following formula:

$$C = TxP^{*}G^{*}L_{\mathrm{d}}^{*}L_{\mathrm{S}}^{*}L_{\mathrm{F}}$$

where
TxP = transmission power (in watts)
G = transmitter and/or receiver gain
L_{d} = distance path loss
L_{S} = slow-fading component (each MS has its own correlated slow-fading process)
L_{F} = fast-fading component (each MS–BS pair has its own correlated frequency-dependent fast-fading process).

All the co-channel and adjacent channel interferers in the whole network area are taken into account. The carrier-to-interference ratio is calculated with the following formula:

$$\mathrm{CIR}_{f_N} = \frac{\mathrm{C}_{f_N}}{\sum_{k} I_{f_N} + ACP1 \times \sum_{l} (I_{f_{N-1}} + I_{f_{N+1}}) + N_0}$$

where
CIR_{f_N} = carrier-to-interference ratio on carrier frequency f_N
C_{f_N} = signal strength at carrier frequency f_N
$\sum_k I_{f_N}$ = sum of all k interfering signals at carrier frequency f_N
$ACP1$ = adjacent channel protection on the first adjacent carrier frequency
N_0 = receiver noise floor.

After the CIR calculation, radio bursts and blocks for each link are received. As the last step in the simulation loop, all the necessary RRM algorithms are executed (most of the algorithms are executed quite seldom, typically on slow associated control channel (SACCH) frame basis) and the network time (TDMA frame number) is increased. If the network is unsyncronized, every tranceiver site has its own clock (frame offset).

E.4.3 Link-Level Interface

The simulator uses non-averaged burst-level mapping between system and link-level simulators [5–8]. The mapping consists of two phases—first, burst-wise C/I is mapped on to raw bit error rate (in the simulated cases presented here, only two mappings are needed; one for Gaussian minimum shift keying (GMSK) and one for octagonal phase shift keying (8-PSK)). Next, mean and standard deviation of the bursts over the interleaving period are calculated. On the basis of extensive link-level simulations, a frame error probability (FEP) mapping matrix has been created where each matrix element represents estimated FEP for a block having certain mean and standard deviation of errors.

Figure E.3 shows an example of the mapping procedure in the second phase. Figure E.3(a) shows the raw data, i.e. frame error probabilities on the (stdBER, meanBER)-plane. Figure E.3(b) shows the final mapping matrix. In the final matrix, the intensities

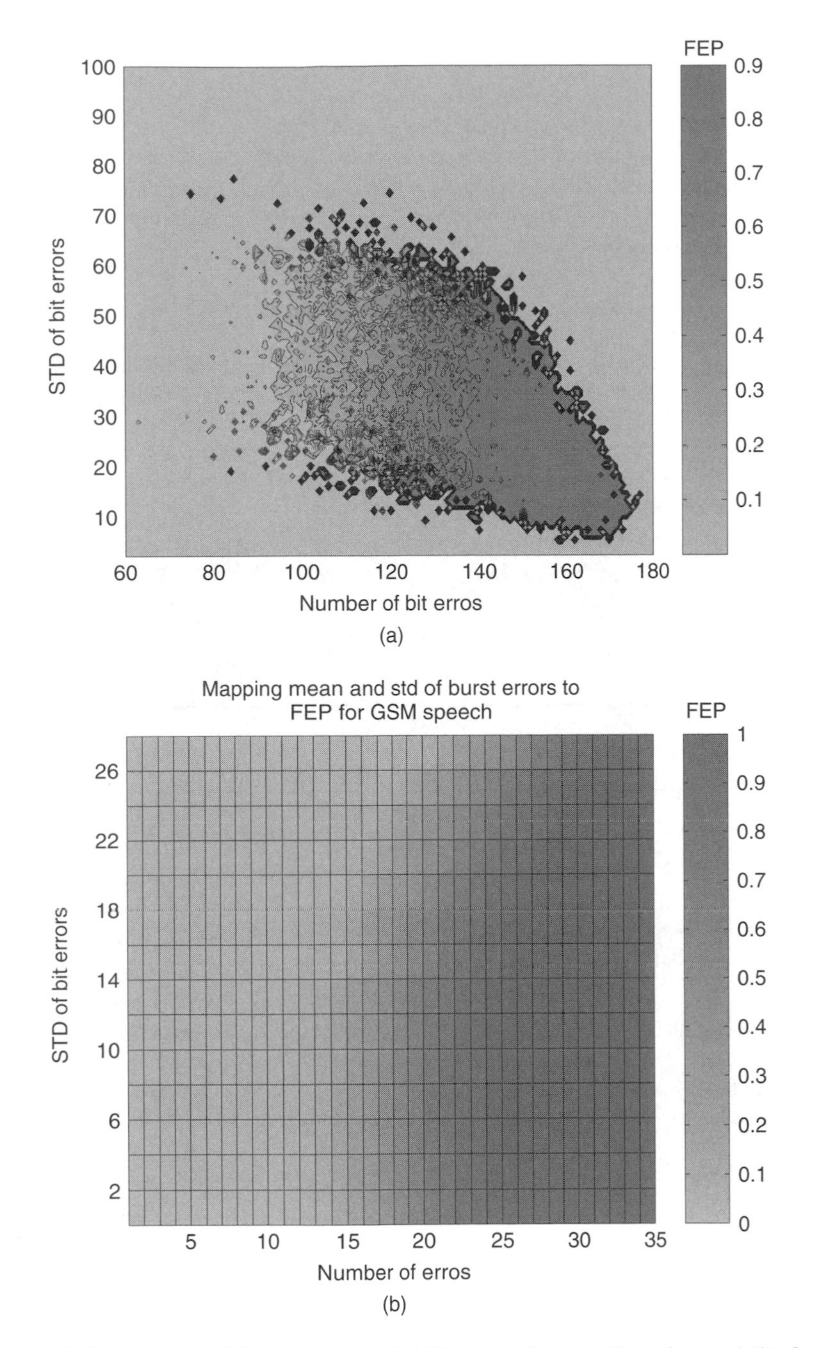

Figure E.3. Example of frame error probability mapping. (a) Raw data and (b) final mapping matrix

Table E.1. Common simulation models

Model	Description	Comment
Cellular layout	Hexagonal grid, three-sector sites	
Path loss model (macro cell)	$L[\text{dB}] = 128.1 + 37.6\log_{10}(R)$	R in km, minimum coupling loss of 70 dB taken into account
Call arrival process	Poisson	
Call length distribution	Exponential	For speech
Channel allocation	Channel with least amount of interference selected	Idle slot uplink interference (non-ideal) measured at BTS
Cell selection	Path loss based	
Inter-cell handover	Path loss based with hysteresis and penalty timer	
Intra-cell handover	If RXLEV good but RXQUAL bad	
BTS antenna diversity	Yes	Simple selection diversity
Synchronization	No	Slot synchronization exists from interference point of view, but each site has its own frame offset

Table E.2. Common simulation parameters

Parameter name	Value	Unit	Comment
Time resolution in simulations	4.615	ms	Every traffic burst is simulated
Cell radius	1000	m	Corresponds to 3 km site-to-site distance
BS antenna height	15	m	No tilting
BS antenna gain	14	dBi	3 dB point
BS antenna beamwidth	65	deg	Horizontal antenna pattern used
BS Tx power (max.)	43	dBm	20 W
MS antenna height	1.5	m	Body loss not taken into account
MS speed	3 or 50	km/h	MS wrapped to other side when network border is reached
Slow fading correlation distance	50	m	
Slow fading standard deviation	6	dB	
Carrier frequency	900	MHz	
Adjacent channel protection	18	dB	First adjacent taken into account

Table E.2. *(continued)*

Parameter name	Value	Unit	Comment
System noise floor	−111	dBm	
Handover margin	3	dB	
Handover interval	10	SACCH frames	Minimum period between inter-cell HOs
DTX silence/talk periods	3.65	s	When DTX is applied, it is used in both directions. When UL is silent, DL is active and vice versa

Table E.3. Some of the most important (E)GPRS simulation parameters

Parameter name	Value	Unit	Comment
RLC window size for GPRS	64	RLC block	Max. for GPRS
RLC window size for EGPRS	384	RLC block	Max. according to multislot class
Downlink TBF establishment delay	240	ms	
RLC acknowledgement delay	220	ms	
Minimum C/I for CS-2	−2 (NH)/7 (FH)	dB	LA related
Minimum C/I for CS-3	3 (NH)/12 (FH)	dB	LA related
Minimum C/I for CS-4	9 (NH)/17 (FH)	dB	LA related
Expiry limit of T3190	5	s	
Expiry limit of T3195	5	s	
N3105max	8		
Cell reselection hysteresis	6	dB	
Cell reselection penalty timer	4	SACCH multiframes	
Number of TRXs/cell	1–3		
Max number of timeslots per TBF	3		
Max number of TBFs per TSL	6		
Total simulation length	200 000	TDMA frames	∼15 min. real time
MS speed	3	km/h	Mobility model according to [9]

of received blocks on the plane are taken into account. Certain (stdBER, meanBER) combinations are very rare or even impossible. If enough samples are received in certain matrix element, it is statistically valid and can be taken into account in the final mappings. Missing points are extrapolated or interpolated from the available data whenever relevant.

E.5 Description of the Basic Cellular Models

Unless otherwise stated, the models in Tables E.1, E.2 and E.3 have been applied to all the simulations in this book.

References

[1] Wacker A., Laiho-Steffens J., Sipilä K., Jäsberg M., 'Static Simulator for Studying WCDMA Radio Network Planning Issues', *IEEE 49th Vehicular Technology Conference*, Vol. 3, 1999, pp. 2436–2440.

[2] Hytönen T., *Optimal Wrap-Around Network Simulation*, Helsinki University of Technology Institute of Mathematics: Research Reports 2001, 2001.

[3] Lugo A., Perez F., Valdez H., 'Investigating the Boundary Effect of a Multimedia TDMA Personal Mobile Communication Network Simulation', *IEEE 54th Vehicular Technology Conference*, Vol. 4, 2001, pp. 2740–2744.

[4] Stroustrup B., *The C++ Programming Language*, Special Edition, Addison-Wesley, Reading, MA, 2000.

[5] Malkamäki E., Ryck F., de Mourot C., Urie A., 'A Method for Combining Radio Link Simulations and System Simulations for a Slow Frequency Hopped Cellular System', *IEEE 44th Vehicular Technology Conference*, Vol. 2, 1994, pp. 1145–1149.

[6] Hämäläinen S., Slanina P., Hartman M., Lappeteläinen A., Holma H., Salonaho O., 'A Novel Interface between Link and System Level Simulations', *Proc. ACTS Mobile Telecommunications Summit*, Aalborg, Denmark, October 1997, pp. 599–604.

[7] Olofsson H., Almgren M., Johansson C., Höök M., Kronestedt F., 'Improved Interface between Link Level and System Level Simulations Applied to GSM', *Proc. ICUPC 1997*, 1997.

[8] Wigard J., Nielsen T. T., Michaelsen P. H., Morgensen P., 'BER and FER Prediction of Control and Traffic Channels for a GSM Type of Interface', *Proc. VTC '98*, 1998, pp. 1588–1592.

[9] Universal Mobile Telecommunications System (UMTS): Selection Procedures for the Choice of Radio Transmission Technologies of the UMTS (UMTS 30.03 Version 3.2.0), ETSI Technical Report 101 112 (1998-04).

Appendix F

Trial Partners

This appendix contains information about network operators who have run field trials in cooperation with Nokia, the employer of all the contributors of this book at the time these trials took place. The following information refers to the trial networks

	Trial partner	Used frequency spectrum in trial (MHz)	Average number of TRXs in trial	Used features during trial
Network 1	Optus/Brisbane	5–8	2.5	FH + PC + DTX
Network 2	Major Chinese city	5–7	4	FH
Network 3	CSL/Hong Kong	6–7	3–4	FH + PC + DTX
Network 4	Major European capital	—	—	—
Network 5	Major Chinese city	5–7	4	FH + PC
Network 6	City in US	4.4	2	FH + PC + DTX
Network 7	Radiolinja/Finland	—	—	E-OTD location method
Network 8	Sonofon/Denmark	—	—	FH + PC
Network 9	Major European metropolis	8–10	3–4	FH + PC + DTX
Network 10	AWS/Several major cities	—	—	—
Network 11	Radiolinja/Estonia	—	—	CI, CI + TA, and CI + TA + RXLEV location methods
Network 12	Sonofon/Denmark	—	—	GPRS
Network 13	Telefonica	—	—	GPRS
Network 14	Cingular Wireless	—	—	AMR, SAIC and network synchronisation
Network 15	AWS	—	—	Link performance

GSM, GPRS and EDGE Performance 2nd Ed. Edited by T. Halonen, J. Romero and J. Melero
 ISBN: 0-470-86694-2

Index

GSM, GPRS and EDGE Performance 2nd Ed. Edited by T. Halonen, J. Romero and J. Melero

ISBN: 0-470-86694-2